THE RAY SOCIETY

INSTITUTED 1844

This volume is 173 of the series

LONDON

2011

Frontispiece. Two lengths from a long panorama of Cambridge drawn by an unknown artist in about 1630. The upper picture shows the area around Cambridge castle; the lower picture is centred on Trinity College. Ray reported *Astragalus glycyphyllos* "*About the Castle hill at Cambridge*" and *Mentha pulegium* "*At the castle end in a bottome over against the furthest house*". Trinity College is given as a site for *Brassica nigra*, *Calystegia sepium* and *Sagina procumbens* in the *Catalogus*, and *Rumex pulcher* is listed from here in the 1685 appendix. Reproduced by kind permission of the Syndics of Cambridge University Library (MS Add. 2655).

John Ray's Cambridge Catalogue (1660)

Translated and edited by

P.H. Oswald

and

C.D. Preston

Ray Society 2011

© The Ray Society

ALL RIGHTS RESERVED

No part of this book may be reproduced, stored in a retrieval system, or transmitted in any form or by any means, including but not restricted to manual, mechanical, electrical, electronic, photocopying or recording, without written permission from the holder or holders of the intellectual property rights.

ISBN 0-903874-43-1
ISBN 978-0903874-43-4

Published in 2011 by the Ray Society, c/o The Natural History Museum, Cromwell Road, London SW7 5BD, U.K.

The Ray Society is a registered charity number 208082

Sold by: Scion Publishing Ltd, The Old Hayloft, Vantage Business Park, Bloxham Road, Banbury, OX16 9UX, U.K.

www.scionpublishing.com

Typeset, printed and bound by
Henry Ling Limited, at the Dorset Press, Dorchester, DT1 1HD

CONTENTS

Foreword vii

Acknowledgements viii

1: Introduction 1

2: John Ray, once Fellow of Trinity College 4
    Introduction 4
    Ray's early life 4
    The University of Cambridge in the mid 17th century 5
    Ray and his colleagues at Trinity College 8
    Ray's botanical studies, *c.* 1651–1662 13
    Samuel Corbyn's records 15
    The Act of Uniformity, 1662: Ray leaves Cambridge 17
    John Ray and Francis Willughby, 1660–1672 18
    A further disruption: Ray leaves Middleton Hall, 1675 20
    Ray's later life 22

3: The *Catalogus* and its appendices – structure and sources 24
    Printing and publishing 24
    Structure of the book 25
    Models for the *Catalogus* 26
    'Preface' 34
    'Explanation' 35
    Main catalogue 39
    'Index of English names' and 'Index of places' 51
    'Etymology' 52
    'Interpretation' 58
    'The more customary headings or divisions of plants' 59
    Authorship of the *Catalogus* 62
    Is the *Catalogus* a Flora of Cambridgeshire? 64
    The 1663 appendix 65
    The 1685 appendix 67

4: Biographical notes on some authors cited by Ray 70
    Classical and mediaeval authors 70
    Modern authors 79

5: Ray's cited sources in public and private libraries 97
    Citation of sources in the *Catalogus* 97
    University and college libraries 98
    Personal book collections 103
    Conclusions 110

6: Identification of Ray's plants 113
    The work of earlier authors 113
    Our approach 113

CONTENTS

Methods for identifying Ray's plants 114
Degrees of certainty in the identification of Ray's plants 115

7: Notes on the translation and editorial methods 117

Translations of John Ray's (1660) *Catalogus plantarum circa Cantabrigiam nascentium* and of the appendices of 1663 and 1685 121

A Catalogue of Plants growing around Cambridge (1660).. 122
    Preface to the Reader 124
    An Explanation of the names of the Authors cited 134
    A Catalogue of Plants growing around Cambridge 146
    An Index of plants of the county of Cambridge, In which the English Names are placed before the Latin in alphabetical order 320
    Index of some places with the plants which grow in them 346
    The Meaning of certain names, or Etymology 356
    Interpretation & Brief explanation of certain terms & more general words .. 414
    The more customary headings or divisions of plants.. 430

Appendix to the Catalogue of Plants Growing around Cambridge (1663).. 436
    Greetings to the Reader 438
    Addenda 440
    Emendanda.. 450

Appendix to the Catalogue of Plants growing around Cambridge, second edition (1685) 454
    Greetings to the Reader 456
    To the Kind Reader 458
    Addenda 460
    Emendanda.. 480

Gazetteer 488

Vocabulary of epithets used to describe plant species 501

Bibliography 530

Index 560

FOREWORD

John Ray was, by any reasonable reckoning, one of the six greatest scientists that Cambridge has produced and probably the greatest scholar that Cambridge has ever sacked. He is remembered as Linnaeus's predecessor in the science of taxonomy, but he had not Linnaeus's gift for publicity. At the height of his international reputation the limit of his ambition was to retreat to his Essex birthplace where his father had been blacksmith, marry his sweetheart and live happily ever after, writing prodigious and learned Latin books on animals and plants from all over the world – which many read at the time and few have read since.

This book is, in effect, a Flora of Cambridgeshire, the first for any English county. Various editions and commentaries exist, but none does justice to its complexity, its discursiveness, its allusiveness, the circumstances of its writing, its vast bibliography or Ray's other works associated with it as appendices or supplements. To cover all these properly calls for a botanist with almost as much erudition as Ray himself, able to find and read the books that Ray used, many of them by what were even then obscure Roman, Byzantine and Renaissance writers. Only now has a pair of authors, Philip Oswald and Chris Preston, arisen who so eminently combine these qualifications.

Ray's career was shaped by the political troubles of his youth. (Those of the 21st century seem trifling by comparison.) He taught himself botany (for there was no one in Cambridge to teach him) and got on his horse and explored. This catalogue was his earliest publication and shows the development of his thought as his life's work was beginning to unfold. In his time the distinction between names and descriptions was emerging: "in some plants a collection of Synonyms can take the place of a description". Part of his book was devoted to trying to show that all plant names were really one-word descriptions. Ray did his best to deal with this problem, but it was left to Linnaeus to solve it (for a time).

This is not an ecological book, though it contains ecological notes on many of the plants; but Ray's researches were to prepare the way for the foundations of scientific ecology two centuries later.

I recommend this book not only to those interested in the British flora and the history of its recording but to anyone wanting to learn more about the development of scientific thinking in the 17th century.

Professor Oliver Rackham, O.B.E., F.B.A.

ACKNOWLEDGEMENTS

In preparing this translation we have used a copy of the *Catalogus* which was given to C.D.P. by Dr David Coombe in 1997; the facsimile pages are taken from this copy. It was the copy owned by Canon Charles Raven when he was preparing his biography of Ray, as he noted in the book when presenting it to T.G. Tutin in 1944. David Coombe was also the first person to draw our attention to the fact that Ewen & Prime's translation of the *Catalogus* did not provide a definitive answer to all the problems posed by the text. Although David died in 1999, long before we contemplated translating and editing the book ourselves, our work is a belated result of both his generosity and his scholarship.

We thank numerous friends and colleagues who have helped us tackle problems which have arisen during the course of the work. Although the project has been a hobby for C.D.P. rather than part of his work at the Biological Records Centre, the initial enthusiasm of the head of B.R.C., Dr Mark Hill, was instrumental in encouraging him to undertake his part of it. Scott Mandelbrote was similarly encouraging when we met him at an early stage of the project, and later he provided useful comments on some of the introductory material. We thank Sam Bosanquet, Dr Marc Botham and Dr Phil Sterling for help with the identification of Ray's moths, Dr Gavin Broad for comments on the more difficult matter of his parasitic insects (and especially for identifying *Orthopelma mediator*), Björn Beckmann and Dr Marco Girardello for translations from German and Italian respectively, Professor Tim Birkhead for his comments on Henry from Heer's outrageous story quoted by Ray in a note in the main catalogue under *Ulmus procera*, Arthur Chater for supplying a copy of Örstan's (2010) paper on Ray's hermaphrodite snails and slugs, Stephanie Ames, Charlotte Coombes, Christine Bartram and Ben Sherwood for help in preparing the figures, Professor James Diggle (through Dr Paul Millett) for explaining the meaning of the Latin phrase *adi sis*, Dr Mike Foley and David Pearman for drawing our attention to Penny's records on the drawings for Gesner's *Historia plantarum* (Zoller, Steinmenn & Schmid 1979), Dr Mark Hill for his comments on Ray's cryptogams and Elizabeth Quarmby Lawrence for her comments on Chapter 5. Prof. em. Dr Gerhard Wagenitz consulted the anonymous *Nomenclator* ascribed to Jung for us in the University of Göttingen library and, when he failed to find the relevant material in it, pursued the matter until he discovered that Ray's abbreviated *Jun.* referred not to Jungius but to Junius.

Our debt to Mrs Gigi Crompton's Flora of Cambridgeshire website is much greater than appears from the text, as it has provided essential background material and led us to the primary source of numerous records. In addition we are most grateful to Gigi for reading and commenting on a draft of the book. While we have been working on Ray's 17th-century Flora of Cambridgeshire, Dr Alan Leslie has been writing a Flora for the 21st century. We have benefited from comparing notes with him on the identity of Ray's plants and from similar discussions with David Pearman, who is working on a revised list of first British plant records.

Our greatest debt is to the collections and staff of the Cambridge University Library. We could not have prepared this work without access to a good library and almost all the books we needed to consult were available in the 'U.L.'. It is difficult to believe that our innumerable requests did not test the patience of the staff of the library's Rare Books Room, but we never detected anything other than their customary friendliness and helpfulness; it is difficult to imagine a better environment in which to carry out such work. We have been made equally welcome on our occasional visits to other

ACKNOWLEDGEMENTS

libraries, namely Department of Plant Sciences, University of Cambridge; Founder's Library, Fitzwilliam Museum, Cambridge; Wren Library, Trinity College, Cambridge; Old Library, St John's College, Cambridge; Archive Centre, King's College, Cambridge; Cambridgeshire Archives, Shire Hall, Cambridge; Cambridgeshire Collection, Central Library, Cambridge; Linnean Society, London; Wellcome Library, London; Manuscripts and Special Collections, University of Nottingham; Department of Plant Sciences, University of Oxford; Old Library, Magdalen College, Oxford; and Merton College, Oxford. For particular help with books and manuscripts we thank the staff of the Manuscripts Room, Cambridge University Library, and Christine Alexander and Gina Murrell (Plant Sciences, Cambridge), Kathryn McKee (St John's College, Cambridge), Mrs Linda Shaw (University of Nottingham), Ruth Bridgen, Adam Green and Jonathan Smith (Trinity College, Cambridge), Tabitha Tuckett (Magdalen College, Oxford) and Julian Reid and Julia Walworth (Merton College, Oxford).

We thank Manuscripts and Special Collections, University of Nottingham, and the Syndics of Cambridge University Library for permission to reproduce the pages of books and documents in Chapters 3 and 5 and The Warden and Fellows of Merton College Oxford for permission to reproduce the map of Gamlingay in the gazetteer; further details of the sources of these figures are provided in the relevant captions. The facsimile pages of the 1663 appendix are taken from the copy in the library of the Linnean Society (R.660 RAY) and those of the 1685 appendix from C.C. Babington's copy in the library of the Department of Plant Sciences, University of Cambridge; we are particularly grateful for permission to reproduce pages from these rare works. Map 1 in the gazetteer is based on a map of Cambridgeshire plotted with the use of the DMAP program written by Dr A.J. Morton.

We express our gratitude to the Ray Society for publishing this book and to the staff of the Dorset Press for their unfailing courtesy while it was going through the press. We are also grateful to Professor Oliver Rackham for his generous foreword. Finally we thank Janet and Susan for their forbearance in putting up with a life dominated by John Ray.

1: INTRODUCTION

The publication of John Ray's *Catalogus plantarum circa Cantabrigiam nascentium* ("A catalogue of plants growing around Cambridge") in 1660 was an important event in the history of British botany. It was the first product of Ray's interest in flowering plants which developed in Cambridge during the 1650s and led to a sequence of botanical books of national and, in some cases, international significance. Ray was greatly influenced by continental botanists such as Caspar and Jean Bauhin, who had begun to move away from the herbalist tradition of the 16th and early 17th centuries, and with his contemporaries he did much to develop the taxonomic approach which dominated 18th-century botany. The *Catalogus* also has a special place in the affections of British botanists as the first County Flora, as a systematic account of the plants of a county has come to be known. Such books are now so familiar to us that it is easy to underestimate the imaginative leap which led Ray to the idea of a critical catalogue of the plants of the county. "Few books of such compass have contained so great a store of information and learning or exerted so great an influence upon the future; no book has so evidently initiated a new era in British botany." (Raven 1950, p. 81).

The significance of the *Catalogus* is not just apparent with hindsight. In his 'Preface to the Reader' Ray hinted that he was planning further works, saying that "if you would be willing to grant us your indulgence, would be favourably inclined to our undertakings & would take these endeavours of whatever kind in good part, you would give us further confidence to undertake greater things". He also urged others to "investigate the spontaneous plants, each of his own area, more diligently so that in this way a complete *Phytologia Britannica* may finally appear from all their contributions". These hints were not lost on the more perceptive of Ray's contemporaries. John Worthington, Master of Jesus College, Cambridge, met Ray (or Wray, as he then was) in June 1661 and on his behalf forwarded a letter from him[1] and a copy of the *Catalogus* to his friend Samuel Hartlib. In his own letter Worthington summarised the contents of the book and picked up on these hints in the 'Preface' (which Ray had perhaps developed in conversation):

> "I believe he hath Thoughts to publish a *Phytologia Britannica*. He hath made good Preparations for such a Work; but he is not hasty to publish it, as he was not to publish this *Phytologia Cantabrigiensis*; he would spend more time about considering it. I wish other Writers would express the like Care, Exactness and Caution about their Books. If others in *England* that have discover'd in their particular Counties any thing remarkable, would candidly impart; such Communications might contribute to its Perfection, and to the Publick Advancement of this part of Learning .... I think there are not many that have attain'd to so great a Knowledge in this part of *Natural Philosophy*; which he is still adding to." (Worthington 1704, pp. 255–256; reprinted in modernised form by Crossley 1847, p. 334).

Ray's *Catalogus* had not only indicated what might be done in the future, it had also demonstrated his competence as a botanist and convinced his contemporaries that they should collaborate with him.

---

[1] An undated letter from John Wray which survives in Hartlib's papers (33/5/7A–8B) and is published on CD (Anon. 2002) is clearly the letter sent by Ray on this occasion; in it he apologises profusely for the delay in thanking Hartlib for his help.

The purpose of this book is to provide an English translation of the *Catalogus*. When we originally planned our work, we thought that all we would need to do would be to write or edit the introductory chapters, translate the text and add the modern scientific names of the species listed by Ray. We soon came to appreciate another of Dr Worthington's perceptive comments to Hartlib: "You will find a great deal put into a little Room." In part this compression is obtained by the frequent but easily comprehensible abbreviation of plant names and synonyms. However, there are also numerous literature abbreviations, only some of which are included in 'An Explanation of the names of the Authors cited'. Ray sometimes appears to assume that his readers have access to a well-stocked library. To the entry for *Cannabis prima sive sativa*, for example, he adds a note saying: "Lake water in which hemp has rotted is so noxious that when drunk it is a very powerful poison to those who are thirsty; on which subject see the notable story told by *Lob. in Adv.*" Those who could not turn to a copy of Pena & de L'Obel's *Stirpium adversaria nova* (1571) must have found this comment rather tantalising. In addition, the *Catalogus* also contains, as would be expected, both classical and contemporary references which the educated 17th-century reader might have taken in his stride but which require explanation today. We soon decided to try to unpack Ray's "little Room" in order to allow the modern reader to gain a greater appreciation of its contents. This we have attempted to do by footnotes to the text, supported by background information in the introductory chapters. The original text can be seen on-line on the Google Books website. We have also added translations of the two appendices to the *Catalogus*, published in 1663 and 1685. These are rare works; most copies of the *Catalogus* lack them and they have been neglected by botanists, who have consequently accepted some erroneous records of plants in the *Catalogus* which were withdrawn in the appendices.

In preparing this book, P.H.O. has translated the text and C.D.P. has taken primary responsibility for preparing the introductory chapters, adding the modern names and compiling the gazetteer and bibliography. Footnotes on matters of translation are by P.H.O.; the other footnotes have been a shared effort, with the notes written individually or in collaboration. We started work towards the end of 2008 and in aiming (unsuccessfully) to complete the text for publication in 2010, to celebrate the 350th anniversary of the *Catalogus*, we have had to work more rapidly than either Ray or Worthington would probably have thought proper.

We must explain why we decided to prepare another translation when Ewen & Prime's translation, *Ray's Flora of Cambridgeshire* (1975), though out of print, is widely available second-hand at very modest prices. Ewen & Prime translated those parts of the book "most likely to interest a field botanist or present day student of the history of botany" (p. iii). They translated the 'Preface' and 'Explanation', and under Ray's 'chosen names'[2] in the main catalogue they provided modern names, reproduced Ray's English text and translated from Latin the habitats and Ray's other observations and numbered notes. They also included versions of Ray's 'Index of places', 'Index of English names' and 'More customary headings or divisions of plants'. They largely succeeded in their aim of providing easy access to the portions of the text of interest to most field botanists – the modern names, habitats and localities of the species listed by Ray. However, it is difficult to believe that anyone interested in assessing Ray's *Catalogus* from a historical perspective could be satisfied with so partial a picture. The absence of the synonyms, in particular, eviscerates the work, depriving it of its historical context and making it impossible to understand the nature of the task confronting

---

[2] We use this term for the first name cited by Ray in his main alphabetical catalogue (see Chapter 3).

1: INTRODUCTION

Ray. The very act of excluding those portions of the book that they thought would be of no interest to the modern reader, including the chapters on the interpretation of technical terms[3] and on etymology, itself introduces so great a bias that it invalidates any attempt to produce a balanced assessment of Ray's work.

We hope that this edition of the *Catalogus* will contribute to a reassessment of Ray's life and career. Any student of Ray must be immensely indebted to Charles Raven's biography, *John Ray naturalist: his life and works* (1942, 1950, republished 1986). Raven's research into Ray's life was meticulous and reliable. He had such command over his material and argued his case so persuasively that after reading his book it is hard to believe that any other interpretation of Ray's career is possible. It is difficult not to be dazzled by his brilliance, but it is important to try to retain an independent perspective. Most successful biographers are likely to identify with their subjects to some extent, but with Raven this identification was exceptionally close (Arber 1943; Walters 1986). Ray is the hero, and "Raven's account of him, in other ways excellent, is too concerned with finding modernity in his concepts to give a balanced estimate of his intellectual position" (Cain 1999b). The only way we can approach Ray for ourselves is through his works. We hope that this translation will contribute towards an increased understanding of Ray at the start of his scientific career and that it will focus attention on the *Catalogus* as a subject for future research. It has been a fascinating book to translate and edit, and we hope that it will be equally interesting to read.

---

[3] Ewen & Prime (p. 6) say: "in this edition every effort has been made to translate the Latin, technical or otherwise, into plain English. There is no point therefore in providing a translation of what in effect has already been translated." This suggests that this chapter, which Ewen & Prime refer to as "a Terminorum" and which we here call the 'Interpretation', is the equivalent of the glossary in a modern Flora, but there is in fact much more to it than this.

2: JOHN RAY, ONCE FELLOW OF TRINITY COLLEGE

JOHN RAY, Master of Arts,
Once Fellow of Trinity-College in Cambridge,
Afterwards
A Member of the Royal Society in London;
And to both these learned bodies
An illustrious Ornament.[1]

Introduction

Canon Charles Raven's book *John Ray naturalist: his life and works* (1942, 1950) provides a full biography of Ray, and there are a number of shorter summaries of his career (e.g. Baldwin 1986; Mandelbrote in ODNB; Stearn 1973, 1981; Walters 1981). Ray's early career is poorly documented and almost all the available information is set out by Raven. As Raven himself predicted (1950, p. xiii), very few additional letters or other manuscripts by Ray have come to light since the publication of the biography. However, some authors have reinterpreted aspects of his life and work (e.g. Sloan 1972; Brooke 2000; McMahon 2000) and there have been many further studies of Cambridge and of the natural sciences in the 17th century. Feingold's (1990a) long essay on Ray's contemporary and friend Isaac Barrow, for example, provides much information which is also applicable to Ray's own time in Cambridge. Morgan's (2004) history of the University between 1546 and 1750 is less directly relevant, but it does much to illuminate the broader scene. In this chapter we have attempted to use a few studies such as these to sketch the background to Ray's work in Cambridge and to provide details of his most important collaborators during his Cambridge years. We have provided only a brief summary of his life after he left Cambridge and especially after the death of his Cambridge friend and patron Francis Willughby. We have not attempted to evaluate the *Catalogus* in the context of the broader intellectual life of the 17th century.

Ray's early life

Ray[2] was born on 29 November 1627 in the hamlet of Black Notley, Essex, the third child of the village blacksmith Roger Ray and his wife Elizabeth. He was named after his paternal grandfather. Ray's elder brother Roger had been born in 1624 but died in 1632; his elder sister Elizabeth was born in 1625 but nothing further is known of her life.

Ray's father, the village blacksmith, must have been "a man of some importance" in the village (Raven 1950, p. 7). His mother was described after her death as "a very religious and good woman, and of great use in her neighbourhood, particularly to her

---

[1] A translation of the Latin epitaph on Ray's memorial at Black Notley, first published in *A general dictionary, historical and critical* (Bernard, Birch & Lockman 1739, p. 695) and later by Boulger (1892) and Baldwin (1986).
[2] Although 17th-century writers were often inconsistent in the spelling of their surnames, Ray seems to have consistently signed himself Wray until he dropped the W in 1670 on the publication of the first book to bear his name, *Catalogus plantarum Angliae*. We follow the usual practice of referring to him as Ray even in the period when he called himself Wray.

neighbours that were lame or sick, among whom she did great good, especially in chirurgical matters" (Derham 1846, p. 37). It seems likely that one or both of the Rectors of Black Notley during Ray's boyhood, Thomas Good (died 1638) and Joseph Plume (Rector from 1638 until his expulsion by Parliament in 1643), spotted Ray's talent and encouraged him to attend the local grammar school at Braintree. This was based in a former chantry chapel in the parish church. Here Ray was certainly helped by the moderate Anglican clergyman and "diligent parochial minister" Samuel Collins, a graduate of Trinity College, Cambridge (Raven 1950, p. 15). Ray himself was admitted to Trinity on 12 May 1644 at the age of 16, a normal age at admission in that period. However, he quickly moved to Catharine Hall (later St Catharine's College) to take advantage of a scholarship for "hopeful poor scholars" which was in the gift of the Vicar of Braintree. Trinity was the largest of the 16 Cambridge colleges, with 277 members in 1641 and 440 in 1651, compared to 102 and 150 respectively at Catharine Hall, a college of middling size (Twigg 1990). Trinity was wealthy, then as now:[3] its average annual income between 1650–1651 and 1659–1660 was £7,517 and its average expenditure £7,119, whereas the annual income of Catharine Hall was only £158, apparently less than the average expenditure of £176.

Ray's story was very typical of those boys of humble origin who obtained places at Oxford or Cambridge in this period and became distinguished clerics in later life. Morgan (2004) reviewed their autobiographies or near-contemporary biographies and concluded that one feature which came through very strongly was the importance attached to education by their parents, relatively uneducated though they were themselves. We have no direct evidence that this was the case with Ray, but it is difficult to see how he could have managed to attend the grammar school unless they were sympathetic to learning. Another crucial feature was the role of networks which "provided the channels through which able boys were forwarded to the universities". For two generations the universities had been educating clergy and sending their graduates into rural parishes. East Anglia was a rich area and particularly well provided with learned clergy who would have links with college tutors (as well as local gentry who might act as patrons to promising boys). Men like Collins who valued their own experience at University would naturally encourage the brightest of their local boys to follow the same path. Provision for scholarships for pupils from specific schools or specific areas had been made by wealthy patrons who might in former times have left their money to the church. Ray benefited at Catharine Hall from a trust established in 1631 by Thomas Hobbes of Gray's Inn, who left cottages and land in Braintree to provide revenues for scholarships at Cambridge (Raven 1950, p. 19).

The University of Cambridge in the mid 17th century

Civil War and Commonwealth: Cambridge in the 1640s and 1650s

Ray entered Cambridge towards the end of a period of sustained growth in the size of the University, as Morgan (2004) relates. The increase had started in Elizabethan times and continued for the first four decades of the 17th century. Although precise figures are impossible to calculate, there are thought to have been fewer than 1,300 resident members in 1564, nearly 2,000 by the 1590s and some 3,000 by 1621 (Morgan 2004). The 1640s was a catastrophic decade, as the effects of the Civil War were felt, the number of students fell sharply and the University was subject to the most drastic of

---

[3] The figures for income and expenditure are calculated from those tabulated by Twigg (1990, Appendix 5); figures are not available for 1654–1655 (Trinity) and 1656–1657 (Catharine Hall).

the purges it suffered in the Civil War and its aftermath. In 1644–1645 the Earl of Manchester enforced Parliament's Solemn League and Covenant and expelled those who would not accept the presbyterian mode of church government that it endorsed. (This was the price Parliament paid for the support of the Scots.) Many College Heads and half the Fellows lost their posts. At Trinity the Master and 47 of the 60 Fellows were ejected and replaced by more politically acceptable substitutes "intruded" into the college. This was followed by a further purge in 1650–1651 when the government demanded assent to the Engagement Oath, which recognised a Commonwealth with no King and no House of Lords. Although this was enforced less rigorously, it cost Manchester his job as Chancellor and Trinity lost a further six Fellows (Twigg 1990).[4]

Rather surprisingly, the Engagement heralded a decade of relative peace in the University. Although there was conflict there which mirrored that in the country as a whole, there was an unexpected degree of coexistence between Parliamentarians and surviving Royalists. Friendships were maintained across political and religious divides and there were some remarkable examples of genuine tolerance for opposing views. One of the most notable of these was the defence of Isaac Barrow, a Royalist Anglican, by the Presbyterian Master of Trinity, Thomas Hill, who told those Fellows who were pressing for his expulsion that "Barrow is a better Man than any of us" (Feingold 1990a). Feingold suggests that the religious and political pluralism in Cambridge might have been a "silent coalition of interest" between Presbyterians and Royalists in response to the growing influence of the Independents and others who were campaigning for wholesale changes to the universities in the 1650s or (in the case of extremists such as the Quakers) for their abolition.

Education and the curriculum

The University syllabus was remarkably flexible. The University statutes, promulgated in 1570, "were deliberately drawn up to avoid putting tutors and students in the kinds of straightjackets which afflict their modern successors ... In the senses in which we use the words today ... there was no syllabus, and the curriculum was what the tutors made of it" (Morgan 2004). Classical studies had become increasingly prominent in the 16th century and mathematics and astronomy were becoming increasingly so in the 17th. Tutors would certainly "indulge their own interests beyond the bounds of the formal curriculum" and Morgan suggests that "sometimes the distinctions between educating pupil-students and the pursuit of a tutor's own intellectual interests must have been blurred". Although in the mediaeval period there was a separation in the content of the Bachelor of Arts (B.A.) and Master of Arts (M.A.) degrees, by the start of the 17th century the entire field of the liberal arts and philosophy would be covered during

---

[4] Botanical historians, whose narratives are centred on Ray, have given much more prominence to the Fellows who, like Ray, suffered as a result of the Act of Uniformity (1662) than to the victims of the earlier purges. Raven (1950, p. 22) recognised the scale of the "wholesale expulsions" of 1644–1645 but argued that, although "no doubt the evictions were drastic and in some cases unjust", "their result, as soon as the newcomers settled in, was definite gain"; he evidently did not number the evicted Fellows amongst the dedicatees of his book, those "who like John Ray sacrificed security & careers for conscience' sake". Raven's judgement was based on the view then current that the University in the years preceding the Civil War was devoted to a "pitifully irrelevant" curriculum (Raven 1950, p. 23), a picture which later historians (especially Feingold 1990a, 1997) have done much to revise. Other accounts have simplified Raven's story to the extent that even as great a scholar as Stearn (1986) could contrast the "tolerant parliamentarian control" of Oxford with "the pernicious Act of Uniformity".

the four years in which the student would study for the B.A., leaving the three years of the M.A. as a period of independent study (Feingold 1990a).

Undergraduates would be taught in University lectures and some colleges provided additional lectures, but they would also be taught by their tutor in his chamber. Tutors had long been a key part of the system, living in close proximity to their students and supervising their studies, and their importance increased in the 16th century as the importance of the colleges grew. The first half of the 17th century was a time when relationships between students and tutors were particularly intimate (Morgan 2004). Most students were tutored by junior Fellows, to allow the senior Fellows time for administrative duties, lecturing, examining and their own research (Feingold 1990a). Tutors would take charge of their students' religious education (it is difficult to exaggerate the importance attached to religion in the mid 17th century), supervise their behaviour, direct their studies and help them manage their financial affairs. Students would arrive at the University with a good knowledge of Latin from school, but the first years of study would be devoted to improving their classical languages, following a course of study laid down by the tutor and reporting daily to him on their progress. Thereafter as they came to study more widely they might be referred to other members of the college for more specialist tuition.

Undergraduates might enter the University as sizars, subsizars, pensioners or fellow-commoners (Westfall 1980). There was a considerable difference in status between the sizars, who were students with some of the duties of college servants, and fellow-commoners, who were students with some of the privileges of Fellows. Sizars were supported by the college; they received their board and lodgings but in exchange they had to serve the Master and senior Fellows, acting as their valets and as waiters at meals in the College Hall. This was the way in which the sons of poor parents entered the University. Subsizars had the same lowly status but did not receive college support. They paid lower fees than pensioners but had to earn the money they needed for their teaching, board and lodgings by making personal arrangements to serve the Fellows or the wealthier undergraduates. Most sizars and subsizars proceeded to a degree, although Westfall (1980) dismisses them as "a plodding group, narrowly vocational in outlook, lower-class youths grimly intent on ecclesiastical preferment". Pensioners paid fees and also had to pay for their accommodation and food. The difference between sizars and pensioners was not as great at Trinity as in some other colleges: they wore the same design of gown and were allowed to share accommodation. Westfall (1980) characterises pensioners as the "merely affluent", in contrast to the fellow-commoners who came from rich families, paid higher fees and were entitled to special privileges, such as eating at high table. Only a minority of fellow-commoners bothered to graduate. Despite these differences of rank and the gowns which distinguished fellow-commoners, all students were educated together. Modern authors make much of the humiliations of life as a subsizar or sizar, but whether these would have had much of an impact on boys brought up in a more hierarchical world is uncertain – except perhaps in cases such as that of Isaac Newton, who came from a wealthy family but entered Trinity as a subsizar in 1661, apparently because his widowed mother refused to pay for his education (Westfall 1980).

One reason for the increasing intimacy between tutors and their students was the growth of the aristocratic element in the University. The growth of numbers in the second half of the 16th century and the first half of the 17th had been caused not only by an increase in the number of students from humble backgrounds, but also by a growing tendency for aristocratic families to send their sons to the University. Wealthy and influential parents would insist that the tutor should look after their son's health and moral welfare as well as his education. It was not unusual for a tutor to travel with one or more of his pupils or to stay with them in the parental home, sometimes by choice in the vacations and sometimes for safety when Cambridge was ravaged by epidemics of disease. The influx of the aristocracy also led to an increasingly diverse curriculum, as men who had no intention of taking a degree saw no reason to restrict their studies to the subjects which were necessary to obtain one.

Ray and his colleagues at Trinity College

Ray spent over two years at Catharine Hall before returning to Trinity on 21 November 1646 as a subsizar and a pupil of James Duport. He was "afterwards much pleased" that he had made the move, "because in *Katherine-Hall* they chiefly addicted themselves to Disputations;[5] but in *Trinity* the politer Arts and Sciences were principally minded and cultivated" (Derham 1760, p. 2). A fellow pupil was Isaac Barrow (1630–1677), who had been admitted to the college as a subsizar in February 1646. He too had come from another college, in his case Peterhouse where he had been a pensioner since 1643 but had left when his uncle (who was also his tutor) was expelled from his fellowship (Walker 1912). Ray and Barrow became good friends and Barrow was later to become first Lucasian Professor of Mathematics, Master of Trinity and Vice-Chancellor of the University. Ray took his B.A. degree in 1648 and was elected a Minor Fellow of Trinity on 8 September 1649. He became a Major Fellow when he took his M.A. in 1651.[6] He was then elected to a series of college posts, documented in the college's Admissions Books (1560–1759, 1645–1659) and held for a year. The first of these were Lectureships in Greek Language and Greek Grammar (1 October 1651), Mathematics (1 October 1653), Humanities (2 October 1655) and Greek Language (1 October 1656). He then became Head Lecturer (*Praelector primarius*, October 1657) and Junior Dean (2 October 1658) and was twice elected Steward (26 December 1659 and 16 December 1660). He was a tutor from 1653 to 1660. He was ordained on 23 December 1660, after the restoration of the episcopacy.

Ray's tutor, James Duport

James Duport (1606–1679) was one of the seven surviving children of John Duport (matriculated 1564, died 1617 or 1618), who was Master of Jesus College, Cambridge, from 1590 and Director of the Second Cambridge Company charged by James I with translating the Apocrypha for the King James Bible, published in 1611. James was elected to a fellowship at Trinity in 1627, received his M.A. in 1630 and in 1639 was elected as Regius Professor of Greek. He was a popular lecturer who would not normally be expected to take students at this stage of his career (and in fact the statutes of the professorship did not allow him to do so, but on appointment he had negotiated a dispensation to permit him to continue to act as a college tutor). Despite his seniority, he was tutor to some 180 students, 20% of the entire intake of the college, between 1645

---

[5] This of course refers to the traditional, formal academic debates rather than to intracollegiate squabbling.

[6] Elections to Major Fellowships followed as a matter of course when a Minor Fellow obtained his M.A. If elections to Major Fellowships were recorded at Trinity during the Commonwealth, the records have not survived, with a few exceptions in 1659 (Innes 1941). However, Ray's receipt of a Major Fellowship in 1651 is confirmed by his entry into the Upper Buttery book in March of that year (see below) and by the increase in his stipend from 10s. to 13s. 4d. recorded in the Senior Bursar's book in the following quarter.

and 1659 (Feingold 1990a). The demand for his services arose from his position as the only remaining tutor at Trinity with overt Royalist and Anglican sympathies. Quite why he took Ray is unclear, but he was presumably unable to resist such a promising student. We know that he was prepared to provide Barrow with free lodgings and waive his tuition fees when his friends were no longer able to support him financially (ODNB).

Duport's "Rules to be observed by young Pupils and Schollers in the University" survive in a contemporary notebook and have been published by Trevelyan (1943).[7] The first section is headed *De Pietate, et cultu Dei, et Religionis artibus* ("About Piety, the worship of God and the practice of Religion"). It begins "Be diligent and constant at Chappell every morning", and later he tells the student to "Come duely to prayers in your Tutors chamber at Night, & that as soon as the Bell hath tolde" and then "Goe immmediately from your Tutors Chamber at night, to your prayers, and retire to your devotions, and reading the scriptures".

Duport was known even as an undergraduate for his knowledge of languages, especially Greek. Feingold (1990a) describes his lifelong devotion to the classics and in particular his passion for Greek poetry. His translation of the biblical Book of Job into Greek verse, *Threnothriambos, sive, Liber Job Graeco carmine redditus* (1637), became an "immediate success" and "established him as a scholar and poet" (Feingold 1990a and in ODNB). He later produced further biblical paraphrases and translated the Book of Common Prayer into Greek. He told his pupils, "Never think you can be a Scholler indeed or well skilled in humane learning, without some competent knowledge of the Greeke tongue." He suggests that "In the course of your studies, use to reade, among the antient classick Authors, the best, & of the best note as Homer, Aristotle, Virgill, Tully [Cicero], Seneca, Plutarch, and the like. ... In reading of heathen Poets, especially Juvenal, & Martiall, suck the hony out of the flower, and passe by the weedes."

A tutor's rules cannot, of course, be taken as an indication of how undergraduates actually behaved. Godliness and learning might not have been as important to many undergraduates as they were to their parents and tutors; it would be surprising if all Duport's students, even in that religious age, were impressed by the need to "Thinke the time of your being in the university an harvest, thinke then you must be as spirituall ants, gathering and laying up food for your soules".

The approved recreations of the period were walking, riding (colleges had their own stables), real tennis, fencing and bowls (Twigg 1996). Duport's advice to "walke often in the fields, and to walke alone, because that will put good thoughts into you, and make you retire into your self, and commune with your owne hearts" is perhaps relevant to Ray's account of how he got interested in botany in the 'Preface' to the *Catalogus*. Similarly, Duport suggests that, rather than swimming, "I ... had rather you should ride out sometimes, and take the aire, so it be moderately, seasonably, and seldome". However, he recommends his charges to "Use Tennis sparingly" and favours bowls "or some like gentle, and handsome and harmlesse exercise". Unauthorised pursuits included hunting, which was prohibited ("without noticeable effect" according to Twigg), and football, "fitter for Clownes[8] then for Schollers" in Duport's view but

---

[7] Although Trevelyan says that he printed only "a few of the more interesting of these Rules", comparison of his paper with the original manuscript (Trinity College O.10A.33) shows that it is a more or less complete transcript, though it omits one rule (presumably by accident) and two Greek phrases (presumably by design). However, Trevelyan does not reproduce the numbering of the rules and so his publication obscures the fact that a page of the notebook has been excised and 27 rules are therefore missing.

[8] At this period a "clowne" was a countryman or peasant (see footnote 732 in the main catalogue).

popular nevertheless. Attendance at bull- and bear-baitings and cockfights was discouraged and at Trinity games of cards and dice were forbidden by the college statutes. Drinking was a serious problem and often led to violence (Twigg 1996).

Although one imagines that Duport's rules envisage a more quiet and godly life than that practised by many of his students, all we know of Ray in later life suggests that he would have been an ideal pupil. His natural inclinations were presumably reinforced by his tutor and by kindred spirits at Trinity. Certainly his statement in the 'Preface' to the *Catalogus* that he is not writing for those who take delight in the tennis court, drinking or gaming is very much a reflection of Duport's views and, incidentally, suggests that there were many in the University who ignored them. (The surviving tutor's rules have nothing to say about "amassing money or procuring men's favour"; perhaps Ray's text here provides a clue to the subject of some of the missing rules.) Duport's suggestion that you should "be not ashamed to profess your ignorance, always remember that one property of a Scholler is to be ζητητικον [zeteticon, "inquiring"]" brings to mind Ray's remarks in the main catalogue such as the one which concludes his long note to *Rapum sylvestre* [495], "We dare not pronounce anything rashly about these things which we have not yet investigated sufficiently."

One theme that runs through Duport's rules is the need to make good use of time. The student is urged to "Keep your set hours in your Study and if at any time you are broken off, make it up at another" and in particular to "be sure you make much use of the morning, for if you loose that, you loose the golden, & pretious time of the day". In reading, students should "observe the most remarkable passages, & note them with a black-lead pen ... transcribe not those sayings and glories at length, for that is tedious, but make short reference to book & page". Duport even discourages chess as "an ingenious play, yet too tedious and time-devouring". Above all, "Goe not a gadding and gossiping from Chamber to Chamber, for that is no recreation, but meere idlenesse and losse of time." The need to make the most of time is a lesson which Ray, with his remarkable productivity in later life, must surely have taken to heart. A phrase in the prayer he wrote after the death of Francis Willughby reinforces this impression: "Help us therefore to work, while we have the light and the day, because the night comes, in which no man can work." (Lankester 1846, pp. 57–58).

Like many of his fellow tutors, Duport arranged for others in the college to teach advanced mathematics and science to his students, and he certainly had reservations about the potential impact in the growth of these subjects on the liberal arts he loved. He has been portrayed as a diehard conservative, but Feingold (1990a) discusses this reputation at some length and concludes that much of it is based on misunderstandings. A tutor who nurtured the talents of Ray and Barrow and regarded them as "The chief of all his Pupils ... to whom he esteemed none of the rest comparable" (Derham 1760, p. 4) cannot have been a blinkered opponent of the new learning. He had some modern scientific works in his library, most of them published or purchased after 1660 according to Feingold (1990a), who suggests that "it would not be far-fetched to muse that Duport allowed his brilliant students – Barrow, Ray and Willughby – to lure him deeper into the new science".

John Nidd, Francis Willughby and Peter Courthope

In his essay on Isaac Barrow, Feingold (1990a) argues that Barrow, Ray, Walter Needham and John Nidd formed "the nucleus of a large and highly active community at Trinity whose members – despite individual tastes – shared many of their colleagues' interests and who met regularly for the purpose of carrying out experiments and scientific discussions". The members of this community, who participated in a "myriad chemical, anatomical, and botanical studies", are discussed by both Raven (1950) and Feingold. John Nidd is the one who features in the *Catalogus*. Nidd was almost certainly the son of a former Fellow of Trinity, Gervase Nidd, a Doctor of Divinity who was Rector of Sundridge, Kent, when he died in 1629. John entered Trinity as a sizar in 1640 and as an undergraduate served as Under Library-Keeper between 1642 and 1645 (Gaskell 1980, p. 229). As such he assisted the full-time Library-Keeper; he was paid wages of £4 per annum and was given board and lodgings (the latter in the Library-Keeper's chamber). The duties were laid down as sweeping the library at least twice a week, dusting the books and furniture and cleaning the windows. However, the 17th-century Under Library-Keepers were "as a group not undistinguished in later life" (Gaskell 1980, p. 78), suggesting that they were selected for more than their ability to dust furniture. The library was not normally available to undergraduates and so they would have had privileged access to the books in return for carrying out their humdrum duties.

Nidd obtained his B.A. in 1646 and his M.A. in 1649; he was a Fellow of Trinity from 1647 until his early death in July 1659.[9] He was an "avid chemist" (Feingold 1990a), but most of the (few) references to him in Ray's books and letters refer to his interests in natural history. In his funeral sermon[10] Ray praised "his excellent learning; particularly his great and exact Skill in the Universal History of Nature, and the Works of GOD". Years later he described the copulation of frogs from "a couple kept on purpose in a Vessel of Water by my Learned and Worthy Friend Mr. *John Nid*, Fellow of *Trinity* College, long since deceased" (Ray 1692, part 2, p. 84). Ray was present when four birds were dissected in Nidd's chamber, birds "whose cases hung up in the cupboards over Mr Nid's portal" (Raven 1950, pp. 47–48); the room must have been an unusual one for a Fellow of Trinity. In another affectionate reminiscence from his Cambridge days, Ray (1686, p. 437) described Nidd as "our special friend & an illustrious botanist". Nidd's library was an important source of material for the *Catalogus* and his role in the authorship is discussed in Chapter 3. Ray included an elegant tribute to his friend in the 'Preface' to the *Catalogus*.

The other two members of Trinity thanked in the 'Preface' are Francis Willughby and Peter Courthope, both fellow-commoners and pupils of James Duport. Willughby[11] (1635–1672), who entered the college in 1653, came from a Royalist family of Middleton Hall, Warwickshire, and Wollaton Hall, Nottinghamshire; he had spent his childhood at Middleton (Welch 1977). Unlike many men of his background, he took his B.A. in 1656 and his M.A. in 1659; Ray described him as "addicted to study" (ODNB). His interests included mathematics as well as the natural sciences; a surviving commonplace book from 1658–1659 refers to chemical experiments, including "Chy. experi. of Mr. Wrays" (Welch 1972). The combination of Willughby's

---

[9] Nidd's name appears consistently in the Upper Buttery book 1658–1660/1 (Trinity College archives) from the start of this book in November 1658 until the week ending 29 July 1659, suggesting that he died in late July. Throughout this period he is marked as 'no co' (not receiving commons, i.e. not dining communally) but with expenses against his name for sizings (food and drink, usually bread and beer, over and above the regular allowance of commons, purchased from the buttery). He was presumably resident but too ill to leave his chamber. Ray warned Courthope in a letter of 3 January 1659 that "if you stay long in the country ... you may never see him in this world any more" (Gunther 1928). His will was proved on 6 September 1659 (Leedham-Green 1986).
[10] The text was available to Derham and summarised by him (1760, pp. 9–11) but has since been lost.
[11] Willughby consistently used this spelling of his name and so we have followed his practice. His son Thomas used both this and Willoughby and the family name later stabilised as the latter (Welch 1972).

gentlemanly status, "remarkable beauty of feature and charm of expression" (Raven 1950, p. 51) and scholarly interests clearly made a great impression on the fellowship. Barrow dedicated his edition of Euclid's *Elements* (1655) to Willughby, a book which remained a bestseller for a century (Feingold 1990a), and Duport dedicated *Homeri gnomologia* (1660), his best known book, to him and three other well-born pupils. The only specific reference to him in the *Catalogus* after the 'Preface' is in the long note to *Rapum sylvestre* in the main catalogue [495], where his observations on the parasites of caterpillars are described. He was one of the early members of the Royal Society, elected in 1661. Ray's friendship with Willughby was to be his lifeline after he left Cambridge in 1662.

Peter Courthope (1639–1725) was born in Cranbrook, Kent, but his grandfather, also Peter, had bought a large estate in Sussex, Danny Park, by the time that the younger Peter entered Trinity in 1655. Ray visited Danny and his letters to Courthope (which survive from 1658 onwards) not only show that they enjoyed a very easy relationship but also suggest that Ray was on friendly terms with the whole family. Courthope left Cambridge before taking a degree; he inherited Danny in 1657 but his payments are recorded in the Trinity Upper Buttery book on occasions until December 1660. Although there is no evidence that he pursued any independent studies of natural history, he and Ray remained in touch. Ray dedicated his zoological work *Synopsis methodica animalium quadrupedum et serpentini generis* (1693) to Courthope and one of his former pupils, Timothy Burrell, "in a letter which shows that the warmth of his affection and the beauty of his Latin had not been impaired by time" (Raven 1950, p. 372). Courthope sent his youngest son, Peter (born 1676), to Trinity in 1694 but he died there the following year; his grave in the college chapel is marked by a fine tombstone.

A chance survival of a facetious oration by Isaac Barrow gives us an insight into Ray and his colleagues at Trinity in the 1650s. The text of Barrow's Latin speech, which was not intended for publication, survived in his papers and was published after his death (Barrow 1687, pp. 123–129; reprinted by Napier 1859, pp. 35–47). It is undated but Whewell (1859) provides convincing arguments for dating it 1654. The following section (given here in our translation) appears towards the end of the speech, after the more serious subjects have been covered.

> "For when, pray, since the foundation of the University, has a bloody curiosity furiously brought about the deaths and cutting-up of so many dogs, fish and birds, through which the nature and functions of the various parts in animals should be made known to you? O most innocent cruelty and easily excusable savagery! What may I say about the history of plants, explored zealously even by your freshmen, who have scarcely struck their roots in the greenswards of the Muses before they are able to distinguish the characters and recount the names of every one of the plants that grow either in the free tracts of the fields or within the fences of gardens, a good part of which *Dioscorides* himself would not know if he were once again among the living?"[12]

Although we know that Ray took part in the dissection of birds and frogs, the references to dissection probably refer mainly to the work of another of Ray's Trinity friends, Walter Needham. Needham left Trinity to become a Fellow of Queens' College in 1655 and published a book based on his dissections, *Disquisitio anatomica de formato foetu*, in 1667 (see Raven 1950, p. 46).

---

[12] It is instructive to see how this oration leads Raven (1950, pp. 27–28) to conclude that "a wider range of subjects than is usually recognised was being studied" before retreating from this view in the face of the evidence to the contrary.

Another insight into Ray's outlook at this time comes from a much more serious source. Although Ray's book *The wisdom of God* was not published until 1691, it incorporates some devotional talks ("common places") that he gave while he was at Trinity over 30 years earlier. The full title of the first issue, which was not retained in the second issue or in later editions (Keynes 1976), is *The wisdom of God manifested in the works of the creation: being the substance of some common places delivered in the chappel of Trinity-College, in Cambridge.* Some of the text written at Trinity is easy to detect. If, as seems likely, the following passage resembles one delivered at Trinity, it constitutes a remarkable attempt to recruit students to the study of natural science (Ray 1691, pp. 124–127).

"Let it not suffice us to be Book-learned, to read what others have written, and to take upon trust more Falshood than Truth: but let us our selves examine things as we have opportunity, and converse with Nature as well as with Books. ... I am sorry to see so little Account made of real *Experimental Philosophy* in this University, and that those ingenious Sciences of the *Mathematicks*, are so much neglected by us: and therefore do earnestly exhort those that are young, especially Gentlemen, to set upon these Studies, and take some pains in them ... .

"But I would not have any man cross his natural Genius or Inclinations, or undertake such methods of Study, as his Parts are not fitted to ... but those who do abound with leisure, or who have a natural Propension and *Genius* inclining them thereto, or those who by the reason of their Strength and Greatness of their Parts, are able to compass and comprehend the whole Latitude of Learning.

"Neither yet need those who are designed to Divinity it self, fear to look into these Studies, or think they will engross their whole time ... ."

Even more remarkable, perhaps, is Ray's suggestion (p. 124) that "It may be ... part of our business and employment in Eternity to contemplate the Works of God".

Ray's botanical studies, *c.* 1651–1662

Ray starts to study Cambridge botany, c. 1651

Ray tells us in his 'Preface' to the *Catalogus* that "after a full period of six years in which we have devoted our efforts to this subject, we have undertaken this little work in earnest, in the preparation and embellishment of which we have spent a period of almost three years". He must therefore have started to study plants in 1650 or 1651. The nature of the illness which necessitated "an immediate rest from more serious studies" and first led him to notice "the variety & elegance of plants" is unknown, but it was later attributed to his "intense application to his studies" (Bernard, Birch & Lockman 1739, p. 692). There is no indication that it had any lasting medical effects (Raven 1950, pp. 61–62).

McMahon (2000) draws attention to the precise information which Ray presents on the timing of the start of his interest in botany. She points out that his readership would have been aware that he started his botanical studies at just the period when the Parliamentary visitors were attempting to impose the Oath of Engagement. She argues that Ray "left the university and returned only after selective enforcement had been abandoned". In an endnote to this sentence she says that "Ray was appointed Greek lecturer and did not return to the university until 1 October 1651 (Raven [1942], p. xvi). Although the Engagement was not repealed until 13 January 1653/4, it was effectively discontinued in the autumn of 1651." McMahon's hypothesis seems inherently unlikely, as it suggests that Ray began to become interested in the Cambridgeshire flora at a time when he had absented himself from the University. Her reference to Raven's biography is to the chronology, which notes (on p. xv) Ray's illness in 1650 and his appointment as Greek Lecturer on 1 October 1651 but does not mention any period of absence. There is in fact evidence that he was present in Cambridge in the summer of 1651. The Upper Buttery book 1649–1654 survives in the Trinity College archives and the entries suggest that Ray was consistently resident from the week ending 22 March 1651 (when his name first appears, bracketed with five others who also received their M.A. in 1651) until the last week covered by the book, 15 September 1654, except for brief periods of absence, the longest being in July 1651 (2 weeks), July 1652 (2 weeks) and August 1653 (3–4 weeks).[13] Although the general thesis in McMahon's paper is convincing (she maintains that Ray was a loyal Anglican rather than a puritan), we cannot accept her interpretation of this particular episode in his career.

The first reference to the preparation of the *Catalogus* apparently comes in a letter from Samuel Hartlib to Robert Boyle sent on 2 February 1658 (1657 old style):

"Mr. Vice-chancellor of *Cambridge* (Dr. *Worthington*) pleases me not a little by writing in his last.—

"We have divers fellows of colleges, who have made excellent progresses in anatomy of several kind of creatures, and have some observations peculiar. They have also much travelled in botanicks, and have got together many hundreds of plants in several gardens here. Some, that are rare, and for which they have been sent to from those, that profess great knowledge. They intend to publish a Phytologia of such herbs as are within ten miles of *Cambridge*, which they have often travelled, and found some not mentioned in other herbals. And for the names of herbs in the Phytologia Britannica, and other herbals, they have found many ertors [errors], which have been propagated from one book to another; and they judge it a good service to restore them to their 'own true names.'" (Boyle 1744, p. 272).

The rather obscure reference to the plants "for which they have been sent to from those, that profess great knowledge" may relate to a note dated 1657 in Hartlib's *Ephemerides*, apparently also based on information from Worthington, "Botanical Garden is already in such request that Lord Lambert sent from some plants out of it" (Webster 2002, p. 151). "Lord" Lambert was the brilliant parliamentarian soldier John Lambert (1619–1684), who was probably educated at Trinity College and was extremely influential in military and government circles between 1653 and 1657; in June 1657 he was forced to resign his offices by Cromwell and retreated to Wimbledon House, where for a time he devoted his energies to his garden (ODNB).

---

[13] Ray's expenditure before March 1651 would have been recorded in the Lower Buttery books but unfortunately these do not survive for this period. The Upper Buttery books have been under-used as a scholarly resource and there is no published guide to their interpretation; we are deeply indebted to Adam Green for generously sharing with us the fruits of his current research into them. Although aspects of their interpretation are still uncertain, the evidence for Ray's residence in 1651 seems clear. A college exit and redit book with entries for this period survives; in theory it too should allow us to identify Ray's absences in term but it was clearly only maintained half-heartedly. (There are frequent cases of exits with no redit and *vice versa*, and some of two successive exits with no intervening redit.) The only absences recorded against Ray's name are on pages which, where the dates include years, record the activities of Fellows in 1648, 1649 and 1650. There are records for Ray of an exit on March 28 with a redit in May (the exact day being difficult to read) and a further exit on August 14; no years are given and the record is clearly too fragmentary to be of any use in tracing his movements.

Studies of plants in gardens

There is little that can be said about Ray's fieldwork for the *Catalogus* other than that which is apparent from the text itself; the sites he mentions are listed in the gazetteer and discussed in Chapter 3. However, he also cultivated plants in a small garden at Trinity. A letter of 9 November 1658 to Courthope in Sussex thanks him for his "paines & care in searching out & sending the small plant growing on yr downes. I received the boxe some-time since, & immediately committed the severall rootes to ye ground. They seemed to me more slender & jejune than wn I duggd them up with you, & I wonder much that you had the good luck to find yem, the tops being quited withered & dryed up" (Gunther 1934; Thompson 1974). Some scraps of evidence about this garden have been pieced together by Raven (1950, pp. 108–110). Most of these are derived from *Historia plantarum* and relate to foreign species, but there are observations in the Cambridge catalogue based on the appearance of two wild plants taken into cultivation, *Plantago coronopus* [152] and *Pulsatilla vulgaris* [475]. When describing the reaction of cats to catmint [363] in the main catalogue, Ray implies that he often transplanted plants from the wild into gardens: "so far we have not been able to observe anything similar in any other plant transplanted from the fields into a garden: for all those that we have transplanted have been untouched by cats, with this one plant excepted". He also describes his failure to grow the hemiparasitic *Melampyrum pratense* [360]. A reference to *Geum × intermedium* in the 1663 appendix [A10] suggests that he brought it into the garden from "*fields somewhere about the Town*", but in retrospect it seems possible that this hybrid arose in the garden, which would explain why Ray failed to remember where it came from. Ray drew up a catalogue of the plants growing in Cambridge gardens which survived in his papers after his death and is cited in 1739 (almost certainly from information provided by Samuel Dale) as *Catalogus Plantarum non domesticarum, quæ aluntur Cantabrigiæ, in hortis Academicorum & Oppidanorum* (Bernard, Birch & Lockman 1739, p. 692). When he sent Willughby the presentation copy of the Cambridge catalogue in February 1660, Ray told him that, in addition to compiling a Phytologia Britannica, "My second design is to make another catalogue, which I will call 'Horti Angliæ'. I intend to write to all the noted gardens, to procure a catalogue of each ... Into this catalogue I shall not admit any that grow wild in England, lest it swell too big." (Lankester 1848, p. 2). There is no evidence that Ray pursued this project.

Fieldwork outside Cambridgeshire

Although most of Ray's botanical fieldwork before 1660 had taken place in Cambridgeshire, he had been on one long botanical tour by the time the *Catalogus* was written. In 1658 he visited Northamptonshire, Warwickshire and Derbyshire in August, then going on to Anglesey, Snowdon, Cader Idris and back to Cambridge *via* Shrewsbury, Worcester and Gloucester, arriving home by mid September (Derham 1760). If he had not noticed them previously, he must have been struck on this tour by the presence, "very frequently and almost everywhere else in England", of plants which he had not recorded in Cambridgeshire, some of which he listed in the 'Preface'.

Samuel Corbyn's records

The Oxford botanist G.C. Druce acquired some manuscripts in Samuel Corbyn's handwriting from another botanist, W. Pamplin, when the latter died in 1899. These included a list entitled "A Catalogue of Plants, First those which growe wild with us about Cambridge, except those mentioned in Text. 20 May, 1657." Druce (1912) published this list (with some obvious errors of transcription and interpretation), but not the list that follows, of plants "such as we have in our gardens and can spare". Corbyn's list has not hitherto been associated with Ray's *Catalogus*, and indeed Druce noted that Ray did not mention him, remarking: "Whether religious differences, which then ran high, was the cause can only be surmised." Corbyn was admitted to Trinity as a pensioner in 1648 and took his M.A. in 1655; he was Chaplain of Trinity from 1655 to 1662 (Venn & Venn 1922–1927; Twigg 1990).[14] It seems inconceivable that Ray, who gave devotional "common places" in the college chapel at this time and was selected to preach more than one funeral sermon, would have excluded a reference to a college chaplain because of religious differences. We have already noted his friendship with the Anglican and Royalist Barrow. Corbyn was on the other side of the religious divide and, like Ray, lost his place in Trinity as a result of the Act of Uniformity; he was a dissenter who in 1672 was licensed to lead a congregation near Green Street, Cambridge (Twigg 1990). In November 1658 Ray wrote to Peter Courthope with news from Cambridge, telling him that "Mr Corbyn hath recovered his ague again, & truly it's but fitting he should have some distemper or other alwaies to keep him doing, & the pipkin[15] going" (Gunther 1934; Thompson 1974), which indicates that he was prepared to joke about the chaplain's idiosyncrasies but scarcely suggests great enmity. In fact, it seems probable that Corbyn was one of those who was associated in at least a minor way with the *Catalogus* and that he had seen the plants that he listed when botanising with Ray and his colleagues or even that some records were supplied by Ray. Most of the species he listed from Cambridgeshire also appear in the *Catalogus*. The only exception is *Anagallis lutea nemorum* [*Lysimachia nemorum*], which is listed by Ray in the 'Index of English names' but not in the main catalogue. The presence of three species in the Corbyn list is particularly significant. Corbyn's "Gnaphalium montanum album Lobelii G[erarde p.] 640, a false figure both in Gerard and Parkinson" is comparable to Ray's "The picture of this plant that is found both in Gerarde and in Parkinson is defective & bad", our translation from his entry for the current *Antennaria dioica* [233]. The other two plants are species for which the *Catalogus* entry constitutes the first published British record. The first is *Tephroseris integrifolia* (*Senecio integrifolius*), listed by both Corbyn and Ray [302] (though Druce erroneously identifies Corbyn's plant as *Senecio paludosus*). The second is *Melampyrum cristatum* [359], listed by Ray as *Melampyrum cristatum flore purpureo* J.B. "In Madingley and Kingston woods, and almost in all woods in this County plentifully ... whence we cannot but wonder that it should not be described or figured by Gerard or Parkinson" and by Corbyn as "Melampyrum cristatum wild in our woods, it is not described in Gerard, Parkinson or Joan. Bauhinum". Although there is a good picture of *Melampyrum cristatum* in volume 3 of Bauhin & Cherler's *Historia plantarum* (1651), *Melampyrum* is omitted in error from the index and this may explain the failure of the Trinity botanists to notice it until after the compilation of Corbyn's list. (*Melampyrum* has been added by hand to the index in the copy in Trinity College library[16]). We can conclude that the relationship of Corbyn's list to the *Catalogus* must be closer than has hitherto been recognised.

---

[14] These dates follow Venn & Venn and differ from those of Ball & Venn (1913), who give 1656 (p. xiv) and 1655 (p. 404) as the start of his chaplaincy and 1660 as the year in which it terminated.
[15] "A small pot for cooking, possibly here for brewing remedies" (Thompson 1974).
[16] Trinity College library Q.18.7.

2: JOHN RAY

The Act of Uniformity, 1662: Ray leaves Cambridge

Ray's decision to leave the University of Cambridge in 1662 was the major turning point of his adult life. He refused to subscribe to the Act of Uniformity (1662), a statute which required clergy to be ordained by a bishop, to swear loyalty to the King, to follow the established order of the church and to declare that the Solemn League and Covenant, an oath of loyalty to the Parliamentary regime (see above), was an unlawful oath. The last of these provisions appears to be the one to which Ray objected. He appears to have had a general reluctance to take oaths and he refused to make the declaration required of him as a clergyman and as a senior member of the University to this effect. The deadline for subscribing to the Act was 24 August 1662, St Bartholemew's Day, and by refusing to take the oath Ray lost his college position.

As McMahon (2000) points out, Ray never left a considered account of his position. Those of his contemporaries who refused to take the oath were largely the more extreme protestants or puritans, a group with which there is no other evidence to associate him (McMahon 2000). The explanation of his position in an account by Samuel Dale published immediately after his death is as follows:

> "upon the 24th of *August* 1662 he quitted his Fellowship aforesaid, because he could not in the Oath of Abjuration Swear that he did believe it was binding to others" (Anon. 1707, pp. 497–498; for the author see Raven 1950, p. xiii).

Presumably "he did not believe" was intended, and this is corrected in a later version of this account:

> "The reason for his quitting his Fellowship was, that though he had never taken the *solemn League and Covenant*, believing it to be an unlawful oath, as he often declared, yet he could not in the oath of Abjuration swear, that he did not believe it to be binding on others." (Bernard, Birch & Lockman 1739, p. 693).

If this truly was Ray's position it was a rather subtle one and unlikely to have been held by many of those who refused to conform to the Act. Another of Ray's friends, James Petiver, transcribed into a notebook in 1711–1712 an account which says that, "altho for some few (& if it was not too presumptuous [I?] would say trivial) reasons, he could not comply with ye Bartholomey Act, … yet he was a strict as well as a pious & exemplary Conformist to ye establish'd Church" (Gunther 1928, p. 11).

Ray had in fact expected to have to leave the college at the Restoration. There were deep disagreements amongst the Trinity fellowship about the most appropriate forms of religious worship, and Ray's sympathies clearly lay with both the religious and the teaching practice of the Protectorate rather than those that were reintroduced with the Restoration;[17] he complained to Courthope in September 1660 that "they have brought all things heer as they were [in] 1641". However, his colleagues were determined to keep him and he responded to this kindness and decided to stay (Mandelbrote in ODNB; Raven 1950, pp. 57–59). Could it be that he came to regret this decision? He seems to have regarded himself as living on borrowed time after 1660. There is no doubt that the decision was a difficult one for Ray, no doubt that he had reasons of conscience for refusing assent to the Act of Uniformity, no doubt that, as he put it at the start of the 1663 appendix to the *Catalogus*, he had unwillingly said his last farewell to his most beloved nurse the bountiful University. However, it is impossible

---

[17] Dr George Seignior of Trinity held contrary views to those of Ray and the account of his career by Fawket (1681) illustrates the divisions in the fellowship at this period. (We owe this reference to Scott Mandelbrote.)

not to wonder whether, as McMahon (2000) suggests, Ray might have had positive reasons for choosing to leave Cambridge in 1662. His clear commitment to natural sciences is apparent from the commonplace later incorporated into *The wisdom of God*, and in particular from the desire to prepare a *Phytologia britannica* which is set out in the 'Preface' to the *Catalogus*. Was this by then a stronger vocation than his University career, one which he could pursue independently? It is impossible too not to wonder what understanding he had reached with Willughby, reflected perhaps in his remark to Courthope, written just before the deadline for compliance with the Act, of his intention to "cast myself upon Providence and good friends" (Gunther 1928, p. 25). Were his views about the Act not only genuinely felt but also a respectable pretext for leaving the college which had meant so much to him since he arrived in 1646, in which he had developed the interests to which he was to devote the rest of his life, and which he could not otherwise have brought himself to leave?

Raven (1950, pp. 59–61) had no difficulty in understanding Ray's motives: "to accept the act was to subscribe to a lie: a man of honour, a teacher concerned with truth, could not thus sacrifice truth to expediency." However, many men of honour, including those who had been and were to remain friends of Ray, clearly did not see things in this light. The total number of expulsions from the University between 1660 and 1662 was about 50, including 22 men removed as a direct result of the Act of Uniformity (one Master, 14 Fellows and seven students). "As in 1650–51 Cambridge chose stability through conformity" (Twigg 1990). Raven himself clearly struggled to understand how many of Ray's colleagues, "a larger number than we, or Ray, might have expected", were prepared to remain. Raven's views were almost certainly tied up with his own position as a pacifist writing during the most desperate years of the Second World War. (The preface to the first edition is dated 1941 and in it he describes the time as one in which "our dreams of the future are black and blank".) We are not as confident as Raven that we fully understand Ray's position on this crucial question.

Though Ray left Cambridge, he did not do so with any feelings of bitterness towards the University, nor did he seek to sever his links completely. His correspondence shows that he visited Cambridge occasionally after his departure from Trinity. When, in 1676, Isaac Barrow, by then Master of Trinity, was raising money to build the Wren Library, Ray donated £10 to the cause (McKitterick 1995), a generous sum for a married man living on an annuity of £60 a year and with an uncertain future. It must have reflected his friendship with Barrow, his continuing gratitude to the college and his devotion to learning. He left a further sum of £5 to the Library in his will for the purchase of books (Raven 1950, p. 481). Nor was he forgotten by the college. The first four busts commissioned for the Library from the sculptor Roubiliac in the 1750s were of Bacon, Newton, Ray and Willughby. Over a century later, in the 1870s, the college refurbished its chapel, introducing stained glass into 15 windows of the choir, each with full-length portraits of eight figures. The portraits range from "Disciples of Christ" to "Worthies of the College" and amongst the latter stands Ray, in black clerical robes and closely examining a fresh specimen of *Caltha palustris*, sharing a window with fellow worthies who include Barrow, Dryden and Newton (Willis & Clark 1886).

John Ray and Francis Willughby, 1660–1672

After the publication of the *Catalogus* Ray wasted no time in starting the fieldwork which would be needed if he was to complete his *Phytologia britannica*. In the summers of 1660 and 1661 he went on long northern tours. The itinerary of the 1660 tour has not survived but we know that Ray travelled with Willughby or, as he put it, he "did accompany Mr. Willughby on his travels" (Gunther 1928, p. 17). They almost certainly visited Yorkshire, Cumbria and the Isle of Man (Raven 1950, pp. 113–116). Raven very plausibly dates the start of Ray's academic partnership with Willughby to this journey. However, in 1661 Ray's travelling companion was Philip Skippon, a former pupil who had been admitted to Trinity in 1655 at the same time as Courthope. Skippon was the son of one of Cromwell's Major Generals, also Philip, who was appointed one of Charles I's judges but prudently failed to turn up for the trial; he died in 1660. The younger Philip was the only one of Ray's Cambridge friends who is known to have had Parliamentary rather than Royalist associations (McMahon 2000), but he was knighted by Charles II in 1675 and was elected M.P. for Dunwich in 1679 (Henning 1983). Ray and Skippon's tour took them north into Scotland, as far as Edinburgh, Stirling and Glasgow. The summer tour of 1662 was to Wales and south-west England, when Skippon joined Ray and Willughby. Ray had confided his anxieties about the trip to Courthope in a letter in which he said he hoped Skippon's company "will not be unacceptable to Mr. Willughby ... I know he would have been a great deal better pleased with yours", but in the event the tour was a great success.

On Ray's return came the melancholy business of his departure from Cambridge, and he spent the winter of 1662–1663 in Suffolk. Willughby had suggested that their next tour should be to the Continent and by November 1662 Ray told Courthope that he had "explicitly obliged myself by promise to accompany him, if he really desire it" (Gunther 1928, p. 33); Ray seems to have treated Willughby at this period with a degree of formality which is conspicuously absent from his relaxed correspondence with Courthope. In April 1663 he was free to "wayte upon Mr. Willughby" and they left Britain with Skippon for their continental tour. (Both Ray and Willughby had tried but failed to persuade Courthope to join them.) They travelled through France, the Low Countries, Germany, Austria and Italy, reaching Naples by the spring of 1664. Here Willughby left the others and travelled to Spain before returning to England by the end of the year. Skippon and Ray continued on to Sicily and Malta, then back through Italy and Switzerland to France, where they made for the great botanical centre of Montpellier. Here they met Martin Lister of St John's College, Cambridge, who became a close friend. The tour was ended only in February 1666 when Louis XIV ordered all Englishmen to leave France within three months. (He had concluded an alliance with the Dutch and declared war on England on 26 January.) Ray completed the last stage of the journey to the Channel in a fish-cart, leaving the more fastidious Skippon and Lister to return in slightly more salubrious fashion. For full details of the continental tour, see Raven (1950, Chapter 5).

Willughby's father died in 1665 and Francis inherited the family properties, living with his mother Cassandra at Middleton Hall. Ray spent the winter of 1666–1667 at Middleton, working with Willughby on their continental collections. Thereafter they spent much of their time together until Willughby's death and they made another botanical tour to south-west England in 1667, reaching Land's End. It was in the winter of 1666–1667 that they prepared the botanical and zoological classifications for John Wilkins' *An essay towards a real character and a philosophical language* (1668). Wilkins (1614–1672) had been Master of Trinity for a year before the Restoration and was clearly popular, as the Fellows petitioned the new King to allow him to remain. However, Charles II felt obliged to honour a promise his father had made to appoint Henry Ferne, and Wilkins was compensated with the first of a series of ecclesiastical appointments which were to culminate in 1668 with the Bishopric of Chester. His book was one of several attempts in the 17th century to devise a universal and 'philosophical' language derived from first principles (Slaughter 1982). Ray and Willughby had to fit the animals and plants into a highly artificial and predetermined structure, an arrangement which Ray described as "manifestly imperfect and ridiculous" when writing to Martin Lister in 1669 (Raven 1950, p. 180: Raven's translation). Wilkins was one of the two Secretaries of the Royal Society and sometimes unofficially regarded as Vice-president (ODNB). Ray was elected a Fellow of the Society in November 1667 and, at Wilkins' instigation, was excused payment of the subscription the following year.

Willughby married Emma Barnard, daughter and co-heiress of Sir Henry Barnard, in January 1668. Ray tactfully absented himself during the summer, when his botanical tour took him to Sheffield (where he stayed with a friend, Francis Jessop, whom he had met at Montpellier), the Craven Pennines and the Lake District. He was back at Middleton by the end of 1668, to join Francis, Emma and their newly born son Francis. By December 1669 he was able to send the text of the second of his books, *Catalogus plantarum Angliae*, to Lister for comment, and the following year he saw it through the press. Another substantial and equally pioneering work, *A collection of English proverbs*, was also published in 1670. Having completed these works, Ray was able to resume his summer tours and in 1671 he visited Yorkshire, the Lake District, Northumberland and Berwickshire.

This turned out to be the last of Ray's botanical tours. Willughby had suffered periods of ill-health from 1669 onwards, but between them he was as active as ever and at the end of 1671 he was thinking of a voyage to the New World to further his zoological research. However, he became dangerously ill in June 1672, forcing Ray to cancel his proposed tour of the West Country. On 3 July 1672 he died, aged only 37.

Ray was one of the five executors of Willughby's will. The education of his sons Francis (born 1668) and Thomas (born 1672) was entrusted to him, though not that of his daughter Cassandra (born 1670). He was also left an annuity of £60 per annum. Until 1675 Ray worked at Middleton to complete the zoological projects on which Willughby had worked. By 1674 he had finished the first of these works, on birds, which was published under Willughby's name as *Ornithologia* in 1676 and illustrated by plates paid for by Emma Willughby. Ray then started work on the next book, eventually published as *De historia piscium* in 1686 (Kusukawa 2000). He also managed to produce several other works in this period including the journal of the European tour on which he had been working when Willughy died, *Observations topographical, moral, & physiological* (1673), bound up with *Catalogus stirpium in exteris regionibus a nobis observatarum*. One result of his work as tutor to the Willughby boys was his *Dictionariolum trilingue* (1675), published, according to its preface, to correct "*some inveterate Errors, especially in the names of Animals and Plants*" and published for "*Children, when they first learn the* Latin *Tongue*".

In June 1673 Ray married Margaret Oakeley, a 20-year-old member of the household at Middleton and the daughter of a local gentleman.

A further disruption: Ray leaves Middleton Hall, 1675

Ray never seems to have established a good relationship with Francis' wife, Emma Willughby. Shortly after Willughby's death, on 17 January 1673, he told Courthope that he was likely to remain at Middleton "at least so long as my old lady lives", so he clearly expected trouble after the death of Willughby's mother, Lady Cassandra. Any prospect of further support for the publication of Ray's joint work with Willughby disappeared with her death in 1675. Ray and his wife had to leave Middleton in 1675.

2: JOHN RAY

They stayed in Warwickshire and then in Essex until 1679, when Ray's mother died and they were able to move into her house at Black Notley.

In August 1676 Emma Willughby remarried. Her new husband, Joshua (later Sir Joshua) Child (1631–1699), was a rich merchant and one of the leading shareholders in the East India Company. The *Oxford Dictionary of National Biography* describes him as "universally disliked and distrusted" (although not, presumably, by Emma) and a man who, to his contemporaries, "appeared ruthless, astute, ostentatious, intelligent, and avaricious". In a later memoir (published by Wood in 1958), her daughter Cassandra tells us that Emma "would never consent to what Sir Josia Child would not approve off". Emma left Middleton to live at Wanstead Abbey, an estate which Child had bought in 1673 and which he was landscaping and extending at enormous cost. She took her three children with her. Cassandra writes of "Mr. Ray refusing to goe with the rest of my mother's family". This may have been what she was told by her mother (she was only six years old at the time), but it is unlikely that Ray was ever invited to accompany the family to Wanstead. Child had Willughby's will set aside, supposedly by convincing the judge that Ray was a nonconformist (an allegation with no basis in fact), and he took the children out of Ray's charge (Raven 1950, p. 483).

It is worth digressing to tell the story of the three Willughby children, as it suggests much about the values imparted to them at Middleton by their father Francis and their tutor, John Ray. Joshua Child sued to obtain control of the Willughby estate in trust for his new stepchildren, then proceeded to milk it to his own advantage. He charged the three children an annual rent of £150–200 to live at Wanstead and removed £1,000 from the estate annually as a management fee. Willughby's elder son Francis soon came to dislike his stepfather and was dissatisfied with his tutor at Wanstead, "who was but a heavy dull person" according to Cassandra (Wood 1958). Child thought that "a home education was best for children born to estates, and much safer for them than to be trusted to Publick Schools or sent to the University". When Francis was no more than 12, in the summer of 1680, he set out as if intending to take his normal morning ride but instead made for Haslingfield, Cambridgeshire, where he arrived at the house of his father's sister, Lettice Wendy, just as she was going to supper. Lettice gave him refuge and found a better tutor, and by 1682 Francis had "made himself a schooler fit for the University" and entered Catharine Hall, Cambridge. His time at Cambridge does not seem to have been entirely happy, for his tutor "had lead him into some extravagancies", but before he left Cambridge he managed to select a more suitable tutor, at Jesus College, for his younger brother Thomas. Without informing either his mother or his stepfather, although Cassandra was in on the secret, he somehow arranged for Thomas to leave Wanstead and to enter the University. After he left Cambridge he moved to London in 1685–1686 to open a court case against Sir Joshua, but he was unable to find a successful response to Child's delaying tactics.

Cassandra was left at Wanstead, but only for a couple of years. By 1687 Francis invited her to settle with him at Wollaton and help him manage the household, to her great delight. "Thus we two young creatures," she wrote, " my self but 17 and he not 19 years of age, went to Wollaton to keep house there together". They travelled from London *via* Aunt Lettice's house in Cambridgeshire. Though Wollaton was in disrepair, Cassandra proved more than capable of managing the house and, when Francis died on his 20th birthday after a short illness, she continued to work with her brother Thomas to rebuild the Hall and redesign the gardens. Her vividly written history of the Willughby family, based on manuscripts at Wollaton, and the journal of her travels in England have gained her an entry in the *Oxford Dictionary of National Biography* in her own right. In 1713 she married a widower, James Brydges, first Duke of Chandos, who had made a fortune (£600,000) as paymaster of the forces in the War of Spanish Succession. Brydges, a more generous if less astute businessman than Child, had built a princely house at Cannons and managed, by his patronage of Handel, to ensure that his memory survives through the Chandos anthems. Needless to say, "the excellent Cassandra" proved to be a much-loved companion to the Duke, a good stepmother to his sons and a capable manager of Cannons, where she died in 1735.

Ray's opinion of Sir Joshua Child is revealed by the brief summary by Derham of a letter (now lost) from Ray dated 24 January 1676 [i.e. 1677, new style] which reads "Mr. Child will treat with bookseller about printing *Hist. Fishes*. He sordidly covet[ous]." (Gunther 1928, p. 136). In the event *Historia piscium* was not published until 1686. In 1691 Ray dedicated his book *The wisdom of God* to Lettice Wendy, "First, because I owe it to the Liberality of your Honoured Brother, that I have this leisure to write any Thing". The other reasons were her "many and signal Favors" to Ray and her abilities as an "able Judge" and "candid Reader" of the work. He could scarcely refer in the dedication to her crucial role in assisting the Willughby children to escape from their stepfather, but he must have been aware it.[18]

Ray's later life

The death of Willughby in 1672 and Ray's departure from Middleton in 1675 severed his main surviving link to his Cambridge years. One of his later books, *Synopsis methodica animalium quadrupedum et serpentini generis* (1693), was dedicated to Peter Courthope and Timothy Burrell, who were by then the last survivors of his early associates (Raven 1950, p. 52). However, he made new friends amongst men of kindred interests; as one of them, Samuel Dale, said, "he was affable, (being not puft up by his learning) and always communicative of any thing he was Master of" (Anon. 1707, p. 497).

John and Margaret Ray lived an unsettled life after leaving Middleton until John Ray's mother died in 1679, at which point they were able to move into her house at Black Notley. Here he settled for the rest of his life, and here his daughters Margaret and Mary (twins, born in 1684), Catharine (1687) and Jane (1689) were born. Somehow Ray managed to fit his wife and daughters, his books (some 1500 at the time of his death, including 350 of folio size[19]) and his plant specimens into the house and to write his remaining books. A work on plant classification, *Methodus plantarum* (1682), was the first to appear while he was at Black Notley, followed by the first two volumes of *Historia plantarum* (1686–1688), a massive work which included descriptions of all the plants then known; a large additional volume was required by 1704. The slim *Fasciculus* of 1688 documented recent additions to the British flora and was a precursor to *Synopsis methodica stirpium Britannicarum* (1690), the first true British Flora, which was based on *Historia plantarum*. A second edition appeared in 1696. In producing these works he was sustained by his friends, including the physician

---

[18] Keynes (1976) refers to a copy of the first issue of *The wisdom of God* in his collection which had belonged to Francis Willughby's mother Cassandra and which he thought (from its red morocco binding) was probably a presentation copy from Ray. This fine copy, now in Cambridge University Library (Keynes F.2.6), bears the signature "Cass: Willughby", but this must be Francis' daughter Cassandra, not his mother Cassandra (who died in 1675).

[19] These figures are derived from the sale catalogue (Feisenberger 1975) and include the actual number of books in multi-volume works sold as a single numbered lot but count together several works sold as one lot on the assumption that they were bound together.

Tancred Robinson, the Braintree apothecary Samuel Dale, who had become interested in natural history under Ray's influence and who was to became his closest local collaborator in these years, and the wealthy London physician and collector Hans Sloane, a constant support to Ray's scientific endeavours.

Other books published during this period were *The wisdom of God manifested in the works of creation* (1691, with a much enlarged second edition in 1692), a combination of biology and theology which reached a wider audience than any of his purely biological works, and *A persuasive to a holy life* (1700), which had much less impact. Another book, *Miscellaneous discourses concerning the dissolution and changes of the world* (1692), dealt with fossils and was rearranged as *Three physico-theological discourses* in the following year. The last zoological work published in his lifetime was *Synopsis methodica animalium quadrupedum et serpentini generis* (1693), but much of Ray's research in his latter years was devoted to Lepidoptera. Even when he was no longer fit enough to undertake fieldwork himself, his friends provided him with material. Thomas Simpson, perhaps Ray's man-servant, and his "little daughters", in particular, were enlisted as collectors of local specimens; the youngest daughter, Jane, appears to have made her first significant catch in Ray's orchard at the age of four and a half (Raven 1950, pp. 394–395).

Not surprisingly, Ray was unable, despite his heroic efforts, to complete a book on invertebrates in his old age. As he had ruefully observed to Edward Lhwyd in 1695, "it were a fitter task for a young & ingenious person, who had the perfection of all his sences & time enough before him" (Gunther 1928, pp. 261–262). He did prepare a brief *Methodus insectorum*, which was published in 1705, but by 1704 his health was failing and towards the end of the year he was unable to undertake any further study (Anon. 1707). By January 1705 he knew that he was dying, and the last and most moving of his surviving letters is addressed to Hans Sloane, "Dear Sir, the best of friends" (Derham 1718, p. 373; Raven 1950, p. 306). He died on 17 January 1705 and was buried in the churchyard at Black Notley. His incomplete notes on invertebrates were published as *Historia insectorum* in 1710.

3: THE *CATALOGUS* AND ITS APPENDICES – STRUCTURE AND SOURCES

Printing and publishing

John Ray sent a copy of *Catalogus plantarum circa Cantabrigiam nascentium* to Francis Willughby on 25 February 1660 (Lankester 1848, pp. 1–2). Willughby was one of Ray's three main collaborators, and Ray clearly sent him a copy as soon as one became available. Indeed, Ray told Willughby that he would have "received [it] a week sooner had not the bookbinder deceived me", thus betraying the impatience of any author waiting for a copy of his first book.

The *Catalogus* is a small book, with pages measuring about 14.7×8.5 cm. Keynes (1976) gives a full bibliographical description. It is known with three different title-pages. Two of these are variants of the title-page translated here, one with two errors in the Latin and the rule across the page in black and the other with these errors corrected and the rule in red. The corrected title-page was clearly intended to replace the incorrect one, but copies of the book are known with the incorrect title-page, with the correct one and with both. The third variant replaces the details of the Cambridge publisher with a London imprint: "Londini: Apud Jo. Martin, Ja. Allestry, Tho. Dicas, ad insigne Campanæ in cœmeterio D. Pauli 1660" ("London: From Jo. Martin, Ja. Allestry, Tho. Dicas, at the sign of the Bell in St Paul's churchyard").

John Field, who printed the *Catalogus*, was a London printer who had been appointed University Printer in 1655, in part because he was regarded as politically reliable. McKitterick (1992), who provides a detailed account of his tenure in office, describes him as "a creature of the government" (p. 312). He was the only printer in Cambridge and worked near the junction of the present Silver Street and Queens' Lane, on the site now occupied by the Master's Lodge of St Catharine's College. Ray's former tutor James Duport was an enthusiastic supporter of the Cambridge printing press (McKitterick 1992, pp. 336–337) and he may have been a useful initial link between Ray and Field. At this period the University Printers were commercially unadventurous and invested their money in guaranteed best-sellers such as bibles and almanacs. Other books were printed on behalf of Cambridge booksellers, who were in effect the publishers. Three booksellers were prominent at this period, William Graves, William Morden and William Nealand. Nealand published the *Catalogus* but he appears to have gone out of business in 1662 (Raven 1950, p. 81) and the 1663 appendix was published by Morden. Samuel Worthington, writing to Hartlib on 10 June 1661, reported that "the most active Booksellers, *Allestree* and his Partners, at the *Bell* in S. *Paul*'s Church-yard, bought all the Impression of one in *Cambridge*, for whom the Book was printed" (Worthington 1704, p. 255; reprinted in modernised form by Crossley 1847, p. 333). Keynes (1976) argues that this cannot be true, as copies with Cambridge title-pages are much commoner than those with the London variant.

There are few misprints in the text of the *Catalogus*, perhaps in part because Ray's handwriting was neat and extremely legible (see Gunther 1934). However, the printing, though adequate, does not seem to be of a very high standard, with uneven spacing and a scattering of poorly inked or even uninked letters, and the typography is inconsistent. Even to the non-expert eye, it is clear that at least two different designs of italic *C* and two different italic *B* letters are used in Ray's common abbreviation *C.B.*, and all four possible combinations occur. (In the 1685 appendix the printer appears to have become so short of italic *B* letters that in many cases he was forced to use roman font.) All the copies of Ray's *Catalogus* that we have examined carefully share the same inconsistencies of typography and the only differences we have noted are attributable to variation in inking. According to Gaskell (1972), there was "a steady and general decline in printing standards throughout Europe during the sixteenth and seventeenth centuries. The rate of decline was not the same everywhere ... French and Dutch printing was notably better than most in the seventeenth century, though even so it was plainly worse in 1700 than it had been 150 years before. Seventeenth-century English printing was abysmally poor, and there were few books listed in Wing [a catalogue of books published in English between 1641 and 1700] that were not set in ill-cast, battered type, clumsily arranged and carelessly printed in brown ink on shabby paper." (It came as no surprise to us to read Gaskell's account after we had started work on the text of the *Catalogus*.) Even by the standards of the time, John Field seems to have been a poor printer. When he and his business partner Henry Hills held a monopoly of domestic bible production in the 1650s he was in constant trouble for the errors in his bibles. As the printer of the "Unrighteous Bible" of 1653 he was responsible for one of the most notorious biblical misprints, "Know ye not that the unrighteous shall inherit the kingdom of God?". Although criticism of his bibles came from potential commercial rivals, there seems to be no doubt that Field produced "abominably erroneous texts", though it was argued that most of the faults were "but slight" and related to commas, full stops, italics and odd letters (McKitterick 1992). However, by the late 1650s Field had achieved a level of competence as a printer. After the Restoration he spent less time in London (where he was probably embarrassed by his former allegiance to Cromwell) and more time in Cambridge.

Structure of the book

The *Catalogus* consists of eight different parts. We refer to the whole volume as the *Catalogus* or, if it has to be distinguished from other catalogues, the Cambridge catalogue. We call the constituent parts the 'Preface', 'Explanation', main catalogue, 'Index of English names', 'Index of places', 'Etymology', 'Interpretation' and 'The more customary headings or divisions of plants'. As an alphabetical catalogue, it needs no index to Ray's chosen names; some of their synonyms are cross-referenced in the main catalogue.

The 'Preface' (16 pp.) and 'Explanation' (12 pp.) are on unnumbered pages. The main catalogue (pp. 1–182) is followed by a second, internal, title-page, and the pagination then starts again for the 'Index of English names' (pp. 1–29), 'Index of places' (pp. 30–35), 'Etymology' (pp. 36–83), 'Interpretation' (pp. 84–99) and 'The more customary headings or divisions of plants' (pp. 100–103). Raven's supposition (1950, p. 81) that it is "probable that the two sections could be bought separately" has been convincingly refuted by Keynes (1951, 1976), who pointed out that no copies of the two parts have been discovered in separate, contemporary bindings and, crucially, that the second title-page is in the middle of a printed section of pages, not an arrangement that would have been adopted for books that were intended to be sold separately. Many of the botanical works Ray cited in the *Catalogus* consist of several *libri*, a word which is always translated as "books" but often means the major internal divisions of a volume and thus has a rather different meaning from the usual modern sense of a discrete object. These *libri* sometimes have their own title-pages and are occasionally paginated separately, and it is possible that this is the model that Ray had in mind when planning his book.[1] The second edition of his much later book *The wisdom of God* (1692) has the same arrangement of two separately paginated parts.

Models for the *Catalogus*

The title *Catalogus plantarum circa Cantabrigiam nascentium* is based on Caspar Bauhin's catalogue of the plants of Basel, *Catalogus plantarum circa Basileam sponte nascentium* (1622), with *sponte* omitted by Ray because he included crops as well as wild plants. Ray's friend and colleague John Nidd had a copy of this book, which he left to the library of Trinity College in the will he drew up in 1658 (see Chapter 5). Raven (1950, pp. 80–81) devoted a paragraph to establishing that Bauhin's book "is obviously the model on which Ray's first book was planned. ... The two books are both small octavo volumes with pages of the same size." The Basel catalogue may well have given Ray the idea of writing a catalogue of Cambridgeshire plants, but it is probably a mistake to look solely at botanical works in considering precursors to the *Catalogus*. Elizabethan authors had written works arranged on a county basis, notably William Camden whose *Britannia* was first published in 1586. Later editions of *Britannia* included county maps by John Norton and John Speed. Norden himself published detailed descriptions of Middlesex and Hertfordshire in the 1590s and Richard Carew's *Survey of Cornwall* appeared in 1602. Charles Webster (2002, p. 427) detected a tendency for natural history to be pursued on a regional basis in the Commonwealth. His examples are mainly drawn from surveys of Ireland following Cromwell's 'pacification', but they also include John Beale's *Herefordshire orchards* (1657).[2]

In addition to Bauhin's catalogue there are three other small octavo volumes, scarcely mentioned in this connection by Raven,[3] to which the Cambridge catalogue appears to owe something of its contents and much of its form. These are the two parts of Thomas Johnson's *Mercurius botanicus* (1634, 1641) and William How's *Phytologia britannica* (1650). The relationship between Johnson's *Mercurius* and How's *Phytologia* was first established by Kew & Powell (1932). The first part of Johnson's work (1634) was itself divided into three parts. It began with a description of an excursion that Johnson and his friends from the Society of Apothecaries had made from London to Bristol and back in 1634, when they spent their days travelling and botanising and their evenings being entertained by local apothecaries, physicians and other worthies. This is followed by a description of the town and waters of Bath. Finally, there is a consolidated list of the plants found on this and earlier journeys, a

---

[1] One of Ray's major sources, Bauhin & Cherler's *Historia plantarum* (1650–1651), has two runs of page numbers in volumes 1 and 3, although volume 2 is continuously paginated. It also has misprinted page numbers, a section in volume 3 where the leaves rather than the pages are numbered, an inconsistent use of roman numerals and words in the titles of the books (*libri*), a chapter structure within the books which sometimes follows a logical pattern but sometimes appears temporarily to depart from it (e.g. in volume 1, book XII, after chapter VIII) and pairs of chapters that sometimes have the same number (e.g. chapter IV in book XXIII).

[2] Gerard Boate's *Irelands naturall history* (1652), published by Hartlib, was in Ray's possession at the time of his death (Feisenberger 1975). We have not come across any evidence that Ray or his colleagues owned a copy of Beale's book. The improvement of fruit trees "assumed outstanding importance in the puritan writings on husbandry" (Webster 2002, p. 477), but the cultivated apple is not included in the *Catalogus*. The omission illustrates the differences between Ray's viewpoint and the outlook of Samuel Hartlib's circle and the other Puritan writers discussed by Webster.

[3] The only exception we have found is a single footnote (Raven 1950, p. 80) which acknowledges that Johnson's *Mercurius* and Ray's *Catalogus* share an alphabetical arrangement.

3: THE *CATALOGUS* AND ITS APPENDICES

departure from Johnson's (1629, 1632) earlier practice of listing separately the plants seen at each of the places visited (see Gilmour 1972). The second part of the *Mercurius* (1641) presents the account of a journey to North Wales, followed by a list of additions to the 1634 list made on this journey or by other authors. Together the two lists, of 708 and 190 species, comprise a complete checklist of the known British plants. Johnson was killed in 1644, fighting in the Civil War, but William How put the lists in the two parts of the *Mercurius* together and published them with some additional records as *Phytologia britannica* (1650). How followed the design and typography of Johnson's lists very closely (see Figures 3.1–3.3).

Bauhin's *Catalogus* (1622) shares many similarities with Johnson's *Mercurius* and How's *Phytologia*. For each species there are a chosen name, synonyms, usually a habitat and sometimes specific localities. Johnson and How differ from Bauhin in consistently adding an English name. (Bauhin gives only six vernacular names and his entire text is in Latin.) In format the English and Swiss works are very different (compare Figures 3.1–3.3 and Figure 3.6). Bauhin lists species in systematic order with his 'genera' separated by rules across the page; the name of each genus is entered as a marginal note and is put in capitals at the start of the first species entry. The rest of the species accounts are all in roman text, with the first line indented, and, where a species account occupies several lines, the appearance is of a uniform and somewhat impenetrable block of print. A few German names introduce some typographical variety as they are in a black-letter font. By contrast Johnson and How have an alphabetical list of species, indent the second and subsequent lines of the species entries, italicise the authors of chosen names and synonyms, give the habitat as a Latin phrase in roman font but italicise the English name of the plant, and also give the specific localities in English in a combination of black letter font and (for place names) italics. This greatly enhances the accessibility of the text. Bauhin's systematic text does not require synonyms to be cross-referenced, but Johnson and How cross-reference synonyms using the abbreviation *v.* or V. for *vide* ("see"). The whole arrangement of Johnson's and How's books, with their combination of Latin and English text and varied typography, is strikingly similar to Ray's *Catalogus* (see Figures 3.4 and 3.5). The main difference is the absence of black-letter text in Ray, who puts the entire English text for his specific localities into italics. Ray may not have known Johnson's *Mercurius botanicus* (which he does not mention in the *Catalogus* and which was not in his library at the time of his death), but he certainly knew How's *Phytologia*. In the 'Preface' Ray refers to the need to compile a complete *Phytologia britannica*, which perhaps suggests that How's work was in his mind as a model. He must surely have been influenced by the *Phytologia* as well as the Basel catalogue in planning the content of his *Catalogus*, and it is impossible to believe that he was not greatly influenced by it in devising its format.

The *Catalogus* is, however, a more ambitious work than that of either Bauhin or How. Bauhin (1622) has a single page of introductory matter (*Ad Medicinæ Studiosos*), his main catalogue, an index (required by any systematic arrangement) and, crammed onto the last of his 128 pages, a key to the abbreviations for authors and errata. How's book is similar in size (134 pages), with more introductory material (10 pages) and single pages devoted to author abbreviations and errata. Bauhin has no notes on species and How has only one, a long note on *Pisa marina Anglica* (pp. 93–94) taken from Johnson (1641, pp. 29–30) and divided into a series of numbered points. There is nothing in either Bauhin's or How's work to match the 'Index of English names', 'Index of places', 'Etymology' and 'Interpretation' in the second part of Ray's *Catalogus*.

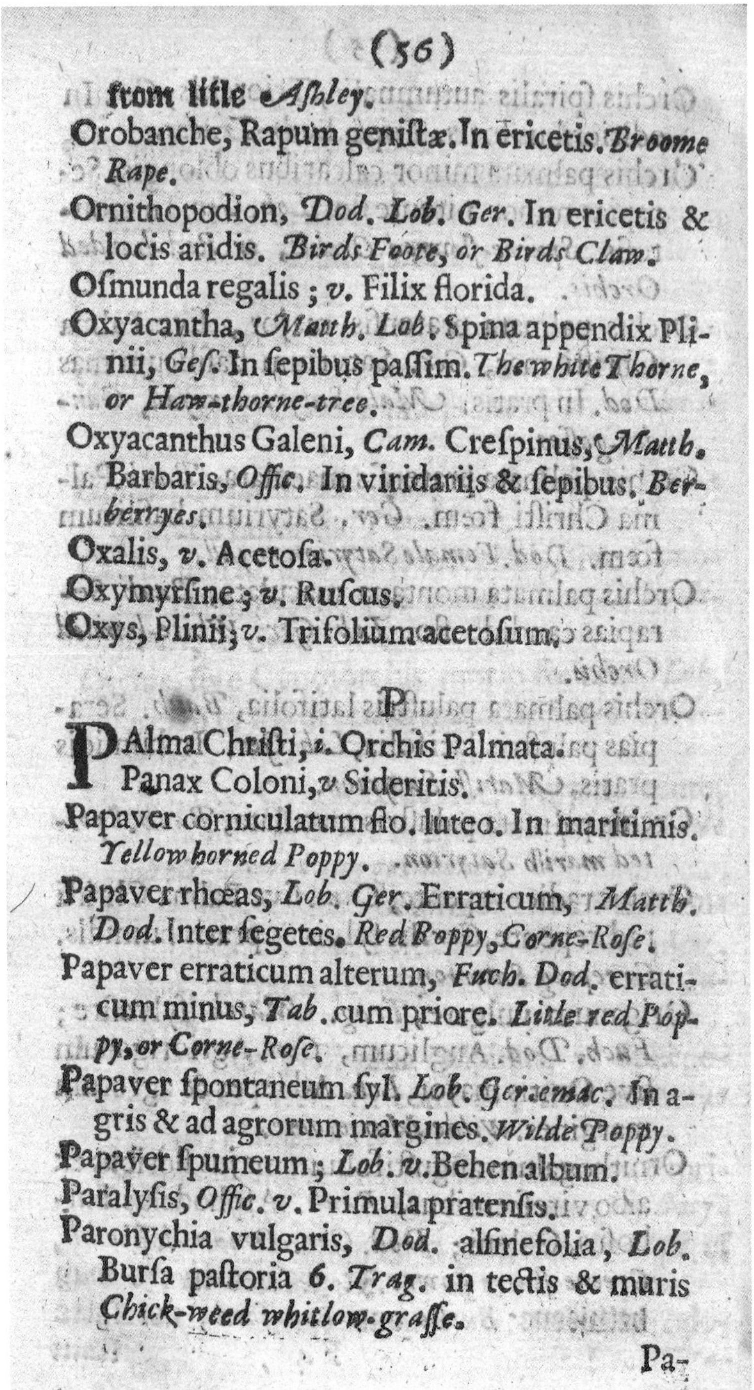

(56)

from litle *Ashley*.
Orobanche, Rapum genistæ. In ericetis. *Broome Rape*.
Ornithopodion, *Dod. Lob. Ger*. In ericetis & locis aridis. *Birds Foote, or Birds Claw*.
Osmunda regalis; *v*. Filix florida.
Oxyacantha, *Matth. Lob*. Spina appendix Plinii, *Ges*. In sepibus passim. *The white Thorne, or Haw-thorne-tree*.
Oxyacanthus Galeni, *Cam*. Crespinus, *Matth*. Barbaris, *Offic*. In viridariis & sepibus. *Berberryes*.
Oxalis, *v*. Acetosa.
Oxymyrsine, *v*. Ruscus.
Oxys, Plinii; *v*. Trifolium acetosum.

Palma Christi, *v*. Orchis Palmata.
Panax Coloni, *v*. Sideritis.
Papaver corniculatum flo. luteo. In maritimis. *Yellow horned Poppy*.
Papaver rhœas, *Lob. Ger*. Erraticum, *Matth. Dod*. Inter segetes. *Red Poppy, Corne-Rose*.
Papaver erraticum alterum, *Fuch. Dod*. erraticum minus, *Tab*. cum priore. *Litle red Poppy, or Corne-Rose*.
Papaver spontaneum syl. *Lob. Ger. æmac*. In agris & ad agrorum margines. *Wilde Poppy*.
Papaver spumeum; *Lob. v*. Behen album.
Paralysis, *Offic. v*. Primula pratensis.
Paronychia vulgaris, *Dod*. alsinefolia, *Lob*. Bursa pastoria 6. *Trag*. in tectis & muris. *Chick-weed whitlow-grasse*.

Pa-

Figure 3.1. Thomas Johnson's *Mercurius botanicus* (1634), p. 56. Reproduced by kind permission of the Syndics of Cambridge University Library (classmark Bb*.12.29).

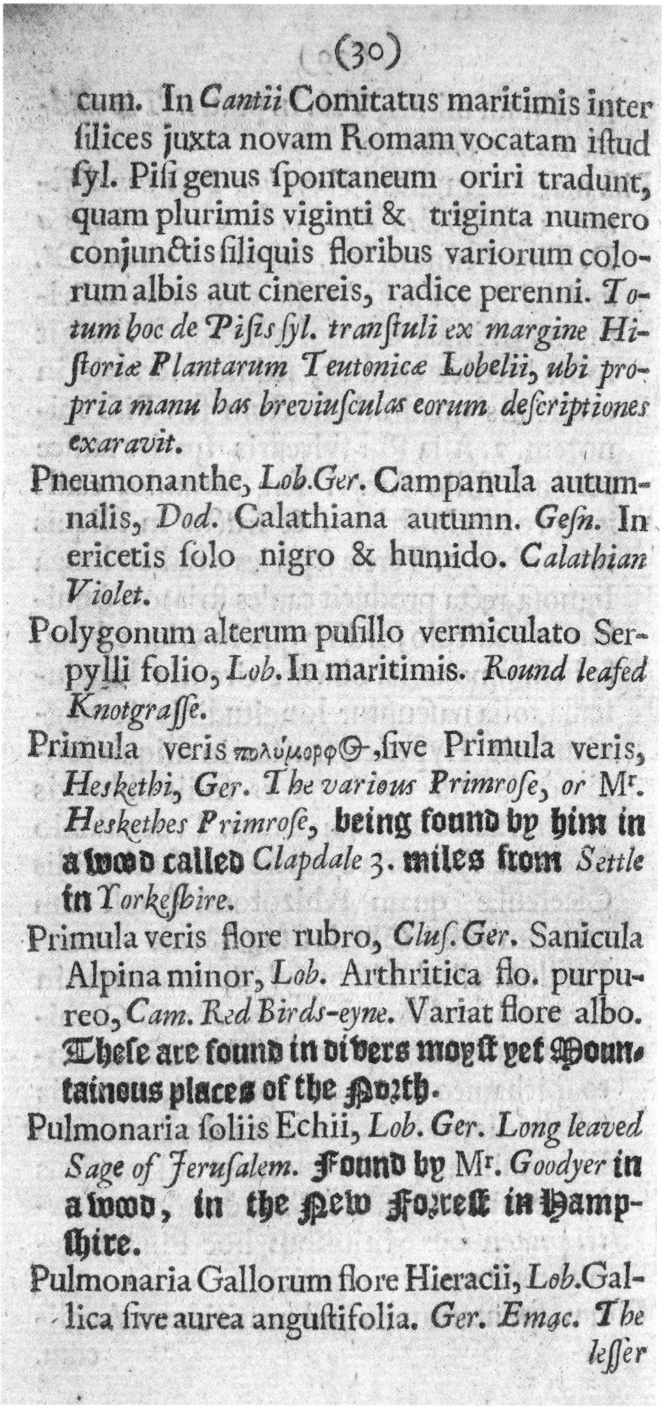

(30)

cum. In *Cantii* Comitatus maritimis inter filices juxta novam Romam vocatam istud syl. Pisi genus spontaneum oriri tradunt, quam plurimis viginti & triginta numero conjunctis siliquis floribus variorum colorum albis aut cinereis, radice perenni. *Totum hoc de Pisis syl. transtuli ex margine Historiæ Plantarum Teutonicæ Lobelii, ubi propria manu has breviusculas eorum descriptiones exaravit.*

Pneumonanthe, *Lob. Ger.* Campanula autumnalis, *Dod.* Calathiana autumn. *Gesn.* In ericetis solo nigro & humido. *Calathian Violet.*

Polygonum alterum pusillo vermiculato Serpylli folio, *Lob.* In maritimis. *Round leafed Knotgrasse.*

Primula veris πολύμορφο-, sive Primula veris, Heskethi, *Ger. The various Primrose,* or Mr. *Heskethes Primrose,* **being found by him in a wood called** *Clapdale* 3. **miles from** *Settle* **in** *Yorkeshire.*

Primula veris flore rubro, *Cluf. Ger.* Sanicula Alpina minor, *Lob.* Arthritica flo. purpureo, *Cam. Red Birds-eyne.* Variat flore albo. **These are found in divers moyst yet Mountainous places of the North.**

Pulmonaria soliis Echii, *Lob. Ger. Long leaved Sage of Jerusalem.* **Found by** Mr. *Goodyer* **in a wood, in the New Forrest in Hampshire.**

Pulmonaria Gallorum flore Hieracii, *Lob.* Gallica sive aurea angustifolia. *Ger. Emac. The lesser*

Figure 3.2. Thomas Johnson's *Mercurii botanici pars altera* (1641), p. 30. Reproduced by kind permission of the Syndics of Cambridge University Library (classmark CCD.47.92).

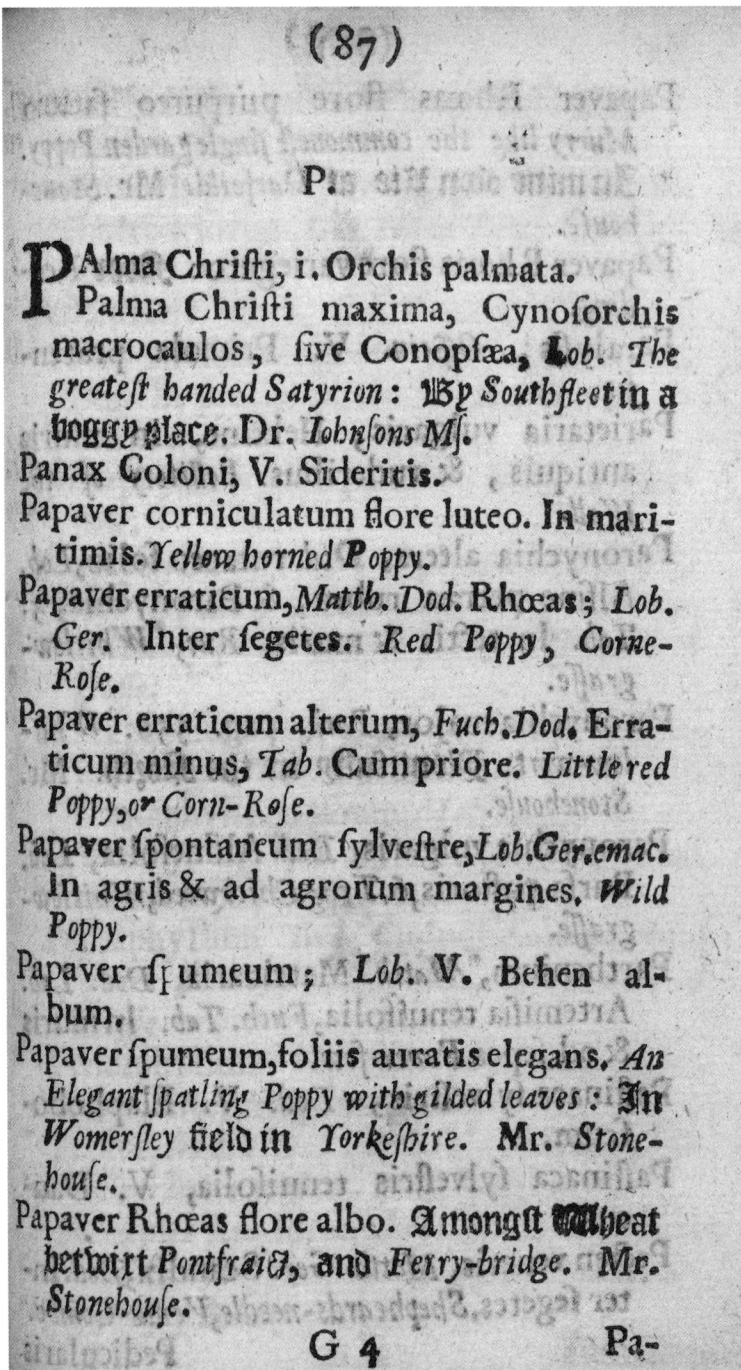

Figure 3.3. William How's *Phytologia britannica* (1650), p. 87. Reproduced by kind permission of the Syndics of Cambridge University Library (classmark CCE.47.18).

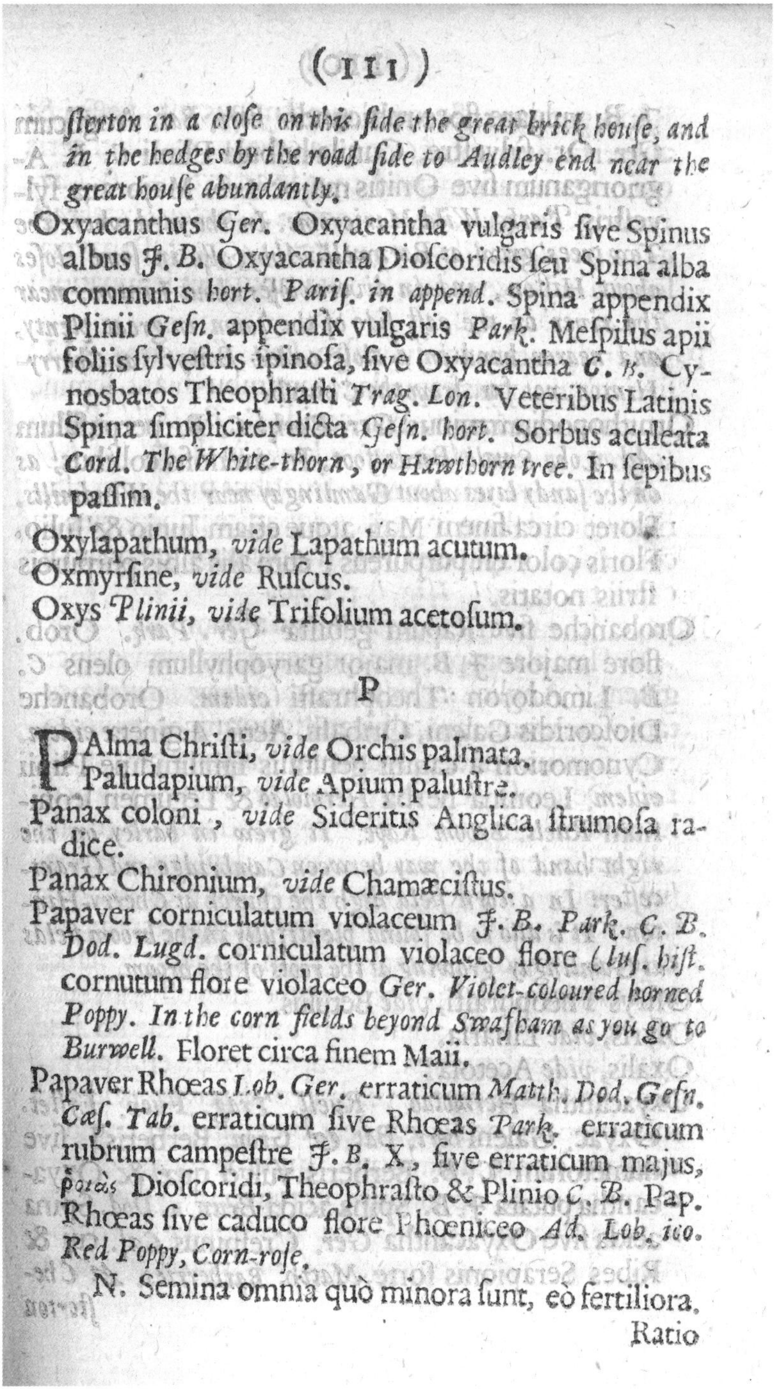

(111)

...sterton in a close on this side the great brick house, and in the hedges by the road side to Audley-end near the great house abundantly.

Oxyacanthus *Ger*. Oxyacantha vulgaris sive Spinus albus *J. B.* Oxyacantha Dioscoridis seu Spina alba communis *hort. Parif. in append.* Spina appendix Plinii *Gesn.* appendix vulgaris *Park.* Mespilus apii foliis sylvestris spinosa, sive Oxyacantha *C. B.* Cynosbatos Theophrasti *Trag. Lon.* Veteribus Latinis Spina simpliciter dicta *Gesn. hort.* Sorbus aculeata *Cord.* The *White-thorn*, or *Hawthorn tree*. In sepibus passim.

Oxylapathum, *vide* Lapathum acutum.
Oxmyrsine, *vide* Ruscus.
Oxys *Plinii*, *vide* Trifolium acetosum.

P

Palma Christi, *vide* Orchis palmata.
Paludapium, *vide* Apium palustre.
Panax coloni, *vide* Sideritis Anglica strumosa radice.
Panax Chironium, *vide* Chamæcistus.
Papaver corniculatum violaceum *J. B. Park. C. B. Dod. Lugd.* corniculatum violaceo flore *Cluf. hist.* cornutum flore violaceo *Ger*. *Violet-coloured horned Poppy*. In the corn fields beyond *Swafham* as you go to *Burwell*. Floret circa finem Maii.
Papaver Rhœas *Lob. Ger.* erraticum *Matth. Dod. Gesn. Cæs. Tab.* erraticum sive Rhœas *Park.* erraticum rubrum campestre *J. B.* X, sive erraticum majus, ῥοιὰς Dioscoridi, Theophrasto & Plinio *C. B.* Pap. Rhœas sive caduco flore Phœniceo *Ad. Lob. ico*. *Red Poppy, Corn-rose*.
    N. Semina omnia quò minora sunt, eò fertiliora. Ratio

Figure 3.4. John Ray's *Catalogus plantarum circa Cantabrigiam nascentium* (1660), p. 111.

(126)

Ang. Verbaſculum odoratum *Fuch.* pratenſe odoratum *C. B.* Φλομίσκος σκιαδὴχῶ χρυσάνϑης μακρόφυλῶ, id eſt, Verbaſculum umbellatum flore aureo, ſtylo longo *P. Renealm. Paigles or Cowſlips.* In paſcuis ubique.

Primula veris elatior pallido flore *Cluſ. hiſt. hort. Hafn.* veris major floribus pallidis ac dilutius coloratis quaſi albidis *Dod.* pratenſis inodora lutea *Ger.* pratenſis inodora luteo-pallida *Lob.* veris caulifera pallido flore inodoro aut vix odoro *J. B.* Verbaſculum pratenſe aut ſylvaticum inodorum *C. B.* Φλομίσκῶ σκιαδῆχος διασκόπιος ἐρυϑρόζωνος, id eſt, Verbaſculum umbellatum floribus in omnem partem diffuſis, zonâ rubrâ ad petiolorum baſin cinctum *P. Renealm. Great Cowſlips or Oxſlips. In Kingſton and Madingley woods abundantly and elſewhere.*

Primula veris vulgaris *Park.* veris minor *Ger.* ſylvarum *Ad. Lob. Lugd.* veris floribus ex ſingularibus pediculis majoribus, ſimplicibus *J. B.* veris pallido flore humilis *Cluſ. hiſt. hort. Hafn.* Aliſma ſylvarum *Col.* Verbaſculum ſylvarum majus ſingulari flore *C. B.* Φλομίσκος ἀσκιάδιος βραχυσιφώνιος, id eſt, Verbaſculum non umbellatum floris tubulo brevi *P. Renealm. Common Primroſe.*

N. In omnibus ferè floribus qui unico colore non conſtant, mediæ eorum partes luteæ erunt, raro pallidæ aut herbaceæ, rariſſimè alio colore imbutæ: Sed quæ varietatem ſuſcipiunt, folia ſunt vel calathi, quæ medias partes circumambiunt. *Marant. method. cogn. ſimpl. lib* 2. *cap.* 15.

Prunella *Ger.* vulgaris *Trag. Park.* vulgaris foliis ſubrotundis *Cluſ.* flore minore vulgaris *J. P.* Brunella major folio non diſſecto *C. B.* Conſolida minor *Matth.* minima *Cam.* Britannica *Guiland. Selfheal.* Variat floris colore cæruleo, carneo, albicante.

Prunus

Figure 3.5. John Ray's *Catalogus plantarum circa Cantabrigiam nascentium* (1660), p. 126.

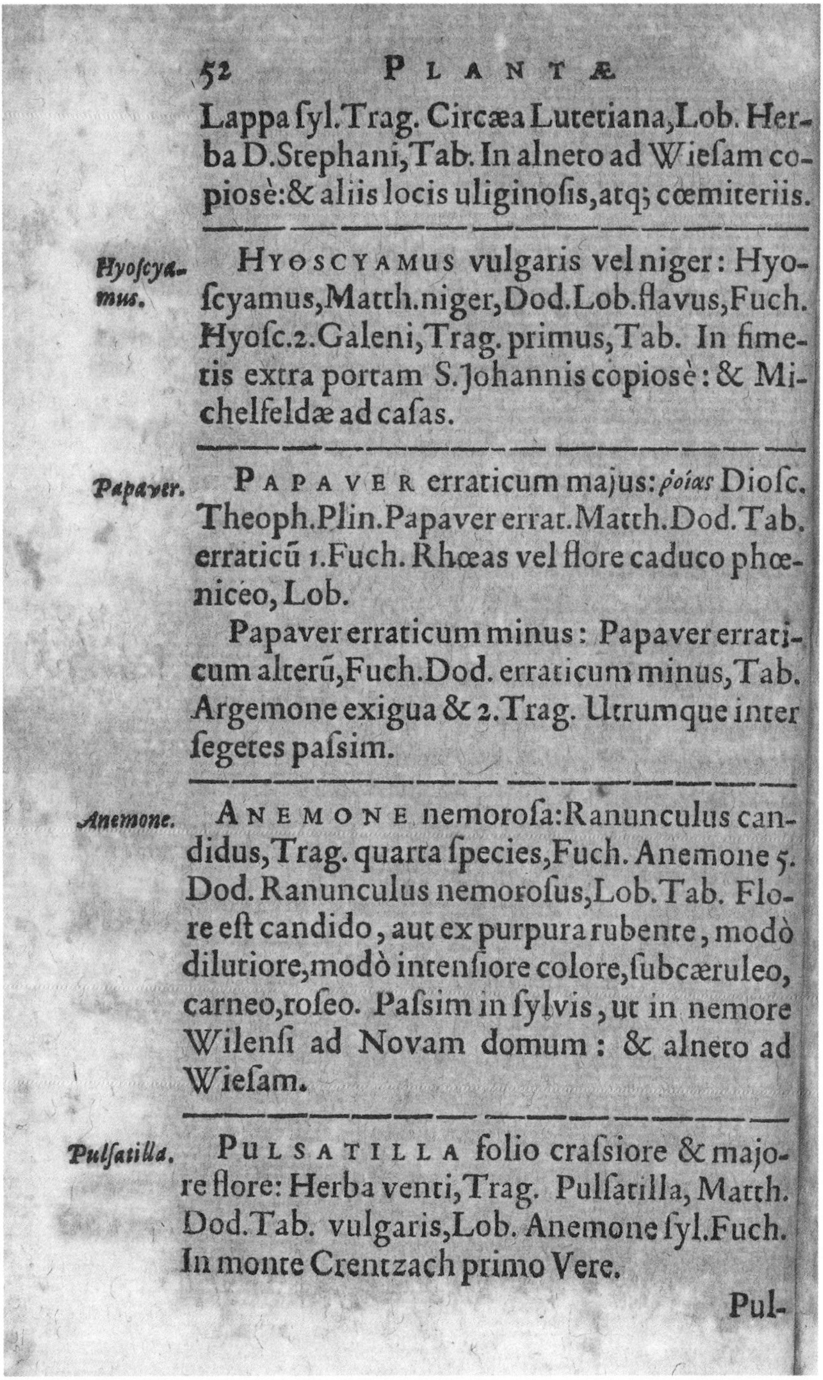

52 PLANTÆ

Lappa syl. Trag. Circæa Lutetiana, Lob. Herba D. Stephani, Tab. In alneto ad Wiesam copiosè: & aliis locis uliginosis, atq; cœmiteriis.

Hyoscyamus. HYOSCYAMUS vulgaris vel niger: Hyoscyamus, Matth. niger, Dod. Lob. flavus, Fuch. Hyosc. 2. Galeni, Trag. primus, Tab. In fimetis extra portam S. Johannis copiosè: & Michelfeldæ ad casas.

Papaver. PAPAVER erraticum majus: *pòias* Diosc. Theoph. Plin. Papaver errat. Matth. Dod. Tab. erraticũ 1. Fuch. Rhœas vel flore caduco phœniceo, Lob.

Papaver erraticum minus: Papaver erraticum alterũ, Fuch. Dod. erraticum minus, Tab. Argemone exigua & 2. Trag. Utrumque inter segetes passim.

Anemone. ANEMONE nemorosa: Ranunculus candidus, Trag. quarta species, Fuch. Anemone 5. Dod. Ranunculus nemorosus, Lob. Tab. Flore est candido, aut ex purpura rubente, modò dilutiore, modò intensiore colore, subcæruleo, carneo, roseo. Passim in sylvis, ut in nemore Wilensi ad Novam domum: & alneto ad Wiesam.

Pulsatilla. PULSATILLA folio crasiore & majore flore: Herba venti, Trag. Pulsatilla, Matth. Dod. Tab. vulgaris, Lob. Anemone syl. Fuch. In monte Crentzach primo Vere.

Pul-

Figure 3.6. Caspar Bauhin's *Catalogus plantarum circa Basileam sponte nascentium* (1622), p. 52. Reproduced by kind permission of the Syndics of Cambridge University Library (classmark F162.d.3.3).

There is one further work which Ray may have drawn upon when planning his *Catalogus*, a list of plants grown in the Oxford Botanic Garden which was published as *Catalogus Horti Botanici Oxoniensis* by Stephens & Browne (1658). The first part of this catalogue is an alphabetical list, not unlike How's *Phytologia* in its arrangement; however, unlike How and later Ray, Stephens & Browne provide page references to the names they quote from Gerarde and Parkinson, their nomenclatural sources. The second part of their catalogue has its own title-page but, unlike the second part of Ray's *Catalogus*, is not separately paginated. It comprises a translation of the preface of the first part into English and a list of English names and their Latin synonyms. Its format is very similar to that of Ray's 'Index of English names', although in this case the Latin rather than the English names are italicised. It is a likely model for Ray's index, and it is even possible that he took the idea of a second part of his *Catalogus*, "For the benefit of beginners", from this work. According to its preface, Stephens & Browne's book was published in the spring of 1658 ("this little book offers itself to thee now in the spring") and we know that Ray had a copy by the time that his book was published in 1660 (Lankester 1848, pp. 1–2).

The component parts of Ray's *Catalogus* are discussed in more detail below.

'Preface'

The 'Preface' is a rather elaborate composition, replete with classical allusions and quotations, in which Ray describes his struggles to learn botany and pays tribute to his collaborators. It differs in style from the rest of the book and reads like the bravura display of a relatively young classics tutor revelling in the exercise of his skills. Raven (1950, p. 81) describes "the dignity and the simplicity of its Latin", but this does not really capture its essence. We have insufficient knowledge of 17th-century literature to put it into context, but we note that, when Ray's friend Martin Lister published three tracts on spiders, terrestrial and freshwater molluscs and marine molluscs in 1678, he introduced them with a much more matter-of-fact introduction "To the reader" (translated by Parker & Harley 1992).

The 'Preface' begins with a detailed explanation of how Ray came to be interested in botany. He then thanks his collaborators, John Nidd, Francis Willughby and Peter Courthope, and sets out the design of the work with admirable clarity. Finally he comments on the uses of the work and urges like-minded University men to follow him in the study of natural history.

An unusual feature of the 'Preface' is that Ray describes his efforts to learn botany from scratch. One would have expected that, as a country boy from Essex, he would have started with a good basic knowledge of plants,[4] especially as his mother was described after Ray's death as "of great use in her neighbourhood, particularly to her neighbours that were lame or sick, among whom she did great good in chirurgical matters", from which she is assumed to have been a herbalist or herb-woman (Raven 1950, p. 9). However, there is no suggestion in the 'Preface' that Ray started with any existing knowledge or that he saw any point in seeking help outside the University (for example, from herb-women in Cambridge). Presumably this is in part because academic botany was regarded as a different subject, with names (for example) that differed from those of the pharmacists, and perhaps in part because of his belief that people should "begin to be wise from their own experience rather than from someone

---

[4] In the main catalogue Ray says that *Salix folio longissimo* [515] is known as "Osiar" in Essex and Cambridgeshire, a rare reference in the *Catalogus* to the plants of his home area.

else's brain" (although this did not preclude him from starting by reading "whole books"). The absence of reference to earlier knowledge may also simply reflect the self-dramatising nature of this part of the 'Preface'. Indeed, Ray describes his struggles to teach himself botany in such magnified terms that it is possible that the reader was expected to realise that he was exaggerating for dramatic effect.

Two other features of the 'Preface' will be discussed. Ray does not mention previous botanists in Cambridgeshire; his omission of earlier records from the main catalogue is discussed below. Secondly, his acknowledgements raise the question of Nidd's role in the authorship of the *Catalogus*, which we discuss later in this chapter.

'Explanation'

The *Explicatio* lists the works of the authors from which Ray's chosen names and synonyms are cited, listing them under the abbreviations that he uses (not altogether consistently) in the main catalogue.

Almost all the authors listed by Ray in his 'Explanation' are cited as the authorities for names in the main catalogue, although Ray almost certainly mentions more of their works than are relevant to the content of his *Catalogus*. The only one who does not appear in the text is Honorius Bellus, although Aetius and Bellonius are cited only indirectly on the authority of *C.B.* (Caspar Bauhin). Sixteen authors of names in the main catalogue are not included in the 'Explanation'. The omissions, with Ray's names (in the genitive or dative) or abbreviations in brackets, are the classical and mediaeval authors Arnald of Villanova (*Arnold. Villanov.*), Oribasius (*Oribas.* and *Oribasii*), Scribonius Largus (*Scribonio*) and Virgil (*Virgilii*) and the modern authors Girolamo Cardano (*Cardani*), Janus Cornarius (*Cornario*), Giovanni Costeo (*Costæo*), Gabriele Falloppio (*Fallopio*), Hieronymus Braunschweig or Jerome of Brunswick (*Hieronymo Brunswicensi*), Adrianus or Hadrianus Junius (*Jun.* and *Jun. nomenc.*), Peter Lauremberg (*P. Lauremberg. in apparatu*), Georg Melich (*Meilichio*), Thomas Moufet (*Moufet. de diæta*), Adriaan van der Spiegel (*Spigel. Isag.*), Valerand Dourez (*Valerando*) and Giovanni da Vigo (*Vigoni*). Very few names are attributed to these authors: of the modern authors Junius is the source of four names, Lauremberg and Moufet the source of two each and all the remainder have just one name attributed to them (see Table 3.1). For further details of these names, see the biographical notes in Chapter 4.

Ray says in his title of the 'Explanation' that its content is derived "principally from *C. Bauhinus*" and indeed much of it is taken *verbatim* from the *Nomina authorum quorum operâ usi sumus* ("Names of the authors whose work we have used") in the introductory part of Bauhin's *Pinax* (1623, unnumbered pages). Bauhin separates these remarks on his sources from his author abbreviations, *Explicatio nominum authorum citatorum* ("Explanation of the names of the cited authors"), which appear later in his introductory matter, whereas Ray integrates the two. Ray also departs from Bauhin in arranging his list by the surnames of the authors rather than by their first names: for example in Bauhin's treatment MATTHIÆ de Lobel is listed under M,[5] whereas Ray lists him under the abbreviation *Lob*.

Ray obviously had to provide his own brief summaries of books which were not listed by Bauhin. It is easy to assume that these are succinct, independent

---

[5] Stephen Jay Gould (1999) once wrote an essay around another bibliography of Caspar Bauhin constructed on the same principle. He shows that it was the usual way of listing authors in the early 17th century and suggests (on the basis of a small sample of works) that it had been replaced by the current system by 1650.

Table 3.1 Sources of the names in the main catalogue. The following modern authors are directly cited as the sources for Ray's names in the main catalogue. The few classical and mediaeval authors to whom names are directly attributed are not included in this table but are given in the biographical notes in Chapter 4. Chosen names are separated into those for which the author is the sole source cited for the name and those for which he is cited with one or more other sources, but the totals for synonyms and queried synonyms include both sole and joint sources. Authors are counted only once per species, so that, if an author is the source of both a chosen name and a synonym of a species, only the chosen name is counted and, if he supplies two synonyms for a single species, only one is counted. Biographical notes on all these authors are provided in Chapter 4.

Authors of works published anonymously are square-bracketed. Ray's abbreviations are square-bracketed if they are cited in the 'Explanation' but not actually used as abbreviations of the primary sources for names. (Some are used when these authors are cited in other circumstances.)

Author	Ray's abbreviation(s), if any	Chosen names: sole source	Chosen names: joint source	Synonyms	Queried synonyms	Total	Chosen names as % of total
Amatus Lusitanus	Amat., Amat. Lusit.			2		2	0
Anguillara, Luigi	Ang., Anguill.		6	35	1	42	14
Barbaro, Ermolao	[Hermol.]		1	4		5	20
Bauhin, Caspar	C.B.	42	54	465	9	570	17
Bauhin, Jean	I.B., J.B.	57	57	427	13	554	21
Besler, Basileus	Eyst., Eystet.		1	8		9	11
Bock, Jerome	Trag.	2	34	109		145	25
Bodaeus, Johannes	Bodæ., I. Bodæ., Io. Bod., Io. Bodæ., Jo. Bod., Jo. Bodæ.			15		15	0
Brasavola, Antonio Musa				1		1	0
Braunschweig, Hieronymus				1		1	0
Brunfels, Otto	Brunf., Brunf. 4.		2	26		28	7
Camerarius, Joachim	Cam.	1	28	33		62	47
Cardano, Girolamo				1		1	0
Cesalpino, Andrea	Cæs.		7	54		61	11
Colonna, Fabio	Col., Column.			19		19	0
Cordus, Euricius	Eric. Cord.			4		4	0
Cordus, Valerius	Cord.		9	38	1	48	19
Cornarius, Janus				1		1	0
Costeo, Giovanni				1		1	0
Daléchamps, Jacques	Dalec., Dalechamp.			4		4	0
[Daléchamps, Jacques]	Lugd.		28	84		112	25
Dodoens, Rembert	Dod.	7	90	125		222	44
Dourez, Valerand				1		1	0
Durante, Castore	Cast.		2	12		14	14
Falloppio, Gabriele				1		1	0
Fragoso, Juan				1		1	0
Fuchs, Leonhart	Fuch.		24	48	1	73	33
Gassendi, Pierre	Gassend., [P. Gass.], P. Gassend.			2		2	0
Gerarde, John	Ger., Gerard.	72	156	179	5	412	55
Gesner, Conrad	Gesn.		15	81	1	97	15
Guilandinus, Melchior	Guiland.			4		4	0
Hofmann, Caspar	C. Hofman., Casp. Hofman.			4		4	0
[How, William]	P.B., Phyt. Brit.	5		4	1	10	50

3: THE *CATALOGUS* AND ITS APPENDICES

Author	Ray's abbreviation(s), if any	Chosen names: sole source	Chosen names: joint source	Synonyms	Queried synonyms	Total	Chosen names as % of total
Imperato, Ferrante	Imp., Imperat.			2		2	0
Johnson, Thomas	Ger. emac., Johnson in Ger., Johnson apud Ger.	40	38	56		134	58
Junius, Adrianus	Iun., Jun.			4		4	0
Laguna, Andreas	Lac.		8	13		21	38
Lauremberg, Peter	Lauremberg., P. Lauremberg.			2		2	0
L'Écluse, Charles de	Clus.	1	14	45	1	61	25
L'Obel, Mathias de	Lob.†	8	80	117		205	43
Lonicer, Adam	Lon.	1	2	19		22	14
Mattioli, Pietro Andrea	Matt., Matth.	2	62	59		124	52
Melich, Georg				1		1	0
Moufet, Thomas				2		2	0
Officinarum‡	Offic., Officin.	1	5	14		20	30
Parkinson, John	Park.	50	106	376	6	538	29
Paulli, Simon	hort. Bat., hort. Bata., hort. Lugd-Bat.	2	1	25		28	11
	hort. Bat. inter indigenas§			3	1	4	0
	hort. Gron., hort. Groning.		1	14		15	7
	hort. Hafn.	2	2	34		38	11
	hort. Par., hort. Paris.			13		13	0
	hort. Pat., hort. Patav.			5		5	0
	hort. Wars.			8		8	0
Pena, Pierre & L'Obel, Mathias de	Ad., Ad. Lob.†	3	25	82		110	25
Plantin, Christopher	Lob. ico., Plant. ico.	3	12	30		45	33
Pona, Giovanni	Pon.			1		1	0
[Ray, John*]		35				35	100
Reneaulme, Paul de	P. Renealm.			6		6	0
Ruel, Jean	Ruel, Ruell.		1	8		9	11
Schröder, Johannes	Schrod., Schroder.			3		3	0
Schwenckfelt, Kaspar	[Schwenck.], Schwenckf.			2		2	0
Spiegel, Adriaan van der	Spigel.	1				1	100
Tabernaemontanus, Jakob Dieter	Tab.	3	42	92	2	139	32
Thal, Johannes	Thal.		8	49		57	14
Turner, William	Tur., Turn.		16	22		38	42
Vigo, Giovanni da				1		1	0

†Although names from *Ad.* and *Ad. Lob.* are attributed to Pena & de L'Obel, those from *Lob. Ad. part alterâ* and *Lob. Ad. in appendix* are attributed to de L'Obel alone. (See explanation of these works in Chapter 4.)
‡"Of apothecaries' shops". We have not included in this table the synonyms attributed to other unpublished sources such as Barbarians (*Barbaris*) or Italians (*Italis* or *Italorum*).
§As *hort. Bat. among the indigenous plants* in our translation of the main catalogue.
*Names newly coined by Ray in the *Catalogus*.

commentaries, but in fact they are almost certainly simply derived from the long titles of the books themselves. Thus Ray says of Bauhin (1623):

"Finally in the year 1623 he published his *Pinax Theatri Botanici*, the most complete work of 40 years, containing the names of about 6000 plants with their synonyms and distinctive characters: Basel, in quarto."

Compare this with the long title:

Πιναξ *Theatri Botanici ... sive Index in Theophrasti, Dioscoridis, Plinii et Botanicorum qui à Seculo scripserunt Opera: plantarum circiter sex millium ab ipsis exhibitarum nomina cum earundem Synonymiis & differentiis Methodicè secundùm earum & genera & species proponens. Opus XL. annorum Hactenus non editum summoperè expetitum & ad auctores intelligendos plurimùm faciens.*
("A List of the Botanical Theatre ... or an Index of Theophrastus, Dioscorides, Pliny and the Botanists who have written Works over the Ages: setting forth the names of about six thousand plants described by them with Synonyms of the same & the distinctions Methodically according to both their genera & their species. A Work of 40 years Hitherto not published, greatly desired & doing very much towards making the authors understood.")

In three of the entries in the 'Explanation' (under *Apul.*, *Ger.* and *Matth.*) Ray mentions the preface of Thomas Johnson's (1633) edition of Gerarde's *Herball*, but he seems to have drawn on it elsewhere without mentioning it. Because Johnson's text is in English rather than Latin and because he too clearly used Bauhin's (1623) *Nomina authorum* as a source, it is not as easy to identify with certainty where Ray was following Johnson as it is where he was using Bauhin's text, but phrases in the entries for the *Hortus* [Garden] of the Bishop of Eichstätt (*Eyst.*) and for Pliny (*Plin.*), for example, are so close to Johnson's that they can scarcely be independent.[6] Ray's main

---

[6] In the former case Johnson wrote: "*Basil Besler* an Apothecarie of Norenberg, *Anno* 1613 set forth the garden of the Bishop of Eyset in Bauaria, the figures being very large, and all curiously cut in brasse, and printed vpon the largest paper: he onely giues the *Synonima's* and descriptions, and diuideth the worke first into foure parts, according to the foure seasons of the yeare; and then againe he subdiuides them, each into three, so that they agree with the moneths, ..." Ray's text is: *Hortus Eystettensis Episcopi in Bavaria à Basilio Beslero Pharmacopœo Norimbergensi editus, 1613. cum magnis fig. æri incisis. In hoc opere exhibentur synonyma tantùm & descriptiones. In 4 partes secundùm 4 anni tempestates dividitur, & singulæ partes iterum in tres subdividuntur secundùm anni menses.* ("The *Hortus* [Garden] of the Bishop of Eichstätt in Bavaria, published by Basilius Besler, a Nuremberg Apothecary, in 1613, with large figures engraved on copper. In this work just synonyms & descriptions are presented. It is divided into 4 parts according to the 4 seasons of the year & each part is again divided into three according to the months of the year.") In the latter case the relevant words in Johnson's long account are: "*Caius Plinius secundus* ... liued in the time of *Vespasian*, and was suffocated by the sulphureous vapours that came from mount Vesuvius, falling at that time on fire; he through ouermuch curiositie to see and finde out the cause thereof approching too nigh, and this was *Anno Domini, 79*. He read and writ exceeding much, though by the iniurie of time wee haue no more than 37. books *de Historia Mundi*, ... he treats of Plants, ... following ... *Dioscorides*, though he neuer make mention of [him] ... ." Ray wrote: *Caius Plinius Secundus. Vixit temporibus Vespasiani: Vesuvii vaporibus sulphureis suffocatus, dum incendii causas rimaretur, interiit, Anno Dom. 79. Plurima scripsit, quæ injuriâ temporum interciderunt: supersunt de Historia mundi libri 37. in quos plurima quæ plantas spectant è Dioscoride transcripsit, nec ejus tamen unquam meminit.* ("He lived at the time of Vespasian; he died in AD 79, suffocated by the sulphurous fumes of Vesuvius while he was investigating the causes of the conflagration. He wrote numerous works which have perished with the ravages of time; but 37 books about the History of the world survive, into

original contributions to the 'Explanation' are his critical evaluations of the work of his English predecessors Johnson (*Ger. emac.*) and Parkinson (*Park.*) – he cites but does not comment on How's *Phytologia britannica* (*P.B.*) – and his fulsome tribute to Jean Bauhin (*J.B.*) and to his great work *Historia plantarum universalis* (Bauhin & Cherler 1650–1651).

Main catalogue

The main catalogue is the core of the book, an alphabetical list of the species known from Cambridgeshire to Ray or (in a few cases) species recorded by earlier authors but not encountered by him. Ray's entries follow a standard format, with his 'chosen name' first, with the source or sources of the name in italics, followed by synonyms in similar format. There is then a short section in English, set in italics, which comprises one or more English names, a brief statement of the habitat and frequency of the species in Cambridgeshire and a list of its localities. The English name is almost always given, but the habitat and localities are often omitted. A brief informal note sometimes follows. Some species have more formal notes, the "Observations" mentioned in the 'Preface', usually prefixed by N. for a single note (see Figure 3.5) or N. 1., 2., 3. etc. for multiple notes.

Ray's 'chosen name' and synonyms

We refer to the first name in the species entry as Ray's 'chosen name'. Drawing up an alphabetical list was not as straightforward for Ray as it would be later for authors such as Crompton & Whitehouse (1983), who published the most recent alphabetical checklist of the Cambridgeshire flora. Crompton & Whitehouse simply had to choose their standard source of names and the order of the alphabetical list then followed automatically. There was no standard list available to Ray, who therefore had to choose which name to apply to each species. This meant that there was an element of taxonomy in his construction of the alphabetical list. He decided for example to select '*Trifolium*' names for the four species that we currently classify in the genus *Medicago*, although he cited J. Bauhin's names beginning with *Medica* for three of the species and could have listed the species under this 'genus' had he chosen to do so. Ray did not always follow the majority of earlier authors in his choice of name, notably in his name for "Horn-beam tree" where he chose *Betulus* Lob. over *Carpinus* of six cited authorities, perhaps because Gerarde's name was *Betulus sive Carpinus*.

There are numerous departures from strict alphabetical order in Ray's list, some of them probably deliberate as they bring together similar species and some for which there seems to be no apparent reason.

Ray does not seem to have regarded his 'chosen name' as a definitive name for the species. He sometimes cross-references synonyms in the alphabetical list to names other than the chosen name – a considerable inconvenience to the user in an alphabetical list. For example his chosen name for the plant we now call *Aegopodium podagraria* was *Angelica sylvestris minor*, but he cross-referenced both the Latin name *Podagraria* and the English names *Ashweed* and *Herb Gerard* to the synonym *Herba Gerardi*. It seems unlikely that he himself had any concept of the 'chosen name' and he may have regarded all names as having equal validity.

which he transcribed many things that relate to plants from Dioscorides, though he never mentioned him.")

Sources of the names

The sources of the chosen names and synonyms are summarised in Table 3.1. The table demonstrates that, as Ray states in the 'Explanation', his main sources are Gerarde, Parkinson and the Bauhin brothers. It is interesting that, despite his critical remarks about Gerarde (1597), he adopts his names and those of *Ger. emac.* (Johnson 1633) as his chosen name much more frequently than he adopts the names given by the Bauhins. He may have chosen them because they were more familiar to his British readers. Biographical notes on all the authors cited by Ray as sources for his names are provided in the next chapter.

In the 'Preface', Ray describes the works of the Bauhin brothers, Gerarde and Parkinson as "now commonly in everyone's hands among us".[7] However, he did not have access to the works of all the authors he cited as the sources of synonyms (see Chapter 5). One or two works seem to be cited as the sources of synonyms more frequently than one would expect. The only possible value of the names from the garden catalogues assembled by Paulli (1653) in *Viridaria varia*, for example, is that they contributed to the plant portraits that Ray aimed to provide when he wrote in his 'Preface' that "in some plants a collection of Synonyms can take the place of a description". The garden catalogues are simply lists, with no descriptions or illustrations, so that reference to them provides no further information about the species in question. Paulli's book was a recent work which happened to be available to him. It is unclear whether he regarded it at the time as a scarce or rare book and so made a point of citing it. Although both Nidd and Willughby owned copies, Nidd's copy in Trinity College library is the only copy now held by a Cambridge college, and no Oxford college library holds it, unless there are copies which are not revealed by the electronic catalogues. It is of course only with hindsight that we can see that it was not cited by many later authors.

Table 3.2. The length (in words) of Ray's chosen names and synonyms.

	Number	Length (words)									Maximum	Mean
		1	2	3	4	5	6	7	8	9	>9		
Chosen names	630	95	225	123	73	50	21	21	11	4	7	15	3.1
Synonyms	2497	156	591	536	456	279	196	109	87	42	45	19	3.9
Total	3127	251	816	659	529	329	217	130	98	46	52	19	3.7

On average, Ray's chosen names are just over three words long (see Table 3.2). Commentators normally write about the "cumbersome polynomials" of pre-Linnean authors, but over half of Ray's chosen names are only one or two words long. However, as the number of species increased (from the more critical examination of European plants or from the exploration of the New World), the length of names was bound to increase, and there are indications that this was already happening in 1660. The willowherbs, for example, are a group of rather similar species, now classified in the genus *Epilobium*, which are unlikely to have names from the classical authors or from folk taxonomy. Ray included them with loosestrifes in the genus *Lysimachia* and

---

[7] One key work which may have been rather scarce is C. Bauhin's *Pinax* (1623). Examination of the relevant electronic catalogues suggests that this is not now held in any Cambridge college library, even the rich collections at St John's and Trinity, and that the only Oxford college with a copy is Magdalen, which acquired it in the library of the 17th-century botanist John Goodyer. Indeed, it is difficult to imagine that the *Pinax*, an unillustrated list of plant names and synonyms, would ever have attracted a wide readership.

his phrase-names for them are 4–7 words long. Ray's synonyms are rather longer than his chosen names (mean length 3.9 words). These statistics are based on the names that Ray cites from modern authors and exclude names attributed solely to classical authors and queried synonyms; in calculating them we have counted numerals and the abbreviation "i.e." – usually (i.) or just i in Ray's text – as single words. We have had to make some arbitrary decisions about which names to include, as names are not always clearly demarcated from the rest of the text.

A more serious problem in 1660 than the length of names would appear to have been the absence of any criteria for deciding on the correct name for a plant, so that every author was free to select any existing name or to coin his own. In the *Catalogus* Ray refrained from introducing new names, except when he refashioned the taxonomy of a group (e.g. *Salix* and *Triticum*) or when he could not trace a species in the existing literature. He devised descriptive phrase-names for 35 species, and these are 3–15 words long (mean 6.7), much longer than the average length of his chosen names from other authors.

Definition of species

Ray comments in the 'Explanation' on Parkinson's tendency to recognise too many species, and he occasionally mentions problems of species delimitation in his own text, for example when discussing the variable *Bursa pastoris* [97] and *Coronopus* [152], the two ivies *Hedera arborea* and *Hedera helix* [270, 271] or the two millefoiles *Millefolium vulgare album* and *Millefolium vulgare flore diluti ruboris* [371, 372]. However, he also includes in separate entries (and therefore presumably by implication treats as species) a number of plants which he later came to recognise as variants of a single species, such as purple- and white-flowered *Jacea segetum* [298, 299], single- and double-flowered *Morsus ranæ* [375, 376] and *Lysimachia purpurea* with two and three leaves in a whorl [347, 348]. We have included information on his subsequent treatment of many of these variants in our footnotes. In the following years Ray gave more consideration to the problem of whether such variants constitute a species, and he discusses the question in detail in his preface to *Catalogus plantarum Angliae* (1670) and in the paper *Of the specifick differences of plants* which he sent to the Royal Society in 1674 (Cain 1999a, b). In the latter he introduced the criterion that variants of common parentage must belong to the same species.

Treatment of habitats and localities

Much of the charm of the *Catalogus* to the modern reader lies in the italicised English text at the end of many of the species accounts, and especially in the localities that Ray cites. Some of these are given in the manner of a modern Flora, such as *Orchis morio mas foliis maculatis* [*Orchis mascula* 398] "*In Madingley and Kingston woods*" and *Potamogeiton perfoliatum* [*Potamogeton perfoliatus* 463] "*In the river Cam plentifully every where*". More evocatively, he describes other sites as he would have come across them during his exploration of the county. Most readers must surely imagine themselves accompanying Ray through the countryside of 17th-century Cambridgeshire when they read entries such as *Pimpinella sylvestris* [*Sanguisorba officinalis* 438] "*in the pastures about Long-Stanton and Cottenham as you ride to Audrey caussey in great plenty*" or *Cirsium Anglicum primum* [*Cirsium dissectum* 137] "*In the first close you pass through, as you go in the foot-way from Cambridge to Cherry-hinton, near a little ditch or gripe that crosseth the close from corner to corner*".

Table 3.3. The number of species for which Ray provides habitat and locality details. For examples of the different classes, see the text.

	No frequency or locality	Frequency only	Sample locality or localities	All localities	Total
No habitat	128	6	6	18	158
Habitat stated	161	75	111	125	472
Total	289	81	117	143	630

There is in fact great variation in Ray's treatment of localities (see Table 3.3). Of the 630 taxa in the main catalogue, only 260 are reported from a more or less precise location. For 117 of these one or more localities are provided as examples, so that *Circaea lutetiana* [136] is described as occurring "*In the lanes about Cherry-hinton, & in many other moist and shady places*" and *Gramen tomentosum & Linogrostis* [*Eriophorum angustifolium* 265] "*On Hinton and Teversham moores and elsewhere*". For the remaining 143 taxa, the sites listed appear to be the only ones known to Ray. These include *Polygonum exiguum* [*Scleranthus annuus* 453] "*About Hoginton where the* Myosuros *growes, & in the sandy grounds about Gamlingay*" and *Erica vulgaris* [*Calluna vulgaris* 184] "*On Gogmagog hills and Newmarket heath*". Almost all these species with a stated locality also have habitat information, although it is omitted for 24 of them.

There are 370 taxa listed in the *Catalogus* without precise localities. Habitats are provided for 236 of them. Of these, 75 also have an indication of frequency in the county, including *Armerius sylvestris* [*Lychnis flos-cuculi* 61] "Frequent in moist meadows" and *Speculum Veneris minus* [*Legousia hybrida* 556] "Common among standing corn". Sometimes this information is provided by reference to the preceding species, as *cum priore*. The remaining 134 taxa lack both habitats and localities, and of these only six have any indication of frequency, so that 128 are simply listed as names without any ecological or distributional information. These include some very common trees, a wide range of herbs and (frustratingly) a high proportion of some of the genera that Ray found taxonomically troublesome, including all four *Lapathum* (i.e. *Rumex*) species [321–324] and four of the eight *Carex* species [243–245, 247–250, 263].

A few localities recur repeatedly throughout the text, and the more notable species from these are listed in the 'Index of places'. The two boulder-clay woods which Ray clearly knew well and which are repeatedly cited together are Madingley Wood, on the outskirts of Cambridge, and Kingston Wood, by the (then) road to Gamlingay. Also frequently paired are two localities for plants of chalk grassland, again one near Cambridge (Gogmagog Hills) and another further afield (Newmarket Heath). Ray listed several species of dry, acidic soils from the Hill of Health in Cambridge and from Gamlingay on the county boundary. Gamlingay was the main site that he knew for plants of acidic peatlands, particularly "*the boggy ground by Sir Roger Burgoynes park, where they dig turfs*".

The numerous sites known to Ray in the city of Cambridge are listed in the gazetteer. It is notable that, although Ray mentions several colleges, "*the back side*" of the colleges, now The Backs, and the two large commons on the north side of the city, now called Coldham's Common and Stourbridge Fair Green, he does not mention by name two sites close to the city which are much cited by later botanists, Coe Fen and Sheep's Green. He did however visit Coe Fen, as he reports *Morsus ranæ* [*Hydrocharis morsus-ranae* 375] "*In the ditches cut out of the river Cam behind Peterhouse*". He often mentions "closes", small enclosed fields, in villages around Cambridge and in the south of the county, such as Barnwell, Chesterton and Linton. The characteristic land-use pattern of closes near the centre of the village and open fields beyond

3: THE *CATALOGUS* AND ITS APPENDICES

is clearly shown on Langdon's map of Gamlingay, part of which is reproduced as Map 2 in our gazetteer. Most closes were pastures and all those mentioned by Ray appear to have been grassland rather than arable land. From one such close, "*a bushy close on the south side of Whitwell a single farm-house beyond Coton in the way to Hardwick*", Ray reports *Helleborus niger hortensis flore viridi* [*Helleborus viridis* 276], *Narcissus pallido-luteus longo calyce* [*Narcissus pseudonarcissus* 386], *Bistorta major* [*Persicaria bistorta* 87] and *Pimpinella sylvestris* [*Sanguisorba officinalis* 438]. *H. viridis* grew "*near unto the house*" and the first three species were almost certainly planted here or garden escapes. Ray's only comment in the *Catalogus* on the problem of distinguishing wild plants from garden escapes is in his 'Preface', where he lists some of the species "which we suspect did not originally spring up naturally but were sown by some chance & by accident in the places which we have mentioned". The fact that his *Catalogus* was not restricted to wild plants presumably reduced the need to decide which of these species to include and which to exclude. Later, in preparing his *Catalogus plantarum Angliae* (1670, 1677), he struggled with the problem of which garden escapes from Cambridgeshire to include. (See our discussion of the 1685 appendix below.) The problem is still as acute today.

Ray's observations on aquatic plants are often based on the River Cam, especially in the vicinity of Cambridge, and on the Fens, especially by Aldreth Causeway ("*Audrey caussey*") and Stretham Ferry, the two main crossing-points of the Ouse north of Cambridge. Several of the aquatic plants he described were new to science or were known only from very inadequate descriptions in earlier works. Ray clearly gave aquatics more attention than earlier authors and was able to take advantage of the rich aquatic flora then present in the county.

The records from the "moors" or "moores" around Cambridge are particularly valuable, as these habitats have been destroyed or transformed beyond recognition by enclosure followed by agricultural improvement in the early 19th century and (in the case of sites nearer Cambridge) have often been obliterated by more recent urban development. Hinton and Teversham Moors are the two main sites Ray mentions, with Trumpington Moor sometimes added; more frequent plants such as *Pinguicula vel Liparis* [*Pinguicula vulgaris* 443] are described as "*On all the moores about Cambridge plentifully*." The species that he reports from the moors are listed in Table 3.4. Although the word "moor" now suggests an acidic habitat, these must have been for the most part calcareous wetlands, fed by streams and springs flowing from the nearby chalk, with drier pastures nearby; they must, however, have had some areas of more acidic habitat. The best surviving evidence of the flora of the spring-fed moors of Cambridgeshire comes from the "detailed and vivid description" of Triplow Heath provided by the manuscripts compiled by G.N. and N. Maynard in the early 19th century and summarised with other records from this site by Crompton (1959). Curiously Ray does not list any plants of Triplow Heath itself, though he lists plants from cornfields and "layes" on its borders. Later records show that this was an area which was almost unbelievably rich by modern Cambridgeshire standards. There was both chalk grassland and some more acidic chalk-heath or heath, supporting "a profusion of *Calluna vulgaris*", and both calcareous and acidic wetlands. The closest surviving vegetation in the county to Ray's moors is almost certainly at Chippenham Fen, a site which was not known to Ray but which was called Chippenham Moor by 19th-century botanists. Several of the species listed by Ray from the moors around Cambridge still survive in the National Nature Reserve at Chippenham, and others survived until the 20th century but now appear to be extinct there; the latter include the species found in the more acidic wetland habitats (see Table 3.4).

Table 3.4. Species reported from the moors about Cambridge in the *Catalogus* and its two appendices.
Species recorded from Chippenham Fen in the late 19th or the 20th century are marked by an obelus (for those not seen since 1969) or an asterisk (for those seen from 1970 onwards). I indicates records taken from the 'Index of places' but not specifically cited from the moors in the main catalogue (though records of *Carex* and *Juncus* spp. in the former are excluded). A indicates records from the 1663 appendix and B records from the 1685 appendix.

Plants of calcareous wetlands

*Anagallis tenella**
Butomus umbellatus† (I‡)
*Carex lepidocarpa**
*Cladium mariscus**
*Epipactis palustris**
Liparis loeselii†
*Parnassia palustris**
*Pinguicula vulgaris**
*Ranunculus lingua**
Sagina nodosa
*Samolus valerandi**
*Schoenus nigricans**
*Valeriana dioica**

Wetland species with a broad ecological range

*Alnus glutinosa**
Carex pendula (A & B, in a ditch "*adjoyning to Teversham moor*")
*Galium palustre**
*Hydrocotyle vulgaris**
*Marchantia polymorpha** (A & B)
Nymphaea alba†
*Potamogeton berchtoldii** and/or *pusillus*
*Potamogeton natans** (I)
*Triglochin palustris**

*Typha latifolia**
*Veronica scutellata**

Plants of more acidic wetlands

Drosera intermedia
Epilobium palustre
Eriophorum angustifolium† (generally calcifuge, but sometimes found in calcareous sites)
*Polytrichum commune**
Utricularia minor† (B)

Plants of drier ground

*Aquilegia vulgaris** ("*In a little thicket*")
*Asperula cynanchica**
Astragalus danicus
*Blackstonia perfoliata**
Erigeron acris
Euphrasia species*
Hippocrepis comosa† (I, A & B, "*on the drier part of Hinton moor*")
Lythrum hyssopifolia (I) (winter-flooded ground)
*Polygala vulgaris**
*Potentilla erecta**
Spiranthes spiralis ("*on the skirts of Teversham moor*")

‡ Reported from Teversham in the 'Index of places', "*About the closes and on the moor*". We have assumed from its usual habitat, base-rich waters and swamps, that it was found on the moor.

Omission of earlier Cambridgeshire records

As noted above, Ray does not mention any earlier Cambridgeshire botanists in the 'Preface'. In the main catalogue the only records he lists from earlier publications are those that he has failed to refind and in most cases regards as errors. These are *Turritis* [602] from Gerarde (1597, p. 213), *Galega* [208] and *Staphylodendron* [561] from Parkinson's *Theatrum botanicum* (1640, pp. 640 and 1417), *Abotanum campestre* [1], *Abrotanum inodorum* [2] and *Portulaca sylvestris* [459] from How's *Phytologia britannica* (1650, pp. 1 and 97) and an unpublished record of *Plantago major paniculâ sparsâ* [449], "*Found by Dr Strachey*". It is extraordinary that the only indication in the *Catalogus* that there had been botanical activity in Cambridgeshire before that of Ray and his colleagues comes from these unconfirmed records. In fact, published records from the county, though few, date back to those of Turner (1538). An outstanding omission from the *Catalogus* is Gerarde's (1597, p. 309) site for *Pulsatilla* [475] at Hildersham, "very plentifully in the pasture or close belonging to the personage house of a small village six miles from Cambridge, called Hildersham; the parsons name that liued at the impression heereof was master *Fuller*, a very kinde and louing man, and willing to shewe vnto any man the saide close, who desired the same". Hildersham was also mentioned by How (1650) as a site for the species and later by Ray himself in *Catalogus plantarum Angliae* (1670). In addition to published records, there were botanists in the county who have left no such tangible record of their activities. In a letter written from Christ's College, Cambridge, on 15 September 1648, for example, Henry Power (1623–1668) told Sir Thomas Browne that he "had simpled in the woods, meadows & Fields ... I have both Gerard, with Johnson's addition, & Parkinson ... I compared also Dodonæus with them" (Keynes 1964). By the 1650s Power was practising as a physician in Halifax, but one wonders what records Ray might have obtained if he had contacted him. Ray's failure to give any attention to his Cambridgeshire predecessors has influenced botanists to the present day, and there is still no adequate account of botany in the county before 1660.

Numbered observations

In a letter written to his friend Samuel Hartlib on 13 July 1661, John Worthington looked forward to Ray's proposed *Phytologia britannica*, the book which was to be published as *Catalogus plantarum Angliae*:

> "I wish him to be as full at least (if not more large) in some pertinent *Observations* upon such and such Plants, as he is in the *Phytologia Cantabrigiensis*; for methinks such portable *Phytologies* as have only the *Latin* and *English* Names (without any choice *Notes*) are but lean and imperfect Things." (Worthington 1704, p. 262; reprinted in modernised form by Crossley 1847, p. 345).

The inclusion of "choice *Notes*" is one of the distinctive features of the Cambridge catalogue. They do not appear in any of its apparent models, Bauhin (1622), Johnson (1634, 1641) or How (1650), with the exception of Johnson's (1641) long discussion of *Pisa marina Anglica*, reproduced by How. Information culled from previous authors had of course been a feature of books about plants since classical times; however the form of Ray's notes, in discrete paragraphs prefixed by N. with multiple notes numbered, does not seem to be taken from his immediate botanical predecessors. If this format was based on an earlier model, rather than devised by Ray himself, it was perhaps taken from a non-botanical work.

Ray's formal notes are easily distinguished by their layout from the brief comments he sometimes makes in the species accounts, such as his remarks on flower colour variation in *Polygala* [*Polygala vulgaris* 451] or the comment from Jean Bauhin on the variation of *Bursa pastoris* [*Capsella bursa-pastoris* 97]. For the botanical observations the distinction is to a large extent a matter of format rather than content, as some of the comments that are included in the initial species entries, such as his description of the flowers of *Stellaria aquatica* [*Callitriche* 562], are certainly more substantial than some of his formal botanical notes, such as the observation that the leaves of *Conyza media* [*Pulicaria dysenterica* 142] smell of soap when squeezed with the fingers. However, non-botanical matter tends to be treated solely in the notes.

There are 125 taxa with associated notes and a total of 204 notes in all. The taxa to which the notes are attached tend to be plants which are of some practical use, particularly crops and trees but also dye-plants, species grown for medical, culinary or horticultural use and wild plants which can be eaten ('food for free'). The nine species which attract the most observations are hazel [*Corylus sylvestris* 154], hemp [*Cannabis prima sive sativa* 104], oak [*Quercus latifolia* 476], wheat [*Triticum* 601] and wormwood [*Absinthium vulgare* 3], all of which have four notes, and flax [*Linum sativum* 336], holly [*Agrifolium* 15], peas [*Pisum arvense* 444] and elm [*Ulmus* 626], which have five each. Some general observations appear to be attached to one of many species to which they could apply, such as the remarks on annual and perennial plants included under the annual weed *Chenopodium album* [*Atriplex sylvestris* 70] or the comment on bulbs under bluebell [*Hyacinthus Anglicus* 293]. An interesting note on the distribution of montane plants follows the entry for *Clinopodium ascendens* [*Calamintha vulgaris* 100], for no apparent reason.

There are 39 observations which are not attributed to any source, 125 based on one or more sources and 40 which consist of a mixture of attributed material and subsequent comment or include original material with a reference to a related text. Some of the unattributed observations simply report what was probably common knowledge, such as the use of holly roots [15] to produce bird-lime, or at least widely held beliefs, such as the efficacy of the water collected in axils of the leaves of teasels [*Dipsacus* 168] for getting rid of warts. However, many others appear to be based on original observations. Some of these are relatively trivial, such as those dealing with the nausea induced by chewing a succulent branch of hemp agrimony [*Eupatorium cannabinum* 191] or the effects of eating saffron [*Crocus sativus* 159] on "the excrement of the belly". However, there are a number of extended notes, and it is a remarkable feature of the *Catalogus* that some of the longest notes are on zoological subjects. In these Ray describes his (or his colleagues') observations on animals in the wild, in captivity or *post mortem*. Ray's enthusiasm for natural history is manifest in these notes, which are perhaps the most vivid passages in the book. They include the account of 'Cuckoospittle', which unusually is a note attached to a synonym, *Papaver spumeum* [after 417], and the observations on parasitised caterpillars under turnip [*Rapum sylvestre* 495], where Ray feels obliged to apologise for the length of the note but excuses it "since we have so far read, seen or heard nothing like it". There are a few long notes about plants, including two, one on tree rings under ash [*Fraxinus vulgaris* 203] and the other on the Doctrine of Signatures under elder [*Sambucus* 519], in which his analytical abilities are particularly evident. These long, original notes presumably include the "very unusual observations" mentioned, perhaps with some pride, by Ray on the title-page of the *Catalogus*.

The notes drawn from other authors cover a wide range of topics, including what would now be regarded as pure botany as well as the practical use of plants for food or building materials and in horticulture and medicine. They are extraordinarily varied in character, ranging from the short and practical note from Lauremberg describing a method of killing large trees with mercury under *Quercus latifolia* [476] to the long anecdotes of Henry from Heer under *Ulmus vulgatissimus* [626]. The sources that Ray draws on are listed in Table 3.5. Some of these books are described in the 'Explanation', such as those by Pliny the Elder and Theophrastus amongst the classical authors and by Camerarius and Jean Bauhin amongst the modern. However, the notes draw upon a much wider range of sources than the primarily botanical books listed in the 'Explanation', including, for example, Lauremberg's horticultural works, Hofmann's (1646) *De medicamentis officinalibus*, Butler's (1634) book on bees and Moufet's (1634) on insects. It is clear that the 'Explanation' lists the authors of Ray's chosen names and synonyms but does not deal with other works cited in the *Catalogus*. Ray also drew extensively on a wider range of works than those listed in the 'Explanation' in his discussion of the meanings of botanical names in the 'Etymology' and of botanical terms in the 'Interpretation'.

We suspect that Ray or his collaborators had access to all the works cited in the notes, as he is meticulous in acknowledging when observations by one author are cited from the works of another; see, for example, the information from the works of Costeo and Goebel taken from Camerarius (1588) in the notes to *Clematis daphnoides minor*

[138] and *Solanum sive Solatrum vulgare* [546] respectively. We are not aware of many instances in the main catalogue where Ray completely omits what was obviously the source of his information, although in the note following *Betonica* [84] he does not mention Pliny as his source for the list of classical authors who wrote about single species.

It is when reading the notes that it becomes clear that Ray was writing at a time of transition between the mediaeval and the modern world views. Ray cites, for example, observations on spontaneous generation, as under *Solanum sive Solatrum vulgare* [546]; his comments there suggest that this theory of the origin of life from inanimate matter had recently received a boost from observations made with the microscope. However, it received no support from the experiments of Francesco Redi (1668, 1671), and Ray was to reject it unequivocally in his later work (Ray 1692, pp. 74–94, 1693, pp. 14–22; cf. Raven 1950, pp. 375–376), although it was not until the 19th century that all scientists abandoned it in the face of the more precise experiments of Louis Pasteur. He describes small experiments by which he has tested the assertions of some of the authors he cites, such as his investigations of the effect of soaking the inner bark of ash in water [203] or of cooking peas with salt [444]. The need to collect a medicine on Midsummer Eve or Midsummer Day is reported without comment under *Artemisia* [62], although Ray does refer the reader here to Bauhin & Cherler (1651), who denounced "these and similar superstitions"; he himself describes as superstitious the belief that the "seeds" (i.e. spores) of bracken, *Filix fæmina* [200], should be collected on particular days.

Medical background to the Catalogus

The botany of the 17th century emerged from a predominantly medical tradition in which plants were used as the main source of drugs. Many of the writers cited in the notes were primarily physicians, and some of the observations refer directly to the medical uses of plants. Ray discusses whether plants and other substances are "hot" and "dry" (see *Ranunculus flammula* [483]) or "cold" (see the note on cuckoo-spittle after species 417). This refers to Galenic medical practice. Galen, basing his views on the Hippocratic corpus,[8] identified hot, cold, dry and wet as the four basic qualities. The body was made up of four humours which reflected these qualities, yellow bile (hot, dry), black bile (dry, cold), blood (moist, hot) and phlegm (moist, cold). Each person had a particular balance of humours; illness was the result of an imbalance which could be corrected by appropriate treatment. The whole body rather than the specific illness was therefore the subject of treatment (often by measures such as blood-letting, purging or sweating) and the constitution of the individual patient had to be taken into account in devising the appropriate treatment. The plants from which drugs were obtained could be classified in terms of the four qualities, which they possessed in one of four degrees, weak, obvious, strong and massive. Dioscorides had listed medicinal plants without any theoretical framework, but Galen was able to classify many of his plants by their qualities, although he was often unable to specify the degree to which they possessed them. Important editors of Dioscorides such as Mattioli were followers of Galen. Gerarde's *Herball* usually includes a section in each chapter on *The temperature*, and this provides information such as "Ribwoort is colde and drie in the seconde degree, as are the Plantaines" (Gerarde 1597, p. 342) or "Betonie is hot and drie in the second degree: it hath force to cut, as *Galen* saith" (Gerarde 1597, p. 578).

---

[8] Galen usually attributed his views to Hippocrates, but "Galen's Hippocrates is to a large extent a hagiographical fiction" (Hankinson 2008).

Table 3.5. Sources cited in the observations in the main catalogue. Both directly and indirectly cited sources are included, as are references to works provided simply as authors of the names of organisms (usually insects) discussed in Ray's otherwise original notes. If Ray identifies his specific sources, or if we have identified them, we give the name of the works of classical authors or a reference to the relevant entry in the bibliography for modern authors; the works cited in the bibliography are not necessarily the editions used by Ray.

Author	No. of notes in which cited	Entry number(s) of species with notes
Aldrovandi (1599–1603)	1	628
Aristotle, *De generatione animalium* and *Historia animalium*	5	128, 193, 502, 615, 622
Bacon (1626)	8	74, 215, *Papaver spumeum* (note to a synonym), 476, 502, 569, 617, 618
Bassus, *Geoponica*	2	142, 541
Bauhin, Caspar in Mattioli (1598)	1	200
Bauhin, Caspar (1623)	2	44, 502
Bauhin, Caspar (1658)	1	291
Bauhin, Jean (1598)	2	199, 336
Bauhin, Jean (Bauhin & Cherler 1650–1651)	18	20, 62, 104, 159, 170, 200, 291, 329, 331, 360, 457, 476, 502, 560, 622, 626 (2 notes), 629
Bennet (1654)	1	19
Bock (as Tragus)	3	85, 200, 215
Bodaeus (1644)	12	62, 104, 131, 142, 342, 362, 444, 456, 529, 541, 548, 603
Brasavola (1540)	1	607
Braun & Hogenberg (1575)	1	209
Browne (1658)	1	601
Butler (1634)	5	20, 155, 167, 476, 615
Caesar, *De bello gallico*	2	226, 565
Camerarius (1588)	19	3 (2 notes), 5, 15, 53, 131, 138, 203, 303 (2 notes), 310, 358, 471, 500, 519, 546, 565, 607, 626
Cesalpino (1583)	1	203
Charleton (1650a)	1	85
Cicero, *De divinatione*	1	474
Columella, *De re rustica*	2	342, 495
Costeo (1578)	2	138, 474
Daléchamps (1586–1587)	1	329
Dioscorides, *De materia medica*	7	63, 154, 164, 200, 203, 361, 362
Dodoens (1583)	2	200, 560
Fernel (1644)	1	62
Galen	2	200, 569†
Gassendi (1658)	5	15, 203 (2 notes), 279, 509
Gerarde (1597)	4	291, 457, 548, 626
Gesner	2	20, 54, 191
Goebel (1565)	1	546
Goropius (1580)	1	336
Heer (1645)	1	626
Helmont (1644)	1	85
Highmore (1651)	1	5
Hippocrates, *De alimento*	1	154
Hofmann (1646)	17	3, 84 (2 notes), 154, 161, 164, 201, 215, 229, 270, 310, 328, 363, 474, 481, 560, 569
Johnson (1633)	2	200, 262
Kentmann (1565)	1	20
Kircher (1658)	3	132, Fungi (introductory note), 546
Krantz (1580)	1	19

3: THE *CATALOGUS* AND ITS APPENDICES

Author	No. of notes in which cited	Entry number(s) of species with notes
Lauremberg (1654a)	14	64, 79, 85, 148, 260, 293, 333, 356, 361, 416, 444 (2 notes), 476, 569
Lauremberg (1654b)	2	19, 159
Linden (1651)	1	84
Maier (1616)	2	336 (2 notes)
Maranta (1559)	7	19, 164, 276, 361, 456, 469, 626
Moufet (1634)	9	54, 154, 168, 333, 495, 502, 518, 521, 628
Moufet (1655)	2	4, 79
Musa [misattributed], *De herba vettonica*	1	84
Parkinson (1640)	4	161, 200, 291, 626
Paulus	1	342
Pena & L'Obel (1571)	6	100, 104, *Papaver spumeum* (note to a synonym), 548, 565, 626
Pena & L'Obel (1605)	1	262
Pliny, *Naturalis historia*	18	6, 20, 62, 84, 154 (2 notes), 159, 200, 203 (2 notes), 226, 303, 336, 456, 502, 565, 603, 622
Porta (1589)	1	628
Scaliger (1557)	1	622
Schröder (1648)	1	148
Schröder (1649)	5	47, 62, 207, 471, 617
Sennert (1643)	1	626
Spiegel (1633)	9	19, 160, 188, 193, 200, 296, 417, 444, 502
Theophrastus, *De causis plantarum*	15	15, 20, 70, 173, 289, 328, 361, 416, 423, 444, 476, 601, 617, 626, 629
Theophrastus, *Historia plantarum*	19	15, 53, 62, 63, 64 (2 notes), 151, 154, 159, 200, 203, 356, 456 (2 notes), 476 (2 notes), 509, 565, 601
Varro, *De re rustica*	1	495
Vitruvius, *De architectura*	1	20

† Cited by Ray as *De medicamentorum facultatibus*, presumably the work *De simplicium medicamentorum facultatibus* which was included in collected editions of Galen's works.

There was quite a gulf between Galen's theory and his medical practice. Measuring the degree of a quality in a plant was difficult, given the absence of any methods of measurement and the fact that the temperatures were not absolute but had to be considered in relation to the patient's state. Many drugs were not applied as 'simples' but in compound prescriptions which had effects that could not be predicted by theory. Galen therefore emphasised the need for empirical study of the effects of drugs and set great store by observation and experience. For helpful introductions to Galen's medicine, from which this brief summary is derived, see Hankinson (2008), Nutton (2004, 2008) and Vogt (2008).

In the early 16th century orthodox Galenists had been assailed by the extremely heterodox figure of Theophrastus Philippus Aureolus Bombast von Hohenheim (*c.* 1493–1541), otherwise known as Paracelsus. The son of a physician, he was appointed Professor of Medicine at Basel in 1527 despite an apparent absence of formal qualifications. He set out to reform medicine just as others were seeking to reform the church, and he regarded the medicines sold by traditional physicians as no better than the indulgences sold by fund-raising churchmen. On his appointment to the Basel chair he had the support of Erasmus, Oecolampadius and other church reformers.

However, his behaviour at Basel soon embarrassed his supporters. He became notorious for delivering his lectures in German rather than in Latin and for his wholesale condemnation of traditional science and medicine. He publicly burnt a traditional medical textbook on a bonfire celebrating the eve of St John's Day in 1527, and he made no secret of his belief that, as he once wrote, his shoelaces knew more than Galen and Avicenna. "Expressions of scientific reserve or uncertainty were entirely alien to the Paracelsus method." (Webster 2008, p. 59). He only lasted a year in the university, leaving in 1528 to spend the rest of his life as a wandering preacher and medical practitioner.

It is almost impossible to summarise Paracelsus' approach to medicine in a single paragraph. "Magic, the kabbalah and their related disciplines were extolled at every opportunity from his earliest to his last literary efforts." (Webster 2008, p. 65). Paracelsus recognised three important 'principles', salt, sulphur and mercury, of which salt (broadly interpreted to include all salts rather than just sodium chloride) was the most important. He regarded diseases as having an external cause (sometimes poisons brought from the stars) rather than resulting from humoral imbalances. He therefore advocated treatments for specific diseases rather than general treatments such as blood-letting. He favoured drugs based on chemicals, including metals, which were treatments based on alchemical practices, rather than the plant-based drugs of the apothecaries. He criticised the apothecaries for the large profits they made from selling traditional drugs and he thought that the foreign medicines recommended by Greek and Arab authors were useless in Germany. He published very little during his lifetime, but his followers published his medical works from 1560 onwards and over 350 editions of his books were published in the next century. The *Dictionary of scientific biography* (DSB) gives a summary of his life and beliefs, and Weeks (1997) and Webster (2008) provide more extended treatments. Paracelsus does not feature directly in Ray's text but Ray does draw on the work of Joseph du Chesne or Quercetanus (*c.* 1544–1609), whose views were similar. He also cites those who developed the Paracelsian approach in later generations, including J.B. van Helmont (1579–1644) and van Helmont's English translator Walter Charleton (1620–1707). Van Helmont's influence was particularly strong in the Commonwealth, "soon eclipsing that of Paracelsus himself; the 'Helmontian' physicians became the most articulate and active opponents of Galenic medicine" (Webster 2002, p. 276). Charleton, regarded by Webster (2002) as "an opportunist and populariser" (p. 91) who "reflected almost every philosophical point of view in turn" (p. 496), later (1654) wrote a treatise based on the work of Gassendi. Daniel Sennert (1572–1637), also cited in the main catalogue, attempted to reconcile the work of Aristotle, Galen and Paracelsus (DSB) whereas Caspar Hofmann (1572–1648) was one of the most diehard defenders of the Galenic tradition (see Chapter 4).

The final medical theory which needs to be mentioned is the Doctrine of Signatures, the view that the appearance of a plant indicated the part of the body that it could be used to treat medically. This probably had its origins in ancient Greek medicine and in traditional folklore, but it was not systematically set out as a theory until the 16th and 17th centuries. Paracelsus developed it, and his belief in treating specific diseases doubtless led to its increasing prominence. Works which played an influential part in its spread included Giambattista della Porta's *Phytognomonica*, first published in Naples in 1588 (see Arber 1986), and Jakob Böhme's *De signatura rerum*, written in 1622 but published posthumously in 1635. Böhme (1575–1624) was a master cobbler of Görlitz, largely self-taught, who believed that the secrets of nature had been revealed to him in 1600 in a mystical vision. His book dealing with signatures was translated into English by J. Ellistone and published in 1651 as *Signatura rerum, or,*

3: THE *CATALOGUS* AND ITS APPENDICES

The signature of all things: shewing the sign and signification of the severall forms and shapes in the creation, and what the beginning, ruin, and cure of every thing is. In it he attempted "to restore a sense of order to the human world by uncovering the concealed design of the divine and natural worlds" (Weeks 1997). The Oxford graduate and botanist William Cole (1626–1662) "carried the doctrine of signatures to as extreme a point as can well be imagined" (Arber 1986) in his book *Adam in Eden, or, Natures Paradise* (1657). One reason for the popularity of the belief in signatures was the view that it was a benevolent God who "hath imprinted upon the Plants, Herbs, and Flowers, as it were in Hieroglyphicks, the very signature of their Vertues" (Turner 1664, introductory pages).[9] Critics of the doctrine had included Dodoens and the French botanist Guy de La Brosse (*c.* 1586–1641), first Intendant of the Jardin du Roi, Paris. Ray's opinion is set out at some length in a note which follows *Sambucus* [519], one of the best known passages in the *Catalogus*.

'Index of English names' and 'Index of places'

The second half of the *Catalogus* begins with two sections, the 'Index of English names' and the 'Index of places'. They are the most straightforward parts of the book and present few problems of interpretation. The English names are listed in alphabetical order of 'genus' and 'species', so that the Archangells are all listed under A and all the Hawkweeds under H, but the order of the names is not reversed as it would be in a modern index. This arrangement, in combination with the design of the book, has one unfortunate consequence. Some of the emphasised letters at the start of the alphabetical sections are misleading, so that D, for example, starts with a large letter E for "English Daffodill" and M with a large G for "Great bastard Madder".

As in the main catalogue, the degree of inconsistency in the 'Index of English names' surprises the modern reader. There are some departures from alphabetical order, so that "Daisie" comes between "Dandelion" and "Danewort" and O starts with "Oneberry", followed by "Oak-tree" and "Oats". The numerous minor differences between the names in this section and those in the main catalogue are not surprising. However, some of the Latin names are Ray's synonyms rather than his chosen names. There are also both English and Latin names that do not appear in the main catalogue, although most of them can be identified with species listed in it. The three exceptions are species that are completely missing from the main catalogue: *Phyllitis* [*Asplenium scolopendrium*] is listed here and in the 'Index of places', as well as in the 1663 appendix [A29], whereas both *Anagallis lutea nemorum* [*Lysimachia nemorum* L.] and *Pirus sylvestris* [*Pyrus pyraster* (L.) Burgsd.] appear nowhere else in the *Catalogus* or its appendices.

The 'Index of places' provides the field botanist with an indication of the most interesting sites to visit. It lists the rarer species of 12 areas in the county, most of them localities mentioned repeatedly in the main catalogue. As in the 'Index of English names', the names are not always Ray's chosen names in the main catalogue. (These inconsistencies are discussed in our footnotes.) The closest precedents to this index in the English botanical literature are Thomas Johnson's lists of plants seen on the excursions of the Society of Apothecaries and in particular his Flora of Hampstead Heath, the first British Local (as opposed to County) Flora (Johnson 1629; Gilmour 1972, pp. 3, 41–43, 65–68); there is, however, no evidence to suggest that Ray knew of these publications in 1660. Ray's 'Index of places' presumably proved useful, since, in his

---

[9] It is not perhaps surprising that the Doctrine of Signatures proved to be more popular than the theory of the 15th-century Friar Nicholas of Poland, who thought that God had implanted special healing virtues in revolting things and therefore urged his patients to eat snakes, lizards and frogs (Siraisi 1990).

later Cambridge Flora, *Plantae Cantabrigienses* (1763), Thomas Martyn included similar lists in a section entitled *Herbationes Cantabrigienses* (Walters 1981, p. 39). His successor as Professor of Botany, John Stevens Henslow, issued printed lists of plants found on his field excursions (Walters & Stow 2001, pp. 72–73).

'Etymology'

The last three sections of the *Catalogus* are the 'Etymology', 'Interpretation' and 'The more customary headings or divisions of plants', treatments of etymology, terminology and classification respectively. The 'Etymology' and the 'Interpretation' appear to draw on a rather different (although broadly overlapping) scholarly tradition from that of the main catalogue. They are based largely on classical authors and the more general works of the humanist scholars of the Renaissance rather than on herbals and primarily botanical and medical works. None of the three sections makes any explicit reference to plants in Cambridgeshire. The 'Etymology' is the largest section of the *Catalogus* other than the main catalogue, taking up 48 pages or 16% of the total. Despite this, it has received almost no attention from students of Ray's career. Raven (1950) describes it as "a very complete and scholarly treatise" in a footnote on p. 99; later (pp. 101 and 171) he briefly mentions "the derivations of the names of plants" as giving Ray "scope for much ingenious discourse" and discusses Ray's long entry on *Hyssopus* and his belief that many Greek and Latin words were derived from Hebrew (which he describes as "of course quite illusory": see below). Ewen & Prime (1975, pp. 5–6) paraphrase Raven's comments and then dismiss the 'Etymology' as "ingenious rather than reliable" and, though "not without some merit", "only indirectly botanical", so they do not translate it. It is therefore worth analysing the contents of this section in some detail.

Ray's full title is *Nominum quorundam Ratio, sive Etymologia*, which we have translated as "The Meaning of certain Names, or Etymology". It consists of an alphabetical list of 321 mainly generic names with notes to explain their meaning and possible derivation; of these five are simply cross-references to other entries. There are 21 instances where two or three names relate to the same species or several species, as itemised below. In addition, about 120 further names in various languages are briefly explained within the main entries.[10] Five of Ray's chosen names are explained in other

---

[10] The full list is as follows: *Acetosella* under *Acetosa*; *Agrimonia sylvestris* under *Argentina*; *Anserina* under *Argentina*; *Argentina* under *Potentilla* (as well as in its own entry); *Aristis* under *Holcus*; *Baianæ* under *Faba*; *Basilicum* under *Ocymum*; *Beane* under *Faba*; *Boona* under *Faba*; *Carduus lacteus* & *Mariæ* under *Carduus*; *Cauda equina* under *Equisetum*; *Cervicaria* under *Trachelium*; *Chafe-weed* under *Gnaphalium*; *Chamædrys* under *Trisago*, a name not found in the main catalogue (as well as in its own entry); *Cloves* under *Caryophyllus*; *Columbines* under *Aquilegia*; *Cow-wheat* under *Melampyrum*; *Crista Galli* under *Pedicularis*; *Cudweed* under *Gnaphalium*; *Dropwort* under *Oenanthe*; *Enchanters Nightshade* under *Circæa*; *Eugalacton* under *Glaux*; *Fagotriticum* under *Fegopyron*; *Farfarella* and *Farfarus* under *Farfara*; *Filipendula* under *Oenanthe*; *Flammula* under *Ranunculus*; *Fuga dæmonum* under *Perforata*; *Fusanus* under *Euonymus*; *Genserich* under *Argentina*; *Gramen Parnassi* under *Gramen*; *Graymill* or *Grummell* under *Milium Solis*; *Heraclion* under *Nymphæa*; *Herba benedicta* under *Caryophyllata*; *Herba hæmorrhoidum* under *Ficaria*; *Herba impia* under *Gnaphalium*; *Herba venti* under *Pulsatilla*; *Herbæ capillares* under *Polytrichum*; *Hermupoa* under *Mercurialis*; *Hippolapathum* under *Lapathum*; *Hippuris* under *Equisetum*; *Hydropiper* under *Persicaria*; *Jacea nigra* under *Jacea*; *Labrum Veneris* under *Dipsacus*; *Lantana* under *Viburnum*; *Laurus* under *Laureola*; *Laver* under *Sium*; *Lenticula palustris* or *Lens palustris* under *Lens*; *Leucographis* under *Carduus*; *Melothron* under *Bryonia*; *Muralium* under *Helxine*; *Mustarda* under *Sinapi*; *Nenufar* under *Nenuphar*; *Nipplewort* under *Papillaris*; *Nufar* under *Nenuphar*; *Ocymum*

3: THE *CATALOGUS* AND ITS APPENDICES

entries – *Lithospermum*, *Oxyacanthus*, *Parietaria*, *Parthenium* and *Pilosella* (see footnotes 10 and 13). Many of the entries begin with a Latin name followed by its Greek equivalent. The Latin name is often an adaptation of the Greek, but, where this is not the case, the Greek name is not always explained (though we have sometimes done this in a footnote). The choice of names is more relevant to the main purpose of the book than are the botanical terms chosen for the 'Interpretation', but it is by no means a complete list of the genera of the Cambridgeshire flora as conceived by Ray, though very few of the names that are included are not represented amongst the chosen names or synonyms in the main list.

The entries vary greatly in length, some being short essays while four of them, those for *Anthyllis*, *Aphaca*, *Aracus* or *Arachus* and *Ballote*, merely list the Latin names with their Greek equivalents. The length of any particular entry is unrelated to its significance in the flora of Cambridgeshire; for example Ray lists 12 species of *Alsine* and eight of *Anagallis* in the main catalogue, but neither of the entries for these two genera in the 'Etymology' even fills two lines. Αιρα [Aera] and *Lolium* both relate to *Lolium temulentum* [340]; *Ballote* and *Marrubium* to *Ballota nigra* [77], but see also below for *Marrubium*; *Berberis* and *Crespinus* both relate to *Berberis vulgaris* [413], though Ray's chosen name for this, *Oxyacantha*, is discussed only under the former; *Capnos* and *Fumaria* both relate to *Fumaria officinalis* [204]; *Caprifolium* and *Periclymenum* to *Lonicera periclymenum* [432]; *Chamædrys* and *Trisago* correctly to Wall Germander, *Teucrium chamaedrys* L., but at least the former with the epithet *sylvestris* was also applied to *Veronica chamaedrys* [126]; *Consolida* and *Symphytum* both relate to *Symphytum officinale* [147]; *Dens leonis* and *Hedypnois* to *Taraxacum* species [167]; *Dulcamara* or *Amara dulcis* and *Salicastrum* to *Solanum dulcamara*

garyophyllatum under *Ocymum*; *Ononis* under *Anonis*; *Orchides* (plural of *Orchis*) under *Satyrium* (in addition to the entry for *Orchis*); *Orobanche* under *Limodoron*; *Oxalis* under *Rumex* (as well as under *Acetosa*, in its Greek form); *Oxyacanthus* and *Oxyacantha* under *Berberis*; *Panax coloni*, so named by Gerarde (1597) and translated as *Clown's All-heal*, under *Panax*; *Papaver spumeum* under *Papaver*; *Parietaria* under *Helxine*; *Perdicium* under *Helxine*; *Philanthropos* under *Aparine*; *Philetæria* under *Polemonium* or *Polemonia*; *Pilosella* under *Myosotis*; *Potentilla* under *Argentina* (as well as in its own entry); *Primula veris* under *Paralysis*; *Pruinnen* under *Prunus*; *Pulicaria* under *Conyza*; *Radix* under *Raphanus* or *Rhaphanus*; *Ranunculus flammeus* under *Ranunculus*; *Rapum Genistæ* under *Limodoron*; *Rhœas* under *Papaver*; *Ropalon* under *Nymphæa*; *Rosa canina* under *Rosa*; *Sanamunda* under *Caryophyllata*; *Sanguisorba* under *Pimpinella*; *Scrophularia* under *Ficaria* (as well as in its own entry); *Senecio* under *Erigeron* (as well as in its own entry); *Sorrell* under *Acetosa*; *Struthium* under *Saponaria*; *Tetragonium* under *Euonymus*; *Tithymalus characias* under *Tithymalus*; *Tomentaria* under *Gnaphalium*, *Tutsan* under *Androsæmum*; *Urceolaria* under *Helxine*; *Veratrum* under *Helleborus* (as well as in its own entry); *Vitex* under *Elæagnus*; *Vitrearia* under *Helxine*; Ἁγνος [Agnos] under *Elæagnus*; Ἁγρωστις [Agrostis] under *Gramen*; Ἀκαλυφη [Acalyphe] under *Urtica*; Ἀρνογλωσσον [Arnoglosson] under *Plantago*; Βηχιον [Bechion] under *Tussilago*; Βληχων [Blechon] under *Pulegium*; Ἑρμου βοτανιον [Hermu botanion] under *Mercurialis*; Ἡριγερων [Erigeron] under *Senecio* (as well as in the entry for *Erigeron*); Ἱεροβοτανη [Hierobotane] under *Verbena*; Ἰτεα [Itea] under *Salix*; Κισσος [Cissos] under *Hedera*; Κνιδη [Cnide] under *Urtica*; Κυαμοι [Cyamoi] under *Faba*; Κωνειον [Coneion] under *Cicuta*; Λιθοσπερμον [Lithospermon] under *Milium Solis*; Μαλαχη [Malache] under *Malva*; Μελανορριζον [Melanorrhizon] under *Veratrum*; Μηκων [Mecon] under *Papaver*; Μυκης [Myces] under *Fungus*; Ξυρις [Xyris] under *Spatula fætida*; Ὀξαλις [Oxalis] under *Acetosa* (as well as under *Rumex*, in its Latin form); Πηγανον [Peganon] under *Ruta*; Προυνος [Prunos] or Προυνη [Prune] under *Prunus*; Πτερις [Pteris] under *Filix*; Πυρος [Pyros] and various Latin and Greek names for kinds of bread under *Triticum*; Ῥοδον [Rhodon] under *Rosa*; Συμφυτον [Symphyton] under *Consolida*; Φλομος [Phlomos] under *Verbascum*; Ψωρα [Psora] under *Scabiosa*; Ὠλεσικαρπος [Olesicarpos] under *Salix*.

53

[547]; *Erigeron* and *Senecio* to *Erigeron acris* [144] and to some *Senecio* species [300, 534, 535]; *Galeopsis* and *Lamium* to species of *Lamium* [25, 317, 319], *Lamiastrum* [318] and *Stachys* [209]; *Lampsana* and *Papillaris* to *Lapsana communis* [320]; *Lapathum* and *Rumex* to various *Rumex* species [321–323]; *Marrubium* and *Prassium* to *Marrubium vulgare* [357], but see also above for *Marrubium*; *Spiræa Theophrasti* and *Viburnum* to *Viburnum lantana* [612]; and *Trachelium* and *Uvularia* to *Campanula trachelium* [572]. *Nasturtium* is a name used in the main catalogue for several of the cruciferes, one being *Cardamine* [106], which also has an entry, and similarly *Eupatorium* is used for several species including *Agrimonia* [16]. There are three separate entries each for names for *Arctium* [79], under *Bardana*, *Lappa* and *Personata*; for *Potentilla anserina* [60], under *Anserina* (listed only as a synonym), *Argentina* and *Potentilla*; and for *Primula veris* [467], under *Dodecatheon* and *Paralysis* as well as its own name (though this last also included *P. elatior* [468] and *P. vulgaris* [469] in Ray's time). Under *Lupulus*, i.e. *Humulus lupulus* [342], only the synonym *Salictarius* is explained.

Sixty-seven of the plant names chosen for explanation are not the first words of Ray's chosen names (nearly a quarter of the 277 entries corresponding to entries in the main catalogue), and indeed these seem to go some way towards explaining his choice of synonyms to include alphabetically in his main catalogue. For example, A in the 'Etymology' includes *Acetosella* (under *Acetosa*), *Achillea*, *Acinos*, *Alectorolophus*, *Alleluia*, *Amellus*, *Androsæmum* and *Avellana*, all of which are listed in A in the main catalogue also but only as cross-references to chosen names elsewhere.[11] There are nine other such synonyms with entries in A of the main catalogue but none of these features in the 'Etymology', except that there is an entry there for *Conferva*, for which *Alga aquatilis capillacea*, listed among the nine, is a synonym.

As in Ray's 'Interpretation', there is a strong bias towards classical sources and one gains the impression that he regarded any plant name for which he could not find one as inferior. Eight names are described as "modern" (*recens vocabulum* or *recens nomen*)[12] and *Saponaria* as "recently formed" (*nuper factum*). About *Argentina* or *Potentilla*, i.e. *Potentilla anserina* [60], Ray writes: "It has not been related by anyone by what name the Greeks & Latins call this herb." *Milium Solis* is "a barbarous name", Ray thinks *Sinapi* is "foreign" (*exoticum*), *Tanacetum* has been "called thus by more recent authors", and some people are "of the opinion" that *Tormentilla* too was "named thus by more recent authors" (*à recentioribus*). On *Alliaria* Ray comments: "Thankfully the name is not found in botanical works." Among the modern names Ray says that *Bardana* is taken from Spanish, *Becabunga* from German, *Bedeguar*, *Behen*, *Esula*, *Nenuphar* and *Taraxacon* from or via Arabic, *Berula* and *Bugula* from French (though the former of these may not be correct) and *Bella donna*, *Branca ursina*, *Favagello*, *Galega* and *Pulsatilla* from Italian, while *Herba Gerardi* and *Spergula* are perhaps from Brabantian and *Rogga* is "a Belgian name"; *Atriplex* "seems to be a made-up name" from the Greek Ἀτραφαξις [Atraphaxis]; *Barbarea*, like "very many other plants", has been named by country folk after a saint. None of the principal names are said to be of English origin, but 11 English names are explained within other entries (see footnote 10 above). Ray cites Martini (1623, 1655) for the theory that *Farfara* – listed in addition to *Tussilago*, i.e. *Tussilago farfara* [603] – "is perhaps

---

[11] One would expect *Anacampseros* to be similarly treated but it is listed in the main catalogue only under *Telephium*.

[12] These are *Campanula*, *Dens leonis*, *Pimpinella*, *Sanicula*, *Scabiosa*, *Spergula*, *Succisa* and *Veronica*.

3: THE *CATALOGUS* AND ITS APPENDICES

a word imported into Italy by the Phoenicians"; this is on the basis of a highly implausible suggestion that the name was derived from the Biblical River Pharpar (Φαρφαρ in the Greek Septuagint version).

Eighty-six initial words of Ray's chosen names are not included as main entries in the 'Etymology'. He may have regarded some of these as too obvious to explain, such as *Millefolium*, *Pentaphyllum*, *Perfoliata* and *Trifolium*, and a few are clearly diminutives of listed words, for example *Genistella* from *Genista* and *Hederula* from *Hedera* (but another diminutive, *Acetosella*, is mentioned as such under *Acetosa*). There is no obvious reason for the selection of some names and not others: among the trees *Acer*, *Alnus*, *Betula*, *Fraxinus*, *Juniperus*, *Populus*, *Salix* and *Taxus*, for example, are included, but *Betulus* (i.e. *Carpinus*), *Malus*, *Quercus* and *Ulmus* are omitted; among the shrubs *Corylus*, *Euonymus*, *Ligustrum* and *Rosa* are listed but not *Cornus*, *Rhamnus* and *Ribes*; among wild grasses, sedges and rushes *Gramen* (including *Agrostis* as Ἀγρωστις) and *Juncus* but not *Arundo*, *Avena* and *Festuca*; and among crop plants *Cannabis*, *Crocus* (i.e. saffron), *Faba*, *Fegopyron* (i.e. *Fagopyrum*), *Lupulus* (i.e. *Humulus*), *Pastinaca*, *Pisum*, *Secale* and *Triticum* but not *Hordeum*, *Glastum* (i.e. *Isatis*), *Glycyrrhiza* and *Linum*; the herbs *Buglossum*, *Chamædrys* and *Herba Gerardi* are included but not the comparable *Cynoglossum*, *Chamæpitys* and *Herba Paris*. Among the omissions that appear surprising because the names seem to demand interpretation are *Bistorta*, *Cynocrambe*, *Hipposelinum*, *Leucoium*, *Luteola*, *Myosurus*, *Ophioglossum*, *Ornithopodium*, *Polygala*, *Polygonum*, *Potamogeiton*, *Pulmonaria*, *Saxifraga*, *Sophia*, *Trichomanes*, *Turritis*, *Ulmaria* and *Viorna*.[13]

By contrast, only three names in the 'Etymology' are completely absent from the main catalogue, *Astragalus*, *Meum* and *Trisago*,[14] but this is not a true indication of the mismatch between the two parts of the book because several more names that are given their proper usage in the 'Etymology' appear as synonyms of other plant names in the main catalogue. Examples are *Aconitum*, properly species of the genus still so called but used by Cordus for Herb Paris, *Paris quadrifolia* [277]; *Chamædrys*, properly the labiate Wall Germander, *Teucrium chamaedrys*, but also used for Germander Speedwell, *Veronica chamaedrys* [126], as mentioned above; *Elæagnus*, probably originally a species of *Salix* (see footnote 164 in the 'Etymology'), but used, again by Cordus, for Bog-myrtle, *Myrica gale* [173]; *Eruca*, Garden Rocket, *Eruca vesicaria* subsp. *sativa* (Mill.) Thell., but used by some authors for *Barbarea*, *Rorippa* and even *Reseda* species; *Hyssopus*, the southern European aromatic labiate undershrub and pot-herb Hyssop, *Hyssopus officinalis* L., but used for *Helianthemum nummularium* [125] and *Melampyrum pratense* [360]; *Osyris* (or *Osiris*), probably *Osyris alba* L., a switch-like shrub of the Sandalwood family and another southern European species, but used also for *Linaria vulgaris* [335]; *Ruta*, properly species of rue, *Ruta*, including the evergreen herb *R. graveolens* L., but used, as *Ruta muraria*,

---

[13] The remaining 45 omissions are *Althæa*, *Arum*, *Baccharis*, *Bursa pastoris*, *Caltha*, *Camelina*, *Cicutaria*, *Cotula*, *Cratæogonon* (see footnote 226 in the main catalogue), *Cuscuta*, *Elatine*, *Gladiolus*, *Gratiola*, *Helleborine*, *Jacobæa* (but see the entry for *Artemisia*), *Lathyrus*, *Linaria*, *Lithospermum* (but see the entry for *Milium Solis*), *Militaris*, *Mollugo*, *Morsus Diaboli*, *Morsus ranæ*, *Myrrhis*, *Ophrys*, *Oxyacanthus* (but see the entry for *Berberis*), *Parietaria* (but see the entry for *Helxine*), *Parthenium* (but see the entry for *Matricaria*), *Pecten Veneris*, *Perchpier*, *Pilosella* (but see the entry for *Myosotis*), *Pinguicula*, *Pseudomelanthium*, *Radix cava*, *Rapistrum*, *Rapunculus*, *Rorella*, *Ros solis*, *Sagina*, *Sedum*, *Sesamoides*, *Sison*, *Spartum*, *Speculum Veneris*, *Tripolium* and *Ustilago*.

[14] *Trisago* is a synonym for *Chamædrys*, which does have an entry in the main catalogue, and appears, with *s* doubled and in the genitive case, *Trissaginis*, as part of Ray's chosen name for *Veronica agrestis* [22].

for the fern Wall-rue, *Asplenium ruta-muraria* [11], and, as *Ruta pratensis*, for two species of meadow-rue, *Thalictrum* [567, 568]; and *Veratrum*, the continental liliaceous mountain perennials Black False-helleborine, *Veratrum nigrum* L., and White False-helleborine, *V. album* L., but used, yet again by Cordus, for *Helleborus viridis* [276]. In the case of *Melampyrum* itself Ray includes two species in the main catalogue, Crested Cow-wheat, *Melampyrum cristatum* [359], and Common Cow-wheat, *M. pratense* [360], but the entry in the 'Etymology' clearly relates to a species that he did not record in Cambridgeshire, Field Cow-wheat, *M. arvense* L. One suspects that some of the entries, for example those for *Astragalus*, *Dodecatheon* (apparently the humble Cowslip) and *Elæagnus* and especially that for *Hyssopus*, were included mainly because Ray had an appropriate classical quotation, historical or mythical allusion or etymological theory relating to them.

As in other parts of the book, Ray does not hesitate to digress if he has what he considers something interesting to mention. The principal cases here are lists of "plants found by heroes and gods and dignified by their names" and of others named after saints under *Artemisia*; Pliny's story of "a pruner who fell from a tall tree with nearly all his bones broken" but "recovered with scarcely believable speed", under *Conferva* (various filamentous algae); details of the twelve gods of the ancients, under *Dodecatheon* (see above); an elaborate explanation of the various theories to account for the Pythagoreans' supposed horror of beans, under *Faba*; a detailed account of the "very pleasing sight" of the antics of a cat when it finds catmint, under *Nepeta*; a suggestion under *Ranunculus*, "in passing" (*obiter*), that "it is advantageous for gardeners to foster a great army of frogs in their gardens", with observations on frogs' diet "as we have frequently seen by dissecting their stomachs"; another "memorable story" from Pliny under *Rosa* explaining why the dog-rose, *Rosa canina*, is so called; a wholly irrelevant story about a stolen pig and two Romans, ending with a jokey punch-line, under *Scrophularia*, with no more excuse than that one of the Romans "was surnamed Scrofa"; and a description of the "four kinds of bread" that were "made from wheat among the ancients", under *Triticum*. An extraordinarily long digression under *Hyssopus* (a loan-word in Greek from Hebrew) seeks to show that many Greek and Latin words, most of them nothing to do with botany, were derived from Hebrew and other Semitic languages (see Figure 3.7); the fact that *Serpyllum* (Thyme) is a Latin word made from the Greek Ἑρπυλλος [Herpyllos] "by the rough breathing being changed into s" is the pretext for a list of 16 similar cases, only one of them at all botanical; similarly the statement that *Viscus* (Mistletoe) is from the Greek Ἰξος [Ixos] "of the same meaning, with the smooth breathing ... changed into v," leads on to a list of nine further instances of this linguistic phenomenon. The last two of these lists are founded on sound etymology, but the theory that Greek and Latin were related to Hebrew and other Semitic languages, generally accepted in the 16th and 17th centuries, no longer has scholarly support. Angelo Canini (or Angelus Caninius, 1521–1557), one of the greatest linguists of the 16th century, cited by Ray in his long etymological entry on *Hyssopus*, seems to have taken it for granted. It "was strongly held by Joseph Mead [1586–1638], the famous scholar who influenced Henry More and the Cambridge Platonists" (Raven 1950, p. 171); he had in his study a book "containing the Hebrew *Radices*, &c. with *Greek*, *Latin* and *English* words derived from many of them" (Worthington 1664, p. vi). Ray's tutor at Cambridge, James Duport, also expounded "the affinity of Greek to Hebrew" in his *Homeri gnomologia* of 1660 (Feingold 1990a). Ray probably accepted the popular theory that Hebrew was "the mother of all languages" (though he did not use this phrase); if so this would help to explain his uncharacteristic sarcasm in his entry on *Prunus* in the 'Etymology' about

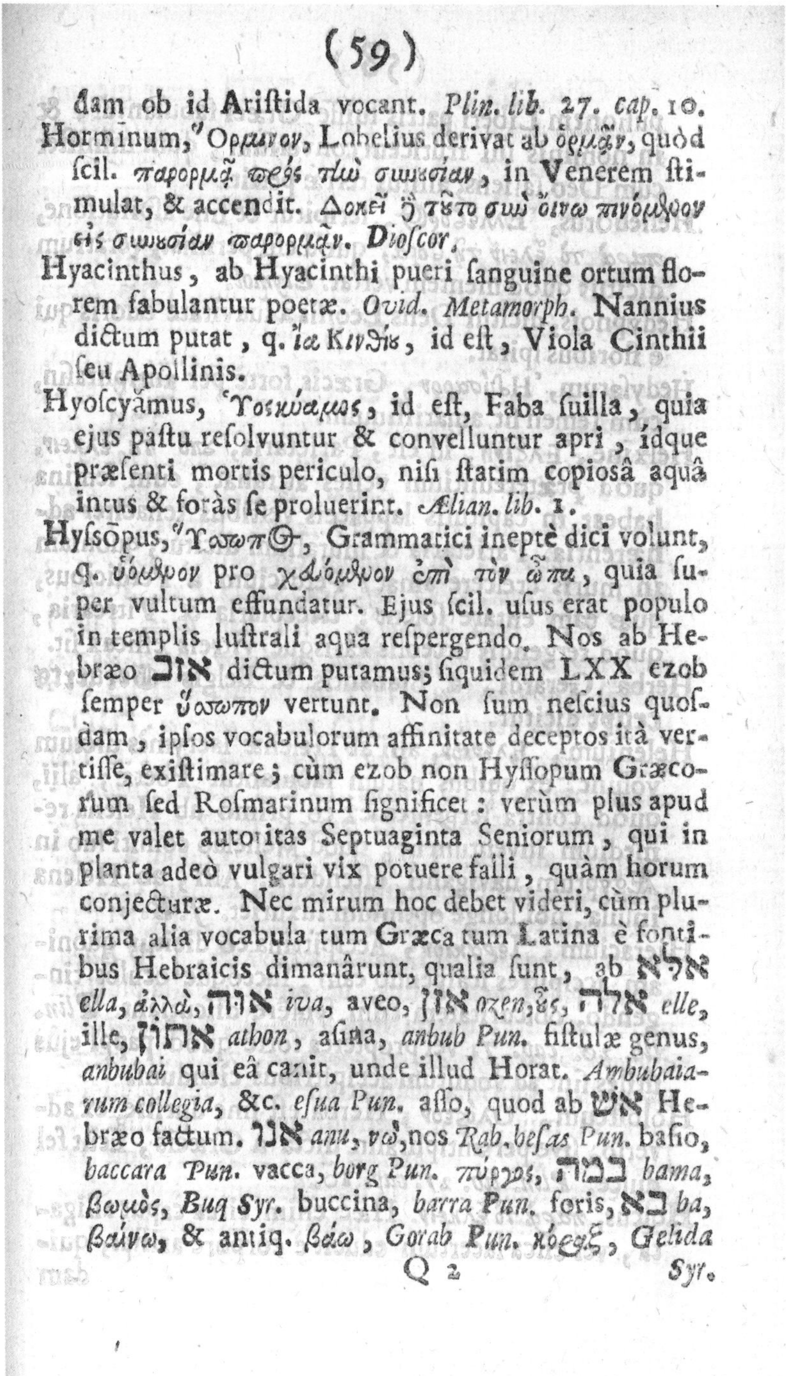

(59)

dam ob id Ariſtida vocant. *Plin. lib.* 27. *cap.* 10.
Horminum, Ὅρμινον, Lobelius derivat ab ὁρμᾶν, quòd ſcil. παρορμᾷ περὶ τὴν συνουσίαν, in Venerem ſtimulat, & accendit. Δοκεῖ ἢ τῦτο σὺν οἴνω πινόμενον εἰς συνουσίαν παρορμᾷν. *Dioſcor.*
Hyacinthus, ab Hyacinthi pueri ſanguine ortum florem fabulantur poetæ. *Ovid. Metamorph.* Nannius dictum putat, q. ἴα Κινθις, id eſt, Viola Cinthii ſeu Apollinis.
Hyoſcyămus, Ὑοσκύαμος, id eſt, Faba ſuilla, quia ejus paſtu reſolvuntur & convelluntur apri, idque præſenti mortis periculo, niſi ſtatim copioſâ aquâ intus & foràs ſe proluerint. *Ælian. lib.* 1.
Hyſſōpus, Ὕσσωπος, Grammatici ineptè dici volunt, q. ὑόμβρον pro χαλόμβρον ἐπὶ τὸν ὦπα, quia ſuper vultum effundatur. Ejus ſcil. uſus erat populo in templis luſtrali aquâ reſpergendo. Nos ab Hebræo אזב dictum putamus; ſiquidem LXX ezob ſemper ὕσσωπον vertunt. Non ſum neſcius quoſdam, ipſos vocabulorum affinitate deceptos ità vertiſſe, exiſtimare; cùm ezob non Hyſſopum Græcorum ſed Roſmarinum ſignificet: verùm plus apud me valet autoritas Septuaginta Seniorum, qui in planta adeò vulgari vix potuere falli, quàm horum conjecturæ. Nec mirum hoc debet videri, cùm plurima alia vocabula tum Græca tum Latina è fontibus Hebraicis dimanârunt, qualia ſunt, ab אלא ella, ἀλλά, איוה iva, aveo, אזן ozen, ἒς, אלה elle, ille, אתון athon, aſina, anbub *Pun.* fiſtulæ genus, anbubai qui eâ canit, unde illud Horat. *Ambubaiarum collegia,* &c. eſua *Pun.* aſſo, quod ab אש Hebræo factum. אנו anu, νώ, nos *Rab. beſas Pun.* baſio, baccara *Pun.* vacca, borg *Pun.* πύργος, במה bama, βωμὸς, Buq *Syr.* buccina, barra *Pun.* foris, בא ba, βαίνω, & antiq. βάω, Gorab *Pun.* κόραξ, Gelida

Q 2 *Syr.*

Figure 3.7. The page containing the beginning of Ray's long digression under *Hyſſopus* about the supposed derivation of Latin and Greek words from Hebrew and other Semitic languages. See also footnote 249 in Ray's 'Etymology'.

the Dutchman Goropius, who believed that his native dialect Brabantic was the original language of Adam and Eve.

There is little evidence of any original thinking in Ray's 'Etymology' and many of his suggested derivations of plant names are attributed to earlier writers. Among those that feature prominently are the classical authors Theophrastus, Dioscorides and Pliny the Elder and among more recent ones the Bauhin brothers (as elsewhere in the *Catalogus*), Mathias de L'Obel (Lobelius) and Caspar Hoffmann. Ray cites for the first time in the 'Etymology' Matthias Martini (1572–1630), the author of an enormous etymological dictionary published in 1623 and in an enlarged edition in 1655, which is clearly a major source, though not always acknowledged; for example 14 of the 24 entries on the first four pages include phrases identical or similar to those in Martini's dictionary, though Ray's first mention of it is in the entry on *Carduus* eight pages later. It seems that he often blended material taken from Martini with information from other sources.

Generally the etymology of the Greek names of which many of the Latin generic names are adaptations is sound, but some of the speculation about the origin of Latin words is far-fetched, examples being *Acer*, *Eruca*, *Lantana* (under *Viburnum*), *Lens*, *Lolium*, *Veratrum* (under *Helleborus* and in its own entry), *Verbena* and *Veronica*. Impossible or improbable derivations of Greek words include those for Ἀκαλυφη [Acalyphe] (under *Urtica*), Ἐυπατωριον [Eupatorion], Ἰτεα [Itea] (under *Salix*), Σογχος [Sonchos] and Φλομος [Phlomos] (under *Verbascum*).

At a more trivial level, Ray's 'Etymology' demonstrates once again his inability to arrange entries in alphabetical order: for example *Chondrilla* comes between *Chamædrys* and *Chamæmelum*, *Fragaria* and *Fraxinus* follow *Fungus*, and there is a sequence of six names that runs *Ligustrum*, *Lichen*, *Limodoron*, *Limonium*, *Lupulus*, *Lolium*!

'Interpretation'

No one reading the 'Interpretation' in isolation could guess that it appeared in a catalogue of Cambridgeshire plants, or indeed in a Flora of any area north of the Alps. It discusses and defines general botanical terms, such as root, stem and leaf, and some more specific terms. Many of the examples illustrating the latter are drawn from crop plants and fruits, and some of the terms are specific to single species. No less than four of the terms relate to the flowers or fruits of the pomegranate (*Balaustium*, *Cicus*, *Cytinus* and *Malicorium*) and a further three refer specifically to date palms (*Dactylus*, *Palmula* and *Spadix*). The other crops and fruits mentioned in the 'Interpretation' include apples, barley, beans, cherries, chickpeas, grapes, hazelnuts, lentils, millet, olives, peas, plums, sweet chestnuts, walnuts and wheat. These are all familiar species in the Mediterranean world and, although some are also staples of northern European agriculture, others would certainly not have been grown in Cambridgeshire in 1660. Although plants used for food figure prominently in the 'Interpretation', there is relatively little emphasis on medical plants in this section.

As in the notes to the main catalogue and the 'Etymology', Ray draws upon a much wider range of sources in the 'Interpretation' than those listed at the start of the *Catalogus* in the 'Explanation'. These sources include Columella and Varro, whose works constitute two of the main surviving treatments of Roman agriculture, as well as a range of classical authors whose writing is more general in its scope. The latter inevitably include Pliny, whose *Natural history* is cited in the 'Explanation' and who is much quoted elsewhere in the *Catalogus*, as well as Celsus, Caelius Aurelianus, Aulus

3: THE *CATALOGUS* AND ITS APPENDICES

Gellius, Macrobius and Scribonius. The numerous classical sources explain in part the emphasis on the agricultural species of the Mediterranean region. Only rarely are the examples cited drawn from the cultivated or wild plants of Cambridgeshire or more generally from those of Temperate (rather than Mediterranean) regions of Europe, although this is the case for one or two definitions such as that of *Spica*.

Ray's modern sources include Schröder, cited in the 'Explanation', Spiegel, not listed in the 'Explanation' but familiar from the notes in the main catalogue, the scholars Julius Caesar Scaliger and his son Joseph Justus Scaliger, who had produced an annotated edition of the works of Varro, and the lexicographer Ambrogio Calepino, who has not hitherto been mentioned in the *Catalogus*.

In preparing the 'Interpretation', Ray had access to the unpublished manuscript of Joachim Jung's *Isagoge phytoscopica*, which he abbreviates to *Jung.*; this was eventually published posthumously in 1678. As he explained in the 'Interpretation' under *caulis*, Jung's work was lent to him by Samuel Hartlib (*c*. 1600–1662). Hartlib, the son of a Polish father and English mother, had lived in London since 1628 and conducted an extraordinarily wide-ranging international correspondence, employing copyists and translators to ensure as wide a circulation as possible of ideas, manuscripts and books. He was particularly active in this role of 'intelligencer' in the Protectorate (Webster 2002). Ray's *Catalogus* is the first work to draw on the *Isagoge*, and it is therefore cited by students of Jung's work (e.g. Kangro 1968) in contexts where one would certainly not expect a British County Flora to be mentioned. Raven (1950, p. 106) argues that the absence of any reference to Jung in the 'Explanation' suggests that Ray did not receive this manuscript until his work on the *Catalogus* was nearly completed. This may be true, but the argument is not convincing as it is clear that the 'Explanation' includes only a subset of the works referred to by Ray in the *Catalogus*.[15] Raven (*loc. cit*.) also suggests that the references to Jung in this 'Interpretation' seem to be "last-moment additions", but we cannot agree. Although Jung's definitions sometimes appear at the end of an entry, crucial words such as *Flos*, *Folium* and *Radix* as well as several other less important terms are defined solely by reference to his work. Some of Jung's definitions are clearly superior to those from Ray's other sources, such as that for *Umbella*, although occasionally they are perhaps too abstract to be immediately comprehensible: see, for example, *Folium*.[16] The references to *Jung.* need to be distinguished from those to the *Nomenclator* of Junius, another author omitted from the 'Explanation' (see Chapter 4). This is frequently cited in the 'Interpretation' as *Jun.* and it is not clear whether earlier students of the *Catalogus* have realised that these similar abbreviations refer to two separate authors. Despite the many inconsistencies elsewhere in the *Catalogus*, Ray applies these abbreviations accurately throughout this section.

'The more customary headings or divisions of plants'

In this brief section Ray presents a classification of "perfect plants", in which group he includes ferns but excludes bryophytes, lichens and fungi. He recognises eight groups

---

[15] A letter from Hartlib to Boyle, quoted in Chapter 2, shows that he was aware of the Cambridge project by February 1658. The *Catalogus* was probably finished in the autumn of 1659. The long note on the parasitism of caterpillars under *Rapum sylvestre* [495] in the main catalogue includes observations made in August 1658, and in his 'Preface' Ray says that the book was almost complete by the time that Nidd died (July 1659). Hartlib's complete correspondence, now published (Anon. 2002), throws no further light on the question.

[16] For a discussion of Jung's definitions, see Stearn (1992).

of trees, two groups of shrubs and 21 groups of herbs (numbered 1–22, but with 18 omitted). As Raven (1950, p. 108) points out, the classification set out by Ray is based on that of Jean Bauhin in *Historia plantarum* (Bauhin & Cherler 1650–1651) but it departs from it in numerous ways. It is possible to compare the two classifications, but, as Ray lists only a few examples of the plants in each taxonomic group, the comparison cannot be exact.

Ray's treatment provides a more rigid separation of trees, shrubs and herbs than that adopted by Bauhin. His eight groups of trees correspond with the species in the first 15 books (*libri*) of *Historia plantarum*. The first three of the groups, *Pomiferæ* (pome-bearing), *Pruniferæ* (plum-bearing) and *Nuciferæ* (nut-bearing), are equivalent to Bauhin's first three books, although Ray includes the palm and the olive in the *Pruniferæ* whereas Bauhin treats the palm as nut-bearing and the olive in a later group of berry-bearing spiny trees and shrubs. Ray also includes the chestnut, *Castanea*, in the nut-bearers whereas Bauhin includes it with the oaks and beeches in the acorn-bearing species. Bauhin's next three books cover aromatic trees and shrubs including the bay laurel, cloves and cinnamon (IV), berry-bearing trees, shrubs and dwarf shrubs (V) and berry-bearing, spiny trees and shrubs (VI). Ray includes examples of the first two of these groups in his fourth group, *Bacciferæ*, the berry-bearing trees, and adds *Juniperus*, which Bauhin treats with the conifers, and *Morus*, which Bauhin rather curiously treats as pome-bearing. However, he excludes other genera which are included in his later group of spiny shrubs. Ray then recognises two groups, the acorn-bearing *Glandiferæ* (5) and the *Coniferæ* (6), equivalent to Bauhin's books VII and IX. He apparently disregards Bauhin's short Book X, *De arbores et fructibus scopariis* ("About broom-like trees and shrubs"), which includes *Erica* and *Tamarix*. This part of the classification ends with the *Siliquosæ*, a group of leguminous trees equivalent to those treated in Bauhin's books XI and XII, and the remaining trees, a miscellaneous group of "woodland trees" which is directly comparable to Bauhin's book VIII, *De arboribus sylvestribus*.

Ray's next two groups are the thorny and thornless shrubs. The examples illustrating his first group of shrubs all come from Bauhin's book VI, the berry-bearing spiny trees and shrubs (mostly, in fact, shrubs). The non-spiny shrubs are taken from a range of Bauhin's books, namely V, the berry-bearing, non-thorny shrubs (*Frangula, Ligustrum*), VIII, the woodland trees (*Vitex*), XI, leguminous trees, shrubs and sub-shrubs (*Genista*) and XV, climbers (*Jasminum*). Ray has therefore extracted the shrubs from Bauhin's books V and VIII, treating the trees from these books in the groups discussed above and the shrubs here. Ray then remarks that shrubs "can also be divided otherwise into *flowering*, *fruiting*, *climbing*, &c.", though this curious distinction between flowering and fruiting is not explained. (It may, perhaps, refer to the uses of the plants.) By contrast Bauhin, who has to set out a classification which includes all the species known to him, devotes three further books to the shrubs, including *Cistus* (plus *Helianthemum* and *Tuberaria*) in book XIII, *Rosa, Rubus* and *Capparis* in book XIV and climbers such as *Clematis, Hedera, Humulus, Jasminum, Lonicera* and *Vitis* as well as *Calystegia* and the non-spiny *Ribes* species in book XV.

Ray starts the section on herbs with a statement of the near-impossibility of the task. The modern reader can see that his difficulty arises from the use of just one or two criteria to define the groups. As Ray points out, a plant may, for example, both be climbing and have bell-shaped flowers. Ray then goes on to list 21 groups of herbs. Of these, 15 are equivalent to the plants covered in single books in Bauhin's treatment. These are Ray's group 1, Bulbous (equivalent to book XIX), 3, Umbelliferous (XXVII), 4, Whorled (XXVIII), 7, Corymb-bearing (XXVI), 8, Pappus-flowered (XXIV), 9,

3: THE *CATALOGUS* AND ITS APPENDICES

Capitate (XXV), 10, Bell-shaped (XX), 12, Round-leaved (XXXII), 13, Vein-leaved (XXXI), 14, Star-like (XXXVI), 16, Succulent (XXXV), 17, Grass-leaved (XVIII), 20, Aquatic (XXXVIII), 21, Marine (XXXIX) and 22, Rock-dwelling (XXXVII). Bauhin tends in *Historia plantarum* to start the books with the typical members of the group and then sometimes adds at the end some additional species which do not fit well but which were presumably placed there because they had to be accommodated somewhere. Ray takes all his examples for these groups from the typical members of Bauhin's groups.

There are six of Ray's groups that do not appear to be directly equivalent to those of Bauhin. However, his group 11, *Coronariæ* (garland-making), which to a modern botanist appears to be defined by a singularly inappropriate criterion, is probably based on Bauhin's Liber XXIX, *Coronariæ semen in vasculis et calyculis gerentes et quæ florum foliorumque, ac totius adeò Plantæ, forma nonnihil accedunt* ("Garland-makers bearing seed in capsules and calyces and plants that somewhat resemble them in the shape of their flowers and leaves and even of the whole Plant"). Bauhin's first chapter in this book, *De Betonica coronaria sive cariophilleo flore in genere* ("About Betony for garland-making or a carnation-like flower in general"), deals with *Dianthus* and Ray's only examples are from this genus as now understood. However, Bauhin goes on to cover other genera including *Alsine* (see Ray's main catalogue for the range of species covered by this genus in the 17th century), *Anagallis*, *Centaurium*, *Hypericum*, *Nummularia* (i.e. *Lysimachia nummularia* [388]), *Polygala* and *Polygonum*. Although no modern botanist would regard this as constituting a natural grouping, it is at least broadly defined by morphological criteria intrinsic to the plants rather than by their use for garland-making. This suggested interpretation differs from that of Raven (1950, p. 108), who considered that Ray omitted Bauhin's *Coronariæ* from his classification; however, if our view is correct, it cannot be denied that Ray abbreviated Bauhin's group to such an extent that it was rendered incomprehensible to the reader of the *Catalogus*. Ray's *Coronariæ* may also have been intended to include Bauhin's next book, Liber XXX, *Coronariæ etiam pleræque* ("Also for the most part garland-makers"), which includes *Anemone*, *Geranium*, *Papaver*, *Ranunculus* and *Verbascum*, none of them cited as examples of any group in Ray's text.

Another of Ray's groups, *Scandentes* (6), the climbing herbs, is an amalgamation of three of Bauhin's books, but Bauhin's taxonomy is reflected in Ray's text where he lists examples of the three types separately, those which intertwine themselves with other plants (book XV), those which are pome-bearing (XVI) and the leguminous climbers (XVII).

This leaves four of Ray's groups with no real equivalent in Bauhin's taxonomy. His *Tuberosæ* (group 2), tuberous herbs, are exemplified by *Asphodelus* and *Battata* from Bauhin's bulbous plants (book XIX) and *Scrophularia* from the rather miscellaneous species added at the end of his second book of *Coronariæ* (XXX). Ray's *Spicatæ* (5), spiked herbs, can, he says, "almost all be referred to the *verticillatæ*" and one of his examples, *Veronica*, comes from Bauhin's *Vertillatæ* (XXVIII) whereas the other is *Lysimachia*, which Bauhin includes in one of his books of *Herbae siliquosæ* (XXI), which are primarily crucifers. Ray's other two groups are *Cerealia, fruges* (15), cereals and fruits of the earth, and *Oleraceæ* (19), vegetables. The cereals are an obviously problematic group even in his own taxonomy as they include two components, the grain-plants, which are also cited later amongst *Graminifoliæ* (17), grass-leaved herbs, and the pulses, most of which have already been cited amongst the leguminous climbers (6) since they possess tendrils. Bauhin deals with the true cereals in a long book (XVIII) covering grasses, sedges and rushes and with the pulses in an equally long book (XVII) covering all the leguminous herbs. Ray's treatment of *Cerealia, fruges* may owe something to Parkinson (1640), who recognises 17 "Classes or Tribes" including *Legumina*, Pulses (11), *Cerealia*, Cornes (12) and *Gramina, Iunci & Arundines*, Grasses, Rushes and Reedes (13). Ray's other group is the vegetables (19), which takes its examples from a range of Bauhin's books including *Brassica* from the chapter which deals primarily with crucifers (XXI), *Beta* and *Spinachia* from book XXIII, *Mollientes et Holeraceæ pleræque* ("Mainly Softening plants and Vegetables"), which includes Malvaceae and Amaranthaceae *sensu lato* (including Chenopodiaceae), and *Endivia* from one of the books on Asteraceae (XXIV).

Bauhin's books of herbs that have no equivalent in Ray's system are the two which cover crucifers, which he splits into those with longer pods (XXI) and those with shorter, broader pods (XXII), the *Mollientes* (XXIII) mentioned above, the book dealing with plants which modern botanists place in the Boraginaceae (XXXIII) and finally his book XXXIV, *Plantæ malignæ et venenatæ* ("Malign and poisonous plants").

It is not easy to see why Ray did not simply report Bauhin's classification, which would have been backed up by the detailed treatment in *Historia plantarum*. He gives so few examples that it is impossible to know where many Cambridgeshire genera should be placed in his system, and indeed some do not appear to be accommodated at all. Few of the changes that he makes to Bauhin's system improve it, whether judged as an attempt to provide a better artificial system or (anachronistically) as an approach to a more natural system. It is difficult to disagree with Raven's conclusion (1950, p. 108) that, in drawing up this classification, "as yet Ray had given little attention to taxonomy. ... He had not yet sufficiently studied either species or structure to be ready for the problems of classification." Judging by the title that Ray gave to this section and the qualifications he made in the text, he may well have realised this himself. The title-page of the *Catalogus* lists all the sections in the second half of the book except this one. "It is interesting to follow Ray's taxonomic work from its beginning in 1660 to its end after the turn of the century ... What we can see in that development is the entire history of taxonomy." (Slaughter 1982, p. 62). Although Slaughter's view fails to do justice to more recent taxonomic developments, it nevertheless emphasises the extent to which Ray's thinking about taxonomy developed after the publication of the few pages on classification at the end of the *Catalogus*. Ray's work on classification above the species level is also discussed by Breidbach & Ghiselin (2006), Raven (1950, chapters 8 and 11), Sloan (1972) and Stearn (1986).

Authorship of the *Catalogus*

The *Catalogus* was published anonymously, although there is no suggestion that there was any other attempt to conceal the identity of the author(s). As far as we know, Ray did not discuss his motives for omitting his name in any of his later works or in his surviving correspondence. Worthington, in the letters we have cited above, treats the book as Ray's work, but the Hampshire botanist John Goodyer, who bought his copy in May 1660 for 2*s*. 6*d*., noted inside the cover that it was "written by Mr John Nid, who is dead [and] Mr John Wray of Trinitie Colledge in [Oxford deleted] Cambridge" (Gunther 1922, 1928). Another note at the end of the book lists Nidd, Ray, Courthope, Willughby and Ray's friend Thomas Pockley, "All of Trinitie Colledge in Cambridge", and adds details of Courthope's and Willughby's seats. This information (as well as an unrelated note on an apothecary, Thomas Thornton) is attributed to John Snagge, "an Apothecarie of Petworth", and John Mapletoft. Mapletoft was a Fellow of Trinity

3: THE *CATALOGUS* AND ITS APPENDICES

College until 1658, when he left to become tutor to Joceline, son of the Earl of Northumberland, at Petworth (ODNB); he received a minor legacy in Nidd's will (see Chapter 5). He is the most likely source of the information about the authorship. The use of the first person plural throughout the 'Preface' might seem to be evidence for multiple authorship, but the 'Preface' includes some material which is personal and can only have applied to one author, notably the first few words which describe how the writer had been "very ill" and "unoccupied and at leisure". However, it seems very likely that Ray, Nidd and their younger associates Courthope and Willughby were a team who worked on the *Catalogus* just as a team of botanists works on the compilation of a County Flora today, with one or more taking the lead and others making a range of contributions, including simply joining field meetings to help with recording. But can Nidd be regarded not just as a member of the team but as a co-author of the work?

Ray's clearest statement of Nidd's role in the production of the *Catalogus* is in his warm tribute to his friend in the 'Preface', where he writes of "*Master John Nid, ... whose part in this enterprise was outstanding & whose industry was most valuable, and to whom the Reader is chiefly indebted for the observations woven into the work. He died wept for by many good men, but by none more wept for than by us*. However, since this little book was almost complete before his death, we apply ourselves vigorously to the work, ...".

Raven (1950, p. 82) does not quote this passage although he does refer to it, commenting that the "suggestion that Nidd was responsible for joint authorship is not borne out by this paragraph". He argues that Nidd's "influence hardly appears except in the note on the rings of trees (*C.C.* p. 57),[17] where Ray rejects his views, and in the insertion of four species of *Veronica* (*C.C.* pp. 10–11),[18] which Ray in his Appendix of 1663 says was due to Nidd". Here Raven appears to come close to concealing significant evidence from the reader. Ray's comment in the 'Preface' that "*the Reader* is chiefly indebted" to Nidd "for the observations woven into the work" surely suggests that the compilation of the numbered notes was mainly his work. This is consistent with what we know of him, as he was clearly a bibliophile (see Chapter 5). One might even speculate that the confused citation of a few works in the observations (see footnotes 133 and 214 in the main catalogue) might have arisen because Ray was working from Nidd's notes rather than compiling the observations himself. Walter Charleton's book *Spiritus gorgonicus* (1650) is cited as *De lithiasi* in both the list of books that Nidd appended to his will (Leedham-Green 1986) and in the main catalogue under *Betula* [85]. Ray's remarks in the 'Preface' and this evidence suggest that Nidd helped to compile the quotations from other authors in the notes, although they leave open the question of the authorship of those notes which are based on books that Nidd did not own himself (such as Butler's book on bees *The feminine monarchie* (1634), which was in Ray's library but not in Nidd's or Willughby's). The longer notes with original observations are almost certainly written by Ray, as in them he mentions both Nidd (in the note on tree rings mentioned above) and Willughby (in the long description of the parasites of caterpillars in the second note under *Rapum sylvestre* [495]) in the third person.

Another approach to the question of authorship is to look at the periods of residence at Trinity of the protagonists. Evidence is available from the Upper Buttery

---

[17] This refers to a note after *Fraxinus* [203].

[18] The species are those that we have numbered 35–38; see *Emendanda* of the 1663 appendix for Ray's later comments.

book 1658–1660/1 in the college archive, which starts in November 1658 and continues beyond the date of publication of the *Catalogus* (February 1660) until the end of March 1661. (The book for the period immediately preceding this has not survived.) Ray was continually present from November 1658 until July 1660 with the exception of three weeks when he is recorded as having received half commons and may have been absent for a few days (weeks ending 21 January, 1 July and 23 September 1659). There is no evidence that he took a summer break in 1659, as he had in the early 1650s and in 1658 (see Chapter 2) and was to do in 1660, 1661 and 1662. Nidd's death in July 1659 might have prevented him from going away, or Ray might have remained in Cambridge to take advantage of a final summer of fieldwork, as Raven (1950, p. 113) suggested. Nidd, as we have seen (Chapter 2, footnote 9), was continuously resident until his death but very ill. Willughby appears only to have been resident for two periods of more than a fortnight, from the week ending 29 April 1659 (when he received half commons, suggesting that he arrived during that week) until 8 July 1659 and again for the weeks ending 2 September 1659 to 20 January 1660.[19] Courthope was resident for three different periods of more than a fortnight, from the end of May to the week ending 24 June 1659 and perhaps (after a week's absence) until the week ending 8 or 15 July 1659, in September and early October 1659 and from the start of December 1659 until April 1660 (except for a brief absence of between one and two weeks at the end of January). Both Willughby and Courthope would therefore have been able to take part in fieldwork in the early summer of 1659 and Willughby would have been able to help with any writing in the autumn and winter, when Courthope was present more briefly. However, Ray appears to have been the only one to have been able to undertake long and sustained periods of work in Cambridge in the 15 months before the publication of the book.

Webster (2002) has suggested that the *Catalogus* was published anonymously to emphasise the collaborative nature of the work. It is impossible to know what Ray would have written if he had been forced to state the authorship on the title-page. It is quite possible that he would have included Nidd as a co-author, as he was notably generous in acknowledging his colleagues. A later work, the *Fasciculus* of 1688, is charmingly attributed to "John Ray and his friends". However, there is no doubt that Ray was the main author of the *Catalogus* and that the contributions for which Nidd was specifically acknowledged are not the most significant botanically.

Is the *Catalogus* a Flora of Cambridgeshire?

The county of Cambridgeshire, like its neighbours immediately to the west, dates from the late Saxon period, when the administrative system of the West Saxons was imposed on the area which had been Mercia as England became unified under the successors of Alfred the Great (Williamson 2000). It was probably created soon after the capture of Cambridge by Edward the Elder in about 917 (Haslam 1984). Its extent and boundaries would, therefore, have been familiar to Ray and his contemporaries. Although Ray's *Catalogus* is traditionally regarded as the first in the long and continuing sequence of British County Floras, it is not immediately apparent that it is a Flora of the county of Cambridgeshire, as opposed to "a catalogue of plants growing around Cambridge", the literal translation of the title. However, a closer examination of the text confirms that it is indeed the first County Flora. In the 'Preface' Ray describes

---

[19] The interpretation of the Buttery book records for fellow-commoners seems to be less straightforward than it is for Fellows, but the longer periods of residence are easy to detect.

his desire "that we should publish a *Catalogue* of all the plants that have been found by us in *Cambridgeshire*" ("ut *Catalogum* omnium quæ nobis inventæ sunt in *agro Cantabrigiensi* plantarum in lucem emitteremus"). From mediaeval times *ager* was often applied to a county; it is repeated in *Index plantarum agri Cantabrigiensis* on the title-page at the start of the second part of the *Catalogus* and in the entry for *Filix mas* [199] in the main catalogue Ray refers to an observation from Shropshire as *in agro Salopiensi*.

The localities visited by Ray, mapped in the gazetteer, show that he achieved a good coverage of the county and usually stopped at its borders. One or two entries confirm that he was aware of the county boundary, notably that for *Pimpinella saxifraga hircina major* [*Pimpinella major* 441], which he reports in English from "*the woods about St. George Hatley, and many other woods on the borders of Cambridgeshire towards Bedfordshire*". When dealing with two apparently erroneous records from How (1650), those of *Abrotanum campestre* and *A. inodorum* (both *Artemisia campestris* [1, 2]), he says in the 1663 appendix: "*I beleeve neither of them is to be found on Newmarket heath, at least that part of it which is in Cambridge-shire*." This not only shows that he is conscious that the county boundary crosses this large site but also implies that he knows the Cambridgeshire side much better than the Suffolk part. However, the county boundary runs through Newmarket town and Ray later wondered whether the site near Newmarket where he recorded *Sesamoides Salamanticum magnum* [*Silene otites* 539] might have been in Suffolk (Ray 1695, column 418); this was also one of his sites for *Thalictrum minus* [*Thalictrum minus* 568] and *Trifolium echinatum arvense* [*Medicago minima* 582]. This site is indeed in Suffolk, as is a locality for *Mentastrum spicatum flore longiore candicante* [*Mentha spicata* B63] near Exning in the 1685 appendix. There are three other records which certainly lie outside the county boundary. One of the sites for *Oxyacantha* [*Berberis vulgaris* 413], "*in the hedges by the road side to Audley end near the great house*", is in Essex and one of those for *Orchis sphegodes sive fucum referens* [*Ophrys apifera* 408], "*a close behinde the Bell Inn at Haverill*", is in Suffolk. The apparently erroneous *Portulaca sylvestris* [459], reported in How's *Phytologia britannica* (1650) from "*About Ramsey mear in the foot pathes by the rape mills*", is in Huntingdonshire. Ray comments on the last that "*We have not as yet searched that place for plants*". Some maps of the period show Ramsey Mere in the Isle of Ely rather than Huntingdonshire, which perhaps explains Ray's treatment of this record (see our gazetteer). Despite these records from sites outside Cambridgeshire, we have no reason to doubt that Ray's intention was, as he stated, to produce a catalogue of the plants of Cambridgeshire. The records from the Newmarket area and Ramsey Mere can be attributed to an understandable confusion over county boundaries, and the record of *Ophrys apifera* from Haverhill was perhaps included because Ray was reluctant not to publish a record of what was obviously a striking population of hundreds of plants. We can offer no explanation for the inclusion of the record of *Berberis vulgaris* at Audley End, which was perhaps simply an error; even current plant recorders occasionally make such geographical errors despite the detailed maps now available.

The 1663 appendix

The 1663 appendix is a slim publication with 14 unnumbered pages which are the same size as those in the *Catalogus*. Keynes (1976) gives a full bibliographical description. In it Ray published (anonymously) an additional 42 Cambridgeshire taxa in the *Addenda* and made some corrections to the main catalogue in the *Emendanda*.

Keynes suggests that the two leaves of *Emendanda* were meant to be bound after the *Catalogus* with the title-page and *Addenda* following.

The circumstances of its publication are set out in the introductory "Greetings to the Reader". Ray had (perhaps to his surprise) been pressed to remain at Trinity after the Restoration (Raven 1950, p. 59), but he told Peter Courthope in July 1661 that "I cannot look upon my selfe as settled heer" (Gunther 1934; Thompson 1974). He decided in 1662 not to subscribe to the Act of Uniformity and thus had to leave the University. He left Cambridgeshire in August 1662, and the 1663 appendix was prepared while he was living at Friston, Suffolk, in early 1663. In a letter of 16 February 1663 Ray wrote to Courthope: "I am intending this spring before I goe over [to the Continent with Willughby], to print a sheet by way of appendix to Catal. Cantabr: wch shall contain some addenda and emendanda." (Gunther 1934; Thompson 1974). He asked Courthope in this letter for details of any errors he had noticed and repeated the request in a letter of 19 March 1663.

Ray gives a detailed report on his recent botanical finds in three of the surviving letters to Courthope, thus demonstrating how closely Courthope (and by implication Willughby) was involved in the fieldwork for the *Catalogus*. In July 1661 he says: "We have this year made a more narrow search into the countrey about Cambridge for plants & have discovered in all about 26 that are not in our Catalogue, some such as I had not seen before, nor are mentioned to grow wild in England." (Gunther 1934; Thompson 1974). He then lists them; they are (using the modern names to which we assign them in the translation of this appendix and their taxon numbers) *Asplenium scolopendrium* [A29], *Blechnum spicant* [A23], *Carduus crispus* [A8], *C. tenuiflorus* [A9], *Carex pendula* [A16], *Cerastium glomeratum* [A3], *Clinopodium calamintha* [A7], *Fontinalis antipyretica* (as *Fontalis minor lucens*, an identification corrected in the published appendix) [A25], *Galium tricornutum* [A5], *Gnaphalium sylvaticum* [A15], *Helleborus foetidus* [A18], *Herminium monorchis* [A27], *Hieracium* sp. (as *Pulmonaria Gallorum*) [A32], *Hypericum hirsutum* [A22], *Hypochaeris maculata* (as *Hieracii species* but presumably the plant listed as *Hieracium montanum caule aphyllo non ramoso flore pallidiore* in the appendix) [A21], *Knautia arvensis* (as *Scabiosa vulgaris flore pleno*) [A38], *Lathyrus nissolia* [A11], *Limosella aquatica* [A31], *Malva moschata* [A1], *Marchantia polymorpha* (as *Hepatica stellata* and *Hepatica umbellata*) [A19, A20], *Ranunculus acris* [A35], *R. sardous* [A34], *Ruscus aculeatus* [A37], *Solidago virgaurea* [A42] and two kinds of *Thymus* (as *Serpylli duo genera*) [A40]. On 28 April 1662 he told Courthope of the discovery of *Thuidium tamariscinum* (as *Muscus pennatus filicinus*) [A24], which he had first seen in Sussex (Gunther 1928). In the third letter, written in May 1662, Ray reports: "I have been out again in pursuit of plants as farre as Gamlingay. There I discovered some that I have elsewhere found in England, other that I never saw before." The former were *Arabidopsis thaliana* [A29], *Moehringia trinervia* [A4], *Montia fontana* [A2], *Veronica montana* (as "*Chamadrys spuria altera* of yr countrey") [A12], two species of sedge and *Gnaphalium montanum flo. suaverubente*, a plant which was not reported in the appendix and may have been just a colour variant of *Antennaria dioica* [233]. Ray also mentions another species that does not appear in the appendix, "one of those sorts of Ferne, wch we observed about Danny [Courthope's seat in Sussex], caule glabro fere [with an almost glabrous stem]". The species that he had never seen before were *Ranunculus parviflorus* [A33] and *Trifolium subterraneum* [A41]. "I perceive," Ray concluded from this rich haul of spring-flowering species, "formerly we were too negligent & remisse in going out at severall seasons, & thereby came to misse of divers plants." (Gunther 1934; Thompson 1974).

The 1685 appendix

The 1685 appendix is the same size as the *Catalogus* and the 1663 appendix. It is considerably longer than the first edition (1663), consisting of 30 unnumbered pages. Keynes (1976) again gives a bibliographical description. The appendix was printed by John Hayes, who had replaced John Field as University Printer in 1669 (McKitterick 1992).

The appendix has no author on the title-page but the introductory remarks "To the Kind Reader" describe Peter Dent's role in its production. Dent contributed records to the 1663 appendix and appears to have been the most active naturalist in the town after Ray left in 1662. In *Historia plantarum* (1686, p. 856) Ray describes him as "a Cambridge doctor & pharmacist, a distinguished Botanist & old friend of ours"; he died in 1689 (Raven 1950, p. 54). Dent compiled the 1685 appendix from three main sources – Ray's 1663 appendix, his subsequent *Catalogus plantarum Angliae* (1670, 1677) and Dent's own records, many of which had been reported to Ray in earlier years and published previously in one or both editions of the English catalogue.

Ray's Catalogus plantarum Angliae *(1670, 1677)*

Ray had begun to think of preparing a *Phytologia britannica* before the publication of the *Catalogus* (see Chapter 1). He travelled widely in both Britain and Europe in the 1660s. He might have been forestalled by Christopher Merrett's *Pinax rerum naturalium britannicarum* (1666[20]), but this proved to be a slight volume and is remembered in Ray's phrase as "Dr. Merrett's bungling Pinax", a remarkably dismissive comment by the normally charitable Ray, albeit one made in a private letter to Martin Lister (Gunther 1928, p. 112). In 1670 Ray's much more substantial and authoritative *Catalogus plantarum Angliae, et insularum adjacentium* ("Catalogue of the plants of England, and of adjacent islands") was published. "England" is the natural translation but it should be understood in the sense of earlier generations, who were not as particular as we are in distinguishing England from Wales and Scotland and would often use the term in a broad as well as a narrow sense.

In preparing the English catalogue, Ray tried hard to improve the treatment of some of the groups which are clearly inadequately covered in the Cambridge catalogue. In June 1667 he wrote to Martin Lister: "I could wish you would take a little pains this summer about grasses, that so we might compare notes, for I would fain clear and complete their history." (Gunther 1928, p. 112). In his preface to the English catalogue (1670) he pointed out that "in plants that are obvious & common everywhere, which we tread underfoot daily, we generally experience the greatest difficulty"; "the accounts of Grasses, for example, of Hawkweeds, of Oraches and of Docks suffer from great confusion & obscurity."[21]

The structure of the English catalogue is that of the earlier Cambridge catalogue. It is an alphabetical list with synonyms, habitats and localities (in English and italicised) followed by informal remarks or more formal observations like those in the 1660 *Catalogus*. In addition to providing for England a Flora similar to that which he had

---

[20] The first edition of 1666 is rare, almost certainly because the stock was lost in the Great Fire of London; a revised edition was published in 1667 (Henrey 1975).

[21] "historiæ *Graminum*, v.g. *Hieraciorum, Atriplicum, Lapathorum* magna confusione & obscuritate laborant". Both translated passages are on the third, unnumbered page of the preface.

written earlier for Cambridgeshire, Ray also intended the English catalogue to update the Cambridge catalogue itself. He records in its preface that he had been asked on a visit to Cambridge for a new edition of the Cambridge catalogue (Raven 1950, p. 156); instead of producing this, he prefixed species known from Cambridgeshire with C in the English catalogue.

Ray received records from Cambridgeshire for inclusion in the English catalogue from Dent. He included several in 1670, sometimes credited to Dent by name, such as *Lychnis sylvestris annua angustifolia flore rubente* [*Silene gallica* B61], and others such as *Caryophyllus Virgineus* [*Dianthus deltoides* B21] which are not attributed to him as recorder. He also included records from Cambridgeshire from other correspondents, notably Martin Lister, a friend who had entered St John's College, Cambridge, to study medicine in 1655, graduated in 1658 and remained at the college until 1669 (except for three years spent in Montpellier between 1663 and 1666). He is now remembered for his zoological work (Parker & Harley 1992).

In the second edition (1677) Ray listed in the unnumbered preliminary pages of the book nine plants "observed by *Master Dent* in Cambridgeshire & omitted from the Catalogue" (our translation). These included five plants which were excluded as garden remnants or as rejects from gardens – one true species, *Ornithogalum vulgare* [*Ornithogalum pyrenaicum* B72], one hybrid, the daffodil *Narcissus medio-luteus vulgaris* [*Narcissus poeticus* × *N. tazetta* B69], and three variants of common species, the bullaces *Prunus sylvestris fructu albo* and *fructu nigro majore* [*Prunus domestica* subsp. *insititia* B79, B80] and the plantain *Plantago rosea* [a monstrous form of *Plantago major* B78]. The other exclusions were a variant of *Cardamine pratensis*, *Cardamine flore multiplic[i]* [B15], which Ray did not consider a distinct species, *Convolvulus minimus* [B28], which he thought might not be distinct from *Convolvulus arvensis*, and two species which he excluded without stating his reasons, *Elleborine minor flore albo* [B42] and *Lichen sylvestris calic[e] striato*.

Sources of Dent's material in the 1685 appendix

Dent based the 1685 appendix on the 1663 text, but he added 60 species to the *Addenda* and 18 additional entries to the *Emendanda*, as well as revising some of the 20 *Emendanda* entries that he inherited from the first edition. The additional plants include eight of the nine taxa excluded by Ray from the second edition of *Catalogus plantarum Angliae*. There seems to be every justification for including the garden escapes, as Ray included such plants in the 1660 *Catalogus* even though he later excluded them from the English catalogue.

In the introductory text, "To the Kind Reader", "nearly forty" of the additions are said to be Dent's records, "added to which the contributions of certain others made sixty". It is possible to identify some of the records as Dent's as they are attributed to him in the English catalogue or Ray's later works, but Dent's other sources are unknown. One obvious possibility is Martin Lister (see above). Dent added descriptions of several of the additional species, taking them from Ray's English catalogue. One description in the *Addenda*, of *Trifolium subterraneum*, is based on Morison's *Plantarum historiae universalis Oxoniensis* (1680), as stated by Dent. Very few of the descriptions appear to be Dent's own work, although he may have been responsible for some comments on *Dianthus deltoides* and *Hypericum hirsutum* (see footnotes 31 and 71 in *Addenda* to the 1685 appendix). This has important implications for assessing the 1685 records, as the descriptions were normally not drawn up from Cambridgeshire material and cannot therefore be used as direct evidence of identity. All we can say is that Dent (or "certain others") must have identified the Cambridgeshire plants from *Catalogus plantarum Angliae* and (presumably) other books and clearly considered that Ray's descriptions applied to them. However, Dent may have sent Ray the specimen of *Silene gallica* from a cornfield near the Devil's Ditch that is the one localised Cambridgeshire specimen in Ray's herbarium (Walters 1981), as he discovered it at this site, and he certainly sent Ray specimens of fish and wildfowl (Raven 1950, pp. 339–340). Ray also received Cambridgeshire material from others, and his comments on Dent's *Tithymalus segetum longifolius* in *Historia plantarum* (1686, p. 868) are based in part on dried material sent to him from Cambridgeshire by Samuel Dale. Dent's additional *Emendanda* are also taken from the English catalogue, as the introductory "To the Kind Reader" suggests.

Rather surprisingly, Dent did not include in the 1685 appendix all the Cambridgeshire records previously published in the English catalogue. The most obvious omissions are *Carex pulicaris* L., published in 1670 (p. 148) as *Gramen cyperoides pulicare* "*In the Fenny-meadows near Quoy-water*", and *Littorella uniflora* (L.) Asch., reported as *Holosteum minimum palustre capitulis longissimis filamentis donatis* (p. 169) from "*Hinton-moor near Cambridge*". In the two editions of the English catalogue there are at least a dozen more vascular plant species that are not in the 1685 appendix but are marked C (as occurring in Cambridgeshire) without a precise locality; these include six grasses, a group which had been inadequately treated in 1660. The 1685 appendix includes no bryophytes, lichens or fungi, and perhaps Dent deliberately decided not to cover these, even though he is credited with the record of *Hepatica stellata* [A19] in the 1663 appendix. He excludes *Lichen sylvestris calic[e] striato*, previously reported to Ray for the 1677 English catalogue, and ignores the four *Fungus* species added to the Cambridgeshire list in *Catalogus plantarum Angliae*. Ray provided localised Cambridgeshire records credited to Martin Lister for two of these fungi, including *Fungus pezicæ Plinii* Col., *Cup-Mushrome*, which "*sprang out of the clefts of the ground, in that dry year of the Sickness continued in Cambridge 1666*".

Authorship of the 1685 appendix

The 1685 appendix is sometimes credited to Ray, although neither his biographer nor his bibliographer regarded it as his work. According to Keynes (1976), "Canon Raven cannot discover that Ray himself had any hand in it." There is certainly no evidence that Ray played a direct part in its composition or even commented on a draft. Dent is known to have corresponded with Ray in the 1670s (see above), and it seems strange that he did not involve Ray in preparing the second edition of the appendix, but this seems to have been the case. The description of the compilation of the 1685 appendix in the introduction, "To the Kind Reader", includes a full account of its sources, is entirely consistent with its contents and makes no mention of Ray's direct involvement. However, a substantial proportion of the text is by Ray: it is a second edition of a work by Ray which incorporates the material from the first edition of 1663 and much material from Ray's *Catalogus plantarum Angliae*. It seems reasonable to regard it as the joint work of Ray and Dent.

4: BIOGRAPHICAL NOTES ON SOME AUTHORS CITED BY RAY

Classical and mediaeval authors

The following biographical notes cover some of the classical and mediaeval authors referred to in the text of the *Catalogus*. Authors included in this first section are those born before AD 1450. We have included all the authors mentioned in the main catalogue and many of those mentioned elsewhere in the text; briefer notes on others are provided in our footnotes to Ray's text. Where Ray lists classical plant names attributed to these authors he usually cites a later work in which the classical plant is identified. These identifications are sometimes unconvincing, especially when northern European botanists were attempting to match southern European species in the classical literature with plants they knew in their own countries.

Numbers in square brackets refer to the species entries in the main catalogue. The authorities on which we have drawn, and some sources of further information, are cited at the end of each account. The main reference works are abbreviated as BNP for the English edition of *Brill's New Pauly* (Cancik & Sneider 2002–2010), DSB for *Dictionary of Scientific Biography* (Gillispie 1970–1980) and OCD for *The Oxford Classical Dictionary* (Hornblower & Spawforth 2003). The two columns on a page of BNP are numbered separately but are cited here as if they were page numbers.

Aegineta. Paulus Aegineta or Paul of Aegina was a seventh-century Byzantine Greek physician who is known to have practised in Alexandria in AD 642. His only surviving work, a medical encyclopaedia in seven volumes which was compiled as a synopsis for practising physicians, exists in numerous manuscripts. The seventh volume, on the properties of medicines, is based on Dioscorides and mentions some 600 plants. His work was first printed in Venice in 1528 and there are numerous 16th-century editions (see Rice 1980). Ray cites three of his names from the works of C. Bauhin [340, 412, 421] and also cites him under *Betonica* in the 'Etymology'. BNP vol. 10, pp. 635–636; DSB vol. 10, pp. 417–419; OCD p. 1128.

Aëtius. Aëtius Amidenus or Aëtius of Amida may have been a Byzantine physician; his precise dates are unknown but he appears to have lived in the early sixth century AD. Like Paulus Aegineta he compiled a medical encyclopaedia, in this case with 16 books divided into four groups: hence its usual name, *Tetrabibli*. It was first printed in Venice in 1534. The word "acne" is derived from a corrupt rendering of ἀκμη [acme] in copies of his work. Ray cites two of his names in the main catalogue, Ἀιρα [Aera, 340] and *Orobanche* [412], both also attributed to Aegineta and other authors and both taken from C. Bauhin. He is also mentioned under *Scabiosa* in the 'Etymology'. BNP vol. 1, p. 275; DSB vol. 1, pp. 68–69; OCD pp. 30–31.

Apuleius Platonicus. Apuleius Platonicus (sometimes known as Apuleius Barbarus or Pseudo-Apuleius) is the name given to the author of a herbal, a compilation based on classical sources which was well-known in the Middle Ages. Numerous manuscript copies survive in western Europe (Howald & Sigerist 1927). That in the British Museum was translated from the Anglo-Saxon by Cockayne (1864, 1961) and a Latin version in the Bodleian Library, Oxford, was reproduced in facsimile by Gunther (1925). This was the first illustrated herbal to be printed, but the illustrations are exceptionally crude. A facsimile of the first printed edition of 1481 (Anon.

4: BIOGRAPHICAL NOTES

1979) includes useful introductory material by E. Caprotti and W.T. Stearn. Ray cites only one name from Apuleius Platonicus directly, under *Absinthium vulgare* [3], but he also refers to the interpretation of the herbal's names by other authors. Anderson (1977); Arber (1986).

Aristotle. The philosopher Aristotle or Aristoteles (384–322 BC) came from Stagira in Chalcidice, Greece; he was a student of Plato and became tutor to the young Alexander the Great. His personal studies of zoology were made at Assus in Asia Minor and on Lesbos; his friend Theophrastus (q.v.) concentrated on botany. Ray cites his works *Historia animalium* and *De generatione animalium* in his note on *Chelidonium maius* [128], as well as names from Aristotle listed by other authors such as C. Bauhin; his zoological observations are also mentioned under *Rosa sylvestris* [502] and *Vicia* [615]. BNP vol. 1, pp. 1136–1147; DSB vol. 1, pp. 250–281; OCD pp. 165–169.

Arnald of Villanova. Arnald or Arnaldus de Villanova (*c.* 1240–1311), a Catalan, studied medicine at Montpellier. He became physician to the Kings of Aragon, taught medicine at Montpellier, translated Arabic medical works into Latin and produced editions of Hippocrates and Galen. However, he became increasingly concerned with theology as his life progressed, and his religious views attracted papal disapproval. A number of editions of his works were published between the late 15th and the late 16th centuries. Ray cites three of his names as synonyms of *Polygonum mas vulgare* [454]. DSB vol. 1, pp. 289–291.

Asclepiades. Asclepiades of Bithynia was a successful physician in Rome in the early first century BC (though his dates are uncertain); his followers included Themison (q.v.). He was primarily responsible for introducing Greek medicine to the Roman world and specialised in non-invasive treatments such as massage and bathing; his opponents accused him of spoiling his patients (e.g. by the over-prescription of wine). Ray includes a work on Mayweed by Asclepiades in a note under *Betonica* [84] which lists classical authors who wrote about plants; this work was mentioned by Pliny (see footnote 123 in the main catalogue) but does not survive. BNP vol. 2, pp. 96–98; DSB vol. 1, pp. 314–315; Green (1955); OCD p. 187.

Athenaeus. A Greek of Naucratis in Egypt, Athenaeus lived at the end of the second and the beginning of the third centuries AD. In his *Deipnosophistae* ("Scholars at a Banquet") 29 guests converse on a wide range of topics at a banquet lasting several days. The work is an immense storehouse of information, chiefly on matters connected with dining, but also containing remarks on music, songs, dances, games, courtesans and luxury; nearly 800 writers and 2500 separate works are referred to. It was first printed in Venice in 1514. Ray cites his work under *Lens* [328] in the main catalogue and under *Raphanus*, *Ruta* and *Urtica* in the 'Etymology'. BNP vol. 1, pp. 240–243; OCD p. 202.

Aurelianus. Virtually nothing is known of the life of the physician Caelius Aurelianus, who came from Sicca Veneria in modern Tunisia and flourished about AD 400. His surviving medical works were first printed in 1529, and in them are preserved the writings of other authors, notably Soranus of Ephesus. He is included by Ray in the classical authors listed under *Capreolus* in the 'Interpretation'. BNP vol. 2, pp. 894–895; Bynum & Bynum (2007) vol. 2, pp. 291–292.

Avicenna. Avicenna is the westernised form of the name of the scholar Abū ʿAlī al-Husayn Ibn ʿAbdallāh Ibn Sīnā (AD 980–1037), who was born at Afshana near Bukhara, Uzbekistan; his native language was Persian. He was well educated in Bukhara and his extraordinary intellect and remarkable memory were apparent at an early age. In his remarkably self-congratulatory autobiography he says that distinguished physicians began to read medicine under him when he was only sixteen (Gohlman 1974). He wrote about 270 works, but he is best known for his philosophical work *Al-Shiftā'* (*The cure [of ignorance]*, also often but misleadingly called *The book of healing*) and his medical encyclopaedia *Al-Quārūn* (*The canon*). He was able to write rapidly, even when travelling on horseback or when imprisoned, citing his sources from memory and sometimes continuing to work through the night. He died at Hamadhan, Iran; some have suggested that he was weakened by his remarkable sexual voracity and died of colic, others that he was poisoned by his servants. *The canon* was extremely influential in mediaeval Arabic and western medicine; it had been translated into Latin by the 12th century and was first printed in 1473 in Milan. Ray mentions two names from his works, *Eupatorium* cited indirectly from C. Bauhin under *Agrimonia* [16] in the main catalogue and *Taraxacon* in the 'Etymology'. Bynum & Bynum (2007) vol. 5, pp. 1156–1159; DSB vol. 15, pp. 494–501.

Bassus. At the end of the sixth or the beginning of the seventh century Cassianus Bassus compiled from earlier writers a collection of agricultural literature, parts of which were revised as *Geoponica*, prepared in about AD 950 by an unknown editor for the Emperor Constantine VII Porphyrogenitus. This contains very little original material; it was based on authors such as Cato, Columella and Varro, though these sources are not cited. An abridged Latin translation was first published in 1538. Ray cites *Geoponica* from Johannes Bodaeus' edition of Theophrastus [142, 148, 541]. BNP vol. 5, pp. 780–783; Greene (1983a).

Caesar. Gaius Julius Caesar (100–44 BC) was a Roman of patrician family who by his brilliant military talents, ruthless political skill, bribery and force rose to the head of the Roman state. He was appointed dictator for life and deified shortly before his assassination in 44 BC. Ray's three references to him in the main catalogue [226, 490, 565] include the famous mention of Woad in *De bello gallico*, Caesar's account of the campaign in which he completed the Roman conquest of Gaul (58–51 BC) and invaded Britain (55–54 BC). BNP vol. 1, pp. 900–916; OCD pp. 623, 780–782.

Calepino. Ambrogio Calepino or Ambrosius Calepinus (*c*. 1435–*c*. 1510), born near Bergamo, entered the order of Augustinian hermits in 1458. His Latin dictionary was first published in 1502 with the aim of returning the language to its pure, classical form. It was immensely successful, with 24 editions published before 1520 and many thereafter. Other languages were added to the Latin and "the name of Calepino became synonymous with dictionary". It frequently appears in the lists of books owned by members of Cambridge University in the 16th century (Jardine 1975) and is one of the sources cited by Ray in the 'Interpretation', under *Bulbus* and *Termes*. Bietenholz (1985–1987), vol. 1, p. 244.

Cato. Marcus Porcius Cato (234–149 BC) was a prominent politician in Rome in the second century BC and the foremost orator of his age. He is regarded as the founder of Roman prose literature, but the only work to survive intact is the disorganised *De agricultura*, which gives advice to the owner of a medium-sized estate and was first printed in Venice in 1472 (Brown 1980a). It includes the encomium of Cabbage mentioned by Ray in a note under *Betonica* [84]. However, Cato's text survives only in a very corrupt state (Hooper & Ash 1967). Dalby (1998) has suggested in his translation of *De agricultura* that the section on Cabbage is "an unintelligent memorandum by an anonymous enthusiast" which differs stylistically from Cato's work and which must have been inserted into his text (where it appears after an apparently genuine shorter section on the medicinal uses of Cabbage). BNP vol. 3, pp. 20–23; OCD pp. 1224–1225.

4: BIOGRAPHICAL NOTES

Celsus. Aulus Cornelius Celsus, who lived in Rome in the first century AD, was the author of an encyclopaedia which covered numerous subjects, but only the eight books covering medicine, *De medicina*, survive. Whether or not he himself was a physician has been debated in an "intense ... scholarly quarrel" (DSB). His work was rediscovered in the Middle Ages and first printed in 1478. *Uva Taminia* of Celsus is a synonym of *Viorna* [621] in the main catalogue; he is also cited under *Alica* in the 'Interpretation'. BNP vol. 3, pp. 74–75; Bynum & Bynum (2007), vol. 2, pp. 314–315; DSB vol. 3, pp. 174–175; OCD pp. 392–393.

Cicero. Marcus Tullius Cicero (106–43 BC), the son of a rich Roman, became famous for his oratory and was elected Consul in 63. His career suffered as the Republic declined under Pompey and Caesar, but in this period he wrote the works on rhetoric and philosophy which were to become immensely influential in the Middle Ages and Renaissance. He was murdered on the accession to power of Augustus. Many authors modelled their style on his prose and Ray's 'Preface' includes several Ciceronian phrases. Cicero's work *De divinatione*, a dialogue about divination which refutes and even ridicules many beliefs held in Roman times, is cited in a note on *Pulegium* [474] in the main catalogue and under *Faba* in the 'Etymology'. Cicero is also mentioned under *Capreolus* in Ray's 'Interpretation'. BNP vol. 3, pp. 317–327; OCD pp. 1558–1564.

Columella. Lucius Junius Moderatus Columella was a Roman author from Cadiz who lived in the first century AD. His main work is a comprehensive description of Roman agriculture, *De re rustica*, in 12 volumes, which was regarded as authoritative even in antiquity. There are printed editions from 1472 onwards (Brown 1976). The text of another work, *De arboribus*, was combined with *De re rustica* in late antiquity and is thought to represent part of a separate work by Columella, though its attribution "remains controversial" (Armendáriz 2003). Ray refers to Columella under two agricultural crops in the main catalogue, *Rapum sylvestre* [495] and *Triticum* [595], and there are also references to his work in the 'Interpretation' and 'Etymology'. BNP vol. 3, pp. 584–585.

Damocrates. Damocrates Servilius, a Greek physician, wrote prescriptions in the first century AD; some survive in the works of Galen. There is a single very indirect reference to him in the main catalogue, under *Convolvulus major* [148]. BNP vol. 4, p. 64.

Diocles. Diocles of Carystus, Euboea, may have been a contemporary of Aristotle but his dates are very uncertain; he was known to the Greeks as a "second Hippocrates". Some of his writings on medical subjects survive as quotations by other authors and they cover a wide range of subjects, including a treatise on roots with medicinal effects, *Rhizotomicon*. Pliny mentioned a work on Turnip which Ray includes in a list of classical works on plants in a note under *Betonica* [84], but this does not survive. Bynum & Bynum (2007), vol. 2, p. 418; DSB vol. 4, pp. 105–107; BNP vol. 4, pp. 424–426; Eijk (2000); OCD p. 470.

Dioscorides. Pedanius Dioscorides of Anazarbus near Tarsus lived in the first century AD. He travelled widely in pursuit of information on the medical uses of plants, animals and minerals. His five-volume work known as *De materia medica* is thought to have been written between AD 65 and 75 and is characterised by an empirical approach to treatment. It describes the use of 550–700 plants in medicine (estimates vary) and is "by far the largest pharmaceutical guide in antiquity" (DSB). The plants are described in detail, and some early manuscripts are illustrated, notably the Vienna Codex of *c.* AD 512. A Latin text was first printed in 1478 and the Greek text in 1499. His work remained a standard authority until the 17th century, and several of the works cited by Ray are commentaries upon it, including Amatus Lusitanus (1554, 1558), Barbaro (1530), Cordus (1561), Gesner (1541), Laguna (1554, 1555), Ruel (1516) and most notably Mattioli's work, first published in 1554 and known from numerous subsequent editions (e.g. 1563b, 1568, 1571, 1598). Interpretations of his names occur throughout Ray's main catalogue, almost always attributed to a named source, although there is no such attribution for *Smilax* of Dioscorides, a synonym of *Taxus* [565]. "The botanist of to-day may find it difficult to avoid a feeling of impatience at the amount of time and energy devoted, in earlier times, to the elucidation of Dioscorides." (Arber 1986, p. 11). Riddle (1980) provides a detailed account of the mediaeval and Renaissance translations and commentaries and notes the curious absence of interest in Dioscorides in the British Isles in the 16th century. The Hampshire botanist John Goodyer (1592–1664) translated *De materia medica*, but his version was not published until 1934; it "served the English-speaking world as the only rather wobbly version of Dioscorides" until the recent translation by Beck (2005). Anderson (1977); Arber (1986); BNP vol. 10, pp. 670–672; DSB vol. 4, pp. 119–123; Greene (1983a); OCD pp. 483–484; Stannard (1969); Stearn in Anon. (1979).

Empiricus. Marcellus Empiricus, also known as Marcellus Burdigalensis as he may have lived in Bordeaux, is the author of a treatise *De medicamentis*, completed after AD 408, which is a compilation of the medicines used at the time. It was first printed in 1536. Ray cites his names from Johannes Bodaeus' edition of Theophrastus in the main catalogue, *Magia* for *Anagallis arvensis* [42] and *Vitrum* for *Isatis tinctoria* [226]. BNP vol. 8, pp. 300–301.

Erasistratus. The physician Erasistratos of Ceos came from a medical family and was probably born in the late fourth century BC and died in the mid third century. He probably practised in Alexandria and is one of only two anatomists in the ancient world who are known to have dissected human bodies. He wrote treatises on various subjects, including pharmacology; some are quoted verbatim by Galen (q.v.). Ray mentions a work by him about Loosetrife in a note under *Betonica* [84], information he must have derived from Pliny (cf. Garofalo 1988, pp. 161–162). BNP vol. 5, pp. 13–15; Bynum & Bynum (2007) vol. 3, pp. 637–639; DSB vol. 4, pp. 382–386; OCD pp. 552–553.

Galen. Galen of Pergamum (AD 129/130–199/217), latinised as Galenus and sometimes erroneously known as Claudius Galenus, was a Greek doctor and philosopher who after travelling widely settled in Rome, where he remained in imperial service until his death. He was a brilliant diagnostician and a prolific though prolix author whose works cover a wide range of medical topics including anatomy, physiology, surgery and preventative medicine, though his pharmacology was largely based on existing sources. In mediaeval times his works formed the basis of the medical curriculum (see Chapter 3). The first Greek edition of his works was printed in Venice in 1525, but with their publication Renaisssance scholarship began to call his authority into question. Ray makes numerous references to his names in the main catalogue, always providing a modern source for the interpretation. He also cites his *De medicamentorum facultatibus* directly in a note under *Tithymalus helioscopius* [569] and draws on his work in the 'Etymology'. BNP vol. 5, pp. 654–662; DSB vol. 5, pp. 227–237; OCD pp. 621–622.

Gellius. Aulus Gellius (born *c*. AD 125) spent most of his life in Rome. His *Noctes Atticae* ("Attic Nights"), published in about 180, includes notes on a wide range of topics; 19 of the 20 books survive and his work was highly regarded during the Renaissance. Many of his quotations are drawn from works which are otherwise lost.

4: BIOGRAPHICAL NOTES

Ray drew on his work in his account of *Faba* in the 'Etymology' and under *Spadix* and *Termes* in the 'Interpretation'. BNP vol. 5, pp. 727–728; OCD pp. 627–628.

Hippocrates. Hippocrates of Cos (*c.* 460–370 BC) was the most famous doctor in antiquity, but little is known about him. There are about 70 works in the Hippocratic corpus, but none can be ascribed to him with confidence and many of the stories that were told about him in ancient and mediaeval times are probably or certainly fictitious. He is best known through the accounts of Galen (q.v.). As with the names of Dioscorides and Galen, Ray cites a modern source when he includes the names of Hippocrates in synonymy in the main catalogue. He also includes in a note under *Betonica* [84] a reference derived from Pliny to a work by Hippocrates on Ptisan. BNP vol. 6, pp. 354–363; DSB vol. 6, pp. 418–431; OCD pp. 710–711.

Homer. By tradition Homer was the author of the two great epic poems of ancient Greece, the *Iliad* and the *Odyssey*, which date from the eighth century BC but tell the story of the Trojan War and its aftermath in the 13th century BC. In the eighth century writing had only been practised for a few decades, and it is uncertain whether Homer was an illiterate bard or whether he was literate but drew upon an existing oral poetic tradition. The first printed edition was published in Florence in 1488. His name appears in its Latin form, Homerus, in Ray's text. Ray lists two Homeric names, Ἀχερωις [Acheroïs: 456] and Ἐλειοδρεπτον [Eleiodrepton: 53], in the synonymy of species in the main catalogue, as well as making other references to his work in the main catalogue and the 'Etymology'. BNP vol. 6, pp. 450–463; OCD pp. 718–720.

Horace. Quintus Horatius Flaccus (65–8 BC) was well-educated but he lost his inheritance in 42 BC when he fought on the losing side in the civil wars. He was, however, later able to devote himself to poetry thanks to the patronage of the fabulously wealthy Maecenas. His major works are the *Odes*, *Epodes*, *Satires* (in Latin *Sermones*) and *Epistles*, with two shorter ones, *Ars poetica* and *Carmen saeculare*. Though admired in antiquity, Horace was forgotten in the Middle Ages and only rediscovered in the Renaissance; his work was first printed in 1470. He is quoted twice by Ray in non-botanical contexts in the 'Preface' and there is a brief reference to the *Satires* in the long entry under *Hyssopus* in the 'Etymology'. BNP vol. 6, pp. 486–494; OCD pp. 724–727.

Isidore. Isidore (*c.* 560–636), Bishop of Seville from about 600, was called by the historian Montalembert "le dernier savant du monde ancien". He came from a Hispano-Roman family and was the first Christian writer to try to compile a summa of universal knowledge, set down in his most important work, the incomplete *Etymologiae*, also known as the *Origines*, a huge compilation posthumously edited into 20 volumes. He was a careful compiler, much more critical than Pliny, and his work "was one of the main routes by which classical learning was transmitted to the Middle Ages" (OCD). The DSB account perhaps judges him against too modern a yardstick when it comments that "he wrote nothing original ... made no new observations ... and discovered nothing". Ray cites his work under *Cortex* in the 'Interpretation' and under *Laureola* in the 'Etymology'. BNP vol. 6, pp. 963–964; DSB vol. 7, pp. 27–28; Gerli (2003); OCD p. 768.

Juba. Juba II (*c.* 50 BC–AD 23), son of Juba I of Mauretania, was only a child when his father was defeated in battle by Julius Caesar (46 BC) and as such he was led captive in Caesar's triumph in Rome. He was educated there, granted citizenship and returned to Mauretania in 25 BC to rule as a client king of Augustus, marrying Cleopatra Semele, the daughter of Anthony and Cleopatra. Deeply learned, he wrote many books which, though now lost, were quoted by Pliny the Elder, Plutarch and others. (They included a monograph describing methods for capturing elephants.) He organised the first proper exploration of the Canary Islands. Ray mentions him in a note under *Betonica* [84] in connection with his work on *Euphorbium*, said by Pliny to have been named by Juba after his physician Euphorbus. BNP vol. 4, pp. 910–912, and vol. 6, pp. 1205–1206; OCD p. 799.

Juvenal. The angry *Satires* of the poet Decimus Junius Juvenalis were written in Rome, probably in the early second century AD. Virtually nothing is known of his life. Ray quotes from the *Satires* towards the end of the 'Preface'. BNP vol. 6, pp. 1146–1148; OCD pp. 804–805.

Macrobius. Virtually nothing is known about Ambrosius Theodosius Macrobius; even the order of his names differs in different manuscripts. He probably lived in North Africa in the early fifth century AD (perhaps *c.* 410). The three surviving works attributed to him include *Saturnalia*, a dialogue in seven books covering a multitude of subjects both serious and entertaining. It is cited by Ray in the 'Interpretation'. BNP vol. 8, pp. 101–103; DSB vol. 9, pp. 1–2; Davis (1969); OCD pp. 906–907.

Matthaeus Sylvaticus. Matthaeus Sylvaticus is the latinised form of the name of Matteo Silvatico, an Italian physician and a member of the Salerno school of medicine who died *c.* 1342. He travelled extensively and grew some of the plants he collected in a botanic garden in Salerno. He compiled the popular medical manual *Liber pandectarum medicinae* (or *Opus pandectarum medicinae*), which was first printed in 1474. A single synonym in Ray's main catalogue, *Pes asini* under *Alliaria* [17], is taken from this source. Greene (1983a).

Mela. Pomponius Mela, of Tingentera in southern Spain, was the author of *De chorographia*, written as three books in AD 43–44 and perhaps designed to celebrate Claudius' invasion of Britain. They are the earliest surviving geographical works in Latin. His work was well-known in the Middle Ages and first printed in Milan in 1471. Ray mentions him once, under Woad, *Glastum sativum* [226]. BNP vol. 11, pp. 582–584; OCD p. 1218.

Moschion. Little is known of this Greek physician, who practised in the first century AD. He is one of the authors of monographs, in this case on Radish, cited from Pliny in a note under *Betonica* [84]. BNP vol. 9, p. 227.

Musa. Antonius Musa was physician to the Emperor Augustus at the end of the first century BC and achieved fame by curing him of a grave illness by prescribing cold water and lettuce leaves. A work entitled *De herba vettonica* (or *betonica*) came to be attributed to him and was copied, translated into numerous languages (including Anglo-Saxon) and later printed (from 1481 onwards) with the work of Apuleius Platonicus (Anon. 1979). Ray devotes a note to it under *Betonica* [84]. However, the attribution is regarded as false, and the work on *Betonica* appears to date from the fourth century AD. For Antonius Musa's brother Euphorbus, see the entry for Juba above. BNP vol. 1, pp. 807–808; OCD p. 116.

Oribasius. The Greek doctor Oribasius (AD *c.* 320–400) was personal physician to the Emperor Julian. His vast compilation of medical writings, *Collectiones medicae*, was part of Julian's energetic programme to revive the classical past. Ray cites the names *Aegilops* [14], *Aera* [340] and *Orobanche* [412] from his and other ancient works. BNP vol. 10, pp. 203–205; OCD pp. 1074–1075.

Ovid. Publius Ovidius Naso (43 BC–AD 17) was educated in Rome and rapidly became the city's leading poet. However, he was banished by Augustus to Tomis on the Black Sea (in modern Romania) in 8 AD, for reasons unknown. His best known works are the *Heroides*, *Amores*, *Ars amatoria* and *Metamorphoses*; the miseries of his exile are dealt with in the *Tristia* and *Epistulae ex Ponto* ("Letters from

Pontus"). His work was influential in the Middle Ages and the Renaissance and remains a source of inspiration to this day. Ray quotes from *Epistulae ex Ponto* in the 'Preface' and three times from *Metamorphoses* and once from a less well-known work, *Remedia amoris*, in the 'Etymology'. BNP vol. 10, pp. 398–306; OCD pp. 1084–1087.

Phaenias. "The naturalist Phanias" is cited by Ray in a note under *Betonica* [84] as the author of a work on Nettle. Ray must have taken the reference from Pliny (see footnote 118 in the main catalogue), who was presumably referring to Phaenias or Phaeneas of Eresos on Lesbos (approximately 375–300 BC). Phaenias, a pupil of Aristotle and a friend of Theophrastus, wrote a work on plants from which only a few quotations survive; these suggest that it was a more popular work than that of Theophrastus with a concentration on practical uses. BNP vol. 10, pp. 901–902; OCD p. 1152.

Pliny the Elder. Gaius Plinius Secundus (AD 23/4–79), born at Como, was a hard-working, patriotic batchelor of inexhaustible curiosity and an author whose energy and diligence "astonished his nephew [Pliny the Younger] ... and still amaze today" (OCD). His *Natural history* in 37 books, an encyclopaedia of all contemporary knowledge, is an immense compilation of the work of hundreds of authors. Much of our knowledge of medicine and natural history in the ancient world comes from Pliny, as many of his sources do not survive (though, when they do, they show that Pliny often transcribed them inaccurately). Books 12–19 deal with botany, agriculture and horticulture and 20–27 with medicines from botanical sources. Pliny was killed while investigating the eruption of Vesuvius. His work was well-known in mediaeval times and first printed in Venice in 1469 (Nauert 1980); an English translation by Philemon Holland was published in 1601. Pliny was an uncritical compiler, and Sir Thomas Browne (1646) remarked that "there is scarce a popular errour passant in our dayes, which is not either directly expressed, or diductively contained in this worke, which being in the hands of most men, hath proved a powerfull occasion of their propogation". There are numerous references to Pliny in Ray's text. His names in the main catalogue are usually taken from cited authorities, but no modern source is cited for *Sertula campana*, *Olus sylvestre* and *Rubigo*, listed as synonyms of *Melilotus vulgaris* [361], *Rapistrum arvorum* [490] and *Ustilago* [629] respectively. Pliny is also the source for some of the observations in the main catalogue (see Table 3.5) and an important source of information in the 'Etymology'. Anderson (1977); BNP vol. 11, pp. 383–390; DSB vol. 11, pp. 38–40; OCD pp. 1197–1198.

Plutarch. Plutarch (AD *c*. 46–*c*. 120) was a Greek, born Plutarchos at Chaeronea, Boeotia, who became a Roman citizen as Mestrius Plutarchus. An advocate of a partnership between Greece and Rome, he taught philosophy in Rome and came to know many distinguished Romans. He was also a devoutly religious man who became one of the two permanent priests of the shrine at Delphi. He wrote prolifically in Greek on many subjects, mainly biographical and philosophical. His *Symposiaca* or *Quaestiones convivales* ("Table talk") make up part of his *Ethica* or *Moralia*, modern names for a collection which makes up one of his two great works; book 3 forms part of Volume VIII in the Loeb edition. He is quoted under *Faba*, *Narcissus*, *Rosa* and *Ruta* in Ray's 'Etymology'. BNP vol. 11, pp. 410–425; OCD pp. 1200–1201.

Porphyrius. Porphyrius of Tyre or Porphyry (AD 234–*c*. 305) studied in Athens and Rome. A prolific Neoplatonic philosopher, he wrote commentaries on Aristotle and Plato and other works on a wide range of subjects including history, music and religion. Ray cites his work only once, under *Termes* in the 'Interpretation'. BNP vol. 11, pp. 646–652; OCD pp. 1226–1227.

Scribonius. Scribonius Largus (*c.* AD 1–50), a Roman physician, accompanied Claudius in his invasion of Britain. His only surviving work is the book of prescriptions *De compositione medicamentorum* or *Compositiones medicae*, which Ray cites under *Volva* in the 'Interpretation'; it was first published by Ruel in 1528. Scribonius is cited with Fallopius (see Falloppio below) as one of two authors of *Altercum*, a synonym of *Hyoscyamus niger* [294]. *Urceolaris Scribonii* also appears as a synonym of *Parietaria* [418], but this is cited indirectly from Camerarius (q.v. below). BNP vol. 13, p. 118; Bynum & Bynum (2007), vol. 5, pp. 1124–1125; OCD p. 1370.

Solinus. Gaius Julius Solinus wrote a geographical work, *Collectanea rerum memorabilium* ("Collections of curiosities"), in the third or fourth century AD. It was largely based, without acknowledgement, on the work of Pliny the Elder and Pomponius Mela. His text was well known in the Middle Ages, for example to authors such as Bede and Isidore, and numerous manuscript copies survive; the first printed edition is undated but was published before 1473 in Rome. Ray cites Solinus under *Acer* in the 'Etymology'. BNP vol. 13, pp. 616–618; OCD p. 786.

Themison. Themison of Laodicea studied medicine as a pupil of Asclepiades (q.v.) and probably practised as a physician in Rome in the latter half of the first century BC. Ray cites (in a note under *Betonica* [84]) his work on Plantain, which was mentioned by Pliny but does not survive (Tecusan 2004). BNP vol. 14, p. 426; OCD p. 1497.

Theophrastus. Theophrastus (*c.* 372–*c.* 287 BC) of Lesbos was a close associate of Aristotle (q.v.). Although Theophrastus is credited with 225 works on many different subjects, only his botanical works survive intact. They provide the counterpart of Aristotle's zoological texts and are regarded as the starting point of scientific botany. *Historia plantarum*, often known as *Enquiry into plants*, deals with trees, shrubs and herbaceous plants and describes some 550 species in 11 books. *De causis plantarum* ("About the causes of plants") deals in six books with subjects such as reproduction, growth, environmental effects, the differing niches of different species, the cultivation of plants, diseases, death, flavours and odours. Theophrastus' ecological insight can still impress modern botanists, but classical scholars say remarkably little about his botany in recent biographical accounts. Ray cites his names throughout the main catalogue, quoting them from numerous sources. Two names, *Milos* and *Erysine*, are cited without modern authorities as synonyms of *Taxus* [565] and *Ustilago* [629] respectively. Ray also draws some of his observations from Theophrastus' works. There is strong evidence (set out, for example, in footnotes 21, 372, 549, 608 and 799 in the main catalogue) that he used the translation of Theophrastus into Latin made by Theodoros Gaza (Γαζης [Gazes] in Greek: *c.* 1400–*c.* 1475), a Greek who fled to Italy when his native Thessaloniki was captured by the Turks in 1430 and translated both Aristotle and Theophrastus. His edition of Theophrastus (1529) includes both *De causis plantarum* and *Historia plantarum* and was certainly held in Cambridge University Library in Ray's time (see Chapter 5); Bodaeus reprinted the translation of *Historia plantarum* in his 1644 edition of that work. BNP vol. 14, pp. 508–517; DSB vol. 13, pp. 328–334; Greene (1983a); Hünemörder (2009); OCD pp. 1504–1505.

Varro. Marcus Terentius Varro (116–27 BC) was famously described by Quintilian as *vir Romanorum eruditissimus*. Although he was a prolific author, only one of his 74 titles survives intact, *De re rustica* or *Rerum rusticarum*, a treatise on farming in three books written as a dialogue. It was first published as a printed book in Venice in 1472 with Cato's *De agricultura* and Columella's *De re rustica* (Brown 1980b). Ray cites it in a note following *Rapum sylvestre* [495]. There is another reference to *De re rustica* in the 'Etymology' under *Ocymum* and Ray here also quotes, under *Baptisecula*, from Varro's *De lingua latina*. This is the only other work by Varro which is known as more than fragments, but only six of the original 25 books survive in an incomplete form as a single manuscript, three of them dealing with etymology. Three other references to Varro in Ray's 'Etymology', under *Raphanus*, *Rapum* and *Triticum*, are cited without a source but also come from *De lingua latina*. BNP vol. 15, pp. 209–226; DSB vol. 13, pp. 588–589; OCD pp. 1504–1505; Taylor (1996).

Virgil. In the *Catalogus* Ray makes use of material from the three major works of the Roman poet Publius Vergilius Maro (70–19 BC), the *Aeneid*, the *Eclogues* and the *Georgics* (a didactic poem about farming in four books dedicated to his patron Maecenas). Ray usually cites plant names from his works indirectly from modern sources, but *Ulmus aeria* ("The lofty elm") is given as a synonym of one of the elms in the main catalogue, *Ulmus vulgatissimus folio lato scabro* [626], without any supporting authority. Another name from Virgil, *Amellus*, is discussed under two different species [101, 129] and as *Amellus montanus* under a third [144]. Greene (1983a); OCD pp. 1602–1607.

Vitruvius. Vitruvius was a Roman architect and military engineer in the first century BC who served in the army under Julius Caesar and later as an engineer in Rome in the reign of Augustus. He is often called Marcus Vitruvius Pollio, but the additional names are not authentic; classical authors refer to him as simply Vitruvius. His treatise *De architectura*, in 10 books, written *c.* 30 BC, deals with engineering as well as architecture as currently understood. It was first published as a printed book in Rome in about 1485. Ray cites book 2, on building materials, in a note under Alder, *Alnus* [20], and also mentions him under Woad, *Glastum sativum* [226]. BNP vol. 15, pp. 480–484; DSB vol. 15, pp. 514–521; OCD pp. 1609–1610.

Modern authors

The late 15th-century and 16th- and 17th-century authors described below are those cited as the sources of Ray's chosen plant names and synonyms. Details of many of these authors are provided by Ray in his 'Explanation', and we also include one author (Belli) who is listed in the 'Explanation' but not mentioned in the main catalogue. Many authors of the period latinised their names; the following alphabetical list of authors normally gives the name in the author's native language, except in the case of authors such as Amatus Lusitanus where the Latin form is very different and much more familiar. The plant names attributed to these authors are enumerated in Table 3.1 and any observations based on their works are listed in Table 3.5. If Ray attributes only one or two plant names to an author, we have listed these names in the following notes. Details of modern authors who are mentioned by Ray in the text of the *Catalogus* but not as the sources of his plant names are provided in our footnotes.

Information for these entries has been taken from the references cited; DSB indicates *Dictionary of Scientific Biography* (Gillispie 1970–1980) and ODNB *Oxford Dictionary of National Biography* (Matthew & Harrison 2004).

Aldrovandi. Ulisse Aldrovandi (1522–1605) of Bologna, the son of a nobleman, studied mathematics, law and philosophy before he acquired a wide-ranging interest in natural history. In 1551 he climbed Monte Baldo with his fellow botanists Anguillara (q.v.) and Luigi Alpago, and later as a professor at Bologna he stimulated much interest in natural history by his lectures and expeditions. He was a pioneer in making a scientific collection for research purposes and his herbarium survives at the botanic garden in Bologna, which he founded in the teeth of much opposition. His largely posthumous work on natural history contains much that is valuable mixed with a large proportion of fables (e.g. about the cockatrice or basilisk and the original 'cock and bull story'). Ray cites *Holosteum Dioscoridis* as a synonym of *Coronŏpus Ruellii* [153] on the basis of Aldrovandi in the work of J. Bauhin. Arber (1986); Greene (1983b); DSB vol. 1, 108–110.

Amatus Lusitanus. This is the *nom de plume* of João Rodigues (1511–1568), a Portuguese physician of Jewish origin who studied medicine at Salamanca Univerity. He practised in Antwerp (1533–1540) and thereafter in Italy until he left for Thessaloniki in 1559, driven out by the papal prosecution of marranos (Jews who outwardly conformed to Christian practice). He is better known for his medical case histories (published in seven *centuria*) than for his botanical work, which consisted of commentaries on Dioscorides. The first of these was published in Antwerp in 1536 (his first book and the only one published under his own name) and the full work was initially published in Venice in 1553; Ray cites an edition published in Strasbourg in 1554. Amatus criticised Mattioli (q.v.), who was sufficiently enraged to publish a short book attacking him, *Apologia adversus Amathum Lusitanum cum censura in ejusdem enarrationes* (1558), which is mentioned in Ray's account of Mattioli's works in the 'Explanation'. Mattioli's hostility presumably contributed to Amatus' insecurity in Italy. DSB vol. 8, 554–555.

Anguillara. Luigi Squalermo (1512–1570) took the name Anguillara from his Italian birthplace. He is described in modern texts as the first director of the botanic garden in Padua, although Ray follows Caspar Bauhin (1623) in describing him as the third curator. The garden was founded in 1545 by the Venetian Senate and Anguillara worked there from his appointment in 1546 until he moved to Ferrara in 1561. He travelled widely in southern Europe, North Africa and the Near East. His book on materia medica, *Semplici*, was first printed in Venice in 1561 and was translated into Latin by Caspar Bauhin in 1593. He was highly regarded as 'a commentator of unusual insight' on Dioscorides (Arber 1986) and his names are frequently cited by Ray. DSB vol. 1, p. 167; Greene (1983b); Legré (1899, 1901); Riddle (1980).

Barbaro. Ermolao Barbaro or Hermolaus Barbarus (1453/4–1493) was the son of a Venetian diplomat who studied in Verona, Rome and Padua. He himself followed a political and diplomatic career, but he became famous for his commentaries on classical texts, especially his *Castigationes Plinianae*, initially published in Rome in 1492–1493. This corrected innumerable copyists' errors in existing editions and was "immediately saluted as the most authoritative discussion of Pliny's *Historia naturalis* then available". He died of plague at an early age, "a great intellect, cut off in his prime". His Latin translation of Dioscorides and his commentary (*Corollarium*) on it were probably written in 1481–1482 but were first published, posthumously, in 1516. Bietenholz (1985–1987), vol. 1, pp. 91–92; Greene (1983b); Riddle (1980).

Bauhin, C. Caspar or Gaspard Bauhin (1560–1624), Casparus Bauhinus (*C.B.*) in Ray's *Catalogus*, was the son of Jean Bauhin of Basel and brother of the botanist Jean (q.v.). He was taught anatomy by his father and botany by his much older brother. He pursued his anatomical studies at Basel, Padua, Bologna, Paris and Montpellier between 1572 and 1580. In 1582 he was appointed to the Professorship of Greek(!) in Basel, but he also taught anatomy in the winter and led botanical

4: BIOGRAPHICAL NOTES

expeditions in the summer; in 1589 a special chair of Anatomy and Botany was created for him. He did not share the reluctance of his brother to publish and wrote several influential anatomical textbooks. His botanical works are listed by Ray in the 'Explanation'. Much the most important is the Πιναξ (*Pinax*) of 1623, which lists all the plants known to him in a systematic sequence, with numerous synonyms (but no other information about them); it is the most useful single compilation of pre-Linnean plant names and is often cited by Linnaeus in *Species plantarum*. It is one of the two sources from which Ray says he took synonyms from works that he had not himself seen. Bauhin intended to produce a more descriptive work following the sequence of the *Pinax*, but only one volume of this, *Theatri botanici sive Historiae plantarum ... Liber primus* (1658), was published and that long after his death by his son Jean Caspar. Bauhin's Flora *Catalogus plantarum circa Basileam sponte nascentium* (1622) is a work of only local interest but it provided one of the models for Ray's own *Catalogus* (see Chapter 3). Arber (1986); DSB vol. 1, pp. 522–525.

Bauhin, J. Jean Bauhin (1541–1613) was born in Basel, the eldest son of a protestant physician, also called Jean, who had left his native France because of religious persecution. His initials *I.B.* or *J.B.* and those of his brother Caspar (q.v.), *C.B.*, occur throughout the *Catalogus* and Ray includes a remarkable tribute to his work in the 'Explanation'. Jean Bauhin was educated at Basel and Montpellier, and much of his career (from 1571 onwards) was spent in Württemberg as physician to Duke Frederick, ruler of the principality of Württemberg-Montbéliard. He became interested in botany as a young man and was befriended by Conrad Gesner of Zurich (q.v.), with whom he travelled to the Alps in 1561. In the next few years he broadened his experience of plants in the field, meeting Fuchs (q.v.) at Tübingen, studying under Guillaume Rondelet while he was at Montpellier and visiting northern Italy and the Apennines. Once he was established at Württemburg he created botanic gardens at Montbéliard and Stuttgart. The two botanical books that he published during his lifetime are listed by Ray in the 'Explanation' and cover very restricted topics; he also published medical works including a description of the baths at Bol which includes much material on apples, pears and fossils. In about 1600 he began to compile his *Historia plantarum universalis*, a detailed account of over 5,000 plant species with many illustrations. Jean-Henri Cherler, a physician who had married Bauhin's youngest daughter, collaborated with him during this period. The manuscript was virtually complete at the time of Bauhin's death, when it passed to Dominic Chabrey of Yverdon, yet another physician, who published a summary in 1619. The whole work did not appear until long after Bauhin's death, in 1650–1651, when Franz Ludwig von Graffenried of Yverdon paid for the publication. Cherler was named as co-author, "an honor he scarcely deserved" (DSB). Arber (1986); DSB vol. 1, pp. 525–527.

Belli. Onorio or Honorius Bellus (died 1604) was a Venetian doctor from Vicenza who lived at Cydonia (now Chania) in Crete from 1583 to 1599, during the period when the island was ruled by the Venetians. He corresponded with de L'Écluse (q.v.), who appended his letters to his *Rariorum plantarum historia* (1601). Dizionario biografico degli Italiani, vol. 7, pp. 671–672 (1965); Stearn (1984).

Belon. Pierre or Petrus Bellonius (*c.* 1517–1564), born at Le Mans into an obscure family, became an apothecary and was then able, through the patronage of the Bishop of Le Mans, to study under Valerius Cordus (q.v.) at Wittenberg. He then "travelled in the Near East between 1546 and 1549 and survived all the many hazards of his journeys only to be murdered in Paris" (Stearn 1984). By then he had published the books on fish by which he is best remembered; based on original descriptions, they entitle him to be considered "the originator of comparative anatomy" (DSB). His botanical books are listed by Ray in the 'Explanation', and the title-page of the original French edition of *Les observations* (1553) is reproduced by Stearn (1984). Backer *et al*. (1993); DSB vol. 1, pp. 595–596; Kusukawa (2007); Legré (1901); Voet (1980–1983), vol. 1, pp. 270–276.

Besler. The Nuremberg apothecary Basileus Besler (1561–1629) kept his own garden and natural history collection, but it is as the author of *Hortus Eystettensis* (1613), written for Johann Konrad von Gemmingen, Bishop of Eichstätt, that he is remembered. This superbly illustrated account of the plants in the formal gardens at Willibaldsburg of the Prince-Bishops of Eichstätt has illustrations arranged in the sequence in which the plants flower through the season. It was published in three versions, two (luxurious and less luxurious) with text and one without; it is clearly the last that is mentioned in Ray's 'Explanation' under *Eyst*. For a facsimile and commentary see Littger, Lorenz & Menghini (2006), and for another book about the book see Barker (1994). Killy & Vierhaus (2001–2006), vol. 1, p. 494.

Bock. Hieronymus Tragus is the latinised name of the German botanist Jerome Bock (1498–1554). Like Brunfels, he was a follower of Luther who worked as a preacher and at times as a physician. He was encouraged by Brunfels to write his *New Kreütter Bůch*, first published (without illustrations) in 1539. Unlike Brunfels' work, the importance of the *Kreütterbuch* (the name of later editions) lies in its text, written "in the plain, racy German of the people" and displaying a "sturdy independence of mind" in his critical approach to the classical authors (Arber 1986). His precise descriptions of about 700 species were based on the careful observation of plants in the field and in cultivation and were as revolutionary as the illustrations in Brunfels' *Herbarum vivae eicones*. An illustrated edition was published in 1546 and a Latin translation in 1552. Anderson (1977); Arber (1986); DSB vol. 2, pp. 218–220.

Bodaeus. Johannes (or Joannes) Bodaeus à Stapel (or van Stapel), sometimes known as Stapelius, studied medicine in Leiden. He died in 1636 before completing his edition of Theophrastus' *Historia plantarum*, which was prepared for publication by his father, Egbert Bodaeus, in 1644. This large volume gives Theophrastus' Greek text accompanied by Gaza's (1529) Latin translation with detailed commentaries, descriptions of further plants and numerous figures by Bodaeus; there are also observations by Julius Caesar Scaliger (see footnote 67 in Ray's 'Interpretation'), notes by Robertus Constantinus and a full index. One of the plants that Bodaeus added was a succulent South African carrion-flower which he called *Fritillaria crassa promontorii bonae spei* (p. 335) but which was later included in the genus *Stapelia* (Apocynaceae), named by Linnaeus in his honour. Aa (1854); Greene (1983a).

Brasavola. Antonio Musa Brasavola (or Brassavola) (1500–1555) of Ferrara, latinised to Bras(s)avolus, was an Italian physician. His book on medicinal plants, *Examen omnium simplicium*, first published in 1536, was very popular in the 16th century and is cited by Ray in the main catalogue. Like the *Botanologicon* of Euricius Cordus, it is set out as a dialogue. Ray includes one synonym from Dioscorides on the authority of Brasavola, under *Pimpinella vulgaris, sive minor* [439], and also cites in the main catalogue [607] a work on syrup (Brasavola 1540) which he knew through Camerarius (1588). Greene (1983b); Hirsch (1884–1888), vol. 1, p. 561.

Braunschweig. Hieronymus Braunschweig or Brunschwig (*c*. 1450–*c*. 1512) of Strasbourg was a surgeon and pharmacist. He published an illustrated work on surgery, *Chirurgia*, in 1497, but his pharmaceutical works were of greater originality and importance as they contained detailed and illustrated accounts of distillation techniques. *Liber de arte distillandi, de simplicibus* was published in 1500 and

Liber de arte distillandi, de compositis in 1507; the latter contained a pharmacopoeia for the poor, *Thesaurus pauperum*, which was reprinted in the later version of his work, *Grosses Buch der Destillation*, first published in 1512. This became one of the most popular medical works of the 16th century and appeared in numerous editions. Ray refers to it in citing Braunschweig's name *Hæmorrhoidum herba* as a synonym of *Chelidonium minus* [129]. Anderson (1977); DSB vol. 2, pp. 546–547; Killy & Vierhaus (2001–2006), vol. 2, p. 186.

Brunfels. Otto Brunfels (*c.* 1489–1534), latinised as Otho Brunfelsius, is usually coupled with Bock (Tragus) and Fuchs as one of the three German "fathers of botany" by those who investigate the paternity of academic disciplines. He attended university in his home town of Mainz and spent a period as a Carthusian monk in Strasbourg before adopting Lutheran views and leaving the monastery in 1521. He then spent time as a pastor, teacher, writer of theological works and finally as a physician. His *Herbarum vivae eicones* was published in three parts in 1530, 1531 and (posthumously) 1536. It describes and illustrates about 230 species growing in the vicinity of Strasbourg. Its importance lies not in its text, a compilation of quotations from earlier authors, but in its illustrations. These were drawn from life by the artist Hans Weiditz and set completely new standards for botanical drawings. A German adaptation of the work, *Contrafayt Kreüterbuch*, was first published in two parts, in 1532 and 1537. Anderson (1977); Arber (1986); DSB vol. 2, pp. 535–538; Greene (1983a).

Camerarius. Joachim Camerarius the Younger (1534–1598) was the son of the classical scholar Joachim Kammermeister (1500–1574), who under his adopted name Camerarius published editions of numerous classical texts (see Bietenholz, 1985–1987, vol. 1, pp. 247–248). The younger Joachim trained as a physician in Germany and Italy before practising in Nuremberg, where he established a botanic garden. His *Hortus medicus et philosophicus* was published with Thal's *Sylvia Hercynia* in 1588 and is often cited by Ray. He also published a new edition of Mattioli's commentary on Dioscorides (Mattioli 1586, 1590). Arber (1986); Killy & Vierhaus (2001–2006), vol. 2, p. 286.

Cardano. Girolamo Cardano or Hieronymus Cardanus (1501–1576) of Pavia in the Duchy of Milan was the illegitimate son of a friend of Leonardo da Vinci; his mother was "ignorant and irascible" (DSB) and his childhood mistreatment presumably explains his highly unstable personality in later life. He is best known for his contribution to the development of mathematics, but he also taught medicine and wrote 200 works on a wide range of subjects. These included two encyclopaedias of natural history, *De subtilitate libri XXI*, first published in Nuremberg in 1550, and *De rerum varietate*, first published in Basel in 1557. The two works have been described as "a mine of facts, both real and imaginary" (DSB). There were several later editions of both works, each of which contains a 'book' dealing with plants. According to Egerton (in Greene 1983b, p. 1016) his "original and pointed remarks" include the observation that species of plants are to be looked for in nature, not in Dioscorides. His autobiography, *De vita propria liber*, was translated into English by Stoner (1931). Ray cites Cardano with a full reference to *De rerum varietate* as an author of *Platanus*, a synonym of *Acer major* [5]. DSB vol. 3, pp. 64–67; Kessler (1994).

Cesalpino. Andrea Cesalpino or Andreas Caesalpinus (1524/5–1603) studied at Pisa, where he became professor of medicine and director of the botanic garden; he moved to Rome in 1592. He wrote books on philosophy and medicine but his fame rests on his botanical text *De plantis libri XVI* (1583). This began with the first modern statement of the principles of botany. He classified plants on Aristotelian principles, based on morphological characters rather than unessential or accidental features such as their medicinal properties. His work was valued most highly by later authors such as Jungius and Linnaeus who had a strong interest in the theoretical side of the subject. Three of the references to Cesalpino in the main catalogue are clearly indirect citations [203, 271, 495]. Although his names are cited in many other places in the main catalogue and under *Polypodium* in the 'Etymology', there is no evidence to suggest that Ray had first-hand knowledge of his work in 1660 (see Chapter 5). Arber (1986); DSB vol. 15, pp. 80–81; Dizionario biografico degli Italiani, vol. 24, pp. 122–125 (1980); Greene (1983b).

Clusius, see **L'Écluse**.

Colonna. Fabio Colonna or Fabius Columna (1567–1640[1]) was born into an ancient Roman family. He was trained in law but was also skilled in Greek and Latin, mathematics and music. He suffered badly from epilepsy, which he treated fairly effectively with *Valeriana officinalis*. His first major work was *Phytobasanos* (1592), which includes a critical discussion of Dioscorides' *Phu* (*Valeriana*) and 25 other plants of classical authors as well as illustrations of a few hitherto unknown southern European species. In *Ecphrasis* he provides detailed descriptions and figures of over 200 plant species. The first part was published in 1606 but the work did not become widely known until the publication of the second part in 1616; even then it did not provide him with much financial return. It was Colonna who suggested the use of the word petal (*petalum*) in its modern sense, although he did not use it in this sense himself. Ray did not use the term in the *Catalogus* either but he was later to become the first person to employ the word in its botanical sense (Stearn 1992). Colonna became interested in palaeontology and minerals in later life, but little is known of his last few years. Anderson (1977); Dizionario biografico degli Italiani, vol. 27, pp. 286–288 (1982); Greene (1983b).

Cordus, E. Euricius Cordus (1486–1535), of Simtshausen in Hesse, did not decide to follow a medical career until 1519. In 1527 he was appointed Professor of Medicine at the University of Marburg, where he laid out a botanic garden and was the first German botanist to organise field excursions. He left for Bremen in 1533 and published his main botanical work, *Botanologicon*, in 1534. This takes the form of a conversation between Cordus and a few friends and is much concerned with exposing the errors of apothecaries who sold German plants erroneously labelled with names from Dioscorides and Galen. Bietenholz (1985–87), vol. 1, pp. 339; DSB vol. 3, pp. 412–413; Greene (1983a).

Cordus, V. Valerius Cordus (1515–1544), son of Euricius (q.v.), was a botanist and pharmacist of outstanding gifts and, by all accounts, a most attractive personality. His works were published posthumously after his tragically early death. He was educated at the University of Marburg, where his father was Professor of Medicine, and he subsequently trained at Leipzig and taught at the University of Wittenberg. He was well-travelled, did much fieldwork in Germany, discovered numerous new species and died when visting Italy in a quest to see the plants about which the ancients had written. His pharmaceutical work, *Dispensatorium*, was first published in Nuremberg in 1546. (He had left a copy of the manuscript in the city in 1542.) *Annotationes in Dioscoridis libros V* was first published in 1549 from notes made at his lectures in Wittenberg and Gesner helped publish his major botanical work,

---

[1] The date of death is given as 25 July 1640 in *Dizionario biografico degli Italiani* but other sources (e.g. Greene 1983b and Jarvis 2007) give it as 1650.

Historiae stirpium libri IV (1561), and the subsequent *Stirpium descriptionis libri quinti* (1563) from Cordus' manuscripts. The last two were notable for the precision of his plant descriptions, based on living material. Gesner added various items to the 1561 volume, including his work on German gardens which is cited by Ray as *Gesn. hort.* and an account of the plant he called *Tulipa Turcarum*, the first description and European illustration of a tulip (Pavord 1999). Arber (1986); Backer *et al.* (1993); DSB vol. 3, pp. 413–415; Greene (1983a); Riddle (1980).

Cornarius. Johann Hainpol or Hagenbut (*c.* 1500–1558), born in Zwickau, Saxony, adopted the name Janus Cornarius in about 1520. In his twenties he travelled round Europe, even visiting England, before living in Basel and then returning to Germany. He practised as a physician, publishing *Universae rei medicinae* ἐπιγραφη [epigraphe] in 1529, but his important work was as a prolific editor and translator of Greek texts, work which he was encouraged to pursue by Erasmus. His critical edition of Hippocrates, based on a comparative study of the available manuscripts, was particularly significant, but he also edited and translated texts by Aegineta, Aëtius, Dioscorides, Galen and various non-medical authors including the ecclesiastics Basil and Chrysostom. Ray cites him as the author of *Ellychnion Tarsicum Galeni*, a synonym of *Fungi arborei ad ellychnia* [206]. We have not traced the source of this name. Bietenholz (1985–87), vol. 1, pp. 339–340; Hirsch (1884–1888), vol. 3, pp. 14–15; Riddle (1980).

Costeo. Giovanni Costeo or Costaeus (1528–1603) was born in Lodi, northern Italy, of a noble family. Nothing is known about his youth and studies before the publication of his first medical work in 1565. From 1570 he taught medicine at the University of Turin and was physician to Duke Emanuele Filiberto; after the death of the duke in 1580 he moved to Bologna. He studied plants in the vicinity of Padua and Venice and corresponded with Aldrovandi (q.v.). His main work, *De universali stirpium natura libri duo* (1578), was modelled on Theophrastus. Ray cites from Camerarius (1588) a passage in the first of the two books in a note after *Clematis daphnoides minor* [138] and relates two "unbelievable" statements of his, without a specific reference, in a note after *Pulegium* [474]. He also cites him indirectly from Hofmann (1646) as one of three authors of *Ben sive Behen album Arabum*, a synonym of *Pastinaca sativa Dioscoridis* [423]. Dizionario biografico degli Italiani, vol. 30, pp. 403–405 (1984).

Daléchamps. Jacques Daléchamps or d'Aléchamps (1513–1588) of Caen, whose latinised his name as Dalechampius, studied under Guillaume Rondelet at Montpellier, where he obtained a degree and a doctorate in medicine. He moved to Lyon in 1552 and was the main author of the massive herbal *Historia generalis plantarum* (1586–1587), the most complete botanical compilation of the period. Many of its figures were taken from the works of Dodoens, Fuchs and Mattioli. Daléchamps was assisted in the work by the young Jean Bauhin and by Jean Desmoulins (Molinaeus). The book was published anonymously and was often cited (as by Ray) as *Lugd.* from its place of publication, Lugdunum [Lyon]. Daléchamps also translated the works of Theophrastus into Latin and produced translations or commentaries on other classical authors. Arber (1986); DSB vol. 3, pp. 533–534.

Dieter, see **Tabernaemontanus.**

Dodoens. Rembert Dodoens (1517–1585), latinised as Dodonaeus, was born in Leeuwarden, was brought up in Mechelen (now Malines in Belgium) and studied medicine at the University of Louvain, graduating in 1535. In the next decade he travelled widely in Italy, Germany and France, before returning to Mechelen. His career as an author of herbals is closely linked to that of his younger contemporaries de L'Écluse and de L'Obel. Dodoens' first major botanical work, *Cruijde Boeck* (1554), was published in Flemish; most of the illustrations were based on those of Fuchs. This was soon translated into French by de L'Écluse as *Histoire des plantes* (1557), with additions by Dodoens, and later from French into English by Henry Lyte (1578). In 1582 Dodoens was appointed to the Faculty of Medicine at the University of Leiden. His collected works, *Stirpium historiae pemptades sex*, were published in 1583 by the great Antwerp publisher Christopher Plantin; most of the illustrations had already been used in earlier works by Dodoens, de L'Écluse and de L'Obel. Anderson (1977); Arber (1986); Backer *et al.* (1993); DSB vol. 4, pp. 138–140; Greene (1983b); Kusukawa (2007); Voet (1980–1983), vol. 2, pp. 821–834; Wille (1993).

Dourez. Valerand Dourez had "une grande passion pour la science des végétaux" (Legré 1904) and was on friendly terms with some of the leading botanists of the 16th century (including Jean Bauhin, Daléchamps, Gesner, de L'Écluse, de L'Obel and Pena), but he published nothing himself and he would be forgotten today had his name not been preserved as the specific epithet of *Samolus valerandi*. Born in Lille, he became a pharmacist in Lyon and was therefore able to botanise in the Alps. He also spent time in Montpellier and in 1564–1565 he visited Venice, the east coast of the Adriatic, Greece, Crete and Syria. Dourez knew Thomas Penny and gave him an account of gall-flies which was eventually published in 1634 in *Insectorum ... theatrum* compiled by Moufet (Raven 1947). The exact date of Dourez's death is not known, but it was between 1571 and 1575; Jean Bauhin inherited his collections and papers. The scattered references to Dourez in the botanical works of his friends were collected together by Legré (1904) in the only adequate account of his life that we have been able to trace. Ray's only reference to Dourez is under *Anagallis aquatica rotundifolia* [39], where he lists two synonyms from Bauhin & Cherler (1651, vol. 3(2), p. 792), *Samolus Plinii quorundam* Valerando ("Pliny's *Samolus* of some [authors] according to Valerandus") and *Samolus Valerandi*.

Durante. Castore Durante (1529–1590) came from Perugia, where he studied medicine. His first two medical works, published in 1565 and 1566, proved to be popular and financially very rewarding, as did a later book, *Tresoro della sanità*, published in 1586. In the 1580s he became *Archiatra* (principal personal physician) to Pope Sixtus V. His main botanical work, *Herbario nuovo* (1585), contains 965 woodcuts, some of which show plants set in an appropriate landscape. Anderson (1977); Dizionario biografico degli Italiani, vol. 30, pp. 403–405 (1993).

Falloppio. Gabriele Falloppio (1523–1562), latinised as Fallopius, studied medicine at the University of Modena and later under Brasavola (q.v.) in Padua. It was as Professor of Anatomy at Padua, a chair he held from 1551, that he became famous for his anatomical work. He published only one medical work during his lifetime, *Observationes anatomicae* (1561), which amongst many other important studies included his description of the fallopian tubes (though according to DSB "tube" is a mistranslation of *tuba*, which refers to their trumpet-like shape). His lecture notes were published posthumously in the 1570s in a number of books, and his collected works (*Opera omnia*) were first published in Venice in 1584. They include commentaries on Dioscorides. Ray cites him with Scribonius (q.v. above) as one of two authors of *Altercum*, a synonym of *Hyoscyamus niger* [294]. DSB vol. 4, pp. 519–521; Riddle (1980).

Fragoso. Juan Fragoso or Joannes Fragosus of Toledo (died 1597) was physician and surgeon to Philip II of Spain. In addition to works on surgery, he wrote a book on the fragrant spices, fruits and simples of the Indies. The synonym *Indicum* for Woad,

4: BIOGRAPHICAL NOTES

Glastum sativum [226], which Ray cites is almost certainly taken from Bauhin (1623, p. 113) rather than directly from Fragoso's work. Hirsch (1884–1888), vol. 5, p. 324.

Fuchs. Leonhart Fuchs (1501–1566) of Wemding, a Lutheran physician, was appointed Professor of Medicine at Tübingen in 1535. His great herbal, *De historia stirpium*, published as a folio volume in Basel in 1542, was just one product of an active career as a practising physician, an author of some medical works and editor and translator of others, a participant in numerous controversies and a university administrator. Fuchs' illustrations rival or surpass those of Brunfels (q.v.). He also covers many more species, as he was able to incorporate those described by Tragus and Gesner. Fuchs expresses his delight in field botany, and Arber, after studying his work, concluded that it "was indeed a labour of love". Smaller versions of the illustrations were prepared to illustrate the cheaper, octavo version of the work, first published in 1545, and these smaller versions, in particular, were subsequently copied in numerous works by other authors. Anderson (1977); Arber (1986); DSB vol. 15, pp. 160–162.

Gassendi. The French astronomer and philosopher Pierre Gassend or Gassendi (1592–1655), latinised as Petrus Gassendus, was a follower of Galileo and a supporter of new astronomical theories, although not one who was prepared to risk trouble by incautiously promoting unorthodox opinions. The first two volumes of his collected works, published by his friends after his death, contained his philosophical masterpiece *Syntagma philosophicum*. Ray's references in the main catalogue are to the book *De plantis* in the second of these volumes (Gassendi 1658, pp. 144–192). Most relate to trees and shrubs – *Agrifolium* [15] and palm in a note after it, *Fraxinus* [203] and an observation about tree rings in a note after it, *Prunus sylvestris* [471] and, after *Salix folio utrinque glauco, viminibus albidioribus* [509], a note about trees with denser and looser textures. A reference to the growth of roots is a general statement by Gassendi which Ray chose to put as a note after *Hieracium longiùs radicatum* [279] and another to Theophrastus in Gassendi's work is given under *Secale cereale* [530]. Gassendi is also cited under *Agrifolium* and *Pisum* in the 'Etymology'. DSB vol. 5, pp. 284–289; Jones (1981).

Gerarde. John Gerard or Gerarde (c. 1545–1612) of Nantwich, Cheshire, was a herbalist and gardener; he was appointed curator of the physic garden of the College of Physicians in London in 1586 and published a list of the plants in his own garden in 1596. His *Herball* of 1597 is perhaps the best-known plant book in the English language and one of the most controversial. When revising the book in 1633, Thomas Johnson (q.v.) stated that Gerarde had based the text on an English translation of Dodoens' *Stirpium historiae pemptades sex* (1583) prepared for the publisher in about 1584 by the physician Robert Priest, who died in 1596 or 1597. Gerarde rearranged the species into a sequence following de L'Obel, and most of the plates were obtained from Frankfurt and had previously been used to illustrate the works of Tabernaemontanus. Gerarde's work was criticised by his contemporaries (for example his erstwhile friend de L'Obel said he corrected many mistakes for the publisher before Gerarde took umbrage), by Johnson in his 1633 edition and by Ray (see the 'Explanation'). Raven's assessment (1942, 1950) that "few writers have obtained a great reputation less deservedly" has been much cited; he set out a more detailed criticism in 1947. For a partial defence, see Stearn's account in the DSB and, from a Cambridgeshire perspective, Oswald (1997). In view of Ray's criticisms, it is a surprise to find that he drew far more on Gerarde than on any other author for his chosen names in the main catalogue (see Chapter 3). Anderson (1977); Arber (1986);

87

DSB vol. 5, pp. 361–363; Henrey (1975); Kew & Powell (1932); Raven (1947); ODNB vol. 21, pp. 941–942.

Gesner. Conrad or Konrad Ges(s)ner (1516–1565) of Zurich was the godson of Huldrych Zwingli, a leader of the Reformation in Switzerland, but as a young man he lost interest in the theological studies at which he was aimed from childhood. He qualified as a doctor and in addition to medicine he had a strong interest in the classical languages and in natural history including palaeontology. Arber (1986) describes him as "more remarkable for encyclopaedic versatility than for critical scholarship; he has been called the Pliny of his age". One of the most amiable of men, he maintained a vast network of friends and correspondents all over Europe, unaffected by the fierce personal, political and religious controversies of the age. His most remarkable work was perhaps *Bibliotheca universalis*, a bibliography of all Greek, Latin and Hebrew publications, but his main contribution to natural history was *Historia animalium*, a massive four-volume work of over 4,000 pages published in 1551–1558 with a fifth volume on snakes published posthumously in 1587. He was also instrumental in publishing the works of Valerius Cordus after the latter's untimely death, but Gesner's own death from plague prevented him from completing his *Opera botanica*, for which he had accumulated 1,500 drawings. His immediate successors, lacking his ability to cope with such a mass of material, were unable to take the work through to completion. Although some of the drawings were used by Camerarius, it was not until the mid 18th century that a proportion of the material was published under Gesner's name and not until the 20th century that all the surviving pictures were published, by Zoller *et al.* (1972–1980). Ray does not always cite a specific source for names attributed to Gesner. Most of those for which he does cite a source are taken from *De hortis Germaniae* (*Gesn. hort.*), a book that Gesner added to Valerius Cordus' edition of Dioscorides (Cordus 1561) which he prepared for the press. Only a few are attributed to his *Catalogus plantarum* (*Gesn. cat.*, 1542) and his book on collecting plants, *De stirpium collectione* (*Gesn. coll.* or *Gesn. in coll.*, 1587). Arber (1986); DSB vol. 5, pp. 378–379; Fischer *et al.* (1967); Greene (1983b); Serrai (1980); Wellisch (1984); Zoller (1965).

Guilandinus. According to the DSB, "very little is known about [the] remarkable scholar, polemicist, able botanist, and traveler" Melchior Wieland, latinised to Guilandinus (*c.* 1520–1589). Born in Königsberg, he had studied there and in Rome and travelled to Sicily, Greece and the Middle East before he succeeded Anguillara (q.v.) at the botanical garden at Padua. He was a friend of Falloppio and an enemy of Mattioli. DSB vol. 14, pp. 335–336; Riddle (1980).

Hofmann. Caspar Hofmann (1572–1648), the son of a blacksmith, was born in Gotha and studied at several European universities, including Padua, before receiving his doctorate in 1605 at Basel, where he would have studied under Caspar Bauhin. In 1607 he took up a post as Professor of Medicine at Altorp bei Nürnberg, which he held until his death. He was one of the most devoted adherents to traditional medical orthodoxy, continuing in a changing world to defend the teachings of Galen with dogged determination; William Harvey failed to convince him of the truth of his theory of the circulation of the blood. Hofmann wrote several books, but his planned edition of Galen's work was never published. Ray takes only a few synonyms but numerous notes from his *De medicamentis officinalibus* (1646). Allgemeine Deutsche Biographie (1875–1912) vol. 12, p. 635; Hirsch (1884–1888), vol. 3, p. 248.

How. William How (1620–1656), a Londoner, studied medicine at Oxford. After fighting for the King in the Civil War he practised as a doctor in London and developed his botanical interests in his spare time. His slim *Phytologia britannica* (1650) was based on Johnson's *Mercurius* (see Chapter 3), to which he added records from Johnson's unpublished manuscripts and numerous other records from his own correspondents. It was published anonymously, but Merrett (1666) disclosed its authorship. Raven (1947, p. 299), with his characteristic lack of charity to Ray's English predecessors, describes it as "a very hasty and defective piece of work". Nevertheless, Ray cites several of How's names as chosen names or synonyms, as well as listing three localised records from the borders of Cambridgeshire which he suspected were erroneous. As usual he omits any mention of the Cambridgeshire records he had confirmed, such as that of *Teucrium scordium* [528] from the Isle of Ely. How, who had married Mathias de L'Obel's daughter or granddaughter Elizabeth, edited and published de L'Obel's *Stirpium illustrationes* (1655) shortly before his early death. His interleaved and annotated copy of *Phytologia britannica* was acquired by John Goodyer and is now at Magdalen College, Oxford. Gunther (1922); Kew & Powell (1932); ODNB vol. 28, p. 308; Raven (1947).

Imperato. Ferrante Imperato or Imperatus (1525–1621) was a herbalist from Naples and the author of *Dell'historia naturale* (1599); the details of his life are "obscure and controversial". He amassed a vast collection of objects in his personal natural history museum, which became known throughout Europe. It included a herbarium in 80 volumes of which only a single volume of 442 specimens and perhaps an additional 170 unbound sheets are believed to survive. Ray cites his names as synonyms of *Conferva Plinii* [140] and *Fungus pulverulentus dictus crepitus lupi* [207]. Dizionario biografico degli Italiani, vol. 62, pp. 286–290 (2004); Natale & Cellinese (2009).

Johnson. The attractive personality and gift for friendship of Thomas Johnson (1595/1600–1644) shine through his published works. His accounts of his excursions with his fellow apothecaries also provide a vivid picture of botanising in the first half of the 17th century. Johnson's best-known work was his revision of Gerarde's *Herball* (1633, reprinted in 1636), completed in haste at the request of the publisher in a successful effort to pre-empt Parkinson's *Theatrum botanicum*. The original plates were no longer available and were replaced by a set from a leading European printing house, Plantin of Antwerp. Many of these had illustrated Dodoens' work and were therefore particularly appropriate as Gerarde had used a translation of his text. Johnson rewrote some of Gerarde's chapters and commented on possible errors in other parts of his text, carefully distinguishing his own work from the original by special symbols and generously acknowledging the help that he received from his friends; there are, however, a few occasions when his alterations were misguided and more cases where he left confusing chapters alone. The standard abbreviation for the revison, *Ger. emac.*, presumably originated from Johnson's own phrase *Gerardus à me emaculatus, & auctus* ("Gerarde, by me de-spotted & augmented": Johnson 1634; cf. Kew & Powell 1932, p. 48), but its widespread adoption doubtless expresses the contemporary view of the merits of his work. Johnson went on to publish, in *Mercurius botanicus* (1634, 1641), a list of British wild plants, with synonyms, English names and localities for the rarer species. Typographically this is very similar to Ray's *Catalogus*, except for the localities for species which are in a black-letter font (see Chapter 3). In 1641 Johnson announced his intention of compiling, with his friend John Goodyer, what would have been an illustrated British Flora, but the Civil War intervened and he died in 1644 from wounds received while fighting for the King at the siege of Basing House, Hampshire. DSB 7, pp. 146–148; Gilmour (1972); Gunther (1922); Kew & Powell (1932); Raven (1947); ODNB vol. 30, pp. 323–325.

Junius. The Dutch physician and humanist Adrianus or Hadrianus Junius or Adriaen de Jonghe (1511–1575) was the author of numerous books including the first book devoted solely to fungi, a brief monograph of the stinkhorn which still bears the name he coined for it, *Phallus*. Ray's references are to his *Nomenclator*, a dictionary first published in Antwerp in 1567; there were several later editions and the book survives in many older libraries in Cambridge and elsewhere. There are synonyms of four species in the main catalogue [91, 162, 167, 418] attributed to Junius and he is cited more frequently in the 'Interpretation'. It is impossible to say which edition Ray used, but an edition of 1602 is present in Trinity College library and is cited in our bibliography (Junius 1602). We can tell that Ray did not use the English edition (Junius 1585) because he says of *Acinus* in the 'Interpretation' *à bacca distinguitur, quæ rariùs crescit, ut in olea, lauro, &c*. The English edition, unusually, omits olive as an example, having simply *ut in lauro*. Ainsworth (1976); Harderwijk (1860); Rademaker (1968); Voet (1980–1983), vol. 3, pp. 1271–1281.

Laguna. Andreas Laguna or Andres de Laguna (1499–1560) of Segovia, the son of a physician, studied at Salamanca and (as a pupil of Jean Ruel) at Paris before taking his doctorate at Toledo. He later "led a wandering life" (Voet), travelling widely in Europe and practising as a physician. He published a Latin commentary on Dioscorides in 1554. His Spanish translation of Dioscorides, first published in 1555, was less influential that than of Mattioli, but it was of great importance in the Iberian peninsula and went through numerous editions. Backer *et al*. (1993); Hirsch (1884–1888), vol. 3, p. 587; Riddle (1980); Voet (1980–1983), vol. 3, pp. 1311–1312.

Lauremberg. Peter Lauremberg (1585–1639) attended university in his home town of Rostock. After a study tour which took him to France and the Low Countries he held a series of professorships, of philosophy at Montauban (1611), physics and mathematics at Hamburg (1614) and poetry at Rostock (1624). As this suggests, his interests were wide-ranging and included the natural sciences. His most popular work was written in German, despite its title *Accera philologica* (1633), and comprised a collection of anecdotes drawn in part from Greek and Roman history with the moral formulated as a proverb. Ray quotes from Lauremberg's *Horticultura* (1654a) at some length in his 'Preface' and cites it also in numerous observations; not unexpectedly, he refers less frequently to *Apparatus plantarius* (1632, 1654b) as this deals only with bulbous and tuberous plants, but it is the source of a long descriptive synonym (translated in our text) of *Crocus* [159] and of another long synonym of *Hyacinthus Anglicus* [293], as well as of observations about garlic and saffron. Killy & Vierhaus (2001–2006), vol. 6, p. 276.

L'Écluse. Charles de L'Écluse or Clusius (1526–1609) was born in Arras (then in Flanders, now in France) into a "rich and respectable" family, though they came to suffer in de L'Écluse's lifetime for their Protestant views. Like Dodoens he graduated from the University of Louvain, in his case with a law degree. He became interested in botany in 1551 while studying at Montpellier and living with the great botanical teacher Guillaume Rondelet. He subsequently travelled widely in Europe, developing a wide network of friends and correspondents and publishing original accounts of the plants he had observed in Spain (1576) and Austro-Hungary (1583). He first visited England in 1571, when he botanised with de L'Obel, and Raven (1947, p. 168) even reports that in 1579 he "dashed over from Vienna for a brief visit in the autumn". His translations of botanical works, from Flemish, Portuguese and Spanish into French and Latin, are as important as his original works and often incorporate his own observations. Many of them are listed by Ray in the 'Explanation', notably his French edition (1557) of Dodoens' *Cruijde Boeck*.

4: BIOGRAPHICAL NOTES

Clusius succeeded Dodoens at the University of Leiden in 1593 and designed the first botanical garden there. His *magnum opus* was *Rariorum plantarum historia* (1601), which includes both his Spanish and Austro-Hungarian books. Arber (1986); Backer *et al.* (1993); DSB vol. 8, pp. 120–121; Egmont, Hoftijzer & Visser (2007); Hunger (1927); Legré (1897); Voet (1980 1983), vol. 2, pp. 669–677; Wille (1993).

L'Obel. Mathias de l'Obel, Lobel or Lobelius (1538–1616) was younger than his contemporaries Dodoens and de L'Écluse, with whom his name is always linked. Born in Lille (then in Flanders, now in France), he studied at Pisa, travelled in western and central Europe and in 1565 matriculated at Montpellier. Although he was less amiable than de L'Écluse and less of a linguist (his Latin has often been criticised), he soon became a favourite of the Professor of Botany, Guillaume Rondelet, a *bon viveur* and inspirational teacher, and struck up a friendship with a Provençal student Pierre Pena. After the death of Rondelet, Pena and de L'Obel moved to England, arriving in late 1566 or 1567, and published an account of the plants they had seen on their European travels as *Stirpium adversaria nova*. The title-page of this book is dated 1570 but the colophon is dated January 1571, so it was almost certainly published early in 1571. The species were arranged in an original classification which he followed in all his later works and which was adopted by some later authors including Gerarde (1597). De L'Obel left England in 1571, becoming physician to William the Silent, but returned in 1590, working in London for much of the remainder of his life as a garden superintendent and physician; his interest in field botany continued into his old age. The later history of Pena & de L'Obel's sparsely illustrated work is outlined by Ray in the 'Explanation'. It was reissued in 1576 together with a longer work by de L'Obel alone, *Stirpium observationes*, which was copiously illustrated by woodcuts prepared for Dodoens but never used. This edition was published by Plantin and was much more successful than the earlier work published in London. In 1581 Plantin also published a Flemish translation, *Kruydtboeck*. Pena & de L'Obel's *Stirpium adversaria* was published for a third time in 1605, again with a separate work by de L'Obel, *Adversariorum altera pars*. Meanwhile Plantin (1581, 1591) had issued a volume of illustrations which are often attributed to de L'Obel, *Plantarum seu stirpium icones* (see Plantin below). Ray's abbreviations *Ad.* (for *Adversaria*) and *Ad. Lob.* refer to Pena & de L'Obel's work and *Lob.* alone refers to *Stirpium observationes*, but, as most species have the same name in both works, which indeed share a common index, the distinction is rather slight. *Lob. ico.* and *Plant. ico.* (see Plantin below) relate to the Plantin volume. Ray refers specifically to *Adversariorum altera pars* (1605) under *Gramen pratense paniculatum molle* [261] and *Orchis lilifolius minor sabuletorum Zelandiæ & Bataviæ* [397]. Parkinson (q.v.) made use of some of de L'Obel's unpublished material in *Theatrum botanicum* (1640) and How (q.v.) published further material under de L'Obel's name as *Stirpium illustrationes* (1655). Arber (1986); Backer *et al.* (1993); DSB vol. 8, pp. 435–436; Gunther (1922); Louis (1980); ODNB vol. 34, pp. 203–204; Voet (1980–1983), vol. 3, pp. 1405–1416; Wille (1993).

Lonicer. Adam Lonicer, Lonitzer or Lonicerus (1528–1586) was educated at Marburg, where his father was a professor and where he himself became Professor of Mathematics in 1553. The following year he obtained a medical degree from the university and became the municipal physician at Frankfurt, a post he held until his death. He married the daughter of a Frankfurt printer whose books included herbals. Lonicer's own publications included a revision of Rösslin's *Kreuterbuch*, which had originally been published in 1533 and is an amalgam of fact and fancy characteristic of its period. It appeared in numerous later editions (see Belkin & Caley 1978 and footnote 39 in Ray's 'Explanation'). Lonicer's Latin edition was first published in 1551. He is commemorated in the Linnaean genus *Lonicera*. Anderson (1977); Arber (1986); DSB vol. 8, pp. 483–484; Killy & Vierhaus (2001–2006), vol. 6, p. 485.

Mattioli. Pietro Andrea (or Pierandrea) Mattioli or Petrus Andreas Matthiolus (1501–1577), the son of a physician, was born in Siena. He graduated as a medical student from the University of Padua in 1523 and practised medicine in Italy and, for a time, in Prague, combining his medical career with an interest in natural history. His great commentary on Dioscorides originated as a relatively modest translation of the work into Italian, published in Venice in 1544. This was followed by a rewritten and much more successful Italian edition originally published in Venice in 1548, and then by the enlarged and illustrated Latin edition of 1554. It is based on Ruel's translation. Mattioli's commentary was intended as a practical guide for physicians. He added to the plants of Dioscorides all the other species known to him personally or from his various correspondents, inserting them into the text as annotations wherever seemed most appropriate. He included synonyms, original descriptions and many excellent original plates. After 1554 the successive editions of the work became known throughout Europe, and it was reprinted "practically without interruption until the eighteenth century" (DSB) and translated into Czech, French and German. The finest edition was the fourth Latin edition of 1565, which has woodcuts occupying almost a full page (Stannard 1969). Some of the editions are cited in Ray's 'Explanation'. Mattioli easily took offence and his enemies included Anguillara, Amatus Lusitanus and Guilandinus. The plant genus *Matthiola* was named by Robert Brown in honour of Mattioli. Anderson (1977); Arber (1986); DSB vol. 9, pp. 178–180; Dizionario biografico degli Italiani, vol. 72, pp. 308–312 (2009); Ferri (1997); Greene (1983b); Riddle (1980); Stannard (1969).

Melich. Georg Melich (*fl.* 1573–1595) is cited (like Costeo) from Hofmann (1646) as one of three authors of *Ben sive Behen album Arabum*, a synonym of *Pastinaca sativa Dioscoridis* [423]. *Meilichio* is an error for *Melichio* and the name is found in Melich's *Dispensatorium medicum* (1601), p. 8.

Moufet. Thomas Moufet, Muffet, Moffett or Moffet (1553–1604) was born in London and studied classics and medicine at two Cambridge colleges. At Trinity he became friends with the naturalist Thomas Penny (*c.* 1530–1589), a Fellow of the college, and at Gonville Hall he studied medicine under John Caius. After receiving his M.A. in 1576 he lived for a period in continental Europe. He pursued his medical studies in Basel, where he became a convinced proponent of Paracelsian medicine; he then worked as a physician in Frankfurt and visited Italy, where the culture of silkworms perhaps stimulated his great interest in entomology. After returning to England in 1580 he developed a very successful medical practice, initially in Ipswich and then in London; he was also elected M.P. for Wilton in 1597, thanks to the patronage of the Earl of Pembroke. Although Moufet published medical works and a long poem about silkworms during his lifetime, his two best-known books were published posthumously. Both are repeatedly cited by Ray. Moufet compiled *Insectorum ... theatrum* from the unpublished papers of Edward Wotton, Conrad Gesner and Thomas Penny (all bequeathed to him by Penny), to which he added his own material. The book was eventually seen through the press by Sir Theodore Mayerne in 1634 and an English translation, *The theater of insects*, appeared as the second volume of Edward Topsell's *The history of four-footed beasts and serpents* in 1658. (Allen (2010) suggests that the story of its publication may be more complex than this.) The manuscript survives in the British Museum. *Healths improvement* was written

4: BIOGRAPHICAL NOTES

in English for a lay readership; it did not appear until 1655, when it was edited for publication by the physician Christopher Bennet. It also includes observations on British birds. It is cited (as *de diæta*, "about diet") as the source of synonyms for two *Triticum* species [595, 597] as well as in Ray's notes. Moufet's interest in entomology led Bristowe (1946) to suggest that his daughter Patience was the original Little Miss Muffet, but Opie & Opie (1997) treat this theory with caution. DSB vol. 9, pp. 440–441; ODNB vol. 38, pp. 503–504 (Moffet) and vol. 43, pp. 597–598 (Penny); Raven (1947).

Parkinson. John Parkinson (1566/7–1650) was, like his younger contemporary Thomas Johnson, a London apothecary. He was also a very keen gardener and he occupies a special place in the history of horticulture as the author of the first important book devoted to gardening in England, *Paradisi in sole Paradisus terrestris* (1629). This describes over a thousand species, early cultivars and monstrosities. Less generally known today is his herbal, *Theatrum botanicum* (1640), an even more substantial work which aimed to deal with all the remaining plants. It was compiled from Bauhin's *Pinax* (1623) and many other European sources. It is the names from *Theatrum botanicum* that are so often cited by Ray. There are a few references in Parkinson to plants from Cambridgeshire localities (e.g. *Galega officinalis* [208] and *Staphylea pinnata* [561], neither of which Ray could refind). Anderson (1977); Gunther (1922); Henrey (1975); Kew & Powell (1932); Raven (1947); ODNB vol. 42, pp. 799–800.

Paulli. Simon Paulli or Paullus (1603–1680) was Professor of Medicine in Copenhagen and author of a Danish herbal. Ray cites numerous synonyms from another of his books, *Viridaria varia* (1653). The book is a series of lists of plants growing in some of the major European gardens or (in two cases) in the wild nearby. Although it is continuously paginated, some of these have long titles on separate 'title-pages' as well as short titles at the top of the first page of text; others simply have the short titles. Nidd's copy of this rare book is in the library of Trinity College, Cambridge (Q.14.18), dated by him 1657 with the price 3s. 3d. It is interleaved with blank pages between the catalogues, on which Nidd has written some notes. As Ray cites the individual catalogues rather than the book as a whole, these are listed below (with the long titles where these are available). Ray's usual abbreviation is given in square brackets. Bech (1979–1984), vol. 11, pp. 181–183; Eriksson (2004).

Catalogus plantarum horti regii Hafniae. Pp. 1–80. [*hort. Hafn.*]

Catalogue des plantes qui sont de present cultivees au Jardin du Roy de Paris, depuis deux ans & demy qu'il est dressè. 1636. Pp. 81–194. [*hort. Par., hort. Paris.*]. *Appendix earum plantarum, quae in proxime praecedenti indice aut catalago [sic], omissae sunt.* Pp. 195–201. [*hort. Paris. in append.*]

Catalogus plantarum tam exoticarum quam indigenarum; quae anno M.DC.LI. in hortis regiis Warsaviae, nasci observatae sunt. Pp. 203–247. [*hort. Wars.*] *Catalogus I. Plantarum horti regii suburbani, Warsaviensis.* Pp. 205–247. *Catalogus alter plantarum horti regii sub arce regia.* Pp. 248–287. *Catalogus tertius, plantarum indigenarum, quae in locis paludosis, pratensibus, arenosis & sylvis, circa Warsaviam nascuntur.* Pp. 288–324.

Catalogus plantarum horti medici Oxoniensis, eas alphabetico ordine accurate exhibens. 1648. Pp. 325–394. [Not cited by Ray.]

Catalogus plantarum horti gymnasii Patavini. Quibus auctior erat anno 1642. Pp. 395–471. [*hort. Pat.*]

Catalogus plantarum horti academici Lugduno-Batavi, quibus is instructus erat annis 1642 & 1649. Pp. 473–560. [*hort. Bat.*] *Index plantarum indigenarum, quae in locis paludosis, pratensibus, arenosis, & sylvestribus prope Lugdunum in Batavis nascuntur*. Pp. 561–578. [*hort. Bat. inter indigenas*]. *Appendix plantarum quae horto publico Academiae Lugd-Batav. accesserunt. Anno 1641*. Pp. 579–591.

Catalogus plantarum horti Gröningensis ordine alphabetico conscriptus, et editus anno 1646. Pp. 593–706. [*hort. Gron.*]

Catalogus seminum exoticorum. Pp. 707–730. [Not cited by Ray.]

M. Guilielmi Laurembergi Botanotheca … Editio secunda. Pp. 731–799. [Not cited by Ray.]

Pena. Pierre Pena (*fl*. 1558–1604), born at Jouques in Provence, travelled widely in Europe and met several of the notable botanists of the age before matriculating at Montpellier in 1565. De L'Obel (q.v.) arrived within a few weeks and, if they had not already met, they soon struck up a close friendship. They left Montpellier for England after Guillaume Rondelet's death in 1566. It has usually been assumed that de L'Obel played the leading part in writing their joint book *Stirpium adversaria nova*, published in 1571, but Legré (1897) presents an alternative interpretation. When they returned to Europe, Pena gave up botany to pursue what became a highly successful medical career in France. Legré (1897).

Plantin. Christopher Plantin (*c*. 1520–1589) was born near Tours in France; his father was a valet and his mother soon died of plague. He apprenticed himself to a printer in Caen and then moved to Antwerp in 1548 or 1549, where he initially worked as a bookbinder. However, he turned to printing and publishing in 1555 after he was attacked by a party of drunkards and his injuries left him too physically weak to continue as a bookbinder. He became one of the most important publishers of learned books in the 16th century; "his success was essentially based on his zest and capacity for work, honesty, sound intelligence and clear-sightedness" (Voet 1969–1972, vol. 1, p. 127). He was important botanically as the publisher of many of the works of Dodoens, de L'Écluse and de L'Obel. In the course of these publications he built up a collection of over 3,000 woodblocks and copperplates which survive in the Plantin-Moretus Museum in Antwerp. Although his importance is as a publisher rather than an author, it is now considered that the collection of illustrations published as *Plantarum seu stirpium icones* in 1581 and reissued as *Icones stirpium* in 1591 should be ascribed to him and not to de L'Obel (Stafleu & Cowan 1976–1988, vol. 3, p. 132). Ray cites it under *Lob.* as *Lob. ico.* in the 'Explanation' and usually uses this abbreviation in the main catalogue, but he attributes three synonyms of *Linaria vulgaris* [335] to it as *Plant. ico.* Clair (1960); Kusukawa (2007); Voet (1969–1972, 1980–1983, 1993).

Pona. We know little more than Ray did about Giovanni Pona (*fl*. 1595–1608), an apothecary of Verona. His friend de L'Écluse (1601) published his account of a botanical excursion to Monte Baldo, a peak of 2218 m above Lake Garda in the Italian Alps. Ray takes a single synonym, of *Onobrychis* [394], from this work.

Reneaulme. The son and grandson of scholars, Paul de Reneaulme or Renealmus of Blois (*c*. 1560–1624) received a doctorate from Avignon in 1590 and thereafter practised as a physician. He was involved in medical controversy over his advocacy of chemical treatments, but it is difficult to agree with Blunt's (1950) view that he is "mainly remembered for his irascibility". To botanists he is chiefly known for his illustrated book *Specimen historiae plantarum* (1611), from which Ray takes several synonyms. Anon. (1866).

Ruel. Little is known of the life of Jean Ruel (1474–1537), latinised as Joannes Ruellius. He studied medicine in Paris and became physician to Francis I of France.

4: BIOGRAPHICAL NOTES

He was a family man, but after the death of his wife he was ordained, became a Canon at Notre-Dame in 1526 and devoted himself entirely to study. Ray mentions his two botanical books. *De medicinali materia* (1516) is a translation of Dioscorides which was used by Mattioli in his own commentary on Dioscorides and was so popular that it eventually appeared in over 20 different editions. In *De natura stirpium* Ruel provided a wide-ranging treatment of botany which was particularly important in developing the descriptive terminology we now use in plant descriptions. It was first published in Paris in 1536, although Ray mentions the Basel edition of 1537 in his 'Explanation'. Arber (1986); Bietenholz (1985–87), vol. 1, p. 415; DSB vols 4, p. 121 and 11, pp. 594–595; Greene (1983b); Riddle (1980).

Schröder. Johannes Schröder (1600–1664) was a German physician who studied at universities in Germany, Denmark, France and Italy and then served with the Swedish army during the Thirty Years War before settling in Frankfurt in 1635. His "widely read textbook" *Pharmacopëia medico-chymica*, first published in Ulm in 1641, is cited by Ray in his 'Explanation' and is the source of three synonyms in the main catalogue [207, 378, 471] and several formal notes and informal observations. Ray also quotes from his *Quercetanus redivivus* (1648) in a note about *Convolvulus major* [148]. Killy & Vierhaus (2001–2006), vol. 9, p. 151.

Schwenckfeld. Kaspar Schwenckfeld (1563–1609) of Greiffenberg (then in the Hapsburg territories, now in Poland) studied medicine and botany under Caspar Bauhin in Basel before returning to his native Silesia to practice as a physician. His *Stirpium et fossilium Silesiae catalogus* ("Catalogue of the plants and fossils of Silesia") was first published in 1600 and Ray takes two synonyms from it, for *Fumaria* [204] and *Trachelium minus* [573]. Hirsch (1884–1888), vol. 5, p. 324.

Spiegel. Adriaan van der Spiegel (1578–1625) appears in Ray's *Catalogus* as Spigelius. He was born in Brussels into a successful medical family and studied in Padua, subsequently practising as a physician in Italy, Germany, Moravia and from 1616 as Professor of Anatomy in Padua. He studied botany in his early years and his *Isagoges in rem herbariam libri duo* was first published in Padua in 1606 and appeared in several later editions. Ray cites this work as the source of the name *Lysimachia purpurea trifolia caule hexagono* [348], but much more frequently in his observations. Spiegel's best-known work was a book on anatomy, *De humani corporis fabrica*, published posthumously in 1627. DSB vol. 12, pp. 577–578.

Tabernaemontanus. Jakob Dieter (*c*. 1525–*c*. 1590) is usually known as Jakob Theodor of Bergzabern or by the Latin name that he chose for himself, Tabernaemontanus. He initially practised as a pharmacist in Hornbach and then, after studying medicine at Montpellier, as a physician elsewhere in Germany. He was a pupil of Brunfels and Bock, and like them was a protestant. His *Neuw Kreuterbuch* (1588–1591) was very popular and went through numerous printings until the final edition was published in 1731. It was fully illustrated, with plates taken from a range of sources; they were published separately as *Eicones plantarum* in 1590 and later used to illustrate Gerarde's *Herball* (1597). Arber (1986); Killy & Vierhaus (2001–2006), vol. 9, p. 716.

Thal. Johannes Thal or Thalius (1542–1583) became interested in plants when he was still at school in Ilfeld. He left the University of Jena without a degree but soon started working as a doctor. In 1572 he became the city doctor at Stolberg and personal physician to the Count. He lived near the Harz mountains and was keenly interested in both the history and natural history of the area. In 1581 he moved to Nordhausen but he died in 1583 as the result of injuries sustained in a carriage accident. His Flora of the Harz, *Sylvia Hercynia* (1588), was posthumously published by his friend

Camerarius with his own *Hortus medicus et philosophicus*. It has been described by Greene (1905) as "the earliest local flora", though Italian accounts of Monte Baldo antedate it (Stannard 1969). Thal's name is commemorated in *Arabidopsis thaliana* (L.) Heynh., which he illustrated as *Pilosella siliquata* (the name that Ray chose for the plant in the 1663 appendix [A30]) and which Linnaeus named as *Arabis thaliana* in his honour. Allgemeine Deutsche Biographie (1875–1912) vol. 37, pp. 642–643; Jarvis (2007), Stafleu & Cowan (1976–1988), vol. 6, pp. 231–232.

Theodor of Bergzabern, see **Tabernaemontanus**.

Tragus, see **Bock**.

Turner. William Turner (1509/10–1568) of Morpeth, Northumberland, if not the first British botanist, was certainly the first to leave a substantial legacy in print. A naturalist from childhood, he attended Pembroke Hall, Cambridge, and at university became a committed Protestant. His religious views led to periods of exile under Henry VIII and Mary, as a result of which he became familiar with the central European and Italian as well as the British flora and struck up a great friendship with Gesner (q.v.). His first botanical books, *Libellus de re herbaria novus* (1538) and *The names of herbes* (1548), both contain brief references to plants in Cambridgeshire. His major botanical work was *A new herball*, published in three parts in London in 1551 and in Cologne in 1562 and 1568. The third part includes a new edition of the first part and a reissue of the second. The illustrations in all three parts were largely copied from those in Fuchs' *De historia stirpium* (1542). In addition to his botanical books, Turner was a religious controversialist whose works were intemperate even by the standards of Tudor England, and he was almost as ill-disposed towards the moderate protestants of the Elizabethan settlement as he had been to their Catholic predecessors. Turner was "respected but not well-known" to later botanists (Raven 1947, p. 135). British botanists seem not to have had access to his works, doubtless because under Mary all his books had been prohibited and ordered to be destroyed and the two later parts of his herbal were published abroad. Raven concludes that Parkinson almost certainly did not know the herbal and that Johnson knew only the first part. Ray mentions only the first part of *A new herball* in the 'Explanation' even though the 1568 edition was present in both the University Library and Trinity College library in the 1650s (see Chapter 5). Ray's omission of any mention of Turner's Cambridge plants is a conspicuous example of his practice of not citing records of his predecessors for species he himself had recorded. Turner was also neglected by continental authors, presumably because he wrote all but his first book in English. Anderson (1977); Britten, Jackson & Stearn (1965); Chapman, McCombie & Wessencraft (1995); Chapman & Tweddle (1995); DSB vol. 13, pp. 501–502; Henrey (1975); Jones (1988); Rydén, Helander & Olsson (1999); Raven (1947); ODNB vol. 55, pp. 674–677.

Valerand, see **Dourez**.

Vigo. Giovanni da Vigo (1450–1525) was an Italian surgeon who served Julius II when, as a cardinal, he was captain of the papal armies in Umbria and after he became pope. His work *Practica in arte chirurgica copiosa* was first published in Rome in 1514; an amplified version, *Practica in arte chirurgica compendiosa*, appeared in 1517. Ray gives him as the author of *Lingua passerina*, a synonym of *Polygonum mas vulgare* [454]. (*Lingua passerina* is also listed earlier in the synonymy of that species as one of the three names attributed to Arnald of Villanova). This name appears in Latin editions of his *Opera in chyrurgia* (e.g. Vigo 1531) and in an English translation (Vigo 1571); in the latter "knotgras" is given as an English name. DSB vol. 14, pp. 27–28.

5: RAY'S CITED SOURCES IN PUBLIC AND PRIVATE LIBRARIES

A book such as the *Catalogus* is likely to reflect, to at least some degree, the research facilities available to the author when he wrote it. The wide range of literature cited by Ray in the *Catalogus* raises two obvious questions. How many of these books did he (or his colleagues) actually see, as opposed to citing from other sources? And how did they get access to the works they consulted? These topics are explored in this chapter. We have concentrated almost exclusively on the modern authors cited by Ray. The works of the classical authors were almost certainly too widely available to provide much information on the whereabouts of Ray's source material.

Citation of sources in the *Catalogus*

We distinguish direct citations of works in the *Catalogus* from indirect citations. The direct citations are those where Ray provides information which is attributed to a specific source. The clearest examples are from the notes, such as the note in the main catalogue to *Absinthium vulgare* [3]: "Wormwood resists putrefaction remarkably & either drives away or kills animals generated by putrefaction. *Casp[arus] Hofman[nus], De medicam[entis] officinal[ibus], book 2, chapter 2, section 19.*" Ray is not always so specific about his sources, as in the 'Etymology', where a typical entry reads: "Erica Ἐρικη [Erice] or Ἐρεικη [Ereice] from ἐρεικω [ereico] I break, because it acts powerfully in breaking up bladder and kidney stones. *Martin[i].*" This is a direct citation from the first or second edition of Martini's *Lexicon philologicum* (1623, 1655). Sometimes the direct citations are included in discussions in the text, as under *Gnaphalium montanum album* [233] where Ray provides direct citations of five works in this comment: "The picture of this plant that is found both in Gerarde [1597] and in Parkinson [1640] is defective & bad, in as much as for anyone comparing it either with the plant itself or with de L'Obel's figures in his *Adversaria* [Pena & L'Obel 1571, 1576: the picture appears in both editions] & *pictures* [Plantin 1581, 1591] it will [not?] be good enough, and *Jean Bauhin's* [Bauhin & Cherler 1650–1651] picture is no more satisfactory."

Indirect citations are those where Ray takes his information on the work of an author from a secondary source. A good example is one of the notes on *Fraxinus vulgaris* [203]: "The inner bark of this tree makes the water in which it has been soaked blue, which they give to hens when they are ravaged by disease. *Cæsalp[inus] apud Cam[erarium] in hort[o].*" Here the indirect citation is of Cesalpino (1583) and the direct citation that of Camerarius (1588), cited as *Cam. hort.* in the 'Explanation'. We also count as indirect citations all the works listed in the 'Explanation' except those that are directly cited elsewhere in the *Catalogus*, as Ray states that the 'Explanation' is principally derived from C. Bauhin. This takes care of almost all the authors of Ray's chosen names and synonyms, who are almost all cited in the 'Explanation'. (The few exceptions are also classed as indirect citations.) The distinction between direct and indirect citations is fairly straightforward in the main catalogue but less straightforward in the 'Etymology', where Ray's practice seems to be much less rigorous. It is clear that in some entries in the 'Etymology' he cites a source but takes the information from an uncited secondary source, and in others he provides no clue to the source of his information.

The books that are cited directly and indirectly by Ray are distinguished in our bibliography. These assessments are only provisional, both because of the difficulties outlined in the previous paragraph and because we have not always attempted to identify specific sources in cases where Ray gives simply the name of an author, especially when that author is as prolific as, say, Gesner. Our working hypothesis is that Ray saw those books which he cited directly but that he did not have access to those works which he cited indirectly, such as Cesalpino (1583) in the above example. The books listed in the 'Explanation' are less straightforward, as Ray may have seen them but have had no reason to cite them directly. We now need to test our working hypothesis against the information that we have about the libraries and personal book collections available to Ray.

University and college libraries

Cambridge University Library

Cambridge University had owned a small collection of books since the 14th century, and in the early 15th century a special room was built to house them. By 1529 the library had grown to a collection of over 500 books. Its history in the 16th and 17th centuries was marked by periods of activity which alternated with periods of remarkable stagnation and decline, chronicled by Oates (1986) in a detailed narrative spiced with the driest of librarian's humour. The middle of the 16th century was a low period, in which some volumes were probably discarded as outdated and others taken by scholars who failed to return them; by 1557 only 175 volumes remained and in 1568 the library was emptied of its books and converted into a lecture room. (Oxford lost its entire library at this time and sold the library furniture as useless.) The library was restored from 1574 under the leadership of Andrew Perne, and the first printed catalogue appeared in this year. With this revival it began to attract significant donations of books. It received an up-to-date collection of some 270 medical works in 1594, a bequest from Thomas Lorkyn (c. 1528–1591), Regius Professor of Physic (Sayle 1921). Another sizeable gift, 87 duplicates from the library of John, Lord Lumley (c. 1534–1609), was received in 1598 (Oates 1986; Jayne & Johnson 1956). The latter included some of Thomas Cranmer's books; Cranmer's possessions were confiscated when he was imprisoned on the accession of Mary Tudor and his personal library had been incorporated by 1594 into the Lumley library. By the time these donations were incorporated into the University Library, it had grown to some 950 volumes.

This period of growth was followed by 30 years of extreme stagnation, when there were almost no additions to the library and the University did not even manage to spend the money given to it for books. This came to an end with the appointment in 1629 as University Librarian of Abraham Whelock, a noted scholar of Anglo-Saxon and Arabic. However, the massive increase in the size of the library which happened towards the end of his career was not, or was only indirectly, a result of his librarianship but was rather a consequence of the turbulence of the times. In 1641 the Archbishop of Canterbury, William Laud, was imprisoned in the Tower, in 1642 the army occupied Lambeth Palace and in 1643 Parliament abolished the episcopacy. The Palace library was donated to Cambridge in 1647 and 10,000 manuscripts, books and pamphlets arrived a couple of years later. A donor was found to pay for the new bookcases, the library was rearranged to integrate this huge collection and a catalogue of the whole library, arranged by author, was produced. Donations of money for books from Alexander Ross (£50) and Richard Foxton (£40) were spent in 1656, and the books bought with these funds included some medical and scientific works.

5: RAY'S CITED SOURCES IN LIBRARIES

These were incorporated into the new author catalogue, which was completed by about 1658.[1] This therefore provides an accurate record of the books available to Ray in the University Library when he was writing the *Catalogus* at the end of the 1650s. The Library Donors' Book[2] provides details of the provenance of many of the works in the University Library proper, as opposed to the Lambeth Palace library, though it has to be used with some caution (Oates 1986).

Examination of the 1658 author catalogue shows that most of the books in the University Library which are relevant to the *Catalogus* were published before 1600. These include works on Dioscorides by Amatus Lusitanus (1558), Barbaro (1530) and Marcellus Virgilius (1529), Gaza's translation of Theophrastus (1529) and Scaliger's commentary on his *De causis plantarum* (1566). Other herbals in the library were Dodoens' *De stirpium historia* (1559), Gesner's *Historia plantarum* (1541), Matthaeus Sylvaticus' *Opus pandectarum* (1499), Tabernaemontanus' *Neuw Kreuterbuch* (1588–1591) and Turner's *Herball* (1568). Several of the herbals were from Lorkyn's bequest, including the works by Amatus Lusitanus, Dodoens, Gesner, Matthaeus Sylvaticus and Turner. The Marcellus Virgilius was originally Cranmer's copy and came from the Lumley library; Ray provides information from this work in the 'Etymology' but *via* Bauhin & Cherler (1650–1651) rather than directly. Tabernaemontanus' *Neuw Kreuterbuch* is a rare example in the library of a botanical book in a modern European language; it had been given to the library in 1657 by a London surgeon, Thomas Holler, and is listed in the Donors' Book as "Tabernaemontanus his Herball in high duch".[3] Other relevant 16th-century works in the library were Brasavola's *Examen omnium syruporum* (1540), Cardano's *De subtilitate* (1553) and *De rerum varietate* (1557), Daléchamps' edition of Pliny (1587), Dodoens' *Purgantium ... historia* (1574), Goropius' *Opera* (1580), Krantz's *Saxonia* (1580), Orta's *Aromatum, et simplicium ... historia* (1567) and Ruel's *De natura stirpium* (1537).

The relevant 17th-century works in the library were few. It did contain Aldrovandi's "Historia de Avibus. Franc: 1610" (two of the three volumes of the *Ornithologia* were published in Frankfurt in 1610), Bacon's *Sylva Sylvarum* (1626), a 1628 edition of Gilbert's *De magnete*, Butler's *The feminine monarchie* (but the second edition of 1623 rather than the third edition used by Ray), Dodoens' *Stirpium historiae pemptades* (1616), Parkinson's *Paradisus* (1629) and the works of van Helmont (1655) and Sennert (1650). The most significant of the 17th-century publications was undoubtedly Bauhin & Cherler's *Historia plantarum* (1650–1651), bought (as were van Helmont's and Sennert's works) by Vice-Chancellor Dillingham with money from the Ross bequest in 1656.[4]

One other library should be mentioned here, that of Richard Holdsworth. Holdsworth, who was Master of Emmanuel College until he was deposed for his

---

[1] Cambridge University Library, MS Mm.4.1.

[2] *Catalogus librorum quos habet Bibliotheca Publica Academ. Cantabrig:*, Cambridge University Library, MS Oo.7.52. This lists, amongst many other donations, the books bought by Vice-Chancellor Theophilus Dillingham with the Ross and Foxton bequests, but we have failed to find his draft expenditure account, which according to Oates (1986, pp. 277–278) details for each book the sum that he paid and the bookseller who supplied it.

[3] MS Oo.7.52, p. 16; the entry in the author catalogue (Mm.4.1) is similar.

[4] Some works which are directly or indirectly cited in the *Catalogus* were acquired by the library soon after 1660, including Besler's *Hortus Eystettensis* (1613), Gassendi's *Opera omnia* (1658) and de L'Obel's *Plantarum seu stirpium historia* (1576) (Oates 1986, pp. 390 and 409–410). The copy of Daléchamps (1586–1587) currently in the library also arrived in this period, bequeathed with the rest of Bishop John Hacket's library in 1670.

Royalist views in 1643, died in 1649. His will made complex provision for his library, which had just under 10,000 printed books, the largest collection then in private hands. The library was at Emmanuel College but at some stage in the 1650s it appears to have been put into the care of Thomas Buck, a University official with the post of Esquire Bedell and "a man of considerable substance" (Oates 1986). The fate of Holdsworth's library was not decided until 1664 when, after negotiations, legal proceedings and arbitration, it was awarded to the University and effectively replaced the Lambeth library which had to be returned to the Palace whence it came. Holdsworth's library was catalogued in 1664. We have assumed that it was not available to Ray.

Trinity College library

In the absence of an adequate University Library in the early decades of the 17th century, the college libraries had increased in importance. In 1600 the Trinity College library had contained some 324 volumes. This was the average size of a college library in the 16th century, and until the last decades of the century they tended not to grow in size because new books were bought to replace old ones, which were then discarded or added to a lending collection. However, from the late 16th century books tended to be added to the existing stock and this, coupled with the expansion of learning, meant that "the college libraries grew steadily larger, and at the same time increased in importance as the need for books began to exceed the capacity of individuals to own enough of them privately". Fortunately the history of the Trinity library is very well documented and the records have been summarised in Gaskell's monograph *Trinity College library: the first 150 years* (1980), from which this summary and much of the following information are taken.

The 324 volumes in the Trinity library in 1600 included just five classified as medical or mathematical. By the time the library was catalogued in 1640 there were 1,900 volumes, including 106 medical and mathematical volumes.[5] Most of the additions were bought with money donated for the purpose, although some books were donated directly and a few were bought with college funds. The scientific (medical and mathematical) books present in about 1645 are listed by Gaskell. They include the works of Dioscorides and Galen; there is no specific reference to Theophrastus but the ten volumes of the works of Johannes van Meurs listed (as "Meursii vol. 10") might have included his edition of Theophrastus, which is listed in the more detailed catalogue compiled in 1667.[6] The library also included some of the great herbals of the 16th century, Fuchs' *De historia stirpium* (1542), Mattioli's commentary on Discorides (1563b) and Camerarius' 'epitome' of it (Mattioli 1586), Turner's *Herball* (1568) and "*Gerhard*s *Herbal 1(-2)*", which Gaskell interprets as Gerarde's *Herball* of 1597, "presumably bound in two volumes". (Two volumes are indeed listed in the 1667 catalogue, but one wonders whether they might have been the original edition of 1597 and Johnson's of 1633.) In addition to the medical and mathematical volumes there were a few natural history works classified as *Philosophi*, including Aldrovandi's *Ornithologia*. The source of many of these works can be traced through the magnificent manuscript volume in the Trinity archives known as the *Memoriale*,[7] which was started in 1612–1614 and lists the donors of books or money to buy books (with, in the case of the latter, a list of the books purchased). It begins with a retrospective listing of

---

[5] The 1640 catalogue is Trinity College Add. MS a.103.
[6] Trinity College Add. MS a.101.
[7] Trinity College R.17.8.

5: RAY'S CITED SOURCES IN LIBRARIES

earlier donations and continues to provide detailed lists of books acquired until 1640; after this the entries peter out (perhaps a symptom of the disruption to the life of the college caused by the Civil War). The dates of the donations are not usually stated but they have been estimated by Gaskell (1980). Gerarde's *Herball*, for example, was donated in 1633, the year of publication of Johnson's 'emaculated' edition, which might suggest that the donor, John Furtho, a Doctor of Medicine, had bought the new edition and was disposing of the old, but as he also gave eight other volumes at the same time (including Fuchs' herbal) the date is probably just a coincidence.

Few relevant works were added to the library between 1640 and 1667 except for those known to have been bequeathed by Ray's friend John Nidd (see below). There are two important exceptions, both scientific works published in the 1650s and added to the Trinity library before 1667; we have not traced their dates of accession. One is Bauhin & Cherler's three-volume *Historia plantarum* (1650–1651), one of Ray's fundamental sources. The copy still present in the library was almost certainly consulted by Ray and his colleagues in the 1650s. It has clearly been used as a working copy, as is shown by the additions to the indexes of all three volumes of some names omitted from them in error. These include *Fumaria vulgaris* [*Fumaria officinalis* 204], *Hedera terrestris* [*Glechoma hederacea* 272], *Ladanum segetum* [*Galeopsis angustifolia* 315], *Lantana vulgo alijs Viburnum* [*Viburnum lantana* 612], *Laureola* [*Daphne laureola* 327] and *Melampyrum* [359–360]. These annotations appear to have been made by John Nidd, as the handwriting closely resembles that of the notes in his copy of Lonicer (1551) and in his interleaved copy of Paulli (1653). Some small plant specimens have also been pressed between the pages. It seems very likely that this book was bought in the 1650s for Ray and his colleagues. The other relevant volumes added to the library between 1640 and 1667 are the collected works of Pierre Gassendi. There are a number of references to the second volume of Gassendi's works (Gassendi 1658) in Ray's *Catalogus*.

In addition to the scientific works, the 1640 catalogue of the Trinity College library lists some of the more general works cited in the *Catalogus*. These include two of the more surprising books to be mentioned by Ray, and their presence in the Trinity library presumably explains their inclusion as sources in the *Catalogus*. One is indicated by a brief entry in the 1640 catalogue, "Krantii rerum German. Histo.", but this fortunately has a more detailed entry in the 1667 catalogue, where three volumes by Albert Krantz are listed, including *Saxonia*, cited under *Allium sylvestre* [19] by Ray. All three volumes had been bought with money from Sir Michael Stanhope in about 1625 (*Memoriale*, p. 105, as "Krantzii opera fol:"; cf. Gaskell 1980, p. 239). *Saxonia* was also in the University Library, but the Trinity copy seems more likely to be the book consulted for the *Catalogus*. The other is Braun & Hogenberg's *Civitates orbis terrarum*, listed in the *Memoriale* (p. 96) as "Georgij Bruin vol. 5". These five volumes[8] were a luxury purchase, bought with Mercator's *Cosmographica* and Camden's *Britannia* with £20 donated by William Smyth in 1609 (cf. Gaskell 1980, p. 238; £20 in 1609 had the equivalent purchasing power of £2,732 in 2009). Volume 2 (1575) was cited by Ray (see footnote 309 in the main catalogue). Both the Krantz and the Braun & Hogenberg volumes are still present in the Trinity library.

Works of reference in the Trinity library listed in the 1640 catalogue include the *Suda* (as *Suidas*, cited in the 'Explanation'), Calepino's *Dictionarium*, the *Etymologicon Magnum*, the works of Goropius Becanus and Martini's *Lexicon philologicum* (which must at this date have been the first edition), all authorities cited (or, in the case of Goropius, derided) in the 'Etymology'.

---

[8] The sixth and final volume was not published until 1618.

St John's College library

The other college library which currently has rich holdings of 16th- and 17th-century scientific works is Trinity's neighbour, St John's. Would Ray have been able to consult this library? St John's library regulations[9] of 1651 start by stating: "That no person whatsoeuer belonging to this Colledge vnder the degree of a Master of Arts (except hee bee ffellow or ffellowcommoner) shall bee permitted to study in the Library. Neyther shall any, whither straunger, or of the Colledge, vnder the degree before mentioned (except the before excepted) bee admitted to view the Library vnlesse by the appointment of the Master, or in his absence, of the President, or that some one of the ffellowes goe along with him, and there abide with the party brought in by him vntil his departure thence." (Scott 1903). A further order, made in 1654, notes that "the fore mencioned Orders for preserving the Bookes in the Library haue bin by experience found ineffectuall" and therefore "neither of the Library Keepers shall lend any booke to any ffelow or Schollar of this Colledge, or to any of other Colledges whatsoeuer without leaue of the Master and Seniors first desired and granted". In 1662 this rule was again declared to have been ineffectual.[10] This suggests that Ray would have been able to consult the books at St John's College if he had chosen to do so. We do know that Thomas Smith of Christ's College studied manuscripts at St John's College, as he happened to overhear a Quaker sermon when "returning home wearied" after an afternoon's work there in August 1659 (Smith 1659) and was so provoked that he challenged the Quakers to a public disputation, which led to a "celebrated wrangle" (Oates 1986, p. 301). In the 18th century evidence from the borrowing registers of other colleges, notably Pembroke, suggests that Fellows did borrow books from colleges other than their own (E. Quarmby Lawrence, *in litt.*).

The holdings of the St John's library are known from a college manuscript which lists the books present in 1634, ordered by class-mark and shelf, and then adds later accessions up to 1690.[11] As we do not know whether Ray actually used this library, there is little point in listing the books at St John's which we know were available to him in the University or Trinity College libraries or those listed below which were owned by his collaborator John Nidd. There are in fact few books cited in the *Catalogus* that were present in the St John's library in 1660 but not represented in these other collections. The most significant is *Historia generalis plantarum* by Daléchamps (1586–1587), published anonymously and listed in the St John's catalogue under the printer's name as "Rovilii historia plantaru[m] 2 vol". This was one of many medical books bequeathed to St John's by Dr John Collins in 1634. The works of de L'Écluse were well represented at St John's;[12] curiously they were apparently

---

[9] A version of these regulations was published by Gaskell (1980, p. 259) from a manuscript at Trinity College. He assumed that they applied to the Trinity library; we are grateful to Elizabeth Quarmby Lawrence for drawing our attention to the fact that they are actually a copy of the St John's regulations.

[10] St John's College archives D94.372 (E. Quarmby Lawrence, *in litt.*).

[11] St John's College manuscript U.3; a transcript of the manuscript with useful introductory notes is available and these notes help to identify the books listed up to folio 70 *recto* as present in the library by 1660.

[12] The catalogue lists from Collins' bequest "Clusii historia plantaru[m] 2. vol", which correspond to two volumes from Collins currently in the library, *Rariorum plantarum historia* (1601) and, bound together, *Exoticorum libri decem* (1605) and *Curae posteriores* (1611). It also lists separately "Clusii Stirpes Hisp. & Pannonicu[m] 2 vol", which must be *Rariorum aliquot stirpium per Hispanias* (1576) and *Rariorum aliquot stirpium, per Pannoniam* (1583), although neither appears to be in the library

5: RAY'S CITED SOURCES IN LIBRARIES

absent from the other two institutional libraries and from Nidd's personal collection. The only other relevant works that we have not traced in the other three collections are Caspar Bauhin's edition of Mattioli (1598), Jean Bauhin's *Historia novi et admirabilis fontis balneique Bollensis* (1598), including Book 4 which is cited by Ray but missing from some copies of the book, Cordus' *Annotationes in Pedacii Dioscoridis Anazarbei De medica materia* (1561), Pena & de L'Obel's *Dilucidae simplicium medicamenorum* [sic] *explicationes, & Stirpium adversaria* (1605)[13] and Canini's *Hellenismos* (1555).

Personal book collections

It is clear from all that we know of the institutional libraries discussed above that they would have contained some scientific works that would have been useful to Ray, and which he cited directly in the *Catalogus*, and one, Bauhin & Cherler's *Historia plantarum*, which was essential to him. However, they contained only a small proportion of the works cited directly in the *Catalogus* and they would have been completely inadequate as working libraries for anyone investigating the Cambridgeshire flora in the 1650s.

In the absence of an adequate institutional library, Ray would have had to rely on his own botanical books and those of his friends. The "willingness to open one's private library to friends" was an "important part of early modern book culture" and is described by Feingold (1990b) with reference to the library of Ray's friend Isaac Barrow.[14] Fortunately, and remarkably, a list made by John Nidd of the books in his library survives in the Vice-Chancellor's Probate Inventories in the University of Cambridge archives[15] and has been published by Leedham-Green (1986). Nidd made his will[16] on 6 December 1658 when he was "weak in body", leaving "my Gerards Herball" to Ray,[17] "Forestus his workes" to John Mapletoft and "Sennertus his workes" to Thomas Pockley, all Fellows of Trinity (see Figure 5.1). These were all books already held in the Trinity College library (Gaskell 1980). He left copies of Origen and Epictetus to "Mr Sharp, Minister of Sundridge".[18] His other books were itemised in two lists annexed to the will, one of 126 works left to the college library and the other of 151 left to William Lynnett, another Fellow of Trinity (see Figure 5.2). Nidd's library list is important as it was drawn up before 1660; it is the only list of books in a private library that we can be sure would have been available to Ray when he was preparing the *Catalogus*.

---

today (both were reprinted in the 1601 volume), and "Clusius de rariorib[us] stirpibus", which we can only suggest might have been another copy of *Rariorum plantarum historia* but which again matches nothing currently in the library.

[13] From Collins' bequest and misbound so that the first part of the book, by Rondelet, is bound within this second part.

[14] However one of Duport's rules for his pupils was "Lend not your books out of your Studie unlesse it be for a short time, and to a speciall Friend whom you knowe will make good use of them, and restore them faithfully" (Trevelyan 1943).

[15] Cambridge University Library, University Archives VCCt. Invs 14.

[16] Cambridge University Library, University Archives VCCt. [Original] Wills 14.

[17] Probably the 1636 printing of Johnson's revision, which was in Ray's library at the time of his death (Feisenberger 1975).

[18] Previous authors, following Venn & Venn (1922–1927), have suggested that John Nidd was probably the son of Gervase Nidd. Gervase had been Rector of Sundridge from 1615 until his death in 1629 and this bequest strongly supports this suggestion. Samuel Sharpe was a Cambridge graduate (B.A. 1626, M.A. 1629) and is known to have been Rector of Sundridge in 1645; he was buried there in 1680 (Venn & Venn 1922–1927).

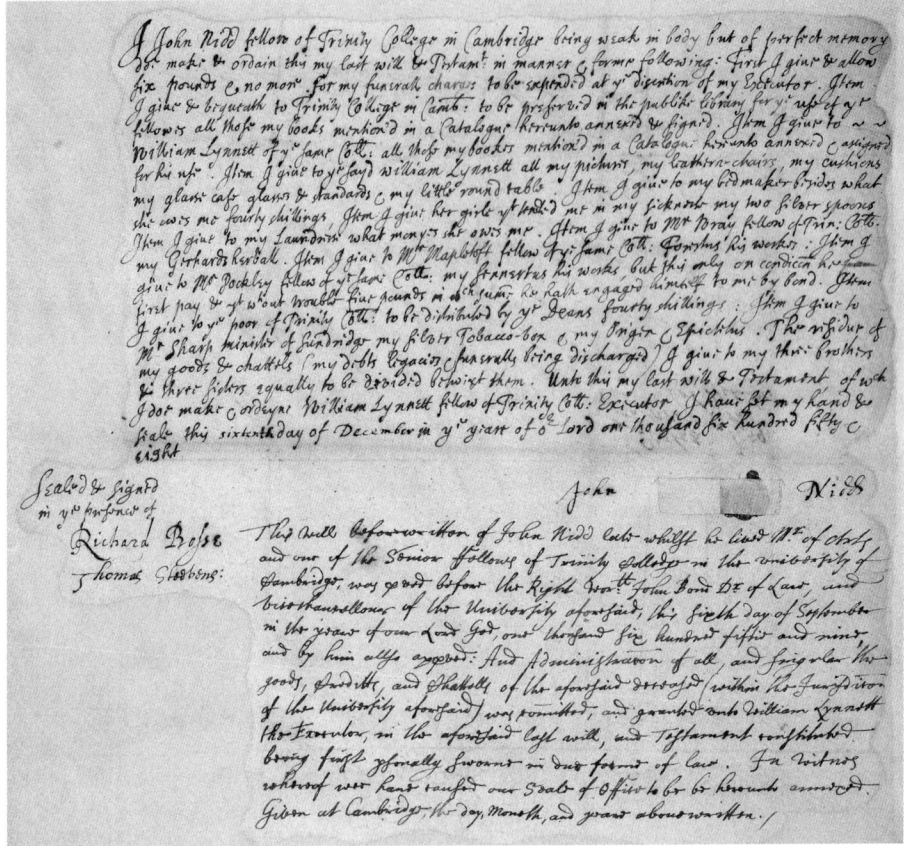

Figure 5.1. John Nidd's will, 16 December 1658. Cambridge University Archives VCCt. [Original] Wills 14, reproduced by kind permission of the Syndics of Cambridge University Library.

The list of books left by Nidd to Trinity College starts with seven botanical works (were these together on his shelves?). These were, in chronological order, Pena & de L'Obel's *Stirpium adversaria* (1571), Plantin's *Icones stirpium* (1591), attributed to Plantin rather than to de L'Obel as in Ray's 'Explanation', three books by Caspar Bauhin, *Phytopinax* (1596), *Prodromus* (1620) and *Pinax* (1623), and Parkinson's *Paradisus* (1629) and *Theatrum botanicum* (1640). Bauhin's *Pinax* and Parkinson's *Theatrum botanicum* were, with Gerarde's *Herball*, the really key books used as nomenclatural sources in the *Catalogus*, along with one work which Nidd did not possess, Bauhin & Cherler's *Historia plantarum* (1650–1651).

Nidd's collection includes a few other botanical works cited in the *Catalogus*. These include Valerius Cordus' *Dispensatorium* (1651), Lauremberg's *Apparatus plantarius* (his *Horticultura* is not listed in this or in any of the other contemporary library catalogues which we have examined because the two were bound together[19]) and Paulli's

---

[19] This was not only the case with Nidd's 1632 edition but also with Willughby's 1654 edition, as described later in this chapter, and with Goodyer's copy of the same edition, bought in the year of publication and now in the Old Library, Magdalen College, Oxford (cf. Gunther 1922).

5: RAY'S CITED SOURCES IN LIBRARIES

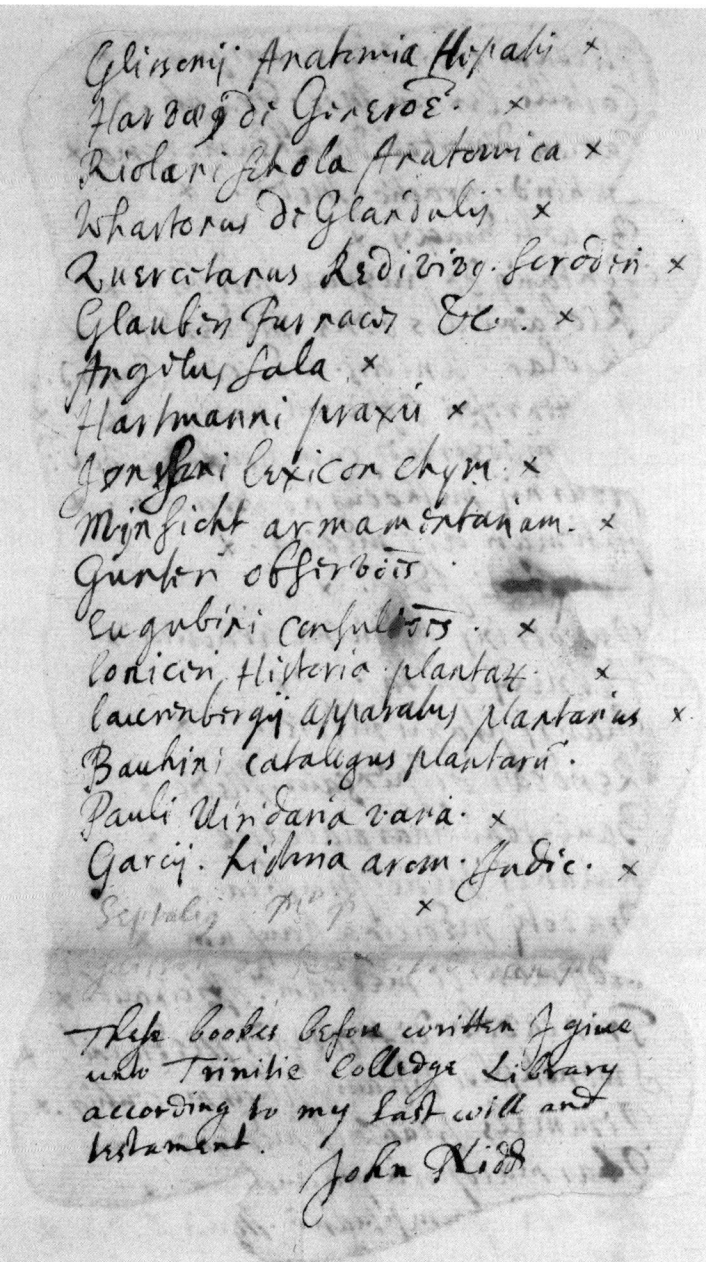

Figure 5.2. John Nidd's inventory of his books, prepared in conjunction with his will (Figure 5.1). This folio shows the end of the list of books left to Trinity College. Books mentioned in the *Catalogus* include Schröder's *Quercetanus redivivus* (fifth from top) and the last four listed in ink, Lauremberg's *Apparatus plantarius*, Bauhin's *Catalogus plantarum circa Basileam sponte nascentium*, Paulli's *Viridaria varia* and Garcia de Orta's *Aromatum, et simplicium aliquot medicamentorum apud Indos nascentium historia*. Reproduced by kind permission of the Syndics of Cambridge University Library (Cambridge University Archives VCCt. Invs. 14).

Viridaria varia (1653). Of particular interest is the presence in his list of "Bauhini catalogus plantarum", which can only be Caspar Bauhin's *Catalogus plantarum circa Basileam sponte nascentium* (1622), one of the models for Ray's *Catalogus* and perhaps the copy still present in the Trinity library (though it lacks any inscription by Nidd). The medical works cited in the *Catalogus* are very well represented and include Bennet's *Theatri Tabidorum vestibulum* (1654),[20] Charleton's *Spiritus gorgonicus* (1650), Fernel's works,[21] Hofmann's *De medicamentis officinalibus* (1646), van der Linden's *De scriptis medicis* (1651),[22] and Schröder's *Quercetanus redivivus* (1648) and *Pharmacopëia* (1649). Other books owned by Nidd that appear in the *Catalogus* included Moufet's *Insectorum ... theatrum* (1634) and Primrose's *De vulgi ... erroribus* (1638). Virtually all the books in Nidd's collection which were cited in the *Catalogus* were left to Trinity rather than to William Lynnett, the exceptions being a folio edition of Pliny's *Natural history* (in Latin), Junius' *Nomenclator* and "Suidas latine". Some of the books in the Wren Library can still be identified as Nidd's, as he wrote in them not only his name but also the year when he bought them and sometimes the price that he paid. The rich representation of mid 17th-century books relevant to the *Catalogus* in Nidd's library is a striking contrast to the paucity of relevant material in the University and Trinity College libraries in his lifetime.[23]

Ray would of course have had his own books. A letter to Courthope of 9 November 1658 suggests that he was an active though canny purchaser in his Cambridge years. "I lately saw heer at our Booksellers a book of one Gulielmus Piso in folio wch he calls Historia naturalis & medica utrivsque India. they ask heer unreasonably deare for it if you returne by London I pray search it out and cheapen it." (Gunther 1934; Thompson 1974).[24] However, our only knowledge of his library comes from the sale catalogue drawn up after his death (published in facsimile by Feisenberger 1975). All the private book lists that we have were drawn up after the death of the owners; Ray outlived his closest Cambridge colleagues, so the list of his books is therefore the most distant from the period during which the *Catalogus* was written. It is impossible to say which of the books published before 1660 were in his possession when he was writing the *Catalogus* and which were bought in the next 45 years – impossible to know, too, whether he disposed of or lost any books in the intervening years. He

---

[20] The book listed by Leedham-Green (1986) as "Benne[..]dicti Tabidorum theatrum" must be this; it is listed as "Xtophori Benedicti Theatrum Tabidorum" in the 1667 Trinity catalogue as "in ye Presse", as were some of the other medical books on Nidd's list.

[21] This reference to "Fernelij opera" presumably refers to Fernel's *Universa medicina*; Nidd's copy survives in the Trinity College library and is an edition published in Leiden in 1645. The 1658 University Library catalogue also lists "Jo: Fernelius Opera. 2 vol. Francof. 1581" and Willughby owned "Fernelius's works"; these may also be references to Fernel's *Universa medicina*, which appeared in several editions.

[22] On the assumption that this is "Vander Lindanus de simplis medicis" of Leedham-Green (1986).

[23] One problem is assessing the college libraries is that their smaller works may not have been catalogued as assiduously as the large folio volumes, and Elizabeth Quarmby Lawrence (*in litt*. 2010) considers that books in smaller formats are under-represented in the surviving records. However Nidd made careful provision for his books and it seems reasonable to assume that he did not leave to Trinity books which the library already possessed; furthermore his smaller volumes do appear in the 1667 catalogue.

[24] Our quotation is from Thompson, who differs from Gunther in minor details of spelling and punctuation. Willem Piso's book, published in Amsterdam in 1658, appears in the catalogue of Ray's library, so presumably Courthope was successful (or perhaps Ray eventually decided to pay the asking price in Cambridge). It is immediately clear from the exotic fauna depicted on the engraved title-page that this work is not relevant to European natural history (it deals with South America and the East Indies), suggesting that Ray had money to spare for non-essential books.

clearly did not own Gerarde's *Herball* until he received the copy bequeathed to him by Nidd in 1659, and this suggests that he gave priority to buying books to which he did not have access rather than simply expanding his personal collection. Bauhin & Cherler's *Historia plantarum* is listed in the Ray sale catalogue (although, curiously, as two volumes rather than three); it would be interesting to know when he bought it. Ray of course owned many other standard botanical works including Bauhin's *Prodromus* (1620), *Pinax* (1623) and *Theatrum botanicum* (1658), How's *Phytologia britannica* (1650), de L'Écluse's *Rariorum plantarum historia* (1601, assuming that "Clusius de Plantis, 2 vol. Ant. 1601" of the sale catalogue is this) and the 1586 edition of Mattioli. None of Thomas Johnson's works are included except the reprint of his emaculated Gerarde (1636). Rather remarkably, there is no copy of Parkinson's *Theatrum botanicum* (1640) in the sale catalogue although his *Paradisus* (1629) is included. The sale catalogue was prepared by Ray's friend Samuel Dale (Feisenberger 1975) and it seems reasonable to assume that he would have been invited to take one or two books for himself; perhaps this was one of them. Maranta's *Methodus cognoscendorum simplicium* (1559) and Kircher's *Scrutinium physico-medicum contagiosae luis, quae pestis dicitur*, listed as 'Kircherus De peste' (the 1659 edition, published in Leipzig) are the only copies we have traced in the records of the libraries to which Ray might have had access in the 1650s. Ray also owned Butler's *The feminine monarchie*, but the year of publication is (unusually) omitted from the sale catalogue list, the second edition of van Helmont's *Opuscula medica* (1648) and Scaliger's commentary on *De causis plantarum* of Theophrastus (1566). His reference works included Calepino's *Dictionarium* (1609), Martini's *Lexicon* (1655) and a 1619 edition of "Suidæ Lexicon". However, some of the books that are frequently cited in the *Catalogus* and are not in Nidd's library are missing from Ray's too, including Moufet's *Healths improvement* (1655) and Camerarius' *Hortus medicus et philosophicus* (1588).

A library catalogue from the archives of the Willughby family[25] is believed to list the books in the library at Middleton Hall and was perhaps drawn up at the time of its transfer to Wollaton Hall at the end of the 17th century (Welch 1972). It is at the end of a book originally used by Francis Willughby and many of the books were clearly acquired by him for his library (see Figure 5.3). The library survived at Wollaton until most of the books were sold in 1925; the catalogue of the Christie's sale (15 June 1925) provides full bibliographical details of many works which can be linked to those catalogued in the late 17th century. The catalogue reveals that the books which one would expect to have been bought by Francis Willughby usually bear the signature of his son Thomas, thus confirming their presence in the library in the decades immediately after Francis Willughby's death. Our impression (and we have not carried out a detailed comparison) is that Willughby's library had a much higher proportion of scientific books than Ray's library and fewer theological works.

There is a considerable similarity between the botanical books owned by Nidd, Ray and Willughby. There are relatively few books owned by all three, although these include Caspar Bauhin's *Prodromus* (1620), Parkinson's *Paradisus* (1629), Sennert's *Opera* (1650) and Blockwitz's *Anatomia sambuci* (1650), the last cited only indirectly in the *Catalogus*. Many more were owned by two of the three. Nidd and Willughby

---

[25] University of Nottingham Department of Manuscripts and Special Collections Mi I 17/1. The books are listed in alphabetical order by author and are given shelf numbers of one of two types, a simple letter and number (e.g. F 23) or a more complex combination of a number, letter(s) and number (e.g. 9 F 5 or 4 CC 3). There are also shelf lists for the simple letter and number codes against which the alphabetical book lists can be checked, but not for the more complex codes. This catalogue deserves more detailed study.

Figure 5.3. The catalogue of books in the Willughby library, undated but believed to have been compiled towards the end of the 17th century. Books mentioned in the *Catalogus* include Gilbert's *De magnete* (5), Lauremberg's *Apparatus plantarius* (20), Columella's *De re rustica*, in an edition which as usual also included "Terent[ius] Varro[,] Cato &c" (105), Schröder's *Pharmacopëia* (114) and van der Linden's *De scriptis medicis* (117); also listed are "Fernelius's works" (8) and an unidentified work by Van Helmont (111). Reproduced by kind permission of Manuscript and Special Collections, University of Nottingham (Mi I 17/1).

5: RAY'S CITED SOURCES IN LIBRARIES

(but, ironically, not Ray) had copies of Caspar Bauhin's *Catalogus plantarum circa Basileam sponte nascentium* (1622). They also owned the 1619 edition of Lyte's translation of Dodoens, *A new herbal*, Lauremberg's *Apparatus plantarius* (Nidd owned the 1632 edition and Willughby the 1654 edition; both were bound with *Horticultura*), Parkinson's *Theatrum botanicum* (1640) and Paulli's *Viridaria varia* (1653). Ray and Willughby both owned Bauhin & Cherler's *Historia plantarum* (1650–1651) and How's *Phytologia britannica* (1650). Few of the botanical volumes cited in the *Catalogus* were owned by Willughby alone; they include Bock's or Tragus' *De stirpium ... usitatis nomenclaturis* (1552), Fuchs' (1542) *De historia stirpium*, which we have already seen was in the Trinity library, and a couple of volumes which one suspects were bought by Willughby in Italy in the 1660s, Imperato's *Dell'historia naturale* (1599) and Pona's book on the plants of Monte Baldo (1608).

There is also a broad overlap between the non-botanical works in the library of the three Cambridge colleagues. All owned Cardano's *De rerum varietate*, van der Linden's *De scriptis medicis* (1651) and Moufet's *Insectorum ... theatrum* (1634). Nidd and Willughby had Bennet's *Theatri Tabidorum* (1654) and Schröder's *Pharmacopëia* (1649). More interestingly, works owned by Willughby but not recorded in either Nidd's or Ray's library included Aldrovandi's *Ornithologia* (1599–1603) (although this was also in the Trinity library), Moufet's *Healths improvement* (1655) (catalogued as "Muffets food" in the 17th-century catalogue), Cardano's *De subtilitate*, Gilbert's *De magnete* (the 1633 edition, rather than the first edition of 1600) and Spiegel's *Isagoges* (1633). Also interesting is the presence of Borel's *Observationum microcospicarum* [*sic*] *centuria* (1656), not cited in the *Catalogus* but perhaps the basis of the reference to the "little worms recently detected in vinegar" mentioned under *Absinthium vulgare* [3]: see footnote 3 in the main catalogue.

We have lists of the libraries left by two further contemporaries of Ray at Trinity, Duport and Barrow. Of all Ray's colleagues, Duport was the greatest bibliophile. He left over 2,000 books to Trinity on his death in 1679, the largest such donation the college had ever received. It was too large a collection to be accommodated in the existing library, and the books were not incorporated into the college library until Wren's library building was completed in the 1690s (Gaskell 1980). There is no published list of Duport's books, but they are recorded in a manuscript at Trinity.[26] Although there are few botanical works in his library, it did, surprisingly, include at least one book which does not appear in any of the other documented libraries to which Ray might have had access, Camerarius' *Hortus medicus et philosophicus* (1588), which was presumably bound with Thal's *Sylvia Hercynia* with which it was published. Another botanical work which he owned was Reneaulme's *Specimen historiae plantarum* (1611), although Ray also owned this. Other works cited in the *Catalogus* include the *Schola Salternitana*, "Bacon his Natural History", i.e. *Sylva Sylvarum* (1626), which rather surprisingly does not appear to have been in Nidd's or Ray's libraries although there were copies in all three institutional libraries and Willughby also had one, and "Brown's Urn-Buryal", that is Browne's *Hydriotaphia* (1658), again a book which Willughby also owned. As one would expect, classical authors are well represented in Duport's library, as are reference works. The latter include Canini's *Hellenismos*, Junius' *Nomenclator* (absent from the institutional library catalogues but also present

---

[26] The so-called *Benefactions book* (Add. MS a.106), pp. 332–427 (cf. Gaskell 1980). The works by Camerarius and Reneaulme discussed here are listed on p. 346. Feingold (1990a) suggests that the scientific works in the library were bought or published after 1660, but he does not provide the detailed evidence and we cannot therefore know whether this generalisation applies to works published before 1660.

in Nidd's, Ray's and Barrow's libraries, so clearly on everyone's shelves) and two volumes of Martini's *Lexicon philologicum* (perhaps the first and second editions?).[27]

Barrow could not have played any part in the writing of the *Catalogus* as he had obtained a travelling fellowship from Trinity and left Cambridge in May 1655 for a European tour which lasted until September 1659 (Feingold 1990a). He purportedly sold his books to pay for the tour, but Feingold (1990b) speculates that they may have been bought by a Cambridge friend such as Duport or Worthington, only to be redeemed after his return. Barrow was interested in botany (ODNB), but there are virtually no botanical works in the library of over 1,000 books listed by Feingold (1990b) – nothing, for example, by either of the Bauhins, Gerarde or Parkinson, although he did have Bodaeus' edition of Theophrastus (1644). Perhaps he was able to sell any botanical books that he owned to one of the other Trinity botanists before his departure and never reclaimed them. The general works relevant to the *Catalogus* in his library were almost certainly available elsewhere in Cambridge in the late 1650s.

Conclusions

We know a remarkable amount about the books which Ray might have seen in Cambridge in the 1650s, but our knowledge is very incomplete. Several of the relevant libraries were catalogued after 1660. We have traced catalogues of the libraries of some of his closest associates, but we know nothing about the books owned by Courthope and there must have been several other colleagues with books which he might have consulted. We cannot even be sure that all the works Ray saw were in Cambridge. This means that conclusions about particular titles can only be tentative, but despite many uncertainties the general picture is very clear.

There is strong evidence that most of the books cited directly in the *Catalogus* are likely to have been available to Ray. The few books cited directly which we have not been able to trace in any of the libraries to which Ray might have had access (including St John's) include Bodaeus' edition of Theophrastus (1644), Charleton's *A ternary of paradoxes* (1650), which is a translation of three of van Helmont's works, Heer's *Observationes medicae oppido rarae*, published with *Spadocrene* (1645),[28] Highmore's *The history of generation* (1651) and Maier's *Lusus serius* (1616, 1654). There can be virtually no doubt that Ray knew Bodaeus' Theophrastus[29] and our failure to find it may be attributable to the very meagre information supplied in some of the catalogues, and thus our inability to distinguish it from other editions of Theophrastus, rather than to a genuine absence. A book published in 1644 would be unlikely to have been bequeathed to an institutional library by the late 1650s, so it was most probably specially purchased in the 1650s by the University or Trinity or present in a private

---

[27] Intriguingly, for someone who warned his pupils against wasting their time and in particular claimed that he was "no great friend to going downe the water", adding "some under colour of going a fishing, drop into a blind house and there drink like fishes" (Trevelyan 1943), Duport owned a copy of Walton's *The compleat angler* and another book listed as *Angling improv'd*, doubtless *The experienced angler, or, angling improved*, which was written by the retired Cromwellian soldier Colonel Robert Venables (1612/13–1687), was first published in 1662 and appeared in five editions in the author's lifetime (ODNB).

[28] The 17th-century Willughby catalogue lists this, but the copy offered for sale by Christie's was the 1685 edition.

[29] See in the main catalogue, for example Ray's note 4 to *Cannabis prima sive sativa* [104], note 1 to *Cichoreum sylvestre* [131], note 3 to *Lupulus* [342], note 3 to *Populus alba* [456] and note 2 to *Tussilago* [603] and our related footnotes.

collection. A note following *Acer major* [5] strongly suggests that Ray knew Highmore (1651). The reference to Charleton's *A ternary of paradoxes* (1650) is less certain as it relies on the interpretation of an obscure reference (see footnote 465 in the main catalogue). The books by Charleton and Highmore were published in London and are likely to have been available to Ray. We know that he owned another work by Highmore which appeared in 1651 (*Corporis humani disquisitio anatomica*, published in The Hague) and that Nidd owned another work of Charleton which was published in 1650 (*Spiritus gorgonicus*, published in Leiden); it would be interesting to know if these were sometimes bound with the works that we cannot trace. It is disappointing not to have discovered the whereabouts of the works by Heer and Maier, the source of notes on *Ulmus vulgatissimus folio lato scabro* [626] and *Linum sativum* [336] respectively.

There are three books which appear to be cited directly in the *Catalogus* and which we can trace only in the St John's library. These are *Historia generalis plantarum* by Daléchamps (1586–1587) but published anonymously, Mattioli's works edited by C. Bauhin (1598) and J. Bauhin's *Historia novi et admirabilis fontis balneique Bollensis* (1598). All three were included in Dr Collins' bequest and they contain a 1634 book label identifying them as such. We may have failed to trace Bauhin's Mattioli in other library catalogues because of the difficulty in identifying specific editions of Mattioli's works. Daléchamps' herbal reveals one of the potential problems with our identification of direct citations. The entry for *Sium umbellatum repens* [545] suggests that Ray knew the Daléchamps herbal, as he states: "There is a picture of this in *[Daléchamps' General] Hist[ory] of Plants published in] Lyon, page 1092*." It was only after we failed to find definite evidence that Ray had access to this work that we searched for other possible sources for this information and realised that it could have been derived from Bauhin & Cherler (1651, vol. 3(2), p. 172), *pace* Raven (1950, p. 80). Other statements derived from this work in Ray's 'Etymology' also appear to have come from the same source. All we can conclude is that Ray may have seen Daléchamps' work but need not necessarily have done so.

Some of the works cited only indirectly in the *Catalogus* were available to Ray, but many do not appear in any of the catalogues. These include almost all the continental herbals in modern languages (Dutch, French, German and Italian) except, as described above, Tabernaemontanus' *Neuw Kreuterbuch* (1588–1591). We have also failed to trace any record of numerous 16th- and early 17th-century Latin works, including those of Acosta (1582), Anguillara (1561), Belon (1553a, b, 1589), Besler (1613), Brunfels (1530–1536), Cesalpino (1583), Colonna (1592, 1616),[30] E. Cordus (1534), Durante (1585), Fragoso (1601), Guilandinus (1558a, b, 1608, 1613) and Monardes (1574).[31] Most of these are relatively minor works which are cited by Ray because they appear in Bauhin's *Pinax* (1623), but they nevertheless provide some of his chosen names and synonyms (see Table 3.1). Besler's *Hortus Eystettensis* and Brunfel's *Herbarum vivae eicones* are historically significant, although the importance of both

---

[30] Ray "enquired diligently" for Colonna's books "in Italy, in those cities where they were printed and elsewhere, but could hear no news of them"; however by 1678 he had Colonna (1616), "wch I owe to ye generosity of my honoured friend Chr. Hatton Esquire" (Gunther 1928, p. 162). This was perhaps a loan as it does not appear in the sale catalogue of his library. Rather remarkably, John Goodyer had been able to obtain Colonna's works for his library (Gunther 1922) and information he sent to Ray from Colonna (1616) about *Chamædrys spuria foliis pediculis oblongis insidentibus* [*Veronica montana* A12] is given in the 1663 appendix.

[31] Later editions of the works by Acosta (1582), Belon (1553a) and Monardes (1574) were, however, included in de L'Écluse's *Exoticorum libri decem* (1605) which was in the library at St John's.

lies only in their fine illustrations. The absence of evidence that Ray had access to Cesalpino's *De plantis* (1583) supports Raven's conclusion (1950, p. 80) that he did not at this stage know of its important treatment of plant classification.

We cannot be sure that Ray used the University Library: the only book which Ray appears to cite directly and which we have not traced elsewhere (except in St John's) is Ruel's *De natura stirpium* (1537) and, in view of the uncertainties around individual works, this is insufficient evidence from which to draw definite conclusions. It does, however, seem almost certain that, as one would expect, Ray used the Trinity College library. Whether or not he also used the St John's College library is an open question, although the presence of Collins' medical works would have rendered it useful to him. There is little doubt that personal book collections would have been essential to him when he was writing the *Catalogus*. This has broader implications. Even if there was no evidence to suggest that John Nidd was involved in the preparation of the "Observations" in the *Catalogus* (see Chapter 3), we would have suspected that the research for the work was likely to have had a collaborative element. It would have been natural for Ray to explain to his colleagues why he wanted to borrow their books and to tell them how useful he had found them. In turn they would surely have been drawn into the project and would have reported to him any interesting botanical observations they had found in the course of their reading, observations which (however tangential) he might sometimes have felt obliged to include in his text. It may also be that the relative unimportance of institutional libraries to the more botanical aspects of Ray's work on the *Catalogus* encouraged him to think that he could continue to work as a naturalist outside the University when he was considering his future in 1660 and 1662.

6: IDENTIFICATION OF RAY'S PLANTS

The work of earlier authors

The standard source of information on the identity of Ray's Cambridgeshire plants is C.C. Babington's *Flora of Cambridgeshire* (1860). Babington started each of his species accounts with the names used in earlier Cambridgeshire Floras, including Ray's *Catalogus* and its appendices. This allows most of Ray's chosen names to be matched with Babington's species. Babington does not deal with those species which fall outside the scope of his Flora (cultivated plants, bryophytes, lichens and fungi) and he sometimes cites only a single name in cases where Ray describes more than one variant of a species. Babington's copy of Ray's *Catalogus* and its appendices, preserved in the library of the Department of Plant Sciences, University of Cambridge, shows that he wrote the (then) modern names of the species into the margins of the 1660 *Catalogus* and the 1663 appendix. There are minor discrepancies between these annotations and the published Flora, in one case because his identification was confused by a compilation error in the published Flora (see footnote 716 of the main catalogue) and in another because an obvious error in his annotations (the identification of *Ranunculus hirsutus arvensis flore minimo* in the 1663 appendix [A33] as *Ranunculus fluitans*) was corrected before he went to press. Proctor (1956) cites Ray's bryophyte names in the synonymy of modern species, in a similar manner to Babington.

Almost all Babington's identifications have been accepted by later authors. Raven (1942, 1950) provided critical commentary on some cases where the identification of Ray's plant is doubtful. Perring *et al.* (1964) cite the first record of the vascular plants and bryophytes of Cambridgeshire by author and date (e.g. "Ray, 1660"). Although they do not cite the full name, this brief treatment often allows one to deduce their views on difficult cases. Crompton (2001–2004) lists Ray's chosen names for vascular plants under the equivalent modern taxon, based on Babington (1860) and with some critical comments. Ewen & Prime (1975), unlike Perring *et al.* and Crompton, present Ray's taxa in the order in which they appear in his main catalogue and provide a modern name for almost all of them, based on Babington (1860) and sometimes with insufficient appreciation of changes in species delimitation since his day.

We have discussed the difficulties of interpreting some of Ray's names in earlier papers on *Equisetum* (Oswald & Preston 1998), *Lactuca* (Oswald 2000) and *Potamogeton* (Preston 2010, a paper based on the unpublished work of J.E. Dandy and G. Taylor).

Our approach

Like our predecessors, we have taken Babington (1860) as our starting point. Babington's species concept was often broader than that of modern taxonomists, even though he was a notorious "splitter" in his day (Allen 1986). Some of his "splits" were only mentioned as subsidiary notes in 1860 and other taxa were only segregated as a result of cytological studies in the 20th century. We have reassessed the identity of the species affected by such changes and provided explanatory footnotes. In some cases where we have suspected that there is something amiss with Babington's identifications we have re-examined the evidence in some detail; it is quite possible that we have overlooked other problems. We have also had to identify the species which, for various reasons, were not tackled by Babington. Our nomenclature follows Stace (2010) and (where stated) Sell & Murrell (2006, 2009) for some infraspecific taxa.

Our aim in identifying Ray's plants has been to name the plant which Ray saw in Cambridgeshire. Thus, if Ray reported a plant under a name taken from Gerarde, Parkinson, Jean or Caspar Bauhin or Mattioli, we have tried to provide the modern name of the Cambridgeshire plant rather than the species of the original author. The only exceptions are for the few plants which were reported from Cambridgeshire in earlier literature but not seen by Ray (see Chapter 3); in these cases we have provided the modern name of the species intended by the earlier author.

One would of course expect that in most cases Ray selected a name which was applied by the original author to the same plant as the one he himself knew in Cambridgeshire. However, there must be cases, especially in names drawn from foreign authors, where he misapplied names to the Cambridgeshire plant, and this is perhaps more likely to be true of the synonyms than of the chosen names. We have not attempted to identify such misapplications of earlier names, though we have drawn attention to a few instances where, in retrospect, his names or synonyms are clearly misapplied or where the listed synonyms cover more than one species.

We have in general identified Ray's taxa only to species level, not to subspecies (even if only one subspecies is known from Cambridgeshire or would have been present in Ray's time). Identification to subspecies would in most cases suggest, anachronistically, an undue degree of taxonomic precision. We have cited subspecies in a few cases where they might arguably be treated as species (e.g. *Medicago sativa* subsp. *falcata* [587], *Raphanus raphanistrum* subsp. *raphanistrum* [491]) and named plants to subspecies or variety in cases where Ray or Dent themselves recognised variation within modern species (e.g. *Daucus carota* [166, 423, 424], *Prunus domestica* [B79, B80]).

Methods for identifying Ray's plants

Ray's names can often be identified simply from the evidence in the main catalogue. There may be a continuity of at least some elements in the chosen name, synonyms and vernacular names from 1660 to the present day, as in *Fraxinus vulgaris* (synonym *Fraxinus I, sive excelsior*), Common Ash tree [*Fraxinus excelsior* 203], *Parthenium*, Feverfew [*Tanacetum parthenium* 421] and *Perchpier Anglorum*, Parsly-piert [*Aphanes arvensis sensu lato* 430]. The synonyms often provide an adequate description: even if the identity of the Cambridgeshire plant called by Ray *Acorus palustris* [10] is not immediately apparent, the synonym *Iris palustris lutea* reveals its identity as the modern *Iris pseudacorus*. Even for less obvious cases Ray normally gives sufficient synonyms for these to be identified by using Linnaeus' *Species plantarum* (1753) directly or by matching them to entries in the third edition of Ray's *Synopsis* (1724) and then to *Species plantarum via* Linnaeus' *Flora Anglica* (1759), conveniently reprinted by the Ray Society with the 1724 *Synopsis* (Stearn 1973). Bauhin's *Pinax* (1623) is an invaluable source of information on the more obscure synonyms of vascular plants and Dillenius (1741) has a comprehensive coverage of synonyms of bryophytes and lichens in the earlier literature. In straightforward cases these methods usually suffice, but there are traps at all stages. For example some scientific names are now applied to plants quite unrelated to those listed by Ray and in *Flora Anglica* Linnaeus certainly misinterpreted some of the plants in the 1724 *Synopsis*. The final and most important check is to ask whether the information given for Cambridgeshire by Ray is plausible for the species identified.

There are two possible approaches to the investigation of more complicated problems. One is to look back to the sources of Ray's chosen names and synonyms, which sometimes clarify the identity of the species. The other is to look forward in Ray's

6: IDENTIFICATION OF RAY'S PLANTS

later works, in particular to the *Catalogus plantarum Angliae* (1670, 1677), which identifies plants occurring in Cambridgeshire with an initial C, the first two volumes of *Historia plantarum* (1686, 1688a), which are particularly valuable as they include a description of most species and often a discussion of taxonomic problems, and the first two editions of the *Synopsis* (1690, 1696). The first edition of the *Synopsis* was based on *Historia plantarum*, edited down to make a concise British Flora; some of the additions included in it are described in a small publication, *Fasciculus stirpium Britannicarum* (1688b). In many cases Ray's subsequent treatment clarifies the identification of plants in the Cambridge catalogue, and some of the relevant changes made in the English catalogue are summarised by Dent in the 1685 appendix. However, Ray learnt much after he left Cambridge in 1662 and it is sometimes inappropriate to apply his mature taxonomic concepts to the entries made in the Cambridge catalogue, his first botanical work.

We have been cautious in using the third edition of the *Synopsis* (1724) when investigating the more complex problems as that was edited after Ray's death by J.J. Dillenius. We are aware from his great work on bryophytes, *Historia muscorum* (1741), that Dillenius liked to make a definite identification of all synonyms, even if to us the taxonomic concept of the original author seems to be much too vague to allow a confident identification to a single species. It is clear that he also took this approach on occasion when editing the 1724 *Synopsis*; see, for example, footnotes 22 and 136 in the 1685 appendix dealing with *Aparine semine laeviore* Park. and *Geranium malacoides sive columbinum minimum* Park. C.B.

For an example where Ray's sources have helped us identify the plants of the Cambridge catalogue see our discussion of the identity of *Cardui polyacanthi secunda species* Ger. emac. in footnote 19 in the 1663 appendix. An alternative example, where examination of the earlier sources reveals a confused situation from which we have been unable to draw any definite conclusions, is provided by *Senecio hirsutus viscidus major odoratus* J.B. (see footnote 726 in the main catalogue). Ray's later works have helped to sort out the identity of Ray's larger *Lactuca* species (Oswald 2000, summarised in footnote 394) and of *Saxifraga graminea pusilla, flore parvo herbido et muscoso* (see footnote 718).

Degrees of certainty in the identification of Ray's plants

Many of Ray's species can be identified without a shadow of doubt. This applies especially to distinctive members of the Cambridgeshire flora such as *Anthyllis vulneraria* [49], *Butomus umbellatus* [225], *Pulsatilla vulgaris* [475], *Schoenus nigricans* [309] and *Zannichellia palustris* [466], each the sole member of its genus in the county. Even where there are two similar species in Cambridgeshire, there is often a convincing separation of species such as *Rumex acetosa* and *R. acetosella* [7, 8], *Papaver argemone* and *P. hybridum* [58, 59] and *Sparganium erectum* and *S. emersum* [553, 554]. However, other species can be identified with certainty only because just one of the possibilities is known from the county. Even *Alchimilla* [B3] can be identified, despite the critical nature of the genus *Alchemilla*, because *A. filicaulis* subsp. *vestita* is the only native representative in the county; it persisted into the 20th century in one of the two sites recorded in the 1685 appendix.

Although Ray's plants and those of other early botanists are often identified in publications such as Babington's (1860) *Flora of Cambridgeshire* or (in the national context) in the lists of first British records published by authors such as Clarke (1900) and Druce (1932), few if any authors have discussed the very real problems associated with such identifications. There are a number of cases where Ray might have seen two or more closely allied species but included them both under a single name. In some cases one of the species is so much commoner than the other in the county that we have accepted the usual identification of Ray's plant as the commoner species. We have, for example, identified Ray's *Quercus latifolia* [476] as *Quercus robur*, as this is very much commoner than *Q. petraea* or the hybrid, and his *Fumaria* [204] as *F. officinalis*, as this is much more frequent than the other species of arable land. However, we have, for example, interpreted *Cicutaria palustris* [134] as *Oenanthe aquatica* and/or *O. fluviatilis* and *Oxyacanthus* [414] as *Crataegus laevigata* and *C. monogyna*. In these and other similar cases Babington regarded Ray's plant as just one species. Babington almost certainly chose *Oenanthe aquatica* rather than *O. fluviatilis* because the former was regarded as the normal species and the latter was then a recent segregate, described as a variety by Babington himself in 1843 and raised to specific rank by his friend W.H. Coleman in 1844. The two species of hawthorns were not fully accepted in his day and so he had no difficulty in identifying Ray's plant as "*Crataegus Oxyacantha*". There is obviously a fine and rather arbitrary line between our decisions to identify *Fumaria* as *Fumaria officinalis* but *Oxyacanthus* as *Crataegus laevigata* and *C. monogyna*. For more recent segregates it is more usual to identify Ray's species only to aggregate, and our decision to identify *Nasturtium aquaticum* [387] as *Rorippa nasturtium-aquaticum sensu lato* and *Perchpier Anglorum* [430] as *Aphanes arvensis sensu lato* is likely to surprise few botanists.

There are particular difficulties in identifying species in the modern genera *Carex* (sedges) and *Rumex* (docks), since Ray had not mastered their taxonomy in 1660 and he reports notably fewer species than he is likely to have encountered. These difficulties are described in our footnotes. Ray also described very few grasses but, because he recognised only the most distinctive species, there are only one or two difficult problems of identification in this family.

Raven (1950, p. 90) summarised the identification of Ray's plants in a well-known passage. "There is in fact (and this is a notable tribute both to Ray and to his successors) only one species in the *Catalogue* which defies identification, and that is not unnaturally a Sedge: ... In others, and particularly in the grasses which were then little known, there is occasionally room for doubt; ... Otherwise every entry ... can be given its proper equivalent." These views were echoed by Ewen & Prime (1975, p. 13). There are several reasons why this appraisal is misleading, even if it is taken as referring solely to Ray's vascular plants. It applies to Ray's plants *as identified by Babington in 1860*. We argue above that Babington identified some species with undue precision, even by the species concepts of 1860. Secondly, and perhaps most importantly, species concepts are narrower than they were 150 years ago and some of the species to which Babington attributed Ray's plants are now species aggregates. (This would have been less apparent to Raven than it is to us as he was writing at a time when there was no up-to-date national Flora and before some of the broad species had been subdivided.) Finally, we question Raven's view that the large proportion of Ray's plants that can be identified to a modern species is primarily a tribute to the botanical abilities of Ray and his successors. The compilation of the *Catalogus* in the 1650s must have been a very taxing intellectual exercise, and like Raven we greatly admire the speed with which Ray mastered the subject and the skill that he brought to the task; nevertheless, it must be admitted that the ease with which most of his plants can be identified is at least partly due to the small size of the county, its relatively limited flora and the fact that the plants he recorded are not so very different from those that were known to Babington, and even from those known to modern botanists.

7: NOTES ON THE TRANSLATION AND EDITORIAL METHODS

This new edition of John Ray's first work offers, unlike that of Ewen & Prime (1975), a full English translation, including those parts that they omitted as unnecessary or irrelevant to its main purpose as a catalogue of Cambridgeshire plants. The only exceptions are Ray's 'chosen names'[1] and synonyms from previous authors, which we retain in the original Latin while providing a vocabulary of the epithets that qualify the 'generic' nouns and giving literal translations of some of the more complex phrase-names (polynomials) in our footnotes. Our aim has been to give the reader as clear a picture as is possible in a translation of what Ray originally wrote in Latin rather than, as is frequently the case in the works of both Raven (1950) and Ewen & Prime (1975), a paraphrase that abbreviates some of Ray's more complex phrases and may even leave out important details.[2] While thus attempting to omit nothing, we have often found it desirable, in order to avoid over-literal English, to add words that are not present in the intrinsically more succinct Latin. Where a literal rendering of the Latin seemed too stilted or obscure in English, we have sometimes provided a more natural phrase in the main text while giving the original Latin and a truly literal translation in a footnote.[3]

We recognise that there are many pitfalls in translating 16th- and 17th-century Latin, and it will be surprising if we have succeeded in avoiding all of them. Selosse (2008) gives an example of a pre-Linnean polynomial where ten errors of translation are possible in French for only four Latin words.[4] Fortunately some of the difficulties that he describes in his paper do not apply to English, but his warning of the need for vigilance has not gone unheeded. As mentioned in the previous paragraph, we have excluded Ray's 'chosen names' and synonyms from our translation, so that the constituent parts of Latin polynomials are generally translated separately in our vocabulary,

---

[1] We use this term for the first name cited by Ray in his main alphabetical catalogue (as explained more fully in Chapter 3).

[2] For example, Raven (1950, p. 81) renders the first long and complex sentence of Ray's 'Preface' thus: "I had been ill, physically and mentally, and had to rest from serious study and ride or walk. There was leisure to contemplate by the way what lay constantly before the eyes and were so often trodden thoughtlessly under foot, the various beauty of plants, the cunning craftsmanship of nature." Ewen & Prime (1975, p. 22) translate it thus: "When I was forced, following an illness that affected me both physically and mentally, to rest from more serious studies, and to spend my time in riding and walking, I had leisure in the course of my journeys to contemplate the varied beauty of plants and the cunning craftsmanship of Nature that was constantly before my eyes, and had so often been thoughtlessly trodden underfoot." See Ray's 'Preface' for our translation and footnote 1 there for further comments.

[3] We have not, however, always done this in the case of standard Latin usages such as the accusative and infinitive (where English has a clause beginning with "that"), the ablative absolute (where English often uses a longer clause) and the verb "to be" with an ablative (where English employs the verb "to have"). Examples of each are: *Filicem semine donatum esse asserunt* (literally "They assert a fern to have been derived from seed"); *circumjecto sacco unicuique fasciculo* (literally "with a sack having been thrown to each bundle") and *testibus Theophrasto & Plinio* ("on the evidence of Theophrastus & Pliny", literally "with Theophrastus & Pliny witnesses"); *Floribus est communiter purpureis* (literally "It is commonly with purple flowers").

[4] *Lactuca montana purpurocaerulea major*, for which the classically received translation (*traduction classiquement reçue*) is *Grande laitue de montagne bleu pourpre* but which should, Selosse claims, be translated as *Laitue des lieux pentus pourpre bleue plus grande*.

thus avoiding some of Selosse's pitfalls (*écueils*), and, where we have given a translation in our footnotes, rather than provide an elegant translation, we have tried to show the meaning word by word.

We have not tried to turn Ray's Latin into the English that he himself would have used, except sometimes in the case of individual words where two or more options were available (e.g. "watery" or "moist" rather than "wet" or "damp"[5]); nor have we used his 17th-century spelling except when reproducing parts of his text already in English or when quoting *verbatim* (in double inverted commas) in our footnotes. Ray (or his printer) was not consistent about the use of an ampersand for *et* ("and"), but nevertheless we have followed what he has done in each case.[6] We have sought to be consistent in our translation of the words used by Ray in his habitat notes.[7] There can be some doubt about the most appropriate English translation for plant-parts or for adjectives describing their attributes when terms close to Ray's Latin are used in modern technical botanical descriptions but alternative essentially English words also exist: we have, for example, used "pod"[8] in preference to "siliqua", "husk" rather than "glume", "stalk" or "flower-stalk" rather than "pedicel"[9] and "scalloped" instead of "crenate". Ray did not employ the modern botanical terms "bract", "bracteole", "stipule", "petal",[10] "sepal" and "tepal" or their Latin equivalents but usually *foliolum* ("leaflet"); in translation we have used only "bract" and "petal".[11] He did however use *calyx* (plural *calyces*) and even *calyculus* ("little calyx").

In the main catalogue and the two appendices Ray (and Dent in the 1685 appendix) used roman type for the 'chosen names' and synonyms taken from the works of earlier authors, italic type for the abbreviations for those authors, for the English names and for any locational and habitat notes in English, and principally roman type again for any Latin commentary and for notes following the main entries.[12] In our translation in these parts of this book we have replicated these typefaces irrespective of the language used, but in our footnotes Latin words and those from other foreign languages

---

[5] These last two are words that Ray never uses in his habitat notes in English (though "wet" occurs twice in the 1685 appendix). We have translated *aquosus* as "watery" and *humidus* as "moist". Both words are most frequently used by Ray in the ablative plural after *in*, often qualifying a noun but the former seven times (and once each in the two appendices, in the same passage) and the latter four times with *locis* understood, i.e. *in aquosis* (meaning "in watery places") and *in humidis* (meaning "in moist places", but this is always combined with another adjective such as *uliginosus*, meaning "marshy", though *in humidioribus* occurs alone once in the 1685 appendix).

[6] Sometimes, however, our use of "and" reflects other Latin usages, for example the suffix *-que* after a second or subsequent item.

[7] Thus we have translated *ager* as "field", *arvum* as "ploughed field", *pascuum* as "pasture", *pratum* as "meadow", *sata* (plural) as "sown crops" and *segetes* (again plural) as "standing corn". We have used "everywhere" for Ray's *ubique* but "generally" for his *passim*; in his habitat notes in English Ray used the former (as "every where") four times and the latter only once.

[8] Ray himself used "cod" or "husk" when writing in English, but we have preferred the modern English word.

[9] A complication here is that Ray used *pediculus* indiscriminately for the modern "pedicel", "peduncle" and "petiole", defining the word in his 'Interpretation' as *à quo folium, aut flos, aut fructus dependet* ("that from which a leaf or a flower or a fruit hangs").

[10] However in the 1685 appendix the entries for *Montia fontana* [B6], *Torilis arvensis* [B23], *Moenchia erecta* [B49] and *Juncus squarrosus* [B54] contain the originally Greek word *petalon* in its plural form *petala* or a compound adjective derived from it – *tetrapetalos*, *pentapetalos* or *hexapetalos*.

[11] Thus, for example, we use "petals" loosely for all the parts of the perianth of an orchid flower.

[12] Most names of persons and places within these last two are however italicised.

7: EDITORIAL NOTES

(other than Greek) are italicised and authors' abbreviations, like the rest of the text, are set in roman type, as is conventional in modern works.[13] Translations into English in the footnotes are usually in double inverted commas. We have generally followed Ray in capitalising or not capitalising plant names, though there are cases where there is ambiguity over first words in his phrases that do not come first also in an English translation. We have also followed his hyphenation of English words as faithfully as we can, although again there are some ambiguous cases (where the hyphen comes at the end of a line) and Ray's inconsistent approach to hyphenation sometimes makes the decision whether or not to retain these hyphens rather arbitrary. We have not always reproduced Ray's capitalisation or otherwise of book titles or his punctuation, in the latter case punctuating in the way that we think most acceptable to a modern reader. In Ray's 'Explanation' we have, for greater clarity, italicised book titles and provided English translations in square brackets, as also many of the authors' native names. In his 'Etymology' we have freely used square brackets within the main text to transliterate Greek and to explain meanings; the use of sanserif type is explained in footnote 1 there. In the 'Interpretation' and 'The more customary headings or divisions of plants' we have provided Ray's Latin names for plants, plant-parts and classificatory terms in square brackets after our English versions.

In Ray's and other authors' original text – but not in other contexts (including bibliographical references) – we have retained the ligatures æ and œ (rather than ae and oe), anomalous usage of i and j (particularly ij for ii) and of u and v, and diacritical marks in Latin. Fortunately Ray used few of the abbreviations that can often hamper understanding of Latin texts, but he occasionally used -$\bar{u}$ and -$\bar{a}$ for -*um* and -*am* at the end of a word and a superscript 9 as an abbreviation for the ending -*us*.

We have noted misprinted words in Ray's text in our footnotes, but we have sometimes silently corrected other minor errors and anomalies. The spacing of Ray's text is uneven: we have not tried to replicate this but instead have left a slightly wider space between the various components of each entry in the main catalogue and appendices, including between the plant names listed. At the start of each full entry here we give, in bold type, the current scientific name (with author) of the taxon or taxa that we believe Ray's or Dent's entry represents, preceded by a running entry number.[14] In the case of the synonym entries that cross-refer to the principal ones, the current scientific name (without author) is given in square brackets after the synonym, followed by the relevant entry number which we provide here to assist the reader in finding the entry quickly. For the same reason we give these entry numbers in our footnotes and elsewhere after both pre-Linnaean and modern scientific names when the taxon discussed is included in the main catalogue or the appendices and we cite the author abbreviation only when this is not the case. The current scientific names with their entry numbers are also given in square brackets in the 'Index of English names' and 'Index of places'.

Ray readily used Ancient Greek as well as Latin in his text, both when giving names and when quoting from original sources, and in our footnotes and vocabulary we also sometimes supply the original Greek forms of latinised words or in the former provide relevant quotations from Greek authors. We have omitted accents on Greek words (which are of no significance in the pronunciation of Ancient as opposed to Modern Greek) but we have retained the signs for smooth and rough breathings (*lenis* and *asper*), the latter indicating an aspirated vowel or rho (ῥ). We have also spelled out

---

[13] This results in the typography of plant names quoted from Ray's text being reversed.

[14] In the 1663 appendix the entry numbers are preceded by A and in the 1685 one by B with any equivalent 1663 number following in brackets.

119

in full any Greek words that are abbreviated in Ray's text. Words in Greek characters are followed by a transcription into the Roman alphabet in square brackets. There are several ways of transliterating Greek, none of them wholly satisfactory for all purposes. We have chosen to use that invented by the Romans and thus used by authors such as Pliny the Elder but also through the Middle Ages and up to the present time for scientific plant names. Most of the transliterations are obvious – for example a for alpha (α) and b for beta (β) – but a few are worthy of mention. An initial vowel with a rough breathing is given h before it, the Greek consonant theta (θ) is transliterated as th, kappa (κ) as c, rho (ϱ) as rh when beginning a word or when the second of two rhos together but otherwise as r, phi (ϕ) as ph, chi (χ) as ch, and psi (ψ) as ps. Of the vowels, both epsilon (ε) and eta (η) are transliterated as e, both omicron (o) and omega (ω) as o, and upsilon (υ) on its own as y. When combined in diphthongs, alpha and iota (αι) are rendered as æ (or ae in modern orthography), omicron and iota (οι) as œ (or oe in modern orthography), and epsilon and iota (ει) as ei or simply as i.[15] Upsilon (υ) in combination with other vowels is not represented by y but by u, αυ becoming au, ευ becoming eu, and ου becoming simply u.

Ray used the Hebrew alphabet only in the entry for *Hyssopus* in his 'Etymology', and there without the additional markings used to indicate vowels (see Figure 3.7), but we have thought it preferable to transcribe these words into the Roman alphabet in bold type and to include the vowels. We give a more detailed explanation in footnote 249 in the 'Etymology'.

We have used the time-honoured abbreviations BC and AD rather than the now fashionable equivalents BCE and CE. All dates are given in new style.

---

[15] This is a rare case where authors have been inconsistent, creating, for example, the genera *Cheiranthus* and *Chiropetalum*. In the case of the Greek word Ποταμογειτον, ει has even been transliterated as e, as in the modern *Potamogeton*, which has replaced *Potamogeiton* of Ray's period.

Translations of

John Ray's (1660)

Catalogus plantarum circa Cantabrigiam nascentium

and of the appendices of 1663 and 1685

CATALOGUS PLANTARUM CIRCA CANTABRIGIAM nascentium:

In quo exhibentur
Quotquot hactenus inventæ sunt, quæ vel sponte proveniunt, vel in agris seruntur;

Unà cum
Synonymis selectioribus, locis natalibus & observationibus quibusdam oppidò raris.

Adjiciuntur in gratiam tyronum,
Index Anglico-latinus, Index locorum, Etymologia nominum, & Explicatio quorundam terminorum.

CANTABRIGIÆ:

Excudebat *Joann. Field*, celeberrimæ Academiæ Typographus.
Impensis Gulielmi Nealand, *Bibliopolæ.*
Ann. Dom. 1660.

A CATALOGUE

OF PLANTS

growing

AROUND

CAMBRIDGE:

In which are presented
All those so far found, which
either occur naturally, or
are grown in the fields;

Together with
The more important synonyms, native localities
& various very unusual
observations.

Appended for the benefit of beginners are
An English-Latin Index, an Index of places,
An Etymology of names, & an Explanation
of certain terms.

CAMBRIDGE:

Printed by *John Field,* Printer
to the most renowned University.

At the expense of **William Nealand,** *Bookseller.*
A. D. **1660.**

Præfatio ad Lectorem.

Um nobis plurimùm valetudinariis, corporis animíque causâ, à severioribus studiis subinde quiescendum, & vel equitandum esset vel deambulandum; ferè necessarium fuit, vacuis præsertim & otiosis, quæ continuè ob oculos versabantur, quásque toties pedibus calcavimus, plantarum varietates & elegantias, polydædala artificis *naturæ* opera, obiter nonnunquam contemplari. Queis semel attentiùs spectatis, non sustinuimus ea diutiùs sicco prorsus pede præterire, & omninò contemptim ducere. Splendidus nempe vernantium agrorum ornatus amæno spectaculo nos primùm invitavit & detinuit; figurâ deinde stirpium singularium, colore, totáque adeò externâ specie concinnâ & decorâ mirificè plerumque delectabamur. His dum sæpiùs intenti oculos pascimus, animos exhilaramus, paulatim sese nobis insinuavit *rei herbariæ* studium; magnúmque protinus

in

Preface to the Reader.

WHen it was necessary for us, being very ill, to take an immediate rest from more serious studies for the benefit of our body and mind[1] and either to go riding or to take walks, it was almost unavoidable for us, especially as we were unoccupied and at leisure, sometimes to contemplate along our way the variety & elegance of plants, those richly wrought works of the artificer *nature*,[2] which were continually passing before our eyes and which we so often trod underfoot. Once we had looked at them more attentively, we no longer continued to pass them by entirely dryshod[3] & to reckon them as altogether contemptuous. First, to be sure, the brilliant attire of the fields growing green in spring enticed & detained us with its pleasant spectacle; then we became principally delighted by the shape, the colour and, what is more, the whole harmonious & wonderfully graceful outward appearance of individual plants.[4] While we were frequently intently feasting our eyes[5] on and cheering our mind with these, little by little an interest in *botany*[6] worked its way into us; and straightaway we conceived a great desire & passion to

---

[1] *corporis animíque causâ*: both Raven (1950, p. 81) and Ewen & Prime (1975, p. 22) translate this as "physically and mentally" and interpret it as relating to Ray's illness rather than to his need to rest. McMahon (2000) even cites the phrase when stating her mistaken belief that Ray abandoned his university studies because of an illness of 'body and soul' (see Chapter 2). The phrase occurs (between commas) between "very ill" and "from more serious studies" and so there is indeed an ambiguity within this long, complex opening sentence occupying 11 lines of text, but we believe that the Latin favours our interpretation.

[2] *polydædala artificis naturæ opera*: *polydædalus, -a, -um* is a latinised form of the Homeric adjective πολυδαιδαλος, -ος, -ον, meaning "highly, richly wrought, chiefly of metal work" (Liddell, Scott & Jones 1940), so the phrase seems to be consciously extravagant. In view of Ray's pious belief that the wonders of nature were attributable to divine providence (e.g. Ray 1691) it is rather surprising that he attributes this craftsmanship to nature rather than to God.

[3] *sicco prorsus pede*, literally "with entirely dry foot". The most likely source of the phrase *sicco pede* is the Israelites' miraculous crossing of the Red Sea or of the River Jordan on dry land, yet the actual words are not found in the Vulgate versions of these stories, though "drye shod" was used by Miles Coverdale in 1535 in Joshua 3: 17, Isaiah 11: 15 and Nehemiah 9: 11, while *pede sicco* does occur in the retelling of the Red Sea crossing in the Apocrypha, in Judith 5: 12. Both *sicco pede* and more rarely "dryshod" have been used – mainly with a verb such as "pass over", in biblical commentaries and literary criticism – to mean (as one commentator explains) "without wading in to solve the difficulties". Ray may be using the phrase here either metaphorically or literally, meaning that he now dismounts from his horse or leaves the path to examine plants even if this means getting his boots wet or muddy.

[4] Ray here uses the word *stirps*, originally meaning the stock or stem of a tree but later used as a synonym for *planta*. No distinction will be made in this translation.

[5] *oculos pascimus*, a phrase that Ray seems to have borrowed from Cicero (e.g. *Actio in Verrem*, 2: 5, 26, 65; *Philippics*, 11: 3, 8).

[6] *rei herbariæ*, literally "of the herbal matter/affair". In the course of this preface Ray uses three further words or phrases for botany – the Greek feminine adjective *Botanice*, the Greek feminine noun *Phytologia* (both of these spelled with an initial capital) and the Latin *res plantaria* (literally "the plant matter/affair").

make progress in it. Certainly we were promising ourselves thereafter a huge pleasure, solitary, tranquil & innocent. Soon we are searching the *University*, looking around & diligently seeking some teacher & initiator[7] who may first instruct us &, so to speak, lead us by the hand, hoping to receive that as a great kindness.[8] However, amazing to relate, among so many princes of the educated, among so many luminaries of letters, we found no one who could go before us on the road or carry the torch ahead of us, no one who had deeply drained the cup of *Botany* and scarcely one or two who had even tasted it with the tips of their lips.[9] Then our mind readily recognised how almost insuperable would be the heaps of difficulties obstructing anyone running in this stadium, how tortuous would be the windings of the ways, and finally with how great an expenditure of time & sweat even a mediocre proficiency in that discipline would be achieved, of which a full & accurate knowledge could have been obtained with the leadership & help of an expert namer and inductor[10] with the greatest advantage and almost no trouble to both. What were we to do in this misfortune? Surely we should not permit this same passion to cool & be diverted elsewhere? By no means. We were really ashamed and disgusted how often we perceived that this principal part of *natural Philosophy and history*, so useful & all but essential to common life, should alone be languishing, neglected & untended, when all the other sciences & noble disciplines were flourishing & being celebrated among us. So, having plenty of leisure if not of ability, we were thinking that there was no way to do some good for this *University* more suited to our moderate ability[11] lying open to us than if we should try to repair this defect as far as was in our power & should cultivate *Phytology*, so neglected & passed over by others. We therefore gave ourselves up without restraint to our talent, which was ready enough for these studies, having obtained of course a rare happiness in that which would have enabled us both to meet our own desire and at the same time to serve the convenience of others. And, spurred on more strongly by the weight of these arguments, even though lacking any outside help, we easily endured all the toil & weariness, and we tenaciously apply ourselves to our undertaking, partly by seeking out everywhere the various kinds of *simples* and by stocking our little gardens with them, and especially by frequently visiting and wandering

---

[7] *mystagogum*, accusative of the Greek word μυσταγωγος, as used by Cicero, again in *Actio in Verrem* (2: 4, 59, 132).

[8] *ingentis id beneficii loco*, a further phrase that may be modelled on Cicero.

[9] *primoribus labris*, another Ciceronian phrase (e.g. *De oratore*, 1: 19, 87; *Oratio pro M. Caelio*, 12: 28).

[10] For some reason Ray here switches into Greek for two words, και εισηγητου, though the etymological equivalent in Latin, *et inductoris*, seems quite adequate.

[11] *mediocritati nostræ*, literally "to our mediocrity".

through the fields surrounding *Cambridge*, so that we might not appear to despise the treasures that *nature* freely offers and venture forth in foreign parts like Lynceans[12] in an inappropriate desire for novelty and be blind to what is near at home. Finally, with unwearied effort, we have overcome most of the difficulties by our own resources[13] and, having marched through rough roads & rocks, have achieved what we wanted. Certainly it was necessary for us at the beginning to read whole books & to compare the plants newly found in the fields with the pictures &, if by good fortune any similarity between them was manifest at first sight, also to attend to the descriptions. At last, having become more experienced with practice, when by chance we lit upon an unknown plant, we first considered to which tribe & family it might belong or could properly be assigned (since, after we had been occupied with *Phytology* for some considerable time, it was natural for us to investigate it by similarity & by characters in which it might agree with its congeners) & we searched for it therein first of all, in this way now relieved of much effort. From this time a vehement desire took hold of us to advance the studies of others too, if any might equally be seized by a love of *botany*, lest they, being less patient of labour, might tire, discouraged by the long series of difficulties, and we carefully consider in what way we may be able to serve them most appropriately. For, since everyone may be excessively assiduous with his own interests and may perhaps favour them more than is right (& we likewise, we do not deny it), we were being impelled by the will of our nature to advance our own interests. And, since we have always valued highly the intimacy and friendship of more talented men, nothing seemed more conducive to procuring this than that we ourselves should (may the word be pardoned) seduce them into our interests, nothing more effective in obliging & rendering most of them compliant than that we should publish[14] a *Catalogue* of all the plants that have been found by us in *Cambridgeshire*.[15] Certainly, because a likeness of interests is the firmest bond of fellowship and friendship.[16]

---

[12] *in extraneis lyncei evadamus*. The Lynceans were the members of the first scientific society in Europe, the Accademia dei Lyncei, founded by Federico Cesi, Prince of Acquasparta, in 1603. They were particularly interested in scientific novelties and took their name from Lynceus, a Messenian hero and one of the Argonauts, who was renowned, like the lynx, for the sharpness of his sight; Cicero used the word to mean sharp-sighted. Ray mentions in his 'Explanation' that Fabius Columna was a Lyncean (see *Col.*).
[13] *proprio Marte*, literally "by [our] own Mars" (the god of war).
[14] *in lucem emitteremus*, literally "into the light we should send forth".
[15] *in agro Cantabrigiensi*, literally "in the Cantabrigian field". From mediaeval times *ager* was often applied to a county.
[16] This use of *quippe quod* is another Ciceronian trait.

Clearly we find benefit for us all in the sharing of interests,
But let us all hold fast faithfully each to his own.[17]

Having carefully considered these matters for a long time, finally, after a full period of six years in which we have devoted our efforts to this subject, we have undertaken this little work in earnest, in the preparation and embellishment of which we have spent a period of almost three years. Meanwhile there has died[18] that most friendly & inseparable companion of ours, a Man deserving most highly from the literary World,[19] of old-fashioned fidelity & sincerity, of singular guilelessness & candour of character, of honesty & innocence of life, but not ostentatious in the common courtesy & modesty of his manners, *Master John Nid*,[20] most deserving senior Fellow of the College *of the Holy and Undivided Trinity* in *Cambridge*,[21] whose part in this enterprise was outstanding & whose industry was most valuable, and to whom *the Reader* is chiefly indebted for the observations woven into the work. *He died wept for by many good men, but by none more wept for than by us.*[22] However, since this little book was almost complete[23] before his death, we apply ourselves vigorously to the work, even though heavily dispirited by such a disaster, though not completely discouraged, lest the University be defrauded of a benefit however small & our friends complain that they have been deceived by our promises & so that we may build a lasting monument to the memory of that dearest life,[24] & at last, having, with the help of the good God, successfully completed all our labours, we present it to the press.[25]

---

[17] *Scilicet ingeniis aliqua est concordia junctis: Et servat studii fœdera quisque sui.*
The Latin, from Ovid's *Letters from Pontus*, Book 2, 5 (to Salanus): 59–60, is in his usual verse form of an elegiac couplet consisting of a dactylic hexameter followed by a dactylic pentameter, which we have rendered in the same metre in English above. A more literal version is:
"Of course there is some harmony in united interests, and let each keep faith with his own interest."

[18] *temporis fatis concessit*, literally "to time's fates has yielded". The use of *fatum* in a similar way by classical authors suggests that this was a death from a natural cause.

[19] *de Republica literaria*, literally "from the literary State/Republic". Ray seems here to be adapting a phrase of Cicero's used by him in relation to the Roman state or republic.

[20] *D.* is the standard abbreviation for *Dominus*, used for a Master of Arts. The usual spelling of his name, which we have adopted in this book, was "Nidd"; Raven (1950, p. 44) suggests that the signature "Jo. Nid" in his books "may be due to the Latin form Nidus".

[21] *apud Cantabrigienses*, literally "among the Cantabrigians".

[22] *Multis ille bonis flebilis occidit, Nulli flebilior quàm nobis.*
With the exception of the final word this is a quotation from the lament for Horace's and Virgil's friend Quintilius in the former's first book of *Odes*, 24: 9–10, where the second line ends *quam tibi, Vergili* ("than by you, Virgil"), thus maintaining the metre spoiled by Ray's alteration.

[23] *ad umbilicum penè perductus est*, literally "to the umbilicus almost was brought". *Umbilicus*, as well as meaning a navel, was also the word used by the Romans for the end of a roller on which a manuscript was rolled; thus Horace wrote in *Epode* 14: 7–8 *inceptos ... iambos ad umbilicum adducere* ("to bring to a conclusion the iambics [that I have] begun").

[24] *charissimi capitis*, literally "of the dearest head".

[25] In classical Latin *prelum* was a wine-press or olive-press.

PREFACE

Now, since it is honourable to acknowledge those through whom you have succeeded, those most noble Young Men *Master Francis Willughby & Master Peter Courthope*, Knights, illustrious by the distinction of their birth, by the loftiness of their character, by the pleasantness of their manners and by their fidelity and virtue, not merely very experienced in *botany*[26] but deeply engaged in every kind of science,[27] our most honourable friends, must not be left unnoticed by us,[28] nor would we wish to be ungrateful & arrogant. We freely & frankly acknowledge that we have used their services frequently & that we have been helped by them in no small measure in compiling this little work. It remains for the easier understanding of our intention & of the material communicated in the following *Catalogue, dear Reader*, that we should reveal to you in a few words at what target we have particularly been aiming, what scheme for arranging the plants we have used[29] and what we have achieved in *it*. And as to what pertains to our general aim, that was to illustrate the Divine glory in investigating the works of nature or creation. Next to this comes the fame of our *Alma Mater*, the *University of Cambridge*, which must be diminished in the world outside if scholarship in this field should be defective among us. Finally its usefulness to the common life & the considerable advantage which may overflow from it to people generally.[30] Throughout we have followed no other scheme than one which suits an index, namely by arranging the names of the plants in Alphabetical order. In the case of the individual plants we have in fact observed the following rules.

1. We have always employed the nomenclature of the four authors who are now commonly in everyone's hands among us,[31] viz. *J. Bauhin, C. Bauhin, Gerarde and Parkinson*, when it could be found.

2. We have ignored the names of almost all other authors, unless they contained something characteristic, distinctive & emphatic or, being now commonly accepted, are used constantly in everyone's speech or are said to be those used by the Ancients, so that in some plants a collection of Synonyms can take the place of a description. Generally, when

---

[26] Ewen & Prime (1975, p. 24) say "though they have no special knowledge of Botany", but, though *duntaxat* alone can mean "exactly", *non duntaxat* means "not merely" rather than "not exactly".

[27] The Latin *literae* can mean "literature", "letters", "science" or "culture", but "science" seems the best translation here.

[28] *non sunt à nobis silentio transmittendi*, literally "are not by us to silence fit to be transmitted"; this phrase may derive from the works of Tacitus.

[29] *quâ methodo usi sumus*. The Greek philosophical term μεθοδος [methodos, "a system"] was adopted into Latin in post-classical times. Ray used the title *Methodus plantarum* for a later work published in 1682.

[30] *ad plures*, literally "to more" (plural).

[31] *qui vulgò jam apud nos omnium manibus teruntur*, literally "who [i.e. whose books] commonly now among us by everyone's hands are rubbed".

the Authors themselves were not to hand, we have taken the Synonyms from the Pinax of C. Bauhin or the Historia of J. Bauhin.

3. When several authors signify some plant by the same name but with epithets added to distinguish it, differing[32] according as seemed appropriate to each, we have placed the name in first position and begun it with a capital letter so that it can be told apart from the epithets. In the second case & subsequently we have omitted the name, just giving the epithets, as for example in Abrotanum, first, beginning *Abrotanum* with a capital letter, we write *Abrotanum campestre* C. B., then, with Abrotanum omitted & understood, we put *sylvestre vel quartum* Trag., beginning with a small letter, & so on subsequently until we come to another name, as here, for example, *Artemisia leptophyllos* Gesn. hort., which we again distinguish with a capital; and, lest any confusion or obscurity may arise therefrom, we very rarely leave anything more than the generic noun alone to be understood in this way.

4. We have signified places *in English*, since this both was easier for us & would be clearer for the Reader, and knowledge of particular places would not be of any use to foreigners who do not know *English*.

5. We have also given descriptions of some plants which either have not been published by others or at least have been described in an obscure or confused way & (as Aristotle is supposed to have written about his own books on Natural Science) are "published and unpublished".[33]

6. Finally, we have added Observations in as many cases as possible, not chosen randomly & without careful choice whencesoever, but either when confirmed by our own experience or when depending on the credibility & witness of the weightiest authors or when in themselves sufficiently credible and probable; whence, unless the result may deceive our hope, *the Reader* may derive some profit or delight.

To this work, which we principally intended, we have appended also some other things: first, for the benefit of beginners, an Index of plants, in which we have put English names before the Latin ones; secondly, an Index of some of the places lying close to *Cambridge*, with the spontaneous plants that grow therein which are not met with everywhere; thirdly, the Etymologies of & reasons for certain Names; lastly and finally, an interpretation of terms & more general words with the principal & most widely used headings[34] & divisions of plants.

These, *dear Reader*, are the tasks that we have performed for you in this little work, in which, granted that we have applied the highest diligence, yet, being conscious of our frailty, we do not doubt that

---

[32] *aliis atque aliis*, literally "with some and with others".
[33] Ray here quotes Aristotle's original words in Greek, ἐκδεδομεναι και μη ἐκδεδομεναι.
[34] *capitibus*, literally "with the heads".

somewhere we have made a mistake[35] & that very many things in this little book remain to be corrected, even though we have now gone through it several times to polish it.[36] If you would be willing to grant us your indulgence, would be favourably inclined to our undertakings & would take these endeavours of whatever kind in good part, you would give us further confidence[37] to undertake greater things; what is more we promise as well that we shall be more careful not to submit to your judgement & public censure anything which *much time and much erasure have* not *pruned*,[38] for we have allowed ourselves to hasten the completion of our present efforts[39] so that we may be able as soon as possible to revive & rekindle[40] the almost extinct & lifeless study of *Botany*.

What is more we hope that another advantage will accrue to the world of literature from this enterprise of ours, undoubtedly that others, in emulation of us, may investigate the spontaneous plants, each of his own area, more diligently so that in this way a complete *Phytologia Britannica* may finally appear from all their contributions.[41] And as far as *the county of Cambridge* is concerned,[42] we do not doubt that very many other plants, especially those that are met with very frequently and almost everywhere else in England, also grow wild somewhere in it, even though we have not yet come upon them; such are *Digitalis purpurea* [Foxglove], *Scorodonia* [Wood Sage], *Ruscus* [Butcher's Broom], *Dryopteris* [a fern], *Erica tenuifolia* [Bell Heather], etc.,[43] which we hope we shall also find very soon. There are also some among those that we have listed, for example *Juglans* [Walnut], *Betulus* [Hornbeam], *Helleborus* [Green Hellebore], *Rhaphanus rusticanus* [Horse-radish], *Papaver* [Poppy], *Bistorta* [Bistort], etc., which we suspect did not

---

[35] *ratio nos fugiat*, literally "reason may escape us".

[36] *licet jam aliquoties ad limam à nobis revocatus sit*, literally "granted that now several times to the file [for polishing] by us it has been recalled". Ray seems to have made a rare grammatical error here, since *revocatus* is masculine while *libellum* ("little book") is neuter; he may have thought that he had written *in hoc libro* (masculine) rather than *in hoc libello*.

[37] *animos nobis addes*, literally "courage/confidence [plural] to us you would add".

[38] Ray here quotes from Horace's *Ars Poetica*: "quod non *multa dies et multa litura coercuit*". All but the last word (*atque*) of the hexameter of Horace's line 293 is included, in italics, but the first two words, which end line 292, are printed in roman text.

[39] The Latin here, *nam in præsentiarum ideo festinandum duximus*, is obscure, though the general meaning seems clear.

[40] Ray uses, in the infinitive, a rare Greek verb that occurs in the works of Euripides and Aristotle, ἀναζωπυρειν. He was perhaps following the advice about letter-writing set out by his tutor James Duport "now & then to intermingle some odd, old word, so it be sparingly, & when you write to your Equalls" (Trevelyan 1943).

[41] *omnibus suis numeris*, literally "from all their own numbers".

[42] *Et quod ad Cantabrigiensem comitatum attinet*, literally "And what to the Cantabrigian county pertains".

[43] This list includes several plants that are indeed not native or are unusually rare in Cambridgeshire, but not Red Campion, *Silene dioica* (L.) Clairv. The identity of Ray's *Dryopteris* is unclear.

originally spring up naturally but were sown by some chance & by accident in the places which[44] we have mentioned.[45] In addition we ask that students should not immediately become angry & attribute the crime of deceit to us if by ill fortune they fail to find some of the species listed by us in the places mentioned; for that must frequently happen, since many are annual & by some chance do not sow themselves again, others are destroyed by beasts of burden & wild animals and others are even completely eradicated & carried off by root-collectors.

We have given the plan of the work and, so that the *Reader* may know, have completed those things that we think are of importance either to him or to us; it remains that we should vehemently exhort *University men* to whom God has graciously given leisure and granted talent & intelligence, without indeed neglecting their other studies but just interrupting them for a little while, to accustom themselves to examine nature itself & to compile a comprehensive account of its creatures[46] so that they may begin to be wise from their own experience rather than from someone else's brain and in particular may strive to read the leaves of plants & interpret the characters stamped on flowers & seeds. It is not the case that anyone should fear that this study will be fruitless & useless, for, to use the words of *P. Lauremberg*: "nothing is contained in this orbit of the world's system which may either bless human life with greater advantage & success or offer fuller delight both to the mind and to the body, the servant of the mind, than the rich store of plants and the most copious and exceedingly varied stock of things growing from the earth;" &, a little later: "I say that from plants a man receives almost everything the use of which he needs for life, whether it may be living simply or pleasantly or luxuriously. Undoubtedly human frailty has need of[47] *food, drink, medicines, clothing, housing, furniture, ships* and *the delights of the senses and the mind*. None of these is something which plants do not bestow on us from their store, none of them something which we do not hold as a possession received from them," as he demonstrates excellently by running through them one by one.[48] However,

---

[44] Ray has made another small grammatical error here: *quæ* is the neuter accusative plural of the relative pronoun but *loci* ("places") are masculine.

[45] This list demonstrates Ray's sound appreciation of which plants that he found were likely not to be native. Hornbeam in Cambridgeshire is on the very edge of its native range in Britain, but the trees at Abington would not have been native ones. The poppy that Ray lists is probably Opium Poppy, *Papaver somniferum* L., which is not native in Britain, although he calls it Wild Poppy.

[46] *& amplissimum creaturarum volumen evolvere*, literally "and to unwind a very large roll of the creatures".

[47] *opus habet*, literally "has work [for]".

[48] Peter Lauremberg (Petrus Laurembergius: 1585–1639) was a professor in botany at the Universities of Hamburg and Rostock and published two books in 1632, *Apparatus plantarius* and *Horticultura*, with second editions in 1654. This quotation is from page 7 of the Foreword (*Præloquium*) to

we may concede that, as human affairs are now, studies of this kind do not greatly contribute to the accumulation of wealth or the procurement of the favour of other men, but yet we know of no occupation either worthier or more delightful for an honourable man than to contemplate the most glorious works of *nature* and so to honour both the infinite wisdom & the goodness of the Divine creator. We are not however recommending anyone[49] to give himself up & devote himself wholly to these studies but to pay respect to them in passing, within reason, and sometimes to be diverted by them[50] for the sake of his mind, so that he may learn something thoroughly while exercising and not allow any part of his life to slip away wholly empty. We do not doubt that the pursuit of plants will be pleasurable to studious young men, since we know several noble and well-born students of *Trinity College* for whom that occupation offered both exercise for the body and delight for the mind. At the same time, however, we know also that not everyone is captivated by the sight of flowers or of meadows growing green in spring, or, if they are captivated, there is something else that delights them more. Some take delight in the tennis court, some in drinking, gaming, amassing money or procuring men's favour,[51] and they all show themselves very diligent merchants for these goods. We are not writing *Phytologies* for these because there is something else which they do, but we are preparing a hundred banquet dishes for the Pythagoreans, dedicated, to be sure, to true Philosophy, whose *heart the Titan fashioned with benign nature and from finer clay*[52] and for whom it is not as important to be thoroughly acquainted with the opinions of authors as to contemplate the very nature of things with their own eyes & to drink in its voice with their ears, who would prefer to know the truth rather than much that is uncertain & things prepared in right measure for use rather than for show. And so, principally for the use of these and in accordance with God's glory, we dedicate both this present little work & all our studies.

---

the latter. Ray cites both of them in the main catalogue, the former four times (twice each in his species accounts and in his notes) and the latter 14 times (all of them in his notes), but, like many of the authors cited for reasons other than as sources of plant names, Lauremberg is not included in those listed in his 'Explanation'.

[49] *Nec tamen cuiquam autores sumus*, literally "Neither however to anyone proposers are we".
[50] Ray actually wrote *eóque* ("and by it").
[51] *Alii sphæristerio, alii potando, ludendo, pecunias cumulando, hominum favorem conciliando delectantur.*
[52] "quibus *natura benigna Et meliore luto finxit præcordia Titan.*" The end of one dactylic hexameter and the whole of another are taken from lines 34–35 of Juvenal's *Satire* 14, but Ray has written *natura benigna* instead of Juvenal's *arte benigna* ("with benign art"), thus spoiling the metre; "quibus" ("whose" or, literally, "for whom") actually forms part of line 34 even though Ray does not print it in italics. The Titan is Prometheus, who, according to the Greek myth, made the human race from clay.

Explicatio nominum Autorum
citatorum, è *C. Bauhino*
præcipué.

AD. & Ad. Lob. Adversariorum opus à Petro Pena & Matthia de Lobel, Londini Ann. 1570. editum, additis circiter 284. figuris: postmodum Anno 1576. Antverpiæ novus titulus præfixus, epistola ad Reginam omissa: cui operi Lobelius observationes suas cum novis figuris junxit. Tandem Anno 1605. iterum Londini edidit Lobelius Adversaria, unà cum parte secunda, & Pharmacopœâ Rondeletii.

Ægin. Paulus Ægineta.

Aet. Aetius.

Amat. Amatus Lusitanus, qui in libros quinque Dioscoridis enarrationes edidit, additis diversarum linguarum nominibus, Argentinæ 1554. in 4.

Ang. Aloysius Anguillara horti Patavini tertius in ordine præfectus, de plantis suam sententiam diversis communicavit: opusculum in partes 14. divisum, operâ Johannis Marinelli Italicè prodiit (additis duabus figuris Chamæleontis & Sedi arborescentis) Venetiis 1561. in 8.

Apul. Apuleius Platonicus, non antiquus ille Madaurensis, sed alius recentior, quem Johnsonus in præfat. ad Gerardum à se emaculatum sexcentis retro annis aut circiter vixisse conjectatur, & opusculum suum de herbarum virtutibus primùm Græcè scripsisse. Idem Johnsonus observavit, in duobus libris manuscriptis à se visis autorem hunc Platonem Apoliensem nominatum. Qui plura scire desiderat, adeat Johnsonum loco citato.

An Explanation of the names of the Authors cited,[1] principally from *C. Bauhinus*.

A D. & Ad. Lob. A work of *Adversaria* [Disputations] by Petrus Pena & Matthias de Lobel [Mathias de L'Obel]: London, in the year 1570; later published, with about 284 figures added, in the year 1576 at Antwerp, new title-page prefixed, letter to the Queen omitted: to this work Lobelius [de L'Obel] added his own observations with the new figures. Finally, in the year 1605 Lobelius published the *Adversaria* again in London, together with a second part & the *Pharmacopœa* of Rondeletius.

Ægin. Paulus Ægineta [Paul of Aegina].

Aet. Aetius [Aëtius of Amida].

Amat. Amatus Lusitanus [João Rodigues], who published commentaries on the five books of Dioscorides, with names in various languages added: Strasbourg, 1554, in quarto.

Ang. Aloysius Anguillara [Luigi Squalermo], the third curator in succession of the garden at Padua, published his opinion about various plants: a small work came out in Italian, divided into 14 parts, through the agency of Johannes Marinellus (with two figures added of a *Chamæleon* & a tree-like *Sedum*): Venice, 1561, in octavo.

Apul. Apuleius Platonicus, not that ancient author of Madaura[2] but another more recent one, whom Johnsonus [Thomas Johnson], in his preface to the edition of Gerardus [John Gerarde] corrected by himself, surmises lived six hundred years ago or thereabouts & wrote his small work about the virtues of herbs first in Greek.[3] The same Johnsonus observed that in two manuscript books seen by him this author was called Plato Apoliensis. Anyone who wants to know more should consult Johnsonus in the place cited.[4]

Honorius Bellus Vicentinus [Onorio Belli of Vicenza], a Cretan physician: letters of his about plants survive as additions to the *Historia* of Clusius.[5] [See also *Pon*.]

Petrus Bellonius [Pierre Belon] wrote three books of *Observationes*, which, when rendered into Latin by C. Clusius [Charles de L'Écluse] with 10 figures of plants added, came out in Antwerp, 1589, octavo; a tract about the neglected cultivation of plants, which the same Clusius translated & published at the same time as his books on exotics,[6] a book about coniferous, resin-producing & evergreen-leaved trees with 8 pictures of them: Paris, 1553, quarto; finally a book about the medicine of funerals & medicaments having the power to preserve the corpse, which came out in the same place.

---

[1] Biographical notes on all the authors cited in Ray's 'Explanation' and details of some of their works are provided in Chapter 4.

[2] Lucius Apuleius Platonicus was a Roman writer born at Madaura, in Numidia, in about AD 130, whose best-known work, *The Golden Ass*, a satirical and fantastic moral romance, has remained popular to this day.

[3] Printed versions of this work appeared from the late 15th century onwards and one was cited by Bauhin (1623), one of Ray's major sources.

[4] Johnson (1633, pp. 3–4, unnumbered, of his preface headed *To The Reader*). Ray refers to this preface only three times but he seems to have drawn upon it elsewhere in writing his 'Explanation', though, as Johnson, like Ray, clearly used Bauhin's (1623) *Nomina Authorum* as one of his sources, there are many doubtful cases. See also footnotes 27, 29, 45 and 49.

[5] L'Écluse (1601).

[6] L'Écluse (1605).

By Antonius [Antonio] Musa Brassavolus [Brasavola] of Ferrara, an investigation of all the simples of which use is made in pharmacies: Lyons, 1556,[7] 16mo.

Brunf. Otho Brunfelsius [Otto Brunfels] in folio. *Brunf. 4.* Brunfelsius in quarto. He was the first of the Germans to publish living portraits of plants. His Latin history came out in three volumes, the first in 1530, the second in 1531 & the third posthumously in 1536, at Strasbourg, with 238 figures but sixteen were presented twice: altogether 288 are described with & without figures. The same author wrote also in German in folio with the same figures & in quarto with small figures added: Strasbourg, 1539.[8]

C. B. Casparus Bauhinus in his *Pinax* [List], *C. B. Phyt.* The same author in his *Phytopinax* [Plant-list]. C. Bauhinus [Caspar Bauhin], doctor & Professor at Basel, presented 2460 names of plants with some synonyms & 164 descriptions of new plants and attached eight figures: Basel, 1596, in quarto.

The same author published in a corrected form all the works of P. Andreas Matthiolus [Pietro Andrea Mattioli], not previously printed in Germany, with the addition of 330 figures & 50 new plants first described by himself, with synonyms: Frankfurt, 1598, in folio.[9]

The same author published his *Animadversiones in Historiam generalem* printed in Lyons [Observations on the General history of Daléchamps], in which he showed that about 400 figures in that work had been included & repeated two or three times: Frankfurt, 1600,[10] in octavo.

The same author corrected and for the most part enlarged the German history of Jacobus Theodorus Tabernamontanus [*Neuw Kreuterbuch* of Jakob Theodor of Bergzabern]; he added several new figures, described some plants & appended synonyms to the first part: Frankfurt, 1613, in folio.

The same author's *Prodromus* [Forerunner] of his *Theatrum Botanicum* [Theatre of Botany], in which about 600 plants were first described by himself, with 140 new figures added, was published in Frankfurt in 1620 in quarto.

The same author's *Catalogus plantarum circa Basileam sponte nascentium* [Catalogue of plants growing naturally around Basel], with synonyms of the same & their native localities: Basel, 1628,[11] in octavo.

Finally in the year 1623 he published his *Pinax Theatri Botanici* [List for a Theatre of Botany], the most complete work of 40 years, containing the names of about 6000 plants with their synonyms & distinctive characters: Basel, in quarto.

Many years after his death, that is in the year 1658, his son, Jo[annes] Casp[arus] Bauhinus [Jean Caspar Bauhin], published the first volume of *Theatrum Botanicum sive Historiæ plantarum* [The Theatre of Botany or Histories of plants], which is a twelfth part of the whole work, as is clear from the *Pinax*.

C. Hofman. By Casparus Hofmannus [Caspar Hofmann], M.D. & distinguished public Professor in the University of Altdorf, *De Medicamentis Officinalibus* [Concerning Pharmaceutical Medicines], both simples and composite ones, the work of thirty years: Paris, 1646, in quarto.

Cæs. Andreas Cæsalpinus [Andrea Cesalpino] Aretinus [of Arezzo], Professor in the University of Pisa, wrote 16 books about plants, with no figures added; they came

---

[7] This work by Antonio Musa Brasavola (1500–1555) was first published in Rome in 1536.

[8] The German *Contrafayt Kreüterbuch* was first published as two folio volumes in 1532 and 1537 respectively; these and the 1539 quarto edition are cited in our bibliography.

[9] Mattioli (1598) in our bibliography.

[10] This is an error for 1601.

[11] This is a misprint for 1622; it is correctly dated in Bauhin (1623).

out in Florence in 1583. The same author's appendix to these books: Rome, 1603, in quarto.

Cam. & Cam. hort. Camerarius in his *Hortus. Cam. ep.* Camerarius in his *Epitome* of Matthiolus. Joachimus Camerarius [Joachim Kammermeister the Younger] corrected again the history of Matthiolus [Mattioli], first translated into German by Handschius[12] & printed with Matthiolus' figures, and expanded it with 123 new pictures & notes.[13] The same author produced an *Epitome* [Abridgement] of Matthiolus with 1003 figures: Frankfurt, 1586, in quarto.[14] The same author wrote *Hortus Medicus & Philosophicus* [A Medical & Philosophical Garden] and added 56 figures: Frankfurt, 1[5]88, in quarto.

Cast. Castor Durantes [Castore Durante], a Roman physician, published in Italian a new herbarium [entitled *Herbario nuovo*] with figures of plants which grow in the whole of Europe and India both East & West, with their properties described in Latin verse; he followed alphabetical order & appended 879 figures of plants: Rome, 1585, in folio.

Clus. Clus. hist. Clusius in his *Historia. Clus. pan.* Clusius in his *Historia Pannonica* [Hungarian History]. Carolus Clusius [Charles de L'Écluse] made a French translation of Dodonæus' [Rembert Dodoens'] history written in his vernacular language [Flemish] & added a supplement about fragrant spices, with figures which came out separately in octavo: in the year 1560 in folio.[15]

The same author condensed into an *Epitome* [Abridgement] the dialogues about fragrant spices & simples by Garcia ab Horto [de Orta] compiled in Portuguese, translated them into Latin and illustrated them with short annotations & 30 figures of plants: Antwerp, 1574, octavo.[16]

The same author translated from Spanish the book by Nicolaus Monardes about the medicines of the West Indies and illustrated it with notes & 9 figures: in the year 1574, octavo.

The same author made a Latin translation from Spanish of the *History* of Christophorus [Cristóbal] Acosta, condensed it into an epitome [abridgement], illustrated it with Scholia [learned comments] & published it in the year 1582.

The same author wrote two books about plants observed throughout Spain, with 230 figures added: Antwerp, 1576, octavo. Later, summoned by the Emperor Maximilian the Second, he travelled through Austria, Hungary & the adjoining provinces, compiled 4 books about the simples observed there and added 353 figures: Antwerp, 1583, in octavo. However he later compiled a history, as much Spanish as Hungarian, bound up with 6 other books added, & published it under the title of *Historia Plantarum rariorum* [A History of rarer Plants] with 1135 figures added: Antwerp, 1601, in folio.

The same author wrote ten books of exotics, of which the first six contain various material about animals & plants, with 194 pictures of plants; also 2 books of Garcia, 1 book of Chr[istophorus] Acosta and 1 book of Monardes, which previously came out separately; also 3 books of Nic[olaus] Monardes about the gem Bezoar, about *Scorzonera* and about Iron & Snow, in the year 1605, in folio.

The same author's later efforts were published after his death, with 27 figures added: Antwerp, 1611, in quarto.

---

[12] Mattioli (1563a).

[13] The first edition of Camerarius' translation is cited in the bibliography as Mattioli (1590) but there were several later editions. See also *Matth*.

[14] Mattioli (1586).

[15] See footnote 23.

[16] Orta (1574).

Col. Fabius Columna [Fabio Colonna], a Neapolitan from the Roman family of the Columnæ, a very learned man & one of refined judgement, a Lyncean,[17] published a *Phytobasanos* [Enquiry into plants] with 34 figures: Naples, 1592; also two parts about lesser known plants, of which the first was printed with 161 figures in Rome in the year 1606 and also the second with 44 figures in the same place in 1616.

Cord. & Cord. hist. Cordus in his *Historia. Cord. in Diosc.* Cordus in his notes on Dioscorides. *Cord. in disp.* Cordus in his *Dispensatorium. Cord. schol.* The scholiast of Cordus, who is thought to be Æmilius. Valerius Cordus wrote *Annotationes in Dioscoridem* [Notes on Dioscorides]; also 4 books about the history of plants, with 272 figures from Tragus [Jerome Bock] & several new ones from Gesnerus [Conrad Gesner] added; also a *Sylva* of observations, which were all published at the same time in folio under Gesnerus' supervision in Strasbourg in 1561; also a *Dispensatorium* [a work prescribing the composition of remedies], frequently reprinted.[18]

Dalechamp. in Plin. Dalechampius [Jacques Daléchamps] with notes on Plinius [Pliny the Elder].[19] *Dalechamp. in Diosc.* with figures added to the Dioscorides of Ruellius[20] [see *Ruel.*].

Diosc. & Dioscor. Pedacius Dioscorides Anazarbæus: according to Suidas[21] he lived at the time of Antony with Cleopatra.

Dod. Dodonæus [Rembert Dodoens] in a Latin edition in folio. *Dod. gall.* Dodonæus in the French edition of Clusius [Charles de L'Écluse]. *Dod. ut.* Dodonæus in both editions, both French and Latin. *Dod. Belg.* that is in his Flemish edition, the last. Rembertus Dodonæus, a physician of Mechelen, wrote a history of plants in Flemish,[22] which, when translated into French by Clusius, came out in Antwerp in 1559 with some additions.[23] See Clusius. In the year 1552 he published a history of fruits in octavo in Latin[24] & a little later a history of flowers & purgatives in octavo.[25] He published these same two bound together, divided into 30 books, in Antwerp in 1583 in folio with 1305 figures. This history was reprinted again in Flemish in the year 1608, with someone's[26] lengthy annotations and with 36 further new figures added. It came out similarly in Latin in the year 1616, corrected by the author, with a few alterations & with the figures which are included in the Flemish edition added with a description of each.

---

[17] A member of the first European scientific society, the Accademia dei Lyncei: see footnote 12 in Ray's 'Preface'.

[18] An edition published at Leiden is cited as Cordus (1651) in the bibliography, although the copies in Cambridge University Library and Trinity College, Cambridge, have an additional title-page dated 1652.

[19] Daléchamps (1587) is cited in our bibliography, though there were later editions.

[20] Amatus Lusitanus (1558).

[21] The massive Byzantine historical encyclopedia *Suda* dates from the 10th century and was formerly attributed to an author, Suidas. An edition edited by Portus (1619) is cited in our bibliography. Dioscorides is now believed to have lived in the first century AD rather than in the first century BC.

[22] The first edition was published in 1554; the last Flemish edition for which Rembert Dodoens (1517–1585) was responsible was published in 1563 but Flemish editions continued to be published until 1644 (Pritzel 1871–1877; Arber 1986; Opsomer 1978).

[23] Dodoens (1557): the date of publication is given here by Ray as 1559 but as 1560 under *Clus.* earlier in this 'Explanation', but it is actually 1557 (Pritzel 1871–1877; Arber 1986; Opsomer 1978). Ray does not mention the English translation by Henry Lyte based on this French edition (Dodoens 1578, 1619).

[24] Dodoens (1568).

[25] Dodoens (1574).

[26] Françoys van Ravelingen.

EXPLANATION

Eyst. The *Hortus* [Garden] of the Bishop of Eichstätt [*Eystettensis*] in Bavaria, published by Basilius Besler, a Nuremberg Apothecary, in 1613, with large figures engraved on copper. In this work just synonyms & descriptions are presented. It is divided into 4 parts according to the 4 seasons of the year & each part is again divided into three according to the months of the year.[27]

Eric. Cord. Ericius [or Euricius] Cordus, father of Valerius [see *Cord.*], a Physician of Siemershausen, published his *Botanologicum*, set out as a dialogue, in Cologne in 1534 in quarto.

Joannes Fragosus [Juan Fragoso], Physician & surgeon of the King of Spain, wrote a history in Spanish of certain fragrant spices, fruits & simples brought to Europe from the two Indies. Israel Spachius, a Strasbourg physician, published it: Strasbourg, 1610,[28] octavo.

Fuch. Fuchsius in his Latin history in folio. Leonhardus Fuchsius [Leonhart Fuchs], a German physician, a learned & diligent botanist, the rival of Tragus [Bock], published his Latin history with 516 figures at Basel in 1542 in folio. His pictures were published separately in Basel in 1549 in octavo.

Gal. & Galen. Galenus [Galen], easily the foremost of all Physicians after Hippocrates.

Ger. The English history of plants [*The Herball*] of Jo[hannes] Gerardus [John Gerarde], with the figures of Tabernamontanus [Jakob Theodor of Bergzabern] himself and also some of Lobelius [Mathias de L'Obel] & in addition 16 new ones added: London, 1597, folio. Gerardus (as Johnsonus [Thomas Johnson] relates in his *Preface to the Reader*[29]) was born at Nantwich in the county of Cheshire, whence he came to London and there gave his attention to Surgery, in which he was so successful that he was made Master in that discipline. He had moderate or no skill in languages. The *Pemptades* of Dodonæus [Dodoens] make up the principal part of his work, translated into English by a certain Doctor *Priest*, which Gerardus transcribed almost unchanged. To conceal this he changed Dodonæus' arrangement to that of Lobelius, as is to be seen in the Grasses & the Orchids. To this translation of Dodonæus he added as well some plants from Clusius [Charles de L'Écluse], others from the *Adversaria* [Disputations of Pierre Pena & Mathias de L'Obel] and about 14 from his own observations not previously published; but he also appropriated the pictures of Tabernamontanus [Jakob Theodor of Bergzabern], which he often ignorantly confused & transposed. Anyone who may wish for more about him should consult Johnsonus in the place recommended.

Ger. emac. Gerardus corrected by that very learned man Thomas Johnsonus [Johnson], citizen & apothecary of London, whom the University of Oxford honoured very deservedly with the status & title of Doctorate for his remarkable knowledge of the whole of medicine but especially of botany. This work was printed in London twice, first in 1633 and again in 1636.

Gesn. Gesnerus. *Gesn. hort.* Gesnerus in his book about the gardens of Germany. *Gesn. Col.* Gesnerus in his little book on the collecting of plants. *Gesn. cat.* Gesnerus in his four-language catalogue of plants. Conradus Gesnerus [Conrad Gesner], a

---

[27] There is only a very brief mention of this book in Bauhin (1623) and much of Ray's entry seems to be derived from Johnson's (1633) entry on p. 9 (unnumbered) of his preface headed *To The Reader* (see footnote 4).

[28] This work was first published in Madrid in 1572. All the details given by Ray fit editions published in Strasbourg in 1600 and 1601; we cannot trace an edition published in 1610 and therefore suggest that 1610 is a misprint for 1601.

[29] Johnson (1633, p. 10, unnumbered, of his preface headed *To The Reader*: see footnote 4).

German, published a history of plants and their properties from Dioscorides, Ægineta, Theophrastus, Plinius [Pliny the Elder] and more recent Greek writers together with an arrangement of the elements: Venice, 1541, 16mo; also a four-language Catalogue of plants: Zürich, 1542, quarto; also *De Lunariis et noctu lucentibus* [About Lunar studies and objects shining at night], with 5 figures, and a description of Mons Fractus or Pilatus [Pilatusberg] of the people of Lucerne: in the same place, 1552,[30] quarto; also a book about the gardens of Germany with an appendix and supplement to Cordus' history, in which are 5 figures with the works of Cordus: Strasbourg, 1561, folio;[31] also *De lapidibus, fossilibus et gemmis* [About stones, fossils and gems], with 8 figures of marine plants: Zürich, 1565, octavo. The same author's medical letters published through Casparus Wolfius [Caspar Wolf], with an assertion added about the first *Aconitus*, with 3 figures: Zürich, 1577, quarto; also a book about the collecting of plants, published through Wolfius: Zürich, 1587, octavo.

Guiland. Melchior Guilandinus [Wieland], the fourth curator in succession of the garden at Padua, published a book which is entitled *Theon sive Apologia adversus Matthiolum* [Theon or A Defence against Mattioli]: Padua, 1558, quarto; also some Letters about plants, in the same place; also a Commentary on three chapters of C. Plinius [Pliny the Elder] about papyrus, which is entitled *Papyrus*.[32] The same author's conjectural synonyms of plants, with a catalogue of the garden at Padua for the year 1591, published through the agency of Jo[hannes] [S]chenckius in Frankfurt, 1600, octavo.

Hermol. in corol. Hermolaus Barbarus [Ermolao Barbaro] published 5 books of *Corollaria* [Annotations] on Dioscorides: Cologne, 1530, folio. The same author published *Castigationes* [Criticisms] of the *Historia naturalis* [Natural history] of C. Plinius [Pliny the Elder]: Basel, 1534, quarto.

Hort. Bat. Catalogue of the Leyden Garden.[33]

Hort. Hafn. [Catalogue] of the Copenhagen Garden in Denmark.

Hort. Gron. [Catalogue] of the Groningen Garden in Confederate Belgium [the Netherlands].

Hort. Par. & Paris. [Catalogue] of the Paris Garden.

Hort. Pat. [Catalogue] of the Padua Garden.

Hort. Wars. [Catalogue] of the Warsaw Garden.

All collected & published together through the agency of Master[34] Simon Paulus under the title *Viridaria varia* [Various pleasure-gardens]: Copenhagen, 1653, 16mo.

J. B. Joannes Bauhinus [Jean Bauhin], brother of Casparus [Caspar], Chief Physician of the Duke of Württemberg, a man of exceptional erudition, the highest fidelity, boundless scholarship and mature judgement, highly versed in all the writings of Botanists both ancient and more recent, most learned in every kind of the more humane & serious literature, in a word the Coryphæus[35] of Botanists, besides some short works such as a little book about plants named after Gods, published in Basel in 1591 in octavo, a little book about the kinds of *Absinthium*, published in Montbéliard in 1593 in octavo, & a catalogue of the plants growing around the bath at Bol, with 60 figures of apples & 40 of pears added, with a history of the Bath at Bol, published in

---

[30] This appears to be an error for 1555.
[31] Cordus (1561) in our bibliography.
[32] This was published in three editions, in Venice (1572), Lausanne (1576) and Amberg (1613); the last is cited in the bibliography.
[33] Ray sometimes uses the abbreviation *hort. Lugd-Bat*.
[34] D. is the standard abbreviation for *Dominus*, used for a Master of Arts.
[35] The chief member of a group, originally the Greek word for the leader of the chorus in Attic drama.

EXPLANATION

Montbéliard in 1598 in quarto, besides these, I say, began his *Historia plantarum universalis* [Universal history of plants], but, forestalled by his death, did not fully complete it. Dominicus Chabræus [Dominic Chabrey], Master of Arts of Geneva, revised & enlarged this, but Franciscus Lud[ovicus] from Graffenried [Franz Ludwig von Graffenried], Master of Arts of Gerzensee, etc., made it public. The work is the most perfect of all that have so far seen the light, containing almost everything worthy of knowledge & observation that occurs in the work of either the ancients or more recent authors, together with Synonyms, equivalents, supplementary material, etc., divided into three volumes; if he himself had lived & supervised this edition, there is no doubt that he would have arranged some things in it more suitably and improved others & added very many, so that he would have been able to fulfil the duty of all Botanists, "not to seem but to be",[36] as the Title-page prefixed to the work has it.

Ferrantes Imperatus, an Apothecary of Naples, published 28 books of natural history with figures of stones, corals, sponges, etc., but also 33 of plants & fruits: Naples, 1599, in folio.

Jo. Bodæ. Joannes Bodæus à Stapel, a physician of Amsterdam, in commentaries or notes on Theophrastus' Historia, published in Amsterdam, together with their Greek context, in folio, 1644. Very many pictures of rarer plants are included in this work.

Lac. Andreas Lacuna [Laguna]. His commentaries on Dioscorides with figures, written in Spanish, survive: Salamanca, 1552, folio;[37] also Annotations on Dioscorides, in which Ruellius' version is criticised: Lyons, 1554, 16mo.

Lob. Lobelius in his *Observationes*. *Lob. ico.* Lobelius in the illustrations published separately [*Stirpium icones*]. Matthias de Lobel [de L'Obel], in addition to the work of *Adversaria* [Disputations] mentioned earlier [see *Ad. & Ad. Lob.*], published his *Plantarum seu stirpium historia* in Flemish [*Kruydtboeck*], in which he included both the *Adversaria* & the *Observationes*. The pictures from this work, 2116 in number, printed separately, came out in long form in Antwerp in 1581.[38] In addition he planned another great work, the title of which was to be *Stirpium Illustrationes* [Illustrations of Plants], but his death interrupted it.

Lon. Adamus Lonicerus [Adam Lonicer]: he enlarged the *Herbarium* of Eucharius Roslin [Rösslin], his predecessor in his post at Frankfurt, written in German in quarto, & published it the next year, 1569; afterwards he removed Eucharius' name and prefixed his own & published it with about 833 figures in the year 1582 at Frankfurt.[39]

Lugd. Historia generalis plantarum [A general history of plants], in 18 books, begun by Dalechampius [Jacques Daléchamps] and compiled & published by Joannes Molinæus [Jean Desmoulins], who did not append his own name, was printed in Lyons in the year 1587. In this 2686 figures are presented, but about 400 are repeated twice or three times.

---

[36] οὐ δοκειν ἀλλ' εἶναι.

[37] We have not traced an edition as early as 1552 and this date appears to be an error; the Latin commentaries were first published in 1554 and the Spanish translation in 1555 (see Riddle 1980). A 1563 edition of the latter is cited in our bibliography.

[38] Although it is often attributed to Mathias de L'Obel, it is now thought that this book should be attributed to the publisher as Plantin (1581).

[39] This oversimplifies a complex history of publication. The *Kreutterbůch* of E. Rösslin was first published in 1533 (see Arber 1986). By 1660 there had been numerous editions, revised by Theodor Dorsten and then by Adam Lonicer (Lonicerus), the son-in-law of the publisher Christian Egenolph, including the edition cited here (Rösslin 1569). It was also published in Latin (Lonicer 1551). It continued to be published for over a century after 1660 and the final edition (1783) may have been the last book to report as fact the belief that barnacle geese emerged from barnacles, a view that had been refuted by Albertus Magnus 500 years before (Arber 1986).

Matth. Petrus Andreas Matthiolus [Pietro Andrea Mattioli]. His commentaries in Italian with 957 large figures were published in Venice in 1568 and the same author's work in Italian with small figures in 1563. The same author's German edition produced by Camerarius;[40] an edition of the same author corrected by C. Bauhinus;[41] the same author's epitome [summary], published in the year 1571;[42] the same author's epitome corrected by Camerarius;[43] the same author's letters with commentaries, published at Frankfurt, with a defence against Amatus Lusitanus.[44] Concerning Matthiolus, see Johnsonus in his preface, where he is frequently praised.[45] He incurs great criticism from Lobelius [Matthias de L'Obel] and others.

P. B. Phytologia Britannica [A British Phytology], published by Master [William] How: London, 1650, octavo.

P. Gassend. Petrus Gassendus [Pierre Gassend or Gassendi] in his natural Philosophy [*Syntagma philosophicum*] has a general treatment of plants.

P. Renealm. Paulus Renealmus [Paul de Reneaulme] of Blois published a sample of a history of plants with 43 figures printed from copper plates in Paris, 1611, quarto.

Pandect. Pandectarius: Matthæus Sylvaticus [Matteo Silvatico] wrote a work of *Pandectæ* [*Opus pandectarum*]; it came out in Venice in 1499, folio.[46]

Park. Joannes Parkinsonius [John Parkinson] of London, Royal Apothecary, in the year 1629 published his *Paradisus terrestris* [Earthly paradise] in English, in which he treats at length the history of all the flowers & also the fruit trees, vegetables & choicer shrubs that are grown in gardens. Later, in the year 1640, he issued his *Theatrum Botanicum* [Theatre of Botany], likewise in English, in which he used a new method and divided plants into various classes according to their various qualities or properties. This work encompasses more species than any other history which exists today. In most respects it follows C. Bauhinus [Caspar Bauhin], even where he multiplies entities unnecessarily. I do not criticise the man's industry, but I frequently question his judgement & accuracy[47].

Plin. Caius Plinius Secundus [Pliny the Elder]. He lived at the time of Vespasianus [the Emperor Vespasian]; he died in AD 79, suffocated by the sulphurous fumes of Vesuvius while he was investigating the causes of the conflagration. He wrote numerous works which have perished with the ravages of time; but 37 books *De Historia mundi* [About the History of the world] survive,[48] into which he transcribed many things that relate to plants from Dioscorides, though he never mentioned him.[49]

---

[40] Mattioli (1590); see *Cam. & Cam. hort*.

[41] Mattioli (1598); see *C.B.*

[42] This must refer to Mattioli's *Compendium* (1571).

[43] Mattioli (1586); see *Cam. & Cam. hort*.

[44] These are also included in Bauhin's edition of Mattioli (Mattioli 1598). See also *Amat*.

[45] Johnson (1633, p. 6, unnumbered, of his preface headed *To The Reader*: see footnote 4); he applauds Matthiolus' "957 large and very faire figures" but, far from frequently praising him, goes on to say that "he is iustly reprehended by some" for various defects which he takes seven lines to describe.

[46] This is one of many printed editions of this work, the first of which was published in 1474.

[47] ἀκριβειαν [acribeian] (accusative singular).

[48] Numerous printed editions of this work had been published by 1660, some with titles beginning with *Historiae mundi* and others with *Naturalis historiae*. See Daléchamps (1587) in the bibliography for one of the former; we cite this one as Ray sometimes quotes this author *in Plinium*.

[49] Bauhin (1623) has no entry for Pliny and several of Ray's phrases suggest that he used as a source Johnson's (1633) account of him on p. 3 (unnumbered) of his preface headed *To The Reader* (see footnote 4).

EXPLANATION

Pon. Jo[hannes] [Giovanni] Pona, an Apothecary of Verona, wrote a catalogue of the simples growing on Mons Baldus [Monte Baldo] & added descriptions of some of them with 16 figures.[50] Clusius [Charles de L'Écluse] added it to his own history of rarer plants.[51] The same author [Pona] published it in enlarged form & added descriptions of some plants observed by Honorius Bellus [Onorio Belli] in Crete with 38 figures: Basel, 1608, quarto. The same book, enlarged by very many other exotics received from the well stocked garden of the Most Noble Nicolas Contarenus, a Venetian Senator, with 91 figures added, translated by his son Franciscus [Francesco] Pona M.D. into the Italian language, came out in Venice in 1617, quarto.

Ruel. Joannes Ruellius [Jean Ruel]: he translated Dioscorides into Latin;[52] he wrote three books about the nature of plants: Basel, 1537,[53] in folio.

Schwenck. Casparus Schwenckfeldius [Kaspar Schwenckfeld] wrote a catalogue of the plants & fossils of Silesia: Leipzig, 1601, quarto.

Schrod. Schroder's *Pharmacopœa medicochymica* [A medicochemical guide for pharmacy].[54]

Tab. The German history of Jacobus Theodorus Tabernamontanus [*Neuw Kreuterbuch* of Jakob Theodor of Bergzabern], published in three parts with 2087 figures in Frankfurt, 1588, folio. The same corrected, enlarged & with very many descriptions & figures of plants & medicaments by C. Bauhinus [Caspar Bauhin] in the year 1613, folio. The pictures from the same came out in long form at Frankfurt in 1590 with simply the Latin & German names.

Thal. Sylva Harcynia[55] [of the Harz Region] of Jo[hannes] Thalius [Thal], a Physician of Northaus, or *Catalogus plantarum in montibus & locis vicinis sponte nascentium* [A catalogue of the plants growing naturally in the mountains & adjoining areas], with descriptions of very many of them, together with the *Hortus* of Camerarius [see *Cam. & Cam. hort.*], with 9 figures added, published in Frankfurt, 1588, quarto.

Theoph. & Theophr. Theophrastus: *Theophr. hist.* Theophrastus in his *Historia plantarum* [A history of plants]. *Theophrast. de caus.* Theophrastus in his books *De causis plantarum* [About the causes of plants]. Theophrastus was the successor of Aristoteles [Aristotle] in his school; he lived about 322 years before the birth of Christ; he wrote 10 books about the history of plants, of which nine have come down into our hands, and 8 books about the causes of plants, of which only six have done so.

Trag. Hieronymus Tragus [Jerome Bock]. He wrote a history, which came out frequently in German in Strasbourg[56] &, translated into Latin by David Kyberus [Kyber], was printed in quarto in Strasbourg in 1552, with 567 pictures, albeit up to 800 plants

---

[50] It is not clear whether this was published separately before it was included by L'Écluse (1601) in his book. Pritzel (1872–1877) cites an edition published in Verona in 1595 but comments *Editionem Veronensem anni 1595 ubique frustra quaesivi* ("I have searched everywhere in vain for the Veronese edition of the year 1595") and we can find no evidence for its presence in European libraries with on-line catalogues.

[51] L'Écluse (1601). See also *Clus. hist.*

[52] The first edition of this translation was published in 1516 and is included in our bibliography; Pritzel lists numerous other editions published between 1516 and 1552.

[53] This had originally been published in Paris the previous year.

[54] Johannes Schröder's *Pharmacopëia* was first published in Ulm in 1641; it was quickly followed by a number of later editions including the 1649 edition cited in our bibliography and one 'englished' by William Rowland and published in London in 1669 as *The compleat chymical dispensatory*.

[55] *Harcynia* is an error for *Hercynia*, presumably copied by Ray from Bauhin (1623) though he has *Harcynica*.

[56] The first edition, published in 1539, is cited in the bibliography.

are described. A German edition was corrected by Melchior Sebitzius & Nicolas Agerius: 1595, folio.

Tur. Gulielmus Turnerus [William Turner] M.D., a man of sound learning & judgement, published a history of plants in English in the year 1551, in which for the most part he used the figures of Fuchsius [Fuchs]. He gave the names in Latin, Greek, English, German and French. He followed the order of the Latin alphabet.

Explicatio &c.

Tab. Jacobi Theodori Tabernamontani historia Germanica tribus partibus edita cum figuris 2087. Francofurti 1588. fol. Idem emaculatus, auctus, & plantarum descriptionibus & figuris ac medicamentis plurimis à C. Bauhino anno 1613 fol. Ejusdem icones cum nudo nomine Latino & Germanico Francofurti 1590. in longâ formâ prodière.

Thal. Jo. Thalii Medici Northusani Sylva Harcynia, sive Catalogus plantarum in montibus & locis vicinis sponte nascentium, cum plurimarum descriptionibus, cum Camerarii horto, 9. figuris additis, editus Francofurti 1588. 4.

Theoph. & Theophr. Theophrastus: *Theophr. hist.* Theophrastus in historia plantarum. *Theophr. de cauſ.* Theophrastus in libris de causis plantarum. Fuit Theophrastus Aristotelis in schola successor; vixit circiter 322. annos ante Christum natum: scripsit de historia plantarum lib. 10. quorum novem; de causis plantarum libros 8, quorum sex tantùm ad manus nostras pervenerunt.

Trag. Hieronymus Tragus. Scripsit ille historiam, quæ sæpiùs Germanicè Argentinæ in folio prodiit, & per Davidem Kyberum Latinè reddita, cum iconibus 567. licèt ad 800. describantur, Argentinæ 1552. excusa est in 4. Editio Germanica à Melchiore Sebitzio & Nicolao Agerio emaculata est. 1595. fol.

Tur. Gulielmus Turnerus M. D. vir solidæ eruditionis & judicii, anno 1551. emisit plantarum historiam Anglicè, in qua figuras Fuchsii plerunque adhibuit. Nomina expressit Latinè, Græcè, Anglicè, Germanicè, Gallicè. Ordinem Alphabeti Latini secutus est.

CATALOGUS PLANTARUM CIRCA CANTABRIGIAM nascentium.

A

Abrotanum campestre *C. B. Ger. Park. Tab.* sylvestre vel 4 *Trag.* Artemisia leptophyllos *Gesn. hort.* tenuifolia 3 *Cluf.* tenuifolia sive leptophyllos, alius Abrotanum sylvestre *J. B.* Ambrosia altera *Matth. Lugd.* tenuifolia *Lob. Wild or field Sothernwood.*

Abrotanum inodorum *Ger. Park. Lob. Lugd.* latifolium *Tab.* II, sive latifolium inodorum *C. B.* Artemisia 2 tenuifolia *Cluf. in post. hist.* quoad figuram. Artemisiæ tenuifoliæ simile si n n idem, Abrot. sc. inodorum Lobelii *J. B.* Absinthium inodorum & insipidum *Dod. Unsavory Sothernwood.* Both said to be found on Newmarket heath by Mr Sare. P. B. We have searched diligently, but can as yet find neither there.

Absinthium vulgare *Park. Trag. Lob. Gesn.* vulgare majus *J. B. hort. Warf.* commune *Turn. Cam. Gesn. hort.* rusticum *Thal. Apul.* latifolium sive Ponticum *Ger.* Ponticum seu Romanum officinarum seu Dioscoridis *C. B. Common Wormwood.*

N. 1. Hujus plantæ ob amaritudinem eximiæ radices

A CATALOGUE
OF PLANTS
growing
AROUND
CAMBRIDGE.

A

1 *Artemisia campestris* L.

Abrotanum campestre *C.B. Ger. Park. Tab.* sylvestre vel 4 *Trag.* Artemisia leptophyllos *Gesn. hort.* tenuifolia 3 *Clus.* tenuifolia sive leptophyllos, aliis Abrotanum sylvestre *J.B.* Ambrosia altera *Matth. Lugd.* tenuifolia *Lob.* Wild or field Sothernwood.

2 *Artemisia campestris* L.

Abrotanum inodorum *Ger. Park. Lob. Lugd.* latifolium *Tab.* II, sive latifolium inodorum *C.B.* Artemisia 2 tenuifolia *Clus. in post. hist.*,[1] as far as the figure goes. Like Artemisia tenuifolia if not the same, Abrot[anum] sc[ilicet] inodorum Lobelii *J.B.* Absinthium inodorum & insipidum *Dod.* Unsavory Sothernwood. *Both said to be found on Newmarket heath by Mr Sare*. P.B. *We have searched diligently, but can as yet find neither there.*[2]

3 *Artemisia absinthium* L.

Absinthium vulgare *Park. Trag. Lob. Gesn.* vulgare majus *J.B. hort. Wars.* commune *Turn. Cam. Gesn. hort.* rusticum *Thal. Apul.* latifolium sive Ponticum *Ger.* Ponticum seu Romanum officinarum seu Dioscoridis *C.B.* Common Wormwood.

    N. 1. The roots of this plant, which are exceptional for their bitterness, are sweet & almost tasteless with a little salt. *Camer. in hort.*

    2. It leaves an unpleasant taste if it is drawn through boiling water. *The same author in the same place.*

    3. Wormwood resists putrefaction remarkably & either drives away or kills animals generated by putrefaction. *Casp[arus] Hofman[nus], De medicam[entis] officinal[ibus], book 2, chapter 2, section 19.*

    4. If those who travel through the countryside to search for or collect plants happen to come upon a beer which is sour or has an unpleasant taste, they can improve it by infusing some common wormwood & make it more pleasant to the palate & the stomach and less injurious; for the bitterness removes the acidity

---

[1] Carolus Clusius [Charles de L'Écluse] in his later history: see the final item under *Clus.* in Ray's 'Explanation'.

[2] The letters P.B. stand for William How's (1650) *Phytologia britannica*. See also *Emendanda* of the 1663 and 1685 appendices, where Ray asks the reader to expunge both species because he believes that "neither of them is to be found on Newmarket heath, at least that part of it which is in Cambridge-shire".

wonderfully, even more so than Sugar. Hence, generally, not sweet things but bitter ones like hops, wormwood etc. are boiled with beer so that it may not turn sour quickly. What is more, the little worms recently detected in vinegar³ are killed by bitter things.

4 *Onopordum acanthium* L.

Acanthium, *Matth. Dod. ut. Ad. Lob. Cam.* vulgare *Park. Tab.* Spina alba sylvestris Fuchsio *J.B.* Carduus sylvestris *Cæs.* Onopordon Athenæi *Ang. Gesn. hort.* Bedeguar Arabum *C. Hofman[nus], De medicam[entis] officina[libus], book 2, chapter 50, section 5*; he asserts in the same place that the leaves of this plant are not inferior to musk in their very ageeable scent. *The common Cotten-thistle.* By roads. It commonly has purple flowers, sometimes white ones.

N. When its roots first begin to produce leaves they are very nutritious & restore strength, either boiled in broth or baked in an oven in meat pies in the manner of artichokes. *Moufet, writing about diet.*⁴

5 *Acer pseudoplatanus* L.

Acer major *Cord. hist. Gesn. hort. Ad. Lob.* majus *Ger. emac.* majus latifolium, falsely called Sycomorus *Park.* majus, for many falsely Platanus *J.B.* montanum candidum *C.B.* Platanus *Trag. Lon. Cardani, De varietate rerum, book 6, chapter 23*.⁵ Sycomorus Ruellii *Gesn. hort. The great Maple, commonly, yet falsly, the Sycomore tree.*

N. 1. The shoots of this & of the lesser maple gush forth with milk like a fig-tree, but not with pungent milk, in fact with no evident taste. *Cam. in hort.*⁶

2. It is clearly established from skilful dissection of the seed of this tree or of the ash that, at least in most plants, the seed is nothing other than the small leaves that emerge first, suitably folded or rolled up by nature, together with the bud of the root, in an appropriate portion of nourishment for the tender plant and a suitable covering. *V[ide] histor[iam] generationis Highmori Anglicè conscript[am].*⁷

---

³ The nematode *Turbatrix aceti* (Müller), or Vinegar Eel, was described by Pierre Borel in his *Observationum microcospicarum* [sic] *centuria* (1656), on pp. 7–8 as 'Observatio prima. De Vermibus aceti', and soon became known to British authors (Power 1664; Hooke 1665). "Once present in practically all vinegar", it was the first free-living nematode to be described, although parasitic species had been known since classical times (Thorne 1961). Borel's book is likely to have been the source of Ray's knowledge, although Rossi (2000) refers to a study by Fontana, completed in 1646, which we have not traced.

⁴ This is a reference to *Healths improvement*, written by Thomas Moufet (or Muffet, Moffett or Moffet, 1553–1604) and published posthumously in 1655. See also footnotes 26 and 76, the entries on *Avena nigra* [75] and *Triticum spicâ muticâ* [595 and 597], Ray's second note on *Bardana major* [79] and Chapter 4.

⁵ Girolamo Cardano or Hieronymus Cardanus (1501–1576) is best known for his contribution to the development of mathematics, but he wrote books on a wide range of subjects. His *De rerum varietate*, an encyclopaedia of natural history, was published in Basel in 1557 and is "a mine of facts, both real and imaginary" (DSB, vol. 3, pp. 64–67). There were several later editions. See also footnote 526.

⁶ The larger *Acer* with the milky latex described by Camerarius (1588) in his *Hortus* must be *A. platanoides* L. rather than the non-lactiferous *A. pseudoplatanus*, and indeed the leaf illustrated as *Acer maior* in his earlier work, *De plantis epitome … Matthioli* (Mattioli 1586), is clearly *A. platanoides*.

⁷ "See [Nathaniel] Highmore's *The history of generation*, written in English." This was published in London in 1651 and was the first serious study of the germination of seeds since that of Theophrastus (*c.* 372 to *c.* 287 BC); it also contains the first reference in English to the use of the microscope (DSB, vol. 6, pp. 386–388).

MAIN CATALOGUE

6 *Acer campestre* L.
Acer minus *Ger. emac.* minus & vulgare *Park.* vulgare minori folio *J.B.* IV, sive campestre & minus *C.B.* Opulus *Turn. Gesn. hort. Cæs.* The lesser or common Maple.

N. Maple second only to cedar in the elegance & fineness of the articles made from it. *Plin[y], book 16, chapter 15.*[8] What is more even today in the Northern parts of England elegant vessels from this wood are turned on a lathe which are so thin that they are even translucent.[9]

7 *Rumex acetosa* L.
Acetosa vulgaris *Park.* I, sive pratensis *C.B.* Lapathum 4. for Dioscorides, sylvestre for Pliny *of the same author.* Lapathi sylvestris 4 genus Dioscoridis[10] *Lugd. Trag. Fuch.* Oxalis *Trag. Matth. Fuch. Dod. Ad.* vulgaris folio longo *J.B.* Ox[alis] seu Acetosa *Ger.* Common Sorrell.

8 *Rumex acetosella* L.
Acetosa arvensis lanceolata *C.B. hort. Groning.* Oxalis parva auriculata repens *J.B.* minima *Trag.* minima seu Leporis *hort. Paris.* tenuifolia sinuata vervecina *Lob. ico.* ovina *Tab.* Sheeps Sorrell. *In sandy & drier places, as about Gamlingay, and on Newmarket heath.*

Acetosella, *see* Trifolium acetosum.[11] [*Oxalis acetosella* 578]

Achillea, *see* Millefolium. [*Achillea millefolium* 371]

Acinos, *see* Clinopodium. [*Clinopodium vulgare* 139]

9 *Clinopodium acinos* (L.) Kuntze
Acinos Anglica *Clus.* Anglica Clusii *Park.* Ac[inos] 4, falsely called Polium montanum by the herb-selling women of London[12] *Ger. emac.* Clinopodium Ocymi facie alterum *C.B.* Small stone Basill. *In the ploughed lands on the borders of Gogmagog hills and Newmarket heath.*

10 *Iris pseudacorus* L.
Acorus palustris *Lob.* nostras *Ruel.* II, sive adulterinus[13] *C.B.* palustris, sive Pseudoiris & Iris lutea palustris *Park.* Iris palustris lutea *Ger. Tab.* palustris lutea,

---

[8] Pliny's *Natural history*, XVI. 66 in the Loeb Classical Library edition, the text of which has *citro* ("to citrus") rather than *cedro* ("to cedar").

[9] *nunc dierum ... ut etiam lumen transmittant*, literally "now of days ... that even light they may transmit".

[10] "Dioscorides' fourth kind of *Lapathum sylvestre*".

[11] The name *Acetosella* is not actually cited in the synonymy of *Trifolium acetosum vulgare* [578]; it is mentioned in the 'Etymology' under *Acetosa*, but without any indication of the species to which it relates.

[12] *mulierculis botanopolis Londinensibus*, literally "for the little women herb-selling of London". In his revision of Gerarde's *Herball* Thomas Johnson (1633, p. 676) wrote: "I haue sometimes seene it brought to Cheapside market, where the herbe women called it Poley mountaine, some it may bee that haue taken it for *Polium montanum* misinforming them. *Clusius* first tooke notice of this plant, and called it *Acinos Anglicum*, finding it growing in Kent, *Anno,* 1581." Johnson and his fellow apothecaries, whose prices were notoriously high, disliked the Cheapside herb-sellers, but Johnson nevertheless seems to have taken a keen interest in the plants that they sold (Kew & Powell 1932, pp. 70–71).

[13] i.e. counterfeit: this epithet and the names *Pseudoiris* and *Pseudacorus* were used for this waterside plant because the leaves were sometimes substituted for those of the pleasantly scented Sweet Flag, *Acorus calamus* L., which was used for strewing on house floors. See also footnotes 264 and 412.

sive Acorus adulterinus *J.B.* Pseudoiris *Dod.* Gladiolus aquaticus *Amat. Lusit.* Butomon *Clus. hist.* Yellow water Flower de luce. *In many places down the river Cam, and in the fenns in the Isle of Ely.*

11 *Asplenium ruta-muraria* L.
Adianthum album *Tab. Cam.* Ruta muraria *C.B. I.B. Dod. Ger.* muraria sive Salvia vitæ *Park.* Salvia vitæ *Ad. Lob. Lugd.* Paronychia *Matth.* Saxifraga sive Empetron *Fuch.* Wall-rue, Tentwort. *On the walls of Hinton church.* It likes unworked rocks and dies on vertical brick surfaces.[14] *Cam. in hort.*

12 *Polytrichum commune* Hedw.
Adianthum aureum majus *Ger.* aphyllon majus *Thal.* Polytrichum aureum majus *Park. C.B.* Apuleii majus quibusdam *I.B.* It is also called Muscus capillaris *in Gerarde.* Great golden Maiden-hair, or Goldi-locks. *On Hinton moor in the watery places.*[15]

13 unidentifiable mosses[16]
Adianthum aureum minus *Tab. Lugd.* aphyllum minus *Thal.* Polytrichum minus *Trag.* Apuleii minus *Fuch.* Apuleii minus quorundam *I.B.* aureum minus *C.B. Park.* Muscus capillúsve aureus minor *Lob.* Muscus capillaris, sive Adianthum aureum minus *Ger.* Musci capillaris altera species omnino minor & humilior *Dod.* Little Goldi-locks, golden Maiden-hair. *On walls generally.*

14 *Avena fatua* L.
Ægilops bromoides *Ger. emac.* bromoides Belgarum *Park. Lob.* Æg[ilops] quibusdam aristis recurvis, sive Avena pilosa *I.B.* Gramen festucæ XIII, sive Festuca utriculis lanugine pubescentibus *C.B.* Ægil[ops] of Diosc[orides], Theoph[rastus], Gal[en], Oribas[ius] & Aegin[eta] *for the same author.* Bearded wild Oats, or Dutch Haver-grasse. *Among standing corn.*[17]

15 *Ilex aquifolium* L.
Agrifolium *Ger. Dod. Cam. Cæs.* Agrifolium sive Aquifolium *Park.* Aquifolium *Ruel. Gesn. hort. Lob. Lugd. Tab.* Aquif[olium] sive Agrifolium vulgò *J.B.* Ilex V, sive aculeata baccifera folio sinuato *C.B.* Ἀγρια [Agria] of Theophrastus, Hist[oria], [book] 3, [chapter] 4, & Aquifolium of Pliny, book 16, chapter 21,[18]

---

[14] This seems the most likely meaning of *Amat saxa viva, coctis lateribus immoritur*, literally "It loves living stones, on baked sides it dies". Virgil has *vivoque sedilia saxo* ("seats in the living stone") in *Aeneid*, 1: 167 and Propertius *cocto aggere* ("on a brick rampart") in *Elegies*, 3. 11. 22.

[15] Ray (1670, p. 8) says "*In boggy places frequently, in most Countries [sic] of England. In Cambridgeshire we once found it on Hinton-moor*", suggesting that it was rare at this site.

[16] Martyn (1763, p. 24) and Proctor (1956) identify this as *Funaria hygrometrica* Hedw. but Proctor admits that Ray's comment in *Catalogus plantarum Angliae* (1670, p. 12), *Hujus duas aut tres varietates observavimus* ("We have observed two or three varieties of this"), suggests that it "might have included several of our small acrocarps". We do not believe that Ray's concept here can be equated with a single modern species, as shown by the fact that Dillenius (1741) regards *Adianthum aureum minus* Tab. as the species now known as *Polytrichum juniperinum* Hedw. and several of Ray's other synonyms as the completely different *Funaria hygrometrica*.

[17] *Inter segetes*. The primary meaning of *seges* (singular) is "a cornfield" but it is also used of the standing crop, which seems to be the meaning here after *inter* ("among"). Ray uses "corn" and "cornfields" in his English passages when referring to cereal crops in general.

[18] Ἀγρια [Agria] can only be the feminine of the adjective ἄγριος [agrios], meaning "wild", so ἀρια [aria] must have been intended, but this was the holm (= holly) oak, *Quercus ilex* L., not holly. In Theophrastus' most important work, Περι φυτων ιστορια (usually called, as by Ray, *Historia*

MAIN CATALOGUE

of the same author. Ilex veterum *Ruel.* Paliurus *[P.] Gassend., volume 2, p. 148. Holly. About Stitchworth.*

N. 1. Later in its lifetime this is represented by a leaf that is by no means fringed nor prickly but only pointed at the tip.[19]

2. Country people prepare bird-lime from the outer layer removed from the roots.[20]

3. If roses are inserted into the young saplings they produce greenish flowers, as some very skilled gardeners have proved. *Cam. in hort.*

4. In evergreens the shedding and new growth of the leaves occur in turn, for the same ones do not last for ever but some shoot forth and others fall off; and they seem to do this particularly frequently around the summer solstice, in some species even after the rise of Arcturus. *Theophr. hist. plant., book 1, chapter 15.*[21] The palm is to be excluded here, which never sheds its leaves once they are produced but keeps them for ever. *P. Gassend., volume 2.*

5. Almost all the same trees produce their fruit late & slowly. *Theophr. de caus., book 1, chapter 21.*[22]

16 *Agrimonia eupatoria* L.

Agrimonia *Ger. Officin[arum].* vulgaris *Park.* Ag[rimonia] sive Eupatorium *J.B.* Eupat[orium] veterum sive Agrimonia *C.B.* Eupat[orium] Avicennæ *Io. Bodæ. in Theoph.*[23] Eup[atorium] of Dioscorides & other Greek authors *Lob.* Eup[atorium] of Greek authors & of Avicenna & Eupatoria of Pliny *C.B. Common Agrimony.*

Alectorolophos, *see* Pedicularis. [*Pedicularis sylvatica* 427]

Alga aquatilis capillacea, *see* Conferva Plinii. [unidentifiable filamentous algae 140]

Alleluia, *see* Trifolium acetosum. [*Oxalis acetosella* 578]

---

plantarum or in English *Enquiry into plants*, III. iv. 4 in the Loeb Classical Library edition), he says that this tree produces its fruit "when winter is beginning": cp. footnote 22. The second reference, to Pliny's *Natural history*, probably relates to XVI. 90 in the Loeb edition, but there are several other references to *aquifolium* in Book XVI.

[19] The phrase *per ætatem* (literally "through [its] age") is presumably intended to have the meaning given here, but the passive verb *redditur* is obscure; it could mean "is restored". See also Ray's more detailed note under *Hedera arborea* [270].

[20] *radicalib*⁹: the superscript 9 represents the Latin ending *-us*; see Chapter 7.

[21] Ray's text is clearly derived from Theodorus Gaza's (1529, part 1, p. 29) Latin translation of Theophrastus' *Historia plantarum* (I. ix. 7 in the Loeb edition), though the original text says that whether evergreens ever do this after the rise of Arcturus is a matter for enquiry. Arcturus is a star in the constellation Boötes, the third brightest in the night sky, which becomes visible before dawn in mid September. Theodorus Gaza (Θεοδωρος Γαζης, c. 1400–c. 1475) was a Greek humanist who translated the works of Aristotle and Theophrastus on natural science into Latin.

[22] The chapters of Theophrastus' Περι φυτων αιτιων, usually known by its Latin title *De causis plantarum* ("About the causes of plants"), are numbered differently in modern editions; this refers to I. 17. 6 in the Loeb edition, where he says: Των γαρ αειφυλλων απανθ' (ως ειπειν) οψικαρπα ("For the 'always-leaved' all (so to speak) [are] late-fruiting.")

[23] *Jo. Bodæ.* of Ray's 'Explanation', that is the commentary of Johannes Bodaeus à (or van) Stapel on Theophrastus' *Historia plantarum*, published in Amsterdam in 1644, on p. 887.2 of which he argues that Avicenna's *Eupatorium* is *Agrimonia*. For the 10th- to 11th-century Persian scholar Avicenna (Ibn Sīnā) see Chapter 4.

17 *Alliaria petiolata* (M. Bieb.) Cavara & Grande
Alliaria *Ger. Park. C.B. I.B. Trag. Matth.* Thlaspi cornutum *Lon. Trag.* Pes asini *Pandect. Iack by the hedge, Sawce-alone. Upon the bankes of ditches.*

18 *Allium ursinum* L.
Allium sylvestre latifolium *C.B. Dod.* ursinum *Ger. Fuch. Matth.* ursinum Plinii *Cord. Cæs.* ursinum bifolium vernum sylvaticum *I.B.* ursinum latifolium, sive Moly Hippocraticum[24] *Ad. Lob. Ramsons. At Ditton, on a little woody hillock near the rivers side.*

N. This plant when applied externally raises blisters on the skin (which is common to it with mustard and many other herbs), but in fact when taken internally it does not affect the inward parts in the same way, because these are clothed with tougher and less sensitive skin & besides they are covered with mucous fluid. Also the acid fluid of the stomach & other things with which it is mixed weaken & disperse its strength.

19 *Allium vineale* L.
Allium sylvestre *Park. Ger. emac.* sylvestre tenuifolium *Lob. hort. Bat.* Ophioscorodon *Matth. Crow-garlick. On Iesus Colledge wall, nigh the gate which opens out of the rode into Garlick fair; also about a gravell pit near the foot way leading from Christs Colledge to Cherry-Hinton, and in many other places.*

N. 1. It is certain that milk & especially butter will be infected with the odour & taste of this plant if this kind of garlic is growing mixed with the grass where cows are grazing. *P. Laurembergius*[25] *in Apparat[us] plantar[ius], book 1, chapter 2, also* reports that garlic given to hens & other fowl with their food brings out its taste & odour in their flesh & eggs. Finally *Christoph[er] Bennett*, a London physician, *in Theat. tabidorum exerc. 29*,[26] asserts that weeping ulcers[27] in the limbs of sick people, when lanced or quarterised, in men of thinner build[28] give off the odour of any garlic that they have eaten[29] three or four hours after its ingestion.

---

[24] "or Hippocrates' Moly". Hippocrates of Cos (*c.* 460–370 BC) was an ancient Greek physician often referred to as the father of medicine (see Chapter 4). *Moly* (Greek μωλυ) is described by Homer (*Odyssey*, 10: 302–306) as the magic herb with a black root and a flower like milk that was given to Odysseus by Hermes to nullify the effects of the drug that the sorceress Circe was planning to use to turn him into a pig. The identity of the plant subsequently called moly in the ancient world is a complex question discussed at length by Stannard (1962). Some early modern botanists used the name in a generic sense; for example Gerarde (1597, pp. 143–145) entitled a chapter "Of Moly, or the Sorcerers Garlick", describing five garden species but not including *Allium ursinum*. Later Linnaeus adopted the name as an epithet for the yellow-flowered *A. moly* L.

[25] Peter Lauremberg (Petrus Laurembergius: 1585–1639) was a professor in botany at the Universities of Hamburg and Rostock and published two books in 1632, *Apparatus plantarius* and *Horticultura*, with second editions in 1654. See also Ray's 'Preface' and footnote 48 in it.

[26] Christopher Bennet (1617–1655), an Oxford graduate, took his M.D. degree at Catharine Hall, Cambridge, in 1646 and was elected a Fellow of the College of Physicians in 1649 (Munk 1878; ODNB, vol. 5, p. 99). He was the author of *Theatri Tabidorum vestibulum* (1654), an important early scientific study of consumption, from which he himself died the next year at the age of 38. He also edited Moufet's *Healths improvement* (1655), another work referred to in Ray's notes (see footnote 4).

[27] *fonticulos*, literally "little fountains".

[28] This seems the most likely meaning of *in rarioris texturæ hominibus*. Bennet's own phrase is *in rarioris contexturæ individuis*.

[29] *allii intùs sumpti*, literally "of garlic taken within".

MAIN CATALOGUE

N. 2. About Garlic in general.[30] The juice of Garlic when drunk not only upsets a man but even kills him. *Spigel[ius in] Isag[oges], book 2, chapter 16.*[31] Maggots in his flesh generated from excessive ingestion of garlic ended the life of the Emperor Arnulphus.[32] *Cranz., hist. Sax., book 3, chapter 3.*[33]

3. Garlics smell more strongly when they begin to grow up than when recently planted. *Marant[a], Meth[odus] cogn[oscendorum] simp[licium], book 2, chapter 3.*[34]

20 *Alnus glutinosa* (L.) Gaertn.
Alnus *Ger. Matth. Dod. Turn. Lob.* vulgaris *Park. Clus. hist.* vulgaris, κληθρα [clethra] of Theophrastus *I.B.* conifera *C. Hofman[nus], lib[er] de medicam[entis] officin[alibus].* rotundifolia glutinosa viridis *C.B.* Common Alder. At Fulborn in the moory places, and at Willborham.

N. 1. Let pieces of the wood of this tree of whatever size you wish be put into those large bronze vessels in which hops are roasted to make beer and, when the hops have been fully roasted, let that same wood be taken out & covered with sand or gravel right there in the store-rooms[35] for a period of three years; when it is taken out it will appear to be hardened into stone, from which the whetstones that carpenters use for sharpening their iron tools are chiefly made.[36] Gesner relates this from Master Kentman,[37] *I.B.*

2. The wood of this tree placed under the ground in the foundations of buildings lasts for ever in moist soil. *See Vitruv[ius] where he deals with wooden*

---

[30] *De Allio in genere*. Many of the observations in Ray's three notes on garlic must relate to the cultivated species, *Allium sativum* L.

[31] *Isagoges in rem herbarium* by Adrianus Spigelius (Adriaan van der Spiegel) was published in Padua in 1606; a second edition was published posthumously in Leiden in 1633.

[32] Arnulf (*c*. 850–899), a direct descendant of Charlemagne, was Duke of Carinthia and became King of the East Franks when he deposed his uncle, Charles the Fat, in 887. He was crowned Carolingian Emperor in Rome in 896 but died in Germany after a long illness.

[33] Albert Krantz (*c*. 1448–1517; Ray omits the t in error) occupied a succession of ecclesiatical and academic posts in Rostock and Hamburg, finally becoming Dean of Hamburg Cathedral. This reference is to *Saxonia*, one of four important historical works published posthumously from his papers. It was first published in 1520 but a later edition (Krantz 1580) is cited in our bibliography and, with a 1621 edition, is held in Trinity College library.

[34] The Italian physician and botanist Bartolomeo Maranta's *Methodus cognoscendorum simplicium* ("A method for understanding simples") was published in three volumes in Venice in 1559.

[35] *in ipsis cellariis*, literally "in the store-rooms themselves", "in the very store-rooms".

[36] *è quo potissimum cotes fiunt quibus ad ferramenta acuenda mensarii utuntur*: *cotes* is the plural of the rare word *cos* (meaning "whetstone"); *mensarius* (someone associated with *mensa*, a table or counter) usually means "money-changer" or "banker" and the only other meaning we have encountered, "A Servitour, or he that serveth at the table" (Gouldman 1669), is almost equally unlikely in this context, but Ray (1675, p. 85) writes that some "Trades and Manufactures are denominated from the subject or matter about which they are conversant" and gives 28 examples all ending in *-arius*.

[37] *Recitat Gesnerus ex D. Kentmanno*. *D*. is the standard abbreviation for *Dominus*, used for a Master of Arts. Johannes Kentmann's *Nomenclaturae* (or *Catalogus*) *rerum fossilium*, the first published catalogue of a mineral collection, was one of eight works published in a single volume entitled *De omni rerum fossilium genere, gemmis, lapidibus, metallis, et huiusmodi* in Zürich in 1565 together with Conrad Gesner's *De rerum fossilium, lapidum et gemmarum maximè, figuris et similitudinibus liber*, cited by Ray as *De lapidibus, fossilibus et gemmis* (see *Gesn.* in Ray's 'Explanation'). Kentman's catalogue (pp. 39–40) is the original source of this note, where it is prefaced by the words *Ex alno etiam fit lapis artificiose hoc modo*. ("From alder also stone is made artificially in this way.")

building material. Plin[y], book 16, chapter 37. Theoph. de causis, book 3, chapter 17.[38]

3. Bees for half of the year, namely from Aries to Libra,[39] feed so much on the buds of the alder that cuts & lines gnawed by them may be seen in its expanded leaves. *Butler, in his treatise about bees, chapter 6, number 34.*[40]

21 *Myosoton aquaticum* (L.) Moench
Alsine aquatica major *C.B.* aquatica sive palustris major *Park.* palustris *Tab. Ger. Great Marsh Chickweed, On the banks of the River Cam plentifully, and in the osier holts, &^{c}.*

22 *Veronica agrestis* L.
Alsine foliis Trissaginis[41] *Lob. Ger. Park.* chamædryodes *hort. Bat. among the indigenous plants.* III, sive chamædryfolia flosculis pediculis oblongis insidentibus *C.B.* serrato folio glabro *I.B.* flore cæruleo interdum lacteo aut dilutè purpurascente[42] *Cam. hort. Germander-chickweed.* In ploughed fields.

23 *Veronica arvensis* L.
Alsine foliis veronicæ *Tab. Ger.* foliis subrotundis veronicæ *Park.* II, sive veronicæ foliis flosculis cauliculis adhærentibus[43] *C.B.* serrato folio hirsutiore, floribus & loculis cauliculis adhærentibus[44] *I.B. Speedwell-chickweed.* In ploughed fields & on roofs.

24 *Veronica hederifolia* L.
Alsine hederacea *Tab. Ger.* I, sive hederulæ folio *C.B.* hederulæ folio minor *Park.* hirsuta *Cam.* Alsines genus Fuchsio folio hederulæ hirsuto[45] *I.B.* Morsus gallinæ

---

[38] Vitruvius, who served Julius and Augustus Caesar (see Chapter 4), wrote in his *De architectura*, 2. 9. 10: *Itaque in palustribus locis infra fundamenta ædificiorum palationibus crebre fixa, recipiens in se quod minus habet in corpore liquoris, permanet inmortalis ad æternitatem et sustinet inmania pondera structuræ et sine vitiis conservat.* ("And so in marshy places, being frequently fixed as stakes driven in below the foundations of buildings, absorbing into itself liquid that it has less of in [its own] mass, it [alder] remains immortal for ever and supports huge weights of construction and maintains [them] without defects.") The reference to Pliny's *Natural history* relates to XVI. 219 in the Loeb edition, where he says: *Contra adacta in terram in palustribus alnus aeterna onerisque quantilibet patiens.* ("On the other hand alder driven into the ground in marshy places lasts for ever and bears a load however great.") Chapter 17 of book 3 of Theophrastus' *De causis plantarum*, corresponding to III. 11. 5–6 in the Loeb edition, is about planting slips of trees in wet and dry conditions and so seems only loosely related to this topic.

[39] i.e. from late March to late October.

[40] *The Feminine Monarchie; Or, A treatise concerning bees, and the dve ordering of them* (in the third edition subtitled *or The Histori of Bees. Written out of Experience*) by Charles Butler was first published in 1609, with a second edition in 1623 and a third in 1634 (the last using the author's own more or less phonetic orthography, which it is not possible to reproduce fully here but is reminiscent of the Initial Teaching Alphabet of the 1960s). The statement reported by Ray is based on an addition to paragraph 34 of Chapter 6 on p. 109 in this third edition: "They gather on the Alder-buds half a yeer^e together; [even from *Aries* to *Libra*:] insomuch, that the leavs, when they spred, doo^e ther^{e}by appeer^e ragged."

[41] "with the leaves of *Trissago*", i.e. of Wall Germander, *Teucrium chamaedrys* L., hence some of the other names listed by Ray. See also the entry for *Trisago* (spelled thus) in Ray's 'Etymology'.

[42] "with a blue flower, sometimes milk-white or pale purplish".

[43] "2nd, or with the leaves of speedwell [and] with the florets clasping the stems".

[44] "with a hairier serrate leaf [and] with the florets & capsules clasping the stems".

[45] "A kind of *Alsine* in Fuchs with a hairy leaf of a little ivy".

folio hederulæ *Lob.* Lappago Plinii *Ang.* Als[ine] chamædryodes repens *hort. Bat. Ivy-chickweed, or smal Henbit.* In fields & on their margins, especially among sown crops.

25 *Lamium amplexicaule* L.

Alsine hederula[46] altera *Ger.* hederulæ folio major *Park.* Morsus gallinæ folio hederulæ alter *Lob.* gallinæ perfoliatus *hort. Bat.* Ballote crispa major, an minor? *Lugd.* Oculus cati sive Ballote crispa *hort. Par.* Galeopsis sive urtica iners folio caulem ambiente *I.B.* Lamium folio caulem ambiente majus, an minus? *C.B.* Urticæ fætidæ secunda species[47] *Dod. gal. Great Henbit.* In ploughed fields.

26 *Cerastium fontanum* Baumg.

Alsine hirsuta myosotis *Ad.* spuria 3 *Dod.* hispida 1 *Thal.* Auricula muris[48] quibusdam *Dod. gal.* muris quorundam, flore parvo, vasculo tenui longo *I.B.,* who assigns to the same chapter Auricula muris[48] 1, 2, & 3 of C.B. in his Phytopinax, i.e. Alsine hirsuta arvensis 1, 2, & 3 of the same author in his Pinax. *Mouseare-Chickweed.* Generally.

27 *Stellaria alsine* Grimm

Alsine longifolia uliginosis proveniens locis *I.B.* aquatica media, an palustris minor folio oblongo? *C.B.* fontana *Tab. Ger.* comes close to our plant. *Small long-leaved Water-chickweed.* On the boggy grounds beyond Gamlingay.

28 *Stellaria media* (L.) Vill.

Alsine major *Dod. Ger.* maxima *Lob. ico. Park.* VIII. sive altissima nemorum *C.B. Great Chickweed.* By hedges. The individual flowers of this & the following species consist of five petals so deeply divided to the base that they might seem to be ten to anyone looking at them somewhat carelessly.[49]

29 *Stellaria media* (L.) Vill.

Alsine media *Dod. Ger.* XI. sive media *C.B.* vulgaris sive Morsus gallinæ *I.B.* Als[ine] morsus gallinæ fœmina *Gesn. hort.* Hippia minor *Cord. hist. Middle Chickweed.* By hedges generally.

30 *Arenaria serpyllifolia* L. *sensu lato*

Alsine minor multicaulis *C.B.* minima *I.B. Dod. Lob. Ger. Fine Chickweed.* On walls & rubble and in drier places generally. The flower of this species consists of five undivided petals.

---

[46] i.e. "little ivy".

[47] "The second kind of stinking nettle".

[48] Literally "little ear of a mouse", i.e. "mouse-ear".

[49] See also *Paronychia vulgaris* [*Erophila verna sensu lato* 419], where Ray likens the four deeply divided petals to those of the next plant, *Alsine media.* In *Emendanda* of the 1685 appendix Ray states that this plant is the same as the next, *Alsine media.* Both Babington (1860) and Ewen & Prime (1975) apparently overlooked this entry and identified *Alsine major* as *Myosoton aquaticum* [21], which (even without the evidence of the 1685 appendix) is very unlikely in view of the stated habitat, "By hedges". Raven (1950, p. 88) was more astute and suggested that the entry for *Alsine major* might well refer to "the large form of *Stellaria media* or the kindred species *S. neglecta,* Weihe". These two species were not separated in Ray's time, but there are very few records of *S. neglecta* in Cambridgeshire so we believe that the plant was indeed *S. media,* which is "very variable and abundant" in the county (Perring *et al.* 1964).

31 *Sagina nodosa* (L.) Fenzl
Alsine palustris foliis tenuissimis, sive Saxifraga palustris alsine folia *Ger. emac.* Saxifraga palustris Anglica *Park*. *Small water Saxifrage, upon Teversham and Hinton moors near the watery places.* It bears a white flower which is very large for the size of the small plant.

32 *Minuartia hybrida* (Vill.) Schischk.
Alsine tenuifolia *I.B.* but not found in the works of other botanists. *Narrow-leaved Chickweed. In the cornfields on the borders of Triplow heath, and in divers other places.* We found it flowering in May. While it is in flower it is similar to Linum catharticum in outward appearance.

33 *Althaea officinalis* L.
Althæa *Dod. Lob. Matth.* vulgaris *Park*. I, sive Dioscoridis & Plinii *C.B.* Alth[æa] sive Bismalva *I.B.* Hibiscus of Theophrastus, Alth[æa] of Dioscorides & Hibiscum of Maro[50] *for the same author.* For the Greeks Δενδροµαλαχη & Ἀναδενδροµαλαχη[51] *for the same author.* Al[thæa] Ibiscus *Ger.* Al[thæa] sive Malva palustris *Cord. hist.* Al[thæa] sive Malvaviscus *Ang. Marsh Mallow. In many ditches about Wisbich.*

34 *Asperugo procumbens* L.
Alysson Germanicum echioides *Lob. Lugd. hort. Hafn. hort. Ludg-Bat.* Aparine major Plinii *Ger. Tab.* Borago minor sylvestris *Col. Park.*, who makes more than one mistake[52] in its description. Buglossum sylvestre caulibus procumbentibus *C.B.* Cynoglossa topiaria sive repens *C.B. Phyt.* Cynoglossa fortè topiaria Plinio, Echium lappulatum quibusdam[53] *I.B.*, volume 3, book 33, chapter 7, page 590, & Cynoglossa topiaria fortè Plinio *of the same author in the same book, chapter 10, page 601*, where there is a better description & a better picture, which may be substituted in the previous place. *Small wilde Buglosse, Great Goose-grasse, German Madwort. In the hedge of a lane leading out of the fields to Newmarket, near the church which stands by the Kings house, and in the close adjoyning to the Churchyard.*[54]

Ambrosia campestris repens *C.B. see* Coronopus Ruellii. [*Lepidium coronopus* 153]

Ambrosia tenuifolia *Lob. see* Abrotanum inodorum. [*Artemisia campestris* 2]

Amellus Virgilii,[55] *see* Caltha palustris, Chelidonium minus, & Conyza cærulea acris. [*Caltha palustris* 101, *Ficaria verna* 129 and *Erigeron acris* 144]

---

[50] *Maronis*, i.e. of Virgil, whose full name was Publius Vergilius Maro. One of the poet's two references to *hibiscum* is in *Eclogues*, 10: 71, *dum sedet et gracili fiscellam texit hibisco*, "while he sits and twines a basket of slender mallow"; the species concerned is perhaps *Althaea cannabina* L. rather than *A. officinalis*.

[51] *Dendromalache* ("Tree-mallow") and *Anadendromalache* when transliterated.

[52] *non semel errat*, literally "not [only] once errs".

[53] "Perhaps the Hound's-tongues of ornamental gardening for Pliny, Viper's-bugloss with small burs for some [authors]".

[54] In his contribution to the 1695 edition of *Camden's Britannia*, Ray (1695, column 416) wrote: "*It once grew plentifully at Newmarket*, (vid. Cat. Cant.) *but being an annual plant, I hear it is now lost there: possibly it may agpear* [sic] *again hereafter.*" *Asperugo procumbens* is an introduction in Britain and this may be a more pertinent factor in explaining its failure to persist.

[55] "Virgil's *amellus*", described by the poet in *Georgics*, 4: 271–278, is almost certainly *Aster amellus* L. and is definitely neither of the first two and unlikely to be the third species cross-referenced here, since he describes its flowers as golden but surrounded by violet "leaves" (*foliis*, i.e. by ray florets).

MAIN CATALOGUE

35 *Veronica beccabunga* L.

Anagallis aquatica I, sive major folio subrotundo *C.B.* aquat[ica] folio rotundiore major *I.B.* aquat[ica] folio maximo *Thal.* aquat[ica] sive Becabunga major *Park.* Berula[56] sive Anag[allis] aquat[ica] *Tab.* Anagallidis sive Becabungæ varietas major *Ger. The greater round-leaved Brooklime.* In rivulets.[57]

36 *Veronica beccabunga* L.

Anagallis aquatica II, sive aquat[ica] minor folio subrotundo *C.B.* aquat[ica] *Dod.* aquat[ica] flore cæruleo, folio rotundiore minor *I.B.* aquat[ica] sive Becabunga Germanorum *Ad. Lob. Park.* Anagallidis sive Becabungæ primæ varietas minor[58] *Ger. emac. Common Brooklime.* With the preceding.

37 *Veronica anagallis-aquatica* L. *sensu lato*[59]

Anagallis aquatica III, sive major folio oblongo *C.B.* aquat[ica] major *Ger.* aquat[ica] folio oblongo crenato *Park.* aquat[ica] flore cæruleo, folio oblongo major *I.B.* Berula major *Tab. The greater long-leaved Brooklime.* With the two preceding.

38 *Veronica anagallis-aquatica* L. *sensu lato*[60]

Anagallis aquatica IV, sive minor folio oblongo *C.B.* aquat[ica] flore purpurascente folio oblongo minor *I.B.* Anagallidis aquaticæ majoris varietas minor[61] *Ger. emac.* This has a round stem, closely preceding an angled one. *The lesser long-leaved Brooklime.* All these species grow generally, in marshy places & bubbling springs.

39 *Samolus valerandi* L.

Anagallis aquat[ica] rotundifolia *Ger.* aquat[ica] tertia *Lob.* aquat[ica] VI, sive rotundifolia non crenata *C.B.* aquat[ica] tertia Lobelii folio subrotundo non crenato *Park.* Samolus Plinii quorundam *Valerando.*[62] Samolus Valerandi *I.B. Round leaved water Pimpernell. On Teversham, and Hinton, and Trumpington moors in the ditches, and by the water-courses.*

40 *Veronica scutellata* L.

Anagallis aquatica angustifolia *I.B.* aquat[ica] VIII, sive angustifolia scutellata *C.B. Narrow-leaved Brooklime.* We have long been in doubt whether this may be the

---

[56] See the entry on *Berula* and the footnote on it in Ray's 'Etymology'.

[57] In *Emendanda* of the 1663 and 1685 appendices Ray asked for this larger variety to be expunged.

[58] "A smaller variety of the first *Anagallis* (pimpernel) or *Becabunga*". *Beccabunga* (usually spelled thus) is derived "ultimately from Low German *beckbunge*, from *beck*, stream, and *bunge*, of disputed origin" (Gilbert-Carter 1964); hence *Germanorum* in the previous synonym.

[59] In *Emendanda* of the 1663 and 1685 appendices Ray asked for this species to be expunged. However, the flower colours specified in the Bauhin synonyms for this and the following species suggest that he might have known both the segregate species recognised today, *V. anagallis-aquatica sensu stricto* (with blue flowers) and *V. catenata* Pennell (with pink flowers). The matter is complicated by the occasional presence of a blue-flowered hybrid, and by the fact that some authorities do not regard flower colour as a reliable guide to the identification of the species in Cambridgeshire (Preston 2008).

[60] This may be *Veronica catenata* Pennell, as Ewen & Prime (1975) suggested (see footnote 59).

[61] "A smaller variety of the greater water *Anagallis* (pimpernel)".

[62] "Pliny's *Samolus* of some [authors] according to *Valerandus*". The reference to Pliny relates to XXIV. 104 of his *Natural history* in the Loeb edition, where he describes *Samolus* as a herb growing in moist places.

same as de L'Obel's fourth Anagallis aquatica, but his picture does not correspond to it, for our plant does not bear its flowers & seeds at the apex of branchlets, but the spikes of flowers burst forth from the axils of the leaves, nor are the seed-vessels round as they are portrayed in de L'Obel's picture. However, these things notwithstanding, perhaps it may be the same, for both in this species & especially in Anagallis longifolia we have observed that the seed-vessels often degenerate to a round shape.[63] *Found on Teversham moor near Fulborn, and in a close, compassed about with a great ditch and hedge in the open field within a quarter of a mile of Barnwell.*

41 *Anagallis arvensis* L.[64]

Anagallis fœmina *Matth. Dod. Ger. Trag. Fuch.* cærulea foemina *I.B.* terrestris II., sive cæruleo flore *C.B. Park.* annua flore cæruleo minor *hort. Hafn. Female Pimpernell. In the corn on the left hand of the way leading to Histon, a little beyond the first closes.*

42 *Anagallis arvensis* L. subsp. *arvensis*

Anagallis mas *Ger.* terrestris mas *Park. Thal.* flore phœniceo *Park.* flore phœniceo annua *hort. Hafn.* phœnicea mas *I.B.* terrestris I, sive phœniceo flore *C.B.* Corchorus Cratevæ of Theophrastus & Nicander *Ang.* Magia of Marcellus Empiricus *Io. Bodæ. in Theophr. hist.* Corcorus of Theophrastus, Nicander and Pliny *for the same author. Male red Pimpernell.*

43 *Lithospermum arvense* L.

Anchusa degener facie milii solis[65] *Lob. Ger. emac.* arvensis minor *Tab.* Lithospermum nigrum flore albo, semine Echii *I.B.* VII, sive arvense radice rubra *C.B.* arvense radice rubente *Park. Bastard Gromwell, or Salfern. In fields among standing corn, & elsewhere.*

Androsæmum hirsutum, *see* Ascyrum [*Hypericum tetrapterum* 65]

44 *Anemone nemorosa* L.

Anemone nemorum *Ger.* sylvestris V, sive nemorosa flore majore *C.B.* Ranunculus nemorosus *Lob.* nemorosus albus simplex *Park.* albus, sylvis & umbrosis gaudens, ac Aprili florens *Turn.* Ran[unculus] phragmites albus & purpureus

---

[63] This is a gall caused by the larvae of a weevil, *Gymnetron villosulum* Gyllenhal. In addition to *V. scutellata*, it parasitises *V. anagallis-aquatica sensu lato* [37, 38] and *V. beccabunga* [36] (Redfern & Shirley 2002). The name *Anagallis longifolia* does not appear elsewhere in the *Catalogus*, nor is it listed in Bauhin's *Pinax* (1623), but Ray must be referring to one or both of the *Anagallis aquatica* species for which he includes "long-leaved" in his English names, i.e. *Veronica anagallis-aquatica sensu lato* [37, 38]. In Cambridgeshire the gall is indeed particularly conspicuous on the plants in this aggregate, as Ray suggests.

[64] Ray (1670, 1677) retained separate entries for this blue-flowered plant and the following, more familiar red-flowered taxon, and by 1677 he knew no more British localities for the blue-flowered plant than the one cited here. In *Historia plantarum* (1688a, p. 1024) he still kept them separate, though noting that they differed only in flower colour; by then he was able to report that in Italy and France the blue-flowered plant was obvious everywhere and no less frequent than the scarlet-flowered one. There are now known to be two blue-flowered variants in Britain, subsp. *foemina* (Mill.) Schinz & Thell. and subsp. *arvensis* forma *azurea* Hyl., but they are more or less indistinguishable macroscopically.

[65] "Fake/Bastard Alkanet with the appearance of millet of the sun", i.e. like *Lithospermum officinale* [339].

vernus *I.B.*[66] *Single white Wood Crowfoot, Wood Anemonie or Wind-flower. In Madingley and Kingston woods plentifully.*

N. It has been observed that all anemones produce single stems, sometimes two or three of them, without branches,[67] unless perhaps it may happen otherwise by a trick of nature to one or other of the branchlets springing forth among the upper leaves surrounding the top of the stem. *C.B. in pin., page 175.*

45 *Angelica sylvestris* L.
Angelica sylvestris *Ger. Park. Dod. Matth.* sylvestris major *C.B.* sylv[estris] magna vulgatior *I.B.* minor aquatica *Tab. Wild Angelica. By the rivers side above the mills,* & generally elsewhere in moist & marshy places.

46 *Aegopodium podagraria* L.
Angelica sylvestris minor *hort. Hafn.* erratica *Thal.* minor sive erratica *C.B.* sylv[estris] repens *I.B.* sylv[estris] communis *Cam.* Herba Gerardi *Dod. Ger.* Podagraria *Ad. Lob.* vulgaris *Park.* Ægopodium *Tab. hort. Wars. Herbe Gerard, Goutwort, or Ashweed.*[68]

47 *Ononis spinosa* L.
Anonis sive Ononis *Matth. Tab.* spinosa flore purpureo *C.B. Park.* An[onis] sive Resta bovis *Ger.* An[onis] sive Resta bovis vulgaris purpurea & alba spinosa *I.B.* German horsemen call it Urinalis or Urinaria because a potion of it can be given to horses that are unable to urinate. *The same author. Rest-harrow, Cammock, petty Whinn. Found with a white flower in Huntington rode, not far from Howes bowling-green.*

N. Before it produces spines it is pickled with salt for food. *Schrod[er], Pharmac[opœa] medico-chym[ica].*

48 *Ononis repens* L.
Anonis non spinosa purpurea *Ger.* non spinosa flore purpureo *Park. I.B.* non spinosa sive mitior prima *Clus.* IV, sive spinis carens purpurea *C.B.* perpetuò lævis *Gesn. Purple Rest-harrow without prickles. In the borders of corn fields, and in the pastures in diverse places.*

Anserina Tragi, *see* Argentina. [*Potentilla anserina* 60]

49 *Anthyllis vulneraria* L.
Anthyllis leguminosa *Ger.* legum[inosa] vulgaris *Park.* legum[inosa] Belgarum *Lob.* Lotus latifolia Dalechampii *Lugd.* Loto affinis II, sive L[oto] aff[inis] vulneraria pratensis *C.B.* Vulneraria rustica *Gesn. I.B. Kidney-vetch, Ladies finger. About Gogmagog hills, in the closes about Chesterton, and almost in all chalkie or dry and gravelly pasture grounds.*

50 *Chaenorhinum minus* (L.) Lange
Antirrhinum minimum *I.B. Thal. Cam.* minimum repens *Ger. emac.* VI, sive arvense minus *C.B.* sylvestre minimum *Park.* Os leonis in vineis[69] *Cæs.*

---

[66] The last two names mean literally "White [Buttercup] delighting in woods & shady [places] and flowering in April of [William] Turner" and "Buttercup growing in hedges white & purple of spring of J[ean] B[auhin]".

[67] Literally "endowed with no branches".

[68] The names *Herba Gerardi*, *Podagraria* (from the Greek ποδαγρα [podagra]), "Herbe Gerard" and "Goutwort" all refer to gout, for the treatment of which this herb was formerly much valued. The 10th-century St Gerard of Brogne was invoked to cure the disease.

[69] "Lion's mouth in the vineyards".

Little Snapdragon or Calves-snout. In the corn fields, and in the rodes about Gogmagog hills, and Newmarket heath, and Cherry-Hinton, &c.

Aparine major Plinii, *see* Alysson Germanicum echioides. [*Asperugo procumbens* 34]

51 *Galium aparine* L.

Aparine *Ger. I.B.* vulgaris *Park. C.B.* aspera *Thal.* Lappago of Pliny, book 26, chapter 10.[70] *I.B.* Lap[pago] aspera *Lob.* Philanthropon of Dioscorides & Pliny *C.B. & I[o]. Bodæ. in Theophr. hist.* Philetærius & Philistion of Hippocrates & Galen *of the same author.* Clivers or Goosegrasse.

Aparine lævis palustris *Thalii, see* Gallium album. [*Galium palustre* 210]

52 *Lathyrus aphaca* L.

Aphaca *Lob. Park. Ger. emac.* flore luteo Dioscoridis, Galeni & Plinii *Cæs.* Vicia quæ Pitine Angullaræ latâ siliquâ flore luteo[71] *I.B.* lutea foliis convolvuli minoris *C.B.* Orobanche leguminum *Lugd. Yellow Vetchling.* Frequent in ploughed fields.

53 *Apium graveolens* L.

Apium palustre seù officinarum *C.B.* vulgare ingratius *I.B.* palustre seu Heleoselinum veteribus *of the same author.* Ἐλειοδρεπτον [Eleiodrepton] in Homer, Il. β'.[72] *of the same author.* Paludapium *Ad. Tab.* Ap[ium] Officinarum sive Paludap[ium] *Park.* Eleoselinum *Lob. Tur. Dod.* Eleos[elinum] seu Paludapium *Ger. Smallage.* In divers ditches about the Town, as at Spittle-house end, &c.

N. Certain old seeds, such as of Apium, Nasturtium etc., grow faster than fresh ones, as Theophrastus also noted.[73] *Cam. hort.*

Apium risûs *Lob. see* Ranunculus aquaticus.[74]

Apollinaris, *see* Hyoscyamus. [*Hyoscyamus niger* 294]

54 *Aquilegia vulgaris* L.

Aquilegia sylvestris *C.B.* vulgaris simplex *Park.* flore simplici *I.B.* flore caeruleo *Ger.* Aquilina *Matth. Ad. Lob. Tab.* Isopyron Dioscoridis *Col. Clus.* Columbina,

---

[70] Pliny's *Natural history*, XXVI. 102 in the Loeb edition, but there is a more convincing account of this species in XXVII. 32: see the entry for *Aparine* in Ray's 'Etymology'.

[71] "The vetch which [is called] *pitine* at Angullara, with a broad pod [and] yellow flower". *Pitine* is one of several Italian names for *Lathyrus aphaca*. Angullara Sabazia is a town in the Campagna Romana 30 km from Rome.

[72] The reference is to Homer's *Iliad*, 2: 776, where the horses of Achilles' men stand idle, "eating lotus and parsley of the marsh", because of their leader's wrath against Agamemnon; but the correct reading is ἑλεοθρεπτον τε σελινον (*eleothrepton te selinon*, literally "and marsh-grown parsley"). *Heleoselinum* or *Eleoselinum* (from Greek) and *Paludapium* (the Latin equivalent) also mean "marsh-parsley".

[73] *Historia plantarum*, VII. 1. 6 (in the Loeb edition). Theophrastus lists σελινον [selinon = parsley or, according to some, celery], τευτλιον [teutlion = beet], καρδαμον [cardamon = cress], θυμβρα [thymbra = savory], κοριαννον [coriannon = coriander] and ὀριγανον [origanon = marjoram]. Most of these are not native in Britain and Ray mentions only two indigenous plants, celery, *Apium graveolens* [53], and water-cress, *Nasturtium officinale sensu lato* [387], perhaps interpreting Theophrastus' καρδαμον as the latter.

[74] Ray does not cite the name *Ranunculus aquaticus* in the synonyms of any of his chosen names. Examination of L'Obel (1576, p. 382) shows that *Apium risûs* is a synonym of Mathias de L'Obel's *Ranunculus palustris rotundiore folio*, which Ray calls *Ranunculus palustris rotundifolius* Ger. [*Ranunculus sceleratus* 478]. It is thus a different species from Ray's *Ranunculus aquatilis* [479]. See also footnotes 558 and 611.

because the parts of the flower recall the curved neck & beak of a dove. *Io. Bod[æ]. in Theoph[r]. hist.*[75] Columbines. *In a little thicket at the hither end of Teversham moore, and in other places.*

N. The leaves of this plant swarm with lice in the month of June because of its exceptional sweetness, as Gesner says. *Moufet., De Insect[is], book 2, chapter 23.*[76]

55 *Vicia cracca* L.
Aracus *Tab.* Galega sylvestris Dod[onæ]i *Thal.* Vicia multiflora *C.B.* sylvestris multiflora sive spicata *Park.* multiflora nemorensis perennis seu dumetorum, *I.B. Tufted Vetches. Birds-tares.* Generally in meadows & hedges.

56 *Vicia sativa* L. subsp. *nigra* (L.) Ehrh. and/or subsp. *segetalis* (Thuill.) Gaudin[77]
Aracus sive Cracca major *Lob. Ger. Park.* Vicia semine rotundo nigro *C.B.* vulgaris sylvestris semine parvo & nigro frugum *I.B. Strangle-tare or wild Vetch.*

57 *Vicia hirsuta* (L.) Gray
Aracus sive Cracca minor *Park.* Ar[acus] sive Crac[ca] minima *Ad. Lob.* Vicia sylvestris sive Cracca minima *Ger. emac.* parva sive Crac[ca] minor cum multis siliquis hirsutis *I.B.* segetum cum siliquis plurimis hirsutis *C.B. Small wild Tare, Tinetare.* Frequent among standing corn.[78]

58 *Papaver argemone* L.
Argemone capitulo longiore *C.B. Lob. Ger. Park.* capitulo longiore spinoso *I.B.* Papaver erraticum alterum *Cæs.* rhœas capitulo seminis hirsuto longiore *hort. Bat. Long rough-headed bastard Poppy.*

59 *Papaver hybridum* L.
Argemone capitulo rotundiore *Park.* capit[ulo] breviore *C.B.* capit[ulo] breviore hispido *I.B.* capit[ulo] torulo *Ger.* capit[ulo] torulis canulato *Lob.* Papaver Rhœas capitulo seminis hirsuto breviore *hort. Bat. Round rough-headed bastard Poppy. Both common among the corn in many places, as near the old chappell at Sturbridge faire place.*

60 *Potentilla anserina* L.
Argentina *Dod. Lob. Ger.* Potentilla *Park. C.B. Matt[h].* Potentilla seu Argentina *I.B.* Anserina *Trag.* Tanacetum agreste *Brunf. Wild Tansy, Silverweed.*

---

[75] Johannes Bodaeus' (1644) commentary on Theophrastus' *Historia plantarum*, in a passage about Columbine in column 2 on p. 717 (misprinted as 517) explaining that it is named *Aquilegia* because the flowers have points like eagles' talons and *Columbina* for the reason given by Ray. See also the entry for *Aquilegia* in Ray's 'Etymology'.

[76] *Insectorum sive minimorum animalium theatrum*, compiled by Thomas Moufet (or Muffet, Moffett or Moffet, 1553–1604) and published posthumously in London in 1634, was the first book on insects published in England; it remained the standard work until replaced by Ray's own posthumous *Historia insectorum* in 1710 (Salmon 2000; Thomson 2000). The cited Chapter is entitled 'De Pediculis Brutorum & Plantarum' ("About lice of beasts & plants"); the latter are presumably greenfly. See also footnote 4 and Chapter 4.

[77] Ray lists the cultivated *Vicia sativa* separately as *Vicia Ger.* [615]. The footnote to that entry discusses the problem of allocating subsp. *segetalis* to one or other of these taxa.

[78] Ray accidentally included this species twice in the *Catalogus*. See his comments under *Vicia parva ...* in *Emendanda* of the 1663 and 1685 appendices and under *Aracus sive Cracca minor* in the latter.

61 *Silene flos-cuculi* (L.) Clairv.

Armerius sylvestris *Dod.* pratensis *Ger.* Caryophyllus simplex laciniatus VII, i.e. Car[yophyllus] pratensis flore laciniato simplici, sive Flos cuculi *C.B.* Flos cuculi pratensis *Trag.* Flos cuculi, Odontis quibusdam *I.B.* Odontitis Plinii *Gesn.* simplici flore *Clus. hist. Cuckcw-flower*,[79] *Wild Williams, Meadow-pinke*. Frequent in moist meadows. It flowers in May & June.

62 *Artemisia vulgaris* L.

Artemisia *Trag. Ang. Matth. Cord. in Diosc.* vulgaris *I.B. Park. Tur. Clus. hist. Thal.* vulgaris major *C.B.* Art[emisia] mater herbarum *Lob.* Art[emisia] mater herbarum 1 & 2 *Ger.* Art[emisia] rubra & alba *Tab. hort. Wars.*; for 1 & 2 *Ger.* or rubra & alba *Tab.* do not differ otherwise than in the colour of the stem & flowers, on the evidence of *I.B., C.B.* & *Johnson* in his revision of Gerarde.[80] *Mugwort*. In fields & on their margins, in various places.

N. 1. It is said that a traveller who has Artemisia fastened to him does not experience tiredness. *Plin[y], book 26, chapter 15*.[81]

2. There are those who recommend for epilepsy the charcoal dug up from under its root on John the Baptist's day.[82] Moreover they both provide it in powder form and hang it from the neck. *Schroder[i] Pharmac[opœa], book 4, chapter 6*. Concerning this charcoal, see *Fernelius in his advice about an epileptic boy, Io[annes] Bodæus in Theophrast[us'] Hist[oria], book 9, chapter 18, & J[ean] Bauhin*.[83]

Artemisia tenuifolia, *see* Abrotanum campestre. [*Artemisia campestris* 1]

---

[79] A misprint for *Cuckow-flower*.

[80] "*testibus I.Bº. C.Bº. & Iohnsono* apud Gerardum", i.e. "with, [as] witnesses, J. Bauhin, C. Bauhin & Johnson in the work of Gerarde". Thomas Johnson (1633, p. 1103) added to Gerarde's (1597, p. 945) account of his "second kinde": "I know not how this differeth from the former, but only in the colour of the stalk and floures, which are red or purplish; whereas the former is more whitish." Much importance was formerly attached to the red-stemmed variants of this and other species used medicinally (Cronk 1989).

[81] Pliny's *Natural history*, XXVI. 150 in the Loeb edition.

[82] Midsummer Day, 24 June. For other beliefs relating to Midsummer Eve and Midsummer Day see footnotes 281, 379 and 721.

[83] Jean François Fernel (*c*. 1497–1558), latinised as Ioannis or Joannis Fernelius, was the leading French physician of the 16th century (DSB, vol. 4, pp. 584–586). He published a number of influential works during his lifetime, but *Consiliorum medicinalium liber* was compiled from his papers and published posthumously, initially in Paris in 1582 (Sherrington 1946). The arrangement of material in this work differs from edition to edition and in the editions that we have seen the passage about *Artemisia* does not appear in the advice (*Consilium*) about the epileptic boy; thus, in the edition cited in the bibliography (Fernel 1644), which is held in the library of Trinity College, Cambridge, the *Artemisia* text appears on p. 12 in *Consilium I* (*Epileptico præscriptum*), which is sometimes printed independently of the *Consiliorum medicinalium liber*, whereas *Consilium IX* concerns the epileptic boy (*Pro puero octo annos nato Epilepsia jamdiu correpto, cum abolita fere memoria & ratiocinatione*), and is on pp. 21–22. Bodaeus (1644, p. 1152.1; see footnote 23) quotes from Fernel about enclosing the charcoal in linen and hanging it around the neck. At first sight it is surprising that Ray includes these beliefs in view of his promise in his 'Preface' to add "Observations" only when confirmed by his "own experience or when depending on the credibility & witness of the weightiest authors or when in themselves sufficiently credible and probable". However his reference to Jean Bauhin may have been intended to draw the reader's attention to the question posed by Bauhin & Cherler (1651, vol. 3(1), p. 189): *Sed quis veræ religionis sincerus cultor, his & similibus superstitionibus fidem haberet, quibus seductor ille spiritus, Cacodæmon, nobiles Dei creaturas consparcatas*

MAIN CATALOGUE

63 *Arum maculatum* L.
Arum *I.B.* vulgare *Ger.* vulgare maculatum & non maculatum *Park.* Arum II, & III, sive maculatum maculis candidis vel nigris, & vulgare non maculatum *C.B.* Dracunculus minor Dioscoridis *Fuch.* Dracontium parvum Dioscoridis *Cæs. Wake-Robin, Cuckow-pint.* By hedges.

N. According to *Theophr[astus] & Dioscorides* the roots & leaves of Arum formerly provided a pleasant enough food, but they were cooked in vinegar first to remove their pungency. *Theophr. 7. hist. 11*.[84]

64 *Phragmites australis* (Cav.) Trin. ex Steud.
Arundo vallatoria *Ger. Lob.* palustris *Matth.* vulgaris palustris *I.B.* inodora I, that is vulgaris or phragmites of Dioscorides, καλαμος χαρακια[85] of Theophrastus *C.B.* Harundo vulgaris sive vallatoria *Park. Common Reed.*

N. 1. In the reed & all reedy plants the roots are large. *Theophr. hist. book 1, chapter 10*.[86]

2. Anyone who wants to free a reed from wetness should harvest it in the water when the moon is waning; it will become completely dry within three years. *P. Lauremberg, Horticult[ura], book 1, chapter 30, number 4*.[87]

3. As to how a reed-pipe was made from reeds in ancient times & how reeds were prepared for that operation, see within *Theophrast[us'] Hist[oria], book 4, chapter 12*.[88]

65 *Hypericum tetrapterum* Fr.
Ascyron *Ger.* vulgare *Park.* Hypericum Ascyron dictum caule quadrangulo[89] *I.B.* non perforatum Ascyron dictum *hort. Hafn.* Ascyron dictum IV, sive Androsæmon hirsutum *C.B.* Hyp[ericum] in dumetis nascens[90] *Trag. Ordinary S^t. Peters-wort. Neare water courses.*

---

decipere, deceptas tandem in ignem æternum abducere solet. ("But what sincere adherent of true religion would have faith in these and similar superstitions, with which that seductive spirit, Cacodaemon [the Evil Genius, a term used for Socrates in one of Aristophanes' plays but here clearly the Devil], is wont to deceive God's noble baptised [literally "besprinkled"] creatures [and] finally, having deceived [them], to lead [them] away into everlasting fire?")

[84] VII. xii. 2 of Theophrastus' *Historia plantarum* in the Loeb edition.

[85] This can be transliterated as *calamos characia*, but χαρακιας [characias], meaning "*of or fit for a stake, pale or palisade*" (Liddell, Scott & Jones 1940), is the correct form, as also in Pliny. Theophrastus wrote in his *Historia plantarum* (IV. xi. 1 in the Loeb edition) that there were two forms of the species of reed not used for making pipes (probably *Phragmites australis* (Cav.) Trin. ex Steud.), a strong one, χαρακιας, and a slender weak one, πλοκιμος [plocimos], suitable for weaving. See also footnote 502 in Ray's 'Etymology'.

[86] I. vi. 10 in the Loeb edition.

[87] *Qui volet arundine liberare stagnum, is eam demetat intra aquas lunâ decrescente*, literally "Who will wish to liberate a pool from a reed, that [man] may harvest it among the waters with the moon decreasing". For Peter Lauremberg see footnote 25 and see also his note under *Gramen pratense paniculatum minus* [260] for another popular belief from the same source about harvesting according to the phases of the moon.

[88] IV. xi. 2–7 of Theophrastus' *Historia plantarum* in the Loeb edition.

[89] "St John's-wort called *Ascyron* with a square stem".

[90] "St John's-wort springing forth in thickets".

66 *Hypericum elodes* L.
Ascyrum supinum villosum palustre *C.B. Park*. Asc[yrum] 2, sive supinum ελωδης[91] Clusii *Ger. emac*. *Round-leaved S'. Peters-wort*. *In a boggy ground beyond Gamlingay, near Sir Roger Burgoynes park*,[92] *where they dig turfs*.

67 *Galium odoratum* (L.) Scop.
Asperula *Ger. Lob. Lugd.* odorata *Dod. Clus. hist.* II, sive Rubeola montana odora *C.B.* Asperula aut Aspergula odorata *Park*. Hepatica stellata *Tab*. Rubiis accedens Asperula quibusdam, aut Hepatica stellaris[93] *I.B.* Matrisylva *Trag*. *Woodroof*.

Athanasia vulgaris, *see* Tanacetum. [*Tanacetum vulgare* 563]

68 *Atriplex patula* L.
Atriplex sylvestris angustifolia *Park. Ger. emac.* sylv[estris] IX, sive angusto oblongo folio *C.B.* sylv[estris] Polygoni aut Helxines folio *Lob*. sylv[estris] Polygoni folio *Cam*. vulgaris angustifolia cum folliculis[94] *I.B. Narrow-leaved wild Orrache. On rubble & beside roads.*

69 *Chenopodium vulvaria* L.
Atriplex olida *Ger.* olida, sive sylvestris fœtida *Park*. sylv[estris] VIII, sive fœtida *C.B.* pusilla, olida, hircina, vulvaria vocata, garum olens[95] *Lob. ico.* fœtida *I.B.* canina & Blitum fœtidum *Trag*. Garosmum *Dod. Cord. hist.* Conyza tertia Dioscoridis[96] *Cæs*. *Stinking Orrache or Notchweed. Under the wall that joynes to Peter-house Tennis-court, and at the Tennis-court end and backside, and in several other places.*

70 *Chenopodium album* L.
Atriplex sylvestris *I.B.*[97] sylv[estris] vulgaris *Ger. emac.* sylv[estris] vulgatior sinuata major *Park*. sylv[estris] I, sive folio sinuato candicante[98] *C.B.* simeteria major *Thal*. *Common wild Orrache. On footpaths*[99] *& rubble*.

N. Annual plants, according to the size of their frame, seek out more fertile ground than perennials or those producing a second growth, and those producing a second growth more fertile ground than evergreens, & these more fertile ground

---

[91] The rough breathing over the initial epsilon of this word (*helodes* when transliterated, meaning "of marshes") is invisible in our copy of Ray's book but can be seen in more heavily inked copies.
[92] See the gazetteer.
[93] "*Asperula* resembling the madders for some [authors], or star-like liverwort".
[94] "with follicles", dry fruits, dehiscent in modern usage, though *Atriplex* has indehiscent fruits; see Ray's 'Interpretation of terms', where *Folliculus* is translated as "A cod or huske".
[95] "called *vulvaria*, smelling [like] fish-sauce". The name *vulvaria* (from *vulva*, a wrapper, covering or womb, probably from the plant's use for gynaecological problems rather than with reference to its fruits) was first used for this species by Castore Durante in his *Herbario nuovo* of 1585; it was later adopted as the name of a genus of fungi. *Garum*, also called *liquamen*, was a fish-sauce condiment originating in Greece which was popular in ancient Rome; though the finished product was apparently mild and subtle in flavour, its production created such unpleasant smells that it was relegated to the outskirts of cities.
[96] "Dioscorides' third *Conyza*".
[97] See also Ray's remarks about this species following his comments on *Atriplex sylvestris altera* [71] in *Emendanda* of the 1685 appendix.
[98] "or with a sinuous leaf becoming pure white".
[99] *In simetis*. The standard Latin word for a narrow way, footpath or lane is *semita*; we have been unable to trace the variant *simeta* elsewhere, but it must be the basis of the epithet *simeterius, -a, -um* in the synonym cited for this species.

than those plants of which the stems or trunks but not the leaves withstand the winter.[100] Consequently you will find few annual plants in mountains and barren places but very many among standing corn in more fertile fields & on footpaths. Compare *Theophr. de causis, book 2, chapters 5 and 6.*[101]

71 *Atriplex prostrata* Boucher ex DC.[102]

Atriplex sylvestris altera *C.B. Ger. emac.* sylv[estris] vulgatior sinuata minor *Park. The lesser wild white Orrache.* With the preceding.

72 *Chenopodium murale* L.

Atriplex sylvestris latifolia *Lob.* sylv[estris] III, sive latifolia *C.B.* sylv[estris] latifolia, sive Pes anserinus *Park. Ger. emac.* Atr[iplex] dicta Pes anserinus *I.B.* Pes anserinus *Fuch. Dod. Lon. Lugd. Cam.* Chenopodium primum *Tab.* Solanum 4 venenosum & furiosum *Trag. The same author* reports that learned little old women are accustomed to attribute the death of pigs & their own suffering to its effect.[103] *Goosefoot or Sowbane.* On footpaths.

73 *Chenopodium rubrum* L.

Atriplex sylvestris latifolia altera *Ger. emac.* sylv[estris] IV, sive latifolia acutiore folio *C.B.* Atr[iplex] dicta Pes anserinus alter sive ramosior *I.B.* Blitum 5, & Atriplicis sylvestris genus *Trag. The other Goosefoot-Orrache.* With the preceding.

Avellana sylvestris, *see* Corylus sylvestris. [*Corylus avellana* 154]

74 *Avena sativa* L.

Avena *Matth. Fuch. Dod. Tur.* vulgaris sive alba *C.B.* alba *I.B.* vesca *Lob. Ger. Park.* sativa & I *Trag.* mitior & sativa *Cæs. Common or manured Oates.*

N. The awns of this & the following when dry at maturity are beautifully twisted, & when sprinkled with liquid they untwist prettily. See *Verulam, Hist[oria] nat[uralis], cent[ury] 5, exp[eriment] 494.*[104]

75 *Avena sativa* L.

Avena nigra *C.B. I.B. Dod.* sylvestrior nigra *Cæs.* This does not appear in the works of Gerarde or Parkinson, although it is grown more frequently in England than the preceding. *Black Oates.*[105] *[Thomas] Moufet* praises this species as causing less

---

[100] *Plantæ annuæ pro corporis mole terram fæcundiorem annicaulibus seu restibilibus expetunt, restibiles semper virentibus, & hæ iis quarum caules aut caudices sed non folia hyemem ferunt.* This passage is obscure, the sentence construction is elliptical, but it seems likely that the adjective *restibilis* describes plants rather than, as is usual, land, so meaning "producing a second growth, springing up after being cut down", rather than "that is sown or planted each year without a break" (OLD).

[101] Chapters 5 and 6 of book 2 of Theophrastus' *De causis plantarum* correspond to II. 4. 1–12 in the Loeb edition; these are about soils and contain (e.g. in 4. 2 and 4. 10) statements that relate rather loosely to the content of this note.

[102] This identification is supported by Ray's description of the plant in *Emendanda* of the 1685 appendix (q.v.).

[103] Literally "accustomed to call ... from [its] effect".

[104] This is Sir Francis Bacon's posthumous *Sylva Sylvarum: or a naturall historie in ten centuries*. The first edition is dated 1626 (though with an engraved title-page dated 1627) and is written in English. Gibson (1950) lists numerous editions and variants published in the next 60 years. Bacon had been created Baron Verulam of Verulam in 1618 and Viscount St Alban in 1621.

[105] Black oats are named after the colour of the grains, though these are dark brown rather than truly black. Most are variants of *Avena sativa*, which was almost certainly the plant Ray knew in

flatulence & being of greater food value than other species of oats. *Moufet, in his book about diet, written in English.*[106]

Auricula muris, *see* Alsine hirsuta & Pilosella. [*Cerastium fontanum* 26 and *Pilosella officinarum* 437]

76 *Cerastium arvense* L.
Auricula muris pulchro flore albo *I.B. volume 3, part 2, book 29, chapter 80, page 360.* We have not found this in the works of other authors[107] even though it is very frequent with us. *Mouse-eare with a large white flower. Upon the hill of health, on Newmarket heath among the bushes, & in the Devills ditch plentifully, as also on the bankes by London-rode side between Trumpington, and Hawkston, and almost on every dry bank about Cambridge.* It flowers around the end of April & through almost all of May.

B

B Accharis Monspeliensium, *see* Conyza major. [*Inula conyzae* 141]

77 *Ballota nigra* L.
Ballote *Matth. Fuch. Dod. Tur. Cæs. Lob. Lugd. Lac.* Marrubium nigrum *Ger. emac.* nigrum, sive Ballote *I.B.* nigrum fætidum Ballote dictum *Park.* nigrum IV, sive nigrum fætidum, Ballote Dioscoridis *C.B. Stinking Horehound.* By hedges.

Ballote crispa, *see* Alsine hederula altera. [*Lamium amplexicaule* 25]

Baptisecula, *see* Cyanus minor vulgaris. [*Centaurea cyanus* 162]

78 *Barbarea vulgaris* W.T. Aiton
Barbarea *I.B. Dod. Lob. Tab. Ger.* Barb[area] flore simplici *Park.* Herba S. Barbaræ *Fuch. Trag.* Eruca VII, sive lutea latifolia, sive Barbarea *C.B.* Nasturtium hybernum *Thal.* palustre *Gesn. hort.* latifolium hyemale *hort. Gron.* Pseudobunias *Lob. Dod. gal. Winter-cresses or Rocket. On dry banks and walls near water-courses.*

79 *Arctium lappa* L. and/or *A. minus* (Hill) Bernh.[108]
Bardana major *Ger.* vulgaris major *Park.* Personata *Lon. Ad. Lob. Lugd.* Personata sive Lappa major aut Bardana *I.B.* Lappa I, sive major, Arcium Dioscoridis *C.B.* Arcium *Ang. Dod. gal. Cord. in Diosc.* Arction personata *Thal. Great Burredock, Clot-burr.*

Cambridgeshire, but some are variants of *A. strigosa* Schreb., which was formerly cultivated in northern and western Britain (Chater 1993; Findlay 1956; Letts 2000).

[106] i.e. *Healths improvement* (1655): see footnote 4.

[107] Ray appears to have overlooked *Caryophyllus holosteus arvensis hirsutus flore majore* C.B., which he later adopted as the name for this species, and both *Caryophyllus holosteus* Ger. and *holosteus arvensis hirsutus* Park., which he then cited in synonymy (Ray 1696, p. 208).

[108] There is no reason to suppose that Ray had separated these species by 1660, and Babington's attribution of this entry to *Arctium pubens* Bab., a segregate of *A. minus*, appears to be unjustifiably precise. This identification is in fact given with a question mark in Babington's annotated copy of Ray (1660) in the library of the Department of Plant Sciences, University of Cambridge. *A. nemorosum* Lej., the third British species recognised by Stace (2010), is not known from Cambridgeshire. See also footnotes 24 and 25 of the 1685 appendix.

N. The more tender stem of this plant before the Burs are produced, when the outer covering has been removed, may be recommended in salads either raw or cooked. *P. Lauremberg*,[109] *Horticult[ura], book 2, chapter 7, notes 3 & 5.*

2. Even its roots, dug while still tender in the month of April, then pickled with pepper & salt like young green artichokes & eaten, offer healthy nourishment. *Moufet, in his book about diet, written in English.*[106]

Becabunga, *see* Anagallis aquatica. [*Veronica beccabunga* 35, 36]

Bechion, *see* Tussilago. [*Tussilago farfara* 603]

Bedeguar *Officinarum, see* Rosa sylvestris aut canina inodora, & Acanthium. [*Rosa canina* 502 and *Onopordum acanthium* 4]

80 *Silene vulgaris* (Moench) Garcke
Behen album *Officin[arum]. Ger.* Been album officinarum *I.B.* Papaver spumeum *Lob.* spumeum sive Ben album vulgò *Park.* Lychnis sylvestris, quæ Ben album vulgò *C.B.* Herba articularis *Tab.* Polemonium *Lugd. Thal. Spatling Poppy, White-bottle.* Frequent among sown crops.

Bella donna, *see* Solanum lethale. [*Atropa belladonna* 548]

81 *Leucanthemum vulgare* Lam.
Bellis major *I.B. Ger. Dod. Matth. Trag. Fuch. Lob.* major vulgaris sive sylvestris *Park.* I, sive major sylv[estris] caule folioso *C.B.* Bellium majus *Hermolai Barbari.* Buphthalmum majus *Lon.* Consolida media vulnerariorum *Lob.* The great wilde white Daisie, of some Ox-eye. Very common in meadows & pastures.

Bellis lutea major, *see* Chrysanthemum segetum. [*Glebionis segetum* 130]

82 *Bellis perennis* L.
Bellis minor sylvestris simplex *Park.* minor sylvestris spontanea *I.B.* Bellis of Pliny, book 26, chapter 5, & Bellius of the same author, book 21, chapter 8.[110] *Cæs. I.B.* minor 4. sive sylvestris & 5. sive sylvestris altera *Ger.* VII, sive sylvestris minor *C.B. Common wilde Daisies.* In meadows & pastures.

Berberis officinarum, *see* Oxyacantha. [*Berberis vulgaris* 413]

Berula, *see* Anagallis aquatica. [*Veronica beccabunga* 35]

83 *Scrophularia auriculata* L.
Betonica aquatica *Ger. Dod. Lob. Lugd.* aquat[ica] major *Park.* Scrophularia III, sive aquatica major *C.B.* maxima radice fibrosa *I.B.* Terpentaria *D. I. Dilsii* à Terebinthinæ dote simili.[111] *Water Betony.*

N. The odour, sap, shape of the flowers, colour etc. show that this plant possesses the same properties as Figwort.[112] What is more it so much recalls that plant

---

[109] For Peter Lauremberg see footnote 25.
[110] Pliny's *Natural history*, XXVI. 26 and XXI. 49 respectively in the Loeb edition. The description in the former is convincing, but *Bellio* [sic] is described as yellow (*luteus*).
[111] "*Terpentaria* of D. I. Dilsius, from [its] property similar to [that of] the Terebinth (or turpentine-tree, *Pistacia terebinthus* L.)"; the phrase is taken from Bauhin & Cherler (1651, vol. 3(2), p. 422). Both *Scrophularia auriculata* and *Pistachia terebinthus* were used by Paracelsian physicians, e.g. Daniel Sennert (1572–1637), to treat ailments of the eye, but we have been unable to trace Dilsius (presumably a man named Dils and *D.* may again be the standard abbreviation for *Dominus*, used for a Master of Arts).
[112] *Scrophularia*, i.e. *Scrophularia nodosa* 529.

in its outward appearance that they could be distinguished with difficulty other than by their roots & the shape of their leaves.

84 *Betonica officinalis* L.
Betonica *Ger. Trag. Dod. Tur. Lob. Lugd.* I, sive purpurea *C.B.* vulgaris flore purpureo *Park.* vulg[aris] purpurea *I.B.* minor flore purpureo *hort. Hafn. & Wars.* Vetonica *Cord.* Cestron & Psychotrophon of Dioscorides & Pliny *I.B.* Wood Betonie. In wooded areas & thickets.

N. Antonius Musa, the Emperor Augustus' physician, wrote a remarkable book about this plant; as a result it began to have such worth that it was said in an Italian proverb: *Vende la tonica, e compra la Betonica.*[113] The same people, when about to praise someone, say that he has more virtues than Betony. C. *Hofman[nus], De medicam[entis] officinal[ibus].*

2. In passing, let us make some observations about other authors:[114] because Hippocrates wrote a whole volume in praise of Ptisan[115] made from barley; Marcion the Greek about Radish;[116] Cato about Cabbage;[117] the naturalist Phanias about Nettle;[118] Themison about Plantain;[119] King Juba about the herb Euphorbium;[120] Erasistratus about Loosestrife;[121] Diocles about Turnip;[122] Asclepiades about Mayweed;[123] and among more recent authors Jean Du Choul

---

[113] "Sell the tonic and buy Betony." The pun is lost in translation. The treatise *De herba vettonica*, falsely ascribed to Musa, was often coupled with the work of Apuleius Platonicus and survives in several manuscripts (see Chapter 4).

[114] The nine references to works by ancient authors which follow must have been taken from Pliny's *Natural history*, as in most cases the works themselves do not survive. The relevant text from Pliny is translated in the following footnotes, with the appropriate reference to the Loeb edition. For biographical details of the authors and further details of some of the works see Chapter 4.

[115] "Hippocrates, among the most famous in the knowledge of medicine, has devoted a whole book to its praise." (Pliny, XVIII. 75). Ptisan or tisane is a medicinal infusion of sweetened barley water, derived ultimately from Latin *ptisana* or *tisana* (as in Pliny's text), from Greek πτισανη, from πτισσειν ("to crush") (OED). For Hippocrates see footnote 24 and Chapter 4.

[116] Marcion is an error for Moschion. "Also Moschion the Greek wrote a whole volume about the radish." (Pliny, XIX. 87).

[117] "... and Cato has celebrated it no more frugally, whose opinion it is appropriate to expound even more diligently than his [Pythagoras' opinion], so that it may be known what medicine the Roman people used for 600 years." (Pliny, XX. 78). An account of Cato's opinions about cabbages follows in XX. 80–83.

[118] "The naturalist Phanias has celebrated its praises, ..." (Pliny, XXII. 35).

[119] "The physician Themiso too has praised a common plant, the plantain, as if its discoverer, having published a volume about it." (Pliny, XXV. 80).

[120] "In the age of our fathers King Juba also discovered [a plant] which he called euphorbea from the name of his physician. ... But Juba's volume about this plant is still extant and [is] a clear commendation." (Pliny, XXV. 77–78). The classical *euphorbea* or *euphorbeum* was a shrubby member of the genus, probably *Euphorbia resinifera* Berg, discovered on Mount Atlas by Juba II of Mauretania. His physician was called Euphorbus. In Ray's time *Tithymalus* was the usual name for the species later included by Linnaeus in the genus that he chose to call *Euphorbia*.

[121] "Lysimachus too discovered [a plant] which retains its name from him, celebrated by Erasistratus." (Pliny, XXV. 72).

[122] "The turnip too has medicinal power. ... Diocles spoke of them with high praise, even maintaining that sexual desire is stimulated by them, ..." (Pliny, XX. 18–19).

[123] "Mayweed was celebrated with high praise by Asclepiades." (Pliny, XXII. 53). This plant is called *Anthemis* by Pliny, a name not otherwise used by Ray in the main catalogue or in the 1663 appendix even as a synonym; Pliny says it is also called *chamaemelon* (Greek for "ground-apple") because it

has written a history of the varieties of Oak;[124] Martin Blockwitz about Elder;[125] Henry Crolach about the cultivation of Woad & its preparation for dyeing wools.[126] See *Vander Linden de scrip. med.*[127]

3. The root of Betonica purges upwards and downwards in the manner of hellebore. *C. Hofman[nus], De med[icamentis] offic[inalibus].*

Betonica Pauli, *see* Veronica. [*Veronica serpyllifolia* 610]

85 *Betula pubescens* Ehrh.[128]

Betula *Ger. Park. C.B. & of others*. Betulla of Pliny & σημυδα [semyda] of Theophrastus[129] *I.B. The Birch-tree. About Sir Francis Russells house at Chippenham.*[130]

N. 1. At the beginning of spring before the leaves have appeared, it pours forth a sweet sap if wounded, which shepherds afflicted with thirst are very often accustomed to drink in the woods. We too have been refreshed more than once by that fluid, when we were wandering through vast woods herborising,[131] says *Tragus*.

2. On the strength & effectiveness of this fluid in treating the stone of the bladder & kidneys and painful discharge of urine, which indeed experience witnesses that it possesses to a remarkable degree,[132] see *Helmont & Charleton, De lithiasi, section 2, chapter 3, page 179.*[133]

---

has the smell of an apple, which suggests Chamomile, *Chamaemelum nobile* (L.) All., but he then says that there are three kinds.

[124] *Jo. Du-choul de varia Quercûs historia*: This was published in Lyon in 1555 and is thought to have been the first treatise written on oaks.

[125] Martin Blockwitz's *Anatomia sambuci* was published in Leipzig in 1631, with a further edition published in London in 1650; a version in English, *Anatomia sambuci: or, The anatomie of the elder*, was published in London in 1655.

[126] *Isatis Herba: De cultura herbae isatidis* of Henricus Crolachius was published in Zurich in 1563.

[127] Jan Antonides van der Linden's *De scriptis medicis libri duo*, the second enlarged edition of which was published in Amsterdam in 1651. Only the three works of more recent authors are cited in this bibliography of medical texts.

[128] *Betula pubescens* Ehrh. and *B. pendula* Roth were not separated by British botanists until the 19th century. Babington (1860) recorded *B. pubescens* from Chippenham, Gamlingay and Newmarket Heath but *B. pendula* only as a planted tree on Newmarket Heath, so it seems likely that Ray's plant was *B. pubescens*.

[129] See the entry for *Betula* or *Betulla* in Ray's 'Etymology'. In Theophrastus' *Historia plantarum* (III. XIV. 4 in the Loeb edition) the leaves of σημυδα are likened to those of walnut and the wood is said to be used only for making walking-sticks. The Loeb edition identifies the tree as Judas-tree, *Cercis siliquastrum* L., elsewhere (I. XI. 2) apparently called κερκις [cercis: see also footnote 573], but Gaza (1529, part 1, pp. 105–106; see footnote 21) equated it with Pliny's *betula*. It is difficult to see how Theophrastus could have regarded the leaves of either tree as like those of walnut.

[130] See the gazetteer.

[131] *herbarum gratiâ*, literally "for the sake of herbs".

[132] *quam quidem experientiâ teste insignem obtinet*, literally "which indeed, with experience [as] a witness, remarkable it possesses".

[133] This appears to refer to two separate sources. Jean Baptiste van Helmont's *De lithiasi* ("Concerning the disease of the stone [λιθιασις in Greek]") is one of four *Opuscula medica inaudita* by him published in Cologne in 1644, with four further editions published, together with his *Ortus medicinae*, in Amsterdam from 1648 to 1655. English versions of three of van Helmont's works were "translated, illustrated, and ampliated" by the "translator and paraphrast" Walter Charleton and published in London in 1650, but we have not located any translation of this work and neither of the published

3. Birches and other wild trees[134] lopped when the moon has waned sprout again more slowly. *P. Lauremberg*,[135] *Horticult[ura], book 1, chapter 30, note 4.*

86 *Carpinus betulus* L.
Betulus *Lob*. Bet[ulus] sive Carpinus *Ger*. Carpinus *Matth. Dod. Lob. Ad. Tab. Cam*. Ostrys sive Ostrya I, ulmo similis, fructu in umbilicis foliaceis[136] *C.B.* Ostrys sive Ostrya Theophrasti *Park*. Ostrys Theophrasti, Fagulus herbariorum *Clus. hist*. Fagus sepium *I.B. Gesn*. Ulmus Atinia *Lugd*. *Horn-beam tree, Horse-beech, or Horn-beech*. *In the hedges on this side of Abington.*

Bifolium, *see* Ophrys. [*Neottia ovata* 396]

87 *Persicaria bistorta* (L.) Samp.
Bistorta major *Ger*. major rugosioribus foliis *I.B.* II, sive radice minùs intorta *C.B.* major vulgaris *Park*. Colubrina minor *Trag*. *Snakeweed the middle sort. In a bushy close on the South of Whitwell, a farmhouse beyond Coton in the way to Hardwick.* Johnson, in his revision of Gerarde, in the description of this Bistort departs in a different direction from other writers when he attributes to it narrower & flatter leaves than those of Bistort with a more twisted root,[137] to which he attributes broader & more wrinkled leaves: see & compare.[138]

88 *Chenopodium polyspermum* L.
Blitum album minus *C.B. I.B.* album sylvestre minus *Park*. sylvestre album *Gesn*. sylv[estre] minus *Thal*. minus album *Ger. emac*. minus vulgare repens *hort. Gron*. *The lesser wild white Blite. Found in Kingston wood after it had been new felled, near Gamlingay rode side.*

89 *Chenopodium polyspermum* L.[139]
Blitum rubrum minus *I.B. C.B. Ger. emac*. rubrum sylvestre minus *Park*. rubrum supinum *Lob. ico*. *The small wild red Blite*. In various places, *as in some Osier*

---

bibliographies of Charleton's works (Manning 1926; Fleitmann 1986) includes one. The section, chapter and page references are to Charleton's work on the same subject, *Spiritus gorgonicus*, published in Leiden in 1650 and described by DSB as "a Helmontian exercise".

[134] *aliæque arbores sylvestres*: The usual meaning of *sylvestris* in post-classical Latin is "wild" but it may here have its original meaning, "of woods" or "woodland".

[135] For Peter Lauremberg see footnote 25.

[136] "similar to elm, with [its] fruits in leafy clusters (literally navels)".

[137] *...quàm Bistortæ radice magìs intorta.*

[138] The case is not as simple as Ray's words suggest. Gerarde (1597, pp. 322–323) included three species of bistort, *Bistorta major* ("The great Bistort" or "Snakeweede"), *Bistorta minor* ("The small Bistort" or "Small Snakeweede") and *Bistorta Latifolia* ("Broade leafed Snakeweede"). It was to Gerarde's account of this third species that Thomas Johnson (1633, p. 399) mysteriously added this sentence: "It differs from the first onely in that the root is somewhat more twined in, and the leaues broader and more crumpled." Ray implies that Johnson stated that *Bistorta major* had "narrower & flatter leaves", but this is not actually the case: he retained Gerarde's description of them as "long leaues much like Patience [*Rumex patientia* L.], but smaller & more wrinkled or crumpled", adding a new figure, to replace two of Gerarde's three, showing a plant with a twisted root and much broader leaves than those of *Bistorta minor* (*Persicaria vivipara* (L.) Ronse Decr.) but somewhat more similar to Gerarde's figure of his third species than to that of his first.

[139] This was identified as *Amaranthus blitum* L. by Babington (1860) and as *A. albus* L. by Perring *et al*. (1964). Ewen & Prime identify both Ray's *Blitum album minus* and his *Blitum rubrum minus* as *Amaranthus albus*. All *Amaranthus* species are introduced to Britain and the habitats described by Ray are not those in which they would be expected to occur. *A. albus* is a New World species and 1660 would be a very early date for such an introduction to be recorded in the wild; the first American

holts by the river: also in a ditch on the back side of S. Iohns Colledge in a close on the north of the back-gate.

90 *Chenopodium bonus-henricus* L.
Bonus Henricus *I.B. Ger. Tab. Trag. Matth.* Tota bona *Lob. Dod. Thal.* Spinachiæ facie Rumicis 3 genus[140] *Fuch.* Spinachia sylvestris sive Bonus Henricus *hort. Hafn.* Lapathum unctuosum *Park.* sylvestre latifolium I, sive unctuosum *C.B.* Common or English Mercury or All-good.[141]

N. The shoots[142] of this plant, its young stems & even its newest leaves, put into boiling water & cooked for a quarter of an hour, then eaten with butter & salt, provide a pleasant & healthy food not unlike the common asparagus.

Borago sylvestris tenuifolia *I.B. see* Echium alterum. [*Echium vulgare* 172]

Branca ursina Germanica, *see* Sphondylium. [*Heracleum sphondylium* 560]

Bromos sterilis, *see* Festuca. [*Anisantha sterilis* 196]

91 *Bryonia dioica* Jacq.
Bryonia alba *Dod. Lugd. Thal. Ger.* alba vulgaris *Park.* baccis rubris *Cam. hort.* I aspera, sive alba baccis rubris *C.B.* Vitis alba *Matth. Fuch. Lac.* alba sive Bryonia *I.B.* Uva Taminia Apuleii *Iun. nomenc.*[143] White Bryony. By hedges generally.

92 *Tamus communis* L.
Bryonia nigra *Ad. Ger.* sylvestris nigra *Park.* IV lævis, sive nigra racemosa *C.B.* Vitis nigra *Matth.* I *Tab.* nigra quibusdam sive Tamus Plinii folio Cyclamini[144] *I.B.* Melothron Theophrasti *for the same author.* Tamus *Dod.* Sigillum B[eatæ] Mariæ *Officin[arum].*[145] Black bryony. By hedges & in thickets.

Buglossum sylvestre caulibus procumbentibus *C.B. see* Alysson Germanicum echioides. [*Asperugo procumbens* 34]

---

species reliably recorded as naturalised in Britain was *Anaphalis margaritacea* (L.) Benth., which was well established by 1709. *Amaranthus* species were not recorded in the wild in Britain until the middle of the 19th century (Preston, Pearman & Dines 2002). In *Emendanda* of the 1663 and 1685 appendices Ray concluded that *Blitum album minus* was a shade form of "the common red Blite" and there seems to be no reason to question his later view that this and the preceding plant were the same species; we therefore list both as *Chenopodium polyspermum*.

[140] "The third kind of Dock with the appearance of Spinach."

[141] Interestingly, Ray does not give "Good King Henry" as one of this plant's English names, even though Gerarde (1597) and Johnson (1633) say that "in Cambridgeshire it is called Good king Harry"; but Ray's text was written during the Cromwellian Protectorate. The name *Bonus Henricus* in fact derives from a German goblin, Heinrich or Heinz, rather than from any English king (Oswald 1992).

[142] *asparagi*, the plural of *asparagus*, a word that in Ray's time and earlier was not used exclusively for the vegetable of that name; see also footnote 260 and Ray's third note on *Lupulus* [342] (including footnotes 429 and 430).

[143] "Apuleius' taminian grape": The *taminia uva* of Pliny (*Natural history*, XXIII. 17, etc.) and other Roman authors was "The fruit of black bryony" and *tamnus* (*Natural history*, XXI. 86) was probably "Black bryony, *Tamus communis*" (OLD). The author referred to here, who evidently interpreted the name differently, is presumably the Apuleius Platonicus discussed under *Apul.* in Ray's 'Explanation'. *Iun. nomenc.* is *Nomenclator* of Hadrianus or Adrianus Junius, of which there were several editions (see Chapter 4).

[144] "Black [Bryony] for some [authors] or Pliny's *Tamus* with the leaf of a Cyclamen".

[145] "The Blessed [Virgin] Mary's image *of the Pharmacies*". See also Ray's note 3 on *Lupulus* [342].

93 *Anchusa arvensis* (L.) M. Bieb.
   Buglossa sylv[estris] minor *Ger. emac*. Buglossum sylvestre II, sive sylvestre minus *C.B.* sylv[estre] minus *Park*. erraticum asperum echioides *Ad*. Echium Fuchsii, sive Borago sylvestris *I.B. Small wild Buglosse. On dry banks, and in the borders of plowed fields in many places.*

94 *Helminthotheca echioides* (L.) Holub
   Buglossum luteum *Ger. Cam*. echioides luteum hieracio cognatum *Lob*. Hieracium capitulis cardui benedicti maximum, Buglossum echioides quibusdam *I.B.* Hieracium montanum hirsutum I, sive echioides capitulis cardui benedicti *C.B.* Lingua bovis *Park. Lang de boeuf.*[146] *On the margins of fields.*

95 *Ajuga reptans* L.
   Bugula *Ger. Dod. Lob. Clus*. vulgaris *Park*. Consolida media *Matth. Fuch. Tab.* media pratensis cærulea *C.B.* media, quibusdam Bugula *J.B.* Prunella cærulea *Trag. Bugle. Very frequent in moist meadows.* Bugula is called δισκαυλις [discaulis][147] by *P. Renealmus*[148] because it produces stems of two kinds, creeping ones, evidently almost terete, & erect ones which bear the flowers, which are square in cross-section.[149]

   Bulbocodion, *see* Narcissus. [*Narcissus pseudonarcissus* 386]

96 *Conopodium majus* (Gouan) Loret
   Bulbocastanum *J.B.* minus *C.B. Ger. emac*. Nucula terrestris minor *Park. The lesser Earthnut. In the pastures and Broom fields about Gamlingay, and in Sir Roger Burgoynes park there plentifully.* It flowers around the end of May.

   Bunias sylvestris, *see* Napus sylvestris. [*Brassica napus* and *B. rapa* 385]

97 *Capsella bursa-pastoris* (L.) Medik.
   Bursa pastoris *J.B. Ger*. past[oris] major *Trag*. past[oris] major vulgaris *Park*. past[oris] foliis erucæ *Cæs*. past[oris] major folio sinuato *C.B.* Thlaspi fatuum *Gesn. hort*. Thlaspi species apud antiquos *Cæs*. Herba Cancri *Cast. Shepherdspurse, Pickpurse, or Caseweed.* This plant plays extraordinary tricks[150] in its size & the incision of its leaves so that anyone might suspect that there are different species *J.B.*

98 *Teesdalia nudicaulis* (L.) W.T. Aiton
   Bursa pastoris minor *Park. Tab*. minima *Ger*. foliis sinuatis III, sive B[ursa] p[astoris] media, & IV, sive minor foliis incisis & Thalii *C.B.* B[ursa] p[astoris] parva folio glabro spisso *J.B.*[151] *The lesser Shepherds purse. In a sandy lay near the windmills beyond Gamlingay toward Sandy; and in many other sandy grounds thereabouts.*

   Butomos Theophrasti, *see* Gladiolus palustris. [*Butomus umbellatus* 225]

---

[146] i.e. "Oxtongue", from the shape and wartiness of the leaves; this is also the meaning of *Buglossum* (from the Greek βουγλωσσον or βουγλωσσος, probably, like our bugloss, a boraginaceous plant).

[147] This is the word that Ray presumably intended, meaning "two-stemmed", but it looks like διοσοκαυλις (diosocaulis) though poorly printed.

[148] Paul de Reneaulme's *Specimen historiae plantarum* was published in Paris in 1611: see *P. Renealm*. in Ray's 'Explanation'.

[149] See also Ray's note on *Sanicula sive Diapensia* [*Sanicula europaea* 520].

[150] *mirè ludit*, literally "plays marvellously".

[151] See also Ray's further comments about the synonymy of this species in *Emendanda* of the 1663 and 1685 appendices.

MAIN CATALOGUE

C

99 *Mentha arvensis* L.
Calamintha aquatica *Ger. emac.* arvensis verticillata *C.B.* arvensis verticillata sive aquatica Belgarum *Lobelio Park.* Mentha arvensis verticillata hirsuta *J.B.* Water or Field Calamint with whorled coronets. Where waters have flooded during the winter in various places.

100 *Clinopodium ascendens* (Jord.) Samp.
Calamintha vulgaris *Park.* vulg[aris] officinarum *Ger. emac.* montana vulg[aris] *Lob. Tab.* vulg[aris] vel officinarum Germaniæ *C.B.* flore magno vulgaris *J.B.* Nepeta montana *Cord. in Diosc.* Common Calamint. By the rode side near Linton. On a bank near the hedge in a close by the high way side adjoyning to S. Johns Colledge walks.

 N. It is especially worthy of note that the same plants that grow on the ridges of mountains of warm regions spring up in the plains, forests and well-wooded & low-lying sites of northern regions, no doubt because the colder sites in the warm regions & the warmer ones in the cold regions are in some way congenial to them by the appropriate composition of the atmosphere, *Lob[elius]* in *the preface to Adv[ersaria]*.

Calcitrapa, *see* Carduus stellatus. [*Centaurea calcitrapa* 112]

101 *Caltha palustris* L.
Caltha palustris *J.B.* palust[ris] vulgaris simplex *Park.* IX, sive palustris flore simplici *C.B.* palust[ris] major *Ger.* Chelidonia palustris *Cord.* Tussilago altera *Matth.* Chamæleuce *Ang. Tur.* Chrysanthemum *Lon.* Supposedly Amellus of Virgil[152] *Thal.* Populago *Tab.* Marsh Marigold. In moist meadows.

102 *Erysimum cheiranthoides* L.
Camelina *Ger.* Camelina sive Myagrum alterum amarum *Park.* Myagrum siliquâ longâ *C.B.* Myagro affinis planta siliquis longis *J.B.* Viola lutea sylvestris *Trag.* Erysimum for Galen & Theophrastus *J.B.* Treacle-Wormseed. In the Osier holts about the bridge at Ely abundantly, and in all the other Osier grounds by the rivers side there.

Campanula, *see* Trachelium.[153]

103 *Campanula rotundifolia* L.
Campanula rotundifolia *Ger.* minor rotundifolia *Lob. Lugd.* minor sylvestris rotundifolia *Park.* minor Alpina rotundioribus imis foliis *Clus. hist. hort. Hafn.* minor rotundifolia vulgaris *C.B.* parva Anguillaræ Cantabrica *J.B.* Rapunculus sylvestris minor *Tab.* The lesser round-leaved Bell-flower. In moorish & drier places.[154]

---

[152] *Amellus Virgilii putatus*. See footnote 55.

[153] There are two *Trachelium* species in the *Catalogus*, *Trachelium majus* [*Campanula trachelium* 572] and *Trachelium minus* [*Campanula glomerata* 573].

[154] *In montosis & siccioribus*, strictly "In mountainous & drier [places]"; but of course the former do not exist in Cambridgeshire, while in his English habitat descriptions Ray frequently refers to "*moors*" or "*moores*" and, in one case each, to "*the moory places*" at Fulborne and "*a moorish place*" near Gamlingay. There is some evidence for this usage in mediaeval and later Latin, as P. Levins' English–Latin dictionary *Manipulus Vocabulorum* (1578) translates "A MOORE, heath" as "montana, orum" (Wheatley 1867). (This remarkable dictionary arranges words in order of the last syllable, "being a

104 *Cannabis sativa* L.
Cannabis prima sive sativa *C.B.* mas & fæmina *I.B. Ger. emac.* sativa mas & fæmina *Park.* mas fertilis & fæmina sterilis *hort. Wars. hort. Bat.* Hempe the male and female, or Winter and Summer Hempe. About Ely it is sown.

N. 1. The male produces only seed, the female only flowers;[155] both are produced from the same seed, but no one can tell (says *Jean Bauhin*) in what way it may come about[156] that one is fruitful and the other sterile, and so this is to be classified among the miracles of nature. The same thing is seen in Mercury, in Dog's Mercury and also in the wild Hops that grow in willow-beds.[157]

2. More observant farmers have noticed that a greater quantity of male hemp comes up in richer soil but of the female in more barren soil and also that, where hemp grows too thickly so that the seeds deprive one another of nourishment,[158] the female grows abundantly and, where it grows more sparsely, the male; and on that account it is very probable that the male avoids the one from no other cause than because of the abundance of nourishment and the female the other[159] because of a want of it, because this plant, like the hop also, is very hungry for nourishment and demands a rich soil & one frequently restored with dung so that it may mature & flourish successfully.

3. Lake water in which hemp has rotted is so noxious[160] that when drunk it is a very powerful poison to those who are thirsty; on which subject see the notable story told by *Lob[elius] in Adv[ersaria].*[161]

---

trade not of any man afore attempted, or by the other Dictionaries", so that "moore" is grouped with "floore", "stoore" and "snoore".)

[155] Early authors believed that the seed-producing female plants of this and the other species listed here were male and *vice versa*.

[156] *qua ratione fit*, literally "by what method it may happen".

[157] *Idem observatur in Mercuriali, in Cynocrambe, atque in Lupulis salictariis. Mercurialis* is Annual Mercury, *Mercurialis annua* L., at least the female plant of which William Turner found growing in the garden of King's Hall (incorporated by Henry VIII in 1547 into Ray's college, Trinity) but which was first recorded wild in Cambridgeshire in the 1685 appendix [B64]. Turner (1538, folio B.iij. *recto*) wrote: *Mas haud temere apud nos reperitur. uidi tamen cantabrigie in horto aulæ regiæ, unde radice[m] in nostru[m] transtulimus, hortum.* ("The male [actually the female] is not easily found with us; however I saw [it] in Cambridge in the garden of King's Hall, whence we transplanted a root into our garden.") See also Ray's entries for *Cynocrambe mas & fæmina* and *Mercurialis montana spicata* [163]. Pliny described a wild hop growing in willow-beds in his *Natural history* (XXI. 86 in the Loeb edition), calling it *lupus salictarius. Lupus* primarily means a wolf but was also used as a name for the wolf fish, a spider and the hop, later more usually called *lupulus* (the diminutive, as in the scientific name *Humulus lupulus* L.).

[158] *ità ut grana sese mutuò alimento defraudent*, literally "thus that the seeds mutually from nourishment may defraud themselves".

[159] *hanc ... illam*, translated here as "the one ... the other" but strictly "the latter ... the former"; Ray, however, seems to have them the wrong way round.

[160] *tantopere viru nocet*, literally "so much with poison injures". Rope- and cloth-making required the retting of hemp (soaking in water for a period of about 20 days to loosen the fibres from the other components of the stem), sometimes in specially constructed ponds but sometimes in natural waterbodies. In Ray's time much *Cannabis* was grown on the light soils in the Waveney valley, Norfolk, especially around Diss. See Godwin (1967) and Peglar (1993).

[161] This occurs at the end of Pena & L'Obel's (1571) account of *Cannabis*, on pp. 226–227, where, after the statement quoted by Ray, they write: *... nullis enim antidotis quantumuis tempestiuè datis occurri potuit fato Marquisii Arboii ... eiúsq; vxori & quamplurimis generosis, qui ... imprudenter*

4. Some of the more recent authors write that female hemp causes hens to be more productive of eggs. Moreover I have learned from experience that, if hens are fed on hemp seed in a hard winter (when they produce eggs more rarely), they lay eggs much more frequently, but afterwards they are rendered sterile as if overwhelmed by the quantity of rich food. Small birds also, if it is provided for them in their food, are so fattened by it that it either kills them or takes away their eagerness to sing. *Io. Bodæ. in Theoph[r]. hist.*, book 8, chapter 7, page 945.[162]

105 *Galeopsis tetrahit* L. *sensu lato*[163]

Cannabis spuria *Park.* Ger. spuria facie urticæ *Lob. hort. Wars.* sylvestris quorundam Urticæ inerti affinis *I.B.* sylvestris *in Tragus* because of the shape of the seed. Urtica fatua aculeata I, sive aculeata foliis serratis *J.B.*[164] *Wild Hempe.* It varies in the colour of the flower. In ploughed fields & at the margins of fields.

Cannabina aquatica, *see* Eupatorium cannabinum.[165]

Capnos, *see* Fumaria. [*Fumaria officinalis* 204]

Caput gallinaceum Belgarum, *see* Onobrychis. [*Onobrychis viciifolia* 394]

Caprifolium, *see* Periclymenum. [*Lonicera periclymenum* 432]

106 *Cardamine pratensis* L.

Cardamine *Ger.* altera *Lob.* Nasturtium pratense vel tertium *Trag.* pratense majus siue Cardamine latifolia *Park.* prat[ense] I, sive prat[ense] magno flore *C.B.* Flos cuculi *Dod.* Sisymbrium aquaticum alterum *Matth.* Iberis Fuchsii, sive Nasturtium pratense sylvestre *I.B. Ladies-smock, Cuckow-flower.* In meadows. It flowers in April and May.

107 *Cirsium acaule* Scop.

Carduus acaulis *Lob.* Chamæleon sive Carlina IV, acaulos minore purpureo flore *C.B.* Carl[ina] ac[aulos] minor purpureo flore *Ger. emac.* Carl[ina] ac[aulos] Septentrionalium *Park.* Carl[ina] minor flore purpurascente *Clus.* Chamæleon parvus *Lon.* exiguus Tragi *I.B.* Septentrionalium exiguus *Col. Dwarfe Carline-thistle. Upon the levell near the new Pesthouses,* and elsewhere generally.

---

fontanam lympidam, sed occultis mœandris & subterraneis cuniculis indidem in fontem illapsam hausissent. ("... for the fate of Marquis Arboius ... and his wife and very many other eminent people could not be averted by any antidotes, albeit promptly administered, when ... they had imprudently drunk from a limpid fountain but from water that had flowed into the fountain from the same [i.e. from a lake contaminated by hemp-retting] through hidden windings and subterranean passages.")

[162] Ray's text is based very closely on Bodaeus' (1644, p. 945.1; see footnote 23) commentary on Theophrastus' *Historia plantarum* (part of VIII. vii in the Loeb edition, but with little relationship to Theophrastus' text), but he corrects a grammatical error.

[163] Two very similar species, *Galeopsis tetrahit* L. and *G. bifida* Boenn., occur in Cambridgeshire.

[164] The four preceding synonyms may be translated as "False [Hemp] with the appearance of nettle", "Wild [Hemp] of some authors related to stingless Nettle", "Wild [Hemp]" and "Foolish prickly Nettle I, or prickly [Nettle] with toothed leaves".

[165] Ray cites *Cannabina aquatica folio tripartitò diviso* as a synonym of *Eupatorium cannabinum fœmina* [*Bidens tripartita* 189] and *Cannabina aquatica folio non diviso* as a synonym of *Eupatorium aquaticum folio integro* [*Bidens cernua* 190] but no such synonym under *Eupatorium cannabinum* itself [*Eupatorium cannabinum* 191].

108 *Silybum marianum* (L.) Gaertn.
Carduus lacteus *Matth. Lugd.* lacteus flore purpureo *hort. Wars.* Mariæ *Ger. Trag. Fuch.* Mariæ vulgaris *Park.* Marianus sive lacteis maculis notatus *I.B.* Silybum *Ang. Ad. Lob.* Silybum sive Carduus lacteus I, i.e. albis maculis notatus vulgaris *C.B.* Leucographis Plinii *Lugd. Milke-thistle or Ladies thistle.*

109 *Cirsium vulgare* (Savi) Ten.
Carduus lanceatus *Ger. Tab.* lanceatus latifolius sive major *Park.* Carduus spinosissimus VI, sive lanceatus latifolius *C.B. The greater Speare-thistle.*

110 *Cirsium vulgare* (Savi) Ten.
Carduus lanceatus angustifolius *Park.* spinosissimus VII, sive lanceatus angustifolius, which is described as species 4 in his Prodr[omus] *C.B. The lesser Speare-thistle.*[166]

111 *Carduus nutans* L.
Carduus nutans *I.B.* Cirsion 3 tota sua stirpe magìs spinosum *Dod. lat.*[167] *Thistle with a bending head.* Generally, in uncultivated places and by roads with the other species. We are not entirely certain whether this is the nodding thistle[168] of *I.B.*, even though it corresponds to his description in most respects; however it can aptly be called nodding thistle since it alone of all those that grow naturally with us has a nodding head, whence it will easily be able to be distinguished from the others, nor is there any need for further description. However the leaves are greener and the flower larger and coloured with a deeper purple than any other of the thistles that are seen with us.[169] We have seen it also with a white flower but more rarely.

112 *Centaurea calcitrapa* L.
Carduus stellatus *Ger. Dod. ut.* stellatus sive Calcitrapa *I.B. Lob.* stellatus sive Calcitrapa vulgaris *Park.* muricatus, commonly called Calcitrapa *Clus. hort. Hafn.* stellatus III. i.e. foliis papaveris erratici *C.B.* Tribulus *Ruel.* Spinastella *Tab. Star-thistle.* Alongside roads. It does not produce spines except only on the flower-heads. Indeed it does have very minute spines at the tips of the leaves, but almost harmless ones.

113 *Cirsium arvense* (L.) Scop.
Carduus vulgatissimus viarum *Ger.* ceanothos, sive viarum et vinearum repens *Park.* vinearum repens folio sonchi *C.B.* An Carduus serpens lævicaulis *I.B? Common Way-thistle.* Everywhere along roads and in ploughed fields excessively frequent.

114 *Cirsium palustre* (L.) Scop.
Carduus caule crispo *I.B.* palustris *C.B.Park.* An Polyacanthos *Ger. Tab? The Marsh Thistle.* The descriptions of thistles are so obscure and confused that we cannot

---

[166] In *Emendanda* of the 1663 and 1685 appendices Ray asked for this variant to be removed, as he did not think it differed from the preceding one.

[167] i.e. *Dodonæus editione Latina in folio* ("Dodonaeus in a Latin edition in folio"): see *Dod.* in Ray's 'Explanation'.

[168] *carduus nutans*.

[169] See also Ray's further comments in *Emendanda* of the 1663 and 1685 appendices. In his later works, Ray (e.g. 1670, 1690) used *Carduus nutans* J.B. as his chosen name for this species without expressing any reservations, and Linnaeus (1753) retained *Carduus nutans* of J. Bauhin as a synonym of his own *Carduus nutans*.

decide anything for certain. This thistle certainly often raises itself to the height of a man, with a wavy-margined reddish stem, and is frequent also in marshy places.

115 *Cirsium eriophorum* (L.) Scop.
Carduus capite tomentoso *I.B.*, where there is a full and accurate description. eriocephalus *Ger. emac.* tomentosus et Corona fratrum *Lob.* sphærocephalus V, sive capite rotundo tomentoso *C.B. Woolly-headed Thistle. In many closes about Madingley, Childerley, Kingston, &^{c}*.

Carlina acaulis Septentrionalium, *see* Carduus acaulis. [*Cirsium acaule* 107]

116 *Carlina vulgaris* L.
Carlina sylvestris quibusdam, aliis Atractylis *I.B.* sylvestris major *Ger. Park.* Cnicus V, sive sylvestris spinosior *C.B. The Great wild Carline-thistle. On Gogmagog hills and Newmarket heath in great plenty.*

Carpinus *Matth. see* Betulus. [*Carpinus betulus* 86]

117 *Geum urbanum* L.
Caryophyllata *Ger. Dod.* vulgaris *Park. C.B.* vulgaris flore parvo luteo *I.B.* Geum *Turn.* urbanum *Gesn. hort.* Herba Benedicta *Brunf.* Commonly Sanamunda, Avancia *for Barbarians.*[170] *Avens, Herbe Bennet. By hedges.*

Caryophyllus, *see* Armeria and Holosteum.[171] [*Silene flos-cuculi* 61, *Stellaria holostea* 286 and *S. graminea* 287]

118 *Turgenia latifolia* (L.) Hoffm.
Caucalis altera purpurascens foliis latioribus et crassioribus semine etiam majore[172] *Ad. Lob.* Lappula canaria latifolia sive Caucalis *I.B.* An Caucalis Monspeliaca echinato magno fructu *C.B.?* We confess that we do not know whether this plant is described by *Gerarde* or *Parkinson* or which species of *Caucalis* in their works it fits.[173] *Purple-flower'd bastard Parsley. Among standing corn, as by the*

---

[170] *Sanamunda vulgò, Barbaris Avancia*. The former name, perhaps meaning something like "world's-health", is now the specific epithet of *Giardia sanamunda* (All.) C. Gerber (*Thymelaea sanamunda* All.) and the latter is presumably a latinisation of "avens", so named by the "intellectually uncultivated" (Marchant & Charles 1955): a barbarian (Latin *barbarus* from Greek βαρβαρος) was originally a non-Greek, then, for example in Cicero's Latin, a foreigner unfamiliar with Greek or Roman culture, so that the word became a term of opprobrium; see also footnote 379.

[171] Ray's name for the former is actually *Armerius sylvestris*. He cites names beginning with *Caryophyllus* under both *Holosteum vernum flore majore* [*Stellaria holostea* 286] and *Holostei Ruellii diversitas* [*Stellaria graminea* 287].

[172] "The second purplish *Caucalis* with broader and thicker leaves [and] with an even larger seed". In his contribution to the 1695 edition of *Camden's Britannia*, Ray (1695, column 417) reported this species from Cambridgeshire as *Caucalis arvensis latifolia echinata* C.B., citing the same localities and adding: "*This is a beautiful Plant, and we have seldom found it in other Counties.*"

[173] Ray's uncertainty is understandable: Gerarde (1597, p. 868) thought that his third "bastard Parsley" (which he did not illustrate) grew "in our pastures in most places of England", but Johnson (1633, p. 1023) wrote that he had not heard that this species, which he illustrated and called "Hogs Parsley" (*Caucalis Peucedanij folio*), "growes wilde with vs" but that "Bastard Parsley with red floures" (*Caucalis Apij folijs flore rubro*), which Gerarde said was "a stranger in England", had been found "in the corne fields on the hilles about Bathe, by M^r. Bowles". Parkinson (1640, pp. 919–922) used the same two figures as Johnson but named them differently and seemed unsure how many of his 13 species were found "in our owne Land".

foot-way side to Cherry-hinton church and in many other places, as in the plowed fields near the windmill in the way to Comberton.

119 *Torilis japonica* (Houtt.) DC.

Caucalis minor flosculis rubentibus *Ger. emac*. minor flore rubente *Park*. VII, sive semine aspero flosculis rubentibus *C.B.* An Anthriscus quorundam semine aspero hispido *I.B.*[174] *Hedge-Parsley*. It varies considerably in size: *see Ger[arde].*[175] Alongside hedges and even among standing corn.

120 *Torilis nodosa* (L.) Gaertn.

Caucalis nodosa echinato semine *C.B. Ger. emac. Park*. nodosa echinato semine Anthrisco hispido affinis, si non ejus varietas[176] *J.B. Knotted Parsley, Hedge-hog Parsley*. By the margins of fields.

121 *Caucalis platycarpos* L.

Caucalis tenuifolia flosculis subrubentibus. From a simple long white root come ridged stems, sometimes scarcely a foot long but sometimes even more than a cubit in height, clothed in rather scattered hairs; finely divided leaves adorn these, which are clothed with scattered hairs only on the underside and that on the ribs or veins alone, possessed of a slightly astringent taste but no obvious smell. From the branches of the stems arise three rays, on each of which is an umbel of flowers that turn purple from white;[177] these are succeeded by 'sea-urchins' or burs weighing half an ounce that are channelled and muricate[178] with rather rigid spines, which are green & not purplish and are divided into two parts at maturity. *Fine leaved bastard Parsley with reddish or blush-coloured flowers*. It seems to be Caucalis albis floribus *Ger*. & vulgaris albis floribus *Park*. However *C. B[auhin]us*[179] calls Ger[arde's] Caucalis albis floribus Caucalis arvensis echinatus magno flore,[180] although this plant of ours has a small flower. Certainly it differs very little from Lappula canaria flore minore sive tenuifolia[181] *J.B.*, that is only in that the spines of that plant along the ribs of the 'sea-urchins' are dark purple but those of this one of ours green. *In the corn about Kingston wood, and elsewhere*.

---

[174] "Is it *J. Bauhin's Anthriscus* with a rough bristly seed of some [authors]?"

[175] Johnson (1633, p. 1023) says: "There is a bigger and lesser variety or sort of this plant, for you shall find it growing to the height of two cubits, with leaues and all the vpper parts answerable, and you may againe obserue it not to exceed the height of halfe a foot."

[176] "Knotted [*Caucalis*] with a prickly seed related to *Anthriscus hispidus*, if not a variety of it": by "bristly *Anthriscus*" presumably Jean Bauhin (Bauhin & Cherler 1651, vol. 3(2), p. 83) meant the preceding species, his name for which is translated in footnote 174.

[177] *umbella florum ex albo purpurascentium*, literally "an umbel of flowers out of white purplish/becoming purple". According to Stearn (1992, p. 245) the expected meaning of this phrase would be "... purplish tinged with white", but a more literal interpretation seems more suited to umbelliferous flowers.

[178] The spiny fruits are described as *echini seu lappæ*: for the botanical usage of the former see Ray's 'Interpretation of terms'. Muricate means rough with short hard points like the shell of *murex*, the marine mollusc that was used in ancient times to produce Tyrian or Phoenician purple (Stearn 1992, pp. 230–232 and 450).

[179] A superscript 9 after the initials *C.B.* represents the Latin nominative masculine ending *-us*; see Chapter 7.

[180] "Echinate [i.e. prickly like a sea-urchin] field bur-parsley [or bastard parsley] with a large flower".

[181] Literally "Dog little-bur with a smaller flower or fine-leaved". Raven's (1950, p. 85) translation of *canaria* as "whitish" is incorrect.

Cauda equina, *see* Equisetum.[182]

122 *Blackstonia perfoliata* (L.) Huds.
Centaurium luteum perfoliatum *C.B. Cam. Park.* parvum luteum *Ger.* Perfoliata Achillea *Gesn. hort.* Perfoliatum Centaurium luteum *J.B.* Those who distinguish between Centaurium luteum with a branched & with an unbranched root are reproved *in J.B., book 3, page 355*. Gentianellæ amaræ secunda species called chlora from the yellowish colour of the flower[183] *P. Renealm. Yellow Centory. On Hinton moor, and elsewhere*.

123 *Centaurium erythraea* Rafn
Centaurium minus *C.B. Dod. Trag. Fuch. Matth. Tur. Lugd. Tab.* minus vulgare *Park.* minus flore purpureo & albo *J.B.* parvum *Ger. Lob. Ad.* Gentianellæ amaræ prima species Erythræa *of P. Renealmus*, who maintains that this little plant is not a centaury. *Small purple Centory. About Hinton and Teversham, & many other places*. It varies in the purple & white colour of the flower. It is so highly esteemed that common people scarcely ever make medicinal wine or that prepared with wormwood[184] of which it does not constitute a part. *Schroder[i] Pharmac[opœa] medico-chym[ica]*.

124 *Chaerophyllum temulum* L.
Cerifolium sylvestre *Tab. Ger.* Chærophyllum sylvestre *C.B.* Anthriscus Plinii quibusdam semine longo cicutariæ vel chærephylli[185] *J.B.* Daucus sepiarius *Gesn. Common wild Chervill. By hedges*. The stem filled with pith does not have a cavity.

Cervicaria minor, *see* Trachelium minus. [*Campanula glomerata* 573]

Chamæacte, *see* Ebulus. [*Sambucus ebulus* 170]

125 *Helianthemum nummularium* (L.) Mill.
Chamæcistus vulgaris flore luteo *C.B.* Helianthemum *Lob.* vulgare *Park.* vulgare flore luteo *J.B.* Helianthemum flore aureo sive Panax Chironium[186] *hort. Hafn.* Helianthos *Ad. Lob. hort. Paris*. Panax Chironium sive Flos solis *Matth. Lugd.* Hyssopus campestris *Trag.* Gratia Dei *Dod. gal.* Consolida aurea chirurgis *Cord. Schol. Dwarf Cistus, little Sun-flower. All over Gogmagog hills, and Newmarket heath*. The description of Chamæcistus serpilli folio nigricans & hirsutus, flore aureo odorato *J.B.* agrees well with this our common Chamæcistus, &

---

[182] *Cauda equina* ("horse's tail") does not appear in the synonymy of any of Ray's *Equisetum* species, but is given, as if it was simply a translation of Equisetum and Ἱππουρις [Hippuris], in the relevant entry in the 'Etymology'.

[183] "The second species of *Gentianella amara*" χλωρα *dicta à subluteo floris colore*. The adjective χλωρος, -α, -ον [chloros, -a, -on] is, like other Greek colour words, imprecise in meaning but can mean yellowish-green. Linnaeus named this species *Gentiana perfoliata* in 1753 but *Chlora perfoliata* in 1767.

[184] *vinum medicatum seu absinthites*, the final word being adapted from the Greek adjective ἀψινθιτης [apsinthites] which was used with οἶνος [oenos] ("wine"); it is here given in the nominative when it should be in the accusative.

[185] "Pliny's *Anthriscus* of some [authors] with the long seed of hemlock or chervil".

[186] "Chiron's Allheal": Chiron, the best known of the mythical centaurs of Greece and the mentor of Aesculapius, god of medicine, of Jason and of Achilles, was renowned as a herbalist; παναξ [panax], like the current word "panacea", means "all-healer". See also footnote 290.

better indeed than that of Chamæcistus vulgaris flore luteo;[187] but we are inclined to attribute these names to this plant principally on the authority of [Mathias] de L'Obel, who in his Adv[ersaria] asserts that Helianthemum vulgare luteum, the Synonyms of which we have given, grows generally in England. Instead of Helianthemum Anglicum luteum Gerarde inserts a picture of the fourth Cistus humilis of *Clus. hist.*, completely erroneously as it seems to us since Cistus humilis 4 of *Clus[ius]* has leaves that are wholly whitish but Helianthemum Anglicum luteum of *Ger[arde]*, which is common with us, has green ones. In so confused a matter it is difficult to decide anything certain.[188]

Chamæcissus, *see* Hedera terrestris. [*Glechoma hederacea* 272]

Chamædaphne, *see* Clematis daphnoides. [*Vinca minor* 138]

126 *Veronica chamaedrys* L.
Chamædrys sylvestris *Ger. Clus.* spuria sylvestris *Park.* spuria latifolia sive foemina *J.B.* XV, sive spuria minor rotundifolia *C.B.* Chamædryoides *Lob. ico.* Teucrium pratense *Lob.* Verbenaca supina mas *Dod. gall.* Wild Germander. Generally, in meadows & pastures.

Chamæleon exiguus *Tragi*, *see* Carduus acaulis. [*Cirsium acaule* 107]

Chamæleuce *Ang. see* Caltha palustris. [*Caltha palustris* 101]

Chamæmelum fœtidum *J.B. see* Cotula fœtida. [*Anthemis cotula* 155]

Chamæmelum inodorum, *see* Cotula non fœtida. [*Anthemis arvensis* 156]

127 *Ajuga chamaepitys* (L.) Schreb.
Chamæpitys vulgaris *Park.* vulgaris odorata flore luteo *J.B.* III, i.e. lutea vulgaris, sive folio trifido *C.B.* mas *Ger.* Ajuga sive chamæpitys mas Dioscoridis *Ad. Lob.* Aj[uga] luteo flore *Clus. hist.* Peristerona Catrevæ *Ang.* Iva arthritica *Offic[inarum].* Common Ground-pine. *On the layes about the borders of Triplow heath.* It flowers through almost the whole summer.

128 *Chelidonium majus* L.
Chelidonium maius *Ger.* majus vulgare *Park. C.B.* Chelidonia *J.B.* Hirundinaria *Ad. Lob.* Cœli donum Chymicis *Jo. Bodæ. in Theophr. hist.*[189] On the rubble of walls & in rough shady places. *The greater Celandine.*

N. Aristotle taught long ago that the eyes of nestling swallows that have been gouged out are often restored, not with the help of this herb or of a pebble,[190] but

---

[187] These two names mean literally "Ground-cistus with the leaf of thyme, blackish & hairy, with a scented yellow flower" and "Common ground-cistus with a yellow flower".

[188] Ray's problem was to match the single species that he found in Cambridgeshire, which in Britain is a very distinctive plant, with the great range of taxa found in mainland Europe. Proctor & Heywood (1968) in *Flora Europaea* recognise eight subspecies of *H. nummularium* and characters of leaf pubescence are important in distinguishing them. Mathias de L'Obel (1538–1616) was born at Lille, studied in Montpellier and spent extended periods on the continent before coming to live in England; hence, presumably, Ray's reliance on his judgement.

[189] The last two names mean "Swallow-plant" and "Heaven's gift to Apothecaries".

[190] *nec hujus herbæ nec lapilli ope*. Ray seems to assume that his readers will be familiar with the longstanding belief about swallows that, as Gerarde (1597, p. 911) puts it, "with this herbe the dams restore sight to their yoong ones when their eies be out". Gerarde attributes the name "Celandine", a corruption of *Chelidonium*, derived in turn from the Greek word for a swallow, *chelidon* (χελιδων), to this belief, though he too condemns it as "vaine and false", citing (on p. 912) Aristotle's "sixt booke of the historie of liuing creatures", where he claimed that blinded young swallows recover their sight

MAIN CATALOGUE

only by the strength implanted in them by nature. *Hist[oria] animal[ium], book 6, chapter 5, & De generatione, book 4, chapter 6.*

129 *Ficaria verna* Huds.

Chelidonium minus *Matth. Fuch. Dod. Ger. Park.* Chelidonia minor II, sive rotundifolia minor *C.B.* Scrophularia minor sive Chelidonium minus vulgò dictum *J.B.* Ficaria & Scrophularia minor *Brunf.* Hæmorrhoidum herba, *Hieronymo Brunswicensi in apodixe Germanica.*[191] Favagello *Cæs.* Amellus flos *of Virgil's Georg[ic] 4.*[192] *Cord. in Diosc.* In meadows & by hedges. *Pilewort, the lesser Celandine.* It rejoices in shade.

Chelidonia palustris, *see* Caltha palustris. [*Caltha palustris* 101]

Chenopodium, *see* Pes anserinus & Atriplex latifolia.[193]

130 *Glebionis segetum* (L.) Fourr.

Chrysanthemum segetum *Lob. Ger. emac.* segetum nostras *Park.* folio minùs secto glauco *J.B.* Bellis major lutea *Trag.* Bellis lutea I, sive foliis profundè incisis major *C.B.* fortè, an Chrysanthemum foliis matricariæ *ejusdem*?[194] *Corn Marigold.* Among the corn, but rarely near to Cambridge. Between Hoginton and Huntingdon rode in great plenty.

131 *Cichorium intybus* L.

Cichoreum sylvestre *Ger. emac. Park.* flore cæruleo perenne *hort. Hafn.* II, i.e. sylvestre sive officinarum *C.B.* Seris picris of Dioscorides, Cichoria of Theophrastus, Intybum erraticum of Pliny *of the same author.* Seris sylvestris *Ang. Lob.* Seris picris *Lob. ico.* Cichorea *Trag.* Intybum sylvestre angustifolium Matthioli *Lugd.* Cichorium sylvestre & sativum *J.B.*, who asserts that wild Succory does not differ from the planted form except through cultivation & that the planted form becomes wild if it is not cultivated. *Wild Succory.* By the margins of fields. It varies in the flower being blue & white.

N. 1. The flower of this plant if buried in an ant-hill changes its blue colour to red. *J[o]. Bodæ. in Theophrast. hist.*[195]

---

naturally. Pliny refers to the swallow-stone (*chelidonius lapillus*) fabled to be brought from the seashore by swallows to give sight to their young (OED). See also the entry for *Chelidonium* in Ray's 'Etymology'.

[191] "The herb of piles, in Hieronymus Brunswicensis [Jerome of Brunswick] in [his] German treatise", i.e. in Hieronymus Braunschweig's *Das buch zu distillieren* (1519). This is the German version of his seminal work entitled in Latin *Liber de arte distillandi* (1500), the third part of which provides a list of plant remedies made by distillation grouped under the conditions they were used to treat. Supposedly the knobbly tubers of lesser celandine or pilewort resemble piles, and according to the Doctrine of Signatures (see Chapter 3) this suggests that it could be used to cure piles.

[192] *Favagello* is an Italian name for *Ficaria verna* [129] (also *ranuncolo favagello*) and *Sedum telephium* [566] (also *favagello grande*). For Virgil's *amellus* see footnote 55.

[193] Ray's treatment of these names is confusing. *Pes anserinus* is not one of his chosen names but it appears as a synonym under *Atriplex sylvestris latifolia* [*Chenopodium murale* 72]. This is the only one of Ray's *Atriplex* species in which *Chenopodium* appears as a synonym (as *Chenopodium primum*) and *Atriplex latifolia* here presumably refers to this entry. *Pes anserinus* also appears in the alphabetical list as a synonym, cross-referenced to *Atriplex sylvestris* [*Chenopodium album* 70], presumably an error for *Atriplex sylvestris latifolia*.

[194] "Is it perhaps *Chrysanthemum* with the leaves of mayweed *of the same [author]*?"

[195] Bodaeus (1644, p. 855.2; see footnote 23) introduces this statement by saying that "they tell another miracle of nature" about this plant and adds that it is "as if it blushes from terror" of the ants!

2. The seeds of this plant cannot easily be collected unless they are soaked, for in that way the seed-vessels are opened. *Camer. in hort.*

132 *Conium maculatum* L.
Cicuta *Trag. Matth. Dod. Ad. Lob. Ger.* I, sive major *C.B.* vulgaris major *Park.* serotina caule maculoso *hort. Bat. among the indigenous plants.* Cicuta veteribus & neotericis *J.B. Hemlock.*

N. 1. We dissected the stomach of a great bustard or Tarda avis[196] and found it full of hemlock seed, with only four or five grains of corn intermixed, which that bird had neglected in favour of hemlock seed even at harvest time, so much does it delight in hemlock.

2. For what serious & astonishing symptoms a tray of meats cooked with the roots of hemlock and placed before two religious fathers (i.e. monks) caused, see the work of *Kircher about plague,*[197] *section 2.*

133 *Aethusa cynapium* L.
Cicutaria fatua *Lob.* tenuifolia *Ger.* Cicuta minor sive fatua *Park.* minor petroselino similis *C.B.* Petroselini vitium *Trag.* Cicutaria apii folio *I.B. The lesser Hemlock or Fools Parsley.* It comes up of its own accord among vegetables.

134 *Oenanthe aquatica* (L.) Poir. and/or *O. fluviatilis* (Bab.) Coleman[198]
Cicutaria palustris *Lob. Ger.* palustris tenuifolia *Park. C.B.* Phellandrium Plinii *Dod. Lugd.* Phellandrium vel Cicutaria aquatica quorundam *I.B. Marsh or water Hemlock. In the river and in great ditches of water almost every where.*

135 *Anthriscus sylvestris* (L.) Hoffm.
Cicutaria vulgaris *I.B.* Myrrhis sylvestris *Fuch. Park.* sylvestris III, sive sylvestris seminibus lævibus *C.B.* Cicutaria alba Lugdunensis, & Cicutaria palustris Camerarii *Ger.* But wrongly, as we think; for this plant is different from the Myrrhis sylvestris of Fuchs both in our own opinion and in that of the Bauhin brothers. *Wild Cicely.* In orchards & generally under trees.

136 *Circaea lutetiana* L.
Circæa Lutetiana *Lob. Ger.* Lutetiana major *Park.* Solanifolia Circæa dicta major *C.B.* Lappa sylvestris *Trag.* Herba D[ivi] Stephani[199] *Tab.* Ocymastrum

---

[196] This large gamebird, *Otis tarda* L., was called *avis tarda* (literally "slow bird") by Pliny, though *tarda* may here be a non-Latin word, taken erroneously as the feminine of *tardus*. Though now extinct as a native in Britain, the great bustard was once one of Cambridgeshire's notable birds and a pair of them are supporters of the county's arms (Bircham & Jordan 2000, Appendix).

[197] Athanasius Kircher's *Scrutinium physico-medicum contagiosae luis, quae pestis dicitur* ("A physico-medical investigation of the contagious pestilence that is called the plague"), published in Rome in 1658; the story appears on pp. 121–122 and tells how the two monks *ex insigni familia* ("from a famous family") ate a dish including what they thought were roots *petroselini* ("of parsley"); one of them plunged into a nearby lake, saying he had become a goose, and the other into a river, shouting that he had become a duck and could not live without water; doctors tried to cure them, but they suffered further symptoms and finally died within three years. See also footnote 302.

[198] Ray's *Cicutaria palustris* has previously been ascribed to *Oenanthe aquatica. O. fluviatilis* was not separated until the 19th century but, as it was formerly frequent in the River Cam and its ditches in Cambridge (Preston 2008), it seems very likely that Ray encountered both species and included them under this name.

[199] "Herb of the Divine [i.e. Saint] Stephen". Compare the use of the title *Divus* for St John the Baptist in the note about *Scrophularia major* [529] below and for St Rupert in the 'Etymology': see footnotes 68 and 218 there.

verrucarium[200] *I.B. Gesn. hort. Enchanters Nightshade. In the lanes about Cherry-hinton, & in many other moist and shady places.*

137 *Cirsium dissectum* (L.) Hill

Cirsium Anglicum primum, & Cirs[ium] Ang[licum] aliud *Park.* Cirs[ium] Angl[icum] *Lob. Clus. Ger. emac.* Cris[ium][201] Ang[licum] radice hellebori nigri modo fibrosâ, folio longo[202] *I.B.* Cirs[ium] Pannonicum primum Clusii folio non laciniato, and likewise Cirs[ium] Britannicum Clusii repens *for the same author.* Carduus Cirsium dictus IV, V, & VIII, *C.B. The English soft or gentle Thistle, Melancholy Thistle, for the most part single headed. In the first close you pass through, as you go in the foot-way from Cambridge to Cherry-hinton, near a little ditch or gripe that crosseth the close from corner to corner.* It does not escape us that some excellent botanists have separated this plant into several species and have designated it by different names, as their titles given by us above show, when it is only one (as it seems to us). In this matter, since we have seen even Jean Bauhin uncertain, we have followed the opinion of Johnson in his revision of Gerarde, principally because his description exactly fits this plant of ours.

138 *Vinca minor* L.

Clematis daphnoides minor *I.B. C.B. Dod. Lob.* Vinca pervinca *Brunf. & Offic[inarum].* minor *Ger.* vulgaris *Park.* Chamædaphne altera Dioscoridis *Brunf.* 4. Vinca pervinca of Plin[y], book 21, chapter 11, and Chamædaphne Græcorum of the same author in the same place[203] *I.B. Periwincle. In the next field to Cherry-hinton church through which the foot-path from Cambridge lies, under the bushes in the corner next to the church, near the stile into the lane passing by the church-yard.*

   N. We have frequently seen the leaves kept in the mouth check a flow of blood bursting out of the nostrils. *Costæus in his book about the differences between plants*[204] in *Camer. in hort.*[205]

Clematis latifolia dentata *I.B. see* Viorna. [*Clematis vitalba* 621]

139 *Clinopodium vulgare* L.

Clinopodium majus *Park.* quorundam Origani facie *I.B.* Origano simile *C.B.* Clinopodium & Lectipeda *Lac. in Diosc.* Acinos *Ger. Lob.* Origanum verticillatum *Thal. Great wild Basil.* By hedges in various places.

Clinopodium Ocymi facie, *see* Acinos Angl[ica] *Clusii.* [*Clinopodium acinos* 9]

140 unidentifiable filamentous algae

Conferva Plinii *Lob. Lugd. Ang. Ger. emac.* Linum aquaticum *Imp.* Alga viridis capillaceo folio *C.B.* aquatilis capillacea, sive Conferva Plinii, aliis Linum aquaticum *Park. Hairie River-weed. In stagnant waters.*

---

[200] This presumably means "False-basil that cures warts".

[201] A misprint for *Cirs[ium]*.

[202] Literally "English thistle with the root fibrous in the manner of black hellebore [and] with a long leaf".

[203] Pliny's *Natural history*, XXI. 68 in the Loeb edition. The second name means literally "Ground-bay of the Greeks".

[204] Giovanni Costeo's *De universali stirpium natura libri duo* was published in Turin in 1578. The passage cited is from the first book, *Quo de plantarum essentia, partibus, summis generibus, & differentiis*, p. 82. See also Chapter 4.

[205] i.e. *Cam. hort.* in Ray's 'Explanation', Joachim Camerarius' *Hortus medicus et philosophicus*, published in Frankfurt in 1588.

141 *Inula conyzae* (Griess.) Meikle
Conyza major *Matth.* major Matthioli, Baccharis quibusdam *I.B.* VI, sive major vulgaris *C.B.* mas Theophrasti *Io. Bod[æ]. in Theophr. hist.* Baccharis Monspeliensium *Gesn. hort. Ad. Lob. Lugd. Cam. Ger. Park.* Plowmans Spikenard. Near Abington in the road to Linton, and about Hildersham by the way sides plentifully.

142 *Pulicaria dysenterica* (L.) Bernh.
Conyza media *Matth. Ger. Dod. Ad. Lob.* tertia vulgaris *Clus. pan.* major sive mas *Trag.* aquatica *Thal.* aquatica 3 *Gesn. hort.* palustris I, sive media Asteris flore luteo, vel tertia Doscoridis[206] *C.B.* media Matthioli flore magno luteo, humidis locis proveniens[207] *I.B.* Calaminthæ tertium genus *Fuch.* Middle Fleabane. In watery places.

N. 1. The leaves of this herb when squeezed with the fingers smell of soap, or even of linen cleaned with soap.

2. Where this plant grows is where you should dig if you are looking for water. *Cassian[us] Bassus, book 2, chapter 3, in [Jo.] Bod[æ]. in Theophr., page 264.*[208]

143 *Pulicaria vulgaris* Gaertn.
Conyza minor *Trag. Matth.* minima *Dod. Lob. Ger. emac.* major flore globoso, i.e. palustris II, *C.B.* mediæ minor species flore vix radiato[209] *I.B.* minor grave olens *hort. Hafn.* Pulicaria *Lob. Ger. emac. Dwarf Fleabane.* Beyond the castle in the road to Histon, & in many other watery or moist places of the highways.

144 *Erigeron acris* L.
Conyza cærulea acris *C.B. Ger. emac.* odorata cærulea *Park.* Erigeron 4. *Dod.* Senecio sive Erigeron cæruleus, aliis Conyza cærulea *I.B.* Dentellaria nostra conyzoides quædam herba *Gesn. hort.* Tinctorius flos alter *Trag.* Amellus montana *Col. Blew-flower'd Fleabane or sweet purple Fleabane.* On Hinton moore and in the parke near Linton and in many other places.

145 *Tephroseris palustris* (L.) Fourr.
Conyza foliis laciniatis *Ger. emac.* helenitis foliis laciniatis *Lob. ico. Park.* palustris III, sive aquatica laciniata *C.B. Hoary Fleabane-mullet, or great jagged Fleabane.* In the Fen ditches about Marsh and Chatteresse in the Isle of Ely.

146 *Senecio paludosus* L.
Conyza palustris *Park.* pal[ustris] IV, sive pal[ustris] serratifolia *C.B.* maxima serratifolia *Cam. Thal.* Lingua major Dalechampii *Lugd.* Consolida palustris *Tab.* Solidago Saracenica *Phyt. Brit.* Virgæ aureæ sive solidagini angustifoliæ affinis Lingua avis Dalechampii,[210] *I.B. Marsh Fleabane.* We have found it in many places about the Fens, as by a great ditch side near Stretham ferry, &c.

Consiligo Plinii, *see* Helleborus niger hortensis flore viridi. [*Helleborus viridis* 276]

---

[206] "The first Marsh [Fleabane], or the middle [one] with the yellow flower of *Aster*, or the third of Dioscorides". *Dioscoridis* (genitive) is misprinted here.

[207] "Mattioli's middle [Fleabane] with a large yellow flower, growing in moist places".

[208] Cassianus Bassus compiled from earlier writers a collection of agricultural literature (*Geoponica*) at the end of the 6th or the beginning of the 7th century, parts of which were revised and published by an unknown editor in about AD 950; the original Greek text has been lost. See also footnotes 23, 211 and 736.

[209] "a species smaller than the middle [Fleabane] with the flower scarcely radiate".

[210] Literally "Daléchamps' Bird's-tongue, related to golden-rod or *Solidago angustifolia*".

MAIN CATALOGUE

147 *Symphytum officinale* L.
Consolida major. *Ger. Trag. Cam. Thal.* Symphytum magnum *I.B.* majus vulgare *Park. Matth. Cam.* alum sive alus *Ad Lob.* I, sive Symphytum Consolida major *C.B. Great Comfrey.* In watery places.

Consolida media, *see* Bellis major & Bugula. [*Leucanthemum vulgare* 81 and *Ajuga reptans* 95]

148 *Calystegia sepium* (L.) R. Br.
Convolvulus major *J.B.* major albus, i.e. Smilax lævis I, *C.B.* Volubilis major *Trag. Tab.* Smilax lævis *Matth.* lævis major *Dod. ut.* lævis sive lenis major *Ger. Lob.* Funis arborum Mesues *Trag.* Malacocissus of Damocrates in the work of *Cassian[us] Bassus, Io. Bodæ. in Theophr. hist.*[211] Iasione & Smilax lævis of Dioscorides & Ligustrum of poets *for the same author. Great Bindweed. In many hedges, as in the privet hedge in the fellows garden at Trinity Colledge.*

N. 1. The flowers of bindweeds open at the sight of the sun and close at night. *P. Lauremberg,*[212] *Horticult[ura], book 1, chapter 11.*

2. The flowers & leaves of bindweeds are composed largely of two volatile salts, & hence they wilt & dry out easily.[213] *Quercet. rediv. Pharm. restit. Phytol.,*[214] *chapter 2, note 8.*

149 *Convolvulus arvensis* L.
Convolvulus minor arvensis, i.e. Smilax lævis III, *C.B.* minor albus vulgaris, & minor purpureus *Park.* Volubilis minor *Trag.* Smilax lævis minor *Dod.* lenis minor *Ger.* Helxine cissampelos multis sive convolvulus minor *J.B.* Helx[ine] in segetibus nascens gracilibus & longis foliis *Cord. in Diosc.* Orobanche Theophrasti *Ang. Small Bindweed.* In fields.

150 *Fallopia convolvulus* (L.) Á. Löve
Convolvulus niger *Dod. Park.* minor semine triangulo, i.e. Smilax lævis IV, *C.B.* Convolvulo similis herba præterquam flore[215] *Gesn.* Volubilis nigra *Ger. Tab.* Helxine semine triangulo *J.B.* Helx[ine] cissampelos altera Atriplicis effigie *Lob.* Orobanche Ruellii *Gesn. hort. Black Bindweed.* Among standing corn.

Corchorus, *see* Anagallis mas. [*Anagallis arvensis* subsp. *arvensis* 42]

Cornu cervinum *Officin[arum], see* Coronopus vulgaris. [*Plantago coronopus* 152]

---

[211] Literally "Soft-ivy of Damocrates in Cassianus Bassus' work [quoted by] Johannes Bodaeus [1644] in [his commentary on] Theophrastus' *History [of plants]*". Damocrates (or Democrates) Servilius was a Greek physician at Rome in the 1st century AD, highly praised by both Galen and Pliny. See also footnotes 23, 208 and 736.

[212] For Peter Lauremberg see footnote 25.

[213] ... *plurimum participant de duobus salibus volatilibus, & hinc facilè flaccescunt & exsiccantur*, literally "... very much participate in two volatile salts, & hence easily wilt & are dried out".

[214] This five-word reference is hard to interpret. Joseph du Chesne (or Duchesne, latinised as Josephus Quercetanus) was a French Calvinist and an influential Paracelsian, reputedly physician-in-ordinary to Henry IV; he died in 1609. His *Pharmacopoea dogmaticorum restituta* was first published in Paris in 1607. However, Johann Schröder's edition of his works, *Quercetanus redivivus*, published in Frankfurt in 1648, appears to be the work referred to here. Volume II has a section entitled 'Pharmacopoeæ Phytologicæ', but Chapter 2 of this section is not the source and (like most of the chapters) lacks notes. Chapter 3, however, entitled 'De Potentiis seu facultatibus Vegetabilium & primum de dignotiore earundem', does have notes and Nota VIII, under the second heading 'De Dignotione ex processu generationis' (p. 759), though not referring specifically to *Convolvulus*, appears to be the passage cited here.

[215] "A herb similar to bindweed apart from the flower".

151 *Cornus sanguinea* L.
   Cornus fœmina *Ger. C.B. Park. Lob. Tab.* Virga sanguinea *Matth. Dod.* Cornus fœmina putata, Virga sanguinea *J.B.* Opulus Columellæ *Dod. Gal. Dogberry or Gatter tree.* In thickets & hedges.

   N. Cornus fœmina bears a fruit that is inedible to all animals. *Theophr. hist. plant., book 3, chapter 6.*[216]

152 *Plantago coronopus* L.
   Coronŏpus *Trag. Matth. Fuch. Ang. Tur. Ad. Tab.* I, sive hortensis *C.B.* vulgaris sive Cornu cervinum *Park.* Cornu cervinum *Ger. Lob.* Coron[opus] sive Cornu cervinum vulgò spicâ plantaginis *J.B.* Herba stella sive Cornu cervinum *Dod.* Coronopum aut Coronopodium. *Crow-foot plantain Tur. Buckshorn-plantain. In the sandie grounds about Gamlingay, also about Newmarket, and about Wisbich in the Isle of Ely, and all places thereabout near to the salt water.*

   N. This plant when sown in gardens flourishes with careful cultivation to such an extent & differs so much from the wild form in size & shape as to be regarded as a separate species by some authors.

153 *Lepidium coronopus* (L.) Al-Shehbaz
   Coronŏpus Ruellii *Dod. gal. Ger. Gesn. hort. Lugd.* recta vel repens Ruellii *Park.* Ruellii sive Nasturtium verrucosum *J.B.* repens *Tab. Cam. Lob.* Cornu cervi alterum repens *Dod.* Ambrosia *Matth.* campestris repens *C.B.* Pseudoambrosia, Nasturtium verrucarium quibusdam *Cam. in Matth.* Pes milvinus Columellæ *Dalechampio*, Holosteum Dioscoridis *Ulyssi Aldrovando*[217] *in the work of* J.B. *Swines Cresses.*

154 *Corylus avellana* L.
   Corylus sylvestris *C.B. Trag. Matth. Ger.* Nux Avellana sylvestris *Fuch. Matth.* Corilus sive nux Avellana sylvestris *Park. The Hasell-nut tree.*

   N. 1. Avellana, which was formerly called Abellina from the town of Campania Abella. *Plin[y], book 15, chapter 22.*[218]

   2. The fruits of this tree are called καρυα πλατεα in Hippocrates 2. about diet; *Casp[arus] Hofman[nus], De medicam[entis] officin[alibus]*; καρυα Ποντικα & λεπτοκαρυα in *Dioscor[ides], book 1, chapter 169*, i.e. Pontic & slight or puny, namely as compared with a walnut. In Theophrastus this tree is called Ἡρακλεωτικη καρυα, because it has been brought from Heraclea in Pontus.[219]

---

[216] III. iv. 6 of Theophrastus' *Historia plantarum* in the Loeb edition, where it is stated that the passage is apparently a gloss on the original text.

[217] Literally "Columella's Kite's-foot *for Daléchamps*, Dioscorides' All-bone *for Ulysses Aldrovandus*". For the Roman author Lucius Junius Moderatus Columella, see Chapter 4. Ulisse Aldrovandi (1522–1605) founded the botanic garden at Bologna and was a pioneer in making a scientific collection for research purposes (DSB, vol. 1, pp. 108–110). His largely posthumous work on natural history contains much that is valuable mixed with a large proportion of fables (e.g. about the cockatrice or basilisk and the original 'cock and bull story'). See also footnote 854.

[218] Pliny's *Natural history*, XV. 88 in the Loeb edition, which gives the current name as *Abellana*.

[219] Presumably Hippocrates' book usually called *De alimento* or *On aliment* is meant here; see also footnote 235. The four Greek names can be transliterated as *carya platea* ("broad nuts"), *carya Pontica* ("Pontic nuts", i.e. of the Black Sea), *leptocarya* ("slight nuts") and *Heracleotice carya* ("nut-tree of Heraclea"). The neuter plural καρυα ("nuts") is accented on the first syllable with the final vowel short while the feminine singular καρυα ("nut-tree") is accented on the penultimate syllable with the final vowel long. The last tree is described in Theophrastus' *Historia plantarum*, III. xv. 1–2 in the Loeb edition, besides several shorter mentions, but the epithet is not discussed.

3. An elegant caterpillar, called *Corilaria* by Moufet,[220] feeds on the leaves of this tree, which, when it is now due to be transformed into a chrysalis by the enduring law of nature, constructs a nest for itself from hairs plucked from its own body and arranged crossways with wonderful skill, which also we have observed to be usual for other more densely hairy caterpillars. The chrysalis of this *Corilaria* is spherical, with the upper part dark red or deep reddish-brown and with the lower part or tail adorned with reddish-brown & yellow circles arranged alternately, hairy & shaggy on its back, which occurs in very few other species. After this has passed through the winter, it is changed in the month of May into a moth very similar to the moths of silkworms except that it is shorter & fatter at the shoulders. Certainly there is the same kind of head, eyes and antennae, the same shortness of the wings in both making them useless for flight; the colour also is the same except that this one is marked with more black spots on the upper side of its wings and with a single spot on its back, which is not so for the other. Moreover it is much hairier over its whole body & clothed with thick wool such as is found on poultry chicks. It has six legs tinged with a yellowish green colour.

4. In Moufet's opinion certain insects have received the name of *Julus* as a result of some similarity to the catkins from these trees;[221] when the air is rather humid many of these (as we may note in passing) appear immediately, gleaming at night like glow-worms, their brilliance having arisen from some rather dense and viscid humour with which their bodies are smeared, for that liquid, still shining, clings to the hands of anyone handling them; when dried it ceases to shine, but, when moistened again soon, it regains its former brilliance. These little creatures are more correctly called *Scolopendra* rather than *Julus*.[222]

---

[220] *Corilaria* is clearly named from its food-plant, *Corylus*. The moth was probably a member of the Lymantriidae or the superficially similar Nut-tree Tussock, *Colocasia coryli* (Linnaeus) (Noctuidae), but no species fits the description in its entirety. The description of the imago strongly suggests the Gypsy Moth, *Lymantria dispar* (Linnaeus), which has black markings on pale wings and females which have wings but very restricted powers of flight. (All other possible species have females with wings either absent or fully functioning.) However, the extinct British race of this species was restricted to Fenland and its larval food-plants were *Myrica gale* and *Salix repens* rather than the deciduous trees on which the continental race feeds. The information that Ray gives is a better fit to a species such as the Pale Tussock, *Calliteara pudibunda* (Linnaeus), as Raven (1950, p. 102) suggested, but this has darker forewings and flying females (Heath & Maitland Emmet 1979; photographs on www.ukleps.org). It is possible that the short wings were the result of rearing the species in unsuitable conditions (see footnote 410). Moufet (1634: see footnote 76) illustrates his *Corilaria* caterpillar on a hazel leaf in Book 2, Chapter 3, on p. 188, and describes it as follows: *Corilaria undique cum virore pallet, tribus inter cuneolos maculis nigris exceptis, & cornu illo in ultimo dorso, & quasi ouropygio crescente, quod rosaceum ruborem lætè suscipit. Corili præcipuè folijs insidet, unde Corilariam dicimus. Duas illarum species vidimus; alteram saturatè; alteram pallidiùs viridem.* ("*Corilaria* is pale green all over [literally "everywhere looks pale with greenness"] with the exception of three black spots between the little wedges & that horn at the very end of its back [literally "on the furthest back"] &, so to speak, [its] increasing rump [Greek οὐροπυγιον, uropygion], which luxuriantly/joyfully takes on a rosy redness. It principally inhabits the leaves of *Corilus* [i.e. Hazel], whence we call [it] *Corilaria*. We have seen two kinds of them, one dark [and] the other paler green.")

[221] *Ab harum arborum julis* ... . See also footnote 674.

[222] They are not insects in the modern sense, but myriapods, probably the centipede *Geophilus electricus* (Linnaeus), originally named by Linnaeus *Scolopendra electrica*. *Scolopendra* is the name used for a myriapod by Pliny.

155 *Anthemis cotula* L.
Cotula fætida[223] *Lob. Park. Ger.* alba *Dod.* Chamæmelum fœtidum sive Cotula fœtida *J.B.* Parthenium *Fuch. Ad. Tab.* Parth[enium] leptophyllon Hippocratis *C.B. Maithes or Mayweed.* Among sown crops everywhere.

N. Bees, while they are engaged in honey-making, reject neither fetid flowers nor poisonous ones such as those of *Cotula fœtida*, henbane etc., though they may however neglect many very scented ones such as those of roses, primroses, peas, etc. *Butler in his treatise about bees*,[224] *chapter 6, number 5.*[225]

156 *Anthemis arvensis* L.
Cotula non fœtida *Park. Ger.* Camomilla fatua & 4 *Trag.* Chamæmelum inodorum *C.B.* inodorum sive Cotula non fœtida *I.B. Dogs Camomile.* With the preceding.

157 *Hydrocotyle vulgaris* L.
Cotyledon aquatica *I.B.* aquat[ica] acris Septentrionalium *Lob.* palustris *Ger. Park. Dod.* Ranunculus aquaticus umbilicato folio *Col. hort. Hafn.* aquaticus cotyledonis folio *C.B. Marsh Pennie-wort, White rot. On Hinton and Teversham and Trumpington moors, and the like places abundantly.*

Crassula, *see* Telephium. [*Sedum telephium* 566]

158 *Odontites vernus* (Bellardi) Dumort.
Cratæogonon Euphrosynes facie[226] *Ger.* Euphrasia rubra sive altera *Dod. Lob.* pratensis rubra *C.B.* parva purpurea *J.B. Eybright-Cowwheat.* In meadows, & also among standing corn. The colour of the flower varies, being at times more intensely red, at others more palely so and sometimes even white.

Crista galli, *see* Pedicularis. [*Pedicularis sylvatica* 427 and *Rhinanthus minor* 426]

159 *Crocus sativus* L.
Crocus *J.B. Tur. Trag. Dod. Lob.* sativus *C.B.* sativus autumnalis *Eyst. hort. Paris.* verus sativus autumnalis *Park. in Parad.* Crocus autumnalis sativus Anglicus *hort. Lugd-Bat.* Spica Cilissa poetarum *Jo. Bodæ.* The true crocus never flowering except in Autumn, with the sweetest scent and a purplish colour. *P. Lauremberg*,[227] *in Apparatu[s plantarius]. True or common Saffron.* It is cultivated in fields.

---

[223] This is a less correct form than *fœtida* (with linked *oe*), which is the form used by Pena & L'Obel (1570, p. 344), Plantin (1591, p. 773), Gerarde (1597, p. 617) and Bauhin & Cherler (1651, vol. 3(1), p. 120) and by Ray later in this entry and in cross-references to it elsewhere in the *Catalogus*; only Parkinson (1640, p. 86) of the cited sources has *fætida* (with linked *ae*). We have accepted Babington's (1860) identification of this and the following species, although the absence from Ray's account of two other mayweeds, *Matricaria chamomilla* L. and *Tripleurospermum inodorum* (L.) Sch. Bip., suggests that this is not a group that Ray had completely mastered. The latter is now much the commonest species in Cambridgehire.
[224] *The Feminine Monarchie* (Butler 1634): see footnote 40.
[225] This is an error for *number 50*.
[226] Literally "Mighty-born/Mighty-bearer with the appearance of Eyebright". The generic name (which could be either passive or active in meaning) is not explained in Ray's 'Etymology': it is derived from the adjective κραταιος, -η, -ον [crataeos, -e, -on, "mighty"] and γονη or γονος [gone or gonos, "offspring"]. As explained in the entry for *Euphrasia* and the related footnote in the 'Etymology', Ευφροσυνη [Euphrosyne] is a synonym of *Euphrasia* [192], so the meaning is that the species is like Eyebright, as the first synonym in the main catalogue, *Euphrasia rubra sive altera* ("Red or second Eyebright"), and Ray's English name, "Eybright-Cowwheat", also suggest.
[227] For Peter Lauremberg see footnote 25.

N. 1. What the ancients hand down about the cultivation of saffron differs greatly from the usual practice of our own time and the care which countrymen now apply to looking after it. For they, on the evidence of Theophrastus, *book 6*, and Pliny, *book 21, chapter 6*,[228] used to choose as the places where they would put it those trodden down by the footsteps of men and used to think that the more the soil was trampled the better it would grow.[229] In fact now they think nothing to be so harmful as that the ground where the saffron is planted should be trampled by feet and they keep away animals and men with the greatest devotion by specially built fences. *J.B.*

2. Irish women dye their clothes with saffron, both so that they may be kept free of lice and and so that strength may come to their limbs therefrom. *Lauremberg*,[227] *Apparat[us] plant[arius], page 44*. They say that our sailors are also accustomed to do the same.

3. It seems astonishing that saffron, which they say is made of such delicate parts, stains with its colour not only urine but also the excrement of the belly.

160 *Cruciata laevipes* Opiz

Cruciata *Dod. Ger.* vulgaris *Park.* minor *Lob.* minor & Asperula aurea *Lugd.* Gallium latifolium, Cruciata quibusdam flore luteo *J.B.* Cruciata hirsuta *C.B. Crosswort, or Mugweed*. By hedges & in woods.

N. On the square stem one can always see two leaves placed alternately opposite each other (and so they are called winged) separated by a moderate distance from two others occupying the other two sides from the former pair, and likewise four arising from the four sides, as in *Cruciata Dodonæi & Erythrodanum*, which they call *Rubia tinctorum*;[230] but in fact you will never find three or five or more. *Spigel[ius in] Isag[oges]*,[231] *book 1, chapter 11*.

161 *Cuscuta europaea* L.

Cuscuta major *C.B.* Cuscuta sive Cassutha *Ger.* Cuscuta *Park.* Cassuta sive Cuscuta *J.B. Dodder. Commonly growing upon beans.*

N. This plant is extraordinary, for it has neither leaves nor root: it is devoid of leaves always and of a root after it has grown to maturity. See *C. Hofman[nus], De medicam[entis] officin[alibus], book 2, chapter 78, & Park.*

162 *Centaurea cyanus* L.

Cyanus *J.B. Tur. Gesn. hort.* minor vulgaris *Lob. Ger. Park.* segetum *C.B.* segetum vulgaris *hort. Hafn.* Baptisecula & Pesgruis *Trag.* Flos S. Zachariæ *Jun. nomenc.*[232] It varies in the colour of the flower which may be blue, purple, pink or white. *Blew-bottles*. Among standing corn.

---

[228] *testibus*, literally "with ... [as] witnesses". See Theophrastus' *Historia plantarum*, VI. vi. 10, and Pliny's *Natural history*, XXI. 34 in the Loeb editions.

[229] *arbitrabantúrque quantò magis solum tereretur tantò meliùs provenire*, literally "and they were having the opinion by how much more the soil might be worn/visited, by that much better [the saffron] to shoot forth".

[230] C. Bauhin (1623, p. 333) gives *Erythrodanum* as a Greek equivalent of the Latin *Rubia*, as does Ray (1686, p. 480) in his Chapter '*De Rubia tinctorum seu Erythrodano*'. In his *Historia plantarum* (VII. ix. 3 in the Loeb edition and elsewhere) Theophrastus calls this dye-plant ἐρευθεδανον [ereuthedanon] (see footnote 473).

[231] For Adrianus Spigelius see footnote 31.

[232] For Junius' *Nomenclator* see footnote 143 and for the Flower of St Zacharias the end of the entry on *Artemisia* in Ray's 'Etymology'.

163 *Mercurialis perennis* L.
Cynocrambe mas & fœmina *Ger. emac*. mas & fœmina, sive Mercurialis repens *J.B.* Mercurialis sylvestris Cynocrambe dicta vulgaris mas & fœmina *Park*. Mercurialis III, sive montana testiculata, & Merc[urialis] IV, sive montana spicata *C.B.* Merc[urialis] canina mas & fœmina *Col. Dogs Mercury*. In woods & hedges.

164 *Cynoglossum officinale* L.
Cynoglossum vulgare *J.B.* majus vulgare *C.B. Park. Ger.* Lycopsis *Lac*. Great Hounds-tongue. Everywhere along roads.

N. 1. The bad smell of this plant is doglike. *C. Hofman[nus], De medicam[entis] officin[alibus]*, book 2, chapter 79. Certainly when squeezed with the fingers it gives off the smell of a hot wet dog.

2. Dioscorides wrote that *Cynoglossum* produces no stem, because only after the third year does it grow into a plant. *Marant[a], Method[us] cogn[oscendorum] simpl[icium]*, book 2, chapter 4.[233]

Cynoglossa topiaria, see Alysson Germanicum echioides. [*Asperugo procumbens* 34]

Cynosorchis, *see* Orchis.[234]

165 *Cladium mariscus* (L.) Pohl
Cyperus longus inodorus sylvestris *Ger.* longus inodorus vulgaris *Park*. longus inodorus sylvestris Lobelio *J.B.* longus inodorus Germanicus *C.B. Long bastard Cyperus*. *In the watery places of Hinton moor, and in divers Fen ditches*. We dare not assert too confidently that the plant that we know well is the same as this plant whose synonyms we have given, because Gerarde & Parkinson attribute triangular stems to it, though ours possesses round ones. However we are not much influenced by the authority of these writers, for perhaps they write thus from conjecture & because they thought that it was essential for every *Cyperus* to have triangular stems. Certainly it is either this or it is not yet described.

D

166 *Daucus carota* L. subsp. *carota*
Daucus officinarum *C.B.* agrestis Galeno de alim[ento][235] *of the same author*. vulgaris *Clus. hist*. usitatior officinarum Galliæ, Germaniæ, Italiæ[236] *Lugd*. Pastinaca sylvestris sive Staphylinus Græcorum *J.B.* sylvestris tenuifolia *Ger. Park. Dod.* sylv[estris] tenuifolia Dioscoridis *C.B.* Staphylinus sylvestris *Trag. Cæs. Wild Carrot or Birds nest*. Everywhere in pastures.

---

[233] For Bartolomeo Maranta see footnote 34.

[234] Four of Ray's *Orchis* species have *Cynosorchis* synonyms.

[235] i.e. in the book entitled *Galeni in Hippocratis librum de alimento commentarii IV* ("Galen's four commentaries on Hippocrates' book *On aliment*"). For Hippocrates see footnotes 24 and 219 and Chapter 4.

[236] "The more usual [Carrot] of the apothecaries' shops of France, Germany [and] Italy".

MAIN CATALOGUE

167 *Taraxacum* spp.

Dens leonis[237] *Ger.* vulgaris *Park.* D[ens] l[eonis] latiore folio *C.B.* Aphaca from the class of vegetables of Theophrastus, *[book] 7 [of] hist., 7, 10.*[238] *C.B. Jo. Bod[æ].* Aphace of Pliny, *book 21, chapters 15 & 17,* & Hedypnois of Pliny, *book 28, chapter 8.*[239] *C.B.* Hedypnois sive Dens leonis *Fuch. J.B.* Hedypnois Dalechampii *Lugd.* Hieracium minus, Hedypnois altera Plinii *Gesn. hort.* Heracium majus Dioscoridis *Trag.* Taraxacon *Offic[inarum]* minus *Lon.* Chondrilla altera & Rostrum porcinum *Dod. gal. Thal.* Chondrilla Galeni *Ang.* Urinaria, Gallis pueris Lectiminga[240] *Ad. Lob.* Caput monachi *Jun. nomenc.*[241] *Lob.* Ambubeia *Jun.*[241] *Dandelion.* In pastures everywhere. It flowers for almost the whole year.

N. The name of apiastrum or melissophyllum can not undeservedly be given to this plant, because throughout almost the whole time of the year in which bees apply themselves to their work they expend no little of their energy on its flowers. *Butler in his treatise about bees,*[242] *chapter 6, number 34.*

168 *Dipsacus fullonum* L.

Dipsacus *Lac. Lon. Tab.* sylvestris *Dod. ut. Tur. Gesn. hort. Park. Ger.* sylvestris, aut Virga pastoris major *C.B.* sylv[estris] sive Labrum Veneris *J.B.* Labrum Veneris *Cord. in Diosc. Ad. Lob.* Veneris alterum *Matth.* Carduus fullonius erraticus *Trag. Wild Teasell, Venus bason.* In uncultivated places, & by hedges & roads.

N. 1. Rainwater lying stagnant in the axils of the leaves of this plant is recommended for getting rid of warts if the hands are washed in it several times. And it is perhaps from this that it has acquired the name of Labrum Veneris.[243]

N. 2. In each flower-head of this plant a single grub is not infrequently found, which for that reason you might deservedly call solitary, as described in Moufet, *in Insect[orum] theat[rum], book 2, chapter 20.*[244] The common people of our country are accustomed to hang these, shut up in a goose quill in unequal numbers, as an amulet or remedy on the necks of those suffering from quartan

---

[237] "Lion's tooth", referring to the coarsely-toothed leaves; *dent de lion* in French, hence the English "dandelion".

[238] *Aphaca ex olerum genere*. Theophrastus mentions αφακη [aphace] among other pulses in two passages in his *Historia plantarum* (VIII. 1. 4 and VIII. v. 3 in the Loeb edition), so the identification with Dandelion is puzzling. See also Ray's entry for *Aphaca* [52].

[239] Pliny's *Natural history*, XXI. 89 & 105 and XX. 75 respectively in the Loeb edition. The identification of *Taraxacum* spp. with the latter is the more credible, but "book 28" seems to be an error.

[240] *Rostrum porcinum* means "Pig's snout". *Urinaria, Gallis pueris Lectiminga* means "A urinary plant known to French boys as Bed-wetter". They call it *pisenlit* ("piss-in-the-bed") to this day.

[241] For Junius' *Nomenclator* see footnote 143.

[242] *The Feminine Monarchie* (Butler 1634): see footnote 40. Both plant names, now used in quite different botanical contexts, relate to the honey bee, which is *apis* in Latin and μελισσα [melissa] in Greek.

[243] i.e. "Venus's Basin": see the English names above.

[244] For Moufet's (1634) book see footnote 76. On p. 256 of this chapter, entitled *De Vermibus Fructuum, leguminum, frumentorum, Vitis, Herbarum* ("About the Worms of Fruits, pulses, grain-plants, Vine, Herbs"), Moufet writes that in the head or the stem of Teasel he has observed a small grub, easily found at the beginning of October, with a small head, six blackish legs and ten or eleven segments. First, he explains, it feeds on the spongy pith of the stem and then, if that fails, it dies from lack of nourishment.

fever.²⁴⁵ These little creatures are an amphisbaena²⁴⁶ & they quickly avoid the finger of anyone touching their head by going backwards. They excavate a home for themselves with a little beak in the pith or meditullium of the flower-head, the walls of which are perforated only in the upper part by a single little door where they first crawled in; thanks to this they both enjoy the air & have an unrestrained exit after their metamorphosis. For they feel that this metamorphosis is approaching as the next spring arrives, so they form a woven couch for themselves so that their head placed close to the opening may find the way out open. Meanwhile they pass the winter under cover, safe from rain & cold, until in the following May they emerge in the form either of a fly (which may often be the case) or of a moth (which is more frequent).²⁴⁷ The moth is one of the smallest nocturnal kinds that emerges from a round yellowish-grey chrysalis. Its abdomen is short and quite well formed, ending in a point, and it is endowed both above and below with a shining colour midway between lead and silver. The antennae, head & top of the shoulders are short & black, while the six legs are silver and they end in feet in such a way that the front pair seem to stand on one digit each, the next on two and the last on three, the upper of which digits arise from the shins themselves. The wings are four, of the same colour as the abdomen beneath; they are tinged with the same colour also above but with elegantly variegated black. This insect is either the same as the last of Moufet's (*page 97 of book 1*) Phalæna species,²⁴⁸ which he calls *moth* in English, or not completely different. The fly is of the same kind as that which we shall be describing under Rosa canina²⁴⁹ and only different in size, in which this one is greater; for, although it appears to be distinguished by the colour of its abdomen, which in this one is wholly black, yet if it is extracted from its cocoon before its due time of birth, it is tinged with the same spots & colour as that one. In fact both accord with the second species

---

²⁴⁵ i.e. quartan malaria, infection with *Plasmodium malariae* Feletti & Grassi, in which paroxysms recur every 72 hours (or every fourth day if the day of each paroxysm is reckoned as the first). See also Ray's note on *Linaria vulgaris* [335].

²⁴⁶ Greek ἀμφισβαινα, meaning "going both ways", originally a mythological ant-eating serpent with a head at each end, according to Greek mythology spawned from the blood that dripped from the Gorgon Medusa's head as Perseus flew over the Libyan Desert.

²⁴⁷ The ability of the caterpillar to walk backwards, as well as its larval habitat, indicate that the species seen by Ray was a micro-moth. There are two possibilities, *Endothenia gentianaeana* (Hübner) and *E. marginana* (Haworth), tortricoid moths with larvae that inhabit and overwinter in the hollow receptacles of the flowers of *Dipsacus fullonum*, filling the lower part with frass, and adults which when at rest closely resemble bird-droppings (Bradley, Tremewan & Smith 1979; P. Sterling, *in litt.* 2009). *E. gentianaeana* usually feeds in teasels whereas for *E. marginana* they are one of several host plants. The larvae of another tortricoid moth, *Cochylis roseana* (Haworth), feed on the seeds of teasels but they do not excavate the pith and are not solitary, so they can be ruled out. The "flies" were presumably parasitoid ichneumon wasps, which Raven (1950, p. 103) suggests were "probably *Omorga mutabilis*", i.e. *Campoplex difformis* (Gmelin) (= *Limneria mutabilis* Holmgren); see also Ray's long note 2 under *Rapum sylvestre*.

²⁴⁸ *Hæc papilio vel est eadem ultimæ phalænarum Moufeti p. 97. lib. 1.* ... There is a problem in translating *papilio* (which means either a butterfly or a moth) here, because "*moth*" follows soon after as the English name given by Moufet. Moufet describes his 16th (and last) *Phalæna* as being of a uniform dull colour apart from the blackish eyes and having a long, articulated body, four long, narrow wings, six legs, the rear ones twice as long as the front ones, and slender but long antennae (*cornicula*). Mysteriously, Ewen & Prime (1975) translate this whole sentence thus: "This moth is not the same as the last Phalaena of Moufet."

²⁴⁹ This name is actually listed as a synonym under *Rosa sylvestris* [502].

of three-bristled flies of Moufet, *book 1, chapter 12, page 64*,[250] and they do not seem to be distinguished from it otherwise than that among these a fly endowed with three bristles occurs more rarely & the bristles on these are also shorter in comparison with their bodies than in those which are portrayed in Moufet's plate.

We have observed the same uncertain transmutation also in a certain very small Sepiaria[251] caterpillar which not infrequently emerges as a fly but more often as a moth, & in some others. It is a matter worthy of further examination & one in which those who have plenty of leisure & talent may occupy themselves.

169 *Dipsacus pilosus* L.

Dipsacus minor *Gesn. hort. Thal.* minor sive Virga pastoris *Ger. emac.* secundus capite rotundo *Dalechampii in Plinium.* sylvestris capitulo minore, vel Virga pastoris minor *C.B.* Virga pastoris *Park.* pastoris vulgaris *I.B.* pastoris & Dipsacus fatuus *Cam.* Cardui fullonum 3 genus *Trag.* Dipsacus 3 & Galedragon Plinii *Dod. ut.* Molybdæna Plinii *vulgó. Shepheards Rod.* About Ditton, and in many other places.

Dulcamara, *see* Solanum lignosum. [*Solanum dulcamara* 547]

Dracunculus minor Dioscoridis, *see* Arum. [*Arum maculatum* 63]

E

170 *Sambucus ebulus* L.

EBulus *Trag. Matth. Fuch. Tur. Lob.* Eb[ulus] sive Sambucus humilis *C.B. Ger. Park. Dod.* Eb[ulus] sive Sambucus herbacea *I.B.* Sambucus pumila *Hermolai.* Heliosacte sylvestris aut Euboica Sambucus *Trag.*, who also relates that it can be called Sambucus æstivalis. Chamæacte Dioscoridis *C.B. Danewort or Wallwort.* In Madingley, by a well on the hill not far from the church, in the street: along the balks of the plowed fields next the closes, on the left hand of the horseway to Cherry-hinton, and in Cherry-hinton church-yard, and many other places.

N. It has been found that, if a she-goat is fed with Danewort for a day and a night, its stomach will be purged by the sap:[252] Jo. de Monte in the work of *I.B.*

171 *Echium vulgare* L.

Echium vulgare *C.B. I.B. Park. Ger. emac.* Lycopsis *Cord. in Dioscor.* Anchusa major *Cæs.* sylvestris *Thal. Vipers Buglosse.* In ploughed fields.

172 *Echium vulgare* L.

Echium alterum *Dod.* Lycopsis *C.B.* Lyc[opsis] Anglica *Lob. Park. Ger. Lugd.* Cynoglossa vera *Cast.* Borago sylvestris tenuifolia *I.B. Wall Buglosse.* On walls & in drier fields.

---

[250] Moufet (1634) says here that he has observed only five species of this genus. See also paragraph 4 of Ray's second note on *Rapum sylvestre* [495] and footnote 647.

[251] See footnote 638.

[252] *lacte*, literally "by/with milk", but the word has also been used since classical times for the milky sap of plants. Danewort does not produce latex, but it seems unlikely that real milk is intended here, as suggested by Ewen & Prime's (1975) manifestly incorrect translation, "… will be filled with milk".

For J. Bauhin Onosma *Matth. Lac. Lon. Cast.*, Cynoglossum Dioscoridis *Matth.* & Echium vulgare seem to be not three but one & absolutely the same plant, as he demonstrates in many ways. *Volume 3, book 33, page 587.*[253]

Echium scorpioides arvense, *see* Myosotis scorpioides arvensis.[254] [*Myosotis arvensis* 381]

173 *Myrica gale* L.
Elæagnus Cordi *Lob.* Chamælæagnus *Dod.* Rhus sylvestris Plinii *Dod. Gall.* sylv[estris] altera *Lugd.* IV, sive myrtifolia Belgica *C.B.* Myrtus Brabantica *Ger.* Rhus sylv[estris] sive Myrtus Brabantica aut Anglica *Park.* Gagel Germanorum et Myrtus Brabantica quibusdam *Ad. Lob.* Gale frutex odoratus Septentrionalium *I.B. Sweet Willow, Gaul, Dutch Myrtle. In the fens in the Isle of Ely in many places abundantly.* The flowers appear in May and June and in July and August the seed is fully formed.

N. Scented plants are almost all bitter and not uncommonly even acrid. See *Theophr. de caus., book 6, chapter 13.*[255]

174 *Pastinaca sativa* subsp. *sylvestris* (Mill.) Rouy & E.G. Camus [256]
Elaphoboscum *Matth. Dod. gal. Cast.* erraticum *Tab.* Pastinaca latifolia II, sive sylvestris latifolia *C.B.* latifolia sylvestris *Ger. Park. Dod.* Past[inaca] Germanica sylvestris, quibusdam Elaphoboscum *I.B. Wild Parsnep.* On the margins of fields.

175 *Kickxia elatine* (L.) Dumort.
Elatine folio acuminato *Park.* altera *Dod. Lob. Ger.* fœmina folio anguloso *I.B.* sagittæ folio *hort. Pat. hort. Gron. hort. Lugd-Bat.* II, sive folio acuminato, in basi auriculato, flore luteo *C.B.* Linaria hederulæ folio *Col. Sharp-pointed Fluellin.* Among standing corn.

176 *Kickxia spuria* (L.) Dumort.
Elatine folio subrotundo *C.B. Park.* mas folio subrotundo *I.B.* Veronica fœmina *Matth. Fuch. Dod.* fœmina Fuchsii sive Elatine *Ger. Female Fluellin, or Round-leaved female Fluellin, or Speedwell.* With the preceding.

Eleoselinum, *see* Apium palustre vel Paludapium. [*Apium graveolens* 53]

177 *Inula helenium* L.
Enula campana *Offic[inarum] Cam. Park.* Helenium *Brunf. Matth. Dod. ut. Ad. Lob. Ger. Cam.* vulgare *C.B.* I Dioscoridis *Ang.* Helen[ium] sive Enula campana *I.B.* Panax Chironium Theophrasti *Ang. Cord. Elecampane. In the fields about Madingley in great plenty, and at Barton and many other places.*[257]

---

[253] See also Ray's comments in *Emendanda* of the 1663 and 1685 appendices, in which he suggests that *Echium alterum* is synonymous with the preceding species.

[254] Ray's chosen name is *Myosotis scorpioides hirsuta*.

[255] Theophrastus' *De causis plantarum*, VI. 9. 4 in the Loeb edition, where the translation is: "... in a way the bitter is to a greater extent the origin of fragrance, since it is hard to find any fragrant thing that is not bitter, but many non-sweet things have excellent flavour ... ."

[256] Ray lists the cultivated plant (subsp. *sativa*) as *Pastinaca sativa* [422].

[257] In his contribution to the 1695 edition of *Camden's Britannia*, Ray (1695, column 417) reported the Cambridgeshire localities thus: "*In the pasture-fields about Madingley, Coton, Barton, &c. in great plenty.*"

MAIN CATALOGUE

178 *Equisetum arvense* L.
Equisetum arvense longioribus setis *C.B. Park.* segetale *Ger.* Polygonum fœmina *Fuch. ico.* Hippuris minor *Trag. Dod. Thal.* arvensis major *Tab.* Equis[etum] minus terrestre *I.B. Corne Horse-tail.*

179 *Chara* spp.
Equisetum fœtidum sub aqua repens *C.B. I.B. Park.* 8, seu fœtidum sub aqua repens Bauhini *Ger.* minimum aquis cœnosis innatans, vel sub iis occultatum semper, brevissimis et asperis setis ac caulibus lutosum virus olentibus[258] *Lugd. Stinking water Horse-tail.*

180 *Equisetum fluviatile* L.[259]
Equisetum nudum *Ger.* junceum Tragi *Lugd.* junceum sive nudum *Park.* X, i.e. foliis nudum non ramosum sive junceum *C.B. Naked Horse-tail.* In moist and marshy places. *Jean Bauhin, in volume 3, book 36, chapter 27*, unites this species with those that are possessed of leaves and maintains that it is their shoot,[260] and we too support this opinion, since we have observed some naked stems, others clothed with leaves and yet others only partly naked arising, as far as one could conclude, from the same root.

181 *Hippuris vulgaris* L.
Equisetum palustre brevioribus foliis polyspermon *C.B.* alterum brevioribus foliis *Park.* Polygonum fœmina *Matth. Eric. Cord. Gesn. Dod. Amat. Ang. Cast. Lugd. Tab.* fœmina Dioscoridis *C.B.* Equiseti facie Polygonum fœmina *I.B. Female Horse-tail. In the rivulet that runs by Paper mils, and in that which comes from Trumpington in many places.*

182 *Equisetum* sp.[261]
Equisetum primum *Matth. Lac.* majus *Lob. Ger.* majus palustre *Park.* majus aquaticum *I.B.* palustre longioribus setis *C.B.* Hippuris major *Brunf. Dod. The greater marsh Horse-tail.* By rivulets and in watery places.

183 *Ceratophyllum demersum* L.
Equisetum palustre ramosum aquis immersum, seu Millefolium aquaticum equisetifolium. It has a fibrous root and very weak stems in proportion to the size of the plant which are rounded, jointed, weak, brittle, two feet or more long and branched into several small stems. As in a horsetail, the leaves encircle the joints and are green, brittle, single at first and then divided into two, like horns, and usually each horn is divided again into two more, with certain swellings on the front

---

[258] This last name can be translated as "least [Horsetail] floating in muddy waters or always hidden under them, with very short, rough bristles and stems that smell of muddy slime".

[259] For the reasons for treating the Cambridgeshire plant as *Equisetum fluviatile* rather than *E. hyemale* L. (to which it has been ascribed by earlier authors and to which some of the synonyms undoubtedly refer), see Oswald & Preston (1998); this paper also discusses the identification of the other species listed under *Equisetum* by Ray.

[260] *eorúmque asparagum esse asserit*. The Latin *asparagus* has been used for an asparagus-like sprout or shoot (as well as for asparagus itself) since the time of Pliny (died AD 79) at least. See also footnote 142.

[261] Despite Mattioli's *Equisetum primum* indubitably being *Equisetum telmateia* Ehrh., Oswald & Preston (1998) present strong evidence for rejecting earlier authors' identification of this plant as that species, which is very local in Cambridgeshire and indeed may not have occurred in Ray's time.

edge, as if it were furnished with small teeth. The flowers cling closely to the stems on both sides at the joints in clusters and are mossy. *Horse-tail water Millfoil.* In more sluggish waters almost everywhere: so that we are the more surprised that it appears in the work of no author that we have consulted, or at least that we have not been able to find it despite careful investigation. An Hippuris lacustris foliis mansu arenosis *Gesn*?[262]

184 *Calluna vulgaris* (L.) Hull

Erica vulgaris *Ger. Park. Trag.* prima *Matth. Dod.* folio myricæ I, sive Erica vulgaris glabra *C.B.* myricæ foliis *Ad. Lob.* vulgaris humilis semper virens flore purpureo et albo *I.B. Common heath. On Gogmagog hills and Newmarket heath.*

Erigeron, *see* Senecio. [*Senecio vulgaris* 535]

185 *Rorippa sylvestris* (L.) Besser

Eruca aquatica *Ger. emac. Park. Small water-Rocket. In many places by watercourses and rivers, but especially in the Fens.*

Eruca lutea latifolia, *see* Barbarea. [*Barbarea vulgaris* 78]

186 *Sisymbrium officinale* (L.) Scop.

Erysimum Dioscoridis *Dod. gal.* Dioscoridis Lobelio *Ger.* vulgare *C.B.* Tragi flosculis luteis juxta muros proveniens *J.B.* Irio *Matth. Cord. in Diosc. Lac. Lugd.* Irio sive Erysimum *Ad. Lob. Dod. Cast.* Ir[io] sive Erysimum vulgare *Park.* Sinapi alterum genus sylvestre *Fuch. ico.* Verbena fœm[ina] & Sinapi 7 *Trag.* Verbena *Tur.* recta sive mas *Fuch. Lugd.* Cleome Octavii *Ang. Hedge Mustard. By hedges & walls.*

Erysimum Cereale, *see* Fegopyron. [*Fagopyrum esculentum* 195]

187 *Euphorbia exigua* L.

Esula exigua *Trag. Lob. Ger. emac.* Tithymalus cyparissias III, i.e. Tithymalus sive Esula exigua *C.B.* Tithymallus leptophyllos *Matth. Cam. Lugd. Thal. Park.* Tithymallus minimus angustifolius annuus *J.B.* Peplis minor Dalechampii *Lugd. Dwarfe Spurge, or small annuall Spurge.* Among standing corn.

188 *Euonymus europaeus* L.

Euonymus *Matth. Dod. ut. Ang. Tur. Gesn. hort. Ad. Lob.* vulgaris *Park.* I, sive vulgaris granis rubentibus *C.B.* Fusanus Crescentio *of the same author.* Euon[ymus] Theophrasti *Ger.* Tetragonia Theophrasti *Lugd.* Euon[ymus] multis, aliis Tetragonia *I.B.* Siler Plinii *Cæs.* It flowers in May and as Autumn approaches its quadrangular pods are tinged with purple colour. *Spindle-tree, Prickwood. In Madingley and Kingston woods, and elsewhere.*

N. The branches & berries of this shrub are somewhat quadrangular; that is four lines of grey or russet colour run down the length of the green branches. Notwithstanding this, however, the branches are not so square that what *Spigel[ius]* observes in *Isag[oges]*,[263] book 1, chapter 10, should not still be

---

[262] "Is it the lake Mare's-tail with leaves gritty to chew, of [Conrad] Gesner?" *Mansu*, from *mando* ("I chew"), is a rare part of the verb known as the supine in *-u*, not a misprint for *morsu* as we have previously suggested (Oswald & Preston 1998). Raven (1950, p. 86) mistakenly translates the author as "of Gerard". Ray later realised that the description of *Millefolium aquaticum cornutum* in Bauhin & Cherler (1651, vol. 3(2), p. 776) applied to this plant but that the accompanying illustration did not: see the entries for this plant in *Emendanda* of the 1685 appendix and Ray's *Synopsis* (1690, 1696).

[263] For Adrianus Spigelius see footnote 31.

MAIN CATALOGUE

considered universally true, that undoubtedly all shrubs and trees have round stems.

189 *Bidens tripartita* L.
Eupatorium cannabinum fœmina *Ger.* cannab[inum] fœm[ina] Septentrionalium *Lob.* cannab[inum] chrysanthemum *Tab.* aquaticum flore aureo *hort. Hafn.* aquaticum duorum generum *Park.* II, sive Cannabina aquatica folio tripartitò diviso *C.B.* Hepatorium aquatile *Dod.* Verbena supina *Trag.* Verbesina sive Cannabina aquatica flore minùs pulchro elatior & magìs frequens *I.B.* Water Hempe-Agrimony with a yellow flower. In watery places.

190 *Bidens cernua* L.
Eupatorium aquaticum folio integro *Park.* Eupatorii cannabini fœminæ varietas altera *Ger. emac.* Eup[atorium] III, sive Cannabina aquatica folio non diviso *C.B.* Verbesina pulchriore flore luteo *I.B. in append[ix], book 2. Water-Hempe, or Agrimony with an undivided leafe.* In the ditches cut out of the river behind Peter-house, and in the ditches near the bridge at Audrey causey and in many other places plentifully.

191 *Eupatorium cannabinum* L.
Eupatorium cannabinum *Park. C.B.* cannab[inum] mas *Ger. emac.* vulgare *Matth. Dod. Ludg. Tur.* adulterinum[264] *I.B.* Eup[atorium] as believed by Avicenna *Ang. Gesn. hort. Cam.* Herba S. Kunigundis *Trag.*[265] Trifolium cervinum aquaticum *Gesn. hort. Common Hempe-Agrimony or Dutch Agrimony.* By streams.

N. Anyone who, having squeezed the flower of this plant with his fingers, has applied it to his nostrils & has chewed a sufficiently succulent branch of the plant for some time will, we think, as a result of the nausea arising therefrom, believe Gesner *in one of his letters*,[266] who attributed a strong power of provoking vomiting to the roots of this plant.

Eupatorium Græcorum, *see* Agrimonia Officinarum. [*Agrimonia eupatoria* 16]

Euphrasia cærulea, *see* Myosotis Scorpioides. [*Myosotis scorpioides* 382]

Euphrasia rubra, *see* Cratæogonon Euphrosynes facie. [*Odontites vernus* 158]

192 *Euphrasia* spp.[267]
Euphrasia *J.B. Ger. Lob.* vulgaris *Park.* officinarum *C.B.* alba *Brunf.* Euphragia *Matth. Ang. Cæs.* Ophthalmica sive Ocularia *Eric. Cord. Eyebright.* In pastures, as upon the moors, and Gogmagog hills, and almost in all meadows.

---

[264] i.e. counterfeit, presumably as distinct from *Agrimonia* [*Agrimonia eupatoria* 16] (q.v.), which had also been called *Eupatorium* since classical times. See also footnotes 13 and 412.

[265] Cunigunde or Kunigunde (c. 975–1040) married Henry II, Duke of Bavaria and (from 1014 until his death in 1024) Holy Roman Emperor. Both husband and wife were canonised and are buried in the cathedral that Henry founded at Bamberg in Bavaria. Their current magnificent marble tomb was sculpted between 1499 and 1513, when Tragus (Jerome Bock) was young.

[266] Gesner was a prolific correspondent (see Chapter 4) and his letters were published in various works, including Gesner (1577) and Bauhin (1591). We have not attempted to find the source of this reference.

[267] Only two species of this critical species complex have been recorded in Cambridgeshire, *Euphrasia nemorosa* (Pers.) Wallr. and *E. pseudokerneri* Pugsley, both plants of base-rich grassland, but *E. officinalis* subsp. *anglica* (Pugsley) Silverside and *E. confusa* Pugsley are known from more acidic habitats in Norfolk and may have been present in Cambridgeshire in Ray's day.

F

193 *Vicia faba* L. var. *faba*[268]

Faba major vulgaris *Ad.* major hortensis *Ger.* veterum Græcorum *Jo. Bodæ. in Theophr. hist.* Boona sive Phaselus major *Dod.* Faba, cyamos leguminosa *J.B.*, who attributes this name to all kinds of bean. *Great garden Beans.*

N. All seeds have their origins in that part by which they adhere to plants, and from there the root originally grows. In the bean this is called by Aristotle ὄμμα [omma], i.e. an eye. *Spigel[ius in] Isag[oges]*,[269] book 1, chapter 6. This is true in most seeds, but in fact we are doubtful whether or not it holds in all: take for instance[270] the kernel of the Date, which is situated in a swollen part of the stone around the middle of its length, where a minute 'navel' is to be seen from which germination begins; but whether in fact the stone is attached to the fruit at that point we have not yet discovered; rather we believe the opposite.

194 *Vicia faba* var. *equina* Pers. and/or var. *minor* Peterm.

Faba minor *Ad. Lugd.* minor sylvestris *Park.* minor sive equina *C.B.* Boona sive Phaselus minor *Dod.* *Common Beans, or horse-Beans.* It is sown in fields.

Fabaria, *see* Telephium. [*Sedum telephium* 566]

Farfara, *see* Tussilago, & Populus alba. [*Tussilago farfara* 603 and *Populus alba* and/or *P. alba* × *tremula* 456]

Fagus sepium, *see* Ostrys Theophrasti.[271] [*Carpinus betulus* 86]

195 *Fagopyrum esculentum* Moench

Fegopyron *Dod. Park.* Fagotriticum & Tragopyron *Ger. emac.* Fagotriticum *J.B.* Erysimum Cereale *C.B.* Erysim[um] Theophrasti *Lob.* Ocymum veterum *Trag.* Oc[ymum] Cereale *Clus. pan. Tab.* Frumentum Saracenicum *Matth. Lugd. Cast.* Frumenti genus folio hederæ *Cord. in Diosc.* *Buck-wheat, Beech-wheat, Brank.* It is sown in fields, but rather rarely.

196 *Anisantha sterilis* (L.) Nevski

Festuca & Avena Græca *Lon.* avenacea sterilis elatior *C.B.* Ægilops prima *Matth. Tab.* Æg[ilops] Matthioli fortè *J.B.* Lolii I genus *Trag.* Bromos sterilis *Lob. ico. Ger.* Virgilio Avena sterilis[272] *Trag.* *Wild Oats.* By hedges.

197 *Bromus hordeaceus* L.

Festuca altera *Dod.* Fest[uca] graminea glumis hirsutis *C.B.* Bromos sterilis altera *Lob. Ger.* Lolium *Lon.* Gramen murorum *Dalechamp. Lugd.* Gramen Gros

---

[268] *Vicia faba* is usually divided into three varieties, var. *faba*, with large beans, and var. *equina* and var. *minor*, with successively smaller seeds (Bond 1995; Sell & Murrell 2009).

[269] For Adrianus Spigelius see footnote 31.

[270] *pro instantia sit*, literally "let it be for an example". Four of the words used in the description that follows are defined in Ray's 'Interpretation of terms', i.e. *nucleus* ("kernel"), *Dactylus* ("Date"), *ossiculum* ("stone") and *umbilicus* ("navel", here used in a different sense from either of the two definitions there).

[271] *Ostrys Theophrasti*, like *Fagus sepium*, is listed by Ray as a synonym of *Betulus*.

[272] In *Eclogues*, 5: 37 Virgil wrote of two cornfield weeds *infelix lolium et steriles nascuntur avenae* ("unfruitful darnel and sterile oats spring up") and in *Georgics*, 1: 154 he repeated the line but with *dominantur* ("are dominant") instead of *nascuntur*.

Montbelgardense *J.B. Drank, or small wild Oats*. It is a plague of rye,[273] very rarely found among other corn crops. *J.B.* It should be considered whether the plant that we know is not rather Festuca avenacea sterilis humilior *C.B.*[274]

Filipendula aquatica, *see* Oenanthe. [*Oenanthe fistulosa* 393]

198 *Filipendula vulgaris* Moench

Filipendula *Ger. Dod. Matth. Cord. hist. Tab. J.B.* vulgaris *Park.* vulgaris, an molon Plinii *C.B.* Oenanthe *Fuch. Lac. Lob.* vulgaris *Ad.* of Dioscorides *J.B. Common Dropwort.* On Gogmagog hills, & frequent elsewhere.

199 *Dryopteris filix-mas* (L.) Schott

Filix mas *Ger. Matth. Fuch. Dod. ut. Tur. Lob.* mas vulgaris *Park.* F[ilix] non ramosa III, sive dentata *C.B. Male Fern. About Gamlingay.*

N. 1. We have heard from a reliable source that in Shropshire[275] not a few people are accustomed to use this fern when dried instead of hops for brewing beer.

2. *J.B.* reports *in his history of the fountain and bath of Boll, page 204*,[276] from an account by Master Rantzius,[277] that this fern was hung up in the Julian Forum in small bundles, on which flies gathered in the evening, and that they were thus easily captured when a sack was thrown over each bundle.

200 *Pteridium aquilinum* (L.) Kuhn

Filix fœmina *Cord. Ang. Fuch. Lac. Lob. Matth. Dod. Ger.* foem[ina] vulgaris *Park.* major & prior Trago sive ramosa repens *J.B.* ramosa major pinnulis obtusis non dentatis *C.B.* Avia Columellæ, quia avium pennas refert. *Jo. Bodæ. in Theophr. hist.*[278] *Female Fern, Common Brakes.*

---

[273] *Secales* is presumably an error for *secalis*, the genitive singular of the neuter word *secale*.

[274] This entry perpetuates a confusion by previous authors, including Gerarde and Parkinson, which was eventually sorted out by Ray in his *Catalogus plantarum Angliae* (1670, pp. 111–112). The Cambridgeshire plant is listed there as *Festuca avenacea hirsuta paniculis minùs sparsis*, with the comment *Gramini murali Dalechampii simile si non idem* ["Similar to, if not the same as, Daléchamps' *Gramen murale*"] J.B. Ray remarks that "this grass, than which nothing is more frequent in England on walls and at the margins of fields, corresponds well enough to J. Bauhin's description". We agree with Babington (1860) that it is *Bromus hordeaceus* L. In 1670 Ray separated *Festuca altera* Dod. as another species, which he had not seen outside cornfields and in Britain knew only from Warwickshire and the Isle of Man. He ascribed the synonyms *Festuca graminea glumis hirsutis* C.B. and *Gramen Gros Montbelgardense* J.B. and the English name "Drank" to this plant, which is the current *Bromus secalinus* L. This species is the "plague of rye", and it perhaps evolved as a mimic of this crop. It was first recorded in Cambridgeshire in 1727 (Perring *et al.* 1964).

[275] *à fide digno accepimus*, literally "from a worthy fidelity we have received [that]"; *in agro Salopiensi*, literally "in the Salopian field". From mediaeval times *ager* was often applied to a county; compare footnote 15 in Ray's 'Preface'.

[276] Jean Bauhin's *Historia novi et admirabilis fontis balneique Bollensis* (1598), which describes the springs and baths at the German spa town of Boll (Bad Boll) on the Swabian Alb, is mentioned briefly under *J.B.* in Ray's 'Explanation'. The reference here is to Book 4, which deals with natural history and illustrates some of the fossils found in the rich late Jurassic strata of the area. It is missing from some copies of the work.

[277] *ex narratione D. Rantzii*: *D.* is again the standard abbreviation for *Dominus*, used for a Master of Arts. See also *J.B.* in Ray's 'Explanation', his Note 2 after *Linum sativum* [336] and his entry for *Myosotis scorpioides hirsuta* [381].

[278] Literally "Columella's Bird-plant, because it recalls the feathers/wings of birds". *Avia* can mean "grandmother" or, as a feminine adjective, "untrodden" or "out of the way", but it is evidently here related to *avis* ("bird"). Bodaeus (1644, p. 1167.2; see footnote 23) wrote: "Columella, quod

N. 1. Trag[us], J.B., C.B. *in his synonyms on Matthiolus*, Park[inson] *in Theatrum Bot[anicum]* & Johnson *in Gerard[us] emac[ulatus]* assert that a fern has been derived from seed, in contrast to Theophrastus, Dioscorides, Pliny, Galen, Dodonæus, Spiegel *in Isag[oges]*[279] & others. On this matter consult *J.B., volume 3, book 37, page 726, Johnson in his revision of Ger[arde], page 1129,* and *Park[inson] in Theatr[um Botanicum], pages 1036–1037.*[280] There are many superstitious beliefs about the collection of the seeds of fern, which some assert should be undertaken on the very day of the solstice and others on the eve of St John the Baptist's Day[281] with linen cloths spread out below.

2. Not only fern but also Adianthum, Asplenium and Phyllitis[282] lack a seed. But in fact among other plants Jasmine lacks a fruit, Fig a flower, wild Asparagus & Dodder a leaf and Mistletoe a root.[283]

---

avium pennas referat, aviam vocat. lib. VI. de re rust. cap. XIV." For Lucius Junius Columella see Chapter 4.

[279] For Spiegel's *Isagoges* see footnote 31.

[280] Johnson says: "neere the old plants I haue obserued verie many small yong plants growing, which came by the falling of the seed from those dusty scales: for I beleeue all herbes haue seeds in them-selues to produce their kindes, *Gen[esis] 1. 11 & 12.*" Parkinson says: "... those heads on the backe of them ... falling on the ground doe spring, for whereby else should it be encreased, seeing God in the beginning appointed every herbe and tree, to have seede of their kind (and not of another) within them." On p. 1039 he continues: "... experience also sheweth that they beare seede, although *Theophrastus, Galen, Dioscorides* and *Pliny* following him, say they neither beare flowers nor seede: for if about Midsommer (for then usually it is ripe) you gather the stalkes of Ferne and hang them up on a thread with some faire white Paper or cloth under them, you shall finde a small dust to fall from them which is the seede, and from them doe spring plants of the same kindes ..., for as I said before no herbe growing on earth or in the water (except some with double flowers which are encreased by the roote) but doe beare seede, &c."

[281] i.e. on 23 June, the eve of Midsummer Day. Parkinson (1640, p. 1039) says that the seed so gathered was "superstitiously held by divers, not only Mountebankes and Quacksalvers, but by other learned men (yet it cannot be said but by those that are too superstitiously addicted) to be of some secret hidden vertue, but I cannot finde it expresst what it should be". For other beliefs relating to Midsummer Eve and Midsummer Day see footnotes 82, 379 and 721.

[282] *Adiantum* or *Adianthum* was used in different senses in the 17th century for, as Parkinson (1640, p. 1049) says, "There be divers herbes accounted to be Maidenhaires by divers authors." In the *Catalogus* Ray includes the current *Asplenium ruta-muraria* L. and two mosses; C. Bauhin (1623) excludes the former, treating it as *Ruta muraria*. Both Bauhin (1623) and Parkinson (*loc. cit.*) also include *Asplenium adiantum-nigrum* L., as does Ray in his later works, a treatment followed in the 1685 appendix [B1]. *Asplenium* usually refers at this period to *Asplenium ceterach* L., which was not recorded in Cambridgeshire until 1967. (*Asplenium trichomanes* L. was generally listed as *Trichomanes*, as here in the *Catalogus* [577].) *Phyllitis* included *Asplenium scolopendrium* L., which is reported from Cherry Hinton in 'Index of places' and in the 1663 and 1685 appendices [A29, B75].

[283] *Inter alia verò fructu caret Gelsemimum, flore Ficus, folio Asparagus sylvestris & Cuscuta, radice Viscum*. Most of these plants do not appear in the catalogue, but see Ray's note after his entry for *Cuscuta major* [161]; his entry for *Viscum* [622] does not mention the lack of a root. Most garden plants of the common or cottage jasmine, *Jasminum officinale* L., are very uniform and rarely set seed, suggesting that they may represent a single, self-sterile clone (Green & Miller 2009, p. 5). Parkinson (1629, p. 406) knew that the flowers of his *Iasminum album* "fall away without bearing any fruit at all, that euer I could learne in our Country" but reported that "in the hot Countries where it is naturall, it is said to beare flat fruit, like Lupines", whereas in fact its fruits are purple berries. Ray (1688a, p. 1599) withdrew his claim that jasmine does not fruit, saying (as translated by us): "*In colder regions the flowers wither without any fruit following on, whence some have stated without sufficient care that it is completely bereft of fruit, and we too formerly followed them and reported the same thing in our*

MAIN CATALOGUE

3. The ashes of burnt fern are suitable for making glass in preference to others, no doubt because they are rich in nitrous salt.

Filius ante patrem, *see* Lysimachia siliquosa.[284] [*Epilobium hirsutum* 352]

Flos cuculi, *see* Armerius sylvestris & Cardamine. [*Silene flos-cuculi* 61 and *Cardamine pratensis* 106]

Fœniculum aquaticum stellatum, *see* Stellaria aquatica. [*Callitriche* spp. 562]

Fœniculum aquat[icum], *see* Millefolium aquaticum. [*Hottonia palustris* 365 and *Ranunculus trichophyllus* 366[285]]

201 *Fragaria vesca* L.

Fragaria *Trag. Matth. Fuch. Dod. Lob. Ger.* vulgaris *Park. C.B.* Fragaria ferens fraga rubra[286] *J.B. Common Strawberry.* In woods.

N. In the case of someone who had eaten strawberries in too great a quantity I have seen many particles of them in his chamber-pot, so that his kidneys might be thought to be liquefying. To such an extent are they carried to the kidneys. *C[asparus] Hofman[nus], De medicam[entis] offic[inalibus], book 2, chapter 95.*

202 *Potentilla sterilis* (L.) Garcke

Fragaria minimè vesca *Park.* minimè vesca sive sterilis *Ger. emac.* sylvestris minimè vesca sive sterilis[287] *Lob. ico.* VII, sive sterilis *C.B.* non fragifera, vel non vesca *J.B. Barren Strawberries.* In heaths & thickets.

203 *Fraxinus excelsior* L.

Fraxinus vulgaris *Ger. Park.* I, sive excelsior *C.B.* the first & Βουμελια [Bumelia] of Theophrastus[288] *of the same author.* bumelia for Pliny & Bellonius *of the same author.* Fraxinus vulgaris, Dioscorides' Μελια [Melia], Theophrastus' lofty & tall kind of Melia[289] *J.B. Common Ash tree.* This tree has been made very famous by the commendation of Homer and the spear of Achilles, says Pliny, *in book 16, chapter 13.*[290]

Catalogue of Cambridge plants. But now, being better informed both by Master P. Hermann and by Signor Triumfetti of Rome, we know that it bears fruit, in fact greenish berries containing round *seeds* marked with a small black ring." Raven (1950, p. 110) overlooked this mention of jasmine in the *Catalogus* and argued that the passage in *Historia plantarum* must have referred to an unpublished catalogue of garden plants (see footnote 346).

[284] *Filius ante patrem* is cited by Ray as a synonym of *Lysimachia siliquosa hirsuta magno flore* (q.v. for an explanation of the name), but also as a name for *Tussilago* [*Tussilago farfara* 603].

[285] Ray lists *Fœniculum aquaticum* 2 as a synonym of *Millefolium aquaticum dictum Viola aquatica* [*Hottonia palustris* 365] and *Fœniculum aquaticum* 3 as a synonym of *Millefolium aquaticum ranumculi flore & capitulo* [*Ranunculus trichophyllus* 366].

[286] "Strawberry-plant bearing red strawberries".

[287] Literally "Wild [Strawberry], by no means edible, or sterile".

[288] The reference is to *Historia plantarum*, III. xi. 3–5 in the Loeb edition, where two kinds of ash are distinguished, this one, called by people in Macedonia βουμελιος [bumelios] ("ox-ash"), being described as ὑψηλη [hypsele] and εὐμηκης [eumeces] ("lofty and tall"). The second kind, Manna Ash, *Fraxinus ornus* L., is not British; Theophrastus describes it as a mountain tree that is shorter and less vigorous.

[289] Μελιας [*Melias*] *genus excelsum & procerum Theophrasti. Melias* is a Greek feminine genitive singular.

[290] Pliny's *Natural history*, XVI. 62 in the Loeb edition. The reference is to *Iliad*, 16: 140–144, where Homer describes Achilles' "ponderous, large, strong spear" made from an ash from Mount Pelion, which only he of the Greeks could wield, a gift from the centaur Chiron to Achilles' father Peleus. See also footnote 186.

N. 1. The Greeks related that the leaves of this tree were lethal to beasts of burden, but Pliny *in the passage cited* & [P.] Gassendus *in volume 2, page 148*, contradict this.[291]

2. The inner bark of this tree makes the water in which it has been soaked blue, which they give to hens when they are ravaged by disease. *Cæsalp[inus] apud Cam[erarium] in hort[o]*. We have made a trial of this matter in cold water without success, but in lukewarm water in a bath it produced a pale blue dye.

3. The rings which are seen in the trunks & branches of trees cut crossways show themselves more openly in the wood of this tree than in others. Petrus Gassendus has observed *in volume 2, page 178*, that in trees growing within the tropics these rings are equidistant from each other throughout[292] & have the heartwood in the true centre, as in the Brazilian wood called acanthine, but in fact in the remaining regions, whether located to the South or to the North, they are extended towards the south and are contracted in the part looking towards the pole, so much so that the heartwood is found with an appearance that is always formed eccentrically. Hence many authors have taught that from this position of the rings one can find the south and the more experienced Husbandmen warn those about to transplant trees that they should retain for the newly planted trees the same orientation relative to the sky that they had in their former location.[293]

On the subject of these rings we shall append what we have observed to be worthy of investigation and attention.

1. The age of a tree or branch is discovered from the number of rings, since the number of rings equals the number of years that it has lived, unless the tree has ceased to grow.

2. The inner rings are closer to each other[294] than the outer, both because they are pressed and constrained by the outer ones and because the wood becomes dry and contracts with time; as a result, if two outer rings are closer to each other than two inner ones, it is probable that the tree has suffered an injury during those years in which the outer rings were being formed: whence it follows also that more aged trees or branches have less distant rings. This suggestion is perhaps not universally true; for it is probable that in the trunks of the largest trees & of those which because of their age have now ceased to grow, since the sap rises more slowly and is supplied more scantily and in such a size of trunk an incredible abundance of food would be required to produce thick or deep rings, the rings would be narrower and have less depth.

3. The inner rings, while they are reduced by the passing of the years and by dryness, compress & restrict the spongy heartwood more, until in some trees it is completely concealed. This compression of the heartwood is manifested most evidently in Elder.

---

[291] *contradicentibus Plinio ... & Gassendo ...*, literally "with Pliny ... & Gassendus ... contradicting". Pliny's statement is in XVI. 64 in the Loeb edition.

[292] *omni ex parte æqualiter inter se distant*, literally "from every part equally among themselves are distant".

[293] *& Geoponici peritiores arbores translaturos monent, ut eundem denuò plantatis partium situm respectu plagarum cœli servent quem in priore statione obtinebant*, literally "& the more experienced Husbandmen [from the Greek γεωπονοι, geoponoe] warn [those] about to transplant trees that the same, for [those] anew planted, situation of the parts in respect of the tracts of the sky they should keep which in the former resting-place they were possessing".

[294] *magis ad se invicem appropinquant*, literally "more to themselves reciprocally/by turns approach".

MAIN CATALOGUE

4. The wood between the inner rings is harder than between the outer ones & for the most part is tinged with a deeper colour, certainly never with a paler one; hence the outer part of the wood is called *Alburnum* in Pliny, *book 16, chapter 38*,[295] from its colour.

5. Trees & branches which grow to a height over several years show fewer rings in their upper part than in their lower part; moreover the outer ones, but not the inner, are the same in each;[296] for the outer are superimposed every year on the inner, & as a result the inner, which, as we have said, do not reach the very top, always end in a point and each makes as it were a cone, as can clearly be seen in fir planks cut along the length of the tree. The pecten arboris *for Pliny, book 16, chapter 38,* διαϕυσις [diaphysis] *for Dioscorides* & κτηδων [ctedon] *for the same [author]* & *Theophrastus*, or the course of the lines that runs lengthways in timber and which in vernacular English speech we call *the grain of the tree*, arises from these rings cut along the length of the tree.[297]

6. It is likely that for every tree a limit to growth, as it were, & thus a number of rings has been fixed[298] by nature, which it would not be able to exceed. This is the conjecture of *Master Nid*, with whom I confess that I cannot agree.[299] For I think that, as long as a tree is alive, new rings are added every year, but always narrower & thinner ones until finally, when all its vigour has been spent, the tree dies. For this reason, however, it has not been possible to discover the age of an ancient tree, both because generally the inner wood decays and because the outer rings will be so thin that they will not easily be distinguished & counted.

204 *Fumaria officinalis* L.
Fumaria[300] *Trag. Matth. Fuch. Dod.* vulgaris *Park. J.B.* F[umaria] non bulbosa I, seu officinarum & Dioscoridis *C.B.* officinarum annua *hort. Hafn. hort. Gron.* Fumus terræ *Brunf. Cam. Thal.* Capnos *Ang. Ad. Lob.* Cerefolium felinum *of Schwenckfeld. Fumitory.* In ploughed fields & gardens.

Fungi
Fungi esculenti & non esculenti variæ species. *Mushromes of severall sorts.*[301]

N. A fungus, whatever kind it may be, is always malignant, always furnished with a provision of deadly properties, even if this may not be immediately felt, but, when the fungus is consumed frequently, I do not know what effect it may

---

[295] Pliny's *Natural history*, XVI. 182 in the Loeb edition.

[296] *utrisque communes sunt*, literally "to each of the two are common".

[297] The Latin name literally means "the comb of the tree": in his *Natural history* (XVI. 185 in the Loeb edition) Pliny in fact has this in the plural, *pectines*. The Greek κτηδων [ctedon] is also related to the word for "comb" and means "*line of fissure* in the fibre of wood" (Liddell, Scott & Jones 1940). Theophrastus describes in his *Historia plantarum* (V. 1. 9 in the Loeb edition) how these differ in different trees, while Bodaeus (1644, p. 514.2; see footnote 23) explains that Dioscorides (book 3, Chapter 1) uses the word when describing fungi with gills.

[298] *crescendi finem ... determinatum esse*, literally "an end of growing ... to have been determined".

[299] *Hæc est conjectura D. Nid, cui fateor me non posse assentire. D.* is again the standard abbreviation for *Dominus*, used for a Master of Arts. See also Ray's 'Preface' (including footnote 20) and Chapter 3.

[300] See the entry for *Capnos* in Ray's 'Etymology'.

[301] In addition to the fungi dealt with below, "*Jews-eare*, Fungus Sambucinus" [*Auricularia auricula-judae* (Bull.) Wettst.] is listed in Ray's index of English names and described in note 2 after *Sambucus*.

have with the passing of time through its hidden action in the inner working of the entrails. *Kircher about plague*.[302]

205 *Morchella* sp.[303]

Fungus favaginosus *Park. Lob.* rugosus vel cavernosus *J.B.* porosus *C.B.* Fungorum præcipuum genus *Gesn. in coll. Honey-combe-Mushrome. Beyond Ditton.* It grows in the month of April & up to the end of May.

206 unidentifiable bracket fungi

Fungi arborei ad ellychnia *J.B.* F[ungus] durus arborum sive igniarius *Park.* igniarius *Trag.* igniarius, vulgò esca *Cæs.* F[ungus] noxius III, sive F[ungus] in caudicibus nascens unguis equina figurâ *C.B.* Ellychnion Tarsicum Galeni *for Cornarius.* An F[ungus] lignosus arboreus ad furfures *J.B., volume 3, page 840?*[304] For it is a kind of touchwood *for Cæsalpinus*.[305] *Touchwood*. It is named thus in English in *Parkinson*. Tragus [Jerome Bock] also relates a method of preparing it in the work of *J.B.* They boil these fungi in lye & pound them when dry; then they cook them again in natron[306] so that they may become suitable for catching fire. We have seen it growing on the trunk of a walnut, which is described by *J.B. in volume 3, page 840*.

207 unidentifiable puff-ball fungi[307]

Fungus pulverulentus dictus crepitus lupi[308] *J.B.* Lupi crepitus sive fungus ovatus *Park.* Fungus 3 sive orbicularis *Ger.* orbicularis *Dod.* Fungus glomeratæ rotunditatis *Lob.* Fungus vesicarius, aliis Ovum lupi *Imperat.* F[ungus] noxius XLII, sive rotundus orbicularis *C.B.* Pezicæ Plinii *Cæs.* Fungus Chirurgorum *Schrod. pharmac. Puffe-balls, or Dusty Mushrome. On Gogmagog hilles, Newmarket heath and the Hill of health.*

---

[302] *... nescio quid successu temporis occultæ machinationis in interiore viscerum administratione moliatur. Kircherus de peste.* Literally "... I do not know what, in succession of time of hidden action in the inner working of the entrails, it may cause. *Kircher about plague*." This cautionary note is in contrast to the treatment of Parkinson (1640, pp. 1316–1324), who gives a relatively detailed account of the fungi under the headings "Fungi esculenti. Holsome Mushromes that may be eaten." and "Fungi pernitiosi. Dangerous Mushromes." and who stresses the edibility of fungi such as the morels which Gerarde (1597, p. 1385) had described as "venemous". For Kircher's work see footnote 197.

[303] This entry clearly refers to one of the spring-fruiting ascomycete fungi of the genus *Morchella* (the morels); *M. esculenta* (L.) Pers. is recorded from Cambridgeshire and is perhaps the most likely species to have been seen by Ray.

[304] The three preceding names may be translated as: "The third poisonous fungus, or the fungus growing on tree-trunks in the shape of a horse's hoof *of Caspar Bauhin*. Galen's lampwick [Greek ἐλλυχνιον] of Tarsus *for Cornarius*. Is it the woody fungus of trees with scales *of Jean Bauhin, volume 3, page 840?*"

[305] *Est enim igniarii species Cæsalpino*.

[306] *Decoquunt fungos hosce in lixivio, & siccatos tundunt; rursus nitro coquunt, ...* The OED defines "lye" thus: "Alkalized water, primarily that made by the lixiviation of vegetable ashes, but also applied (esp. with prefixed word as in *soap-lye, soda-lye*) to any strong alkaline solution, esp. one used for the purpose of washing." The on-line version of the OED defines "natron" thus: "Native hydrous sodium carbonate, $Na_2CO_3 \cdot 10H_2O$, crystallizing in the monoclinic system and occurring chiefly in solution and as evaporative residues, e.g. in dried lake beds; mineral salt containing this." See also footnote 808.

[307] Parkinson (1640, p. 1324) described *Fungus ovatus,* or *Orbicularis lupi crepitus* as "of severall sizes, some of the bignesse of a ball or balloone or a childes head more or lesse". It is unfortunate that Ray does not indicate the size of his Puffe-balls, so we do not know whether he saw the Giant Puffball, *Calvatia gigantea* (Batsch) Lloyd, or one or more of the smaller species.

[308] "Dusty/powdery fungus called wolf's fart".

N. Surgeons apply a bit of this fungus to a cut vein to staunch the blood and consolidate the wound. *Schroder[i] Pharmac[opœa], book 4, chapter 411.*

G

208 Galega officinalis L.

Galega *Ad. Lob. Lugd. Ger. I.B.* vulgaris *C.B. Park.* Galega sive Ruta capraria *Park. Matth. Gesn. hort.* Goats Rue. *Reported by Parkinson to have been found growing in the meadows about Linton: we could not find it there, and do suspect that it is not there to be found.*

Galega sylvestris Dod. *see* Aracus *Tab.* [*Vicia cracca* 55]

209 Stachys sylvatica L.

Galeopsis legitima Dioscoridis *Park. Clus. hist.* vera *Ger. emac.* Gal[eopsis] sive Urtica iners magna fœtidissima *I.B.* Lamium maximum sylvaticum fœtidum *C.B.* Urtica Heraclea *Lugd. Trag.* Urtica iners rubea spicata *Hort. Gron. hort. Bat.* among the indigenous plants. Hedge-nettle. Frequent in hedges.

N. The roots of this plant protect the Belgians' sea-walls against the force of the ocean. *Georg Bruin, in Theatr[um] Urbium, in the index under the letter B.*[309]

Galeopsis sive Urtica iners minor folio caulem ambiente *I.B. see* Alsine hederula altera. [*Lamium amplexicaule* 25]

Gallitricum, *see* Horminum sylvestre. [*Salvia verbenaca* 292]

210 Galium palustre L.

Gallium album *Tab. Ger.* palustre *Dod.* IV, sive palustre album *C.B.* Gallii albi sive molluginis vulgatioris varietas minor *Park.* Aparine lævis palustris *Thal.* White flowered Ladies Bedstraw. On Teversham moor, and many other moist places.

211 Galium verum L.

Gallium luteum *Lob. Ger. Park.* arvense luteum *C.B. apud Matthiolum.* I, sive luteum *C.B.* verum Dioscoridis & Galeni *I.B.* Bubonium *Gesn. cat.* Ladies bedstraw, Cheese-rening, in Yorkshire Maidenhair.

Garosmum, *see* Atriplex olida. [*Chenopodium vulvaria* 69]

---

[309] This refers to the six-volume work generally known as Braun & Hogenberg's *Civitates orbis terrarum* (1572–1618) after the title of the first volume. None of the other possible sources we have examined has an index which includes descriptive text. The reference to *Galeopsis* appears under Brielium [Brielle] in the index to volume 2, believed to have been published in 1575. Braun signs the dedication to volume 1 Georgius Braun and that to volume 2 Georgius Bruin. The discrepancy in title is less easy to explain, but all the volumes have different titles; the title of volume 2 is *De praecipuis, totius universi urbibus, liber secundus*, but those of volumes 5 and 6 are *Urbium praecipuarum mundi theatrum quintum* and *Theatrum praecipuarum totius mundi urbium liber sextus* respectively. The work is bibliographically complex as numerous editions appeared without distinguishing colophons or imprints and with the same title-page and plates as the first edition but with much variation in the text. Judging by the holdings in Cambridge, many copies are incomplete. For further details of this lavish work, see the introductions to the modern facsimile volumes (Skelton & Vietor 1965; Füssel 2008). Braun describes the tenacious and hairy roots of the plant "which the Batavians call Helm and some people Galeopsis". The Dutch name identifies it as Marram Grass *Ammophila arenaria* (L.) Link, still often planted to bind mobile dunes. The use of *Stachys sylvatica* or any other species of Lamiaceae for this purpose is most unlikely.

212 *Ulex europaeus* L.
Genista spinosa vulgaris *Ger. emac.* spinosa major longioribus aculeis *C.B.* spinosa major vulgaris sive Scorpius Theophrasti, quod Gaza Nepam transtulit[310] *Park. Furze, Whins, Gorsse.* On Newmarket heath and about Gamlingay.

213 *Cytisus scoparius* (L.) Link
Genista *Ger.* & also Chamægenista Anglica *of the same author* in the opinion of *C.B. & Johnson.* vulgaris sive scoparia *Park.* non spinosa I, sive angulosa & scoparia *C.B.* angulosa trifolia *I.B.,* who in his *volume 1, book 11,* has two descriptions & two pictures of the same, one without a Broomrape *on page 388* and the other with a Broomrape *on page 390;*[311] in fact he himself gives the reason why he has acted thus in the place cited last. Genista quadrato junco prima *Cæs. Common Broom.* In the closes about Gamlingay, and at Stitchworth.

214 *Genista anglica* L.
Genistella aculeata *Park.* aculeata foliosa *I.B.* Genista aculeata *Ger.* spinosa III, sive minor Germanica *C.B.*[312] *Needle-Furze or Petty Whins.* On a heath beyond Gamlingay by Sr Roger Burgoynes park plentifully.

215 *Genista tinctoria* L.
Genistella infectoria *Ad. Lob. Ger.* Genista tinctoria *Dod.* tinct[oria] Germanica *C.B.* tinct[oria] vulgaris *Park. Clus. hist.* Tinctorius flos *I.B. Fuch. Green-weed, Diers-weed, Woodwaxen.* On a bushy common on the north side of Madingley; at Cheveley, &^{c}.

N. All plants used for dyeing or staining are slender in structure & especially useful for loosening or curing obstructions, as can be seen in Saffron, Madder, Kermes grains, the inner bark of Barberry, Turmeric &^{c}. Also Tragus asserts that a decoction of Woad is useful for people with complaints of the liver.[313] The reason

---

[310] This is the full name on Parkinson's (1640, p. 1003) figure (with *Genista* at the beginning and the grammatically correct *quem* instead of *quod*); the latter part means "which Gaza [1529] translated [as] *Nepa*", as explained by Parkinson on p. 1005. Theophrastus' σκορπιος [scorpios] (e.g. *Historia plantarum*, VI. I. 3 in the Loeb edition) was a spiny plant without leaves, not *Ulex europaeus*, a western European species, but probably a species of *Genista* or a related genus. For Theodorus Gaza see footnote 21.

[311] *unam sine Rapo ... , alteram cum Rapo ... . Rapum* literally means "turnip" but here refers to the parasitic *Orobanche rapum-genistae* [469], which has a large underground tuber like a turnip; hence its specific epithet.

[312] See the entry for this species in *Emendanda* of the 1685 appendix, where Ray proposes changes to the synonyms.

[313] *Plantæ omnes tinctoriæ sive infectoriæ tenuium sunt partium* [literally "are of slender parts"], *& ad obstructiones solvendas sive referendas utiles, ut videre est in Croco, Rubia tinctorum, granis Chermes, interiore Berberis cortice, Curcuma &c. Glasti etiam decoctum jecinorosis utile esse Tragus asserit.* Saffron Crocus, *Crocus sativus* L., was grown for its golden yellow styles used especially to colour and flavour food; see also Ray's entry for *Crocus* [159] and the notes following it. Madder, *Rubia tinctorum* L., was grown for the red dye known as rose madder obtained from its roots. Kermes grains were the red scale-insects found on Kermes Oak, *Quercus coccifera* L., once thought to be the tree's *cocci* (i.e. berries), which in former times were used to manufacture a scarlet dye, just as South American cochineal was used after its discovery by the Spaniards. Κόκκινος [kókkinos] is the Modern Greek word for "red" and Kermes (from the Arabic and Persian *girmiz*) is cognate with "crimson" and "carmine". The stems, roots and bark of Barberry, *Berberis vulgaris* L., produced a yellow dye for wool and leather; see also Ray's entry for *Oxyacantha* [413]. The rhizomes of Turmeric, *Curcuma longa* L., were ground into a deep orange-yellow powder widely used in mediaeval Europe as an

is that by colouring minute parts of the other body it cannot reach what is abandoned by the slender parts.[314] *C. Hofman[nus], De medicam[entis] officina[libus], book 2, chapter 196, page 553, & Verulam, Hist[oria] nat[uralis]*,[315] *cent[ury] 7, exp[eriment] 692.*

216 *Gentianella amarella* (L.) Börner

Gentianella fugax autumnalis minor *Ger. emac.* autumnalis centaureæ minoris foliis *Park.* fugax sive annua 4 Clusii flore dilutè purpurascente *I.B.* Gentiana pratensis VII, sive angustifolia autumnalis minor floribus ad latera pilosis *C.B.*[316] *Bastard or dwarf Autumnall Gentian or Fellwort.* Two kinds of this plant are described by authors.

217 *Erodium cicutarium* (L.) L'Hér.

Geranium arvense vel minus *Tab.* cicutæ folio inodorum *Ger.* cicutæ folio minus & supinum *C.B.* moschatum inodorum *Park.* moschatum folio ad myrrhidem accedente minus *I.B.* Myrrhis sive Myrrhida Plinii *Matth. Lob. Field Cranes-bill without sent.* On the banks in the highway between Cambridge and Barnwell, & frequent elsewhere.

218 *Geranium pratense* L.

Geranium batrachoides *Dod. Cam. Thal. Lugd. I.B.* batrachoides sive magnum cæruleum *Ad.* batrachoides majus flore cæruleo *hort. Bata. hort. Hafn.* anemones folio rotundo V, sive Geran[ium] batrach[oides] Gratia Dei Germanorum Lobelio *C.B.* Ger[anium] batr[achoides] flore cæruleo *Park. in parad. Crowfoot-Cranes-bill.* In the hedges about Bigwin closes, about Cherry-hinton and Histon and many other places.

219 *Geranium molle* L. and/or ***G. pusillum*** L.[317]

Geranium columbinum *Tab. Ger.* columb[inum] vulgare *Park.* folio rotundo multùm serrato, sive columbinum *I.B.* folio malvæ rotundo *C.B.* Pes columbinus *Dod. Ad. Lob. Doves-foot or Doves-foot-Cranes-bill.*

220 *Geranium dissectum* L.

Geranium columbinum majus dissectis foliis *Ger. emac.* malacoides laciniatum sive columbinum alterum *Park.* malvæ folio II, sive columbinum tenuiùs laciniatum *C.B. Doves-foot with deeply cut leaves.* In meadows & by hedges generally.

---

alternative to the far more expensive Saffron and so known as Indian Saffron; today it is used as a spice in curries and to impart colour to mustard condiments. *Glastum*, i.e. Woad, *Isatis tinctoria* L., is well known as the source of the blue dye used by the Ancient Britons to paint their bodies, but it continued to be used as a dye plant until Ray's time in the mid 17th century, when it was largely displaced by Indigo, *Indigofera tinctoria* L. and *I. suffruticosa* Mill., imported from the New World (Hurry 1930).

[314] *Ratio est, quia alterius corporis partes minutas tingendo nequit subire quod tenuibus partibus destituitur.* The meaning of this passage is obscure.

[315] For the work of Baron Verulam (Sir Francis Bacon) see footnote 104.

[316] See the entry for this species in *Emendanda* of the 1685 appendix, where Ray proposes changes to the synonyms.

[317] It is not clear whether Ray correctly separated *Geranium molle* and *G. pusillum* in 1660: see the following footnote and footnote 136 in *Emendanda* of the 1685 appendix.

221 *Geranium molle* L. and/or *G. pusillum* L.
Geranium malacoides sive columbinum minimu[m] *Park. C.B. The least Doves-foot.*[318] *On the hill of health,* & elsewhere in more barren places. It varies in the colour of the flower.

222 *Geranium sanguineum* L.
Geranium hæmatodes *Thal. Cam. Park.* Anemones folio rotundo IV, sive sanguineum maximo flore *C.B.* sanguineum sive hæmatodes crassa radice *I.B.* batrachoides minus sive hæmatodes *hort. Par.* gruinum sanguinale hæmatodes supinum radice reptrice *Lob. ico.* Geranii altera species major montana & Ischæmon *Gesn. hort.* Sanguinaria radix *Trag.* Ger[anium] sanguinarium *Ger. Bloudy Cranes-bill. Found on Newmarket heath in the Devils ditch, also in a wood adjoyning to the highway betwixt Stitchworth and Chidley.*

223 *Geranium robertianum* L.
Geranium Robertianum *Ad. Lob. Dod. Cam. Tab. Ger. Park.* Robertianum murale *I.B.* folio cicutæ vel myrrhidis V, sive Robertianum I, *C.B.* Herba Ruperti & Geranium secundum Dioscoridis *Lugd. Herbe Robert.* By hedges.

224 *Geranium lucidum* L.
Geranium saxatile *Ger. emac. Cam. Park.* lucidum *I.B.* anemones folio rotundo XV, sive lucidum saxatile *C.B. Shining or stone Doves-foot. On the bank and in the hedge on the right hand of the lane leading from Cambridge to Chesterton plentifully,* & elsewhere.

Geum, *see* Caryophyllata. [*Geum urbanum* 117]

225 *Butomus umbellatus* L.
Gladiolus palustris *Cord. Ger.* Juncus floridus *I.B. Matth. Park.* floridus major *C.B.* cyperoides floridus paludosus *Lob.* Butomos Theophrasti *Cæs.* Calamogrostis 2 *Trag.* I *Lugd. Water Gladiole, The flowring Rush. In the river between Cambridge and Granchester, and in many rivulets and watery ditches about the countrey.*

226 *Isatis tinctoria* L.
Glastum sativum *Ger. Park. Tur. Ad. Lob. Cam.* Isatis sativa *Trag. Fuch. Dod. ut. Cord. in Diosc. Gesn. hort.* sativa vel latifolia *C.B.* Isatis of the Greeks, Nil of Avicenna, Indicum of the apothecaries' shops *Fragoso.*[319] Isatis sive Glastum sativum *I.B.* The herb Vitrum of the Romans on the evidence of Vitruvius, Mela, Caesar & Marcellus Empiricus, *Jo. Bodæ. in Theophr. hist.*[320] *Woade. Planted about Littleport in the isle of Ely.* I do not know whether the sown plant differs from the wild one except by reason of cultivation. *I.B., volume 2, book 21, page 910.*

---

[318] See the entries for this plant in *Emendanda* of the 1663 and 1685 appendices and the footnote about it in the latter. Although Babington (1860) identified this species as *G. pusillum*, the remark on the variation in flower colour is more appropriate to *G. molle*. Yeo (1992) describes the petals of the former as "lilac" and those of the latter as "purplish pink with a white base, pale pink or white". All these variants of *G. molle* occur in Cambridgeshire (Crompton 2004).

[319] *Isatis Græcorum, Nil Avicennæ, Indicum officinarum* Fragoso. Juan Fragoso was the native Spanish name of the author listed as Joannes Fragosus in Ray's 'Explanation'; see also Chapter 4.

[320] *Vitrum herba Latinis, Vitruvio, Mela, Cæsare & Marcello Empirico testibus*, literally "The herb *Vitrum* for the Latins [i.e. Romans], with Vitruvius, Mela, Caesar & Marcellus Empiricus [as] witnesses". For these authors, see Chapter 4. Ray's sentence is based on Bodaeus' (1644; see footnote 23) account in column 2 of p. 723 (misprinted as 523).

N. In times past the Britons were accustomed to use the sap of this plant to dye their bodies blue. See *Plin[y], book 22, chapter 1,*[321] & *Cæsar in his commentary on the Gallic War, book 5*.

227 *Astragalus danicus* Retz.
Glaux Dioscoridis *Ger.* Hispanica *I.B.* Hispanica Clusii *Park.* Cicer sylvestre III, sive Ciceri sylvestri minori affinis si non idem[322] *C.B.* Anthyllis altera lenti similis *Dod. Dioscorides his Milk-tare, or Clusius his Spanish Milkwort. On the drier part of Hinton moor, and almost all over Newmarket heath and Gogmagog hills.* The short pods of this plant contain a seed which resembles that shape of a heart that is painted in playing cards.

228 *Astragalus glycyphyllos* L.
Glaux vulgaris *Ad. Lob. Clus.* vulgaris leguminosa sive Glycyrrhiza sylvestris *Park.* Glycyrrhiza sylvestris floribus luteopallescentibus *C.B.* Fœnugræcum sylvestre *Trag. Clus.* Fœnum Græcum sylvestre, sive Glycyrrhiza sylvestris quibusdam *I.B.* Hedysarum Glycyrrhizatum *Ger. emac. Wild Liquorice, or Liquorice Hatchet-vetch. About the castle hill at Cambridge: by the lanes side that leads from Cambridge to Cherry-hinton church and elsewhere.*

229 *Glycyrrhiza glabra* L.
Glycyrrhiza vulgaris *Ger. emac.* vulgaris siliquosa *Park.* siliquosa vel Germanica *C.B.* radice repente vulgaris Germanica *I.B.* Dulcis radix *Trag. Tur. Gesn. hort.* Radix Scythica Theophrasti *Jo. Bodæ. in Theophr. hist. Common Liquorice. Planted in good quantity at Elme in the isle of Ely.*

N. Bitterness is found in all leguminous plants, even in Liquorice. *C. Hofman[nus], De medicam[entis] officinal[ibus], book 2, chapter 313*.

230 *Filago vulgaris* Lam.
Gnaphalium minus sive Herba impia[323] *Park.* vulgare majus *C.B.* Filago sive Herba impia *Ger.* Gnaph[alium] Germanicum *I.B.* Centunculus *Tur. The lesser Cudweed or Herb impious. In sandy & more barren places.*

231 *Gnaphalium uliginosum* L.
Gnaphalium vulgare sive 2 *Ger.* medium *C.B.* Gnaphalio vulgari similis *I.B.* Filago minor *Park. Dod. Lob. Common Cudweed. In dry ditches, &ᶜ*.

232 *Filago minima* (Sm.) Pers.
Gnaphalium minimum *I.B.* Filago minor *Ger.* Gnaph[alium] minus repens *C.B.* minimum repens *Park.* Certainly our plant does not creep but is erect, & it corresponds very well to de L'Obel's figure in *Ger.* & *I.B. The least Cudweed. In the sandy grounds about Gamlingay.*

233 *Antennaria dioica* (L.) Gaertn.
Gnaphalium montanum album *Lob. ico. Tab. Ger.* montanum Dalechampii *Lugd.* montanum I, sive montanum flore rotundiore *C.B.* montanum sive Pes cati *Park.* Pes cati *Offic[inarum]*. Pilosella minor *Fuch. Clus.* Pilosella minor quibusdam, aliis Gnaphalii genus *I.B.* Lagopus 2 *Trag.* Auricula muris *Lon.* The picture of this plant that is found both in Gerarde and in Parkinson is defective & bad, in as

---

[321] Pliny's *Natural history*, XXII. 2 in the Loeb edition, where he says that it is *Britannorum coniuges nurusque* ("the wives and daughters-in-law of the Britons") who stain their whole bodies and march naked in religious ceremonies, *Aethiopum colorem imitantes* ("imitating the colour of Ethiopians").

[322] "The third wild Chick-pea, or [one] related to, if not the same as, the smaller wild Chick-pea".

[323] See footnote 437 and the entry for *Gnaphalium* in Ray's 'Etymology'.

much as for anyone comparing it either with the plant itself or with de L'Obel's figures in his *Adversaria & pictures* it will be good enough,[324] and *Jean Bauhin's* picture is no more satisfactory. The best one is contained in de L'Obel's work in his *Adversaria & pictures*.[325] *Mountain Cudweed, Catsfoot*. *On Newmarket heath, on the right hand of the road from Cambridge to Newmarket, about a quarter of a mile from the beacon, and in other places of the heath in great plenty*. It is found in the same place with a purple flower also.

234 *Alopecurus pratensis* L.[326]
Gramen alopecuroides minus *Park. Ger.* typhoides molle *C.B.* Gramen Alopecuroidi accedens & Phalaridi, spicâ longiusculâ, folio lanuginoso *I.B.*, but the leaves of our plant are not so downy. alopecuroides *Lob*. *Small Fox-tail grass*. In pastures.

235 *Alopecurus myosuroides* Huds.
Gramen Alopecurinum minus *Ger.* alopecuroides minus alterum *Lob. ico.* alopecuroides spicâ longâ majus & minus *Park.* typhoides V, sive spica angustiore *C.B.* cum cauda muris purpurascente *I.B. The lesser bastard Fox-tail-grass*. Among standing corn.

236 *Cynosurus cristatus* L.
Gramen alopecuroides minus spicâ asperâ brevi *Park*. We do not know whether this is found in the work of other authors, but it is certainly not the same as Gramen alopecuroides spicâ asperâ *C.B. Small rough-eared bastard Fox-tail-grass*. Very frequent in pastures. It may easily be distinguished from other species by the fact that in this species the utricles or glumes are turned only to one side so that the ear seems semicircular in section.[327]

237 *Alopecurus geniculatus* L.
Gramen aquaticum spicatum *Lob. ico.* aquaticum geniculatum spicatum *C.B.* fluviatile spicatum *Ger*. Gramen cum parva cauda muris, radice non nodosa, repens *I.B. Spike Float-grasse or Water-grasse*. In watery places. *As beyond the castle near the Wind-mill in great plenty.*

238 *Glyceria fluitans* (L.) R. Br.
Gramen aquaticum cum longissima panicula *I.B.* aquis innatans *Lob*. fluviatile *Ger. emac. Tab. Park.* spicatum aquaticum I, sive aquaticum fluitans multiplici spica *C.B.* Agrostis potamios *of the same author*. Gramen anatum *Trag*. *Float-grass*. Everywhere in rivulets & watery ditches.

---

[324] Presumably *non* must be missing before *satis*, so as to make this mean "it will not be good enough".

[325] See *Ad. Lob.* and *Lob.* in Ray's 'Explanation'.

[326] Raven (1950, p. 90) cites this as an example of a plant of Ray's about which there is "room for doubt", saying that in his *Catalogus plantarum Angliae* (1670, pp. 137–138) he replaced it by Gerarde's (1597, pp. 8–9) *Gramen Alopecuroides maius* "with a different list of synonyms". Ray marks this as a Cambridgeshire species, calls it "The most common Fox-tail-grass" and comments: *Mirum de planta adeò obvia & vulgari, tam obscurè & confusè scripsisse Botanicos. Est enim in pratis per totam Angliam frequentissima, & nulli non rustico nota, … .* ("[It is] astonishing that Botanists have written so obscurely & confusedly about so obvious & common a plant; for it is very frequent in meadows throughout the whole of England & known to every countryman, … ."). Raven identifies this as "being evidently the modern *Alopecurus pratensis*".

[327] For Ray's usage of *utriculus*, *gluma* and *spica* see his 'Interpretation of terms'. See also Ray's comment about a further name for this grass, *Gramen cristatum*, in *Emendanda* of the 1663 and 1685 appendices.

239 *Glyceria maxima* (Hartm.) Holmb.
Gramen aquaticum majus *Ger. emac. Lob. ico.* paniculatum aquaticum IV, sive paniculatum aquaticum latifolium *C.B.* majus aquaticum Lobelii *I.B.*, where the picture is unrelated. Gr[amen] harundinaceum aquaticum *Park. Great water Reed-grasse*. It is almost peculiar to this plant that it also sends out roots from the lower joints of the culm.

240 *Phalaris arundinacea* L.
Gramen arundinaceum acerosâ glumâ nostras *Park. Our great Reed-grasse with chaffie heads*. As far as we know it does not exist in the works of other authors[328] except perhaps in de L'Obel in his *Stirpium illustrationes*. *It is found in the river in many places, and in the Fen ditches in the Isle of Ely*.

241 *Dactylis glomerata* L.
Gramen asperum *J.B.* spicatum pratense V, sive spicatum folio aspero *C.B.* pratense Dalechampii *Lugd.* pratense spicâ multiplici rubrâ *Park. Rough Spiked grasse*.

242 *Elytrigia repens* (L.) Desv. ex Nevski
Gramen caninum *Ger.* caninum vulgatius *Park.* caninum arvense, sive Gramen Dioscoridis *C.B.* repens officinarum fortè triticeæ spicæ aliquatenus simile[329] *J.B. Common Quich-grasse*. Authors differ in their descriptions of this plant. Gerarde's figure corresponds better to our grass than Parkinson's; the one described by Jean Bauhin is not ours, & he himself reasonably expresses doubt whether it is Gramen caninum of the apothecaries' shops, which he describes under his published title.[330]

243 *Carex riparia* Curtis
Gramen cyperoides cum paniculis nigris *J.B.* cyperoides latifolium spicâ rufâ sive caule triangulo *C.B.* cyperoides majus latifolium *Park., page* 1266. cyperoides *Ger. Lob. Lugd. Cam. Tab. The great Cyperus-grasse*. It flowers in May in ditches & the shallows of more sluggish rivers.

244 *Carex* sp. or spp.[331]
Gramen cyperoides angustifolium spicâ spadiceo-viridi minus an majus? *C.B.* Gr[amen] cyperoides majus angustifolium *Park*. *The greater narrow leaved*

---

[328] Ray continued to use Parkinson's name for this species in his later works (e.g. 1688a, p. 1280) without giving any synonyms. However, *Gramen arundinaceum spicatum* of C. Bauhin's *Prodromus* (1620, p. 94) and *Pinax* (1623, p. 6) is cited as a synonym of *Phalaris arundinacea* by Linnaeus (1753).

[329] Literally "Creeping [grass] of the apothecaries' shops perhaps to some extent like an ear of wheat".

[330] ... *quod sub titulo allato describit*. This seems the most likely translation: *allatus, -a, -um* (the past participle of the verb *affero*) literally means "brought to". See Ray's revised opinion in *Emendanda* of the 1663 and 1685 appendices.

[331] There are particular difficulties in identifying Ray's *Carex* species as he himself was clearly still struggling to come to terms with the genus in 1660. This and the following species are particularly problematic. Ray suggests *Gramen nigro-luteum vernum* J.B. as a synonym for both. Bauhin & Cherler (1651, vol. 2, p. 494) illustrate two plants under this name which to the modern botanist are clearly different species, although the text suggests that the differences are caused by their habitat. Ray himself later (e.g. 1688a, p. 1293) treated them as separate species. This entry refers to the larger species of wet habitats which Babington (1860) identifies as *Carex elata* All.; we agree that it is probably one of the larger members of the *nigra* group such as *C. elata* and/or *C. acuta* L. but it would be difficult to rule out *C. acutiformis* Ehrh. Parkinson's (1640, p. 1266) illustration of a tall sedge with rather narrow inflorescences does not help us to decide between them. In fact, it seems likely that Ray's concept included more than one modern species.

Cyperus-grasse. This is the same, we think, as Gramen nigro-luteum vernum *J.B.*, which he states is found mainly in moist meadows; there is however a substantial difference in size between this & the one following next.

245 *Carex* sp.[332]
Gramen nigroluteum vernum *J.B.* spicatum foliis caryophylleis *Park.* cristatum & spicatum VI, sive Gr[amen] caryophylli foliis spicâ divulsâ, *C.B.* spicatum foliis vetonicæ *Lob. ico.* Gramen cyperoides spicatum *Ger. emac. Spiked Cyperus-grasse*. In moist meadows.

246 *Bolboschoenus maritimus* (L.) Palla
Gramen cyperoides palustre paniculâ sparsâ *Park.* cyperoides vulgatius aquaticum *I.B.* cyperoides paniculâ sparsâ majus *C.B.* aquaticum cyperoides vulgatius *Ger. Water or Marsh Cyperus-grasse with sparsed heads. In the ditches near Wisbich.*

247 *Carex otrubae* Podp.[333]
Gramen cyperoides palustre majus spicâ compactâ *C.B.* Gr[amen] palustre cyperoides triquetrum spicâ integrâ *I.B.* palustre cyperoides *Ger.* cyperoides palustre majus *Park. The greater Marsh Cyperus-grasse.*

248 *Carex muricata* agg.[334]
Gramen cyperoides palustre minus *Park.* cyperoides parvum *Ger.* cyperoides spicis minoribus minúsq; compactis *C.B. I.B. The lesser Marsh Cyperus-grasse.*

249 *Carex disticha* Huds.[335]
Gramen cyperoides ex monte Ballon, spicâ divulsâ *I.B.* cyperoides palustre majus spicâ divisâ *C.B.* Gramen cyperoides palustre minus *Park. The lesser sparsed Marsh Cyperus-grasse.*

We have also as well two other species of cyperus-grass which we think have not yet been described; and we have not been able to distinguish sufficiently accurately even these ones for which we have given synonyms, since in

---

[332] Babington identified *Gramen nigroluteum vernum* as *Carex caryophyllea* Latourr. but Ray's plant of "moist meadows" is very unlikely to have been *C. caryophyllea*, which, although spring-flowering, grows, as Babington himself states, in "dry places". It also has green leaves (despite its specific epithet) rather than the glaucous-green leaves which one might expect from the synonyms. Two conspicuous omissions from Ray's text (as interpreted by Babington) are *C. flacca* Schreb., frequent in a range of wet and dry, open habitats in Cambridgeshire, and *C. nigra* (L.) Reichard., a less common plant of wet meadows but one which has inflorescences which are initially black, as J. Bauhin's description demands. We suspect that one or both of these species, which have glaucous leaves, must have been the plant(s) that Ray encountered.

[333] Fortunately this plant is very well illustrated by Parkinson (1640, p. 1266) and the plate in Bauhin & Cherler (1651, vol. 2, p. 497), though poorer, is recognisably the same plant, so there is no doubt about the identity of this species.

[334] In Cambridgeshire this aggregate of closely related species comprises *Carex divulsa* Stokes, *C. muricata* L. and *C. spicata* Huds. Parkinson's illustration (1640, p. 1267) is crude and the account by Bauhin & Cherler (1651, vol. 2, p. 497) is unillustrated, so the identification of Ray's plant even to this aggregate is rather conjectural, especially in the absence of any information on its habitat and localities. However, there are no other likely candidates in Subgenus *Vignea*.

[335] J. Bauhin's plant is *Carex divisa* Huds., but this is a predominantly coastal species in England and has never been reliably recorded in Cambridgeshire. Ray (1696, p. 268) retracted his identification of the Cambridgeshire plant with Bauhin's, and we have followed Raven's (1950, p. 90, footnote 3) view that the new description provided in 1696 suggests that Ray's plant was *C. disticha*.

descriptions of grasses there is amazing confusion & discrepancy among authors.[336]

250 *Carex sylvatica* Huds.
Gramen cyperoides sylvarum tenuiùs spicatum *Park. Lob. in Stirpium illustrationes, page 60*. A figure of this is contained in *Parkinson's Theat[rum] Bot[anicum], page 1172*. *Slender-eared wood Cyperus-grasse*. In Madingley wood plentifully. It flowers around the end of April & at the beginning of May.

251 *Luzula campestris* (L.) DC.
Gramen exile hirsutum *Ger.* nemorum hirsutum minus angustifolium *Park.* exile hirsutum cyperoides *Lob. ico.* nemorosum hirsutum III, sive hirsutum capitulis Psyllii *C.B.* Gramen Luzulæ minus *I.B.* & Gramen exile hirsutum cyperoides Lobelii *of the same author*: as he himself thinks & we agree with him. Frequent in meadows. *The lesser narrow-leaved Wood-grasse, Small hairy Wood-grasse*. It flowers around the end of April & the beginning of May.

252 *Luzula multiflora* (Ehrh.) Lej.
Gramen hirsutum majus paniculâ junceâ compactâ. It resembles Gramen hirsutum angustifolium majus *C.B.*, except that the panicle is not as sparse.[337] *Hairy grasse with a compact rush-like panicle*. In the pastures about Gamlingay, near the boggy ground by Sir Roger Burgoynes park, where they dig turfs.

253 *Juncus articulatus* L.
Gramen junceum aquaticum Bauhini *Park.* junceum IX, sive folio articulato aquaticum *C.B.* Gr[amen] aquaticum *Tab. Ger.* aquaticum alterum *Lob. ico.* Juncus foliaceus capsulis triangulis *I.B.* *Rush-grasse with joynted leaves, water Bur-grasse*.

254 *Juncus compressus* Jacq.
Gramen junceum maritimum *Lob. ico. Ger.* junceum X, sive foliis & spica junci *C.B.* junceum maritimum majus *Park.* Juncus parvus cum pericarpiis rotundis *I.B.* *Marsh Rush-grasse*. In meadows & by footpaths. We think that the synonyms that we have given from other authors do not fit the plant that we intend, but the name, picture & description of Jean Bauhin, from whom we have also taken the remaining synonyms & have not chosen them according to our own judgement, correspond excellently.

255 *Juncus acutiflorus* Ehrh. ex Hoffm. and/or *J. subnodulosus* Schrank[338]
Gramen junceum aquaticum magìs sparsâ paniculâ *Park.* *Water Rush-grasse with a sparsed panicle*. *Parkinson* states that he first published this plant, but we suspect

---

[336] One might suppose that these were the two species listed in the 1663 appendix, *Gramen cyperoides spicâ pendulâ longiore* [*Carex pendula* A16] and *Gramen cyperoides angustifolium paniculâ multiplici* [*Carex remota* A17]. However, in a letter to Courthope Ray listed the former as a species he first found in Cambridgeshire in 1661, and in 1662 he reported finding at Gamlingay two species of *Gramen cyperoides* that he had never seen in England before (Gunther 1934; Thompson 1974). The identity of the "two other species" reported here is therefore unclear.

[337] See also Ray's further comment in *Emendanda* of the 1663 and 1685 appendices and the footnotes about it.

[338] Babington (1860) and Ewen & Prime (1975) identify this as *J. acutiflorus*, a species which is now very uncommon in Cambridgeshire although Ray might well have seen it, for example at Gamlingay. The commoner plant in calcareous fens in Cambridgeshire is another large species with sparse panicles, *J. subnodulosus*, which was not separated in Ray's time but which he must surely have encountered. The record in 'Index of places' from Hinton Moor, for example, seems likely to have been this species.

that he secretly took it from *Lobelius'* illustrations of plants,[339] not then published, like very many other grasses the finding of which he appropriates to himself. Apart from the locality this plant seems to differ in absolutely no way or certainly too little from Gramen junceum folio articulato sylvaticum *C.B.*, except perhaps that the leaves in this plant are not jointed.

256 *Juncus bufonius* L.
Gramen junceum *Ger.* junceum 6, sive bufonium *Tab.* junceum parvum sive Holostium Matthioli & Gramen bufonis Flandrorum[340] *Park.* Holostium Matthioli junceum *I.B.* Gram[en] nemorosum glabrum V, sive nemorosum calyculis paleaceis *C.B. Toad-grasse,* because it comes up in places where toads have their home.

257 *Triglochin palustris* L.
Gramen marinum spicatum *Ger. emac.* junceum spicatum, seu triglochin *C.B.* mixtum ex junco & gramine[341] *Thal.* triglochin *I.B.* triglochin Dalechampii *Park. Arrow-headed grasse, Sea Spike-grasse.* On Hinton and Teversham moors plentifully, &^{c}.

258 *Aira caryophyllea* L.
Gramen montanum paniculâ spadiceâ delicatiore C. Bauhini *I.B.* paniculatum montanum I, sive mont[anum] paniculâ spadiceâ delicatiore *C.B. Park. Soft mountain Panicled-grasse.*

259 *Poa pratensis* L. *sensu lato*[342]
Gramen pratense *Ger.* pratense vulgatius majus *Park.* prat[ense] paniculatum majus *C.B. The greater Meadow-grasse.* Everywhere.

260 *Poa annua* L.
Gramen pratense paniculatum minus *C.B.* pratense minimum album *Park.* minimum album *Ger.* paniculatum minus album & rubrum *I.B. The lesser Meadow-grasse.*

N. I knew a nobleman who took care that one part of a meadow was mown when the moon was waxing and the other when it was waning; the former bore twice as much grass in the ensuing months as the latter. The words are those of *P. Lauremberg [in] Horticult[ura], book 1, chapter 30, number 4.*[343]

261 *Holcus lanatus* L.
Gramen pratense paniculatum molle *C.B. Park.* paniculatum molle *Lob. Ad. part alterâ.* Gramen lanatum Dalechampii *Lugd. I.B.* Gr[amen] 6, *chapter 22, book 1. Ger. emac. Soft tufted Meadow-grasse.*

262 *Deschampsia cespitosa* (L.) P. Beauv.
Gramen paniculatum arvense III, sive Gr[amen] segetum altissimum paniculâ sparsâ *C.B.* Gr[amen] segetum paniculâ speciosâ *Park.* Gramen segetale *Ger.* Gramen capillatum *I.B. The fair panicled Corn-grasse.*

---

[339] L'Obel's (1655) *Stirpium illustrationes*: see also *Lob.* in Ray's 'Explanation'.
[340] Literally "Small rushy [grass] or Mattioli's All-bone & Toad's-grass of the people of Flanders".
[341] Literally "[Grass] mixed from rush & grass".
[342] Three closely related species, *Poa angustifolia* L., *P. humilis* Ehrh. ex Hoffm. and *P. pratensis* L., are recognised by Stace (1997, 2010), but Cope & Gray (2009) treat them as intergrading subspecies. All three are recorded from Cambridgeshire.
[343] For Peter Lauremberg see footnote 25. See also the comment in note 2 under *Arundo vallatoria* [64] for another popular belief from the same source about harvesting according to the phases of the moon.

N. From the culms of this grass together with their panicles, tied together one after another in order with a cord or rush, they construct little hurdles, so to speak, called *Bents* in the vernacular, with which in many towns of England they are accustomed to decorate their fireplaces in summertime. Gramen agrorum Lobelii is also applied to this use, according to Johnson in his edition of Gerarde,[344] and likewise Gramen cristatum,[345] according to [Mathias] de L'Obel *[in] Advers[aria], part 2*. Several other species of grass occur with us which we cannot yet determine for certain by what names they are referred to by botanical authors, which perhaps we shall give separately later in another little publication that we are considering.[346]

Gramen Leucanthemum, *see* Holosteum. [*Stellaria holostea* 286]

Gramen secalinum, *see* Hordeum spurium. [*Hordeum murinum* 288]

263 *Carex lepidocarpa* Tausch[347]

Gramen palustre echinatum *J.B. Lob. Ger. emac.* aculeatum Germanicum *Park.* palustre aculeatum vel minus Germanicum *C.B.* Hedge-hog-grass. On the moores about Cambridge.

---

[344] The name cited here is not to be found among Ray's chosen names or synonyms, but Johnson (1633, pp. 5–6), after the account of *Gramen segetale* ("Corne-grass"), illustrates and describes *Gramen harundinaceum* ("Reed-grasse, or Bent"), for which he cites Lobelius' name "in Latine" as *Gramen agrorum latiore, arundinacea, & comosa pannicula*, adding to Gerarde's original text: "Some in English, much agreeable to the Latine name, call these, Windle-strawes. Now I take this last to be the Grasse with which we in London do vsually adorne our chimneys in Sommer time: and we commonly call the bundle of it handsomely made vp for our vse, by the name of Bents." He thus suggests that it is this grass, rather than his *Gramen segetale* listed by Ray in his synonymy, that was used in this way.

[345] This name appears above as a synonym from C. Bauhin for *Gramen nigroluteum vernum* J.B. [*Carex* sp. 245], but Mathias de L'Obel's plant is *Cynosurus cristatus* [236]: see de L'Obel's *Adversariorum altera pars* in Pena & L'Obel (1605, p. 467) and *Gramen alopecuroides minus spicâ asperâ brevi* in *Emendanda* of the 1663 and 1685 appendices.

[346] It seems unlikely that this refers to Ray's next publication, the 1663 appendix, as one would not expect him to be planning this while writing the main *Catalogus* and in any event it does not include any additional grasses (as opposed to the two undescribed species of cyperus-grass or sedge mentioned earlier). By 1660 Ray was considering at least two further works, a catalogue of those plants in English gardens which did not grow wild and a complete *Phytologia britannica* (Lankester 1848, pp. 1–2; Raven 1950, pp. 108–109 and 111). The former seems unlikely to be the "little publication" referred to here, since it would have excluded wild plants (although, as the references to insects in the *Catalogus* show, Ray could insert much material of only marginal relevance into his works). The latter may be more likely because, although one would not expect Ray to describe it as a "little publication", he told Willughby when he sent him the *Catalogus* that he hoped "to bring [it] into as narrow a compass as this book" (Lankester 1848, pp. 1–2). Alternatively, Ray may have had other ideas for books which (like the garden catalogue) never materialised.

[347] Three closely related species in the *Carex flava* group occur in Cambridgeshire. Ray is likely to have seen *C. lepidocarpa*, which is the largest of the three, has the most echinate infructescences and grows in the most base-rich habitats. Babington (1860) assigned Ray's name to this species (which he called *C. flava* L.). *C. demissa* Hornem. has not been recorded from base-rich habitats around Cambridge. *C. oederi* Retz., though previously known from such habitats (Crompton 2001), is a relatively inconspicuous plant with short utricles and which do not project to give the infructescences an echinate appearance. However, the group is a difficult one (these species are often treated as subspecies of *C. viridula* Michx.) and we identify Ray's plant only on the balance of probabilities.

264 *Parnassia palustris* L.
Gramen Parnassi *Dod. Lob. ico. Ger.* Parn[assi] vulgare *Park.* Parn[assi] flore albo simplici *C.B.* Parn[assi] hederaceum recentiorum *Ad. Lob.* Parn[assi] Dodonæo, quibusdam Hepaticus flos *J.B.* Parn[assi] non verum *Gesn. hort.* Hepatica alba *Cord. hist.* Unifolium palustre *Gesn. coll. Grasse of Parnassus. On Hinton and Teversham and Trumpington moores.*

265 *Eriophorum angustifolium* Honck.
Gramen tomentosum & Linogrostis *Tab.* tomentosum II, sive pratense tomentosum paniculâ sparsâ *C.B.* tomentarium *Ger.* Juncus bombycinus *Ad. Lob. ico.* Gnaphalium Tragi, sive juncus bombycinus *J.B.* Linum pratense *Gesn. Coll.* Gramen junceum lanigerum *Thal.* It bears wool in June & July. *Cotton-grasse. On Hinton and Teversham moores and elsewhere.*

266 *Briza media* L.
Gramen tremulum *J.B.* tremulum II, sive majus *C.B.* leporinum & tremulum *Tab.* Phalaris pratensis *Ger.* pratensis minor *Lob. Quaking-grasse, Cow-quakes.* In meadows & pastures.

267 *Phleum pratense* L.
Gramen typhinum maximum *Park.* typhinnm[348] majus *Ger.* Gramen cum cauda muris majoris longâ majus[349] *I.B. The greatest Cats-taile grasse.*

268 *Phleum bertolonii* DC.
Gramen typhinum medium sive vulgatissimum *Park.* Gramen typhinum minus *Ger. The lesser Cats-taile grasse.*

269 *Lythrum hyssopifolia* L.
Gratiola angustifolia *Ger. emac.* angustifolia sive minor *Park.* minor sive Anthyllis aquatica *hort. Paris.* Hyssopifolia *C.B.* aquatica *I.B.* Hyssopoides *Cam. Grasse Poley Cord. Small hedge-Hyssope. In the corn fields and shadowy lanes about Hoginton and Histon, as also in many places about Cambridge.*

Gratiola latifolia cærulea, *see* Lysimachia galericulata. [*Scutellaria galericulata* 345]

H

270 *Hedera helix* L.
HEdera arborea *C.B. Matth. Gesn. hort. Lugd.* arborea sive corymbosa communis *Ad. Lob.* arborescens fertilis *hort. Pat.* nigra *Fuch. Dod. gal. Lac. Lon.* Hed[era] arborea sive nigra corymbosa scandens *Park. Climing or berried Ivie.*

N. This plant has a peculiarity from nature, that its first leaves are incised and diverse but the later & highest ones are round: for in division nature is otherwise inclined to proceed from the simple to the diverse. *C. Hofman[nus], De medicam[entis] officinal[ibus], book 2, chapter 102.* However the same characteristic is seen also in Holly, as we have noted above,[350] but in fact in Smyrnium

---

[348] A misprint for *typhinum*.
[349] Literally "Larger grass with the long tail of a larger mouse".
[350] See Ray's note 1 under *Agrifolium*, the name that Ray gives here for Holly.

MAIN CATALOGUE

creticum above all others & in Sium aquaticum maximum,[351] whose lowest leaves & those that burst forth in early spring are divided into very slender segments but the higher ones are similar to the leaves of garden parsnip.[352]

271 *Hedera helix* L.

Hedera helix *Park.* hel[ix] sive provoluta sterilis *Ad. Lob.* III, sive major sterilis, & IV, sive humi repens *C.B. Barren or creeping [I]vie*. Tragus, Cæsalpinus & Jo[annes] Bauhinus do not accept that these are different species of ivy, but rather they argue that the whole difference between them arises either just because of their age or because of the place where they grow: for from this cause alone, they say, it happens that they differ in form or in fruit. See *I.B., volume 2, book 15, pages 111–112*. C. Bauhin also seems to have doubts about this matter in his *Pin[ax]*, as also Theophrastus before him.[353]

272 *Glechoma hederacea* L.

Hedera terrestris *Ger.* vulgaris *C.B. Park.* Chamæcissus sive Hed[era] terrestris & Hed[era] sylvatica Romanorum *I.B., in append[ix] to] volume 3, page 839*. Chamæcissus antiquorum *Trag. Fuch. Dod. Lob. Cæs.* Malacocissus *Lugd. Ground-Ivie, Alehoof, Tunhoof, Gill go by ground.*

273 *Lemna trisulca* L.

Hederula aquatica *Lob. Ger.* palustris *Tab.* Lenticula aquatica trisulca *I.B. C.B.* Ranunculus hederaceus aquaticus *Park. Water-Ivie, Ivie-leaved Duckes meat. In ditches cut out of the river in many places.*

Hedypnois *Plinii, see* Dens leonis. [*Taraxacum* spp. 167]

Helenium, *see* Enula campana. [*Inula helenium* 177]

Helianthemum, *see* Chamæcistus. [*Helianthemum nummularium* 125]

274 *Epipactis palustris* (L.) Crantz

Helleborine angustifolia *Tab. Ger.* VII, sive montana angustifolia purpurascens *C.B.* Damasonium purpureum sive Helleborine 6 Clusii *I.B.* Helleborine minor flore purpurante *Park.* An Helleborine minor flore albo *of the same author?* An Helleborine angustifolia palustris *C.B? Narrow-leaved wild white Hellebore.* The flower of this plant[354] consists of five petals, which surround something like the

---

[351] *Smyrnium creticum* Mill. (*S. apiifolium* Willd.) is an umbelliter of Greece and Turkey with highly divided (3-ternate) lower leaves and undivided upper stem leaves (Davis 1972). Both lower and upper leaves are illustrated by Bauhin & Cherler (1651, vol. 3(2), p. 125) as *Smyrnium creticum perfoliatum*. The second species is clearly *Sium majus latifolium* [*Sium latifolium* 543] (q.v.), though this exact name is not among Ray's synonyms for it.

[352] *superiora pastinacæ sativæ foliis similia.*

[353] Ray's later works show that he came to accept this view. In *Catalogus plantarum Angliae* (1670, pp. 158–160) he retained the two taxa but noted his agreement with the views of Tragus, Caesalpinus and Bauhin. When he wrote *Historia plantarum* (1688a, p. 1505) he produced a single account for "Climbing, or berried Ivy; also barren, or creeping Ivy". In this he expressed his surprise that Theophrastus had recognised so many species of ivy, whereas Ray said that he himself had so far observed only a single species in all the regions over which he had wandered.

[354] This and the three detailed descriptions of orchids later in the main catalogue – of *Liparis loeselii* [397], *Coeloglossum viride* [404] and *Anacamptis pyramidalis* [406] – are remarkable examples of Ray's ability to describe plants even at this early stage in his career. A comparison of his text with the detailed modern descriptions of Sell & Murrell (1996) and the close-up photographs of Harrap & Harrap (2005) demonstrates both the accuracy of his descriptions and the insight with which he interpreted the complex orchid flower. When reading them, one should remember that the male and

body of a fly or other insect; their whitish inner part is marked with purple lines or streaks and the outer part (if you look at the three outer petals) is tinged with a faint purple which has a little green & white mixed in with it: the two remaining inner petals have the same colour on each part, both inner and outer. That style (or body of an insect) hidden in these petals has a yellow head with a white thorax furrowed within with purple streaks and a white abdomen joined to the thorax by a slender joint acting as a division and so is easily separated from it. The hanging labellum is large, white & as it were fringed on the margin. A single small acute leaf grows by each flower-stalk[355] & at the highest extremity of the stem are three or four more minute ones forming the apex. The upper parts of the stem where the flowers are attached to it, as also the calyces of the flowers, are besprinkled with a sort of flour. The root creeps under the ground, so it propagates itself easily. We have not observed it growing anywhere except in marshy & watery places, so we wonder why it is called montana by *C.B. On Hinton and Teversham moors, and in many places in the Isle of Ely in great plenty.*

275 *Epipactis helleborine* (L.) Crantz
Helleborine latifolia montana *C.B.* Helleborine *Ger. Dod. Ad. Lob. Lugd.* Helleborine Dodonæi *J.B.* an fortè Damasonium flore herbaceo intus nonnihil candicante *of the same author*? An Helleborine flore viridante *Park*? Certainly with such a diversity of species and brevity of the descriptions we scarcely dare to decide anything for certain. *Wild white Hellebore with whitish green flowers. In Kingston wood:* and we have not found it anywhere else than in wooded & shady places.

276 *Helleborus viridis* L.
Helleborus niger hortensis flore viridi *C.B.* niger vulgaris flore viridi vel herbaceo, radice diuturnâ *J.B.* Helleboraster minor flore viridante *Park*. Helleborastrum *Lob. Ger.* Consiligo Plinii *C.B.* Veratrum nigrum *Cord. in Diosc.* nigrum 2 *Clus. Dod. Wild black Hellebore or Beares-foot. In Bigwin closes, and in Whitwell closes where the Narcissus growes, near unto the house.*[356]

N. Veratrum, Scammonea etc. & almost all plants whose root is sought after seek out cool places. *Marant[a], Meth[odus] cogn[oscendorum] simp[licium], book 3, chapter 7.*[357]

---

female organs in orchid flowers are fused into a single structure, the column, which Ray calls the style and sometimes compares to the body of an insect. Ray's records of *Liparis loeselii* and *Anacamptis pyramidalis* in the *Catalogus* are the first reports from the British Isles.

[355] *Singulis florum pediculis singula folia adnascuntur exigua, acuta*, literally "To the single [i.e. each one of the] pedicels of the flowers single [i.e. one each] small, acute leaves are growing on".

[356] See *Narcissus pallido-luteus* [386] and also *Bistorta major* [87] and *Pimpinella sylvestris* [438]. Ray lists *Bistorta* and *Helleborus* in his 'Preface' as plants which "we suspect did not originally spring up naturally but were sown by some chance and by accident in the places which we have mentioned".

[357] The *Veratrum* referred to by Maranta (see footnote 34) is almost certainly not *Helleborus viridis* but one or both of the continental liliaceous mountain perennials Black False-helleborine, *Veratrum nigrum* L., and White False-helleborine, *V. album* L., highly poisonous plants long used in medicine which were perhaps unknown to Ray at this time. Later Ray (1686), discussing *Helleborus*, wrote (pp. 168–169): *Duorum generum est, albus & niger.* ("It is of two kinds, white & black.") *Helleborus albus* (clearly two *Veratrum* species) was, he wrote, called *Veratrum* by the Latins. On pp. 697–698 he described four species of *Helleborus niger*, among them what is clearly *H. viridis*. In view of this history Linnaeus' use of the epithets *nigrum* and *album* for the two *Veratrum* spp. is confusing. Scammonea is *Convolvulus scammonia* L., a native of the eastern Mediterranean region, of which

Helxine, *see* Parietaria. [*Parietaria judaica* 418]

Hepatica alba *Cord. see* Gramen Parnassi. [*Parnassia palustris* 264]

Hepatica stellata *Tab. see* Asperula aut Aspergula. [*Galium odoratum* 67]

Hepatica vulgaris *Officinarum*, *see* Lichen. [*Marchantia polymorpha* and/or *Lunularia cruciata* 331]

Heptaphyllon, *see* Tormentilla. [*Potentilla erecta* 571]

Herba Gerardi, *see* Angelica sylvestris minor. [*Aegopodium podagraria* 46]

277 *Paris quadrifolia* L.
Herba Paris *Matth. Dod. Ger. Park. J.B.* Solanum tetraphyllon *Ad. Lob. Gesn. hort.* bacciferum X, sive quadrifolium bacciferum *C.B.* Aconitum sive Pardalianches monococcon *Cord.* Aster sed non Atticus *Trag.* *Herb Paris, Herb True-love, One-berry.* In Kingston and Eversden woods.

Herba Trinitatis, *see* Viola tricolor. [*Viola arvensis* 620]

278 *Leontodon hispidus* L.
Hieracium caule aphyllo hirsutum *J.B.* dentis leonis folio acuto IX sive asperum flore magno dentis leonis *C.B.* foliis & floribus dentis leonis bulbosi *Lob. ico.* asperum foliis & floribus dentis leonis bulbosi *Park.* dentis leonis folio hirsutum *Ger. emac.* *Dandelion-Hawkweed.* In meadows & pastures.

279 *Hypochaeris radicata* L.
Hieracium longiùs radicatum *Lob. Ger. emac. Park.* macrocaulon junceum sive minus primum Dodonæi *J.B.* dentis leonis folio obtuso I, sive dentis leonis folio obtuso majus *C.B.* Apargia Dalechampii *Lugd.* *Long-rooted Hawkweed.* With the preceding.

N. The roots of plants are not pushed downwards except by the force of heat; but this seems to happen because of the location of the soil that feeds it,[358] just as the flame of a candle moves gradually downwards because its nourishment is lower down. *P. Gassend., volume 2, page 184.*

280 *Arnoseris minima* (L.) Schweigg. & Körte
Hieracium minimum Clusii, Hyoseris Tabernæmontani & Gerardi *Park.* minus folio subrotundo *C.B.* Hyoseris mascula *Ger.* *Clusius his least Hawkweed.* *In a lay near the Windmills beyond Gamlingay, and in the fields thereabout.* It loves sandy places.

281 *Scorzoneroides autumnalis* (L.) Moench.
Hieracium minus præmorsâ radice *Lob. ico. Park.* minus præmorsâ radice sive Fuchsii *J.B.* Chondrillæ folio glabro radice succisà majus *C.B.* minus sive leporinum *Ger.* *Hawkweed with bitten roots, Yellow Devils-bit.*

282 *Crepis capillaris* (L.) Wallr.
Hieracium luteum glabrum sive minùs hirsutum *J.B.* Cichoreum pratense luteum lævius *C.B. Park.* Hedypnois *Dod. Ad. Lob.*, for whom it is also Aphaca Theophrasti. *Smooth Succory-Hawkweed.* It is surprising that botanists have

---

the dried juice, virgin scammony, obtained by incision of the living root, has long been used as a purgative.

[358] *ob solum alimenti situm*: this is rather obscure but seems to mean literally "on account of the soil of food laid down".

written in such a confused & obscure way about a plant that is so common & encountered everywhere, so that it is difficult to decide anything; go, if you will, to[359] *J.B., volume 2, book 24, chapter 5, page 1024*. This plant varies remarkably in size with regard to its location, for in moorish & more barren places it scarcely ever exceeds three inches, but in pastures & lower-lying meadows it even exceeds a cubit.[360] I am afraid that authors may have divided it into too many species. We have sometimes thought that it was Hieracium minus glabrum *C.B.* Consult his *Prod[romus]*,[361] *book 3, chapter 29*.

283 *Crepis foetida* L.

Hieracium minus Cichorei vel potiùs Stœbes folio hirsutum.[362] From a simple woody white root arise several branched hairy stems half a foot tall or taller. The narrow, hairy leaves, laciniate as in Stœbe Salamantica prima[363] of [Charles] de L'Écluse but smaller by far, end in a sharp point. The flowers are yellow, small, quickly passing away; after they close up, they swell as heads into a minutely channelled protuberance, revealing a pappose down from the top, as is to be seen also in Hieracium flore carneo.[364] *Small rough Succory-leaved Hawkweed*. We suspect that this exists in C. Bauhin's work among such a large number of species, but we have not yet found it there.[365]

Hieracium echioides capitulis Cardui benedicti, *see* Buglossum luteum. [*Helminthotheca echioides* 94]

---

[359] *Adi sis*, literally "Go to, if you wish". *Adi* is the singular imperative of *adeo* ("I go to", "I visit") and *sis* an abbreviated form of *si vis* ("if you wish", "if you will", "if you please").

[360] *Crepis capillaris* varies greatly in habit and size in Cambridgeshire, as elsewhere, ranging from plants with numerous ascending stems (var. *capillaris*) to those with a single erect stem (var. *agrestis* (Waldst. & Kit.) Dalla Torre & Sarnth.). For "moorish" as the translation of *montosis* see footnote 154.

[361] C. Bauhin's work of 1620: see paragraph 5 under *C.B.* in Ray's 'Explanation'.

[362] Literally "Lesser Hawkweed with the leaf of Chicory or rather of Stœbe [and] hairy".

[363] Linnaeus (1753) cites this name in the synonymy of Dagger-flower, *Mantisalca salmantica* (L.) Briq. & Cavillier (*Centaurea salmantica* L.), a Mediterranean species which has become a weed in North America and elsewhere.

[364] "Hawkweed with a pink flower", i.e. the southern European Pink Hawk's-beard, *Crepis rubra* L. Parkinson (1640, p. 794) calls it *Hieracium intubaceum flore carneo* and says that it is assuredly the *Hieracium Apulum flore suaverubente* of Columna, C. Bauhin (1623, p. 127) gives the latter as a synonym of his *Hieracium dentis leonis folio, flore suaverubente* and Linnaeus (1753, p. 806) cites Bauhin's name as a synonym of his *Crepis rubra*.

[365] In *Historia plantarum*, Ray (1686, p. 232) recalls that he first found this species in pastures not far from Cambridge and afterwards in other localities including Montpellier, where he had been shown it as *Hieracium Castorei odore* and where he says it even occurs in the wild (*ubi etiam sponte provenit*). The significance of the last observation is not entirely clear, but it appears to suggest that in Cambridgeshire its occurrence was not entirely natural. Ray later (1695, column 418) called the plant "Small rough Succory-hawkweed smelling like Castor" and gave the Cambridgeshire locality as "In the pastures between Cambridge and Grancester, not far from the river." *C. foetida* has a southern distribution in Europe and Preston, Pearman & Hall (2004) treat it as an archaeophyte (ancient introduction) in Britain. Evans (1939) suggested that at least some of the early records of *C. foetida* in Cambridgeshire were misidentifications of *C. vesicaria* L. Perring *et al.* (1964) reported him as suggesting that all the pre-1860 records were "probably erroneous" and added: "This seems likely, but no herbarium specimens are available to support this suggestion." However, there is no evidence that *C. vesicaria* occurred in Britain before 1713 (Clarke 1900; Preston, Pearman & Dines 2002) and it would in any case be surprising if Ray confused these two species: a reassessment of their early history in Britain is clearly needed and this should take into account the evidence from Montpellier.

MAIN CATALOGUE

284 *Picris hieracioides* L.

Hieracium asperum maiore flore in agrorum limitibus *J.B.* pratense asperum *Gesn. coll.* Cichoreum pratense luteum asperum *Park.* Cich[oreum] pratense luteum hirsutie asperum, vel Hieracium hirsutum foliis caulem ambientibus *C.B.* & Cich[oreum] montanum angustifolium hirsutie asperum *of the same author*, in *J.B.*'s & our opinion. Hieracium asperum *Ger. Rough Hawkweed with a large flower.* At the boundaries of fields.

Hierobotane, *see* Verbena. [*Verbena officinalis* 608]

285 *Smyrnium olusatrum* L.

Hipposelinum *Ger. emac.* Hippos[elinum] sive Smyrnium vulgare *Park.* Macerone, quibusdam Smyrnium semine magno, nigro *J.B.* Apium XIII, sive Hipposelinum Theophrasti, vel Smyrnium Dioscoridis *C.B.* Ap[ium] Macedonicum maximum sive Hipposelinum & Olus atrum *hort. Lugd-Bat.*[366] Petroselinum Alexandrinum *Trag.* [Petroselinum] Macedonicum maximum sive Olus atrum *hort. Gron. Alisanders. At Hinton in a hedge, &ᶜ.*

Hippuris, *see* Equisetum.[367]

Hirundinaria, *see* Chelidonium majus. [*Chelidonium majus* 128]

286 *Stellaria holostea* L.

Holosteum vernum flore majore *P.B.* Holostium Ruellii *Lob.* Euphrasia gramen at nodosum *Trag.* Gramen leucanthemum *Ger. Park. Dod.* [Gramen] Fuchsii sive leucanthemum *J.B.* Caryophyllus holosteus arvensis V, sive Car[yophyllus] arvensis glaber flore majore *C.B. Stichwort, with large flowers. In woods & thickets.*

287 *Stellaria graminea* L.

Holostei Ruellii diversitas *Lob. ico.* Gramen leucanthemum alterum *Ger. hort. Bat.* among the indigenous plants. Gramini Fuchsii leucanthemo affinis & similis herba *J.B.* Caryophyllus holosteus arvensis VI, sive arvensis glaber flore minore *C.B.*[368] Gr[amen] leucanthemum minus *Park. The lesser Stichwort. On the sides of the bank between Mepole and Chatteresse, and on many other Fen banks: also in the pastures about Gamlingay.*

288 *Hordeum murinum* L.

Hordeum spurium *Park. Ger. Tab.* spontaneum spurium *Lob.* murinum *Tur. Cæs. Dod. I.B.* an herba Phœnicea Græcorum in Pliny, *book 22, chapter 25, of the same author?* Holcus of Pliny, *book 27, chapter 10.*[369] Aristida aliorum *Ang.* Gramen

---

[366] These two names mean literally "The thirteenth Celery, or Theophrastus' Horse-parsley, or Dioscorides' Smyrna-plant" and "The largest Macedonian Celery or Horse-parsley & Black vegetable". Linnaeus adopted the last name as the specific epithet of this plant, which was formerly eaten as a vegetable; it is now spelled as a single word (as, indeed, it appears in Ray's 'Index of English names').

[367] *Hippuris minor* is cited as a synonym of *Equisetum arvense longioribus setis* [*Equisetum arvense* 178], *Hippuris major* under *Equisetum primum* [*Equisetum* sp. 182] and *Hippuris lacustris foliis mansu arenosis* as a possible synonym of *Equisetum palustre ramosum aquis immersum* [*Ceratophyllum demersum* 183].

[368] These two names mean literally "A herb related & similar to Fuchs' White-flowered Grass" and "The sixth All-bone Field Pink, or Glabrous Field [Pink] with a smaller flower".

[369] In the Loeb edition of Pliny's *Natural history* the two preceding names are to be found in XXII. 135 and XXVII. 90 respectively. The first phrase means "Is it the Phoenician herb of the Greeks?" Pliny writes about this grass: *Est et herba phoenicea appellata a Graecis, a nostris vero hordeum murinum.* ("There is also a herb called *phoenicea* by the Greeks but by our people mouse barley.")

hordeaceum secundum, sive minus & vulgare *C.B.* Secale sylvestre & Gramen secalinum *of Johnson in Ger[arde]*. *Wall-Barley,*[370] *wild Rie, Rie-grasse, Way-bennet.* By footpaths.

289 *Hordeum distichon* L.

Hordeum distichum *Ger. Park. I.B.* distichum, quòd spica binos ordines habeat, Plinio[371] *C.B. Common Barley.* It is sown in fields everywhere.

    N. Cereals that have been sown densely produce an ear sooner than those sown more thinly, because the latter spread their roots more widely & are divided more intricately below because there is enough space, while the others grow upwards straightaway. *Theophr. de caus. plant.*, book 3, chapter 27.[372]

290 *Hordeum distichon* L.[373]

Hordeum distichum minus *Park. Sprit-Barley, Battledoor-Barley.* This has a shorter ear than the preceding species & more compact grains; moreover they say that it is safer from damage by birds because its grains are with greater difficulty plucked from the ear and hardly yet even when ripe. These two kinds do not tolerate the winter with us and so they are sown in spring.

291 *Hordeum vulgare* L.

Hordeum polystichum *I.B. Tur.* polystichum hibernum *C.B.* polystichum vel hibernum *Park. Winter square-Barley, or Winter beare-Barley. Among Wheat.*

    N. Generally it is four-rowed but sometimes it has more rows of grains, so we think that Hordeum hexastichum pulchrum[374] *I.B.* also does not differ from this and moreover that the spring & winter many-rowed barley differ only in the time of sowing and in developments arising therefrom, as will be clear to anyone comparing their descriptions in the works of the Bauhins. Hence, with much less unfortunate rashness,[375] our English writers seem to have handed down what seem to be contradictory statements, since Gerarde asserts that just spring many-rowed barley is commonly sown with us but Parkinson says exclusively the winter one.

---

[370] The epithet *murinum* means "of a mouse", not "of a wall", which would be *murale*. (The German name for this grass is *Maüsegerste* and the French name *Orge des souris*.) This mistake appears to date back to Turner (1548, folio D.v. *verso*), who wrote: "… Phenicea or Hordeum murinum of Plenie [Pliny], is the wal Barley, whiche groweth on mud walles. Phenix Dioscoridis semeth to be the herbe which is called in Ca[m]brig shire Way bent. It is like vnto barlei in the eare". Later Turner (1562, pp. 16 *verso* and 17 *recto*), wrote: "This herbe … is named of Pliny / Phenicea / & he saith that it is called of the latines Hordeu[m] murinu[m] / yᵗ is wall barley. … it groweth comme[n]ly upon mud walles that are lately made." Turner here correctly reports the brief passage from Pliny quoted above, but the association with walls is his own.

[371] "Two-rowed [Barley] for Pliny, because the ear has two rows".

[372] The reference to Chapter 27 is an error for Chapter 26, the final words of which correspond to the end of III. 21. 5 of Theophrastus' *De causis plantarum* in the Loeb edition. Ray's text is derived almost *verbatim* from Gaza's (1529, part 2, p. 175; see footnote 21) Latin translation.

[373] 'Spratt' or 'Battle-door' Barleys are wide, short-eared variants of *H. distichon*. They were described by Linnaeus as *H. zeocriton* and still sometimes recognised as an infraspecific taxon of *H. distichon* or (by those who consider the two common barley species as conspecific) *H. vulgare* L. (see Beaven 1947; Jarvis 2007).

[374] i.e. "the beautiful six-rowed barley".

[375] *temeritate non adeò infelice*, though this seems to have the opposite meaning from what one would expect.

292 *Salvia verbenaca* L.

Horminum sylvestre *Ger. Fuch. Dod.* pratense foliis serratis *C.B.* sylvestre alterum caule purpurascente *C.B. in Phytop.*[376] sylvestre vulgare *Park.* Orvalæ sylvestris species 4 *Dod.* Gallitrichum vulgò *Cæs.* sylvestre vulgò, sive sylvestris Sclaræa flore purpureo cæruleóve magno *I.B.* Sclaræa sylvestris *Tab.* Oculus Christi *Officin[arum].* Salvia agrestis *Brunf.* agr[estis] flore purpureo *Eyst.* Wild Clary. In various places.

293 *Hyacinthoides non-scripta* (L.) Chouard ex Rothm.

Hyacinthus Anglicus *Ger. Ad. Lob.* Anglicus sive Belgicus *I.B.* Dioscoridis *Dod. Lugd.* oblongo flore cæruleus major *C.B.* Hyacinthus longiflorus præcox I, sive dolichanthus major et Anglicus *[P.] Lauremberg*[343] *in Appar[atus plantarius].* Hare-bells or English-Hyacinth. In Madingley & Kingston woods, and in many other places.

N. All bulbous plants mature slowly from seed if you except the onion. *P. Lauremberg,*[343] *Horticult[ura], book 1, chapter 17, note 6.*

Hydropiper, *see* Persicaria. [*Persicaria hydropiper* 435]

Hydrolapathum, *see* Lapathum aquaticum.[377]

Hyssopifolia *C.B., see* Gratiola minor angustifolia. [*Lythrum hyssopifolia* 269]

294 *Hyoscyamus niger* L.

Hyoscyamus niger *Dod. Lob. Ger.* vulgaris *I.B.* niger vel vulgaris *C.B. Park.* Apollinaris *Cord. in Diosc.* Hyosc[yamus] niger sive Apollinaris herba altera Arabum *Lob. ico.* Altercum *Scribonio & Fallopio.* Jusquiamus *Officin[arum].* Henbane.

N. The seed of henbane placed on live coals emits a smoke with a most offensive smell, by inhaling which through a funnel in the nostrils or mouth they expel the little worms that are sometimes generated in the nose or teeth, and they collect them in a basin filled with water and placed underneath[378] so that they may be observed better with the eyes.

295 *Hypericum pulchrum* L.

Hypericum minus erectum *Park.* IV, sive minus erectum *C.B.* pulchrum Tragi *J.B.* 5 sive pulchrum Tragi *Ger. emac.* Small upright St Johns-wort. In Sr Roger Burgoynes parke a little beyond Gamlingay, and in other places.

296 *Hypericum perforatum* L.

Hypericum *Ger. Matth. Fuch. Dod. Lob. Cæs. Lugd. Cam. Tab.* vulgare *C.B. Park.* vulgare, sive Perforata caule rotundo foliis glabris *J.B.* Herba perforata & Hyp[ericum] vulgare *Trag.* Fuga Dæmonum *Barbaris.*[379] St Johns-wort.

---

[376] i.e. *Phytopinax*, Caspar Bauhin's first major work, published at Basel in 1596.

[377] *Hydrolapathum magnum* and *Hydrolapathum majus* are both cited as synonyms of *Lapathum aquaticum I* [*Rumex hydrolapathum* 323], while *Lapathum aquaticum sive Hydrolapathum minus* is *Rumex palustris* Sm. (but see footnote 400).

[378] *in subjecta pelvi aquâ repleto*, literally "in an underlying basin with water filled", but *repleto* must be a mistake for *repleta* since, as indicated by *subjecta*, *pelvis* is a feminine noun.

[379] Literally "Demons' Flight *for Barbarians*": for an explanation of *Barbaris* see footnote 170. This plant was gathered by country folk on St John the Baptist's Day (Midsummer Day, 24 June) as a charm to repel demons. For other beliefs relating to Midsummer Eve and Midsummer Day see footnotes 82, 281 and 721.

N. The flowers of Hypericum, Androsæmum, Ascyrum & Coris[380] are golden, but when they are squeezed with the fingers they give out a blood-red or reddish juice. *Spigel[ius in] Isag[oges]*,[381] *book 1, chapter 38.*
Hypericum caule quadrangulo, *see* Ascyrum. [*Hypericum tetrapterum* 65]

I

297 *Centaurea nigra* L. *sensu lato*[382]
Jacea nigra vulgaris *Park. Ger.* nigra vulgaris capitata & squammosa *J.B.* vulgaris capitulis nigricantibus *hort. Bat.* nigra pratensis latifolia *C.B.* Centaurii pratensis species *Gesn. hort.* Gethia nigra Scaligeri *for the same author.* Scabiosa vulgaris pratensis vel 4 *Trag.* Hyoseris Plinii *Ang.* Cyanus sylvestris angustifolius *Thal. Knapweed, Black Matfellon.*

298 *Centaurea scabiosa* L.
Jacea segetum major purpurea *P.B.* nigra vulgaris laciniata *Park.* major *Ger.* and likewise Scabiosa flore purpureo *of the same author*, as it seems to J.B. & to us also. Ja[cea] major folio multùm laciniato Dentis leonis, majusculis vasculis, floris purpurei[383] *Lob.* Scabiosa major squammatis capitulis *C.B.* and likewise Scabiosa major altera squammatis capitulis, sive Jacea rubra latifolia laciniata *of the same author.* Centaurium collinum Gesneri flore purpureo *J.B.* Flos cyanoides *Dod. Great Matfellon or Knapweed.* Among standing corn.[384]

299 *Centaurea scabiosa* L.
Jacea segetum major foliis dissectis flore albo *P.B.* laciniata alba *Park.* J[acea] 5, sive Jacea flore albo *Ger. emac.* Scabiosa major squammatis capitulis flore albo *C.B.* Centaurium collinum Gesneri flore albo *J.B. Great Matfellon with a white*

---

[380] At this time the first three names were all used (as they are by Ray in this catalogue) for species now included in the genus *Hypericum*. They were clearly thought to be related, as Gerarde (1597, pp. 434–435) and Parkinson (1640, pp. 572–577) treated them in successive chapters and Bauhin (1623, pp. 279–280) actually included them all in *Hypericum*. *Coris* included two southern European species which were treated next to *Hypericum* by Bauhin (1623, p. 280) and other authors, a yellow-flowered plant which became *Hypericum coris* L. and one with pink, purple or blue flowers which is now *Coris monspeliensis* L. in a different family (Primulaceae).

[381] For Spigelius' work see footnote 31.

[382] This aggregate comprises *Centaurea debeauxii* Gren. & Godr., *C. nigra* L. and their hybrids. The two species were not recognised in Ray's time and even today there is no agreement about the most appropriate taxonomic treatment; Stace (1997) treated them as a single taxon.

[383] "Greater Knapweed with the very laciniate leaf of Dandelion, with somewhat larger seed-vessels [and] a purple flower".

[384] *Inter segetes*: see footnote 17. In modern terms this is a surprising habitat for this perennial composite of chalk grassland and roadsides, but it does not seem to be an accidental error, since Ray presumably deliberately chose his principal name, including *segetes*, from an unusual source, How's *Phytologia britannica*, and he repeats the word and the habitat for the white-flowered form of the same species in his next entry. *C. scabiosa* was frequently reported from cornfields in southern England by the authors of a number of county Floras in the 19th century (e.g. Briggs 1880; Druce 1886; Purchas & Ley 1889), and more explicitly by Bromfield (1856) from the "grassy borders of cornfields, and amongst the corn itself": it was clearly able to persist in unintensively farmed arable land with periods of fallow in the time before intensive modern agriculture. See also footnotes 500 and 662.

flower.[385] Among standing corn but more rarely. *As in the corn a little on this side Babram.*

300 *Senecio jacobaea* L.
Jacobæa *Ger.* vulgaris *I.B. Park.* vulgaris laciniata *C.B.* Jac[obæa] Senecio *Cæs. Lob.* Herba S. Jacobi *Fuch. Gesn. Cæs. Common Ragwort, Seggrum.*

301 *Senecio aquaticus* Hill
Jacobæa latifolia *Ger. emac.* latifolia Pannonica prima *Park.* latif[olia] Pannonica 3 *Clus.* Pannonica latifolia *I.B. Broad-leaved Ragwort.* In moist & watery places.

302 *Tephroseris integrifolia* (L.) Holub
Jacobæa montana lanuginosa angustifolia non laciniata *C.B.* Pannonica 2 *Clus.* Pannonica folio non laciniato *I.B.* angustifolia *Ger. emac.* angustifolia Pannonica non laciniata *Park. Mountain Ragwort. On Gogmagog hills and Newmarket heath.*

Illecebra, *see* Sedum minimum. [*Sedum acre* 533]

Intybum erraticum, *see* Cichoreum sylvestre. [*Cichorium intybus* 131]

Irio, *see* Erysimum. [*Sisymbrium officinale* 186]

Iris palustris lutea, *see* Acorus palustris. [*Iris pseudacorus* 10]

303 *Juglans regia* L.
Juglans nux, sive Regia vulgaris *C.B.* Nux juglans *Ger. I.B.* Perhaps the Persian or perhaps the Euboean Nut of Theophrastus, the Royal Nut of Dioscorides & Galen, the Persian & Kingly Nut of Pliny[386] *I.B.* Nux juglans vulgaris *Park. The Wall-nut-tree.*

N. 1. With the shell of this nut wools are dyed, & hair is reddened with the nuts when they first appear. *Plin[y], Hist[oria] nat[uralis], book 15, chapter 22.*[387]

2. There seems to be an antipathy between this plant and the Oak, so that one does not tolerate the other. *Cam. hort.*

3. The middle of the bark of this tree when stripped off, especially when it abounds in sap, after being both dried and ground, induces vomiting, and the catkins perform this still more gently. *The same author.*

304 *Juncus inflexus* L.
Juncus acutus *Ger.* acutus vulgaris *Park. Lob. ico.* acutus paniculâ sparsâ *C.B.* foliaceus *J.B.* forté. Oxyschœnos *Dod. Common sharp Rush or hard Rush.*

---

[385] Ray (1670, p. 176) later realised that this was simply a white-flowered variant of *Centaurea scabiosa* and in a short note listed numerous purple or blue flowers which produced white or flesh-coloured variants, contrasting these with the rarity in the wild of colour variants of plants with yellow or white flowers. He thought that the variation could be environmental in origin, asking whether it might be ascribed to a shortage of nourishment or the unsuitability of the region. His readiness to amalgamate the two colour variants of *Centaurea scabiosa* contrasts with his retention of other variants as separate taxa in *Catalogus plantarum Angliae* in 1670 even though he had by then concluded that they were not specifically distinct (see, for example, footnotes 64, 458 and 825).

[386] *Nux forté Persica forté Euboica Theophrasti, Regia Dioscoridis & Galeni, Persica & Basilica Plinii.* In his *Natural history* (XV. 87 in the Loeb edition) Pliny says that the Greeks called the best kind of walnut *Persicum atque basilicon. Juglans* (= *Jovis glans*) means "Jupiter's acorn".

[387] Pliny's *Natural history*, XV. 87 in the Loeb edition.

305 *Eleocharis palustris* (L.) Roem. & Schult.
Juncus aquaticus minor capitulis equiseti *Lob. ico. Ger. Park.* clavatus Dalechampii *Lugd.* capitulis equiseti major *C.B.* capitulis longis sive clavatus *I.B.* Club-Rush, Aglet-headed Rush.[388] In rivulets. It flowers in the months of May & June & through the whole summer. This rush varies remarkably in size & stature, so much so that we can be excused for doubting whether there are not[389] distinct species, & we are more inclined to the affirmative.

306 *Schoenoplectus lacustris* (L.) Palla[390]
Juncus aquaticus maximus *Ger. Ad. Lob. ico.* lævis maximus *Park.* max[imus] holoschœnos *I.B.* grandis holoschœnos *Gesn. cat. Dod.* lævis holoschœnos II, sive Juncus maximus & Scirpus major *C.B.* Holoschœnos Theophrasti *Lugd.* Mariscus Plinii *C.B.* Bull-rush, or great water-Rush. In waters. Both C.B. and Johnson in his edition of Gerarde make two species of this, as also Park[inson]. We think that both grow with us.

Juncus bombycinus, *see* Gramen tomentosum. [*Eriophorum angustifolium* 265]

Juncus floridus, *see* Gladiolus palustris. [*Butomus umbellatus* 225]

307 *Juncus effusus* L.
Juncus lævis vulgatior paniculâ sparsâ *Lob. ico.* lævis paniculâ sparsâ major *C.B. Park.* oxyschœnos *Ang.* lævis paniculâ arundinaceâ *I.B.* vulgaris alter sive tenuior cineracei coloris, holoschœnos of Pliny, Virgil's Juncus mollis & limosus *Trag.*[391] Common soft Rush. In & next to waters.

308 *Juncus conglomeratus* L.
Juncus lævis paniculâ non sparsâ *C.B.* lævis glomerato flore *Park. Lob.* Juncus Matthioli *I.B.* Round-headed Rush. We have found it in some moist woods.

---

[388] An aglet is a metal or other sheath at the end of a cord or ribbon which prevents it from fraying; the most familiar modern examples are the plastic sheaths on the ends of shoelaces.

[389] *ut meritò dubitemus annon sint* ..., literally "that with reason we may doubt whether [there] may not be...".

[390] In addition to *Juncus lævis maximus* Parkinson (1640, pp. 1191–1192) recognises *Juncus lævis vulgaris*, commenting: "This rush is in all things like the former but much lesser and shorter." The second Cambridgeshire species of *Schoenoplectus*, *S. tabernaemontani* (C.C. Gmel.) Palla, is a shorter more glaucous plant and Ray's comments suggest that he might have seen this as well as *S. lacustris*. However, it was not until 1696 (p. 273) that he recognised *S. tabernaemontani* as a separate species in England, as "*Juncus aquaticus medius* C.B. Park." (though this is not actually Parkinson's name), which Dale had discovered "*In the Sea-ditches at* Bricklesey *and* Mersey-Island". The long interval between the note in the *Catalogus* and Dale's discovery of the species in England throws doubt on any identification of the Cambridgeshire plant as *S. tabernaemontani*. Martyn (1763) made the first explicit record of this species from Cambridgeshire, at Cherry Hinton.

[391] Literally "The second common or more slender [Rush] of a grey colour, Pliny's *Holoschœnus* [and] Virgil's soft & mud-loving Rush, [in] Tragus". The last two epithets are to be found in Virgil's *Eclogues*, 2: 72 and 1: 48 respectively. This is not a polynomial but contains various elements taken from Bock's (1552, pp. 686–687) account of *Juncus vulgaris*, in which he lists three modern species and six reported by Pliny. He does not say how they relate to each other, but here Ray equates the German *Juncus vulgaris alter*, which Tragus describes as grey (*cinereus*), to Pliny's *Holoschoenus*, to which Tragus adds Virgil's epithets in his synonymy. In his *Natural history* (XXI. 113 in the Loeb edition) Pliny describes *Holoschoenus* as *utilissimus ad vitilia ... quia mollis et carnosus est* ("most useful for wickerwork ... because it is soft and fleshy").

MAIN CATALOGUE

309 *Schoenus nigricans* L.
Juncus palustris paniculâ glomeratâ ex rubro nigricante. We think that this Rush is the same as Juncus sylvaticus Tabernamontani described in the work of J. Bauhin, for both the description & the locality agree to this very well. *The marsh soft Rush with a round blackish head. Every where in the watery places of Hinton and Teversham moors; so that he which shall look there cannot doubt what rush we mean.*

310 *Juniperus communis* L.
Juniperus *Ger. Dod. Lob.* vulgaris sive minor *Park. hort. Hafn.* vulgaris baccis parvis purpureis *J.B.* vulgaris fruticosa *C.B.* humilis *Gesn. hort. Juniper-tree. Near Hildersham on a hill, called thence Juniper-hill, plentifully.*[392]

N. 1. The berries of this tree ripen within a three-year period like those of lemon & orange trees. *C. Hofman[nus], De medicam[entis] officin[alibus], book 2, chapter 118.*

N. 2. No kind of wood produces charcoal that keeps fire alive longer than Juniper. *Camer. in hort.*

K

K Nawell Germanorum, *see* Polygonum exiguum *Dod*. [*Scleranthus annuus* 453]

Keiri officinarum, *see* Leucoium luteum sylvestre.[393] [*Erysimum cheiri* 330]

L

L Abrum Veneris, *see* Dipsacus. [*Dipsacus fullonum* 168]

311 *Valerianella locusta* (L.) Laterr.
Lactuca agnina *Ger. Park.* Valeriana campestris inodora major *C.B.* Album olus *Dod.* Locusta altera foliis serratis *I.B. Lambs Lettuce, or Corn-Sallet. In the corn growing on the sandy grounds at Gamlingay not far from the church.*

312 *Lactuca serriola* L. forma *serriola*
Lactuca sylvestris costâ spinosâ *C.B.* sylvestris laciniata *Park.* sylv[estris] foliis dissectis *Ger. emac.* sylv[estris] sive Endivia multis dicta folio laciniato, dorso spinoso *I.B.* sylv[estris] of Dioscorides, Theophrastus & Pliny, which seems to be Thridacine of Galen *C.B.* Seris domestica *Lob. Cut-leaved wild Lettuce. By the margins of fields. As by the way side from Ely to Cambridge, within a mile or two of Ely.*

313 *Lactuca serriola* forma *integrifolia* (S.F. Gray) S.D. Prince & R.D. Carter
Lactuca sylvestris 2. *Ger. emac.* sylvestris odore viroso *C.B.* sylv[estris] endiviæ foliis, odore viroso *Park.* sylv[estris] lato folio, succo viroso *I.B.* Endivia *Trag., page* 268. Thesion Dalechampii *Lugd., page* 564. *Endive-leaved wild Lettuce.* Johnson, in his edition of Gerarde, creates two kinds of wild lettuce with the smell of opium, which he thinks are confused in the works of other writers; & he

---

[392] Ray (1670, pp. 181–182) reports this locality as "*Upon Juniper-hill near Hildersham, and no where else near Cambridge, that I know*".

[393] *Leucoium luteum vulgare* must be intended here.

attributes to the first sort earlier authors' names which we think are more appropriate to his second sort, since we believe that our plant, which is one with undivided leaves & is Gerarde's second one, is the only one described by botanical writers, rather than as he suggests, Gerarde's first.[394]

314 *Lactuca saligna* L.
Lactuca sylvestris laciniata minima, not yet described.[395] *The least cut-leaved wild Lettuce. This was found on a bank and in a ditch by the side of a kind of drove or lane leading from London road to the river, just at the water near a quarter of a mile beyond the spittle-house end.*

315 *Galeopsis angustifolia* Ehrh. ex Hoffm.
Ladanum segetum quorundam flore rubro *I.B.* segetum Plinii & Tetrahit angustifolium aliis *Lugd.* Sideritis arvensis angustifolia rubra *C.B.* arvensis rubra *Park.* Ancoraria *Tur.* Alysson Galeni flore purpureo *Tab.* Sideritis 7 *Ger. emac. Narrow-leaved All-heal or Iron-wort.* Among standing corn. It flowers from the middle of summer into Autumn.

316 *Trifolium arvense* L.
Lagopus vulgaris tenuifolius *hort. Hafn.* vulgaris *Lugd. Park.* trifolius quorundam *I.B.* genuinus antiquorum *Ad. Lob.* Hippocratis *in the same place.* Lagopus sive Pes leporinus *Matth. Fuch. Dod. ut. Lob.* Lagopodium sive Pes leporis *Ger.*[396] Trifolium spicatum VI, i.e. arvense humile spicatum sive Lagopus *C.B.* Λαγωπους [Lagopus] for Dioscorides, λαγοπυρον [lagopyron] for Hippocrates, Lagopus for Pliny *of the same author. Haresfoot or Haresfoot-trefoile.* Among standing corn, *upon the hill of health, and on many lay-grounds.*

317 *Lamium album* L.
Lamium album *Ger. emac.* flore albo *Park.* album non fœtens folio oblongo *C.B.* Lam[ium] sive Archangelica flore albo *Ad. Lob.* Galeopsis sive Urtica iners floribus albis *I.B. White Archangel, Dead Nettle.*

318 *Lamiastrum galeobdolon* (L.) Ehrend. & Polatschek
Lamium luteum *Ger. emac. Park. Lob.* III, sive folio oblongo luteum *C.B.* Urtica iners 3, sive Lamium flore luteo *Dod.* Galeopsis flore luteo vel Urtica Labeo

---

[394] The reference is to Johnson (1633, p. 309), where he describes three large wild lettuces, the second of which is not illustrated but is described as having "all the leaues whole". Ray was of the opinion that this was one of the two plants which he knew in Cambridgeshire and that to which the earlier authors' names cited by him applied correctly. He retracted this view in his *Catalogus plantarum Angliae* (1670, p. 185; 1677, p. 178), and in *Emendanda* of the 1685 appendix and in his *Synopsis* (1690, p. 41; 1696, pp. 69–70) he regarded *Lactuca sylvestris 2* and *Lactuca sylvestris costâ spinosâ* (Johnson's third species) as variants of the same species (*L. serriola* L.) and separate from the plant now called *L. virosa* L. (Johnson's first species). Despite this, all Cambridgeshire authors ascribed *Lactuca sylvestris 2* to *L. virosa* until the matter was sorted out by Oswald (2000) in a detailed discussion of the history of the larger *Lactuca* species in Cambridgeshire.

[395] See, however, Ray's entry for this species in *Emendanda* of the 1685 appendix.

[396] *Lagopus* (i.e. Λαγωπους), with a diminutive form *Lagopodium* (i.e. Λαγωποδιον), and *Pes leporis* or *Pes leporinus* are the Greek and Latin respectively for "foot of a hare", used to describe the softly downy inflorescence of this clover, so the two preceding names mean literally "Hare's-foot or Foot of a hare, *of Mattioli, Fuchs, Dodoens in [his] French & Latin edition and de L'Obel*" and "Little-hare's-foot or Foot of a hare, *of Gerarde*". Hippocrates' name, listed by Liddell, Scott & Jones (1940) as Λαγωπυρος (with omega in the middle and feminine rather than neuter), literally means "hare's-wheat" and relates to this species. For Hippocrates see footnote 24 and Chapter 4.

Cam. Galeopsis sive Urtica iners flore luteo *I.B.* Yellow Archangel. *In Madingley and Kingston woods.*

Lamium maximum sylvaticum fœtidum *C.B. see* Galeopsis legitima Dioscoridis. [*Stachys sylvatica* 209]

319 *Lamium purpureum* L.

Lamium rubrum *Ger. emac.* vulgare flore rubro *Park.* purpureum fœtidum folio subrotundo, sive Galeopsis Dioscoridis *C.B.* Galeopsis sive Urtica iners flore purpurascente majore, folio non maculato *I.B.* Urtica non mordax vulgaris fœtens purpurea *Lob. ico.* Urtica Labeo mas *Brunf.* Small dead Nettle, red Archangel.

320 *Lapsana communis* L.

Lampsana *Lob. Ger. emac. Cam. Dod. Thal.* Papillaris in Prussia *Cam.* Lamp[sana] vulgaris *Park.* Lampsana Dodonæi *I.B.* Sonchus sylvaticus primus *Tab.* Sonchis cognatum olus *Gesn. hort.* Soncho affinis Lampsana domestica *C.B.*[397] Chrysolachanum Plinii *Ruell. Dock-cresses, Nipplewort or Tetterwort.*[398]

Lantana, *see* Viburnum. [*Viburnum lantana* 612]

321 *Rumex* spp.[399]

Lapathum acutum *Ger. emac. Dod. Lob. Trag.* acutum majus *Park.* acutum sive Oxylapathum *J.B.* Lapathum folio acuto plano *C.B.* Oxylapathum of Dioscorides and Pliny *of the same author.* Rumex acutus *Cord.* Sida sylvestris Avicennæ *C.B.* in Phyt. *Sharp-pointed Dock.*

322 *Rumex obtusifolius* L.

Lapathum sylvestre folio minùs acuto *Ger. emac.* sylv[estre] vulgarius *Park.* vulg[arius] folio obtuso *J.B.* sylv[estre] latifolium II, sive Lap[athum] folio subrotundo *C.B.* Rumex agrestis *Cord. Ordinary wild or roundish-leaved wild Dock.*

---

[397] The two preceding names mean "Vegetable related to the sow thistles" and "Cultivated Lampsana related to sow-thistle".

[398] "Tetter" was a name applied to a variety of skin diseases, including eczema and (Ray 1675, p. 41) ringworm.

[399] The difficulties in identifying Ray's *Rumex* species are similar to those presented by his *Carex* species; in both cases he had not mastered the taxonomy of the genus by 1660 and he presents fewer species in the *Catalogus* than he is likely to have encountered in the field. It therefore seems intrinsically unlikely that there will be a precise relationship between his taxa and our modern species. Babington (1860) and Ewen & Prime (1975) identified *Lapathum acutum* as *Rumex conglomeratus* Murray, presumably on the basis of Ray's later view (1686, p. 174; 1690, p. 31; 1696, p. 56) that *Lapathum acutum minimum* appeared to be the same species as this taxon, but his understanding of the genus had advanced so greatly by then that it seems unwise to use his mature views to define his taxonomic concepts in 1660. The *Lapathum acutum* of 1660 seems likely to have covered several taxa including *R. conglomeratus* Murray, *R. sanguineus* L. and *R. crispus* L. Ray's first reasonably convincing account of the genus is in *Synopsis* (1670, pp. 187–189). In this he retained *Lapathum acutum* and *Lapathum sylvestre folio minùs acuto* (renamed *Lapathum vulgare folio obtuso* J.B.) from Cambridgeshire (but noted that they did not appear to him to be specifically distinct, which suggests that he had still not quite got to grips with the genus) and added to the Cambridgeshire list *R. crispus* and *R. conglomeratus*. For further details see *Addenda* and *Emendanda* of the 1685 appendix. One source of confusion may have been Ray's failure to recognise *R. sanguineus*, which is a fairly common plant in Cambridgeshire. The first county record of *R. sanguineus* is given by Babington (1860) and Perring *et al.* (1964) as that of Relhan (1786), who described the red leaf-veins of his plant. This suggests that the ordinary green-veined plant, var. *viridis* (Sibth.) W.D.J. Koch, was previously confused with other species.

323 *Rumex hydrolapathum* Huds.
Lapathum aquaticum I, sive aquaticum folio cubitali *C.B.* aquat[icum] maximum sive Hydrolapathum *J.B.* Hydrolapathum magnum *Ger.* majus *Lob. Lugd. Park.* Rumex palustris *Trag.* *The great water-Dock.*

324 *Rumex conglomeratus* Murray[400]
Lapathum aquaticum sive Hydrolapathum minus *Ger. emac. Park.* aquat[icum] minus *C.B. J.B. The lesser water-Dock.* It has flowers arranged in whorls.

Lappa major, *see* Bardana. [*Arctium lappa* and/or *A. minus* 79]

Lappago Plinii, *see* Aparine. [*Galium aparine* 51]

325 *Lathyrus sylvestris* L.
Lathyrus major latifolius *Ger. emac.* major perennis *Park.* Narbonensis latiore folio *Ad. Lob.* major latifolia flore purpureo speciosior *J.B.* VI, sive latifolius *C.B.* Ervum sativum *Cord.* Cicercula sylv[estris] *Tab.* Clymenum *Matthioli.* *Pease everlasting.* In Madingley wood, and in severall other woods.

326 *Lathyrus pratensis* L.
Lathyrus luteus sylvestris dumetorum *J.B.* sylv[estris] flore luteo *Park. Ger. emac.* sylv[estris] minor flore luteo *hort. Bat.* XIII, sive sylv[estris] luteus foliis viciæ *C.B.* *Tare everlasting.* In thickets & by hedges.

327 *Daphne laureola* L.
Laureola *Park. Ger. Dod.* mas *Matth. hort. Hafn.* semper virens flore viridi, quibusdam Laureola mas *C.B.* semper virens flore luteo *J.B.* Daphnoides sive Laureola *Lob. Tab.* Thymelæa *Trag.* *Dwarfe Lawrell, or Spurge-Lawrell, Lowry.*

328 *Lens culinaris* Medik.
Lens *J.B.* vulgaris *C.B.* minor *Lob. Dod. Lugd. Cam. Ger. Park.* Lenticulæ I genus *Trag.* *Lentills.* It is sown in fields & very often is mixed with barley. To learn how much lentils were valued in times past see the work of *Athenæus,*[401] *book 4, chapter 18.*

N. About this plant I can bear witness from my own experience that in spring it sometimes turns to stone in a field. *C. Hofman[nus], book 2, De medicam[entis] officin[alibus], chapter 114.* For why leguminous plants grow more quickly & exhaust the ground less than grain crops see the work of *Theophr., de caus., book 4, chapter 9.*[402]

329 *Lemna minor* L.
Lens palustris *Ger. Matth. Dod. Lob. J.B. Lac.* palustris sive aquatica vulgaris *Park.* Lenticula palustris vulgaris *C.B.* palustris *Trag. Tab.* Muscus lenticulæ similis Dioscoridis *J.B.* *Duckes-meat.* An inhabitant of stagnant waters. *Spigel[ius], [in]*

---

[400] Note Ray's decision in *Emendanda* of the 1685 appendix that he had misidentified as *Lapathum aquaticum sive Hydrolapathum minus* [*R. palustris* Sm.] what was in fact *Lapathum acutum minimum* [*Rumex conglomeratus*]. He apparently never found the former in Cambridgeshire; the first records for the county in Babington (1860) and Perring *et al.* (1964) are therefore incorrect. See also footnotes 149, 150 and 158 in the 1685 appendix.

[401] *Quanto olim in pretio lentes fuerunt vide apud Athenæum*, literally "For how much formerly in price/value lentils were, see in the work of Athenaeus". For the Greek author Athenaeus see Chapter 4.

[402] Theophrastus' *De causis plantarum*, IV. 8. 1–3 in the Loeb edition.

Isag[oges], book 1, chapter 5, is astonished that nobody before him had observed the roots of this plant arising from the earth.[403]

N. If (as sometimes happens) it is transferred by the flooding of waters from stagnant waters to flowing ones, this plant, when it has clung to the banks, is propagated in an extraordinary way; for it attaches itself to the earth by certain hairs that have grown up underneath it, which seem to be in place of roots, and so, maturing, it finally grows into a plant not dissimilar to Sisymbrium aquaticum.[404] That is what that investigator of the works of nature, Daléchamps, observed not without some surprise. *J.B.*

Lenticula aquatica trisulca, *see* Hederula aquatica. [*Lemna trisulca* 273]

Leucographis *Plinii, see* Carduus lacteus. [*Silybum marianum* 108]

330 *Erysimum cheiri* (L.) Crantz
Leucoium luteum vulgare *C.B.* luteum, vulgò Cheiri flore simplici *J.B.* Viola lutea *Ger. Trag. Fuch. Dod.* Keiri sive Leucoium luteum simplex vulgare *Park.* Cheiri vel Keiri *Offic[inarum]. Gesn. Lon.* Viola lutea of Pliny, & ιον φλογινον [ion phloginon] of Theophrastus *Lugd.*[405] *Wild Cheir, Wall-flower.* On walls generally.

331 *Marchantia polymorpha* L. and/or *Lunularia cruciata* (L.) Lindb.[406]
Lichen *Matth. Fuch. Dod. ut. Tur.* Lichen sive Hepatica vulgaris *Park.* L[ichen] sive Hepatica fontana *J.B.* L[ichen] petræus latifol[ius] sive Hepat[ica] fontana *C.B.* Jecoraria sive Hep[atica] fontana *Trag.* Hepatica terrestris *Ger. Liverwort, or common ground Liverwort.* In moist shady places.

N. The larger & more succulent leaves of this plant have a resinous flavour when fully tasted, the smaller & less fat ones a bitterness.[407] J. Bauhin attributes to this plant a taste that is to some extent aromatic & bitter & in addition somewhat biting. Let us leave to be considered by learned physicians what ought to be con-

---

[403] For Adrianus Spigelius see footnote 31. *Hujus plantæ radices è terra orientes*: this phrase, which is puzzling because the fronds of duckweed float on the surface with their roots in the water below, is presumably explained by Daléchamps' observation recorded in the following note, though in fact he must have been mistaken about the final outcome.

[404] The plant referred to here appears to be *Mentha aquatica sive Sisymbrium* of Bauhin & Cherler (1651, vol. 3(2), p. 223) and Ray [*Mentha aquatica* 362] although the only *Sisymbrium aquaticum* listed by Ray is *Sisymbrium aquaticum alterum* Matth. in the synonymy of *Cardamine* Ger. [*Cardamine pratensis* 106].

[405] There has long been a sharing of names between *Viola* spp. and scented members of Cruciferae, as still today in the vernacular name "Dame's Violet" for *Hesperis matronalis* L. *Leucoium*, now applied to the monocotyledonous snowflakes related to Snowdrop, *Galanthus nivalis* L., is from the Greek λευκοιον [leucoïon], meaning "white violet" and Theophrastus' name cited here from his *Historia plantarum* (VI. VIII. 1) means "flame-coloured violet".

[406] Almost all the names listed by Ray are included by Dillenius (1741) in the synonymy of the species now called *Marchantia polymorpha*, but it is likely that Ray included the other frequent large thallose liverwort in Cambridgeshire, *Lunularia cruciata*, in his concept of this species. For records that certainly refer to *Marchantia polymorpha*, see the entries for *Hepatica stellata* [A19] and *Hepatica umbellata* [A20] and footnote 29 in Addenda of the 1663 appendix. Ray (1686, pp. 125–126) later recognised *Lunularia* as distinct, having been shown it not far from his house in Essex by the Braintree pharmacist Samuel Dale.

[407] *Hujus plantæ folia majora & succulentiora qualitatem resinosam planè gustantibus exprimunt, minora & minùs pinguia amaritudinem,* literally "Of this plant the larger & more succulent leaves squeeze out a resinous property to [those] tasting [it] thoroughly, the smaller & less fat [ones] a bitterness".

cluded from these facts for rightly understanding the properties of this plant, since Hepatica has without good reason been considered cold by the majority of people & it is thought that in intemperance cold things should be avoided.

332 unidentifiable foliose lichens[408]
Lichen arborum *Ger.* arboreus sive Pulmonaria arborea *J.B.* Muscus pulmonarius sive Lichen arborum *Park.* Muscus arboreus VII, sive pulmonarius *C.B.* Pulmonaria *Trag. Matth. Fuch. Dod. ut. Gesn. hort. Lac.* Pulmonaria fungosa *Lugd.* Tree-Lungwort. It grows on trees, especially on Oak.

333 *Ligustrum vulgare* L.
Ligustrum *Ger. J.B.* vulgare *Park. Trag.* Germanicum *C.B.* Cyprus Plinii *book 12, chapter 24*.[409] C. Hofman[nus], De medicam[entis] officin[alibus]. Privet or Prim-print. In the hedges of the closes about Barnwell, and in many hedges and woods.

N. 1. Among all shrubs, trees & herbs there is none that can be fashioned, bent or formed by clipping into so many figures & graceful shapes as privet. *P. Lauremberg,*[343] *Horticult[ura], book 1, chapter 30.*

2. We have frequently found the caterpillar described by Moufet in *Theatr. insect.,*[410] *book 2, chapter 2,* as the noblest of the greens feeding on the leaves of this shrub, about which it is worthy of note that it feels anger easily compared with other caterpillars, which it expresses by twisting and untwisting its body as many times as it can if you touch it even lightly. This caterpillar also, casting its green skin in Autumn, is changed into a chrysalis and rests thus through the whole winter wrapped up as if in bandages, covered by no web or case. The colour of the bandages is dark red. The shape is round and without corners with a distinctive sheath for the proboscis projecting in the upper part. From this fairly hard case there emerges in the following June a large moth, which is pot-bellied and hairy and has its whole body & wings besprinkled with a whitish powder, whence it is agreed that it belongs to the class of Phalænæ or Nocturnæ. The wings are short and useful for walking rather than for flying. It will not be irrelevant here (since no one paying attention can fail to recognise the truth,[411] though it is manifest also in silkworms & all other caterpillars) to point out an error of other authors & especially of *Moufet in Theatr. insect.,*[410] *book 2, chapter 1, final words,* that the caterpillar's head is changed into the tail of the moth: for in all the caterpillars that I have been able to observe quite the opposite happened.

---

[408] The names strictly refer to the conspicuous lichen *Lobaria pulmonaria* (L.) Hoffm. but this appears to have been extremely rare in eastern England even before the advent of serious air pollution in the 19th century. Ray does not mention any other foliose lichens and we therefore interpret this entry as referring to foliose lichens in general.

[409] Pliny's *Natural history*, XII. 109 in the Loeb edition.

[410] *Insectorum sive minimorum animalium theatrum*, compiled by Moufet and published posthumously in London in 1634 (see footnote 76). The larva, pupa and imago described are those of the Privet Hawk-moth, *Sphinx ligustri* (Linnaeus), but Ray's comment that the "wings are short and useful for walking rather than for flying" is not an accurate description of the normal imago. It seems likely that he is describing moths which have been reared in captivity and emerged with deformed wings. This deformity is often believed to result from storage of the pupae in dry conditions (M.S. Botham, pers. comm.) although Friedrich (1986) suggests that it can also occur if the pupae are overcrowded. See also footnotes 220 and 702.

[411] *ubi res neminem attendentem latere potest*, literally "since the matter can be concealed from nobody attending".

2. Hop, and probably all other twining plants, in climbing up poles follow the movement of the sun, that is they are twisted from east through south towards the west and never in the opposite direction, even though you may endeavour to make that happen by turning & guiding them actually by hand.

3. The young shoots[429] or sprouts of this plant, which are tender when they first appear, cooked & prepared in the same way as the common asparagus, are sought out by very many people for food. Incidentally it may be permissible to note here that the ancients used also to eat the shoots of Bryony that come up at its first appearance. *Colum[ella], book 12, chapter 7. Paulus* also wrote about these that *some shoots of Bryony are suitable for the stomach & induce urine, but they give little nourishment, although, when they are cooked, they are more effective in nourishing than the shoots of vegetables. Jo. Bod[æ]. in Theoph[r]. hist.*[430]

343 *Reseda luteola* L.

Luteola *Lob. Ger.* vulgaris *Park.* I, sive Luteola herba salicis folio *C.B.* Lutum herba *Dod. hort. Bat.* Lutea Plinii quibusdam *I.B.* Lutea vel Luteum Vitruvii *Gesn. hort. Cam.* Antirrhinum *Trag. Lugd. Thal.* Pseudostruthium *Matth. Cast. Lugd.* Guadarella vulgò *Cæs.* Catanance *Lon. Tab.* Yellow-weed, Wild Woad. Diersweed. On walls & in rough places.

Lychnis segetum, *see* Pseudomelanthium. [*Agrostemma githago* 472]

344 *Silene latifolia* Poir.

Lychnis sylvestris flore albo *Ger. emac. Clus. Lon. Park.* Coronaria IV, sive sylvestris alba simplex *C.B.* Ocymoides *Gesn. hort. Matth. Lac. Tab. Cast.* album multis *J.B.* Wild white Campion. Alongside roads & by hedges.

Lychnis viscosa flore muscoso, *see* Sesamoides Salamanticum magnum. [*Silene otites* 539]

Lycopsis Anglica *Lob. see* Echium alterum. [*Echium vulgare* 172]

345 *Scutellaria galericulata* L.

Lysimachia galericulata *Lob. Lugd. Clus. hist.* Lys[imachia] non spicata III, sive Lys[imachia] cærulea galericulata, vel Gratiola cærulea *C.B.* Tertianaria[431] *Tab.* Tertianaria, aliis Lys[imachia] galericulata *J.B.* Sideritis cærulea *Thal.* Gratiola latifolia *Ger.* latifolia nostras *Park.* Hooded Willow-herb. *In the ditches by the foot-way side to Cherry-Hinton, and elsewhere.*

346 *Lysimachia vulgaris* L.

Lysimachia lutea *Trag. Fuch. Tur. Ad. Lob. Cæs. Tab. Cam. Ger. I.B.* lutea communis *Clus. hist.* lut[ea] major vulgaris *Park.* lut[ea] I, sive lut[ea] major quæ

---

[429] asparagi: see Ray's note on *Bonus Henricus* [90] (including footnote 142) and footnote 260.

[430] The whole passage from "the ancients" onwards is taken almost *verbatim* from Bodaeus (1644, p. 1187.2; see footnote 23), who cites Paulus but does not include him in his index. It relates to Black Bryony, *Tamus communis* [92], called in Ray's time *Bryonia nigra*. Some people in France still collect the young shoots to eat, as witnessed by P.H.O. near Castres (Tarn) on 23 May 2004. In limestone areas of central France where Spiked Star-of-Bethlehem, *Ornithogalum pyrenaicum* L., is common, its young spears are also collected and served on menus as *asperges des bois*; they were formerly sold as "Bath asparagus" in the market in Bath, as attested by the late Dr David Coombe from his childhood memories.

[431] i.e. a herb to treat tertian malaria, a disease characterised by paroxysms occurring every 48 hours (or every third day if the day of each paroxysm is reckoned as the first). See also the entry for *Sideritis Anglica* [540].

MAIN CATALOGUE

it is commonly called Milium Solis[424] inappropriately. *Gromwell, or Gromill. It ought to be called Graymill* (i.e. Milium griseum) *in English, not, as it commonly becomes, Grummell. It is found in various places, as in a close near Newnham mills.* Gerarde & Parkinson describe two species of Lithospermum growing wild with us; we cannot see adequately whether or how they may be distinguished, but the descriptions & figures of the smaller of the two fit this wild one of ours more nearly.

Lithospermum sylvestre *Tragi, see* Anchusa degener. [*Lithospermum arvense* 43]

340 *Lolium temulentum* L.
Lolium album *Ger. Park.* Lolium et triticum temulentum *Ad. Lob.* L[olium] gramineum spicatum caput tentans *J.B.* L[olium] et Aera Plinii *C.B.* Gramen loliaceum spica longiore *of the same author.* Αιρα [Aera] of Dioscorides, Aristotle, Theophrastus, Galen, Oribasius, Aegineta and Aetius *C.B.* Ζιζανιον [Zizanion] for Constantinus, Zinzania of the Arabs *of the same author. Darnell, Juray, Ray. Among standing corn.*

341 *Lolium perenne* L.
Lolium rubrum *Ger.* rubrum, sive Phœnix *Park.* Phœnix *Matth. Dod. Tur.* Phœnix lolio similis *J.B.* Hordeum murinum *Lob. Tab.* Gramen loliaceum angustiore folio et spicâ *C.B.* Φοινιξ [Phoenix] for Dioscorides, from the purple colour of the seed, herba Phœnicea for the Greeks and Hordeum murinum for our [Roman] authors for Pliny[425] *of the same author. Red Darnell-grasse. By footpaths.*

Lujula, *see* Trifolium acetosum. [*Oxalis acetosella* 578]

Lupi crepitus, *see* Fungus pulverulentus. [unidentifiable puff-ball fungi 207]

342 *Humulus lupulus* L.
Lupulus *J.B. Clus. Trag. Dod. Matth.* salictarius *Ger. Park. Lob.* Vitis Septentrionalium *of the same author.* Lupus salictarius of Pliny, *book 21, chapter 15.*[426] *Fuch. Dod. gal. Gesn. hort. Lac. Lob. Cæs. Ger.* Lupulus sativus et sylvestris *Lob. Trag. Lugd. J.B.*, for the cultivated hop does not differ from the wild except in size. There are, however, two sorts of Hop: the first can be called useless Hop, which is male, with the flower of the female hemp and devoid of seed, the second useful Hop, which is female, with a scaly flower or rather fruit, bearing seed.[427] The first kind is called by Jean Bauhin male Hop but by C.B. female; the second is for J.B. female but for C.B. male.[428] *Hops, the male and female. The male are in some counties of England called Seed Hops, yet improperly, seeing they bear onely flowers. The manured Hops of both sorts are but in few places of this County. The wild grow in many hedges about Chesterton, and divers other places.*

N. 1. The male or useless hops can easily be distinguished from the female soon after the appearance of the long shoots because the males produce much redder shoots and the females greener ones.

---

[424] i.e. "Millet of the Sun".
[425] See the entry for *Hordeum spurium* above and the related footnotes.
[426] Pliny's *Natural history*, XXI. 86 in the Loeb edition.
[427] The Latin words used to describe the kinds of hops mentioned here are, respectively, *sativus, sylvestris*, *Lupulus inutilis* and *Lupulus utilis*.
[428] i.e. J.B.'s *Lupulus mas* is C.B.'s *fœmina* and *vice versa*.

3. Oil of flax has such subtle components that it cannot be kept in earthenware vessels without sweating. *Michael Mayerus in Lusu serio.*[418]

4. A book produced only from flax will last if you use paper burnt to blackness & mixed with thickened oil of flax[419] for the printing ink. *The same author in the same place.*

5. Let him who wishes to know what wrongs and how many torments this herb suffers from men, above all others, see *Goropius Becanus* splendidly & rhetorically describing them *in Hermath[ena], book 3.*[420]

337 *Linum perenne* L.

Linum sylvestre radice perenni, flore cæruleo *hort. Bat.* sylv[estre] I, floribus cæruleis *Ger. emac.* sylv[estre] angustifolium cæruleo flore majore *Park.* sylv[estre] angustif[olium] I, sive flore magno *C.B.*[421] sylv[estre] angustif[olium] album lineis in umbilico purpureis[422] *of J. Bauhin* does not differ from this except in the colour of the flower, as will be clear to anyone consulting *C.B.* and *Ger. emac.*; but in fact we have seen only that form which has a blue flower.[423] Linum sylv[estre] 4. *Clus. pan.* sylv[estre] 5, angustifolium 2 *in the same author's Hist[ory].* Wild blew Flax. On Gogmagog hills, and thereabouts; and in some closes about Cherry-Hinton, abundantly.

338 *Linum catharticum* L.

Linum sylv[estre] catharticum *Ger. emac.* sylv[estre] pusillum candicantibus floribus Camerarii *Johnson[us] apud Ger[ardum].* sylv[estre] angustifolium VII, sive pratense flosculis exiguis *C.B.* Chamælinum Clusii flore albo, sive Linum sylvestre catharticum *Park.* Linocarpos *Thal.* Alsine verna glabra flosculis albis, vel potiùs Linum minimum *J.B.* Alsine verna Dalechampii *Lugd.* Mil-mountain, Dwarf wild Flax. Very common in meadows & pastures, *as in the closes about Chesterton, and Hinton, and Teversham. &^{c}.*

339 *Lithospermum officinale* L.

Lithospermum, sive Milium solis *J.B.* Lith[ospermum] majus erectum *C.B.* minus vulgare *Park.* minus *Ger. Matth. Dod. Gesn. hort. Lac. Lob. Lugd.* Milium Soler to the Mauritanians, because it grows commonly in the Soler mountains and hence

---

[418] i.e. Michael Majerus (or Mayerus or Maier) in his *Lusus serius*, published at Oppenheim and dated 1616, subtitled in its edition in English of 1654 "Serious passe-time: a philosophicall discourse concerning the superiority of creatures under man".

[419] *si papyrum ad nigredinem ustam; & oleo lini incrassato mistam ... utaris*, literally "if paper to burnt blackness & with thickened oil of flax mixed ... you may use". There is a grammatical error here because *ustam* is feminine, to agree, even though oddly, with *nigredinem*, but *mistam* (also feminine) should be *mistum* (neuter) as it must agree with *papyrum*.

[420] Ray's use of the words *luculenter & rhetoricè* ("splendidly & rhetorically") is probably sarcastic: see the entry on *Prunus* in Ray's 'Etymology'. The passage referred to can be found in the posthumous collection of Goropius' works of 1580, in *Hermathena*, book 3, pp. 40–41, labelled in the margin *Linum tormenta varia patitur* ("Flax suffers various torments"); it describes at great length the treatment of flax both in processing and in subsequent uses and includes the comment *Dies me & verba priùs deficiant, quàm omnes iniurias lini queam recensere ...* ("Time [literally "The day"] & words would fail me before I could recount all the torments of flax ...").

[421] See Ray's comment on these names in *Emendanda* of the 1685 appendix.

[422] Literally "Narrow-leaved wild [flax], white with purple lines in the navel". In his 'Interpretation of terms' Ray defines the *umbilicus* of a flower as the middle part whence the style & stamens emerge.

[423] The white-flowered form occurs today on Signal Hill near Babraham not far from Cambridge (Crompton 2001).

Limonium verum Dioscoridis & antiquorum, *see* Plantago aquatica. [*Alisma plantago-aquatica* 445]

334 *Thesium humifusum* DC.
Linaria adulterina[412] *Tab. Ger. emac.* erecta angustifolia XV, sive montana flosculis albicantibus *C.B.* Linariæ similis *J.B.* Linosyris nuperorum *Lob. ico.* Pseudolinaria montana alba *Park.* Anonymos lini folio *Clus.* Anthyllis montana *Lugd.* Bastard Toadflax. On Gogmagog hills, and Newmarket-heath.[413] The very small flowers of this plant consist of five petals spread out in the manner of a star, whitish, ending in a point and somewhat greenish on the underside. Each petal is endowed at its base with two very minute auricles, as it were, on this side and on that. The base of the flower is green, and from it arise five small stamens, which are yellow tinged with green, with a central white style. Two leaflets[414] spring from each little calyx of the flowers.

335 *Linaria vulgaris* Mill.
Linaria vulgaris nostras *Park.* vulgaris lutea *Ger. J.B.* vulgaris lutea flore majore *C.B.* Osyris *Matth. Jo. Bodæ.* Osyris, Linaria & Urinaria *Plant. ico.* Common Toad-flax.

N. Very many of the common people place this herb under the bare soles of their feet inside their boots & shoes to drive away quartan fever.[415]

Lingua bovis, *see* Buglossum luteum. [*Helminthotheca echioides* 94]

Lingua avis Dalechampii, *see* Conyza palustris. [*Senecio paludosus* 146]

Linum aquaticum, *see* Conferva Plinii. [unidentifiable filamentous algae 140]

336 *Linum usitatissimum* L.
Linum sativum *Ger. emac. Park. Dod. C.B.* vulgare cæruleum *Ad. Lob.* Linum *J.B. Matth. Fuch. Tur. Lugd.* Manured Flax.

N. 1. The linen of the ancients was manufactured from the finer Spanish flax, about which *Pliny* writes in *book 19, chapter 1*; it seems not to differ much from the cambric which even today is imported from Spain.[416]

2. If polished flint is heated in oil of flax it acquires an invincible hardness, as a result of which it turns out more suitable for polishing those things which are needed for a marble tablet. *J[ean] B[auhin]* in his Historia fontis & balnei Bollensis from an account of Master Lutz[ius], a pharmacist of Kirch.[417]

---

[412] i.e. counterfeit, supposedly as distinct from the next species [*Linaria vulgaris* 335], but it could more easily be confused with *Linum sylvestre catharticum* [*Linum catharticum* 338], another small, fine-leaved plant that grows in the same habitat. See also footnotes 13 and 264.

[413] Ray (1695, column 418) reported these localities as "On Gogmagog-hills and New-market heath, but scatteringly."

[414] In fact each flower has a bract and two bracteoles.

[415] For quartan malaria see footnote 245.

[416] See Pliny's *Natural history*, XIX. 10 in the Loeb edition. "Cambric" (our translation of Ray's *tela Cameracensis*, literally "web/warp of Cameracum [Cambrai]") is defined by OED as "A kind of fine white linen, originally made at Cambray in Flanders".

[417] See *J.B.* in Ray's 'Explanation', his Note 2 after *Filix mas* [199] and his entry for *Myosotis scorpioides hirsuta* [381].

Dioscoridis[432] *C.B.* Salicaria *Gesn. hort. Common yellow Loose-strife or Willow-herb. In the close where* Cirsium Anglicum[433] *grows near the foot-path, and in the Islet made by the brook Stoure near to the paper mills, and in diverse other places.*

347 *Lythrum salicaria* L.
Lysimachia purpurea *Lob. Cæs. Cam. Tab.* purp[urea] spicata *Ger. Park.* purpurea quibusdam spicata *J.B.* spicata I, sive spicata purpurea fortè Plinii *C.B.* Solidaginis Saracenicæ alia species *Trag. Purple spiked Willow-herb. In the ditches about Histon, and the Fen ditches, &*[c].

348 *Lythrum salicaria* L.[434]
Lysimachia purpurea trifolia caule hexagono. *Spigel[ius in] Isag[oges],*[435] *book 1, chapter 10.* This differs from the above principally in these respects, that in this plant three leaves joined at one point surround the stem, while in that only two; then that the branches in this plant also arise in triple order from the axil of the three leaves, while in that other one only in twos (since the stem is square); finally that in this plant the flowers surround the stem exactly in whorls but not in the preceding one. *Three-leaved purple Loose-strife, or Willowherb.*

349 *Epilobium montanum* L.
Lysimachia siliquosa glabra major *C.B.* campestris *Ger.* siliquosa major *Park.* lævis *J.B. The lesser wild codded Loose-strife or Willowherb.*[436] In moist stony places.

350 *Epilobium tetragonum* L.
Lysimachia siliquosa glabra minor, sive siliquosa IV *C.B.* siliquosa glabra media sive minor *Ger. emac. The lesser smooth-leaved codded Willowherb.* By ditches & rivulets.

351 *Epilobium palustre* L.
Lysimachia siliquosa glabra minor angustifolia *Ger. emac.* siliquosa V, sive siliquosa glabra angustifolia *C.B. The least smooth codded Willowherb.* On *Teversham moor.*

352 *Epilobium hirsutum* L.
Lysimachia siliquosa hirsuta magno flore *C.B.* siliq[uosa] hirsuta majore flore purpureo *J.B.* siliq[uosa] Filius ante patrem[437] *Ad. Lob.* Lysim[achia] sive Salicaria

---

[432] "The first yellow [Loosestrife] or the greater yellow [Loosestrife], which [is that] of Dioscorides".

[433] i.e. *Cirsium dissectum* 137.

[434] In *Catalogus plantarum Angliae* Ray (1670, p. 203) retained a separate entry for this taxon but noted that it was a variety of the preceding species. In his paper *Of the specifick differences of plants*, sent to the Royal Society in 1674, he listed two characters of the stalk which did not constitute specific differences, including "varying its number of Angles, as in Purple Loose-strife" (Cain 1999a).

[435] For Adrianus Spigelius see footnote 31.

[436] This is given the English name "Greater wild codded Willowh[erbe]" in 'Index of English names', which is consistent with '*major*' in the Latin names and suggests that 'lesser' may have been a slip of the pen.

[437] Literally "Podded [Loosestrife], Son before father". "Son afore (or before) the father" has been more commonly applied to Colt's-foot, *Tussilago farfara* [603] (as mentioned by Ray under *Tussilago*), or to Butterbur, *Petasites hybridus* [436], because the flowers appear before the leaves (Grigson 1955, p. 367), or, like "Herbe impious", to Common Cudweed, *Filago vulgaris* [230], because "for the most part those flowers which appeer first are the lowest & basest; and those that come after growe higher, as children seeking to ouergrowe or ouertop their parents, (as many wicked

altera *Trag*. Onagra campestris hirsuta *Cæs*. Antoniana, sive Herba S. Antonii *Gesn*. *Great codded Loose-strife, or Willowherb*.

N. The leaves of this plant when lightly crushed smell very like half-cooked apples,[438] whence it is also called in English *Codlings and cream*.

353 Epilobium parviflorum Schreb.
Lysimachia siliquosa hirsuta parvo flore *C.B.* siliq[uosa] hirsuta flore minore *J.B.* Lys[imachia] 12 *Gerard*. sive sylvatica. Lys[imachia] siliquosa sylvestris hirsuta *Park*. forté. *Hairy small-flowered Loose-strife or Willowherb*. By ditches & rivulets, with the preceding.

M

MAjorana sylvestris *Park. see* Origanum vulgare. [*Origanum vulgare* 410]

354 Malva neglecta Wallr.
Malva sylvestris minor *Park*. sylv[estris] pumila *Ger. Fuch. Dod*. sylv[estris] repens pumila *Lob*. I, sive sylv[estris] folio rotundo *C.B.* flore candido *Cæs*. vulgaris flore minore, folio rotundo *J.B. Dwarfe Mallow, small wild Mallow*. In rubble & by roads.

355 Malva sylvestris L.
Malva vulgaris *Park*. vulg[aris] procerior *Lob*. sylvestris *Ger*. sylv[estris] major *Tab. Thal. Gesn. hort*. II, sive sylvestris folio sinuato *C.B.* vulg[aris] flore majore, folio sinuato *J.B.* The holiest leaf of Pythagoras, in the work of Aelian,[439] book 4, chapter 17. Anguill. Jo. Bodæ. in Theophr. hist. *Common Mallow*.

356 Malus sylvestris (L.) Mill.
Malus sylvestris sive agrestis *J.B.* & of almost all authors. Mala sylvestria quæ & alba & rubra & majora & minora[440] *C.B.* In woods & hedges. *The Crab-tree, or Wilding*.

---

children do" (Gerarde 1597, p. 518); but Dodoens (1578, pp. 74–75) says of *Epilobium hirsutum* that it is "called of some, in Latine Filius ante Patrem, that is to say, the sonne before the father, bycause y' his long huskes in which the seede is cōteined do come forth and waxe great, before that the floure openeth".

[438] ... *odorem pomorum semicoctorum bene imitantur*, literally "... the odour of half-cooked apples well imitate". Grigson (1955, p. 196) claims that this explanation, first put forward by Ray, "has been repeated a hundred times ... by botanists who never crushed the leaves and smelt them" and that the leaves "have no characteristic smell, nor have the flowers"; he proposes instead that "codded" suggested "codlin" and that then, as codlins were "often boiled in milk and then eaten with cream, ... the rosy and white combination in the flowers" may have suggested this and other fruit names. A codling or codlin is a greenish elongated English cooking apple.

[439] *Folium sanctissimum Pythagoræ apud Ælian[um]*. The Pythagoreans, followers of the Greek philosopher and mathematician Pythagoras (*c*. 570–*c*. 495 BC), were reputedly vegetarian but famously are said to have shunned beans; see the entry for *Faba* in Ray's 'Etymology'. Apparently they also regarded mallows (used as a vegetable in classical times) as sacred and so not to be eaten. Claudius Aelianus (AD *c*. 175–*c*. 235), often known just as Aelian, was a Roman author and teacher of rhetoric who spoke Greek so perfectly that he was called μελιγλωσσος [meliglossos], "honey-tongued". Roman-born, he preferred Greek authors and wrote in a slightly archaic Greek himself; the English titles of his surviving works are *On the nature of animals* and *Various history* (an abridged version).

[440] "Wild apples which are both white & red & greater & lesser".

MAIN CATALOGUE

N. 1. The shape of fruits is never straight-sided and does not have any angles. *Theophr. hist. plant., book 3, chapter 8.*[441]

2. In those fruits whose seed is covered with pulp, the moister, larger and fuller is the pulp in the fruit, the smaller the seed inside, & vice versa, and hence it is that wild trees usually have a larger seed and a smaller pulp. *P. Lauremberg,*[343] *Horticult[ura], book 2, chapter 3, number 2.*

357 *Marrubium vulgare* L.
Marrubium album *J.B. Ger.* album vulgare *C.B. Park.* album odorum *Ad.* Prassium *Ang. Officin[arum].* White Horehound. Beside roads. It may be prescribed for the chest & is good for a cough.

358 *Lycopus europaeus* L.
Marrubium aquaticum *Ger. emac. Trag. Dod. Lob.* aquat[icum] vulgare *Park.* aquaticum quorundam *J.B.* palustre *Lon. Dod. gal.* palustre glabrum *C.B.* Verbenæ alterum genus *Cæs.* Sideritis prima Matthioli *Ang.* Water-Horehound.

N. There are those who call this Egyptian herb,[442] because those who pretend that they are Egyptians apply to themselves the blackish colour from this herb like that which Egyptians possess; for the juice of this herb dyes anything with a blackish colour of this sort, which clings so tenaciously that it cannot be washed or wiped off. *Cam. hort.*

Marrubium aquaticum acutum *Ger. see* Sideritis. [*Stachys palustris* 540]

Marrubium nigrum fœtidum, *see* Ballote. [*Ballota nigra* 77]

Matricaria, *see* Parthenium. [*Tanacetum parthenium* 421]

Matrisylva, *see* Periclymenum, & Asperula. [*Lonicera periclymenum* 432 and *Galium odoratum* 67]

Medica, *see* Trifolium.[443]

---

[441] Σχῆμα οὐδὲν περικαρπίων εὐθυγραμμον ἐστι οὐδε γωνιας ἐχει. Curiously the reference is incorrect: this sentence is from Theophrastus' *Historia plantarum*, book 1, Chapter 18 (I. xii. 1 in the Loeb edition).

[442] *Ægyptia herba*, the reference being to gypsies, as in the modern vernacular name for this species, "Gipsywort". "Egyptian" was a former name for a gypsy, apparently because they were thought to originate in Egypt. The word "gypsy" is derived from it and in Shakespeare's time it could still mean true Egyptians, although it soon acquired its modern more restricted usage. Acts of Parliament from 1562 until the 1822 Vagrancy Act (repealed in 1824) set out penalties (including the death penalty in the earlier legislation) for those (in the words of a 1597 Act) "wandering and pretending themselves to be Egipcyans, or wandering in the Habbite Forme or Attyre of counterfayte Egipcians" (Tawney & Power 1951). It is doubtful whether people actually disguised themselves as gypsies, and the Acts might simply have intended to counteract a defence argument that a person born in England or Wales could not be an Egyptian. For further information on this subject, which is too complex to be discussed adequately here, see Fraser (1992), Netzloff (2001) and Sandland (1996). Floras have consistently reported that *Lycopus europaeus* is called gypsywort or "the Egyptians herbe, bycause of the Rogues and runnegates whiche calle themselves Egyptians, do colour themselves blacke with this herbe" in the words of Lyte's translation of Dodoens (1578). However, "the sixteenth century lives in terror of the tramp" (R.H. Tawney, cited by Beier 1974) and its Flora-writers are no more likely than its legislators to have provided an accurate record of gypsy behaviour.

[443] Three of Ray's *Trifolium* species have synonyms beginning with *Medica*, all now in the genus *Medicago* – *Trifolium cochleatum folio cordato maculato* [*M. arabica* 581], *Trifolium echinatum arvense* [*M. minima* 582] and *Trifolium sylvestre luteum siliquâ cornutâ* [*M. sativa* subsp. *falcata* 587].

359 *Melampyrum cristatum* L.
Melampyrum cristatum flore purpureo *J.B. Purple-headed crested Cow-wheat*. In Madingley and Kingston woods, and almost in all woods in this County plentifully, likewise it overspreads all the pasture or common grounds you ride through going out of Madingley to dry Draiton: whence we cannot but wonder that it should not be described or figured by Gerard or Parkinson.[444]

360 *Melampyrum pratense* L.
Melampyrum sylvaticum flore luteo, sive Satureia lutea sylvestris *J.B.* IV, sive luteum latifolium *C.B.* Cratæogonon I, sive album *Ger.* vulgare *Park.* Parietaria sylvestris *Clus. Eyst.* Hyssopus nemorensis & lutea. *Common Cow-wheat. In a wood at Stichworth & in a wood at St George Hatley in great plenty.* The pictures of this plant which are to be found in the works of *Ger[arde]* & *Park[inson]* do not agree well with the descriptions of the same, especially in the number & position of the flowers, as will be obvious to anyone discussing them.

N. This plant does not like to be cultivated in garden plots. The seed must be collected speedily before it shoots out of the capsules & immediately returned to the soil so that it may fall out of its own accord, for it is believed to lose its fertility if it is handled when naked. *J.B., volume 3, book 30, chapter 63, page 440, column 2*. We have sown both species several times in gardens, but they have never germinated.

361 *Melilotus altissimus* Thuill.
Melilotus vulgaris *Park. Lugd.* vera flore luteo *Gesn. hort.* Germanica *Ger. Ad. Lob. Dod. gal.* I, sive Officinarum Germaniæ *C.B.* officin[arum] flore luteo perennis *hort. Hafn.* Trifolium odoratum sive Melilotus vulgaris flore luteo *J.B.* Sertula campana *Plinii*.[445] Trifolium caballinum *Cam. & for the Italians*, because horses seek after this plant very eagerly in their fodder. *Common Melilot. In ploughed fields & by hedges*.

N. 1. When Dioscorides said that melilot possessed a good scent, he should be understood to have been speaking about the dried herb, because when green it has almost no scent, but when it has dried it has a very fragrant one; and this also was not overlooked by Theophrastus *in book 6 [of] De causis plant[arum]*. Sertula campana, he says, smells more strongly when dried. *Marant[a], Meth[odus] cogn[oscendorum] simp[licium], book 2, chapter 5*.[446]

2. There are no greater lovers of floral garlands or more attentive to them under the sky than the inhabitants of Mexico. Not only do they themselves weave chaplets and garlands from flowers at home in such variety and with such elegance that nothing more marvellous could be imagined, but they also present them to their Gods instead of a large votive offering and adorn their annual festivals with them. And this matter may be highly valued by their chieftains, and no doubt from that cause the custom has spread, because we usually see painted representations of them with some flowers held out in both hands, just as with us gloves are

---

[444] *Melampyrum cristatum* has a very restricted distribution in Britain, centred on Cambridgeshire, and so it is not very surprising that Ray was the first British botanist to record it.
[445] Pliny's *Natural history*, XXI. 53 in the Loeb edition.
[446] Theophrastus mentions plants that have more scent when dried, instancing iris and melilot, in his *De causis plantarum* (VI. 14. 8 in the Loeb edition). For Bartolomeo Maranta see footnote 34.

often added to the hands of men portrayed in paintings instead of an ornament. *P. Lauremberg, Horticult[ura], book 2, chapter 6, number 2*.[447]

Among us English the people of Norwich excel in their fondness for and their cultivation of choice flowers: as we have heard from them themselves, they yearly celebrate the Floralia,[448] in which there is a competition for the finest flowers & the victor is awarded a floral garland as the prize.[449] For this reason they boast that their Norwich is worthy to be called either a city in a garden or a garden situated in a city.

Melilotus coronaria sive coronata *Lob*.[450] *see* Trifolium corniculatum.

362 Mentha aquatica L.

Mentha aquatica sive Sisymbrium *J.B. Ger. emac.* rubra *Brunf.* aquat[ica] rubra *Park.* rotundifolia palustris sive aquatica major *C.B.* aquat[ica] sive Sisymbria rubro folio & flore nigræ menthæ, Balsamine officinarum *Lob. ico.* Calamintha aquatica *Gesn. Tab. Water Mint.*

N. 1. Experience shows that the mint that generates spiked flowers rarely produces a seed & even when it does for the most part bears an infertile one.[451] *Jo. Bodæ. in Theoph[r]. hist.*

2. Leaves of mint immersed in it do not permit milk to curdle or to be thickened into cheese. *Diosc., book 3, chapter 41.*

Mentha arvensis verticillata, *see* Calamintha aquatica. [*Mentha arvensis* 99]

363 Nepeta cataria L.

Mentha cattaria *J.B.* felina sive cattaria *Ger.* cattaria vulgaris & major *C.B.* Nepeta major vulgaris *Park. The greater Catmint or Neppe. In all the lanes about Swafham in great plenty: in a lane on the right hand of Barnwell in going thither, which leads down to the moor on which stand the pesthouses: also at Ditton near Bigwin closes by the foot-way side, and in many other places about Cambridge.*

N. 1. The smell of Catmint suggests a cat hunting for mice & is unpleasant to almost everyone. *C. Hofman[nus], De medicam[entis] officinal[ibus], book 2, chapter 146, page 425.* To us the smell of greater Catmint does not seem unpleasant.

2. This mint, when transplanted from the fields into gardens, we have frequently seen bruised & completely destroyed by the mouths of cats and by their whole bodies abandoned to rolling around on it, so much so that the soil compressed & polished by their bodies even reveals the place where they have bruised the plant & crushed it to pieces:[452] and so far we have not been able to observe

---

[447] For Peter Lauremberg see footnote 25. This passage contains some words requiring explanation: *Nova Hispania* ("New Spain") means Mexico; *Domini* ("Lords") clearly refer to gods; *donarium* can be a temple treasury or a gift to a god; *reguli* ("kinglets") seem best translated as "chieftains".

[448] In Roman times this was the festival of the goddess Flora, celebrated on 27 April.

[449] *quibus de florum elegantia certatur, & victori floreum sertum velut præmium porrigitur*, literally "in which about the elegance of flowers it is contended, & to the victor a floral garland as the reward is supplied".

[450] This presumably refers to the synonym *Melilotus coronata* Ad. Lob. cited by Ray under *Trifolium corniculatum primum* [*Lotus corniculatus* 579].

[451] Peppermint, *Mentha* × *piperita* L., the hybrid between *M. aquatica* L. and *M. spicata* L., is a spicate, sterile hybrid, as are other triploid hybrids of *M. spicata*.

[452] *in quo comminutam plantam triverint*, literally "in which they may have bruised the plant crushed to pieces", though this does not seem to accord with the actual sequence of events. Ray also discusses the reaction of cats to this species in 'Etymology', where he cites J. Bauhin's observations.

anything similar in any other plant transplanted from the fields into a garden: for all those that we have transplanted have been untouched by cats, with this one plant excepted, which was always destroyed in the way described unless covered over with thorns, as we have tried at least six times and that in different places. But in fact, if it is protected for some time in a garden, until it can take root & flower, it will remain untouched in future, like those plants also which have arisen from fallen seed; so much so that cats seem to seek after it when it is weak & languishing but not to pay attention to it when it is vigorous. Hence that English rhyme about Catmint:

> *If you set it the cats will eat it;*
> *If you sow it the cats can't know it.*

Mercurialis montana spicata, *see* Cynocrambe. [*Mercurialis perennis* 163]

364 *Stratiotes aloides* L.
Militaris aizoides *Ger.* Stratiotes sive Militaris aizoides *Park. Ad.* Aloe IV, sive palustris *C.B.* Aloe sive Aizoon *J.B. Water-Sengreen, or Fresh-water Souldier.* In a ditch on the left hand as you go to Stretham ferry, and in the river about the same ferry, and in the river and ditches about Audrey causey abundantly.

365 *Hottonia palustris* L.
Millefolium aquaticum dictum Viola aquatica *I.B.* aquat[ica] seu Viola aquatica caule nudo *C.B.* aquat[ica] floridum, seu Viola aquatica *Park.* Viola palustris *Ger.* Fœniculum aquaticum 2 *Tab.* Myriophyllon *Cæs.* equisetifolium fluviatile *Ad. Lob. Water-Violet, or Water-Millefoil.* Frequent in muddy waters.

366 *Ranunculus trichophyllus* Chaix[453]
Millefolium aquaticum ranunculi flore & capitulo *Park.* aquat[icum] flore albo *Clus. hist.* aquat[icum] VI, sive aquaticum foliis abrotani, ranunculi flore & capitulo[454] *C.B.* Millef[olium] sive Maratriphyllum 3 flore & semine ranunculi aquatici hepaticæ facie[455] *Ger. Lob. ico.* Fœniculum aquat[icum] 3 *Tab.* Alga palustris & fluviatilis *Gesn.* Ranunculus aquat[icus] omnino tenuifolius *J.B.* Ranunc[ulus] trichophyllon aquaticus medioluteus *Col. Water Fennell, Crowfoot, or Millefoile.* Very common in waters.

367 *Myriophyllum verticillatum* L.
Millefolium aquaticum minus *Park. J.B.* aquaticum III, sive flosculis ad foliorum nodos[456] *C.B.* aquat[icum] 6, sive Myriophyllum aquat[icum] minus Clusii *Ger. emac. Small water-Yarrow or Millefoile.* In the rivulet Stoure, by the little Islet which it makes above the Paper mills, and in divers other places.

---

[453] As there is no consensus even today about the taxonomy of *Ranunculus* subgenus *Batrachium*, it might seem foolhardy to ascribe Ray's plant with confidence to a modern name. However, Parkinson's account of *Myriophyllum aquaticum Ranunculi flore & capitulo* (1640, p. 1257) is a remarkably convincing description of *Ranunculus trichophyllus* as Stace (2010) understands this species.

[454] "... or water Milfoil with the leaves of southernwood [and] the flower & head of a buttercup/crowfoot", *capitulum* here presumably being used of the crowded stamens and/or carpels; see Ray's 'Interpretation of terms'.

[455] "The third Milfoil or three-leaved Fennel with the flower & seed of a water-crowfoot [and] with the appearance of hepatica".

[456] "The third water [Milfoil] or [that] with flowers at the nodes of the leaves".

MAIN CATALOGUE

368 *Myriophyllum spicatum* L.
Millefolium aquaticum pennatum spicatum *Park. C.B.* pennatum aquaticum *J.B.* Feathered water Millefoil. *In the river about Stretham ferry.*

369 *Utricularia vulgaris* L.
Millefolium palustre galericulatum *Ger. emac.* aquaticum flore luteo galcriculato *J.B. hort. Hafn. Lob. ico. Park.* aquat[icum] XI, sive aquat[icum] lenticulatum *C.B.* Lentibularia & Meon aquaticum *Gesn. coll.* Fœniculum aquaticum galericulatum *Tab.* Hooded water-Millefoile. *In the brook Stoure, by the Islet. It flowers around the end of June & the beginning of July. It is found with difficulty except while it is adorned with a flower. It grows plentifully in many of the great Fen ditches in the Isle of Ely.*

370 *Potamogeton pectinatus* L.
Millefolium tenuifolium *Ger. emac.*, as far as the picture goes, for the description does not correspond to it. Fennell-leaved water Millefoile. *In the river Cam in many places.* With regard to the spike of seeds that it produces, this plant is similar to Fontalis[457] and is more correctly considered a species of that than of Milfoil. We are surprised that, when it is so frequent & ordinary, it should appear nowhere that we have yet seen, either described or depicted, other than in Gerarde.

371 *Achillea millefolium* L.
Millefolium vulgare album *C.B. Park.* terrestre vulgare *Ger.* majus album *hort. Gron. hort. Bat.* Millef[olium] stratiotes pennatum terrestre *J.B.* Stratiotes millefolia *Fuch.* Militaris, sive Millefolium flore albo *Ad.* Achillea *Dod.* Sideritis Achilleos *Hermolai corall.* Common Yarrow, Millefoile. *Everywhere well known.*

372 *Achillea millefolium* L.
Millefolium vulgare flore diluti ruboris *Trag.* flore rubro *Ger.* vulgare rubrum *Park.* purpureum *Tab.* vulgare purpureum minus *C.B.* Red-flowered or purple Millefoile or Yarrow. *These two do not seem to us to differ in species.*[458]

373 *Galium album* Mill.
Mollugo montana sive Gallium *Ger.* montana angustifolia, sive Gallium album latifolium *C.B.* vulgatior *Park.* Rubia sylvatica altera *Gesn.* Matrisylva 2 *Trag.*[459] Great bastard Madder. *In the hedges in many places, especially about Linton.*

374 *Succisa pratensis* Moench
Morsus Diaboli *Trag. Gesn. Ad. Lob. Cam. Tab. Ger.* Diab[oli] vulgaris flore purpureo *Park.* Succisa *Matth. Fuch. Dod. Cast.* Succisa sive Morsus Diaboli *J.B.* Scabiosa folio integro *Cæs.* Jacca nigra *Brunf.* Pycnocomon *Col.* Succisa glabra *C.B.* perhaps. Devills-bit. *Very common in meadows & pastures, especially moister ones.*

Morsus gallinæ, *see* Alsine hederacea. [*Veronica hederifolia* 24]

---

[457] i.e. *Potamogeton*: see *Potamogeiton foliis latis splendentibus* [460], *Potamogeiton longis acutis foliis* [462], *Tribulus aquaticus minor quercûs floribus* [576] and also *Tribulus aquatic[us] minor muscatellæ floribus* [575], the last now segregated as *Groenlandia*.

[458] Ray (1670, 1677) continued to treat these flower colour variants separately while noting that they did not appear to him to be specifically distinct; he finally treated them in a single account in *Historia plantarum* (1686, pp. 345–346).

[459] See, however, Ray's comment about the last two synonyms and his lengthy discussion about the plant in *Emendanda* of the 1663 appendix and also the comments in *Emendanda* of the 1685 appendix.

375 *Hydrocharis morsus-ranae* L.
Morsus ranæ *Ger. Dod. Ad. Lob.* Nymphæa minor sive Morsus ranæ *J.B.* Nymphæa alba minima sive Morsus ranæ *Park.* Nymphæa alba minima *C.B. Frog-bit. In the ditches cut out of the river Cam behind Peterhouse, and infinite other ditches.*

376 *Hydrocharis morsus-ranae* L.
Morsus ranæ flore pleno odoratissimo,[460] not yet described. *In a ditch by the side of Audrey causey, close by the great wooden bridge plentifully.*

Muscipula Salamantica major, *see* Sesamoides Salamanticum magnum. [*Silene otites* 539]

377 *Usnea* spp. or other fruticose lichens[461]
Muscus arboreus *Trag. Dod. Matth.* arboreus villosus *J.B.* quernus *Lob. Ger. Park.* M[uscus] arboreus, Usnea Officinarum *C.B.* Bryon of Dioscorides & Galen, Sphagnos of Pliny[462] *J.B. Tree Mosse.*

Muscus pulmonarius, *see* Lichen arborum. [unidentifiable foliose lichens 332]

Muscus capillaris, *see* Adianthum aureum.[463]

378 unidentifiable mosses[464]
Muscus ex cranio humano *Ger. Park.* ex cranio *J.B.* Usnea Cranii *Schroder. pharmac. The Mosse on a dead mans skull.* This moss is an ingredient in the composition of an unguent of the cabinet, about the uselessness of which see *Primrosius, De vulgi erroribus, Charletonus, Palinodia in Philosophia Epicuro-Gassendo-Charletoniana, & alii.*[465] *This moss springs from a human skull kept in a moist place for some time.*

---

[460] "Frogbit with a very sweet-smelling double flower". Ray (1670, 1677) retained this as a species, citing only this locality. However, it does not appear even as a synonym in *Historia plantarum* (1688a, p. 1320), and in *Synopsis* (1690, p. 207) he says that it seems to be a variety of the plant now called *H. morsus-ranae*. Such *flore pleno* variants appear to be very rare in this species; we have not encountered them and they are not mentioned by Cook & Lüönd (1982) in their generic monograph.

[461] The commonest of the species in the lichen genus *Usnea* in eastern England is now *U. subfloridana* Stirt., but it is not known which species would have been present before the air pollution of the 19th and 20th centuries or whether Ray included lichens in other fruticose genera in this entry. We know of no evidence to support Ewen & Prime's (1975) identification of this species as the moss *Brachythecium rutabulum* (Hedw.) Schimp.

[462] Pliny's *Natural history*, XXIV. 27 in the Loeb edition. Pliny's *sphagnon* and the *bruon* of classical authors are discussed by Scott (1988).

[463] Ray cites this synonym under both *Adianthum aureum majus* Ger. [*Polytrichum commune* 12] and the unidentifiable *Adianthum aureum minus* [13].

[464] It would be fanciful to suppose that *Muscus ex cranio humano* represents a taxon in the modern sense. The physician William Heberden (1710–1801) lectured on *materia medica* in Cambridge in the mid 18th century and a specimen in his collection, at St John's College, Cambridge, is the common pleurocarpous moss *Homalothecium sericeum* (Hedw.) Schimp. "It is unlikely to have been the only moss used" under this name (Belcher & Swale 1998), though, interestingly, Dillenius (1741) attributes Gerarde's name to this species.

[465] "Primrose, *On the errors of the common people* [1638], Charleton, *A Recantation to the Philosophy of Epicurus, Gassendi & Charleton*, and others". The second reference is obscure, but appears to refer to the long discussion of the "unguent" in 'The Translators Supplement' added by Charleton to 'Of the magnetick cure of wounds', one of three works by the Paracelsian physician J.B. van Helmont which he translated in *A ternary of paradoxes* (Charleton 1650b, pp. 93–105). The unguent contained "the Mosse grown on a humane skull 2. ounces" as well as various other ingredients, "into which a splinter of wood, or the weapon stained with the patients blood, is to be immersed", the idea being

MAIN CATALOGUE

379 *Cladonia* spp.
Muscus pyxoides sive pyxidatus *Ger. Park. J.B.* terrestris ramosus XIII, sive pyxiodes terrestris *C.B. Cup or Challice-Mosse. Upon the hill of health,* & elsewhere in various places.

380 unidentifiable mosses[166]
Muscus terrestris vulgaris *Dod. Lob. Ger.* terrest[ris] vulgatissimus *Park.* terrest[ris] repens I, sive vulgatissimus *C.B.* Muscus terrestris & hortensis *J.B. Common Earth-Mosse.*

Myagrum siliquâ longâ, *see* Camelina. [*Erysimum cheiranthoides* 102]

381 *Myosotis arvensis* (L.) Hill
Myosotis scorpioides hirsuta *Park.* scorp[ioides] arvensis hirsuta *Ger. emac.* Echium scorpioides arvense *C.B.* scorpioides minus folio hirsuto *hort. Gron.* scorpioides solisequum flore minore *J.B.* scorpioides minus hirsutum, flowering in the months September, October & August *J.B. in Historia fontis & balnei Bollensis.*[467] Auricula muris cærulea *Tab.* Alsine myosotis sive Auricula muris *Lob. ico. Mouse-ear-Scorpion-grasse, blew Mouse-ear.* In ploughed fields. It generally has blue flowers, but more rarely white.

382 *Myosotis scorpioides* L.
Myosotis scorpioides palustris *Ger. emac.* scorpioides repens *Park.* Euphrasia cærulea *Trag. Tab.* Scorpioides fœmina *Dod. in ico.* Echium palustre *Cord.* scorpioides palustre *C.B.* scorpioides majus *hort. Gron.* scorpioides solisequum aliud flore majore cæruleo & albo *J.B.* Echium aquaticum *hort. Paris.* Heliotropium minus in palustribus *Cæs.* Heliotropii species altera *Gesn. hort.* Leontopodium *Lon. Lugd.* Auricula muris *Matth. Lac. Lugd.* muris glabra *hort. Pat. Water-Scorpion-grasse.* Generally in watery ditches.

N. From the sap of the leaves and the colour & scent of the flowers it seems to us to have the same properties that Borage[468] possesses.

383 *Myosurus minimus* L.
Myosuros *J.B.* Cauda muris *Ger.* Holostium IX, sive Holostio affinis cauda muris[469] *C.B.* Holostium Loniceri, cauda muris vocatum *Park. Mouse-tail. In the corn by the high-way side leading from Hoginton to Huntington rode plentifully.*

"to cure a wound by unction of the instrument of the harme". However, Charleton is concerned to show that the unguent was invented after Paracelsus' death and to establish the original recipe rather than to condemn its uselessness. Ray's reference does not appear to be to Charleton's *Physiologia Epicuro-Gassendo-Charltoniana* (1654). For Gassendus see *P. Gassend.* in Ray's 'Explanation' and for Charleton see footnote 133, which discusses another obscure citation of his work.

[466] Martyn (1763) identifies this as *Brachythecium rutabulum* (Hedw.) Schimp., as, tentatively, does Proctor (1956). This is, nowadays at least, much the commonest Cambridgeshire bryophyte, and Dillenius (1741) included Gerarde's description of *Muscus terrestris vulgaris* in the synonymy of this species, but he also listed *Muscus terrestris vulgaris* Lob. and Dod. in the synonymy of a species of *Sphagnum*. Ray lists only four moss species and it is clearly impossible to equate them all to modern species.

[467] See *J.B.* in Ray's 'Explanation', his Note 2 after *Filix mas* [199] and his Note 2 after *Linum sativum* [336].

[468] *Borago officinalis* L.

[469] Literally "The ninth All-bone or mouse's tail related to All-bone". *Myosurus* or, with the original Greek spelling, *Myosuros* (μυος ουρος), like *Cauda muris*, means "mouse's tail", with reference to this plant's appearance in fruit, with a long spike of achenes borne on an elongated receptacle.

384 *Anthriscus caucalis* M. Bieb.
Myrrhis sylvestris seminibus asperis *C.B.* sylv[estris] Neapolitana atque etiam Anglicana *Park.* Cerefolium 5, sive Myrrhis Æquicolorum nova Columnæ *Ger. emac. Small Hemlock-chervill.* On the Mudwalles in many places about Cambridge and by the wayes sides.

Myrtus Brabantica, *see* Elæagnus Cordi. [*Myrica gale* 173]

N

385 *Brassica rapa* L. and *B. napus* L.[470]
NApus sylvestris *J.B. C.B.* sylv[estris] sive Bunias *Park.* Bunias sylvestris Lobelii *Ger.* Rapum sylvestre *Trag. Wild Navew, in the Fens Cole-seed.*

N. It comes up naturally at the margins of fields, by footpaths & by the banks of the River Cam. It is also sown in very great quantity in the Ely marshes, where, by its very copious production of seed, it offers a not inconsiderable profit to the farmers, for its seed, warmed, crushed by the power of a windmill prepared for that purpose & then pressed, pours forth a great quantity of oil suited to many purposes.[471]

386 *Narcissus pseudonarcissus* L.
Narcissus pallido-luteus longo calyce, sive sylvestris Anglicus *hort. Bat.* magno & oblongo tubo latifolius VIII, sive sylvestris pallidus calyce luteo *C.B.* totus luteus montanus Theophrasti *Lob.* Βολβουϰωδιον [Bolbucodion] of Theophrastus, & Codiaminum of Pliny *C.B.* Bulbocodium vulgatius *J.B.* Codianum vel Codiaminum flore Codii, i.e. campanulæ *Gesn. hort.* Pseudonarcissus Anglicus *Ger.* Anglicus vulgaris *Park. in parad. English wild bastard Daffodill, or common yellow Daffodill. In a bushy close on the south side of Whitwell a single farm-house beyond Coton in the way to Hardwick.* It flowers in March.

387 *Nasturtium officinale* W.T. Aiton *sensu lato*[472]
Nasturtium aquaticum *officinarum.* aquat[icum] vulgare *Park.* aquat[icum] supinum *C.B.* aquat[icum] sive Cratevæ Sium *Ger. emac.* Sium Cratevæ erucæfolium *Lob.* Sisymbrium sive Nasturtium aquaticum *J.B.* Laver odoratum *Eric. Cord. Water-cresses.*

---

[470] Babington (1860) treats this as *Brassica campestris* L. (now *B. rapa* subsp. *campestris* (L.) A.R. Clapham). This is a reasonable interpretation of "Wild Navew", but the plant cultivated for oil is surely *B. napus* L., which is also listed later by Ray as *Rapum sylvestre* [495]. These two *Brassica* species are closely related as *B. napus* is an allotetraploid derived from *B. rapa* and *B. oleracea* L. Ewen & Prime (1975) treat Ray's *Napus sylvestris* as *B. oleracea* L. but this is known from the county only as a relic of cultivation. See also Ray's entry for *Rapum sylvestre* Ger. in *Emendanda* of the 1685 appendix.

[471] There was a great boom in the production of rape oil in England in the 17th century, especially on drained fenland, even though the fenlanders protesting against drainage during the Civil War had dismissed rapeseed as "trash and trumpery" from Holland (Thirsk 1997).

[472] Cytological studies in the mid 20th century established that the single species known to Ray and later authors, including Babington (1860), consists of three closely related taxa, *Nasturtium officinale* W.T. Aiton, *N. microphyllum* (Boenn.) Rchb. and their hybrid *N.* × *sterile* (Airy Shaw) Oefelein. All three occur in Cambridgeshire. It is very unlikely that the "two species" mentioned in Ray's following note refer to the modern taxa.

MAIN CATALOGUE

N. Two species of this plant are suggested, growing commonly, but we have so far not been able either to find them or to distinguish those that we have found.

Nasturtium pratense, *see* Cardamine. [*Cardamine pratensis* 106]

Nasturtium hybernum, *see* Barbarea. [*Barbarea vulgaris* 78]

Nasturtium verrucosum, *see* Coronopus Ruellii. [*Lepidium coronopus* 153]

Nepeta vulgaris, *see* Mentha cattaria. [*Nepeta cataria* 363]

Nigellastrum, *see* Pseudomelanthium. [*Agrostemma githago* 472]

388 *Lysimachia nummularia* L.
Nummularia *Ger.* vulgaris *Park.* major lutea *C.B.* Nummularia sive Centimorbia *J.B.* Ἐρευθεδανον Theophrasti[473] *Anguill. Ad. Lob. C.B.* Numulus, which is falsely read as Mimulus, of Pliny, *book 18, chapter 28*,[474] for certain writers *in the same author. Yellow Money-wort or Herbe Two-pence.*

389 *Anagallis tenella* (L.) L.
Nummularia minor flore purpurascente *C.B. Ger. emac. Park.* Num[mularia] rubra *J.B. Little purplish-flowered Money-wort. On Trumpington, and Hinton, and Teversham moors.*

390 *Nymphaea alba* L.
Nymphæa alba *Ger. J.B.* alba major *C.B.* alba major vulgaris *Park.* Nenuphar album *Brunf. Great white Water-Lilly. In the water on Teversham moor, and many other places.*

391 *Nuphar lutea* (L.) Sm.
Nymphæa lutea *Ger. J.B.* major lutea *C.B. Park.* Nenuphar luteum *Brunf. Great yellow Water-Lilly. In the river Cam almost every where.* The flower of this plant recalls by its odour the distilled spirit or liquor that they commonly call aqua vitæ.[475]

392 *Nymphoides peltata* Kuntze
Nymphæa lutea minor flore fimbriato *J.B.* lutea IV, sive lutea minor flore fimbriato *C.B.* alia minor *Lugd. Small yellow Water-Lilly with a fringed flower. In the river about Stretham ferry, and in many rivers about the Fens in great plenty.*

Nymphæa lutea[476] minima, *see* Mórsus ranæ. [*Hydrocharis morsus-ranae* 375]

O

OCulus Christi, *see* Horminum sylvestre. [*Salvia verbenaca* 292]

Ocymum Cereale, *see* Fegopyrum. [*Fagopyrum esculentum* 195]

Ocymoides, *see* Lychnis sylvestris. [*Silene latifolia* 344]

---

[473] Ἐρευθεδανον [Ereuthedanon] of Theophrastus and other Ancient Greeks was in fact the dye-plant Madder (*Rubia tinctorum* L.). See also footnote 230.

[474] Pliny's *Natural history*, XVIII. 259 in the Loeb edition, where he describes it as the worst weed in a meadow and bearing *siliquam diram* ("a terrible pod"), making the identification of "certain writers" impossible.

[475] Literally "the water of life". A folk name for this plant is "brandy-bottle".

[476] *lutea* here is clearly a slip for *alba*.

Odontitis *Plinii, see* Armerius sylvestris. [*Silene flos-cuculi* 61]

Oenanthe vulgaris *Lob. see* Filipendula. [*Filipendula vulgaris* 198]

393 *Oenanthe fistulosa* L.
Oenanthe aquatica *C.B. Lob.* Filipendula aquatica *Ger.* aquatica vel palustris *Park. Tab.* Oenanthe, sive Filipendula aquatica *J.B.* Laver minus *Lon.* Juncus odoratus aquatilis *Dod. Water Drop-wort.* In moist meadows & by rivulets almost everywhere.

394 *Onobrychis viciifolia* Scop.
Onobrychis *Dod. Clus. Ger.* vulgaris *Park.* foliis viciæ fructu echinato major *C.B.* Caput gallinaceum Belgarum *Ad. Lob.*[477] Polygalon Gesneri *J.B.* Polygala Dioscoridis Dalechampio *Pon. in Baldo.* vera Dalechampii *Lob. Medick-Fitchling, or Cocks-head.* On Gogmagog hills, and the balks all thereabout.

Ononis, *see* Anonis. [*Ononis spinosa* 47]

395 *Ophioglossum vulgatum* L.
Ophioglossum *Ger. Trag. Matth. Dod. Gesn. J.B. Cam.* Ophioglossum, sive Lingua serpentina *Park. Cæs.* Lingua vulneraria *Cord. hist.* Henophyllon *Ad. Lob. Adders-tongue.* In Grantcester meadow abundantly, also in many other pastures and meadows.

396 *Neottia ovata* (L.) Bluff & Fingerh.
Ophrys *Matth.* bifolia *Ger. C.B.* Bifolium *Dod. Lob. Cam.* sylvestre vulgare *Park.* Bifolium majus, sive Ophrys major quibusdam *J.B.* Pseudoorchis sive Bifolium *Dod.* Alisma *Cord. Gesn. hort. Tway-blade.* Frequent in meadows & woods.

397 *Liparis loeselii* (L.) Rich.
Orchis lilifolius minor sabuletorum Zelandiæ & Bataviæ *J.B.* Monorchis IX, sive Chamæorchis lilifolia *C.B.* Chamæorchis latifolia Zelandiæ *Park.* Orchis lilifolius minor sabuletorum Zelandiæ & Bataviæ, bulbo eriophoro similis bulbus[478] *Lob. Ad. in appendix, p. 506. Dwarf Orchies of Zealand.* This plant is unique; it has a bulbous root, green inside if it is dissected, compacted from scales, & quite unlike an orchid, for the fibres emerge from the bulb below, not above & alongside the bulb; & besides it propagates itself by creeping, with the fibres swelling into new bulbs. It puts out two leaves, which are pale green, between which a stem of a hand's-breadth rises, angled and bearing three or four or even more flowers of a washed-out green at the top, consisting of six petals with a style in the middle. The style is green, marked with a white spot at the tip & curved inwards a little as

---

[477] The two preceding names mean literally "Greater [Ass's-bite] with the leaves of vetch [and] a prickly fruit" and "Cock's head of the Belgians". Sainfoin, as this plant is now called, meaning "healthy hay" in French, was imported to Britain as a fodder plant early in the 17th century but already occurred here as a wild plant. When he visited Zurich in 1565 Thomas Penny annotated Gesner's watercolour of this species "prouenit in agro Cantabrigensi in Anglia, circa agrorum margines" (it grows in Cambridgeshire in England, around the margins of fields) (Zoller, Steinmenn & Schmid 1979). The first published record for Britain was that of Gerarde (1597, pp. 1062 and 1064) as "Medick Fitchling, or Cockes head" and "red Fetchling" in various places including "vpon the grassie balkes betweene the landes of corne two miles from Cambridge, neere vnto a water mill towardes London". Today both the wild subsp. *collina* (Jord.) P.D. Sell and the clearly introduced subsp. *decumbens* (Jord.) P.D. Sell can be found in the county, both growing on the Devil's Ditch for example (A.C. Leslie, pers. comm.).

[478] "The smaller lily-flowered orchid of the sandy places of Zealand & the Dutch Netherlands, the bulb [being] like a wool-bearing bulb".

if it fulfils the role of a hood. As we have said, six petals surround this, with one much broader than the rest which takes the place of the labellum. Of the remaining five, three are of almost the same size and those two that you could reasonably call wings are far narrower. The seed vessels are very well depicted in [Mathias] de L'Obel's picture, & in general de L'Obel's picture portrays this plant very well, but the description differs from the picture.[479] *In the watery places of Hinton and Teversham moors but more rarely.*

398 *Orchis mascula* (L.) L.
Orchis morio mas foliis maculatis *Park. C.B.* Cynosorchis morio mas *Ger. emac. Tab.* Orchis major tota purpurea maculoso folio *J.B. The male Fools-stones.*[480] It flowers first of all the orchids that grow wild with us, around the end of April. The colour of the flower is generally purple, more rarely flesh-pink. *In Madingley and Kingston woods*, abundantly.

399 *Anacamptis morio* (L.) R.M. Bateman, Pridgeon & M.W. Chase
Orchis morio fœmina *Park.* morio IV, sive morio fœmina *C.B.* minor purpurea & aliorum colorum cum alis virentibus[481] *J.B.* Cynosorchis morio fœmina *Ger. Lob.* Testiculus morionis fœmina *Dod. Lugd. The female Fools-stones.* It varies greatly in the colour of the flower, but it is always found with wings streaked with green lines. A white colour in the flower is rarer.

400 *Ophrys insectifera* L.
Orchis myodes *Ger.* myodes major an minor *Park.* myodes galea & alis herbidis *J.B.* an potiùs myodes flore majore *of the same author?* Or[chis] major muscam referens *C.B.*,[482] i.e. Orchis Serapias X.[483] *Flie-Orchies.* On the banks of the Devils ditch, and in the closes about Hinton and Teversham. However, it occurs very rarely & it is not possible to find more than two or three plants in the same place.

401 *Dactylorhiza incarnata* (L.) Soó and/or *D. praetermissa* (Druce) Soó[484]
Orchis palmata major mas, sive Palma Christi mas *Park.* palmata I, sive palmata pratensis latifolia longis calcaribus *C.B.*[485] palmata non maculata *J.B.* Palma Christi mas *Ger.* Palma Christi & Serapias mas lævi folio *Lob.* Satyrium

---

[479] The reference is to Pena & L'Obel (1605, p. 506). See footnote 354 for comments on Ray's orchid descriptions.

[480] Ray's vernacular names of orchids including "stones" ("Fools-stones" and "Dogs-stones") refer to testicles, because many species have a pair of underground tubers reminiscent of them. *Orchis* (ὄρχις) is Greek for "testicle". This species is Shakespeare's "long purples", to which "liberal shepherds give a grosser name" (*Hamlet*, Act 4, Scene 7, 171–172). See also footnotes 492, 496 and 498.

[481] Literally "The smaller [Orchid], purple & of other colours, with green wings".

[482] The preceding synonyms mean literally "Greater or lesser mouse-like [Orchid]", "Mouse-like [Orchid] with the helmet & wings grass-green, or perhaps rather mouse-like [Orchid] with a larger flower?" and "Greater Orchid recalling a fly".

[483] Ray departs here from his usual practice, which would be to cite this synonym as *Orchis Serapias X, Orchis muscam referens major* C.B.

[484] These are the two relatively frequent *Dactylorhiza* species in Cambridgeshire with unspotted leaves, although neither is now a plant of woods.

[485] The two preceding names mean "Male palmate Orchid, or Male Christ's-palm" and "The first palmate [Orchid], or broad-leaved meadow palmate [Orchid] with long spurs". The species of the modern genus *Dactylorhiza* were formerly termed "palmate" or "handed" because the tubers are divided into finger-like segments; Mathias de L'Obel's name, adopted by Parkinson and Gerarde, likens them, not to any palm of Palm Sunday, but to the hand of Christ crucified.

basilicum mas *Dod*. *The great male handed Orchies, or male Satyrion royall*. In woods & meadows generally. It is also found with a white flower, but rather rarely.

402 *Dactylorhiza fuchsii* (Druce) Soó
Orchis palmata fœmina, sive Palma Christi fœmina maculato folio *Park*. palmata speciosiore thyrso, folio maculato *J.B*.[486] palmata III, sive palmata pratensis maculata *C.B.* Palma Christi fœmina *Ger*. *The female handed Orchies, or Female Satyrion royall*. Frequent in meadows & moist woods. It has a more washed-out purple flower, more rarely white.

403 *Gymnadenia conopsea* (L.) R. Br. *sensu lato*[487]
Orchis palmata rubella cum longis calcaribus rubellis *J.B.* palmata minor flore rubro *Park*. Serapias minor nitente flore *Ger*. Serapias minor rubello flore nitente, angustifolia, nullis inspersis punctulis[488] *Lob*. Orchis palmata V, sive palmata minor calcaribus oblongis *C.B.* *Red handed Orchies*. In meadows almost everywhere. It too is also found with a white flower, but rather rarely.

404 *Coeloglossum viride* (L.) Hartm.
Orchis palmata flore viridi *Park. C.B. in Prod*. *Handed Orches with a green flower. In some pastures and closes on the Northside of Chesterton plentifully, and in other places.* The descriptions of this plant are too short, so we shall try to describe it more accurately & more fully.[489] From the two rather small tubers of the *root*, which are palmate or rather divided into two legs, as it were, there arises a *stem* a hand's-breadth & sometimes half a foot high, striped in the upper part, bearing washed-out green *leaves* that are oblong and not spotted and supporting at the top a spike of medium length of rather widely spaced *flowers*, of which each sits on a strongly twisted stalk furnished at its base with a single oblong & angular bract. The *flowers* themselves (which, like the whole plant also when it is squeezed with the fingers, give off a rather powerful scent) have, instead of spurs, translucent vesicles[490] that are almost round and divided into two parts by an impressed line, so much so that they somewhat resemble the berries or seeds of grapes. The helmet or hood, as in most orchids, is made up of five petals, which are of almost equal length and of which the two inner ones, which stand up highest, are much narrower than the outer. Instead of a style, the helmet hides something like the body of an insect. The colour of the helmet is green, with the margins & veins of the petals purplish. Below it is a longish labellum, which is of

---

[486] The two preceding names mean "Female palmate Orchid, or Female Christ's-palm with a spotted leaf" and "Palmate [Orchid] with a more handsome spike [and] a spotted leaf".

[487] Two of the segregates of *G. conopsea* formerly recognised as subspecies but treated at specific rank by Stace (2010) occur in Cambridgeshire, *G. conopsea sensu stricto* and *G. densiflora* (Wahlenb.) A. Dietr. The former is generally said to be a plant of chalk grassland and one might expect it to have been the plant seen by Ray in meadows, but specimens collected by J.S. Henslow in 1821 and 1825 at Cherry Hinton, one of the three sites for the species listed by Ray in the 'Index of places', have been identified by P.D. Sell as *G. densiflora* (Crompton 2004), a plant of damp chalk grassland and fens. Ray may well have seen both taxa.

[488] "Lesser *Serapias* with a bright reddish flower, narrow-leaved with no little spots sprinkled on". *Serapias* is a name used by Dioscorides and Pliny for a kind of orchid.

[489] See footnote 354 for comments on Ray's orchid descriptions. This species had only recently been added to the British flora, by How (1650).

[490] These "vesicles" are indeed the spurs of the flowers, which are semi-transparent and almost hemispherical in shape in this species (Sell & Murrell 1996; Harrap & Harrap 2005).

a green colour tinged with purple or dark brown and divided into three segments, of which the central one is by far the shortest. The overall appearance of the flowers is similar to the flowers of Bifolium[491] except that the flowers of Bifolium completely lack those vesicles that are here present instead of spurs & in them the hanging lip is divided into only two little beards. It flowers around the end of May & at the beginning of June.

405 *Neotinea ustulata* (L.) R.M. Bateman, Pridgeon & M.W. Chase

Orchis sive Cynosorchis minor Pannonica *Ger.* militaris Pannonica *Park.* Cynosorchis militaris VI, sive Militaris pratensis humilior *C.B.* Orchis minor flore incarnato *Eyst.* Or[chis] parvis floribus multis punctis notatis, an Clusio Orchis Pannonica quarta? *J.B.* Or[chis] sive Cynosorchis flore purpurascente Pannonica quarta *Clus.*[492] *Little purple-flowered Dogs-stones. All over Gogmagog hills and Newmarket heath. On the Devils ditch bank you may be sure to finde it.*

406 *Anacamptis pyramidalis* (L.) Rich.

Orchis sive Cynosorchis purpurea spicâ congestâ pyramidali.[493] From the twin non-palmate *bulbs* there arises a *stem* of a hand's-breadth & greater, erect and clothed with many leaves, most of them clasping the stem, which are not spotted and of a rather washed-out green; at its top there is a short crowded pyramidal spike of *flowers*. The *flowers* themselves are purple or reddish, not marked with spots, each possessed of a long spur. Their hoods consist of three petals with two wings. The lip, which hangs down, divided into three segments in the lower part, is furnished in the upper part with two whitish processes peculiar to this orchid; these processes, curved into a small hollow or little cavity, somewhat resemble an ear.[494] It is sometimes found with a white flower. *In many places, as in a chalkie close at Hinton near where they burn lime.* It flowers in the month of June, last of all the orchids. Perhaps this plant may have been the same as that which *J. B[auhin]us* calls Orchis parvo flore sive Phœniceo. An Cynosorchis militaris media *C.B?* Cynosorchis militaris rubra *Park?* Orchis flore rubro elegantissimo *Clus.?*[495]

---

[491] Ray's *Ophrys* [*Neottia ovata* 396] (q.v.), for which he gives *Bifolium* as a synonym; the latter is equivalent to the vernacular English name "Twayblade".

[492] The two preceding names mean literally "Orchid with small flowers marked with many spots; can it be the fourth Hungarian Orchid of de L'Écluse?" and "The fourth Hungarian Orchid or Dog's-orchid with a purplish flower". *Cynosorchis* (κυνος ὀρχις) is the Greek equivalent of Ray's "Dogs-stones": see footnote 480.

[493] The omission of an English name is presumably an error; "Purple pyramidall Orchies" is given for this species in the Index of English names. See footnote 354 for comments on Ray's orchid descriptions.

[494] *qui processus in sinum exiguum seu cavernulam flexi auriculam quadantenus exprimunt.* These prominent ridges are extensions of the column and they converge towards the mouth of the spur. They are sometimes called "guide-plates" and are believed to guide the long proboscis of a pollinating insect (normally a butterfly or burnet moth) towards the spur, ensuring that it is in the correct position to trigger the mechanism that releases the orchid's pollinia onto the proboscis (Harrap & Harrap 2005; Lind *et al.* 2007). There is in fact no nectar in the spur and *Anacamptis pyramidalis* achieves pollination by deception.

[495] The four preceding names mean literally "Orchid with a small or Phoenician purple flower", "Perhaps Middle-sized military Dog's-orchid, *of Caspar Bauhin*?", "Perhaps Red military Dog's-orchid, *of Parkinson*?" and "Perhaps Orchid with a very elegant red flower, *of de L'Écluse*?". A superscript 9 after the initials *J.B* represents the Latin nominative masculine ending *-us*; see Chapter 7. For Phoenician purple see footnote 178. Parkinson (1640, p. 1345) actually called his ninth, unillustrated species *Cynosorchis militaris spica rubente conglomerato* or "Round headed red Souldiers cullions" and described the flowers as having "a very small heele behind the backside", thus ruling out the

407 *Platanthera chlorantha* (Custer) Rchb.
Orchis Serapias bifolia vel trifolia minor *Park.* Hermaphroditica *Ger.* alba calcare oblongo *J.B.* Serapias III, sive alba bifolia minor calcare oblongo *C.B.* If anyone may wish to make this the same as Orchis trifolia major, we shall readily assent. *C.B.* Testiculus vulpinus primus[496] *Lob. Butterflie Satyrion. In Madingley and other woods and bushie places.*

408 *Ophrys apifera* Huds.
Orchis sphegodes sive fucum referens *Park.* fuciflora, galeâ & alis purpurascentibus *J.B.* Serapias IX, sive fucum referens colore rubiginoso *C.B.*[497] Orchis sive Testiculus vulpinus 2. sphegodes *Ger. Humble-bee Satyrion. In a close behind the Bell Inn at Haverill there be hundreds of them. In a close near the church at Burrough-green: In severall closes at Cherry-Hinton near the church, and in sundry other places, some say on Gogmagog hills: we met with none there.*

409 *Spiranthes spiralis* (L.) Chevall.
Orchis spiralis alba odorata *J.B.* Triorchis *Ger.* alba odorata minor *C.B. Park.* Testiculus odoratus minimus omnium, flosculis odore Lilii convallium[498] *Lob. & Pen. Ad. Triple Ladies Traces, on the skirts of Teversham moor towards Gogmagog hills.*

Origanum verticillatum, *see* Clinopodium majus. [*Clinopodium vulgare* 139]

410 *Origanum vulgare* L.
Origanum vulgare *Trag. Matth.* vulgare spontaneum *J.B.* vulgare flore rubicundo hort. *Bat.* Anglicum *Ger.* Or[iganum] sylvestre, Cunila bubula Plinii *C.B.* Agrioriganum sive Onitis major *Lob.* Majorana sylvestris *Park. Wild Marjerome. In the yard where the Yew trees grow at Barnwell Abby, also in some closes about Histon, and in divers closes about Linton, near the river on the east side of the town in great plenty, and nearer hand in a close called Bowyers at Cherry-Hinton, not far from the Church.*[499]

411 *Ornithopus perpusillus* L.
Ornithopodium minus *Ger. Park. C.B.* perpusillum *Ad. Lob. Small Birds-foot. On sandy hills, as on the sandy layes about Gamlingay near the Windmills.* It flowers around the end of May and even in June & July. The colour of the flower is purple, with the wings white marked with purple stripes.

412 *Orobanche elatior* Sutton and *O. rapum-genistae* Thuill.[500]
Orobanche sive Rapum genistæ *Ger. Park.* Orob[anche] flore majore *J.B.* major garyophyllum olens *C.B.* Limodoron of Theophrastus *for the same author.*

---

long-spurred *Anacamptis pyramidalis* and rather suggesting the alpine species *Gymnadenia rhellicani* (Teppner & E. Klein) Teppner & E. Klein.

[496] "The first Fox's-testicle": see footnote 480.
[497] The three preceding names mean literally "Wasp Orchid or [Orchid] recalling a drone", "Drone-flowered [Orchid], with purplish helmet & wings" and "The ninth Serapias, or [Serapias] recalling a drone with a rusty colour".
[498] "The smallest scented *Testiculus* [Latin equivalent of Greek *Orchis*] of all, with flowers with the scent of Lily of the valleys [*Convallaria majalis* L.]". See footnote 480.
[499] The plant in the last site was *Clinopodium majus* Park. [*Clinopodium vulgare* 139]: see Ray's comment in *Emendanda* of the 1663 and 1685 appendices and the related footnotes.
[500] The Gamlingay plant, parasitic on broom, is clearly *O. rapum-genistae*. Previous authors (e.g. Babington 1860; Raven 1950; Ewen & Prime 1975) have identified Ray's other plant as *O. elatior* Sutton, the only other large-flowered *Orobanche* in the county, which is a parasite of *Centaurea*

MAIN CATALOGUE

Orobanche of Dioscorides, Galen, Oribasius, Aetius and Aegineta *for the same author.* Cynomorion of Pliny, from the likeness to canine genitalia,[501] *for the same author.* Leonina herba *Hermolao* & Legumen leoninum *Ruell. Broom Rape. It grew in barley on the right hand of the way between Cambridge and Grantcester: In a corn field nigh the church at Cherry-Hinton: It is also to be found plentifully in the broom fields at Gamlingay growing at the roots of the broom.*

Ostrys Theophrasti, *see* Betulus. [*Carpinus betulus* 86]

Osyris, *see* Linaria.[502] [*Linaria vulgaris* 335]

Oxalis, *see* Acetosa.[503]

413 *Berberis vulgaris* L.
Oxyacantha *Hermolao, Ruell. Trag. Fuch. Eystet.* Oxyac[antha] Galeni *hort. Bat. & Gron.* Berberis I, sive dumetorum *C.B.* Berberis vulgò quæ Oxyacantha putata[504] *J.B.* Spina acida *Brun.* 4. *Dod.*[505] Spina acida sive Oxyacantha *Ger.* Crespinus *Cæs.*, for whom it is perhaps also Ribes Serapionis, *Matth. Barberries. At Chesterton in a close on this side the great brick house, and in the hedges by the road side to Audley end near the great house abundantly.*

414 *Crataegus laevigata* (Poir.) DC. and *C. monogyna* Jacq.[506]
Oxyacanthus *Ger.* Oxyacantha vulgaris sive Spinus albus *J.B.* Oxyacantha Dioscoridis seu Spina alba communis *hort. Paris. in append.* Spina appendix Plinii *Gesn.* appendix vulgaris *Park.* Mespilus apii foliis sylvestris spinosa, sive Oxyacantha *C.B.* Cynosbatos Theophrasti *Trag. Lon.* Veteribus Latinis Spina simpliciter dicta *Gesn. hort.*[507] Sorbus aculeata *Cord. The White-thorn, or Hawthorn tree. In hedges generally.*

---

scabiosa L. and still grows at Cherry Hinton in roadside grassland. In unpublished notes, the late Dr D.E. Coombe suggested that the second plant may have been *O. minor* Sm., a smaller and weedier species, as this is much the most likely broomrape to occur in arable fields. However, Ray recorded *C. scabiosa* "Inter segetes" and *"in the corn"* (see *Jacea segetum majus purpurea* [298] and *Jacea segetum major foliis dissectis flore albo* [299]), so it is possible that both *O. elatior* and its host-plant grew in this habitat in the 17th century: see footnote 384. In addition, the early records of *O. minor* in the county were associated with clover crops rather than cereal fields.

[501] Κυνομοριον [Cynomorion] means "Dog's-member", *à canini genitalis similitudine*, since μοριον was used euphemistically like the English "member". Pliny's description is in XXII. 162 of his *Natural history* in the Loeb edition.

[502] Ray cites *Osyris* as a synonym of *Linaria vulgaris nostras*.

[503] Ray cites *Oxalis* as a synonym of *Acetosa vulgaris* [*Rumex acetosa* 7] and *Oxalis parva auriculata repens* as a synonym of *Acetosa arvensis lanceolata* [*Rumex acetosella* 8].

[504] "Commonly *Berberis*, which [is] considered [to be] *Oxyacantha* [i.e. sharp-thorned]".

[505] This phrase may have been taken in its entirety from C. Bauhin (1623, p. 454), where it appears as "*Spica acida*, Brunf. 4. Dod.".

[506] These two hawthorns were not distinguished in Cambridgeshire until the 19th century; even Babington (1860) does not treat them as separate species, although he recognised that *Crataegus monogyna* was "much the more common form".

[507] The two preceding names mean literally "Dog's-bramble of Theophrastus, *of [Jerome] Bock [and Adam] Lonicer*" and "For the ancient Latins [i.e. Romans] *spina* [thorn], simply said, *of [Conrad] Gesner in his book about the gardens of Germany*". The former reference is to Theophrastus' *Historia plantarum* (III. XVIII. 4 in the Loeb edition), where κυνοσβατον [cynosbaton, a neuter form] is said to have a reddish fruit like that of the pomegranate, to be intermediate between a shrub and a tree and to have a spiny leaf. This plant is usually regarded as being a wild rose rather than a hawthorn: see Ray's entry for *Rosa sylvestris* [502].

Oxylapathum, *see* Lapathum acutum. [*Rumex conglomeratus* and *R. sanguineus* 321]

Oxmyrsine,[508] *see* Ruscus. [*Ruscus aculeatus* A37]

Oxys *Plinii*, *see* Trifolium acetosum.[509] [*Oxalis acetosella* 578]

P

PAlma Christi, *see* Orchis palmata. [*Dactylorhiza* spp.[510]]

Paludapium, *see* Apium palustre. [*Apium graveolens* 53]

Panax coloni, *see* Sideritis Anglica strumosa radice. [*Stachys palustris* 540]

Panax Chironium, *see* Chamæcistus. [*Helianthemum nummularium* 125]

415 *Papaver bivalve* (DC.) K.-F. Günther
Papaver corniculatum violaceum *J.B. Park. C.B. Dod. Lugd.* corniculatum violaceo flore *Clus. hist.* cornutum flore violaceo *Ger. Violet-coloured horned Poppy.* In the corn fields beyond Swafham as you go to Burwell. It flowers around the end of May.

416 *Papaver rhoeas* L.[511]
Papaver Rhœas *Lob. Ger.* erraticum *Matth. Dod. Gesn. Cæs. Tab.* erraticum sive Rhœas *Park.* erraticum rubrum campestre *J.B.* X, sive erraticum majus, ῥοιας [rhoeas] for Dioscorides, Theophrastus & Pliny *C.B.* Pap[aver] Rhœas sive caduco flore Phœniceo *Ad. Lob. ico. Red Poppy, Corn-rose.*

N. For all seeds, the smaller they are, the more fertile. There are two reasons:[512] both because that which is small can more easily be produced & made to grow and because it adapts to the quality of the air and adjusts itself to external circumstances more quickly. *P. Lauremberg, Horticult[ura], book 1, chapter 17, number 2.* Compare *Theophrast[us], De causis plant[arum], book 2, chapter 17, & book 4, chapter 1.*[513] The seeds of poppy kept up to ten years are suitable for sowing. *P. Laur[emberg], in the same place, number 8.*

---

[508] The alphabetical position confirms that *Oxmyrsine* is an error for *Oxymyrsine* (i.e. Ὀξυμυρσίνη, a name found in Dioscorides, Greek for "sharp myrtle"). It is a surprising entry, since *Ruscus* [*Ruscus aculeatus* L.] with its synonym *Oxymyrsine* does not appear in the catalogue and indeed it is mentioned in Ray's 'Preface' as a plant that he had not yet come across in the county but which he hoped to find very soon. His hopes were soon realised, as he discovered the species in 1661 (Gunther 1934; Thompson 1974) and reported it from Anglesey Abbey in the 1663 appendix [A37].

[509] Ray actually lists *Oxys Pliniana* as a synonym of *Trifolium acetosum vulgare* [578].

[510] Ray gives *Palma Christi* and *Palma Christi mas* as synonyms of *Orchis palmata major mas* [*Dactylorhiza incarnata* and/or *D. praetermissa* 401] and *Palma Christi fæmina* as a synonym of *Orchis palmata fæmina* [*Dactylorhiza fuchsii* 402].

[511] It is not clear whether *Papaver dubium* L. *sensu lato* (including *P. lecoqii* Lamotte), which is now widespread in Cambridgeshire, was overlooked by Ray or whether, as Raven (1950, p. 245) suggests, it was a much rarer plant in his time. It was first reported from Britain by Ray (1686, p. 856) on the basis of capitula sent to him from Cambridgeshire by Dent and was afterwards found in Ray's neighbourhood by his friend Samuel Dale of Braintree; Ray said that he had not yet seen living and flowering plants himself.

[512] *Ratio gemina*, literally "The reason [is] twin".

[513] For Peter Lauremberg see footnote 25. His text here must be based partly on Gaza's (1529, part 2, p. 103; see footnote 21) Latin translation of Theophrastus' *De causis plantarum* (II. 12. 1 in the Loeb edition).

417 *Papaver somniferum* L.
Papaver spontaneum sylvestre *Lob. Ger. emac.* III, sive vulgare cujus capitula foraminibus hiant, semine in cano ἀγριωτερον [agrioteron] Dioscoridi[514] *C.B. Wild Poppy. It grows in many places on the banks in the Isle of Ely.*

N. It is characteristic of the fruits of all poppies to have an asterisk or star at the top or apex; this is less obvious in the fruit of a horned poppy than in one with a head. *Spigel[ius in] Isag[oges],*[515] *book 1, chapter 40.*

Papaver spumeum *Lob. see* Behen album. [*Silene vulgaris* 80]

N. This plant has been considered to be Papaver spumeum[516] because of the frothy or spittle-like spume (in English called *Woodseare* & *Cuckoo-spittle*) which settles in the folds of the leaves & joints, not infrequently making them white. *Lob.* About the cause & origin of this froth the throng of philosophers vigorously talk nonsense: some, calling it the spittle of the stars, have believed that it rains down from the sky, in the same way as Manna or honeydew,[517] which however are found condensed on very few plants; others that it is exhaled directly from the earth; finally others that is exuded from the plant itself. For we have discovered[518] that it is vomited from the mouth of an insect, a tiny little creature that is always hidden in the middle of this froth. In fact, if you wipe away all the frothy spume from it with a hand or linen cloth, you will see for yourself with your own eyes the same spume abundantly poured out again shortly afterwards from the mouth of the little creature, which will completely bury and hide itself in it again, so that it may hide therein safe from harm from frogs, small birds and other predators while it is still delicate and cannot protect itself by jumping or flight. Moreover this insect has generally almost the appearance of a louse, except that it is shorter in relation to the size of its body, of a yellowish-green colour and with large & protuberant eyes; the rear legs are designed for jumping, whence we readily agree with certain learned & able men who have confidently asserted to us from their own experience that it turns into a Locusta, called in English a *Grasse-hopper*.[519] To admit the truth, so far we ourselves have neither investigated whence it has its origin nor discovered what it finally turns into. Baron Verulam, in his *Natural history, century 5, experiment 497*, states that this foam is found only on hot plants,[520] but we must disagree with this most distinguished man in this

---

[514] Literally "The third [Poppy], or the common [one] whose heads open with holes, the wilder [one] with a rather grey seed, for Dioscorides" (on the assumption that *in cano* is an error for *incano*).

[515] For Adrianus Spigelius see footnote 31.

[516] i.e. frothy poppy.

[517] For the biblical account of manna, the miraculous food sent by God to the Israelites in the wilderness, see *Exodus* 16: 2–36. Belief that honeydew also falls from heaven was quite general: for example, in Norse mythology it falls from an enormous ash, the world tree Yggdrasil. See also Ray's note 2 on *Quercus latifolia* [476] and footnote 606.

[518] *Nobis enim compertum est*, literally "For for us it has been discovered".

[519] The Cuckoo-spit Insect or Common Froghopper, *Philaenus spumarius* (Linnaeus), is abundant in Britain and is actually a true bug (Hemiptera, superfamily Cercopoidea) rather than a grasshopper. The larvae live in froth which they make by blowing air into their urine. Adult froghoppers are free-living and are able to jump prodigious distances for their size; they achieve very rapid take-off velocities by storing energy in their cuticle by slow muscle contraction and then releasing it suddenly (Burrows, Shaw & Sutton 2008).

[520] For the work of Baron Verulam (Sir Francis Bacon) see footnote 104. For the Galenic theory of the qualities of plants see Chapter 3.

matter,[521] since this foam on which the present discussion is centred should rather be considered cold.

Paralysis *officinarum, see* Primula pratensis.[522] [*Primula veris* 467]

418 *Parietaria judaica* L.
Parietaria *Ger. J.B.* vulgaris *Park.* I, sive officinarum & Dioscoridis *C.B.* Helxine *Brunf. Matth. Dod. gal. Fuch. Gesn. hort. Lugd. Cam.*, for whom it is also Urceolaris Scribonii. Vitriola sive Perdicium *Lob. Cæs. Tur.* Muralium *Jun. nomenc.*[523] *Pellitory of the wall.* On rubble & old walls.

419 *Erophila verna* (L.) DC. *sensu lato*[524]
Paronychia vulgaris *Dod. Ger.* vulgaris alsines folio *Park.* alsinefolia *Lob. Lugd. Cam.* Bursa pastoris 6 *Trag.* pastoris foliis integris loculis oblongis II, sive minor loculo oblongo *C.B.* pastoris minima oblongis siliquis, sive verna loculo oblongo *J.B.* Pilosella siliquata minima *Cam. Ad. Thal. Chickweed-Whitlow-grasse.* On the hill of Health, & almost everywhere in drier places. The flower of this plant consists of four white petals divided, as in Alsine media,[525] right to the bottom, so much so that to someone looking at them somewhat carelessly they may seem to be eight.

420 *Saxifraga tridactylites* L.
Paronychia rutaceo folio *Ger. Lob.* foliis incisis *Park.* Sedum foliis laciniatis VI, sive tridactylites tectorum *C.B.* tridactylites tectorum flore albo *J.B.* Alsine petræa rubra & Paronychia 3 *Tab.* Dactyliobotanon alterum *Thal. Rue-Whitlow-grasse.* On walls & roofs & in drier places. *Upon the hill of health abundantly.*

421 *Tanacetum parthenium* (L.) Sch. Bip.
Parthenium *Matth. Dod. gal. Lac. Cord.* Matricaria *Ger. Dod.* Parthenium I, sive Matricaria vulgaris *C.B.* Matricaria vulgaris *Park.* Matricaria vulgò, minus Parthenium *J.B.* Artemisia tenuifolia *Fuch. Tab.* Amaracus for Galen & Aegineta *C.B. Feverfew.* It delights in rubbly places, oftentimes where even a little charcoal, which it happens is easily found among rubble, underlies the fibres of its roots, about which *Cardanus* provides some trifling information in his *Liber de subtil[itate].*[526]

---

[521] *verùm necesse est ut viro summo fidem hac in re denegemus*, literally "but it is necessary that to the highest/very high man we should refuse belief in this matter". Ray's disagreement arises from the view that "All the Poppies are cold, as *Galen* testifieth in his booke of the Faculties of simple medicines" (Johnson 1633, p. 370, misnumbered 400).

[522] Both *Herba Paralysis* Officin[arum] and *Primula pratensis* Lob. are cited by Ray as synonyms of *Primula veris major* [*Primula veris* 467] (q.v. for an explanation of the former).

[523] For Junius' *Nomenclator* see footnote 143.

[524] British botanists currently divide this self-fertilising annual into three segregate species. All three, *Erophila glabrescens* Jord., *E. majuscula* Jord. and *E. verna* (L.) DC., have been recorded in Cambridgeshire and the last is probably the most frequent, but their relative distributions require critical study (Leslie 2007).

[525] i.e. *Stellaria media* [29]. See Ray's comment on *Alsine major* [28].

[526] *De subtilitate*, Girolamo Cardano's (see footnote 5) first encyclopedia, was first published in Nuremberg in 1550, with revised editions published in Basel in 1554 and Leiden in 1559. Cardano had a high reputation as a physician and astrologer, so his claim to be a universal philosopher and polymath was taken seriously at the time.

Parthenium *Fuchsii, see* Cotula fœtida.[527] [*Anthemis cotula* 155]

422 *Pastinaca sativa* L. subsp. ***sativa***[528]
Pastinaca sativa *Trag. Lugd. Gesn. hort. Cam.* latifolia sativa *Ger. Park. C.B. Dod.* sat[iva] latifolia Germanica flore luteo *J.B.* Siser Dioscoridis *Col.* Sisarum sativum magnum *Fuch. Tur.* Elaphoboscum sativum *Tab. Common garden Parsnep.*

423 *Daucus carota* subsp. ***sativus*** (Hoffm.) Arcang.
Pastinaca sativa Dioscoridis *Lob. Cord. in Diosc.* sativa tenuifolia *Ger.* tenuifolia sativa lutea *Park.* tenuifolia sativa radice lutea vel alba *C.B.* sativa, sive Carota lutea & alba *J.B.* Carotta vulgaris radice flavâ *Ad.* Staphylinus sativus & Daucus domesticus Galeno *C.B.* Daucus sativus radice lutea *hort. Bat. hort. Gron.* Ben sive Behen album Arabum *Meilichio, Costæo & C. Hofmanno de medicam[entis] officin[alibus].*[529] *Yellow and white rooted Carrot.*[530]

N. Plants whose roots are put down deep into the ground grow fairly successfully in poor soil, where they imbibe nourishment rather sparingly & slowly as if they were abstemious, as is well known in this Carrot, which delights in sandy soil, & in trees that drive their roots down deep. But in fact those plants whose shorter roots creep along the surface of the ground demand more fertile soil because they live frugally only with difficulty and they become hungry. Consequently for these the surface of the ground, well broken up by the sun, air & showers, supplies copious nourishment. Compare *Theophr. de caus., book 1, chapter 22, & book 2, chapters 5 and 6.*[531]

424 *Daucus carota* subsp. ***sativus*** (Hoffm.) Arcang.
Pastinaca sativa atrorubens *Ger. emac.* sativa altera tenuifolia atrorubens *Park.* tenuifolia sativa radice atrorubente *C.B.* Staphylinus sativus of Dioscorides, Daucus radice nigra of Theophrastus, Pastinaca sativa of Pliny *of the same author.* Carota altera radice sanguinei coloris *Cæs.* Daucus sativus radice atrorubente *hort. Bat. hort. Gron. Red-rooted Carrot.*[530]

---

[527] Ray's chosen name is actually *Cotula fœtida* (with linked *ae*), a less correct spelling (see footnote 223).

[528] Ray lists Wild Parsnep (subsp. *sylvestris*) as a separate species, *Elaphoboscum* [174].

[529] Meilichio is a copying error or misprint for Melichio (cf. Hofmann 1646, p. 156) and Hofmann must be referring to Melich's *Dispensatorium medicum* (1601), p. 8, where the name appears as *Been album*. For Georg Melich and Giovanni Costeo, see Chapter 4.

[530] Carrots can be divided into two main groups, the anthocyanin or eastern variants with purple, yellow or yellowish-orange branched roots and the carotene or western variants with orange or more rarely yellow or white unbranched roots. The western type was derived from the eastern by a long period of selection starting in the 13th or 14th century, summarised by Riggs (1995). This started with plants with purple roots (called "red" by authors before 1700), which were gradually superseded in the 16th century by plants with yellow carrots growing above the ground; in turn these were replaced by orange-red carrots, which arose in the Netherlands in the 17th century and from which our modern carrots were ultimately derived. The variants listed by Ray clearly represent the earlier stages of this sequence. By 1674, in his paper *Of the specifick differences of plants*, Ray implied that yellow was the usual colour, but included the root colour of carrots amongst his examples of characters which should not be used to separate species (Cain 1999a).

[531] *Pastinaca* usually means "Parsnip" but in view of the context is here translated as "Carrot". The cited chapters of Theophrastus' *De causis plantarum* (equivalent to I. 18. 1–5 and II. 4. 1–12 in the Loeb edition) do not mention *Pastinaca*, but in I. 18. 1 and II. 4. 2 and 5 deep-rooted trees, which do better in leaner soils, are contrasted with cereals, vegetables and other annuals, which thrive in rich soils and get their food from near the surface.

Pastinaca sylvestris *Dod. see* Elaphoboscum. [*Pastinaca sativa* subsp. *sativa* var. *sativa* 174]

Pastin[aca] Sylv[estris] tenuifolia, *see* Daucus. [*Daucus carota* subsp. *carota* 166]

425 *Scandix pecten-veneris* L.
Pecten Veneris *J.B. Ad. Lob.* Veneris sive Scandix *Ger.* vulgaris *Park.* Scandix *Dod.* Sc[andix] semine rostrato vulgaris *C.B. Shepherds-needle, Venus-combe.* Among standing corn.

426 *Rhinanthus minor* L.
Pedicularis sive Crista galli lutea *Park.* VI, sive pratensis lutea vel Crista galli *C.B.* Crista galli *Lob. Ger.* galli mas & fœmina *J.B.* Alectorolophos *Ang. Yellow Rattle or Cocks-combe.* Very common in meadows & pastures.

427 *Pedicularis sylvatica* L.
Pedicularis *Lob. Ger. Tab.* V, sive pratensis purpurea *C.B.* pratensis rubra vulgaris *Park.* Pedicularis, quibusdam Crista galli flore rubro *J.B.* Alectorolophos *Dod. gal. Thal.* Fistularia *Dod. Red Rattle, or Lousewort.* In watery places of meadows & by rivulets.

428 *Comarum palustre* L.
Pentaphyllum rubrum palustre *Ger. Lob. ico.* Quinquefolium erectum VI, sive Quinquefolium palustre rubrum *C.B.* Pentaphyllum vel potiùs Heptaphyllum flore rubro *I.B. Purple marsh Cinquefoil. On the boggy ground at Gamlingay near Sir Roger Burgoynes park in great plenty.*

429 *Potentilla reptans* L.
Pentaphyllum vulgare *Ger.* majus luteo flore *Park. hort. Pat.* Pentaph[yllum] seu Quinquefolium vulgare repens *J.B.* Quinquefolium *Matth.* luteo flore I, sive majus repens *C.B.* repens flore luteo *hort. Hafn. Common Cinquefoil, or Five-leaved grasse.* The root of this plant becomes quadrangular when dried, as Theophrastus reports to be a feature of Pentaphyllum.[532]

430 *Aphanes arvensis* L. *sensu lato*[533]
Perchpier Anglorum *Lob. Ger. emac.* Anglorum quibusdam *J.B.* Polygonum selinoides *Park.* Scandix minor *Tab.* Chærophyllum III, sive Chærophyllo nonnihil similis *C.B.* Scandix VI, sive minor Saxifraga *of the same author* in *Phyt.* Alchimilla montana minima *Col. Parsly-piert. On the hill of Health,* & elsewhere among standing corn.

431 *Bupleurum rotundifolium* L.
Perfoliata *Matth. Dod.* vulgaris *Ger. Park.* vulgatissima sive arvensis *C.B.* Perfoliata simpliciter dicta vulgaris annua *J.B.* Perfoliatum vulgatius flore luteo, folio umbilicato *Ad. Lob. Thorow-wax. In the corn on the left hand of the footpath leading to Teversham, over against the first closes.*

---

[532] See Theophrastus' *Historia plantarum*, IX. XIII. 5 in the Loeb edition.

[533] Two closely related species have been recognised since the mid 20th century, *Aphanes arvensis* L. and *A. australis* Rydb., both of which occur in Cambridgeshire. *A. arvensis* is much more frequent in the county, especially in arable habitats, and it is more likely to have been the plant seen by Ray. *A. australis* shows a greater preference for uncultivated land and less fertile, acidic soils, although the ecological requirements of the species overlap.

MAIN CATALOGUE

Perforata, *see* Hypericum vulgare.[534] [*Hypericum perforatum* 296]

432 *Lonicera periclymenum* L.
Periclymenum *Trag. Fuch. Ger.* non perfoliatum Septentrionalium *Ad. Lob.* non perfoliatum Germanicum *C.B.* Caprifolium *Brunf. Gesn. hort.* Periclymenum sive Caprifolium vulgare *Park.* Periclymenum plurimis, sive Caprifolium non perfoliatum *J.B.* Matrisylva *Offic[inarum].* Wood-bine, Honey-suckle.

433 *Persicaria maculosa* Gray
Persicaria mitis maculosa *Lob. Ger.* mitis *J.B.* vulgaris mitis seu maculosa *Park.* mitis maculosa & non maculosa *C.B.* altera *Trag. Matth. Tab.* Persicaria I, Britannica Plinio *Cæs. Dead or spotted Arsmart.* The leaves are frequently marked with a black spot but sometimes lack it.

434 *Persicaria minor* (Huds.) Opiz
Persicaria pusilla repens *Ger. emac. Park.* IV. sive minor *C.B. Small creeping Arsmart. At the end of Water-Beach load, next to the Town, in the water.*

435 *Persicaria hydropiper* (L.) Delabre
Persicaria urens *Lob.* mordax *Thal.* vulgaris acris sive minor *Park.* acris sive Hydropiper *J.B.* urens seu Hydropiper *C.B.* Hydropiper *Matth. Fuch. Dod. Ger. Lugd.* Piper aquaticum Dioscoridi, & Zinziber caninum Avicennæ *book* 2 & Herba muscarum Germanis (because wounds or ulcers smeared with its juice are protected from flies) *Trag. Arsmart. By ditches & rivulets.*

Personata, *see* Bardana. [*Arctium lappa* and/or *A. minus* 79]

Pes anserinus,[535] *see* Atriplex sylvestris. [*Chenopodium murale* 72]

Pes asini, *see* Alliaria. [*Alliaria petiolata* 17]

Pes cati *Officin[arum]*, *see* Gnaphalium montanum album. [*Antennaria dioica* 233]

Pes columbinus, *see* Geranium columbinum. [*Geranium molle* and/or *G. pusillum* 219]

Pes leporis, *see* Lagopus. [*Trifolium arvense* 316]

436 *Petasites hybridus* (L.) P. Gaertn., B. Mey. & Scherb.
Petasites vulgaris *Park.* Petasites *Ger. Lob.* major & vulgaris *C.B.* vulgaris rubens rotundiore folio *J.B.* flore punicante *Thal.* Tussilago major *Matth.* It puts out its flowers earlier than its leaves: *see* Verulam, Hist[oria] nat[uralis],[536] cent[ury] 6, experiment 591. *Butter-Burre, Pestilent-wort.* It gained the same name also with the Germans because of its remarkable efficacy & virtue against plague. *By the river Cam in many places, as a little on this side Grantcester meadows.*

Phalaris pratensis minor *Lob. see* Gramen tremulum. [*Briza media* 266]

Phellandrium, *see* Cicutaria palustris. [*Oenanthe aquatica* and/or *O. fluviatilis* 134]

Phœnix, *see* Lolium rubrum. [*Lolium perenne* 341]

Phu, *see* Valeriana.[537]

---

[534] Ray actually cites *Hypericum vulgare, sive Perforata caule rotundo foliis glabris* and *Herba perforata* as synonyms of his chosen name *Hypericum*.
[535] *Pes anserinus* is actually cited as a synonym of *Atriplex sylvestris latifolia* [72], not *Atriplex sylvestris* [70]. See also footnote 193.
[536] For the work of Baron Verulam (Sir Francis Bacon) see footnote 104.
[537] Ray cites synonyms beginning with *Phu* under both *Valeriana sylvestris major* [*Valeriana officinalis* 605] and *Valeriana sylvestris minor* [*V. dioica* 606].

437 *Pilosella officinarum* (L.) F.W. Schultz & Sch. Bip.
Pilosella repens *Ger.* minor vulgaris repens *Park.* majore flore, sive vulgaris repens *J.B.* Auricula muris minor *Trag.* Pil[osella] major repens hirsuta *C.B.* The common small creeping Mouse-eare.

Pilosella minor Fuchsii, *see* Gnaphalium montanum album. [*Antennaria dioica* 233]

438 *Sanguisorba officinalis* L.
Pimpinella sylvestris *Ger.* major vulgaris *Park.* Italica major *Clus. hist.* major domestica *hort. Bat.* VI, sive sanguisorba major *C.B.* Italorum capitulo purpureo *Cord.* Sanguisorba major flore spadiceo *J.B.* major Italorum *hort. Paris.* Great Burnet. *In the same close at Whitwell, where the* Narcissus Angl[icus]⁵³⁸ *growes. Also in the pastures about Long-Stanton and Cottenham as you ride to Audrey caussey in great plenty.*

439 *Poterium sanguisorba* L. subsp. *sanguisorba*
Pimpinella vulgaris, sive minor *Park.* hortensis *Ger.* VII, sive sanguisorba minor hirsuta, & VIII, sive minor lævis *C.B.* minor hortensis sive odorata *hort. Gron.* & sanguisorba minor hirsuta *of the same author.* Sanguisorba minor *J.B.* Italica minor *Thal.* Teucrium Dioscoridis *Brassavolæ in exam. Simpl. Small or little Burnet. In moorish*⁵³⁹ *meadows & pastures. It flowers in May & June.* Pimp[inella] hirsuta & lævis do not differ in species but only in cultivation.⁵⁴⁰

440 *Pimpinella saxifraga* L. subsp. *saxifraga* var. *saxifraga*
Pimpinella saxifraga major altera *C.B.* saxifraga major nostras *Park.* Tragoselinum minus *Tab.* Bipinellæ, sive Saxifragæ minoris varietas major *Ger.* Saxifragia hircina minor foliis Sanguisorbæ *J.B. The little roundish-leaved Burnet-Saxifrage. Frequent in meadows.* Ger[arde's], Park[inson]'s & C[aspar] B[auhin]'s names that we have brought here do not fit this plant, which *J. B[auhin]us* describes *in volume 3, book 28, chapter 48*. In this place we relate the plant described by J. B[auhinus] there, from which source look for other synonyms.

441 *Pimpinella major* (L.) Huds.
Pimpinella saxifraga hircina major *Park.* saxifraga *Ger.* saxifraga major umbellâ candidâ *C.B.* Saxifraga⁵⁴¹ hircina maior *J.B.* Bipennella saxifraga major *Ad.* Great Burnet-Saxifrage. *In the woods about S*ᵗ*. George Hatley, and many other woods on the borders of Cambridgeshire towards Bedfordshire. It is seldome or never found but among bushes, or in woods.*

442 *Pimpinella saxifraga* subsp. *saxifraga* var. *dissecta* With.
Pimpinella saxifraga minor *C.B.*⁵⁴² saxifraga hircina minor *Lob. Park.* an saxifraga minor nostras *of the same author?* Saxifragia hircina minima, Pimpinella crispa

---

⁵³⁸ See *Narcissus pallido-luteus* [*Narcissus pseudonarcissus* 386] and also *Helleborus niger* [*Helleborus viridis* 276].

⁵³⁹ For "moorish" as the translation of *montosis* see footnote 154.

⁵⁴⁰ *Poterium sanguisorba* is very variable in size, and the cultivated plant Ray knew might have been a genetically distinct variant, perhaps introduced from Europe. However, the robust subsp. *balearicum* (Bourg. ex Nyman) Stace, which was formerly grown for fodder and is now naturalised in Cambridgeshire and elsewhere, is not believed to have been introduced to Britain until 1803.

⁵⁴¹ Bauhin's name is actually *Saxifragia*.

⁵⁴² Ray retained separate entries for the two taxa treated here as *Pimpinella saxifraga major altera* [440] and *Pimpinella saxifraga minor* [442] in his later works. A detailed description of the latter in *Historia plantarum* (1686, p. 445) made it clear that it differed from the former principally in having more divided leaves, with even the basal leaves long, narrow and deeply divided (rather than

Tragi *J.B.* Bipinella sive Saxifraga minor *Ger. emac*. *Small Burnet-Saxifrage*. In moorish[539] places & pastures.

We have observed besides another larger species of Pimpinella also, with a curled leaf, which we suspect is the same as Jean Bauhin's Saxifragia hircina media, but we have not been attentive enough in describing it.[543]

443 *Pinguicula vulgaris* L.
Pinguicula vel Liparis *Gesn. hort. Clus*. Pinguic[ula] Gesneri *J.B.* flore cæruleo *hort. Hafn. & Patav*. Ping[uicula] sive Sanicula Eboracensis *Park. Ger*. Sanicula montana flore calcari donato *C.B*. Cucullata, quibusdam Crias Apuleii *Lugd*. Viola humida aut palustris, Dodecatheon Plinii Gesnero *Cam*. *Butter-wort, Yorke-shire Sanicle*. *On all the moores about Cambridge plentifully*.

444 *Pisum sativum* L.
Pisum arvense flore albo aut roseo, fructu albido aut cinericeo. Pisa vulgaria parva alba sive arvensia *J.B.* & of others. *Field Pease*. With us three species of peas are commonly sown in the fields. 1 Pisum parvum flore & semine albis. *Common white Pease*. 2 Pisum flore roseo, semine subnigro aut fusco. *Common gray Pease*. 3 Pisum flore roseo, semine variegato.[544] *The Maple Pease*.

N. 1. The flowers of all legumes are called papilionaceous[545] because they somewhat resemble a flying butterfly. *Spigel[ius in] Isag[oges]*,[546] *book 1, chapter 12*.

2. The farmers of our country, taught by experience, diligently avoid sowing peas when the wind is blowing from the North;[547] otherwise those which had been white generally degenerate to grey or certainly are eaten into by maggots. *P. Lauremb[erg]*,[513] *Horticult[ura], book 1, chapter 4, number 4*.

3. I have learned from experience that peas with salt added remain hard and do not become soft with any amount of cooking, even if, so to speak, they are cooked over a period of three days, says *Jo. Bodæ. in Theophr. hist*.[548] We too have found that salt added in quantity to water very much retards, but does not completely

---

undivided and suborbicular). In characters of the root, stem, flowers and seeds they were sufficiently matched, so that one would scarcely know the difference between them. A more succinct treatment along the same lines is provided in Ray's *Synopsis* (1690, 1696). The varietal names are taken from Sell & Murrell (2009); P.D. Sell has confirmed the occurrence of both taxa in Cambridgeshire and of the intermediate var. *intercedes* Thell.

[543] Ray later (1688b, p. 18; 1690, p. 70) attributed *Pimpinella saxifraga hircina media* "Herbariis nostratibus" ("for the Herbalists of our country") to *Apium petræum sive montanum album* J.B., which he reported from the Gogmagog hills. He had earlier annotated his own copy of the 1677 *Catalogus plantarum Angliae* to this effect (Crompton 2001), noting that the record was omitted through forgetfulness in his *Historia*. This is the Cambridgeshire rarity *Seseli libanotis* (L.) W.D.J. Koch; it had been described separately from *Pimpinella saxifragia hircina media* by Bauhin & Cherler (1651, vol. 3(2), pp. 111 & 105 respectively).

[544] "Of peas there were several sorts ... and the local varieties of each were innumerable" (Thirsk 1967, writing about British crops in the period 1500–1640). The three Latin phrases mean "Small pea with white flower & seed", "Pea with a pink flower [and] blackish or dark-coloured seed" and "Pea with a pink flower [and] variegated seed".

[545] i.e. butterfly-like.

[546] For Adrianus Spigelius see footnote 31.

[547] ... *studiosè cavent nè pisa seminent spirantibus à Septentrione ventis*, literally "... diligently beware lest they may sow peas with winds blowing from the North".

[548] Ray faithfully reproduces the words of Bodaeus (1644, p. 916.2; see footnote 23) here, including *ut ita dicam* ("so to speak", literally "that thus I may say").

prevent, the softening of old peas or of those dried by age & time, but that it does absolutely no harm to the cooking of green peas.

4. Peas remain good and suitable for sowing for more than a period of three years. *[P.] Lauremb[erg]*,[513] *Horticult[ura], book 1, chapter 17, number 8*.

5. All white things are in general weaker and more fragile than black ones, both in plants and in the case of animals. *Theophr. de caus., book 3, chapter 27*.[549]

445 *Alisma plantago-aquatica* L.
Plantago aquatica *J.B.* aquatica latifolia *C.B.* aquatica major *Ger. Park.* aquatica foliis betæ aut plantaginis, flore gallii albi *Ad. Lob. & ico.* Limonium verum Dioscoridis & Antiquorum *Ang. Great water Plantain.* Very frequent by waters, & also in waters.

446 *Baldellia ranunculoides* (L.) Parl.
Plantago aquatica minor *Park. Tab. Cæs.* aquat[ica] humilis *Ger.* aquat[ica] humilis angustifolia *J.B.* aquat[ica] II, sive aquat[ica] angustifolia *C.B.* aquat[ica] humilis angustifolia, capitulis rotundis echinatis *P.B. Dwarfe water Plantain. In many places of the Fens.*

447 *Plantago major* L.
Plantago latifolia vulgaris *Park.* latifolia sinuata *C.B.* latifolia *Ger.* latifolia glabra vulgaris *hort. Hafn.* major folio glabro non laciniato ut plurimùm[550] *J.B. Great Plantain, or Waybread*.

448 *Plantago media* L.
Plantago incana *Ger.* major incana *Park. Ad. Lob.* latifolia incana *C.B.* media *Fuch. Dod. Gesn.* latifolia hirsuta *hort. Hafn.* major hirsuta, media à nonnullis cognominata *J.B.* Arnoglossa *commonly. Hoary Plantain or Lambes-tongue.*

449 *Plantago major* L.[551]
Plantago major paniculâ sparsâ *J. B.* latifolia VI, sive spicâ multiplici sparsâ *C. B.* paniculâ sparsâ sarothrodes & corematophora, sive scoparia Gemmæ *Lob.*[552] *Besome Plantain or Plantain with spokie tufts. Found once by Dr Strachey*[553] *in the same close where the* Cirsium Anglicum *is said to grow. We could never meet with it there*.

---

[549] *Infirmiora fragilioráque sunt omnia, ut simpliciter loquar* [literally "that simply I may speak"], *alba quàm nigra, tam in plantis quàm in animalium genere* [literally "in the kind/class/race of animals"]. Ray's text is taken *verbatim* from Gaza's (1529, part 2, pp. 176–177; see footnote 21) Latin translation of a statement in Theophrastus' *De causis plantarum*, III. 22. 2 in the Loeb edition.

[550] "… for the most part not laciniate".

[551] Teratological variants of *Plantago major* such as this feature prominently in the early botanical literature (e.g. Johnson 1633, pp. 419–421; Parkinson 1640, pp. 493–498) and were "cherished in gardens" (Parkinson 1640, p. 494; see also Henrey 1986). Another such variant is reported in the 1685 appendix [B78].

[552] The names listed mean literally "Great Plantain with a sparse panicle", "The sixth broad-leaved [Plantain], or [the one] with a sparse multiple spike" and "Broom-like & besom-bearing [Plantain] with a sparse panicle, or Broom [Plantain] of the Bud". The compound adjectives *sarothrodes* (i.e. σαροθρωδης) and *corematophora* (i.e. κορηματοφορα) are Greek, probably specially coined.

[553] This must surely be Robert Strachey of Ashdon, Essex, who entered Gonville and Caius College in 1640 aged 16. He became a Doctor of Medicine in 1658 and was admitted a Candidate of the College of Physicians in 1659. Little is known of his subsequent career but he died in 1705 and was buried in Bishop's Stortford church, Hertfordshire (Munk 1878; Venn 1897; Venn & Venn 1913). His will (Essex Records Office D/ABW 78/280; inspected on-line) shows that he held land in Ashdon and Bishop's Stortford at the time of his death.

MAIN CATALOGUE

450 *Plantago lanceolata* L.
Plantago quinquenervia *Ad. Lob. Ger.* quinquenervia major *Park.* lanceolata *J.B. Trag. Tab.* angustifolia I, sive major angustifolia *C.B.* Lanceola major *Cæs. Ribwort, or Ribwort-plantain.* Two species of this plant are decided to exist by botanists, and we think both occur with us but we have not yet accurately distinguished them.[554]

Platanus *Tragi, see* Acer majus.[555] [*Acer pseudoplatanus* 5]

Platanaria, *see* Sparganium.[556] [*Sparganium* spp. 553, 554]

Podagraria *Lob. see* Herba Gerardi.[557] [*Aegopodium podagraria* 46]

Polyanthemum palustre *Dod. see* Ranunculus aquaticus.[558]

451 *Polygala vulgaris* L.
Polygala *Ger. Park.* recentiorum *Ad. Lob.* II, sive vulgaris *C.B.* Polygalon multis *J.B.* Amarella *Gesn.* Flos Ambarvalis *Dod.* Onobrychis vera Dioscoridis *Lob.* Flos crucis *Gesn. coll. Milk-wort.* It varies in the blue, purple, white, flesh-pink or variegated colour of its flower. In pastures, meadows & moory[539] places.

452 *Hippocrepis comosa* L.
Polygalon Cortusi *J. B.* An Polygala *Ang? Yellow Milk-wort with knotted cods. In the chalk-pit close at Cherry-hinton and in many places on Gogmagog hills & Newmarket heath, as upon the banks of Devills-ditch abundantly, and generally on chalkie & barren grounds.* We do not know whether this plant was described by Gerarde or Parkinson, or by what name it may be called.[559] Actually we are amazed that a plant so obvious in England could have been overlooked by English authors.

453 *Scleranthus annuus* L.
Polygonum exiguum *Dod.* minus polycarpon *Tab.* tertium Dodonæi sive tenuifolium *J.B.* angustissimo, & acuto vel gramineo folio minus repens *C.B.*

---

[554] "There is another lesse kinde of this Rib-wort, which differs not from the last mentioned in any thing but the smallnesse thereof" (Johnson 1633, p. 421). Parkinson (1640, p. 495) also says that this species is "like the first sort, but that it groweth smaller and lower". *Plantago lanceolata* L. is an extremely variable species. Experimental studies of Cambridgeshire populations have shown that some of the variation is phenotypic and the differences between plants in different habitats are reduced when they are cultivated in common garden experiments (Warwick & Briggs 1979). Sell & Murrell (2009) recognise a small variant of dry inland grasslands as var. *pusilla* Baumg.

[555] Ray's chosen name is actually *Acer major* (with an incorrect gender for *Acer*, which is neuter) but he cites *Acer majus* as a synonym.

[556] Ray cites *Platanaria* as a synonym of *Sparganium ramosum* [*Sparganium erectum* 553] and *Platanaria altera* as a synonum of *Sparganium non ramosum* [*S. emersum* 554].

[557] *Herba Gerardi* is not Ray's chosen name but is cited as a synonym of *Angelica sylvestris minor* [*Aegopodium podagraria* 46].

[558] This is one of two synonyms referred by Ray to *Ranunculus aquaticus*, the other being *Apium risûs* Lob. (q.v.). There is no other entry for *R. aquaticus* or either synonym in the catalogue. *Apium risûs* appears to be a synonym of Ray's *Ranunculus palustris rotundifolius* Ger. [*Ranunculus sceleratus* 478]. However, he probably intended to refer *Polyanthemum palustre* to *Ranunculus aquatilis* [*R. aquatilis* 479], as C. Bauhin (1623) cites both *Polyanthemum palustr. aut aquaticum*, Dod. gal. and *Ranunculus aquatilis*, Dod. Ger. as synonyms of his *Ranunculus palustris & aquaticus V. Ranunculus aquaticus, folio rotundo & capillaceo*, a name that Ray does cite under *Ranunculus aquatilis*.

[559] *veniat*, literally "it may come". See, however, *Ferrum equinum Germanicum siliquis in summitate* C.B. in *Addenda* of the 1663 and 1685 appendices and Ray's comment on *Polygalon Cortusi* in *Emendanda* of both.

Germanicum vel *Knawell* Germanorum *Park.* selinoides *Ger.*, but wrongly, as Parkinson rightly warns, for it does not have the leaves of Parsley.[560] *German Knotgrasse or Knawell. About Hoginton where the* Myosuros *growes, & in the sandy grounds about Gamlingay.*

Polygonum fœmina,[561] *see* Equisetum. [*Hippuris vulgaris* 181]

454 *Polygonum aviculare* L. *sensu lato*[562]

Polygonum mas vulgare *Ger.* mas vulgare majus *Park.* latifolium *C.B.* Polyg[onum] sive Centinodia *J.B.* Centinodia *Brunf.* Sanguinaria Plinii *J.B.* Sanguinalis mascula *Gesn. hort.* Corrigiola *Eric. Cord.* Polygonia, Virga pastoris & Lingua passerina *Arnold. Villanov.* Lingua passerina *Vigoni. Common Knot-grasse.* Generally in ploughed fields & beside footpaths.

455 *Polypodium vulgare* L. *sensu lato*[563]

Polypodium *J.B. Ger.* vulgare *C.B. Park.* Herba radioli Apuleii *Lob. Polypody. On Trinity-hall wall by the lanes-side which leades to Garret-hostle-bridge,* & elsewhere in shady places at the roots of trees, & in rocky places or on rubble.

Polytrichum minus, *see* Adianthum aureum minus. [unidentifiable mosses 13]

456 *Populus alba* L. and/or *P. alba* × *tremula* L.[564]

Populus alba *Park. Ger. Tab. Dod. Matth.* alba, λευκη [leuce] Theophrasti *J.B.* Ἀχερωις [Acheroïs] of Homer & Pausanias[565] *of the same author.* Farfarus

---

[560] *nam non habet Selini folia. Selinum* is a latinised form of the Greek σελινον [selinon], which is probably Garden Parsley, *Petroselinum crispum* (Mill.) Nyman ex A.W. Hill. See also footnotes 72 and 73.

[561] *Polygonum fœmina* of numerous authors is cited by Ray as a synonym of *Equisetum palustre brevioribus foliis polyspermon* [*Hippuris vulgaris* 181], but *Polygonum fœmina* Fuch. ico. is cited as a synonym of *Equisetum arvense longioribus setis* [*Equisetum arvense* 178]. The former is almost certainly intended here.

[562] Stace (2010) recognises four species, of which three occur in Cambridgeshire. *Polygonum arenastrum* Boreau and *P. aviculare* L. are frequent and *P. rurivagum* Jord. ex Boreau is much less common. P.D. Sell and J.G. Murrell will recognise more taxa in a future volume of their *Flora of Great Britain and Ireland*.

[563] Cytological studies in the mid 20th century established that this plant, hitherto regarded as a single species, is a complex which includes three species and their hybrids in Britain. Two of these taxa, *Polypodium vulgare sensu stricto* and *P. interjectum* Shivas, are currently found in Cambridgeshire (Barden & Preston 2009). Although the Garret Hostel Lane population survived until 1949, there appears (surprisingly) to be no extant herbarium specimen and so it cannot be identified beyond aggregate level.

[564] Both *Populus alba* and its hybrid with *P. tremula* (*P.* × *canescens* (Aiton) Sm.) are recorded in Cambridgeshire. Ray's plant has hitherto been identified as *P. alba* but Ray did not distinguish *P.* × *canescens* in any of his works, so it is impossible to know whether he saw the species or its hybrid in the county. Gerarde (1597, pp. 1301–1302) and Parkinson (1640, pp. 1409–1412) described a small-leaved tree which some later authors (e.g. Clarke 1900) have identified as *P.* × *canescens*, but Ray (1677, p. 240) did not regard it as specifically distinct. Although references to *abel* in mediaeval documents suggest that white poplars have a long history in East Anglia, it is equally impossible to use them to identify which of these two taxa was present, although ancient clones tend to be *P.* × *canescens* (Rackham 1990, 1994). From his field experience, P.D. Sell (pers. comm.) believes that *P.* × *canescens* has the longer history in the county, occurring for example in areas of former fenland in south Cambridgeshire, and his observations suggest that the plants reported by Ray cannot be identified with confidence as *P. alba* and might well have been *P.* × *canescens*.

[565] This tree name occurs in Homer's *Iliad* (in two identical similes about the fall of a warrior in Book 13: 389 and Book 16: 482). It is derived from Acheron, one of the rivers of Hades, supposedly

antiquorum *Cord. in Diosc.* Farfarus Plauti in *Pœn[ulus]*[566] *J.B.* White Poplar tree. One tree in the hedge of St Johns Colledge bowling-green; At Teversham diverse great trees, many also at Chesterton, and in many other places, especially toward the Fens.

N. 1. It turns over the undersides of the leaves soon after the summer solstice, and from that proof countrymen know that the solstice has passed. *Theophr. hist. plant., book 1, chapter 16.*[567]

2. Pliny, *book 16:*[568] He says that the leaves of Populus alba are rounded when young,[569] which can also be observed even now, but that by contrast the angular leaves of ivy become rounded, which was reported by Theophrastus.[570] *Marant[a], Meth[odus] cogn[oscendorum] simp[licium], book 2, chapter 3.*[571]

3. The upper side of the leaf of this tree is leek-green, the underside white, because the upper is moister, the under drier; for white is the colour in the dry, of which the first subject is earth,[572] that which is evident in ashes, rocks and sands heated by fire or the sun, but green is that from the moist and the dense, whose source we see is water, for in growing dense it becomes green. Consequently the upper side is greener, because, being exposed to the sun, it attracts to itself the moist, which ought to be on the underside. *Jo. Bod[æ]. in Theophr. hist. plant., book 3, chapter 14, page 217.*

457 *Populus tremula* L.

Populus Libyca *Matth. Dod. Lob. Ger. Park.* Libyca Plinii, κερκις Theophrasti[573] *J.B.* tremula *C.B.* The Aspe, or Aspen tree. *At Teversham and Chesterton* with the preceding species, & elsewhere generally in woods & hedges.

---

because Heracles (or Hercules) brought the tree back from the underworld: see Pausanias, 5. 14. 2. (Pausanias was a Greek traveller and geographer of the second century AD.) The commoner Ancient Greek name is λευκη [leuce], from λευκος [leucos], meaning "white".

[566] *Poenulus* ("The Little Carthaginian" or "The Puny Punic") is a comedy by the early Roman playwright Titus Maccius Plautus (*c.* 254–184 BC). The relevant passage describes a highly fanciful combat in which birdlime was shot from slings to kill flying men, the leaves of *farfara* being used to line the slings so that the birdlime did not stick to them. Ray's entry for *farfara* in the 'Etymology' discussed the use of this word for both *Populus* and *Tussilago*.

[567] This statement in Theophrastus' *Historia plantarum* is in Book I. x. 1 in the Loeb edition: "There is a peculiarity special to the olive, lime, elm and abele: their leaves appear to invert the upper surface after the summer solstice, and by this men know that the solstice is past."

[568] Pliny's *Natural history*, XVI. 86 in the Loeb edition.

[569] ... *in juventa circinatæ rotunditatis esse*, literally "in youth to be of a roundness formed into a circle [as with a *circinus*, a pair of compasses]".

[570] In Theophrastus' account (see above) the following two statements immediately precede the one reported in Ray's previous note: "The young leaves ... are round, the old ones angular, ... On the other hand in the ivy, when it is young, the leaves are somewhat angular, but when it is older, they become rounder."

[571] For Bartolomeo Maranta see footnote 34.

[572] *cujus subjectum primum terra*: the most opaque phrase in a generally rather obscure passage, which presupposes the four elements of Aristotelian physics and Galenic medicine, earth, water, fire and air, each with its own pair of qualities from among dry, wet, cold and hot (see Chapter 3). It is taken almost *verbatim* from Bodaeus (1644, p. 217.1, 2; see footnote 23), though the original has *ex* ("from", "out of") rather than *in* ("in") before "ashes, rocks and sands".

[573] i.e. "Libyan [Poplar] of Pliny [or] Cercis of Theophrastus". The reference is to Theophrastus' *Historia plantarum*, III. XIV. 2 in the Loeb edition, rather than I. XI. 2, which seems to relate to Judas-tree, *Cercis siliquastrum* L.

N. The buds[574] of Populus Libyca, especially in spring-time, give off a very pleasant scent when rubbed with the fingers. *J. Bauhin* reports that the buds of Populus nigra are scented and he denies that the same is the case in Libyca. *Gerarde* reports it about nigra but is silent about Libyca.

458 *Populus nigra* L.

Populus nigra *Ger. Park. Dod. Lob. J.B. C.B.* Ἀίγερος [Aegeros] of Dioscorides & Theophrastus[575] *J.B. Black Poplar. In the meadows about Fordham and Soham plentifully.* Theophrastus falsely teaches in several places that White Poplar is changed into Black by a spontaneous metamorphosis. *J.B., volume 1, book 8, page 256.*

459 *Portulaca oleracea* L.[576]

Portulaca sylvestris *Ger. Trag. Dod. Ad. Lob. Park.* sylvestris minor sive spontanea *J.B.* angustifolia sive sylvestris *C.B. Wild Purslane. About Ramsey mear in the foot pathes by the rape mills P.B. We have not as yet searched that place for plants, yet we fear there might be a mistake in the first reporter*; viz. *of* Portulaca sylvestris *in stead of* Portulaca aquatica.

460 *Potamogeton natans* L.

Potamogeiton foliis latis splendentibus *C.B.* altera major *Dod.* rotundiore folio *J.B.* Potamog[eiton] latifolium *Ger.* Fontalis major latifolia vulgaris *Park. Broad-leaved Pondweed.* In stagnant waters & fish-ponds, and sometimes also in rivulets.

461 *Persicaria amphibia* (L.) Delarbre

Potamogeiton angustifolium *Ger.* II, sive salicis folio *C.B.* Potam[ogeiton] seu Fontalis persicariæ foliis *J.B.* Fontalis major longifolia *Park. Narrow-leaved Pondweed.* Those leaves of this plant that float on water, as can be seen on the banks of a river,[577] are smooth & shining, but those that are out of water are hairy, rough & devoid of that same shininess. The same also happens to the whole plant when (as very frequently happens) it is growing out of water: which is to be noted diligently, for otherwise a fall into error will be likely & anyone who looks at it somewhat carelessly will decide that there are beyond doubt two different plants. And hence we suspect that the plant which is given the name Persicaria hirsuta radice perenni in the catalogue of the indigenous plants *of Leiden Botanic Garden* is not different from this Pondweed, which, when it grows out of water, certainly recalls a Persicaria in outward appearance.[578]

---

[574] *oculi sive gemmæ*, literally "eyes or gems": Ray here uses both the words that he gives for buds in his 'Interpretation of terms'.

[575] This is the common name for the tree in Ancient Greek from the *Iliad* and *Odyssey* onwards.

[576] *Portulaca sylvestris* is Common Purslane, *Portulaca oleracea*, a very widespread weed in warm temperate and subtropical climates but known in Britain only as an uncommon introduction. It is currently (2010) very well established between the cobbles in Great Court of Trinity College, but it is indeed very unlikely to have been the plant seen at Ramsey in the mid 17th century. Ray recorded what he thought might be J. Bauhin's *Alsine aquatica surrectior*, describing it as similar to *Portulaca aquatica*, from Gamlingay in *Addenda* of the 1663 appendix, i.e. *Montia fontana* [A2], following this with a more definite identification in the 1685 appendix [B6]. *Alsine rotundifolia sive Portulaca aquatica*, the native *Lythrum portula* [B8], was reported, also from Gamlingay, in the 1685 appendix.

[577] *ut in fluminis ripis videre est*, literally "as on a river's banks it is to see".

[578] This is hardly surprising since it is in fact a species of *Persicaria*. In the second edition of his *Catalogus plantarum Angliae* (1677, p. 241) Ray retained it under *Potamogeiton* but wrote at the end

MAIN CATALOGUE

462 *Potamogeton lucens* L.
Potamogeiton longis acutis foliis *Ger. emac*. Potamogeiton longo serrato folio *C.B.* Fontalis longo serrato folio *Park*. Lapathum fluitans longo serrato folio *J.B.* Lap[athum] XIII, in aquis fluitans, foliis crenatis *C.B. in Phyt*. Lapathi genus sylvestre *Gesn. hort*. Oxylapathum aquaticum Dalechampii *Lugd*. *Long-leaved Pondweed*. To tell the truth,[579] we cannot satisfy ourselves about this plant's names; neither the pictures nor the descriptions correspond to it.[580]

463 *Potamogeton perfoliatus* L.
Potamogeiton perfoliatum *P.B.* altera Dodonæi *Ger. emac. J.B.* IV, sive foliis latis splendentibus *C.B. Small Pondweed, or perfoliate Pondweed. In the river Cam plentifully every where*.

464 *Potamogeton compressus* L.[581]
Potamogeiton ramosum caule compresso, folio graminis canini,[582] not yet described. The branched & strongly flattened *stems* exceed a cubit & sometimes a cubit and a half. The *leaves* are long, narrow, approaching the leaves of couch-grass except that they are of almost the same width throughout & end in a blunt point, arranged alternately, except at the point whence the stalks of the flower-spikes arise, where they are two opposite each other; on each three more significant veins run along the whole length of the leaf. Under each leaf a thin & translucent membrane invests the stem. The flower-spike is short, bearing washed-out green *flowers*, consisting of four petals which are flesh-pink near the style. The *style* ends in two points at most. The *seed* vessel is enclosed by a line that is straight on one side and curved on the other. *Small branched Pondweed with a flat stalk. In the river Cam in many places.*

465 *Potamogeton berchtoldii* Fieb. and/or *P. pusillus* L.[583]
Potamogeiton pusillum gramineo folio, caule rotundo,[584] not yet described. In other respects it differs very little from the preceding species apart from its size, which in this species is ten times smaller, & in the colour of the flowers, which appear whiter & almost flesh-pink when they emerge above the water. *Small grasse-leaved Pondweed. In the rivulet at Hinton moor and many other places.*

---

of the entry: *Et revera Persicariæ species est non Potamogeitonis*. ("And in reality it is a species of *Persicaria*, not of *Potamogeiton*.")

[579] *ut verum fateamur*, literally "so that we may admit a true [thing]".

[580] See also Ray's comments on the synonymy of this species in *Emendanda* of the 1685 appendix.

[581] We have followed J.E. Dandy and G. Taylor's unpublished draft '*British species of* Potamogeton *L*.' in making this identification (Preston 2010). In this they withdraw their earlier view, published in Evans (1939), that Ray's plant was *P. acutifolius* Link and convincingly demonstrate that Ray's description cannot apply to *P. friesii* Rupr., to which it has been ascribed by Babington (1860), Perring *et al*. (1964) and Ewen & Prime (1975).

[582] i.e. "Branched Pondweed with a flattened stem [and] the leaf of Gramen caninum [Common Couch, *Elytrigia repens* 242]".

[583] The taxonomy of these two species in Britain was first clarified by Dandy & Taylor in 1938. They attribute Ray's name to *P. berchtoldii* in their unpublished draft '*British species of* Potamogeton *L*.' but not in any of their published works. The species are superficially very similar and in the absence of herbarium material it is impossible to know whether Ray encountered one or both species. Herbarium specimens suggest that until recent decades *P. berchtoldii* was the commoner plant south of the Fens.

[584] i.e. "Small Pondweed with a grassy leaf [and] a round stem".

466 *Zannichellia palustris* L.
Potamogeito affine gramen aquaticum.[585] It has long, narrow *leaves* and slender, feeble little stems; it rises to a foot's height & more, always immersed in the water. We have not yet observed the *flowers*. The small *pods* or seed-vessels are recurved, with two or three or even four joined at one point, and stick closely to the nodes. *Water-grasse with small crooked cods.* We have found it in stagnant waters. It could be referred to the grasses or to Stellaria aquatica.[586]

Potentilla, *see* Argentina. [*Potentilla anserina* 60]

Prassium, *see* Marrubium album. [*Marrubium vulgare* 357]

467 *Primula veris* L.
Primula veris major *Ger.* veris major floribus odoratis luteis *Dod.* veris odorata flo[re] luteo simplici *J.B.* veris flore flavo elatior *Clus. hist. hort. Hafn.* pratensis *Lob.* umbellifera flore simplici luteo *hort. Bat.* Herba Paralysis *Officin[arum]*[587] *Brunf. Cast.* Arthritica *Gesn. hort.* Alisma pratorum *Col.* Dodecatheon *Ang.* Verbasculum odoratum *Fuch.* pratense odoratum *C.B.* Φλομισκος σκιαδιουχος χρυσανθης μακροστυλος [Phlomiscos sciadiuchos chrysanthes macrostylos], i.e. Verbasculum umbellatum flore aureo, stylo longo[588] *P. Renealm. Paigles or Cowslips.* In pastures everywhere.

468 *Primula elatior* (L.) Hill
Primula veris elatior pallido flore *Clus. hist. hort. Hafn.* veris major floribus pallidis ac dilutiùs coloratis quasi albidis *Dod.* pratensis inodora lutea *Ger.* pratensis inodora luteo-pallida *Lob.* veris caulifera pallido flore inodoro aut vix odoro *J.B.* Verbasculum pratense aut sylvaticum inodorum *C.B.* Φλομισκος σκιαδιουχος διασκοπιος ἐρυθροζωνος [Phlomiscos sciadiuchos diascopios erythrozonos], i.e. Verbasculum umbellatum floribus in omnem partem diffusis, zonâ rubrâ ad petiolorum basin cinctum[589] *P. Renealm. Great Cowslips or Oxslips. In Kingston and Madingley woods abundantly and elsewhere.*

469 *Primula vulgaris* Huds.
Primula veris vulgaris *Park.* veris minor *Ger.* sylvarum *Ad. Lob. Lugd.* veris floribus ex singularibus pediculis majoribus, simplicibus *J.B.* veris pallido flore humilis *Clus. hist. hort. Hafn.* Alisma sylvarum *Col.* Verbasculum sylvarum majus singulari flore *C.B.* Φλομισκος ἀσκιαδιος βραχυσιφωνιος [Phlomiscos asciadios brachysiphonios], i.e. Verbasculum non umbellatum floris tubulo brevi[590] *P. Renealm. Common Primrose.*

---

[585] i.e. "Water-grass related to Pondweed".

[586] q.v., i.e. *Callitriche* spp.

[587] Cowslip was considered beneficial in all paralytic ailments, being recommended by Pliny and also called "Palsywort" in English. The root was called *Radix arthritica* from its use as a cure for muscular rheumatism.

[588] Literally "Umbellate Mulleinlet with a golden flower [and] long style". Only the third of the four compound Greek words here (meaning *"with flower of gold"*) and none of those in the next two entries is listed by Liddell, Scott & Jones (1940), but their meanings can all be derived to match the Latin explanations. Φλομισκος is a diminutive of Φλομος [*Phlomos*], the Greek for "mullein", just as *Verbasculum* is a diminutive of *Verbascum*.

[589] Literally "Umbellate Mulleinlet with flowers spread out in every direction, surrounded by a red zone at the base of the petioles". Since widely spread flowers are an important character for distinguishing hybrids between Primrose and Cowslip, known as False Oxlips, from the true Oxlip, *Primula elatior*, it seems that Renealmus was describing the former.

[590] Literally "Non-umbellate Mulleinlet with the tube of the flower short".

N. In almost all flowers that are not uniformly coloured[591] the central parts will be yellow, rarely pale or herbaceous, very rarely imbued with another colour, but those that undergo variation are the petals or perianths[592] that surround the central parts. *Marant[a], Meth[odus] cogn[oscendorum] simp[licium], book 2, chapter 15.*[593]

470 *Prunella vulgaris* L.

Prunella *Ger.* vulgaris *Trag. Park.* vulgaris foliis subrotundis *Clus.* flore minore vulgaris *J.B.* Brunella major folio non dissecto *C.B.* Consolida minor *Matth.* minima *Cam.* Britannica *Guiland.* *Self-heal.* It varies in the colour of the flower which may be blue, flesh-pink or white.

471 *Prunus spinosa* L.

Prunus sylvestris *C.B. Ger. Park. J.B.* sylvestris fructu minore serotino *P.B.* Prunellum vel Prunulum *Officin[arum].* Acacia Germanica *Schrod. & of some other authors.* Spodias Theophrasti *Dalechampio* in the work of *J. Bauhin, from his manuscript notes in the margin of Theophrastus.* Spinus of Virgil's *Georg[ic] 4*[594] *in [P.] Gassend[i], C. Bauhin & others.* Κοκκυμηλεα αγρια [Coccymelea agria] or Ἀγριοκοκκυμηλεα [Agriococcymelea] for the Greeks, because it bears apples with the roundness & size of a Berry[595] *C.B. Sloe-tree, or Black-thorn.*

N. 1. This is peculiar to the wild prunus that, dropping its petals as soon as showers touch them, the fruit is converted into a sort of oblong & empty body, which in Italy they call *Turcas.*[596] *Cam. in hort.* However this is common to certain other kinds of prunus too.

2. By transplantation & diligent cultivation the wild prunus becomes the garden one and bears larger fruits, which the Germans call *Grosse garden Schlehe*, i.e. *Great garden Sloes. Schroder[i] Pharmac[opœa] medico-chym[ica], book 4, chapter 6.*

472 *Agrostemma githago* L.

Pseudomelanthium *Ger. Matth. J.B.* Lychnis arvensis *Tab.* Segetum major *C.B.* Nigellastrum *Dod.* Lychnoides segetum sive Nigellastrum *Park.* Githago *Trag.* *Cockle.* Among standing corn. The seed of this plant, when inspected with a lens, portrays quite closely a rolled-up hedgehog.[597]

---

[591] *qui unico colore non constant,* literally "which with a single colour do not remain".

[592] *folia sunt vel calathi,* literally "are the leaves or baskets"; *folium* commonly meant "petal" and a subsidiary meaning of *calathus* was "calyx" or "perianth".

[593] For Bartolomeo Maranta see footnote 34. Maranta is discussing which part of any kind of flower it is that determines the colour that one describes it as having and decides that it is the corolla rather than the filaments or, as Pliny calls them, the stamens. Ray's Latin follows Maranta's with three minor differences, none of them affecting the overall meaning. There seems no particular reason for attaching this passage to his account of the Primrose.

[594] This refers to line 145 of Virgil's fourth Georgic, a famous poem about bee-keeping, in a passage about a green-fingered peasant's plot of unclaimed land which he makes productive, cultivating, among other trees, *spinos jam pruna ferentes* ("blackthorns now bearing plums").

[595] *quòd rotunditate & magnitudine Cocci mala ferat.* The initial capital of *Cocci* suggests that it may refer, not to an ordinary berry, but to the supposed berry of the Kermes Oak, *Quercus coccifera* L.: see footnote 314. The two Greek names mean literally "Wild Berry-apple-tree" and "Wild-berry-apple-tree".

[596] possibly "Turkish women" or "female Turks", but the plural seems Spanish rather than Italian (in which one would expect *Turche*). In Nicaragua *turca* (plural *turcas*) is a penis.

[597] *... microscopio inspectum echinum in se convolutum non malè exprimit,* literally "... with a lens inspected, a hedgehog rolled up into itself not badly it expresses". Raven (1950, p. 200) points out that

473 *Achillea ptarmica* L.

Ptarmica *Ger. Matth. Fuch. Lob.* vulgaris *Park.* vulgaris folio oblongo serrato, flore albo *J.B.* Draco aquaticus *Gesn. in coll.* sylvestris sive Ptarmica *Dod.* Dracunculus 2. sive aquaticus serrato folio *C.B. in Phyt.* II, sive pratensis serrato folio *of the same author in Pin.* Pyrethrum sylvestre *Dod. gal.* Tanacetum album seu acutum *Trag.* Sneezwort, bastard Pellitory. *All over the Fens abundantly.*

474 *Mentha pulegium* L.

Pulegium *Brunf. Matth. Trag. Cord. Dod. Cam. J.B.* vulgare *Park. Ang.* regium *Ger. Ad. Lob. Lugd.* latifolium *C.B.* officinarum *hort. Hafn.* Blechon to the Greeks, because, as Dioscorides relates, & Pliny[598] too, as a result of its pungency & bitterness when tasted by a flock of sheep or goats, it evokes bleating from them.[599] *J.B.* Penny-royall, Pudding grasse. *At the castle end in a bottome over against the furthest house, and in many other places about Cambridge.*

N. 1. Pennyroyal applied to the skin not only reddens it but also makes it sore. *C[asparus] Hofman[nus], De medicam[entis] officin[alibus], book 2, chapter 186.*

2. Dry pennyroyal is said to flower on the very shortest day of the winter. *Cicero, De divinat[ione], book 2.* Costaeus relates the same thing, & there is, he says, an example of this phenomenon in the animal kingdom, in the Woodpecker of Mars,[600] whose corpse, hung from the ceiling, has been observed to put forth new feathers in spring-time with the previous ones dropping away. *Both these things are unbelievable.*

Pulicaria, *see* Conyza major[601] [*Pulicaria vulgaris* 143].

Pulmonaria, *see* Lichen arborum. [unidentifiable foliose lichens 332]

475 *Pulsatilla vulgaris* Mill.

Pulsatilla *Matth. Dod. Tab. Lugd.* vulgaris *Park. Ger. Lob.* purpurea *J.B.* I, sive folio crassiore, & majore flore *C.B.* Herba venti *Trag.* Anemone limonia & Samolus Plinii *Dalechampio in Plinium.* Herba Sardoa[602] *Dod. gal.* Purple Pasque flower. *On the left hand of the way over Gogmagog hills leading to Haverill, so soon as you come to the top of the hill.* It does not readily tolerate cultivation except in

---

"Ray calls any lens a microscope" and "never possessed more than a small magnifying glass". This view is supported by the two objects that he mentions in the *Catalogus* which he has examined with a "microscope", seeds of *Agrostemma* and larvae parasitic on Large White caterpillars [see note 2 under *Rapum sylvestre* 495]. Both of these, and especially the opaque seeds of *Agrostemma*, would be more easily observed under a lens than through a microscope. However, see also footnote 742.

[598] Pliny's *Natural history*, XX. 156 in the Loeb edition.

[599] This derivation is based on the assumption that the name is derived from βληχη [bleche], meaning "a bleating", as in Homer's *Odyssey*, 12: 266.

[600] *& hujus rei, inquit, exemplum est in animalium genere, in Pico Martio, ...*, literally "& of this thing, he says, an example is in the kind/class of animals, in the Martian Woodpecker, ...". The woodpecker was sacred to Mars. In his *Natural history* (X. 40 in the Loeb edition) Pliny writes *pici Martio cognomine* ("woodpeckers with Mars's surname") and describes their importance in augury, hence no doubt the presence of dry ones hanging from people's ceilings! Ray (1675, p. 29) gives *Picus Martius* as the Latin translation of "*A wood-pecker*" but Linnaeus chose it as the scientific name for the Black Woodpecker, now *Dryocopus martius* (Linnaeus). For Giovanni Costeo see footnote 204 and Chapter 4.

[601] It seems clear that *major* here is an error for *minor*.

[602] See footnote 611.

a thoroughly dry & sunny location. *Cam. in hort*. But in our experience even in garden-grounds in quite fruitful & rich soil and in a place that is not sunny it grows successfully, indeed rather it thrives, blooms & is luxuriant.

Q

476 *Quercus robur* L.

Quercus latifolia *Park*. vulgaris *Ger*. platyphyllos mas *P. Renealm*. latifolia mas, quæ brevi est pediculo *C.B.* vulgaris brevibus pediculis *J.B.* Balanon Hippocratis *P. Renealm*. *The common Oak tree*.

N. 1. There is no tree that produces so many spurious & excrescent fruits besides its genuine one as the Oak; see *Verulam, Hist[oria] nat[uralis]*,[603] *cent[ury] 7, exp[eriment] 635, & J[ean] Bauhin*. These are enumerated by Theophrastus in *Hist[oria] plant[arum]*, book 3, chapter 8.[604]

2. In addition, we are indebted to the leaves of this tree more than all the rest for the heavenly dew preserved on them, from which bees make honey. *Butler, in his treatise about bees*,[605] *chapter 6, number 40, & Theophr[astus] in his book about honey*.[606] Sometimes Maple & Hazel[607] do the same, but only sparingly. *Butler, in the same place*.

3. We can also note here in passing *from Theophr. de causis, book 2, chapter 14*, that trees that raise themselves up to a greater size generally produce smaller fruit.[608]

4. Finally this too in passing: there is no more effective chemical for killing even the most enormous trees than quicksilver; if you will insert some quantity mixed with oil into a tree drilled right to the heartwood and will close up the hole with wax & chalk, you will see that it will die shortly afterwards. *P. Lauremberg*,[609] *Horticult[ura], book 1, chapter 34, number 5*.

Quinquefolium, *see* Pentaphyllum.[610] [*Potentilla* spp. 448-449]

R

477 *Adoxa moschatellina* L.

Radix cava minima viridi flore *Ger*. Moschatella foliis fumariæ bulbosæ, de qua Cordus *J.B.* Ranunculus nemorosus Moschatella dictus *Park*. Ranunculus

---

[603] For the work of Baron Verulam (Sir Francis Bacon) see footnote 104.
[604] Theophrastus' long description of the various kinds of oak galls is in III. vii. 4–5 in the Loeb edition of *Historia plantarum*.
[605] *The Feminine Monarchie* (Butler 1634): see footnote 40.
[606] We have not traced a work by Theophrastus specifically about honey, but in the paragraph following those about oak galls (*Historia plantarum*, III. vii. 6 in the Loeb edition), after mentioning Hesiod's (line 233) belief that the oak produced honey, he says: "however, the truth appears to be that this honey-like juice comes from the air and settles on this more than on other trees". See also footnote 517.
[607] *Acer minus* & *Corylus*.
[608] Ray's final words here are clearly adapted from Gaza's (1529, part 2, p. 96; see footnote 21) Latin translation of a sentence from Theophrastus' *De causis plantarum*, II. 10. 2 in the Loeb edition.
[609] For Peter Lauremberg see footnote 25.
[610] Ray cites synonyms beginning with *Quinquefolium* under both *Pentaphyllum rubrum palustre* [*Potentilla palustris* 428] and *Pentaphyllum vulgare* [*Potentilla reptans* 429].

nemorosus IV, sive nemorosus Moschatellina dictus *C.B.* Ranunculus minimus Septentrionalium herbido muscoso flore *Lob.* Denticulata *Lugd. The Musk Wood-Crowfoot. In the bank of a hedge by a grove of Elmes on the North-side of Chesterton.*

478 *Ranunculus sceleratus* L.

Ranunculus palustris rotundifolius *Ger.* pal[ustris] Sardonius lævis *Park.* flore minimo *J.B.* pal[ustris] apii folio lævis *C.B.* Sardonius sive Apium palustre *hort. Hafn. & Gron.* Ranunculus strumea and Apiastrum of Pliny, and Sardonia of Dioscorides *C.B.* Herba Sardoa *Guiland.*[611] *Marsh Crowfoot, or Round-leaved Spearewort.* In waters.

479 *Ranunculus aquatilis* L.

Ranunculus aquatilis *Dod. Ger.* aquaticus hepaticæ facie *Lob. Park.* aquatilis albus tenuifolius *J.B.* palustris et aquaticus V, sive aquaticus folio rotundo et capillaceo *C.B. Water Crowfoot.*

Ranunculus aquaticus capillaceus, *see* Millefolium aquaticum VI, *C.B.*[612] [*Ranunculus trichophyllus* 366]

480 *Ranunculus arvensis* L.

Ranunculus arvorum *Park. Lob. Ger.* sylvestris tertius *Dod.* echinatus I, sive arvensis echinatus *C.B.* arvensis echinatus *J.B. Rough-headed Crowfoot, or Corn-Crowfoot.* Among standing corn everywhere.

481 *Ranunculus bulbosus* L.

Ranunculus bulbosus *Ad. Lob. Ger. Park.* tuberosus *Dod. Lugd.* tuberosus major *J.B.* pratensis radice verticilli modo rotundâ[613] *C.B.* lanuginosus I, *Cæs.* primus exiguus *Trag.* & Ran[unculus] sceleratissimus folio nigricante, levitérq; hirsuto, parte sui adversa ranam viridem referente[614] *of the same author. Round-rooted or bulbous Crowfoot.* In meadows & pastures generally.

---

[611] The four synonyms that include words beginning with *Sard-* relate to Sardinia, known in ancient times for a bitter herb which was believed to distort the face of the eater into a grin; this was supposed to be the origin of a sardonic smile or laughter, so called as resembling this effect. The synonym *Apium risûs* (q.v. earlier in the catalogue), the second word meaning "of a laugh", derives from the same belief: presumably the confusion with true *Apium* species – as also indicated by Pliny's brief comment on *Apiastrum* in Sardinia in his *Natural history* (XX. 116 in the Loeb edition) – arose from the shared habitat and a superficial similarity between the leaves of Celery, *Apium graveolens* [53], and *Ranunculus sceleratus*, which is still called "Celery-leaved Buttercup". (Similarly, *Lysimachia vulgaris* [346] and *Lythrum salicaria* [347] were both named *Lysimachia* in Ray's time and are still both called "loosestrife" in English, while *Apium nodiflorum* [545] is now known as "Fool's Watercress" from its superficial resemblance to *Nasturtium officinale sensu lato* [387].) In fact the derivation is bogus, as Odysseus "smiled a right grim and bitter smile" (μειδησε … σαρδανιον [meidese … sardanion]) in Homer's *Odyssey*, 20: 301–302 long before this explanation was invented. According to another tradition *Herba Sardoa* was *Pulsatilla vulgaris* [475] (q.v.). The epithet *sardous* is now applied to a different species of *Ranunculus*, *R. sardous* [A34, B85].

[612] We have assumed that this entry refers to Ray's chosen name *Millefolium aquaticum ranunculi flore & capitulo* [366], as *Millefolium aquat[icum] VI* is cited as a synonym of it although *Ranunculus aquaticus capillaceus* is not.

[613] Literally "Meadow [Crowfoot/Buttercup] with the root rounded in the manner of a spindle-whorl".

[614] Apparently "Most noxious Froglet with a blackish, lightly hairy leaf, with its front part recalling a green frog". The similarity to a frog is not obvious, but Pliny used a diminutive of *rana* (= "a frog"), *ranunculus*, as an equivalent of the Greek βατραχιον [batrachion], a diminutive of βατραχος [batrachos], which has the same meaning.

N. Beggars make their skin sore with this plant so that they may appear wretched. *C[asparus] Hofm[annus], De medic[amentis] offic[inalibus]*, book 2, chapter 132.

482 *Ranunculus lingua* L.
Ranunculus flammeus maior *Ger.* palustris flammeus major *Park.* lanceolatus major *Tab.* longifolius palustris major *C.B.* longo folio maximus, Lingua Plinii *J.B.* Lingua Plinii *Lugd. Great Spearewort. In some ditches at Teversham moor, and abundantly in many great ditches in the Fens in the Isle of Ely.*

483 *Ranunculus flammula* L.
Ranunculus flammeus minor *Ger.* palustris flammeus minor *Park.* longifolius palustris minor *C.B.* flammeus aquatilis angustifolius *Ad. Lob. Lugd.* Flammula ranunculus *Dod.* Ranunculus longifolius, aliis Flammula *J.B.*, who uses the same name for the following species, namely for serratus. Ægolethron of Pliny, *book 21, chapter 13,* according to some authors: Enneaphyllon of the same writer, *book 27, chapter 9,* according to others.[615] *C.B. The lesser Spearewort.*

N. The majority of plants that grow in water possess a remarkably hot quality; a great part of those that grow in watery places is hot & dry and among these some are even caustic.

484 *Ranunculus flammula* L.[616]
Ranunculus flammeus serratus *Ger. Park.* palustris serratus *C.B.* aquatilis angustifolius serratus *Lob.* lanceatus serratus *Tab.* Flammula ranunculus folio serrato *Dod.* Ranunculus flammula folio serrato altera Dodonæi *hort. Hafn. Indented Spearewort.* By waters in various places & in waters.

485 *Ranunculus hederaceus* L.
Ranunculus hederaceus rivulorum se extendens atrâ maculâ notatus *C.B.* aquaticus hederaceus luteus *C.B.*[617] hederaceus Dalechampii *Lugd. Ivy-leaved water-Crowfoot. In the entrance of a little lane leading out of Coton road up to the bowling-green beyond the Castle of Cambridge.*

486 *Ranunculus auricomus* L.
Ranunculus auricomus *Ger. Lob. ico. Thal.* nemorosus dulcis *Park.* sylvestris dulcis Tragi *hort. Hafn.* rotundifolius vernus sylvaticus *J.B.* nemorosus vel sylvaticus folio rotundo *C.B.* Martius *Trag.* Polyanthemon Plinii *Ang. epist.* II. *Sweet wood-Crowfoot, Goldilocks. In wooded places & thickets abundantly.*

Ranunculus nemorosus albus, *see* Anemone nemorum. [*Anemone nemorosa* 44]

---

[615] Pliny (*Natural history*, XXI. 74 and XXVII. 77 in the Loeb edition) took these two names from the Greek αἰγολεθρος (meaning "goat's-bane" and masculine rather than his neuter) and ἐννεαφυλλον (meaning "nine-leaved"). Neither is found in Theophrastus and the attribution of either of them to *Ranunculus flammula* seems implausible: the former is usually thought to be *Rhododendron ponticum* L.!

[616] Ray (1670, 1677) retained separate entries for this and the preceding taxon whilst noting that they were certainly not different species. *Ranunculus flammula* is an extremely variable species in habit and leaf shape, but many of the characters (including leaf serration) are indeed the result of phenotypic plasticity (Padmore 1957).

[617] See Ray's comment about the colour of the flowers in *Emendanda* of the 1663 and 1685 appendices.

487 *Ranunculus repens* L.
Ranunculus pratensis repens *Park.* pratensis repens hirsutus *C.B.* pratensis etiámq; hortensis *Ger.* pratensis reptante cauliculo *Lob.* repens flore luteo simplici *J.B.* Ranun[culus] 3. vulgari similis per omnia, sed dulcis *Tur.* Ranunculi genus humi serpens, non mordens *Cæs.*[618] *Common creeping Crowfoot, or Butter-cups.*

488 *Rorippa amphibia* (L.) Besser
Rhaphanus aquaticus *Ger. Park. Tab.* aquaticus Tabernamontani *J.B.* VI, sive aquaticus foliis in profundas lacinias divisis *C.B.* sylvestris officinarum aquaticus *Lob. ico. Cut-leaved water Radish.* Acrid plants produce & nourish garden lice very much, because they abound in volatile salt.

489 *Armoracia rusticana* P. Gaertn., B. Mey. & Scherb.
Rhaphanus rusticanus *Ger. Lob. Offic[inarum]. C.B.* sylvestris, sive Armoracia multis *J.B.* Thlaspi alterum Dioscoridis *book* 2, *chapter* 186. Rhaphanus sylvestris Dioscoridis *book* 2, *chapter* 138. Rhaphanus sylvestris seu Armoracia Plinii[619] quibusdam *C.B. Horse-Radish. It was found in Magdalen Colledge close, but since hath been most what dug up and carried away.*

490 *Sinapis arvensis* L.
Rapistrum arvorum *Ger. Lob. Park.* Rap[istrum] I, sive flore luteo *C.B.* flore luteo *J.B.* Sinapi agreste 4 *Trag.* Lampsana flore melino *Tab.* Olus sylvestre Plinii *book 19, chapter 8,* by which Julius Caesar's soldiers stayed alive at Dyrrachium.[620] *Ad. Charlock, wild Mustard.* Among standing corn. Sometimes it has rough pods and sometimes smooth ones. This Charlock usually grows taller than the two following ones; it has small, black, round seeds and angled pods, long and not jointed, carried on short stalks.

491 *Raphanus raphanistrum* L. subsp. *raphanistrum*
Rapistrum luteum siliquâ glabrâ articulatâ. From a white, simple, woody *root*, hairy with rather few fibres, arises a stem a foot high & more, branched immediately from the root & rough with rather rigid spinules. The *leaves* are hairy on both sides, laciniate, especially the lower ones & near their source. The *flowers*, which are yellow but not so deeply so as those of the preceding species, in fact rather sulphur-coloured, and consisting of four petals, emerge from a reddish calyx, like those of the yellow Wallflower[621] (for in this kind the calyces do not split open,

---

[618] The two preceding synonyms may be translated as "The third Crowfoot/Buttercup, similar to the common [one] in all respects except [that it is] mild [i.e. not acrid], *in Turner*" and "A kind of Crowfoot/Buttercup creeping on the ground [and] not acrid, *in Cæsalpinus*". Turner (1562, p. 114 *verso*) wrote: "The thyrd kinde is lyke the common sort in all poyntes / sauing that it is so swete / that it is eaten as a sallet herbe about Mentz in Germany … ."

[619] i.e. in Pliny's *Natural history*, XIX. 82 and XX. 22 in the Loeb edition, as probably correctly identified "by some authors".

[620] The name means "The wild vegetable of Pliny", the reference being to his *Natural history*, XIX. 144 in the Loeb edition, where he says that the plant was made famous particularly by the songs and jests of the soldiers at Julius Caesar's triumph, when they taunted him for the parsimony of the rewards that they received, reproaching him in alternating verses (*alternis versibus*) that they had had to live on *lapsana* at Dyrrachium. In XX. 96 Pliny says that *lapsana* is a kind of wild cabbage, a foot high, with hairy leaves, similar to mustard (*sinapi*) but with a whiter flower. The Greek city of Epidamnos was renamed Dyrrachium by the Romans; it is now Durrës, the second city of Albania, Durazzo in Italian. During the civil war of 49–44 BC Caesar and Pompey fought a battle there in 48 BC, in which Caesar was lucky to survive.

[621] *Leucoium luteum* [330].

MAIN CATALOGUE

as in the remainder, when the flower first unfolds). The *pods* follow, carried on rather long rough stalks, and are glabrous, jointed, large and oblong, containing up to six or eight large seeds, but single ones, not in pairs as in the common Charlock. *Yellow Charlock with a smooth joynted cod*. This species does not dif fer from white jointed Charlock except in the colour of the flower.[622]

492 *Sinapis alba* L.

Rapistrum luteum siliquâ hirsutâ articulatâ.[623] The *root* of this species is simple, white & woody and the stem a foot high, branched and hairy. The *leaves* likewise are hairy and more deeply laciniate, especially the lower ones. The *flowers* are yellow, like those of Charlock[624] except that they are carried on longer stalks; *pods* follow these, standing out from the stem, hairy, jointed, ending in a broad, slender & oblong point and containing three or four pairs of seeds. The *seeds* are larger than those of Rapistrum arvorum, at first russet but dark at maturity and a little compressed. *Yellow Charlock, with a rough joynted cod*. Among standing corn. Is it the Rapistrum arvorum *of Ger.?* It has seemed to us worth the trouble to describe these species more accurately because Botanists have written so confusedly & obscurely about these plants that are so common & everywhere at hand that it is difficult to understand what they mean. Anyone who attends diligently to our descriptions & compares the plants themselves with them will easily recognise each from our notes & distinguish it from the other species.

Rapum genistæ, *see* Orobanche. [*Orobanche elatior* and *O. rapum-genistae* 412]

493 *Brassica rapa* L.

Rapum vel Rapa *Ang. Tur. Lac. Gesn. hort.* Rapum majus *Ger.* majus orbiculatum sive turbinatum *Ad. Lob.* sativum rotundum *J.B.* Rapa I, sive sativa rotunda *C.B.* Rapa mas of Theophrastus & Pliny; Rapum of Dioscorides *of the same author. Round Turnep.*[625]

494 *Brassica rapa* L.

Rapum radice oblonga *Ger. emac.* sativum oblongum *J.B.* tereti, rotundâ, oblongâ radice *Ad. Lob.* Rapa II, sive sativa oblonga, seu fœmina *C.B.* Rapa fœmina of Theophrastus & Pliny *of the same author. Long Turnep.*[625]

495 *Brassica napus* L.

Rapum sylvestre *Ger. Matth.* sylvestre non bulbosum *Park. Ad. Lob.* sylvestre Matthioli *J.B.* Rapa sylvestris & Cæsalpini *C.B. Wild Rape or Turnep. Sown in the Isle of Ely.*[626]

N. From the seed of this turnip when crushed & pressed they extract the oil generally called *Rape-oile*.

2. In some plants old seed (says *Varro in book 1, chapter 40*) even changes its nature, for they say that turnips grow from old seed of cabbage & contrariwise

---

[622] The flowers of this subspecies are polymorphic and may be yellow, white (often with darker veins) or pink. Ray here calls the white-flowered form *Rapistrum album articulatum*.

[623] See also Ray's synonyms in *Emendanda* of the 1685 appendix.

[624] *Rapistrum arvorum*.

[625] In his 1674 paper *Of the specifick differences of plants*, Ray used "diversity of figure, observed in Turneps, which are sometimes long, though commonly round–Root" as an example of "Accidents of the Root mistaken for Notes of Specifick difference" (Cain 1996, 1999a).

[626] See another entry for this crop under *Napus sylvestris* J.B. C.B. [385] and also Ray's comments on *Rapum sylvestre* Ger. in *Emendanda* of the 1685 appendix.

275

cabbage from that of turnips, to which Columella also assents, asserting that three-year-old seed of turnips produces cabbage & contrariwise that of cabbage turnips.[627] In fact we have learned that a fairly close relationship exists between these two plants from the viviparous caterpillars that arise on cabbages, which feed off the leaves of turnip no less greedily than off those of their own cabbage, although they disliked most other leaves that we have given them as food. Let us continue the story of these caterpillars, since we have so far read, seen or heard nothing like it,[628] even if at greater length than the plan of our undertaking demands; at least we shall be worthy of the reader's pardon.

The caterpillar feeding on cabbages,[629] if you look at its size, is average among caterpillars,[630] clothed with scattered, short, whitish hairs which are nowhere clumped together. The colour of its body is composed of black, yellow & blue arranged in a variegated pattern, in which the yellow delineates three stripes down the length of the body, namely a middle one on the back and the others on the flanks on opposite sides to each other; the blue & black colours lie between all of these with the black depicted in spots and the blue diffusely. What is more, these black spots protrude over the rest of the body and sprout hairs from each of their centres. The colour of the head also consists of the three components already mentioned, mixed in various combinations,[631] & the head itself is covered with similar hairs. The feet are sixteen in number, arranged in three types: six make up the first type, eight the middle one (as happens in nearly all caterpillars) and two the last one. This is the outward appearance of the viviparous caterpillars.[632] We shut up ten or so of these in a wooden case around the end of August in the year 1658, where, when sated with the leaves of turnip & cabbage over several days, they fixed themselves to the sides of the case but especially to the lid; there seven of them revealed that they were viviparous or vermiparous; for out of their back & sides there burst out very many little creatures of the class of maggots, namely at least thirty from each caterpillar and from some more than sixty, white, hairless, devoid of legs &, when we examined them with a lens,[633] translucent. Immediately after their birth they were opening a weaver's shop and within the space of two hours were completing the weaving of caskets or silken dwellings for themselves, from which at the beginning of the following October they emerged in the form of flies with the head & whole body black on both the upper side and the underside, the legs reddish[634] and the antennae very long in proportion to the animals. The wings were longer than the body, entirely silvery except

---

[627] Marcus Terentius Varro (who died in 27 BC), author of *De re rustica*, was a Roman writer on husbandry and Lucius Junius Moderatus Columella (who flourished around AD 50) was another: see Chapter 4.

[628] *cùm hactenus à nobis nihil simile lectum, visum aut auditum sit*, literally "since so far by us nothing similar has been read, seen or heard".

[629] i.e. that of the Large White butterfly, *Pieris brassicae* (Linnaeus).

[630] *medium inter erucas locum obtinet*, literally "occupies a middle place among caterpillars".

[631] *Capitis quoque color ex tribus prædictis inter se variè permistis constat*, literally "The colour also of the head consists from the three said beforehand among themselves variously mixed".

[632] *Hæc erucarnm* [a misprint for *erucarum*] ζωοτοκων [zoötocon: genitive plural] *externa facies*, literally "This of viviparous caterpillars the external form". The Greek adjective ζωοτοκος, like the Latin *viviparus*, means "giving birth alive".

[633] *cùm microscopio*: see footnote 597.

[634] *cruribus rufescentibns* (a misprint for *rufescentibus*).

that each was marked with a single black spot.[635] The whole size of the body did not exceed the magnitude of smaller ants. The remainder of this kind of caterpillar (for there were three or four that did not produce maggots), after a longer period of time,[636] were changed into angular & humped chrysalises, at first green tinged with yellow and later conspicuous for their bluish-grey colour, which, with winter finished, were changing in the middle of April into white butterflies, endowed with four wings and marked on each of the front wings with two black spots. The viviparous caterpillars described above experienced no metamorphosis and died some days after giving birth. This also is worthy of notice that, if you have observed the said caterpillars when autumn is drawing near, you will find more viviparous ones, but, if when autumn is now almost finished, more changing into chrysalises.

Since we observed these things, that most talented Man & assiduous investigator of nature, Master *Francis Willughby*,[637] has observed an uncertain transformation in various species of caterpillars, e.g. in Ambulo prima or the more densely hairy caterpillar of Moufet, Sepiaria vulgaris, the first of three yellowish species of Mouf[et],[638] etc., in which (& probably in most others) the caterpillars are not always transfomed into chrysalises but sometimes into oblong eggs, in shape like those of ants[639] but in colour for the most part dark, sometimes one, sometimes more, sometimes larger, sometimes smaller, that is in relation to the size of the caterpillars. One can read in Moufet's *Insect[orum] theat[rum], book 1, chapter 10*, something similar observed by Master Knivet;[640] but the things that are

---

[635] *nisi quòd singulæ singulis maculis nigris notarentur*, literally "except that single [ones] were marked with black spots one each"; Latin uses plurals where singulars are natural in English. The "flies" (parasitoid ichneumon wasps) described here were correctly identified by Raven (1950, p. 103, footnote 4) as *Apanteles glomeratus* (Linnaeus), now *Cotesia glomerata*. See also Ray's long second note under *Dipsacus* [168].

[636] *interjecto diuturnioris temporis spatio*, literally "with the space of longer-lasting time interposed".

[637] See Chapter 3.

[638] For Moufet's (1634) book see footnote 76. *Ambulo prima* is the first in the first group of Moufet's caterpillars described and illustrated on pp. 185–191 in book 2, Chapter 3 (entitled *De Erucis hirsutis atque pilosis*, i.e. "About shaggy and hairy Caterpillars"), where two (rather than three) species of *Sepiaria* are also included, the first called *major*. Moufet says of the first group, clearly relating their name to the verb *ambulo* ("I walk"), *Ambulones dicimus, quibus incerta domus & esca* ... ("We call [them] *Ambulones*, for which home & food [are] uncertain ...") and explains that the English call them "*Palmerwormes ... ab erratica nimirum vita*" ("undoubtedly from [their] wandering life") but also "*Bearewormes*" because of their shagginess. Moufet describes his *Sepiaria major* as having a yellowish face except for a paler (the verb being *lilliasceret*, perhaps suggesting somewhat lily-like in colour) triangle in place of its nose; he says that its body has variable white, yellow, red and black spots (the last illustrated in his figure) and that it bristles with yellowish hairs; it lives in hedges (*sepes*, whence no doubt its name), which it denudes of their leaves and where, leaving its ball of rude silk, it pupates in a chestnut-brown case as if in a sarcophagus.

[639] *figurâ qualia sunt formicarum*, literally "in shape such as are of ants".

[640] *Simile quid à D. Knivetto observatum legitur apud Moufetum* ..., literally "Something similar by Master Knivet observed is read in the work of Moufet ...". Moufet (1634, p. 57) calls him *Thomas Kniuettus Anglus, equestris ordinis & singularis doctrimæ* [sic] *vir* ("an Englishman of the body of knights & a man of extraordinary learning"). He then says (in our translation): "The corrupted or somewhat bruised body of a caterpillar is converted into an imperfect chrysalis, then from it there are ejected not a butterfly but three oblong blackish eggs, from which arise common flies or others similar to them. But sometimes, when the chrysalis itself is corrupted, neither a butterfly nor eggs but white grubs come forth through an aperture (sometimes one, not infrequently more), from which afterwards

related there are partly false and partly confused & obscure. We have not yet discovered whether these eggs that we describe as coming from caterpillars may be laid by that species which we have seen or whether the Cabbage White,[641] & so all the rest, may first have brought forth maggots that afterwards were transformed into eggs, intending to investigate the matter as soon as may be possible next summer, God willing; for the time being, looking to the future,[642] it seems to us more probable that, if indeed these eggs become[643] flies in the same way as chrysalises become butterflies, so also the maggots become eggs just as caterpillars become chrysalises. And yet the contrary is not entirely impossible,[644] that is that caterpillars are sometimes changed directly into eggs in the same way as they frequently are into chrysalises.[645] Besides, it should be examined whether the caterpillars that are transformed into eggs in this way may be weaker & thinner than the others; in them, since nature was unable to achieve her original intention & produce a butterfly, she did at least form what she could, a less perfect little creature, namely a fly, so that it would not disappear completely without its own existence.

Moreover, the same very learned Man observed that these eggs (it may be legitimate for us to name them thus for the present while deprived of a more appropriate appellation[646]) are of two kinds, some shorter, dark and so to speak composed of rings and the others oblong, almost transparent and not divided into rings but covered by a kind of skin. The former always produced flesh-flies and the latter flies from Moufet's genus of three-bristled flies.[647] Pondering these things in earnest and comparing the eggs with chrysalises (because so far we have not been able to experiment,[648] since these things have been communicated to us for the first time with autumn now almost finished), we are of this opinion about them. As some caterpillars wrap themselves in a web before they are transformed into chrysalises but others remain entirely naked during this metamorphosis, so too some of the maggots are transformed naked into eggs but others wrap themselves first in a web as is clearly well known in the Cabbage Whites. Of the former kind are those eggs that appear dark and are divided into rings in almost the same way as the tails of the chrysalises that we have said are transformed into flesh-flies. These, as we surmise, are evidently the naked chrysalises (if it may be legitimate

---

[arise] very mean flies." He concludes by saying that his readers should accept these very fine observations of natural history, coming as they do from the forenamed gilded knight. For an account of Sir Thomas Knyvet (or Knyvett) and his uncle Edmund (who is also acknowledged by Moufet, though described by him as Thomas's brother), see Raven (1947, pp. 173–175) and McKitterick (1978). *D.* is again the standard abbreviation for *Dominus*, used for a Master of Arts.

[641] *Brassicaria*, i.e. Large White, *Pieris brassicae* (Linnaeus). Moufet (1634) describes and illustrates the caterpillar and chrysalis in book 2, chapter 3, p. 189. The two common white butterflies are the first and second of Moufet's *Diurnæ Papiliones mediæ* ("Medium-sized Diurnal Butterflies") on p. 103 of book 1, Chapter 14 (which covers all day-flying butterfies and moths).

[642] *interim posterius*, literally "meanwhile later/afterwards".

[643] *se habent ova hæc ad ...*, literally "these eggs have themselves to ...".

[644] *dissimile veri*, literally "unlike the truth".

[645] *eodem modo ... quo frequenter in aurelias solent*, literally "in the same way in which frequently into chrysalises they are accustomed".

[646] *ità in præsentiarum liceat appellare commodiore vocabulo destitutis*, the exact meaning of which is obscure but must be something like the translation above.

[647] *Muscæ carnariæ* (plural) are probably the same as Moufet's (1634, p. 58) *Musca carnivora*. For Moufet's *tripiles muscæ* see Ray's second note on *Dipsacus* [168] and footnote 250.

[648] *experimenta nos deficiunt*, literally "experiments fail us".

to call them so) of flies. Of the latter kind are those that appear oblong and transparent and produce three-bristled flies; for these are not (as we think) themselves the chrysalises of the flies but cocoons[649] in which are concealed chrysalises of the same kind as the former and divided into rings, in the same way as the chrysalis is concealed in a cocoon in silkworms. We dare not pronounce anything rashly about these things which we have not yet investigated sufficiently.[650]

496 *Jasione montana* L.
Rapunculus Scabiosæ capitulo cæruleo *C.B. Park.* Rapuntium alterum leptophyllon capitatum *Column.* Scabiosa ovilla *Cam.* media *Lob. Lugd.* 12, sive minima hirsuta *Ger. emac.* globularis, quam ovinam vocant *J.B.* Rampions with Scabious-like heads, or Hairy sheep Scabious. On the sandy grounds about Gamlingay, & elsewhere.

Rapunculus sylvestris minor *Tab. see* Campanula rotundifolia. [*Campanula rotundifolia* 103]

Regina prati, *see* Ulmaria. [*Filipendula ulmaria* 623]

497 *Reseda lutea* L.
Reseda vulgaris *C.B.* minor seu vulgaris *Park. hort. Paris.* Res[eda] Plinii *Ger. Eyst.* lutea *J.B. Dalec.* Eruca peregrina Italica vel Cantabrica *Ad. Lob.* Sesamoides 1 *Cæs.* Erucago vulgaris *Col.* Res[eda] minor flore subvirente *hort. Bat.* Italian Rocket, Common base wild Rocket. On the dry bankes of some closes beyond Peterhouse not far from the river; on the bankes about Chesterton, and in many places among the corn.

Resta bovis, *see* Anonis. [*Ononis spinosa* 47]

498 *Rhamnus cathartica* L.
Rhamnus catharticus *J.B. C.B.* catharticus sive Spina infectoria *Matth. Lob.* solutivus *Dod. ut. Ger. emac.* Spina cervina vulgò *Gesn. hort.* It flowers in May and the berries ripen in Autumn. Buckthorne or common purging Thorne. In a little thicket beyond Coton, in Madingley wood, Teversham closes, & elsewhere.

---

[649] *folliculi*, plural of *folliculus*, a small sack and thus a pod, shell, cocoon or, in later Latin medical usage, a stomach, bladder or scrotum. See Ray's 'Interpretation of terms'.

[650] *De his nobis nondum satìs perspectis nihil temerè pronunciare audemus*. Literally "About these [things], for us not yet enough investigated, nothing rashly to pronounce we dare." As Raven (1950, p. 104) comments, Ray was "already on the track of the true solution" of the mystery of parasitoid insects, which he had reached by the time he wrote in his *Historia insectorum*, published posthumously in 1710 (p. 114): *Ego vespas ichneumonas uteri sui tubo fistuloso Erucas hasce compungere existimo, earúmque corporibus ova sua immittere* ("I think that the ichneumon wasps prick these Caterpillars with the hollow tube of their ovipositor [literally "womb"] and insert their eggs into their bodies") and *erosa cute foras exeunt, & thecas sibi texunt* ("when they have gnawed away the skin [literally "with the skin gnawed away"], they come out & construct cases for themselves"). Ray was one of several European scientists who independently discovered the parasitoid life-cycle at this time (van Lenteren & Godfray 2005). The first was probably J.J. Swammerdam, with help from the artist O. Marsilius, who described the parasitism of *Pieris* butterflies in a book which was finished in 1679 but not published until 1737–1738. A detailed, illustrated account of branconid wasp parasitoids of aphids was published (in English) by Anthony van Leeuwenhoek in 1702 in *Philosophical Transactions of the Royal Society*. However, the parasitoid life-cycle had long been known to Chinese sericulturists: the earliest known written record of the parasitism of silkworms by tachinid flies is that of Lu Dian in 1096 (Cai, Yan & Li 2005).

N. The berries of this shrub display three kinds of colours, certainly yellow at first, if they are soaked after being picked at the time of harvest, dried and pounded in water & alum; secondly a beautiful green, which in English they call *Sapgreen*, very suitable for painters & dyers, if they are gathered in autumn, when, having reached maturity, they have become black &, after being pounded, are kept in a brazen vessel; thirdly a purplish-red, if they are still remaining on the trees around the feast of St Martin.[651]

Rhus sylvestris, *see* Elæagnus *Cordi*. [*Myrica gale* 173]

499 *Ribes nigrum* L.
Ribes nigrum vulgò dictum folio olente *J.B.* fructu nigro *Park.* 3 *Ger.* sylvestre *Trag. Thal.* sylvestre aquaticum, Sabinæ ferè odore *Gesn. coll.* Ribesium fructu nigro *Dod.* Piper hortense olitoribus Galliæ *Gesn. hort.* Grossularia XI, sive non spinosa fructu nigro *C.B. Black Currans, Squinancy-berries. By the rivers side at Abington.*

500 *Drosera intermedia* Hayne
Rorella sive Ros solis foliis oblongis *J.B. Park.* Ros solis folio oblongo *C.B. Sundew or Rosa solis with long leaves. On Hinton moor, about the watery places plentifully.*

N. In England it has been observed by shepherds that sheep that frequently feed on this plant have had their liver & lungs badly affected & damaged by some putrefaction.[652] *Cam. hort.* Thence in English it is called *Red rot*.

501 *Drosera rotundifolia* L.
Ros solis folio rotundo *Ger. C.B. Park. J.B.* Rorella sive Ros solis *Eyst. Sundew, or Rosa solis with round leaves. Near Gamlingay in a moorish place where they digge turfes, not far from the high-way leading to Bedford, see* Ascyron supinum villosum,[653] &c.

502 *Rosa canina* L.
Rosa sylvestris *Matth. Dod.* syl[vestris] vulgaris canina *Park.* canina spongifera flore purpureo *hort. Bat.* sylvestris vulgaris flore odorato incarnato *C.B.* sylv[estris] alba cum rubore, folio glabro *J.B.* canina inodora *Ger.* Cynosbatos Dioscoridis[654] *Ad.* Rosa canina odorata & sylvestris *Lob.*, who, with C.B., is considering the flower, but in fact Gerarde, when he says it is scentless, is considering the leaves. *The Briar-bush, or Dogges-rose.* In hedges & thickets.

N. Sometimes there grows on the twigs of this shrub a certain softly hairy little ball, named Spongiola by Pliny, which is wrongly called Bedeguar by the Apothecaries' shops, for Bedeguar is a kind of Thistle *J.B. C.B.*[655]

---

[651] 11 November, Martinmas, celebrated as a feast before the pre-Christmas fast of Advent.

[652] Sheep rot is a disease caused by the liver fluke, *Fasciola hepatica* Linnaeus, which in one stage of its life is a parasite in the bodies of small water-snails and so is more prevalent in places where sundews and other plants formerly believed to cause the disease occur.

[653] *Hypericum elodes* [66], also reported from "a boggy ground beyond Gamlingay, near Sir Roger Burgoynes park, where they dig turfs".

[654] See footnote 507.

[655] *Spongiola* literally means "a small sponge". The reference is to Pliny's Natural history, XXV. 18 in the Loeb edition, where he says that the ashes of spongiola mixed with honey are a cure for alopecia. C. Bauhin (1623, p. 483) is the source for Ray's assertion that the name *Bedeguar* was wrongly applied to *Spongiola* and J. Bauhin & Cherler (1651, vol. 3(1), p. 55) show that *Bedeguar* was the

If you cut into this Spongiola, you will find it filled inside with certain small white grubs, as witnessed by *Verulam in Hist[oria] nat[uralis], cent[ury] 6, exp[eriment] 562, Spigel[ius] in Isag[oges], book 1, chapter 10, and Moufet in Insect[orum] theat[rum], book 2, chapter 20*,[656] and finally from actual experience. *Spigel[ius] & Moufet, in the places presently commended, & Aristot[le] in Hist[oria animalium], book 5, chapter 19,* say that beetles are produced from these grubs, whether rightly the Reader will discover from what follows.[657] Those which have lain hidden through the winter in the sponge (for we have kept a number of them) emerged in the month of May of the following year in the form of flies; the shape and proportions of their body corresponded with those that a flying ant has but their size was a little smaller. In these the head and shoulders were black and connected to the abdomen by a thin black thread; the eyes also (which were large in relation to such a small creature) were tinged with the same colour. The abdomen was reddish on its upper part and on the lower side, except where it was marked with spots, was green tinged with yellow,[658] apart from the blackish tip of the tail itself. The spots just mentioned formed two lines under the belly each consisting of five black points. They have six reddish feet and four silvery wings without wing-cases,[659] the lower ones small, the upper ones larger, each marked with a single black spot.[660] Moreover some of these flies are armed with a sting or spike always projecting from their tail and can be referred to Moufet's Seticaudæ, while others lack this absolutely; and in this there is perhaps a difference between the sexes.[661]

503 *Sherardia arvensis* L.
Rubeola arvensis *Cam.* arvensis repens cærulea *C.B.* Rubea minor pratensis cærulea *Park.* parva flore cæruleo se spargens *J.B. Little field-Madder. In the corn next beyond Gogmagog hills, on the left hand of the way leading to Linton, and in many other places in the corn.*

504 *Rubus caesius* L.
Rubus minor fructu cæruleo *J.B.* repens fructu cæsio *C.B. Ger. emac.* humilis et 1 *Trag.* minor, Chamærubus et Humirubus *Park.* Chamæbatos Theophrasti *C.B. Small Bramble, Dewberries. Among standing corn.*[662]

Arabic name for "a kind of thistle" (*Cardui genus*), specifically *Onopordum acanthium* [4]; see also Hofmann's name for this species, *Bedeguar Arabum*, listed by Ray as a synonym for this species.

[656] For the three authors cited see footnotes 104, 31 and 76 respectively.

[657] *Ex his vermiculis cantharides nasci tradunt … an rectè ex sequentibus Lector intelliget.*

[658] *ex luteo virebat*, literally "out of yellow was green".

[659] *… alas quatuor argenteas, anelytras*: so these insects were not beetles, which have elytra (modified, hardened forewings acting as cases for the hindwings).

[660] *… singulas singulis maculis nigris insignes*, literally "… single ones [i.e. each one] distinguished with single [i.e. one each] black spots".

[661] *Seticaudæ* are literally "Bristle-tails". For Moufet's (1634) book see footnote 76. The bedeguar gall or Robin's pincushion is caused by the cynipid gall wasp *Diplolepis rosae* (Linnaeus), but the galls are often invaded by other species: see Redfern & Askew (1998, pp. 21–22 and plate 3, figure 6). The description of the "flies" which emerged from the galls is a good match for the hymenopteran wasp *Orthopelma mediator* (Thunberg), a common parasitoid of *D. rosae*: see Redfern & Askew (1998, pp. 34 and 85 and plate 4, figure 7). The "sting or spike" which Ray describes is the ovipositor of the female, so his suggestion about "a difference between the sexes" is correct. Excellent photographs (from hedgerowmobile) can be found by searching for the insects' scientific names on the internet.

[662] *Inter segetes*: see footnote 17. Though this seems an unlikely habitat for *Rubus caesius*, it appears not to be an error for *Inter sepes* ("Among hedges"), as has been suggested, since in his *Catalogus*

505 *Rubus fruticosus* agg.[663]
Rubus *Ger.* I vulgaris, sive Rubus fructu nigro *C.B. Batos* for the Greeks, *Sentis* for the Latins, *Batina* for Galen, *Rubus* for Dioscorides and Pliny and Tall *Rubus* for Theophrastus[664] *for the same author.* Rubus major vulgaris *Park.* major fructu nigro *J.B.* Morus, sive Rubus[665] *Ang. Bramble-bush, or Black-berry-bush.* In hedges & thickets.

Rumex, *see* Lapathum. [*Rumex* spp. 321-323]

Ruta muraria, *see* Adianthum album. [*Asplenium ruta-muraria* 11]

Ruta pratensis, *see* Thalictrum. [*Thalictrum flavum* 567]

S

506 *Sagittaria sagittifolia* L.
Sagitta major *J.B. Ger. Matth. Dod. Tab.* aquatica major *C.B.* major latifolia *Park.* Pistana Magonis, sive Sagittaria aquatica Plinii *Ad. Lob.* Phleos mas latifolia *Lugd.* Barba sylvana recentioribus *Cæs. Great Arrow-head.* In waters.

507 *Sagittaria sagittifolia* L.
Sagitta minor *Lob. Lugd. Matth. Dod. Tab.* Sagittaria minor angustifolia *Park. J.B.* Sagitta aquatica minor angustifolia *C.B.* Pistana Plinii *Ang.* Sagitta *J.B.* 3 sive Sagittaria minor angustifolia *Ger.*[666] *Small Arrow-head.* With the preceding. They flower in the months of May and June.

Saginæ Spergula, *see* Spergula. [*Spergula arvensis* 557]

Salix
¶ The confusion and obscurity of botanists in describing and distinguishing willows are astonishing; for, while some divide them into arboreal and dwarf, others into broad-leaved and narrow-leaved,[667] with no account taken of the texture & qualities of the leaves, they both separate related species and join unrelated ones. Moreover most of their descriptions are so short & obscure that it is scarcely possible to conclude anything certain from them. We therefore, partly for the reason mentioned and partly because no one as far as we know has mentioned several

---

plantarum Angliae (1670, p. 269) Ray gives the habitat as *Inter segetes, & in sylvis cæduis* ("Among standing corn & in coppices [literally "woods fit for cutting"]) and in his *Synopsis* (1690, p. 228; 1696, p. 309) as *Inter segetes, & in arvis requietis* ("Among standing corn & in fallow fields"). See also footnote 384.

[663] Babington (1860) identifies Ray's *Rubus* as the species now known as *R. ulmifolius* Schott. This sexual diploid species is much the commonest of the microspecies found in the county, but numerous others are recorded and it is more realistic to equate Ray's plant with the aggregate.

[664] Βατος *Graecis, Sentis Latinis,* Βατινα *Galeno, Rubus Dioscoridi et Plinio, Rubus excelsa Theophrasto*. The "Latins" are the Romans. Βατινα [*Batina*] seems to be the neuter plural of an adjective βατινος, -η, -ον, meaning "of a bramble" or "of brambles".

[665] i.e. "Mulberry-tree or Bramble"; the confusion persists to this day in French, in which *mûres* can be either mulberries or blackberries.

[666] As indicated here, Gerarde (1597, pp. 336–337) made three species out of the one recognised today. Ray (1670, 1677) retained these as separate taxa without any comment on their distinctness. However, he later (1686, 1690, 1696) concluded that they did not seem to him to be specifically distinct because (in our translation) "they very often arise indiscriminately in the same waters, whence they seem also to owe their origin to the same seed" (1686, p. 619).

[667] i.e. *arboreæ* and *pumilæ* or *latifoliæ* and *angustifoliæ*.

willows that are frequent with us, have tried to divide them according to our own system more concordantly with nature and to describe more accurately those of our own species which have either not been reported at all or only obscurely, so that anyone who has attended diligently enough to the descriptions may not be able to fail[668] to recognise & distinguish them.

We have ten species of Willow, which we reduce to two chapters or kinds with respect to the texture of their leaves. Let the first kind be Willow with a more compact leaf,[669] called in the vernacular *Willow*, and the second Willow with a laxer & thicker leaf,[670] in the vernacular *Sallow and Osier*. We have five species of each.

Willow with a more compact leaf.

508 *Salix alba* L.[671]

1. Salix folio utrinque glauco, viminibus rubris.[672] An Salix vulgaris rubens *C.B?* An Salix angustifolia purpurea, seu nigra *Park?* This grows into a tall tree. Its catkin emerges, after the appearance of the leaves, from a long stalk clothed in numerous leaves, as a fairly long, graceful, compact spike which is composed of the pericarps arranged like scales & the yellow bracts adjoining the stalks of the pericarps. The pericarps themselves are smaller than in the remaining pappose species, supported on a very short stalk, adorned at their tips while they are still fairly young with two little stigmas[673] that are yellow & rather short, from which, as they soon dry up, the pericarps enlarge until, finally dehiscing with mature seed (if there is any), they reveal a pappus. The appearance of this whole catkin is not badly represented in the picture of the first Willow in the work of *J. B.* These catkins[674] do not usually drop before the summer Solstice. *The common red Willow.*

509 *Salix alba* L.

2. Salix folio utrinque glauco, viminibus albidioribus.[675] Salix *Ger.* maxima fragilis alba hirsuta *J. B.* alba of Theophrastus, Amerina & candida of Pliny *of the*

---

[668] The Latin has a comparative adverb *diligentiùs* (literally "more diligently") and a double negative (*non possit non*), here translated by "not" and "fail".

[669] *Salix folio compactiore*. The meaning of *compactiore*, though clearly the antonym of *laxiore*, which follows, is not obvious.

[670] *Salix folio laxiore & crassiore*.

[671] In *Catalogus plantarum Angliae* (1670, p. 271) and in *Emendanda* of the 1685 appendix Ray said that he did not regard this and the following plant, *Salix folio utrinque glauco, viminibus albidioribus*, as distinct species and *S. alba* is indeed the only willow likely to have been present in the county in the 17th century with leaves which could be described as glaucous on both sides. Babington (1860) treated Ray's plant as *S. alba* var. *vitellina* (L.) Stokes. Perring *et al*. (1964) and Meikle (1984) describe this variety as having bright yellow or orange twigs, but Meikle and Stace (2010) mention a cultivar, 'Britzensis', in which the twigs are brilliant red or orange-red. Stace allows var. *vitellina* to have "yellowish to reddish" twigs and Ray's plant could perhaps be referred to this variety.

[672] "Willow with the leaf glaucous on both sides, with red withies."

[673] *duobus staminulis*, but what follows indicates that the female catkin is being described, so these are evidently not stamens.

[674] Ray has previously used *nucamentum*, a word derived from *nux* (a nut), for "catkin" but here and subsequently he uses *julus*, a word from the Greek ἰουλος [iulos] used by Pliny. It might be thought that the former related to the female and the latter to the male catkin, but here Ray has used both in the same context. In his 'Interpretation of terms' he translates both as "catkin" or "palm[e]" and, while associating the latter with stamens, says that the former is *idem quod Julus* ("the same as *Julus*").

[675] "Willow with the leaf glaucous on both sides, with rather whitish withies." See also *Emendanda* of the 1685 appendix.

same author. vulgaris alba arborescens *C. B.* vulgaris arborescens longis angustis foliis *P. B.* perticalis *Col.* and perhaps also *Cord.* arborea angustifolia alba vulgaris *Park. The common white Willow.* The catkins of this willow correspond in everything except size with the catkins of Salix fragilis soon to be described.

N. 1. Because life is shorter for aquatic plants than for terrestrial ones it is thus for Willow and the others. *Theoph. hist. plant., book 4, chapter 14.*[676]

2. Box, pine and oak grow slowly, willow & poplar quickly,[677] since the former & similar trees have a dense and firm texture and they do not permit nourishment to permeate easily and abound, but the latter and others, that kind, have a looser and laxer texture and they do not impede the diffusion and production of nourishment. *Volume 2, page 186 of the Works of P. Gassend[i].*[678]

510 *Salix fragilis* L.

3. Salix folio lato splendente fragilis.[679] An Salix fragilis *C.B?* Salix spontanea fragilis, Amygdalino folio, non auriculata *J.B?* It grows into a fairly large tree, with the bark of the twigs reddish tinged with yellow and with brittle branchlets (whence it is commonly called in Cambridgeshire[680] *The Crack-Willow*). It has large *leaves*, sometimes four or five inches long, an inch or an inch and a half wide, glabrous, shining, toothed at the edge, of a cheerful green above, rather glaucous beneath, sometimes with auricles at the base, but small ones, though more often without them.[681] The *catkins* consist of yellow mossy filaments or threads & yellow bracts surrounding their bases; of the two little stamens one stands close to the yellow bract. Together these form a loose, sparse, mossy spike, smelling pleasantly and staining the hands of those handling it with a yellow colour; and indeed the heads of the stamens are sprinkled with a sort of yellow pollen.[682]

511 *Salix triandra* L.

4. Salix folio splendente auriculato flexilis.[683] This, as far as we have seen so far, does not grow into a tree. The *bark* of the younger twigs is yellowish-green, sometimes inclining towards redness. It has *leaves* arranged alternately, two or three inches long, half an inch or an inch wide, neatly toothed on the margin, shining above with a cheerful green but somewhat deeper than that of the preceding willow. What is peculiar to this species is that throughout it possesses large roundish auricles on both sides at the base of the leaves, toothed in the same way as the leaves and ending in the angle towards the leaf. The *catkins* are mossy & in one respect only[684] they differ from the catkins of the preceding species, that by each

---

[676] *Historia plantarum*, IV. XIII. 2 in the Loeb edition, where Theophrastus says: "it appears that trees which grow by water are shorter-lived than those which live in dry places: this is true of willow [ἰτέα, itea], abele, elder and black poplar."

[677] *Buxus, pinus, robur lentè crescunt, properè salix & populus.*

[678] The cited text actually appears on p. 187 of Gassendi (1658).

[679] "Brittle Willow with a broad, shining leaf."

[680] *Cantabrigiensi*, dative or ablative of the adjective formed from *Cantabrigia* (Cambridge), probably standing for *agro Cantabrigiensi*; see footnote 15 in Ray's 'Preface'.

[681] *supernè hilari virore prædita, infernè magìs glauca, ad basin interdum auriculis sed minoribus donata, sæpiùs tamen iis destituta*, literally "above furnished with a cheerful greenness, beneath more glaucous, at the base sometimes endowed with ear-lobes, but small [ones], more often however deprived of those".

[682] This time Ray is clearly describing the male catkins.

[683] "Pliant Willow with a shining, eared leaf."

[684] *in eo duntaxat*, literally "in that [thing] only".

yellow bract they have three little stamens or threads & not just two. An Salix folio Amygdalino utrinque virente aurito *C.B.?* An Salix spontanea Amygdalino folio fragilis auriculata *J.B.?*[685] *The round-eared shining Willow.*

512 *Salix purpurea* L.

5. Salix humilior foliis angustis, subcæruleis, ut plurimum sibi invicem oppositis.[686] The outer *bark* of the branches is of a grey colour & as it were polished. The *leaves* are oblong, narrow and not eared, growing in twos at intervals on this side and on that and for the most part opposite, not scalloped or at least only slightly so, of almost the same width throughout and not gradually narrowed as in the rest. The *catkins* of this willow are the smallest of all, whether you may examine their length or their thickness, and composed of yellow stamens or filaments and woolly bracts. These last, rather densely compressed, make the body of the catkin fairly compact & woolly, which is supported on a short stalk furnished with leaves. The tips of the filaments are red at their first appearance but soon degenerate to yellow. An Salix tenuior folio minore utrinque glabro fragilis *J.B.?*[687] *The yellow dwarf Willow.* By the horse-way side to Cherry-Hinton, in the close just by the water which you passe over to go thither.

N. The inner bark of the branches, which is of a beautiful yellow, colours the water in which it is steeped for twelve hours with the same hue, so that the water looks like Spanish wine.

Willow with a laxer leaf.

513 *Salix cinerea* L.

1. Salix folio ex rotundo acuminato auriculata.[688] Salix folio ex rotunditate acuminato *C. B.* [Salix] Caprea latifolia *Tab. Ger.* latifolia oblongior *Park.* latifolia infernè hirsuta *J. B. Common Sallow.* This species is very common in hedges. Its *catkins* burst out of the very buds of the twigs before the appearance of the leaves, supported by a rather short, somewhat thick, woolly stalk on which there are either no bracts or very small ones. The *catkin* itself, which is rather fat, pot-bellied and short for its size, after the shedding of its mossy hair, is composed of pericarps and woolly bracts, one each for each stalk of the pericarps. In this kind the pericarp is carried on a longer stalk and its appearance to some extent resembles the seed-vessel of Scandix[689] except that the beak is far shorter, for at the thicker and more compressed bottom it tapers to a sharp point. When the fruit has ripened, bursting into two little keels, it reveals a pappus or down. While this catkin grows to maturity, the pericarps become much enlarged, but the woolly bracts gradually wither and are concealed. It emerges in early spring before all the others.

514 *Salix caprea* L.

2. Salix folio subrotundo auriculata.[690] This willow grows into a tree if it is permitted to do so. It has almost round *leaves* except that they end in a sharp point;

---

[685] "Is it Willow with an Almond-like leaf [that is] green on both sides [and] eared, of Caspar Bauhin? Is it uncultivated, brittle, eared Willow with a leaf like Almond, of Jean Bauhin?"

[686] "Lower-growing Willow with narrow, bluish leaves, for the most part facing each other alternately."

[687] "Is it the more slender brittle Willow with a smaller leaf [that is] glabrous on both sides, of Jean Bauhin?"

[688] "Eared Willow with a leaf [that is] acuminate from a round [shape]."

[689] This is Ray's synonym for his *Pecten Veneris* [*Scandix pecten-veneris* 425].

[690] "Eared Willow with a roundish leaf."

they are deep green above and glaucous or grey beneath, of a rather loose texture but nevertheless thick. The more slender *twigs* become somewhat reddish where they face the sun but on the part turned away from it are of a green colour inclining towards grey, as is also the bark of the branches. We have not yet observed the *catkins*. Salix latifolia rotunda *C. B.* caprea rotundifolia *Tab. Ger.* latifolia infernè hirsuta *J. B.*, who does not wish to make these two that are mentioned next different species; consult his work.[691] Salix latifolia rotunda *Park. The round-leaved Sallow.*

515 *Salix viminalis* L.
2. Salix folio longissimo.[692] Salix angustis & longissimis foliis crispis subtus albicantibus *J.B.*, in whose work see also a description. *The Osiar. At least so called in Essex and Cambridgeshire. Parkinson and Gerard seem not to understand what sort is so tearmed. That which they give us for the Osiar hath indented leaves.*

516 *Salix* sp.[693]
4. Salix folio longo, & folio & vimine subluteo, non auriculata.[694] And this grows into a tree even more rarely, with slender yellowish twigs and leaves an inch wide and three inches long, acuminate, pale green on the upper side and grey on the under side, supported by fairly long stalks and not eared. We have not yet observed its catkins. *The long leaved yellowish or green Sallow.*

517 *Salix* sp.[693]
5. Salix folio longo non auriculato vimine rubro.[695] This scarcely differs from the preceding willow in any other way than in the colour of the leaves & of the

---

[691] ... *qui has duas proximè dictas non vult specie differre, quem adi*, literally "... who these two next spoken does not wish by species to differ, whom go to [to consult]". See also footnote 359.

[692] "Willow with a very long leaf." The 2 which precedes it is a misprint for 3.

[693] Ray (1670, p. 274) united the two willows that we have numbered 516 and 517 by omitting the latter, a decision reported by Dent in *Emendanda* of the 1685 appendix; the taxon that they represent has not been satisfactorily identified. Babington (1860) listed Ray's names as synonyms of *S. purpurea*, citing Martyn (1763, p. 22) and perhaps following him rather than out of any firm conviction of his own. He also reported J.E. Smith's (1828, p. 182) view that they are *S. vitellina* L. (*S. alba* var. *vitellina* (L.) Stokes). Ewen & Prime (1975) followed Babington and also called both taxa *S. purpurea*. Variable though it undoubtedly is, *S. purpurea* cannot be the identity of a plant with grey undersides to the leaves and catkins like those of *S. cinerea*, especially as Ray described it tolerably well elsewhere in the catalogue [512]. *S. alba* var. *vitellina* usually grows as a tree and has narrower leaves than the plants described here by Ray; in addition Ray must have considered that *Salix folio longo non auriculato vimine rubro* [517] differed from the variant of *S. alba* with coloured twigs which he called *Salix folio utrinque glauco, viminibus rubris* [508] and which approximates to var. *vitellina*. It seems possible that entries 516–517 represent the common hybrid *Salix caprea* × *viminalis* (*S.* × *smithiana* Willd.) or the similar *S. cinerea* × *viminalis* (*S.* × *holosericea* Willd.). Meikle (1984) describes the former as a robust bush or small tree with yellowish or reddish twigs and leaves which measure 6–12 × 1.3–3 cm and are dull green above and grey-tomentellous below, are gradually acuminate and have stipules that are sometimes well-developed but are more often small and caducous. His illustrations of the catkins show a close resemblance to those of the *S. caprea* parent; Ray had not seen these catkins, but they are similar to those of *S. cinerea*. This hybrid was widely planted for basket-making and is known from the side of the Cam in Cambridge and elsewhere. *S.* × *holosericea* is not quite so good a match for Ray's plants as Meikle says it often has well-developed and persistent stipules, but it is also known by the Cam and can be difficult to separate from *S.* × *smithiana*. It would be rash to claim with confidence that Ray's plants were either of these hybrids, but they seem to be the most likely possibilities.

[694] "Earless Willow with a long leaf [and] yellowish leaf & withy."

[695] "Willow with a long earless leaf [and] red withy."

withy, which is more reddish in this species. *The long-leaved red Sallow.* The catkins are in all respects similar to those of Salix Caprea latifolia except that they end in a less sharp point and are less pot-bellied. *All these sorts of Willows are to be found either planted by water-courses, or in the Osiar-holts by the river Cams side.*

Salix rosea *Ger. Park.* We regard it as certain that this is not a species of willow distinct from the rest, for we have observed that clusters of leaves compacted in such a shape that they closely resemble roses sometimes grow at the tops of the twigs in several species of willows.[696] This arrangement of leaves frequently occurs with us on the common white & red willows;[697] we have also seen it elsewhere more than once on sallow[698] & on a kind of dwarf willow, so much so that it is certain that a trick & exuberance of nature, not some specific pattern,[699] produces roselike willows. Anyone who has sedulously searched for these imitations of roses on willows that are green & covered with foliage and has not found them should wait for a little while until the leaves fall, for when the rest of the leaves are falling they still remain at the tops of the twigs & present themselves to anyone to be seen as leaflets imitating roses, which, even though they are dried out & completely lose their natural greenness, nevertheless through the whole of even the most severe winter cling to the twigs tenaciously because they are firmly connected to them without a stalk. From these facts one is at liberty to conclude what should be decided about Salix angustifolia undecima sive humilis capitulo squamoso[700] *C.B.* & about the same capitula attributed in the works of *J.B., Park.* & *Lugd.* to definite species of willow as if they were peculiar to them.

Salvia vitæ, *see* Adianthum album. [*Asplenium ruta-muraria* 11]

518 *Viburnum opulus* L.

Sambucus aquatilis sive palustris *Ger.* aquatica *J.B.* aquatica flore simplici *hort. Paris. C.B.* palustris sive aquatica *Park.* Lycostaphylon fœmina *Cord.* Opulus *Ruell.* Aceris species 3, quæ Platanus aquatica[701] *Ang.* Chamæplatanus *Gesn. hort. Marsh or Water Elder.* It flowers around the beginning of June; the fruits or berries ripen in September. *By the ditch of one of the closes on the back-side of Clare-hall. In many places in moist woods and by water courses.*

N. The Elder caterpillars[702] described by Moufet *in book 2, chapter 2,* are transformed in the period of autumn into round, dark red chrysalises which are

---

[696] These clusters of leaves are Camellia galls, caused by the gall-midges *Rabdophaga strobilina* (Bremi) and the possibly conspecific *R. cinerearum* (Hardy). The former infects willows, including *Salix alba* and *S. purpurea*, and the latter several species of sallows (e.g. *S. cinerea* and *S. caprea*).
[697] i.e. *Salix alba* [508, 509].
[698] *non semel in caprea*, literally "not once in goat [willow]", i.e. *Salix cinerea* [513] and/or *S. caprea* [514].
[699] *non formam aliquam specificam*.
[700] Literally "The eleventh narrow-leaved Willow or low-growing [Willow] with a scaly capitulum".
[701] Literally "The third species of Maple, which [is] water Plane".
[702] *Erucæ Sambucinæ*. On p. 182 of his chapter entitled *De reliquis glabris Erucis* ("About the remaining hairless Caterpillars") Moufet (1634: see footnote 76) describes as *nobilissima* ("noblest") what is clearly the larva of the Privet Hawk-moth, *Sphinx ligustri* (Linnaeus): see Ray's note 2 on *Ligustrum* [333] and footnote 410. He then goes on to describe *Sambucina*, which he says feeds principally *Sambuco roseâ* ("on rosy Elder", *Viburnum opulus* cv. 'Roseum' (var. *roseum* L.), a garden plant in which all the flowers, rather than just the outer ones, are large and sterile, to which the name "Guelder Rose" was originally applied), saying that it does not differ much from it except that it is green all over apart from very white transverse spots and some little milk-white points. It seems that this is in fact the same species, which feeds on *Viburnum opulus* as well as *Ligustrum*.

not wrapped in any web, out of which in the following May emerge rather large moths, which are either male or female. The males, apart from their eyes which are black, are reddish all over[703] and between the shoulders & on the whole body densely hairy & shaggy. They have long, narrow wings with small rounded teeth and medium-sized antennae not extending upwards but turned along their sides, beautifully jointed and toothed. The wings of the females are distinguished by various colours, whitish, black, reddish and a shabby green. The shabby green colour extends over the hairy shoulders and the upper part of the abdomen; the lower belly is stained with a dirty white. In other respects they resemble the males[704] except that they are more swollen-bellied and they have antennae that are not so obviously toothed. They lay green eggs approaching a grain of mustard in size.

Sambucus humilis, *see* Ebulus. [*Sambucus ebulus* 170]

519 *Sambucus nigra* L.
Sambucus *Ger.* vulgaris *Park. J.B.* domestica *Cast.* vulg[aris] fructu nigro *hort. Paris.* I, sive fructu in umbella nigro *C.B. Common Elder.*

N. 1. A plum tree grafted into an Elder produces purging fruits, even though from my own experience[705] it does not readily take hold. *Cam. hort.*

2. This tree brings forth its own peculiar kind of fungus; they call it Judas' ear[706] because of its resemblance and they infer from its signature[707] that it is useful for the ears. And here in passing one may note the foolishness of Alchemists[708] also in this connection, who prate and proclaim so much about the signatures of plants. We indeed, weighing up the matter quite attentively, are moved, by reasoning that is not to be despised, to assert that the signatures impressed by nature on plants are not indications of their natural faculties and powers. 1. Because, out of the plants said to be specific, namely those that are said to be appropriate for some part or limb of the body or for some disease, the number of them that lack signatures is far the greater, as would be easy to demonstrate in conditions of the heart, thorax, head or liver. 2. Because different parts of the same plant, such as leaves, roots, flowers and seeds, exhibit not only different but sometimes even contrary signatures. 3. Because very many plants express the images of natural or artificial things, to which however they have no relationship, such as the flowers of certain orchids that express the images of flies, spiders, frogs, bees, butterflies, &c. (Among these (that I may note it in passing) there exists such a likeness that no

---

[703] *omnibus partibus rufescunt*, literally "in all parts they grow reddish".

[704] *In cæteris cum maribus convenient*, literally "In the rest with the males they agree".

[705] *ut expertus sum*, literally "as I have experienced/tested".

[706] *auriculam* [accusative] *Judæ*, i.e. the fungus *Auricularia auricula-judae* (Bull.) Wettst. Despite the meaning of the Latin name (Judas' ear), this fungus has been known in English as "Jew's Ear" or variants thereof since at least the time of Gerarde (1597, p. 1233), who wrote in his Chapter "Of the Elder tree" of "a certaine excrescence called *Auricula Iudæ*, or Iewes eare". Perhaps there was once some association between this fungus and the tradition that Judas Iscariot hanged himself on an elder tree; the English name may simply be a mistranslation. The fungus can be abundant on dead elder branches but it is not, as Ray states, peculiar to that species.

[707] See our comments on the Doctrine of Signatures in Chapter 3. Later in his life Ray (1691, pp. 85–86) wrote: "As for the Signatures of Plants, … all that I find mentioned and collected by Authors, seem to me to be rather fancied by men, than designed by Nature to signifie or point out any such Vertues or Qualities as they would make us believe, I have elsewhere, I think upon good Grounds, rejected them; and finding no reason as yet to alter my Opinion, I shall not further insist on them."

[708] For a justification of this translation of *Chymicorum* see footnote 843.

one would not recognise it at once nor persuade himself that it could come about that it could ever happen by chance but that it manifestly proves a plan of nature.) The seed-vessels of Shepherd's-purse have the clear likeness of a small bag, those of Buckler Mustard of a shield, those of the common Snapdragon of a calf's head, Bladder heart-like spotted Pease that of a heart, as also the leaves of certain species of Trefoil or Medick, Goat's-beard of a he-goat's beard, the root of pease earth-nuts of a mouse, Flower de luce and corn-flag of a sword, the pods of Fenugreek of horns, &c., to which objects however they should or can by no means be related.[709] 4. Because sometimes parts of certain plants represent those parts of the body to which they are as completely opposed as is possible. Thus the fruit of the Bean of Malacca represents a heart and is nevertheless poisonous; the sap of Spurges is like milk, but no one is so foolish that he would offer that to drink in order to increase milk in wet-nurses; the flesh of Medlars is similar to human excrement in consistency and colour, but it is not suitable for purging the belly; Touchwood or Tree Mushroom growing on the trunks of trees recalls the lungs in shape, colour and its spongy softness and is nevertheless malignant and harmful to the lungs.[710] It would be endless to enumerate every single case. 5. The same parts of the body are affected by different and often contrary diseases that require remedies that are different and varying in properties. 6. Because in different plants parts of the same kind and shape possess different and opposing powers, as is well known in bulbous roots, of which some, such as narcissi, cause vomiting and others, such as tulips, provide a very pleasant food. 7. Neither is the number of signatures so great nor are the signatures which plants bear so obvious or manifest to anyone that they reveal a pointing finger or purposeful plan of nature;[711] but among such a multitude of plants, even if they had come to exist entirely at random & by chance, it would easily have happened that an equal number of signatures would have been found, especially by human cleverness & with the help of imagination.

520 *Sanicula europaea* L.

Sanicula sive Diapensia *Ger. Park. Brunf. Matth. Dod.* Sanic[ula] officinarum *C.B.* mas Fuchsii sive Diapensia *J.B.* quinquefolia officinarum mas *hort. Wars.*

---

[709] The English names chosen for the plants and fungus listed here and in the next list are those likely to have been familiar to Ray, with spelling modernised where required. Ray (1686, pp. 836–837) describes five species with names beginning with *Thlaspi biscutatum*, to the first of which he gives the English name "Buckler Mustard". "The bladder heart like spotted Pease" is Parkinson's (1640, pp. 1377–1378) clumsy English name for his *Pisum cordatum vesicarium*, which is now called "Balloon Vine", "Heart Pea", "Heart-seed" or "Love-in-a-puff"; Ray (1686) describes this, without an English name, on p. 487. Ray's Latin names (with their current scientific names in brackets) are *Bursa pastoris* (*Capsella bursa-pastoris* [97]); *Thlaspi biscutatum* (probably *Biscutella* sp. or spp.); *Antirrhinum vulgare* (*Antirrhinum majus* L.); *Pisum cordatum* (*Cardiospermum halicacabum* L. (Sapindaceae)); *Medica* (*Medicago* spp., notably *M. arabica* [581], which Ray calls "Heart-trefoile, or claver"); *Tragopogon* (*Tragopogon pratensis* [574]); *terræ glandes* (literally "acorns of the earth", *Lathyrus tuberosus* L.); *Iris* (*Iris* spp.); *gladiolus* (*Gladiolus* spp.); *Fænugræcum* (*Trigonella foenum-graecum* L.). As with those in Ray's 'Headings or divisions', the examples chosen show little relationship to the plants that he had found in Cambridgeshire.

[710] Ray's Latin names for the plants and fungus listed here (with their current scientific names in brackets) are *Anacardium* (*Semecarpus anacardium* L. f.), a member of Anacardiaceae from the outer Himalayas related to the cashew-tree, the nuts of which were called "marking nut" from their use to mark clothing before washing; *Tithymalli* (*Euphorbia* spp. [569]); *Mespila* (fruits of *Mespilus germanica* L.); *Fungus igniarius* (a name applied to bracket fungi used as tinder [206]).

[711] *ut digitum aut consilium intendentis naturæ arguant*, literally "that they may make clear a finger or plan of intending nature".

Sideritis 3 Dioscoridis *Col.* It flowers in the month of June. In woods & shady hedges. *Sanicle.*

N. For the French it comes into a proverb that is commonly bandied about, that those who have Bugle[712] & Sanicle do not need a surgeon.

Qui ha du Bugle et du Sanicle
Fait aux chirurgiens la nicle.[713]

521 *Saponaria officinalis* L.
Saponaria *Dod. Lob. Ger.* vulgaris *Park. J.B.* major lævis *C.B.* Herba fullonum recentiorum *Fuchsii ex Ruellio.*[714] Viola agrestis *Trag.* Struthium *Fuch. Thal. Lugd. Sopewort.* It flowers in July & August. *In a place paled about on the right hand of Barnwell, as you go to Sturbridge fair; also about the brick-wall of the great house at Madingley; and on a bank at Cherry hinton by the springs near the chalk-pits.*

N. In the month of August number 12 of the medium-sized moths in *Moufet's Theatr[um] insect[orum],*[715] *page 105,* loves to sit on the flowers of this plant; apart from its very swift flight observed by the same author, it is notable on account of its very long proboscis and the loud noise that it emits almost like the sound of a hornet.

Satyrium basilicum, *see* Orchis palmata. [*Dactylorhiza incarnata* and/or *D. praetermissa* 401]

522 *Saxifraga granulata* L.
Saxifraga alba *Ger. Dod. Fuch. Gesn. Cam. Tab.* alba vulgaris *Park.* rotundifolia alba *C. B.* alba radice granulosa *J. B.* alba tuberosa radice *Clus. hist.* Thelygono affinis herba *Cæs. White Saxifrage. On the hill of Health, and on a common between Histon and Chesterton abundantly.* It flowers around the end of April and at the beginning of May: except while it is flowering you would search for it in vain in drier places, for its leaves are very minute & are sometimes entirely dried up.

523 *Asperula cynanchica* L.[716]
Saxifraga Anglica Occidentalium *Lob. Park.* Anglica[717] alsinefolia *Ger. emac. Pearlwort, Chickweed-Breakstone. Upon Gogmagog hills, Newmarket heath, and the drier part of the moors.* It flowers at the beginning of June & subsequently.

---

[712] *Bugula* [*Ajuga reptans* 95].

[713] "He who has some Bugle and some Sanicle
    "Cocks a snoot at the surgeons."

[714] "Fullers' herb of more recent [botanists], *of Fuchs [taken] from Ruellius.*"

[715] See footnote 76. The moth is the Humming-bird Hawk-moth, *Macroglossum stellatarum* (Linnaeus).

[716] Babington (1860) does not cite Ray's name anywhere in his Flora, but he gives the similar name of the next, quite dissimilar, species (*Saxifraga Anglica facie Seseli pratensis*) twice, once as a synonym of *Sagina procumbens* L. (an obvious slip) and once (correctly) as a synonym of *Silaum silaus.* Raven (1950, p. 90, footnote 2) therefore suggested that Babington may have intended to identify *Saxifraga Anglica Occidentalium* as *Sagina procumbens* and this is confirmed by Babington's annotated copy of the *Catalogus* in the library of the Department of Plant Sciences, Cambridge. Ewen & Prime (1975) also identify this plant as *S. procumbens* and this has a certain plausibility as the English name "Pearlwort" is now used for *Sagina* species. However, all earlier authors appear to have overlooked the significance of the entries under *Saxifraga Anglica Occidentalium* in *Emendanda* of the 1663 and 1685 appendices, which make it clear that the plant that Ray saw was in fact *Asperula cynanchica,* which he at first misidentified. The localities, habitat and flowering time all fit this species very well.

[717] This is an error for *Anglicana*, but see footnote 718.

524 *Silaum silaus* (L.) Schinz & Thell.
Saxifraga Anglica facie Seseli pratensis *Lob. Ger.* Seseli pratense nostras *Park*. Saxifr[aga] umbellifera Anglorum *Lugd*. Saxifr[aga] Anglica foliis fœniculi latioribus, radice nigra, flore candido, semine fœniculi, similis Silao. *J.B.* Hippomarathrum album & Anglicum *Tab*. Daucus seseli pratensis facie *C.B. in Phytop. Common or meadow Saxifrage, or green Saxifrage.* In meadows & pastures.

525 *Sagina procumbens* L.[718]
Saxifraga graminea pusilla, flore parvo herbido et muscoso. *Small grassie Saxifrage. The single flowers occupy individual stalks. It is found in marshy places and sometimes also in drier ones, as upon Newmarket heath, where the grasse is low. If any one desires certainly to know what plant we mean, he may be sure to find it among the stones in the stone-walk in the Fellows garden at Trinity Colledge. We think this plant hath not been yet described.* It can be assigned to the grasses.

Saxifraga palustris alsinefolia, *see* Alsine palustris foliis tenuissimis. [*Sagina nodosa* 31]

526 *Knautia arvensis* (L.) Coult.
Scabiosa major vulgaris *Dod. Lugd. Ger.* major communior *J.B.* vulgaris foliis dissectis *hort. Bat.* major satorum vulgatior *Lob.* pratensis hirsuta, quæ officinarum *C.B.* vulgaris, inter segetes nascens *Clus. Common Scabious.* In sown crops & meadows. It flowers in June and July: moreover it varies in the colour of the flower, namely purple and white.

527 *Scabiosa columbaria* L.
Scabiosa minor sive columbaria *Ger.* minor vulgaris *J.B.* minor campestris *Park*. capitulo globoso minor *C.B.* perhaps. *The lesser field Scabious.* In moorish[719] meadows & drier places, *as upon the hill of Health, &c.*

Scabiosa major squammatis capitulis, *see* Jacea segetum major foliis dissectis.[720] [*Centaurea scabiosa* 298]

Scandix, *see* Pecten Veneris. [*Scandix pecten-veneris* 425]

---

[718] This species was identified by Babington (1860) and Ewen & Prime (1975) as *Sagina apetala*, but, as Raven (1950, p. 90) suggested, the plant of marshy places is likely to have been *S. procumbens*. It is more than probable that Ray also saw and failed to separate *S. apetala* Ard. *sensu lato* (including *S. filicaulis* Jord.), as this grows with *S. procumbens* in dry places and is certainly difficult to separate from it on trampled ground, such as the paths and pavements of Cambridge; however, we believe that this entry refers primarily to *S. procumbens*, a conclusion that earlier authors have perhaps avoided because of their belief that *Saxifraga Anglica Occidentalium* represented that species. Our belief is strengthened by Ray's treatment of the two species in *Historia plantarum* (1688a, p. 1026), where he first recognised *S. apetala* as *Saxifraga Anglica annua Alsinefolia* D. Plot (as it had been described by Plot from Oxfordshire) and explicitly stated that the plant of the Cambridge *Catalogus* was *Saxifraga graminea pusilla flore parvo tetrapetalo*, as he then called the species that we now call *Sagina procumbens*. See also Ray's comment under *Emendanda* of the 1663 and 1685 appendices equating this plant with Johnson's (1633, pp. 567–568) *Saxifraga Anglicana alsinefolia*, which Ray had previously thought related to the last but one species [*Asperula cynanchica* 523], apparently following "the coniecture of *Pena* and *Lobel*" (Johnson, *loc. cit.*).
[719] For "moorish" as the translation of *montosis* see footnote 154.
[720] Ray actually cites this as a synonym under *Jacea segetum major purpurea* [298] rather than making it his chosen name, although he treats the white-flowered variant of this species separately and calls it *Jacea segetum major foliis dissectis flore albo* [299].

Scirpus major, *see* Juncus aquaticus maximus. [*Schoenoplectus lacustris* 306]

Sclaræa sylvestris *Tab. see* Horminum sylvestre. [*Salvia verbenaca* 292]

528 *Teucrium scordium* L.

Scordium *J.B. C.B. Matth. Dod. Ger.* legitimum *Park*. Trixago palustris *Lob. Water Germander*. In many ditches in the Isle of Ely, in the Osier holts about Ely city: Also in a ditch on the left hand of the road leading from Cambridge to Histon, about the midway thither. It flowers in June, July & August, i.e. in the summer months.

529 *Scrophularia nodosa* L.

Scrophularia major *Ger. Lob.* major vulgaris *Park*. vulgaris & major *J.B.* major radice tuberosa *hort. Hafn*. nodosa fœtida *C.B.* Ocymastrum alterum *Trag*. Galeopsis *Dod. Gal. Figwort*. In Madingley and Kingston woods, and many other places. It flowers in July.

    N. What some people report is remarkable, that is that scarcely any other herb can be found that remains green & is preserved from destruction for longer. Some simple women of Germany hang it up on an upper storey on the eve of St John the Baptist's Day[721] and also put it on the walls of houses & place it over the outer doors, in which places it flourishes for a long time, remains green & resprouts immediately, so much so that it is possible to find it green around the feast of the nativity of Christ. *Jo. Bodæ. in Theophr. hist.*

Scrophularia aquatica major, *see* Betonica aquatica. [*Scrophularia auriculata* 83]

Scrophularia minor, *see* Chelidonium minus. [*Ficaria verna* 129]

530 *Secale cereale* L.

Secale *Matth. Lob. Tab. Ger.* vulgatius *Park*. hybernum vel majus *C.B.* Siligo *Brunf. Lon.* veterum, & Secale *Trag*. Olyra *Cord. in Diosc*. Rogga sive Secale Plinii[722] *Dod*. Tipha Cerealis Theophrasti, *book 8, chapter 1*[723] *P. Gassend. Rie*. It is sown in fields.

    N. If all the flowers projecting from the ear of this cereal are stripped off with the hand, within a quarter of an hour it will thrust forth new ones from other small sheaths, and that even if the ear has been picked, a property that is common to wheat also & to certain other species of grasses.

531 *Sedum rupestre* L.[724]

Sedum minus hæmatoides *Ger*. minus flore luteo *J.B.* minus V, sive minus luteum folio acuto *C.B.* Vermicularis & Crassula minor vulgaris, sive Illecebra major

---

[721] *Germaniæ quædam mulierculæ in vigilia D. Joan. Bapt. in tabulato suspendunt*, literally "Of Germany certain little women on the eve of Divine John the Baptist [i.e. on 23 June] on a storey hang [it] up". For the use of *D*. (for *Divus*) for saints rather than the usual *S*. (for *Sanctus*) see also the entry for *Circæa Lutetiana* [136] above and footnotes 68 and 218 in Ray's 'Etymology'. The meaning of *tabulatum* is "flooring", "boarding", "a storey" or, as for example in Virgil's *Georgics*, 2: 361, "a vine-support". For other beliefs relating to Midsummer Eve and Midsummer Day see footnotes 82, 281 and 379.

[722] Rye is described in very derogatory terms, as very inferior and useful only to avert starvation and as very unacceptable to the stomach, in Pliny's *Natural history*, XVIII. 141 in the Loeb edition.

[723] *Tipha* is a latinised form of Theophrastus' τιφη [tiphe], translated as "one-seeded wheat" in the Loeb edition of his *Historia plantarum* and discussed several times among other cereals in the early part of Book VIII. The following reference is to Gassendi (1658, p. 153) where this work of Theophrastus is cited.

[724] Babington (1860) identified Ray's plant on nomenclatural grounds as *Sedum glaucum* Sm., a segregate of *S. reflexum* L., now *S. rupestre* L. He suspected a misidentification, as he knew of no

Park., a picture of which is substituted, namely of the following plant with a white flower. *The ordinary Prickmadam, or Stonecrop.* On walls & roofs generally.

532 *Sedum album* L.
Sedum minus officinarum *Ger.* Vermicularis flore albo *Park.* Sedum minus I, sive minus teretifolium album *C.B.* minus folio longiusculo tereti, flore candido *J.B. White flowered Stonecrop or Prickmadam with round pointed leaves. Very plentifully on many of the thatch'd houses in Chatteresse in the Isle of Ely.*

533 *Sedum acre* L.
Sedum sive sempervivum minimum acre *P.B.* parvum acre flore luteo *J.B.* minus 8 causticum *Clus. hist.* Illecebra *Officin.* minor, sive Sedum 3 Dioscoridis *Park.* Vermicularis sive Illecebra minor acris *Ger.* Sedum minus VIII, seu Sempervivum minus vermiculatum acre *C.B.* Aizoon minus fervidi gustûs *Thal.* Aizoi facie planta, quæ Dactylon 2 vel 3 Plinii *Col.* Telephium of Hippocrates *in his book On the nature of the woman & a 2nd book On the diseases of women.*[725] *Jo. Bodæ. in Theophr. hist.* Wall-pepper, Stonecrop. *On walls & roofs. It flowers in June and July.*

534 *Senecio* ?*sylvaticus* L.[726]
Senecio hirsutus viscidus major odoratus *J.B.* III, sive incanus pinguis *C.B.* fœtidus *Gesn. hort. Lugd. Park.* Erigeron tomentosum alterum *Lob. Ger. emac.*, in whose work (perhaps through the carelessness of the print-workers) the pictures of the first Erigeron tomentosum & Erigeron tomentosum alterum have been transposed, and the latter is put in the place of the former & vice versa. *Cotten Groundsell, or stinking Groundsell. On all the Fen banks almost in the Isle of Ely.*

---

certain locality for *S. reflexum* (an introduced species) in the county and implied that it was very unlikely to have been as frequent in Cambridgeshire in 1660 as Ray's plant clearly was. We share Babington's surprise but we are forced to conclude that Ray's plant was indeed *S. rupestre*. Ewen & Prime (1975) suggested that it might be *S. acre*, despite the fact that this is also listed in the catalogue [533]. Tempting as this is as an escape from the difficulty, it cannot be the solution. Ray's later works continue to separate these two species and in *Historia plantarum* (1686, p. 691) he describes *Sedum minus haematodes* as a plant nine inches high and with flowers in umbels; he says there that it occurs on walls and roofs almost everywhere in England. *Sedum acre* is so different that he describes it in another section of the book (1688a, p. 1041), with other small *Sedum* species, under J. Bauhin's name *Sedum parvum acre flore luteo*.

[725] lib. de natura muliebri, & lib. 2 de morbis mulierum.

[726] Previous authors have identified Ray's plant as *Senecio viscosus* L. on the basis of the names that he cites, and indeed Clarke (1900) gives this as the first British record of this species, but they have apparently disregarded what he says about Johnson's figures. Johnson (1633, pp. 278–279) captioned the two relevant figures as in Ray's text with the English names "Cotton Groundsell" and "The other Cotton Groundsell". The first, which Ray says *should* apply to his plant, shows a plant without ray florets and with the stems not obviously hairy, which seems to be *S. sylvaticus*, whereas the second shows one with ray florets and hairy stems, peduncles and pedicels, which seems to be *S. viscosus*; neither figure is Gerarde's (1597, p. 217) original one of "Cotton Groundsell", though this is more like Johnson's first one. Ray later (1686, p. 290) concluded that previous authors had been wrong in distinguishing two species of *Erigeron tomentosum* and furthermore that Johnson's second figure was misplaced and false (*supposititiam & falsam*), belonging not just to a different species but to a different genus, because it showed ray florets. Only one of the two species appears in Ray's *Synopsis* (1690, p. 50; 1696, p. 83; 1724, p. 178), which Linnaeus (1759, p. 106) identified as his *S. viscosus*. Ray's statement about Johnson's figure strengthens the argument in favour of his Cambridgeshire plant being *S. sylvaticus*, in which the outer florets are less obviously rayed than in *S. viscosus*, but Linnaeus' opinion argues the other way; clearly a definite decision is impossible.

535 *Senecio vulgaris* L.
Senecio vulgaris *Park*. minor vulgaris *C.B.* vulgaris sive Erigeron *J.B.* Erigeron *Ger.* Verbena fœmina *Brunf.* Carduncellus *Cæs. Groundsell; herb Simpson in some countreys.*

Senecio major *Matth. see* Jacobæa. [*Senecio jacobaea* 300]

Serapias, *see* Orchis palmata.[727]

Seriphium Germanicum, *see* Sophia chirurgorum. [*Descurainia sophia* 552]

Seris picris, *see* Cichoreum sylvestre. [*Cichorium intybus* 131]

536 *Thymus polytrichus* A. Kern. ex Borbás[728]
Serpillum hirsutum *Ger. emac.* VIII, i.e. angustifolium hirsutum, sive montanum repens hirsutum *C.B.* Pannonicum 3 *Clus.* angusto lanuginosóque folio *J.B.* Serpilli Pannonici Clusii varietas altera *Park. Hoary wild Tyme. On Gogmagog hills. It flowers in summer.*

537 *Thymus polytrichus* A. Kern. ex Borbás
Serpillum vulgare *Ger. J.B.* vulgare minus *Park. C.B.* vulgare repens *Clus.* sylvestre *Matth.* Saxifranga *Dod. Wild Tyme, Mother of Tyme.* In moorish[729] & drier places.

538 *Serratula tinctoria* L.
Serratula *J.B. C.B. Lob. Matth. Dod.* purpurea *Ger.* vulgaris flore purpureo *Park.* tinctoria, et Jacea aromatica, sive Caryophyllata, on account of the fragrant root, *Tab.* Centauroides vel Centaurium majus sylvestre Germanicum *Thal. Saw-wort.* In meadows & woods.

539 *Silene otites* (L.) Wibel
Sesamoides Salamanticum magnum *Ger. Clus.* Muscipula Salamantica major *Park.* muscoso flore, sive Ocymoides belliforme *J.B.* Lychnis viscosa X, sive viscosa flore muscoso *C.B. The greater Spanish Catchflie. Near the gravell pits as you go to the nearest windmill on the Northside of Newmarket town.*[730] This plant has so far been believed to be foreign and not indigenous to England: moreover we suspect that those who have claimed that the small Sesamoides Salamanticum comes up naturally around *Newmarket* have been deceived, since the Sesamoides that we

---

[727] Synonyms beginning with *Serapias* are given by Ray for *Orchis palmata major mas* [*Dactylorhiza incarnata* and/or *D. praetermissa* 401] and *Orchis palmata rubella cum longis calcaribus rubellis* [*Gymnadenia conopsea* 403]. *Serapias* is also a component of the polynomial names or synonyms of other orchid species.

[728] Ray (1670, 1677) retained separate entries for this and the following taxon, though he noted that they scarcely differed except in the hairiness of the leaves. *Thymus polytrichus* var. *neglectus* (Ronniger) P.D. Sell and var. *britannicus* (Ronniger) P.D. Sell were collected in a mixed population on the Gogmagog Hills by W.T. Stearn in 1932 (specimens in CGE). The former has leaves which are glabrous on the upper sides whereas in the latter they have frequent hairs; var. *britannicus* is clearly a hairier plant when examined closely but it is not conspicuously hairy from a distance. The mite *Acaria thomasi* (Nalepa) converts the inflorescence of infected *Thymus* species into a "woolly ball" (Pigott 1955), but this phenomenon is well described by Bauhin & Cherler (1651, vol. 3(2), p. 269; see footnote 109 of the 1685 appendix), so such plants are less likely to be the basis of this entry.

[729] For "moorish" as the translation of *montosis* see footnote 154.

[730] When reporting this locality in the 1695 edition of *Camden's Britannia*, Ray (1695, column 418) added: "*This place may be in Suffolk*", as indeed it is. See also the entries for *Thalictrum minus* [568] and *Trifolium echinatum arvense* [*Medicago minima* 582] and our gazetteer.

have found there and which we think is the only one to grow there is the large one unless the pictures & descriptions greatly deceive us.

Sideritis arvensis rubra *Park. see* Ladanum segetum. [*Galeopsis angustifolia* 315]

Sideritis altera *Diosc. see* Pimpinella sanguisorba. [*Sanguisorba officinalis* 438[731]]

Sideritis I *Matth. see* Marrubium aquaticum. [*Lycopus europaeus* 358]

540 *Stachys palustris* L.
Sideritis Anglica strumosa radice *Park.* Panax coloni, & Marrubium aquaticum acutum *Ger.*[732] Stachys palustris Gesneri *Cam.* VII, seu palustris fœtida *C.B.* Galeopsis angustifolia fœtida *J.B.* Tertiola *Cæs.*, because it cures tertian fevers.[733] *Clownes All-heale*. By rivers & ditches.

Sideritis tertia Dioscoridis *Col. see* Sanicula. [*Sanicula europaea* 520]

Sigillum B. Mariæ *Offic. see* Bryonia nigra. [*Tamus communis* 92]

Sisymbrium, *see* Mentha aquatica. [*Mentha aquatica* 362]

541 *Brassica nigra* (L.) W.D.J. Koch
Sinapi 2 sive vulgare *Ger. emac.* siliquâ latiusculâ glabrâ, semine russo, sive vulgare *J.B.* The descriptions of Johnson in the work of Gerarde & of Jean Bauhin differ mutually[734] in this respect, that the latter says that the stem is hairy in the lowest part and glabrous in the upper part but the former that it is entirely glabrous throughout.[735] We indeed also have observed in this plant a stem that is sometimes absolutely glabrous but at another time rough with little spines in the lower part. *Ordinary Mustard*. *On the bankes of ditches newly cast up, as about Trinity Colledge walkes, sometimes also among the Corn.*

N. If mustard may be thrown into the water, it cooks meats & vegetables in a short while, as Cassianus Bassus relates *in book 11, chapter 39,* and *Jo[annes] Bodæus* cites *in his Theophr[astus]*.[736]

542 *Sison amomum* L.
Sison sive officinarum Amomum *J.B.* Si[son] quod Amomum officinis nostris *C.B.* vulgare sive Amomum Germanicum *Park.* of Dioscorides, *book 2, chapter 64 C.B.* Ammi parvum *Gesn. hort.* Petroselinum Macedonicum Fuchsii *Ger. Bastard stone Parsley*. By ditches & hedges.

Sium Cratevæ erucæfolium *Lob. see* Nasturtium aquaticum vulgare *Park.* [*Nasturtium officinale sensu lato* 387]

---

[731] *Sideritis altera* Diosc. ("The second Ironwort of Dioscorides") does not appear as a synonym in the *Catalogus*, but [*Pimpinella*] *sanguisorba major* C.B. is included in the synonymy for *Pimpinella sylvestris* Ger. [*Sanguisorba officinalis* 438] and Johnson (1633, p. 1046) shows that both Gerarde and he believed that this was Dioscorides' "second Ironwoort".

[732] The first of these two names and the English name used by Ray were invented by Gerarde (1597) as a result of witnessing an extraordinary self-healing with the plant by a "husbandman": see the entry for *Panax* in Ray's 'Etymology'. "Clown" is used in its original sense of "countryman, rustic or peasant" (OED). Johnson (1633, p. 1005) points out that Gerarde had described the plant twice, also as *Marrubium aquaticum acutum* (on p. 565).

[733] See footnote 431.

[734] *ab invicem*, a misprint for *ad invicem*.

[735] *ubique & omnino glabrum*, literally "everywhere & altogether glabrous".

[736] Bodaeus (1644; see footnote 23) cites Cassianus Bassus on p. 83.1. See also footnotes 208 and 211.

543 *Sium latifolium* L.
Sium majus latifolium *Ger.* maximum latifolium *J.B.* majus latifolium bituminosum *Tab.* Sium Dioscoridis, sive Pastinaca aquatica major *Park.* Great water Parsenep. In the river Cam in some places, and in great watery ditches. The leaves of this plant that emerge first at the beginning of spring are not like those that follow but finely cut like those of Oenanthe aquatica or some kind of Petroselinum[737] or of other similar species.

544 *Berula erecta* (Huds.) Coville
Sium erectum umbellatum sive Pastinaca aquatica *Lob.* Sium *Matth.* odoratum *Trag.* majus angustifolium *Ger.* Sium minus alterum *Park.* medium *J.B.* perhaps.

545 *Apium nodiflorum* (L.) Lag.
Sium umbellatum repens *Ger. emac.* minus sive Pastinaca aquatica minor *Park.* An Sion umbelliferum *J.B.?* There is a picture of this plant in *[Daléchamps' General] Hist[ory of Plants published in] Lyon,*[738] page *1092.* The most common water Parsenep. In watery ditches & rivulets almost everywhere. We confess that it is not clear to us whether this plant has been described by others, but we are certain about it from Johnson in his revision of Gerarde, to whose description this can be added, that no narrow bracts underlie the base of the umbel as in the previous Sium, which can be called in English *the middle water Parsenep.*

Similax[739] lævis, *see* Convolvulus major. [*Calystegia sepium* 148]

546 *Solanum nigrum* L.
Solanum sive Solatrum vulgare *Park.* hortense *Matth. Dod. Lob. Ger.* hortense sive vulgare *J.B.* nigrum *Cord. in Diosc.* bacciferum I, sive officinarum *C.B.* Nightshade. On footpaths & rubble.

N. From the putrefaction of this herb grow green worms with yellow spots intermixed, on whose head can be seen a horn sometimes almost an inch in length.[740] *Severin[us] Gobelius,*[741] *in his treatise about amber* in the work of *Cam. in hort.* There is no species of plant that may not generate from its putrid liquid a worm appropriate to itself, something that the wonder of the microscope has revealed in recent years.[742] *Kircher about plague.*[743]

547 *Solanum dulcamara* L.
Solanum lignosum sive Dulcamara *Park.* bacciferum XII, i.e. scandens seu Dulcamara *C.B.* Amara dulcis *Ger. Tab. Gesn. hort.* Glycypicros sive Amara

---

[737] i.e. *Oenanthe fistulosa* [393] (rather than the current *O. aquatica* [134]), probably Garden Parsley, *Petroselinum crispum* (Mill.) Nyman ex A.W. Hill, and possibly *Sison amomum* [542].

[738] *Hujus icon habetur in hist. Lugd.*, literally "A picture of this is held in *the hist[ory from] Lyon*"; see *Lugd.* in Ray's 'Explanation'.

[739] A misprint for *Smilax*.

[740] This may be a rather inaccurate description of the caterpillar of the Death's-head Hawk-moth, *Acheronia atropos* (Linnaeus), which is variable in colour, although more prominently striped than spotted, and has a prominent horn on its tail (the caterpillars of some related species are called hornworms in North America). It will feed on plants in several genera of Solanaceae, including *Solanum*.

[741] Severinus Gobelius or Severin Goebel (1530–1612) was physician to the Elector of Brandenburg and a correspondent of the Antwerp publisher Christopher Plantin. His book *De succino* ("About amber") was one of seven works published in a single volume entitled *De omni rerum fossilium genere, gemmis, lapidibus, metallis, et huiusmodi* in Zürich in 1565.

[742] *quod microscopii arcanum ultimis hisce temporibus detexit*, literally "which the microscope's secret in these last times has uncovered". See footnote 597.

[743] For Kircher's work see footnotes 197 and 302.

dulcis *J.B.* Dulcamara sive Vitis sylvestris Matthioli & Melothron Theophr. *Lugd.* Cyclaminos altera Dodonæi *J.B.* Vitis sylv[estris] *Matth.* Salicastrum Plinii *Guiland. Cæs.* Solanum perpetuum, quòd caules perennent[744] *Gesn. Woody Nightshade, Bittersweet.* In hedges beside waters. The taste is at first bitter but afterwards becomes sweet.[745]

548 *Atropa belladonna* L.
Solanum lethale *Park. Ger. Dod. Clus. hist.* bacciferum IV, sive melanocerasos *C.B.* Bella Donna *Clus. pan.* and *of the Italians.*[746] Solanum manicum multis, sive Bella Donna *J.B.* Dioscorides' first Solan[um] maniacum & Theophrastus' Mandragora *Jo. Bodæ. in Theophr. hist.*[747] *Deadly Nighshade or Dwale. In the lanes about Fulborn plentifully.*

N. Not even this lethal plant escapes the teeth of terrestrial snails & slugs, but in early spring even its leaves are nibbled by these same creatures. Incidentally, concerning these little animals, it may be appropriate to explain that individual ones of them share equally in both sexes & that they are androgynous; for they are both active & passive in turn, impregnating & at the same time being impregnated,[748] as will be sufficiently agreed by anyone who has separated them while they are mating in spring, even though neither Aristotle nor other Authors on nature,[749] as far as we know, have made any mention of that matter.[750] The berries of this plant are deadly to eat. See *[Mathias] de L'Obel in Adv[ersaria], [Joannes] Bod[aeus] in Theophr[astus'] Hist[oria], book 6, chapter 2, page 586,*[751] *[John] Gerarde* & others.

Solanum tetraphyllum, *see* Herba Paris. [*Paris quadrifolia* 277]

Solidago Saracenica *Ger. see* Conyza palustris. [*Senecio paludosus* 146]

---

[744] "Perpetual Nightshade, because the stems perennate".

[745] *Gustu primùm est amaro, postea dulcescente*, literally "At first it is with bitter taste, afterwards with becoming sweet".

[746] *Bella Donna* is Italian for "Beautiful Lady"; the plant is said to have been formerly used to dilate women's pupils to make them appear more beautiful. See also the entry for *Bella donna* in Ray's 'Etymology'.

[747] "Dioscorides' first maddening Nightshade and Theophrastus' Mandrake". Other passages in Theophrastus' *Historia plantarum* clearly refer to Mandrake, *Mandragora officinarum* L., but that describing μανδραγορας [mandragoras] as being like ναρθηξ [narthex], *Ferula communis* L., in having a hollow stem and as having a black fruit like a grape (VI. II. 9 in the Loeb edition) probably refers to *Atropa belladonna*. Bodaeus (1644; see footnote 23) discusses this on pp. 583–586.

[748] *Vicissim enim agunt & patiuntur, immitunt simul & recipiunt*, literally "For in turn they act & they suffer, they send in & at the same time they receive".

[749] *rei naturalis Scriptores*, literally "of the natural thing Writers".

[750] It seems that Ray's observation here is indeed the first published report of the hermaphrodite nature of slugs and snails (Örstan 2010, where fuller details are given). The concept of an androgynous (ἀνδρογυνος, from the Greek words for "man" and "woman") or hermaphrodite (ἑρμαφροδιτος) human being had been well known since classical times, especially in connection with the myth of Hermaphroditus, son of Hermes (Mercury) and Aphrodite (Venus), who supposedly grew together with the nymph Salmacis while bathing in her fountain and thus combined male and female characteristics. The subject excited much interest in the Renaissance. Caspar Bauhin, who was an anatomist as well as a botanist, wrote one of the most complete treatises on the subject, *De hermaphroditorum natura* (published at Oppenheim in 1614), which covered both people and animals (Long 2006).

[751] Bodaeus (1644, p. 586.2; see footnote 23) says that there have been many cases of lethal poisoning but that it is sufficient to cite one from his own experience.

549 *Sonchus asper* (L.) Hill
Sonchus asper laciniatus & non laciniatus *C.B.* Sonchus laciniatus spinosus *J.B.* Sonchus asperior *Ger.* asper laciniatus *Park. Prickly Sow-thistle.*

550 *Sonchus oleraceus* L.
Sonchus lævis *Ger.* lævis vulgaris *Park.* lævis laciniatus latifolius *C.B.* laciniatus non spinosus *J.B. Hares Lettuce, Unprickly or smooth Sow-thistle.*

551 *Sonchus arvensis* L.
Sonchus repens, multis Hieracium majus *J.B.* 3 lævis altissimus *Clus. pan.* arborescens *Park.* arboresc[ens] alter *Ger. emac.* Hieracium Dentis leonis folio acuto V, sive arborescens palustre *C.B. The greatest Sow-thistle.* Among standing corn & on the boundaries of fields. We are greatly in doubt whether or not the descriptions of all other authors which are attached to the titles that we have provided fit the plant that we mean, but the description of J. Bauhin corresponds exactly.[752]

552 *Descurainia sophia* (L.) Webb ex Prantl
Sophia Chirurgorum *Park. Dod. Lob. Ger.* Nasturtium sylvestre *Fuch. ico.* sylvestre tenuissimè divisum *C.B.* Seriphium Germanicum & Lumbricorum herba *Trag.* Seriph[ium] German[icum] sive Sophia quibusdam *J.B.* Accipitrina *Cæs.* Thalictrum *Tab. Flixe-weed.* At the margins of fields.

553 *Sparganium erectum* L.
Sparganium ramosum *Park. Ger. C.B.* Sparg[anium] quibusdam *J.B.* Platanaria *Dod. Thal.* Butomon Theophrasti *Dod. lat. ico. Ang.* Butomos dissectâ paniculâ, commonly Platanaria, because it has little balls like the little balls of the Plane. *[Jo.] Bodæ. in Theophr. hist.*[753] Phleos fœmina *Lugd. eidem* & Calamogrostis secunda. *Burre-flag, or branched Burre-reed.* Almost everywhere in watery places. It produces its prickly capitula finally in the month of August.

554 *Sparganium emersum* Rehmann
Sparganium non ramosum *Park. C.B.* non ramosum sive latifolium *Ger.* alterum *J.B.* Platanaria altera *Dod. Great water Burre-reed.* With the preceding species.

555 *Nardus stricta* L.
Spartum parvum Lobelio *J.B.* parvum Batavicum & Anglicum *Park.* nostras parvum Lobelii *Ger. emac. in the appendix.* Gramen Sparteum tenuifolium *C.B. Small Matweed.* On a boggy ground beyond Gamlingay near Sir Roger Burgoynes park, where they dig turffs.

Spatula fœtida, *see* Xyris. [*Iris foetidissima* 630]

556 *Legousia hybrida* (L.) Delarbre
Speculum Veneris minus *Ger. emac. Park. Codded Corn-violet. The lesser Venus looking-glasse.* Common among standing corn. There is no mention of it in foreign writers that we know of.[754]

---

[752] *adamussim quadrat*, literally "to a carpenter's rule is square", *ad amussim* being originally two separate words.

[753] *vulgò Platanaria, quia pilulas habet Platani pilulis similes*. The "little balls of the Plane" are the inflorescences of *Platanus orientalis* L. The reference is to Bodaeus (1644, p. 462.2; see footnote 23).

[754] In *Historia plantarum*, Ray (1686, p. 743) still commented on the failure of foreign writers to mention this plant, even though he himself had by then seen it overseas amongst standing corn no less than in England.

MAIN CATALOGUE

557 *Spergula arvensis* L.
Spergula *J.B.* Saginæ Spergula *Lob. Ger. emac.* Sag[ina] Spergula major vulgaris *Park.* Alsine Spergula dicta I, sive major *C.B.* tenuifolia altera, sive terrestris altera τριχοφυλλος [trichophyllos] *Col. Spurry.* In the corn fields about Hoginton towards Madingley: but especially about Gamlingay, in plenty.

558 *Spergularia rubra* (L.) J. & C. Presl
Spergula alsineformis, or 5 of chapter 463 [in book 2]. *Ger. emac.* Sagina Spergula minima *Park.* An Spergula purpurea *J.B?* Alsine Spergulæ facie minor, sive Spergula minor flore subcæruleo *C.B. Chickweed-Spurry.* Around *Gamlingay* in sandy fields.

559 *Spergularia marina* (L.) Besser and/or *S. media* (L.) C. Presl[755]
Spergula marina of Daléchamps, or 4 *of chapter 463* [in book 2]. *Ger. emac.* Sagina Spergula minor *Park.* Alsine Spergula dicta II, sive Spergulæ facie media *C.B. Sea-Spurry.* By the ditches and river at Wisbich.

560 *Heracleum sphondylium* L.
Sphondylium *Ad. Lob. Dod. Ger.* vulgare *Park.* vulgare hirsutum *C.B.* Sph[ondylium] quibusdam, sive Branca ursina Germanica *J.B.* Branca ursina[756] *Trag. Brunf. Cord. in Diosc.* Acanthus Germanica sive vulgaris *Fuch. Cow-Parsnep.* In meadows & pastures.

N. The Poles & Lithuanians are said to cook up a potion from the leaves & seed of Sphondylium with water and added yeast, which takes the place of beer for the poor. *Dod.* in the work of *J.B. Dod[oens]* relates that it takes the place of the hop for the Poles & Lithuanians in producing beer, for just like the hop it is hot and dry. *Casp[arus] Hofman[nus], De medicam[entis] officinal[ibus],* book 2, chapter 40.

Spina alba sylvestris *Fuch. see* Acanthium vulgare. [*Onopordum acanthium* 4]

Spina appendix, *see* Oxyacanthus. [*Crataegus laevigata* and *C. monogyna* 414]

Spina infectoria, *see* Rhamnus. [*Rhamnus catharticus* 498]

Stachys, *see* Sideritis Anglica strumosâ radice. [*Stachys palustris* 540]

Staphylinus sylvestris, *see* Daucus Officinarum. [*Daucus carota* subsp. *carota* 166]

561 *Staphylea pinnata* L.[757]
Staphylodendron *J.B. Ad. Lob. Cam. Tab.* Plinii *book and chapter* 16,[758] *C.B.* Nux vesicaria *Ger. Dod. Trag.* N[ux] v[esicaria] sive Staphylodendron *Park.* Pistacia III, sive sylvestris *C.B. The Bladder Nut-tree, Said to grow at Milton three miles from Cambridge by Parkinson. We could not find it out by search, nor heare of it by enquiry, howbeit we deny not but possibly it may grow there.*

---

[755] These species were not separated in Ray's time and even Babington (1860) separated them with the comment that *S. media* was "probably a distinct species". Both have been recorded in maritime habitats near Wisbech.

[756] See the entry on *Branca ursina* and the footnote on it in Ray's 'Etymology'.

[757] Parkinson (1640, p. 1417) says: "It groweth in many places of this land, both as a hedge bush, and a standard tree at *Ashford* in Kent, and at *Milton*, three miles from *Cambridge*." This sounds implausible to a modern botanist, but Dunn (1905) says: "Much cultivated in England from as early as the seventeenth century, and, from the first, showing a tendency to establish itself outside gardens." It is now very uncommon as a naturalised species.

[758] i.e. Pliny's *Natural history, book 16, Chapter 16,* or XVI. 69 in the Loeb edition, where he describes the tree as growing north of the Alps and bearing pods containing kernels tasting like hazel nuts.

562 *Callitriche* spp.[759]
   Stellaria aquatica *Lob. ico. Park.* Fœniculum aquaticum stellatum *Tab.* Alsine aquatica *Lugd.* Alsine aquis innatans foliis longiusculis *J.B.* Millefolium aquaticum XIII, sive Stellaria aquatica *C.B. Water Star-wort, or Star-like water Fennell, or Star-headed water Chick-weed.* In waters. The minute flowers of this plant, two at each node of the stem, are opposed to each other alternately, depending on two white bracts alternately bent back on themselves, with a white style that is very long in proportion to the flower; the seed-vessels, compressed a little out of the round, follow these, separated into four parts by four quite deeply impressed little lines. At the tops of the seed-vessels some black hairs emerge constituting as it were a tip. To see how absurdly those clumps of leaflets that float on the surface of the water are regarded & described as flowers by Gerarde & Parkinson just look at their works themselves.

Succisa, *see* Morsus Diaboli. [*Succisa pratensis* 374]

Silybum, *see* Carduus lacteus. [*Silybum marianum* 108]

Symphytum, *see* Consolida major. [*Symphytum officinale* 147]

T

TAmus Plinii, *see* Bryonia nigra. [*Tamus communis* 92]

Tanacetum agreste, *Brunf. see* Argentina. [*Potentilla anserina* 60]

563 *Tanacetum vulgare* L.
   Tanacetum *Ger.* vulgare *Park.* odoratum I, sive vulgare luteum *C.B.* vulgare flore luteo *J.B.* Artemisia Dioscoridis *Tab.* monoclonos, sive tenuifolia *Fuch.* Athanasia vulgaris *Lac. Tansie.* At Histon near the ruines of an old Chappell; and in a pasture close on the North side of Ely, within half a mile of the citie.

564 *Verbascum thapsus* L.
   Tapsus barbatus *Offic. Ger.* Verbascum album vulgare, sive Tapsus barbatus communis *Park.* Verbascum mas *Ang. Ad. Cast. Tab.* latius *Dod.* candidum mas *Lon. Lac.* mas latifolium luteum *C.B.* mas, & Candela regia *Lob.* vulgare flore luteo magno, folio maximo *J.B.* Verbascum aut Phlomos vulgaris mas *Lob. Mullein, High-taper, Cowes Lungwort.* By the margins of fields & on walls or rubble.

Taraxacon *Officinarum, see* Dens leonis. [*Taraxacum* spp. 167]

565 *Taxus baccata* L.
   Taxus *Ger. Park. Matth. Dod. Lugd. Cam. C.B. J.B.* Smilax of Dioscorides, *book 4, chapter 80.* Μιλος [Milos] for Theophrastus. *The Yew tree.* In graveyards; *as in the walled yard at Barnwell abby.*

   N. 1. The yew tree is harmless with us. *Cam. hort.* Pena asserts in *Adv[ersaria]* that in England the berries are eaten without any danger & that pigs eat them generally like acorns.

---

[759] Three species in this difficult genus are frequent in Cambridgeshire, *C. obtusangula* Le Gall, *C. platycarpa* Kütz. and *C. stagnalis* Scop. They were not separated by Cambridgeshire botanists until the 20th century and it is only possible to attribute Ray's plant to the genus, although the "quite deeply impressed little lines" between the mericarps suggest that the fruit described by Ray is unlikely to have been that of *C. obtusangula*.

2. The yew has this property, that if beasts of burden have eaten leaves from it they may die but if ruminants do so they may suffer nothing.[760] The fruit is also eaten by some human beings and it is pleasant and harmless. *Theophr. hist. book 3, chapter 10*. See also *Pliny, book 16, chapter 10*.[761] Caesar *in chapter 6 of De bello gallico* relates that Cativulcus, king of the Eburones, killed himself with yew. It is also well known to us that a certain woman departed from life from vomiting that arose from a potion boiled up from it; hence, on the evidence of Pliny, poisons are said to be "toxic", as if it was "taxic" from the tree Taxus.[762]

566 *Sedum telephium* L.
Telephium *Matth. Lob. Tab. Cam.* Tel[ephium] sive Crassula major vulgaris *Park*. Crassula major *Cæs. Thal.* Cras[sula] sive Faba inversa *Ger*. Fabaria *Matth. Guiland. Lugd.* Anacampseros *Gesn. hort.* Anac[ampseros] vulgò Faba crassa *J.B.* Orpine, Livelong. *In a lane at Shelford; also at Burrough-green in a grove on the West side of the Church.*

Tertianaria *Tab. see* Lysimachia galericulata. [*Scutellaria galericulata* 345]

Tetrahit angustifolium, *see* Ladanum segetum. [*Galeopsis angustifolia* 315]

Tetragonia, *see* Euonymus Theophrasti. [*Euonymus europaeus* 188]

Teucrium pratense, *see* Chamædrys sylvestris. [*Veronica chamaedrys* 126]

567 *Thalictrum flavum* L.
Thalictrum sive Thalictrum majus *Ger.* majus vulgare *Park.* I, sive majus siliquâ angulosâ aut striatâ *C.B.* nigrius, caule et semine striato *J.B.Thal.* 5, sive pratense primum elatius, longioribus & magìs atris foliis, & quodammodo splendentibus *Clus*. Ruta pratensis *Gesn. hort.* Barba caprina minor & Thalictrum quorundam *of the same author.* Ruta pratensis herbariorum *Ad. Lob.* Meadow Rue. *In meadows, in moist places.*

568 *Thalictrum minus* L.
Thalictrum minus *Ger. Park. Dod.* VIII, sive minus, *C.B.* minus, sive Rutæ pratensis genus minus semine striato *J.B.* The lesser Meadow Rue. *About Newmarket where the* Sesamoides *grows,*[763] *and also among the corn between that pit and Cambridge road: also about Bartlow and Linton in the chalky grounds.* The scent of this species is much stronger than that of the preceding one.

Thlaspi fatuum, *see* Bursa pastoris. [*Capsella bursa-pastoris* 97]

Tinctorius flos, *see* Conyza cærulea acris & Genistella infectoria. [*Erigeron acris* 144 and *Genista tinctoria* 215]

---

[760] The two types of animals mentioned here are in Latin the plurals *jumenta* and *ruminantia* and in Theophrastus' Greek λοφουρα [lophura], literally "bushy-tailed", and μηρυκαζοντα [merycazonta].
[761] Theophrastus' *Historia plantarum*, III. x. 2, and Pliny's *Natural history*, XVI. 50 in the Loeb editions. The latter says that the male yew has noxious fruit and that, especially in Spain, the poison in them is lethal; the berries of the female yew are not mentioned!
[762] *toxica venena dicta sunt, quasi taxica, Plinio teste, à Taxo arbore*, literally "poisons are called toxic, as if taxic, with Pliny [as] witness, from the Yew tree". The reference is to Pliny's *Natural history*, XVI. 51 in the Loeb edition. See also the entry on *Taxus* in Ray's 'Etymology'.
[763] See the entry for *Sesamoides Salamanticum magnum* [*Silene otites* 539] and footnote 730.

569 *Euphorbia helioscopia* L. and *E. peplus* L.[764]
Tithymalus helioscopius *Fuch. Matth. Lob. Ger. Park. C.B.* helioscopius sive solisequus *J.B.* helioscopius, sive Esula vulgatior *hort. Haf. Sun-Spurge.* In fields. There are two species,[765] the one with scalloped leaves and larger, the stem of which produces five branchlets for the most part from the same point, & at the base of the branchlets five scalloped leaves surround it, the other with the leaf not scalloped & smaller. These species are also distinguished in other ways, certainly by the flower & the seed: look at & compare them.

N. 1. Baron Verulam has observed, not improperly, *in his Nat[ural] hist[ory]*,[766] *cent[ury] 7, exp[eriment] 639*, that all sap poured forth from cut plants like milk is acrid. Lettuce is food but the sap of lettuce is poison, says *C. Hofman in his De medicam[entis] officinalibus], book 2, chapter 67,* & Galen in his De medicamentorum facultatibus.

2. I frankly confess that in no region anywhere and at no time, even when desiring passionately to observe most diligently, have I been able perceive what so many Botanists have published about Sun Spurge, Marigold & very many other plants,[767] that they are turned around daily with the sun. *P. Lauremberg*,[768] *Horticult[ura], book 1, chapter 11, number 7.*

570 *Euphorbia amygdaloides* L.
Tithymalus characias amygdaloides *C.B. Ger. emac.* characias vulgaris *Park.* sylvaticus, toto anno folia retinens *J.B. English Wood-Spurge. In Sr John Cotton's park at Cheveley, near Newmarket.*[769]

Tithymalus leptophyllos, *see* Esula exigua. [*Euphorbia exigua* 187]

571 *Potentilla erecta* (L.) Rauesch.
Tormentilla *Ger. Trag. Matth. Dod. Lac. Lob.* officinarum *J.B.* vulgaris *Park.* sylvestris *C.B.* Heptaphyllum *Fuch. Tur. Gesn. hort.* Consolida rubra *Tab. Tormentill. On all the moors about Cambridge, and elsewhere in pasture grounds.*

572 *Campanula trachelium* L.
Trachelium majus *Ger.* minus flore purpureo *Park. in parad.* VII, sive Campanula vulgatior foliis urticæ, vel major & asperior *C.B.* Campanula major & asperior folio urticæ *J.B.* Uvularia major *Trag. Great Throatwort or Canterbury bells. In Kingston and Madingley woods and sundry other places.*

573 *Campanula glomerata* L.
Trachelium minus *Lob. Dod. gal. Park. Ger. emac.* minus multis *J.B.* minus umbellatum *hort. Haf.* minus flore violaceo *hort. Paris.* minus pratense umbellatum, & Cervicaria minor *Schwenckf.* Trach[elium] sive Campanula hirsuta XVII, aut pratensis flore conglomerato *C.B.* Cervicaria minor *Thal. Tab.* Sphærocephalus *Clus.* Uvularia exigua *Trag. Little Throatwort or Canterbury Bells.* On

---

[764] The species with crenate leaves is *E. helioscopia* and the one with non-crenate leaves *E. peplus*: see *Peplus sive Esula rotunda* [A28, B74] in the 1663 and 1685 appendices and *Tithymalus helioscopius* in their *Emendanda*.
[765] *Duplex est*, literally "It is double".
[766] For the work of Baron Verulam (Sir Francis Bacon) see footnote 104.
[767] *de Tithymalo helioscopio* [the species listed above], *de Calendula* [Pot Marigold, *Calendula officinalis* L.], *& plurimis aliis.*
[768] For Peter Lauremberg see footnote 25.
[769] See the gazetteer.

Gogmagog hills, and Newmarket heath, and in Little Shelford church-yard, & elsewhere on the way in many places.

574 *Tragopogon pratensis* L.
Tragopogon luteum *Ger.* flore luteo *J.B.* luteo flore I, sive pratense luteum majus *C.B.* Barba hirci *Cordi in Diosc.* flore luteo *Cam.* Barbula hirci *Trag. Matth. Goats-beard, Go to bed at noon.* In meadows & pastures. From the roots of this plant cooked in boiling water to that point when they become tender and then prepared with butter in the same way as the garden parsnip[770] a very fine dish is made, for they both have a more delicate flavour and yield a more wholesome juice than parsnips or carrots. The roots are also frequently eaten raw in salads. If their sweet taste had been known to the Apitii,[771] they would never have failed to include them in their salads in preference to all other herbs and roots. This herb is known to the Italians as Sassefrica and Sassifica[772] and among them its use is very frequent.

575 *Groenlandia densa* (L.) Fourr.
Tribulus aquatic$^{9\ 773}$ minor muscatellæ floribus *Ger. emac.* aquaticus minor alter *Park.* Fontalis media lucens *J.B. Small Frogs Lettuce or water Caltrops.* In rivulets.

576 *Potamogeton crispus* L.
Tribulus aquaticus minor quercûs floribus Clusii *Ger. emac.* aquat[icus] minor prior *Park.* Pusillum Fontilapathum *Lob.* Fontinalis crispa *J.B.* Butomos Damocratis *Ang.* Potamogiton οὐλοφυλλον [ulophyllon] seu crispum aut fimbriatum *Thal. hort. Hafn.* VI, i.e. foliis crispis, sive Lactuca ranarum *C.B.* Alga 2 *Trag. Water Caltrops or Frogs Lettuce.* In waters.

577 *Asplenium trichomanes* L.
Trichomanes *Park.* mas *Ger.* Trich[omanes] sive Polytrichum *J.B.* Trich[omanes] sive Polytrichum officinarum *C.B. English Maiden-hair, or as Gerard, The male English Maiden-hair. On some stone walls about the Cathedrall church at Ely, as also on the church walls at Over and Suacy.*

578 *Oxalis acetosella* L.
Trifolium acetosum vulgare *Park. C.B.* Oxys Pliniana *Gesn. hort.* alba *Ger.* Oxys sive Trifolium acidum flore albo et purpurascente *J.B.* Lujula sive Alleluia *Officin[arum]* Panis cuculi *Brunf. Wood-sorrell. In a grove near Burrough-green church; and in the ditches about the park at Cheveley &c.* The pod containing the seed is not round, as *Ger[arde]* & *Park[inson]* state, but pentagonal, as *J.B.* rightly says; if, when it is ripe, you have squeezed it quite strongly in its upper part with your finger, the seeds will jump out from the corners with force.

---

[770] *eodem modo quo pastinacæ sativæ*.

[771] Correctly *Apicii*: one of the family, Marcus Gavius Apicius, was a notorius epicure in Rome under Tiberius who was written about by Pliny, Seneca and others. An anonymous Latin book on cookery, *De re coquinaria*, probably dating from the late fourth or early fifth century and still extant, was later attributed to an otherwise unknown Apicius Caelius and is often referred to as *Apicius*.

[772] The species of *Tragopogon* used as a vegetable is the purple-flowered *T. porrifolius* L., the English name for which, salsify, is derived from the French *salsifis*, believed to be corrupted from the Italian *sassefrica*, which is of unknown origin (OED).

[773] The superscript 9 represents the Latin ending *-us*; see Chapter 7.

579 *Lotus corniculatus* L.
Trifolium corniculatum primum *Dod. Thal.* siliquosum minus *Tab. Ger.* Lotus corniculata glabra minor *J.B.* pentaphyllos siliquis rectis I, i.e. Lotus sive Melilotus pentaphyllos minor glabra *C.B.* Melilotus coronata *Ad. Lob. Birds-foot-trefoile, Small codded Trefoile.* Very common in meadows & pastures.

580 *Lotus pedunculatus* Cav.
Trifolium corniculatum tertium *Dod.* Loti corniculatæ major species *J.B.* Lotus pentaphyllos flore majore luteo splendente *C.B.* Trifolii siliquosi varietas major *Ger. emac. The Greater Birds-foot-trefoile.* In moist meadows & by hedges. *In a wood at St George Hatley, and about Gamlingay and many other places.*

581 *Medicago arabica* (L.) Huds.
Trifolium cochleatum folio cordato maculato *C.B.* cordatum *Ger.* Medica echinata glabra cum maculis nigricantibus et sine illis *J.B.* Medica Arabica Camerarii sive Trifolium cordatum *Park. Heart-trefoile, or claver.*[774] *In the field on the right hand of the lane which leads from Barnwell to the Pesthouses, or the Common called Coldhams, on the green by the lanes side.*

582 *Medicago minima* (L.) Bartal.
Trifolium echinatum arvense *C.B.* Medica echinata minima *J.B.* echinata parva recta *Park., but wrongly, for it is not erect.* pusilla capsulis echinatis *Cam. The smallest Hedgehog-trefoile. In an old gravell-pit in the corn field near Wilborham church; also at Newmarket where the* Sesamoides Salamanticum *grows.*[775]

583 *Trifolium fragiferum* L.
Trifolium fragiferum *Ger. emac. C.B.* fragiferum Frisicum *Clus. hort. Hafn. Park.* vesicarium *of some authors. Strawberry-trefoile. In the borders of plowed fields, and by the foot paths in many places.* That species of Trifolium fragiferum which grows generally with us at the margins of fields differs from Frisicum at least in the fact that its leaves are not obtuse and cordate but a little more acute.

584 *Trifolium campestre* Schreb.
Trifolium luteum lupinum *Ger. emac.* agrarium *Dod.* pratense luteum fœmina flore pulchriore, sive lupulino *J.B.* pratense luteum capitulo lupuli vel agrarium *C.B. Hop-trefoile.* Frequent in fields & among standing corn, *as in the closes near Chesterton church.*

585 *Trifolium dubium* Sibth.
Trifolium lupulinum alterum minus. *The lesser Hop-trefoile.* This of all the trefoils that we have seen is the smallest. Jean Bauhin seems to understand this plant by his Trifolium luteum fœminum flore lupulino minus,[776] which he asserts is common, just as this one is too.

586 *Medicago lupulina* L.
Trifolium luteum minimum *Ger. emac. Park.* luteum I, capitulo breviore *Thal.* Trifolii genus medicæ simile *Gesn. hort.* Trifolium pratense luteum capitulo

---

[774] The prevalent Middle English name was "claver"; the form "clover", apparently retained from Anglo-Saxon in some dialects, was very rare before 1600 and did not prevail much before 1700, after which it became the standard one and "claver" became obsolete (OED).

[775] See the entry for *Sesamoides Salamanticum magnum* [*Silene otites* 539] and footnote 730.

[776] Literally "Yellow female lesser Trefoil with a hop-like flower".

breviore *C.B.* pratense luteum mas, flore minore, semine multo *J.B. Little yellow Trefoile.* In meadows & pastures.

587 *Medicago sativa* subsp. *falcata* (L.) Arcang.
Trifolium sylvestre luteum siliquâ cornutâ, vel Medica frutescens *C.B.* Medica sylvestris *J.B.* frutescens, sive flavo flore Clusii *Park.* Lens major repens *Tab. Yellow Medick, with flat wreathed cods. In many places in the corn fields, as between Linton and Bartlow by the roads side, between Cambridge and Trumpington near the river Cam, about Quoy church, and Wilborham.*

588 *Trifolium repens* L.
Trifolium pratense album *Fuch. Dod. gal. Park. C.B. Cam. Ad.* minus pratense, flore albo sive 2 *Ger.* pratense flore albo minus & fœmina glabrum *J.B. white-flowered Meadow-trefoile.* In meadows & pastures almost everywhere.

589 *Trifolium ochroleucon* Huds.
Trifolium pratense hirsutum majus flore albo-sulphureo, not yet described as far as we know. *From a black root more slender than the smallest finger there arise several round, hairy stems a foot or a cubit tall, bent towards the ground. The leaves are also hairy, divided in three as in the rest of this genus, neither scalloped nor incised, the lower ones smaller & rounder, borne on long, hairy stalks, the upper longer, joined more closely to the stem. The spike of flowers is fairly long, acuminate and congested; its colour is whitish-yellow & approaching sulphur-coloured in older flowers. Where the flower-stalk begins you have two leaves opposite each other,*[777] *while elsewhere on the stems they arise alternately. This plant approaches very closely to Trifolium majus flore albo incanum Clusii J.B. as far as the picture goes. It also does not differ very much from Trifolium pratense album depicted by Fuchs or mas of the same author. However both of these have scalloped leaves, but ours does not likewise, & besides they have leaves which are more frequent at the root & larger while this one of ours has far smaller ones.*[778] *The great white or yellowish Meadow-trefoile. About Cherry-Hinton in many pasture closes and elsewhere.*

590 *Trifolium pratense* L.
Trifolium pratense *Ger.* pratense purpureum *C.B. Trag. Fuch. Ad.* purpureum vulgare *J.B.* pratense purpureum vulgare *Park. Common purple Trefoile or Honeysuckle-trefoile.*

591 *Trifolium medium* L.
Trifolium majus flore purpureo sive 4 *Ger.*[779] majus 3 purpureum Clusio *J.B. Johnson. in Ger.* spicatum II, sive spicâ oblongâ rubrâ *C.B.* montanum majus flore purpureo *Park. Great purple Trefoile, Common Claver-grasse.*[774] J. Bauhin's description does not square in all respects with our Trifolium majus, for it has neither leaves that are minutely toothed on the edges nor calyces of the florets that are hispid. It is sown in fields. *Lately in an enclosed ground near the river Cam, not farre from Newnham by the foot way to Grantcester.*

---

[777] ... *duo habes folia invicem opposita*, literally "... you have two leaves on both sides opposed". Perhaps *habes* ("you have") is a misprint for *habet* ("it has").

[778] *& præterea folia quæ ad radicem frequentiora illis sunt & majora, huic nostro longè minora*, literally "& besides leaves which at the root [are] more frequent are to them & larger, [while] to this our [one they are] much smaller".

[779] See also the discussion about this species in *Emendanda* of the 1685 appendix.

592 *Menyanthes trifoliata* L.

Trifolium palustre *J.B. C.B.* paludosum *Lob. Ger. Park.* majus & fibrinum *Tab.* Lotus palustris *Gesn. hort.* Manianthes palustre Theophrasti *Lugd.* Isopyron *Dod. gal. Gesn. hort. Marsh-trefoile, Buck-beans.* Very frequent in marshy & watery places.

593 *Trifolium striatum* L.

Trifolium dilutè purpureum glomerulis florum oblongis sine pediculis caulibus adnatis.[780] From a white, woody, useless root it sends forth several stems, a hand's-breadth to a hand's-breadth and a half tall, rounded, hairy, pale green & somewhat recurved at each node. Likewise the leaves are woolly, the lower ones sometimes cordate, the upper more acute, very slightly scalloped around the edge, borne on longish stalks, which project from a sort of membranous appendix[781] clothing the joints. Next the dense[782] oblong capitula of pale purple florets, hidden away in a mass, are borne crowded on the nodes as well as at the top of the little stems. The sometimes purplish, striate calyces of the florets, each containing a single largish seed, are furnished with five rough rays arranged like a star on the margin. This plant is annual & varies in size according to the soil, for sometimes in moorish[783] & more barren places it scarcely exceeds three inches but sometimes in meadows it surpasses half a foot. Is it the Trefoil that is knot-flowered or flowering close to the leaves which is said in *P[hytologia] b[ritannica]* not yet to have been described? Is it *J[ean] B[auhin]*'s Trefoil the stems of which bring forth oblong clusters from the nodes?[784] *In all the closes you pass through going from Cambridge to Chesterton church, and in many other places.* It flowers around the end of May and at the beginning of June. No name is commoner for various plants than that of Trifolium, by which we embrace various kinds not only of grasses but also of trees & shrubs. *Spigel[ius in] Isag[oges], book 1, chapter 17.*

594 *Aster tripolium* L.

Tripolium majus et minus *J.B.* vulgare majus et vulgare minus *Ger.* majus sive vulgare et minus *Park.* Tripolium I, sive majus cæruleum, et II, sive minus *C.B. Sea Starwort. In most of the ditches about Wisbich.* It is taller or smaller in relation to the soil.

595 *Triticum aestivum* L.[785]

Triticum spicâ muticâ, et spica et granis albicantibus. Tritici hyberni aristis carentis genus secundum, sive Triticum siligineum *C.B. in Theatro.*[786] Siligo hyberna sive

---

[780] "Pale purple Trefoil with oblong clusters of flowers without stalks [and] adnate to the stems."

[781] i.e. a stipule.

[782] *denso*, apparently a misprint for the neuter plural *densa* agreeing with *capitula*.

[783] For "moorish" as the translation of *montosis* see footnote 154.

[784] *An Trifolium nodiflorum vel juxta folia floridum nondum descriptum* P.B? *An Trifolium cujus caules ex geniculis glomerulos oblongos proferunt* J.B?

[785] Ray's primary distinction is between the beardless wheats (*Triticum spicâ muticâ* ...) and the bearded (*Triticum aristis* ...); he then uses the colour of the ear and grain to subdivide them. The first five variants were almost certainly land races of *Triticum aestivum* (cf. Percival 1948). It was thought that the best flour for bread-making was obtained from a mixture of land races of red and white wheat, so that they were often grown as a mixture (Letts 2000).

[786] "Wheat with a docked ear, with both the ear and the grains whitish. The second kind of winter wheat lacking awns, or White-wheat Wheat *of C[aspar] B[auhin] in [his] Theatrum [Botanicum]*." The adjective *siligineus, -a, -um* is derived from *siligo*, a kind of very white wheat known as "winter wheat".

Sementina Columellæ *for the same author in the same place*. Siligo *Moufet*, writing about diet.[787] Σιλιγνις [Silignis] of the Greeks and Galen *C.B.* and Τρυγις [Trygis] for Hippocrates. *White Wheat, white Lammas*. This is especially sought after for the making of starch;[788] i.e. it is most particularly sought after so that a sticky paste may be made from its starch, which women use to whiten and stiffen their robes and collars and make them firm and flat. *C.B. in Theat[ro]*.

596 Triticum aestivum L.
Triticum spicâ muticâ albicante, granis rufescentibus.[789] *White-eared red Wheat*. It differs from the first only in the colour of the grains.

597 Triticum aestivum L.
Triticum spicâ muticâ, et spicâ et granis rufescentibus. Tritici hyberni aristis carentis genus primum, πυρος [pyros] Veteribus, Robus Columellæ *C.B.*[790] Robus Columellæ *Moufet, writing about diet.*[787] *Kentish-wheat, red Wheat, red Lammas*.

598 Triticum aestivum L.
Triticum aristis donatum spica albicante, granis rufescentibus.[791] *Red Wheat, with a bearded white ear.*

599 Triticum aestivum L.
Triticum aristis donatum granis & spica rufescentibus. Tritici hyberni primum genus cristatum *C.B. in Theat.*[792] *Bearded Kentish Wheat*. Here and there among the preceding kinds.

The kinds of Wheat enumerated so far are sought after not only for bread but also for making meat pies; those that follow are less suitable for this use both because the lump of dough made from them slumps easily from excessive flabbiness and does not conform to the intention of the person moulding it and because it develops cracks during kneading.

600 Triticum turgidum L.[793]
Triticum aristis circumvallatum glumis hirsutis spicâ albicante.[794] *Gray Wheat*. This kind is chosen above all among our countrymen for making grits or a

---

[787] i.e. *Healths improvement* (1655): see footnote 4.

[788] *turbasis*. We have not found *turbasis* in dictionaries of classical or mediaeval Latin, but it is defined as "sterch" in 16th- and 17th-century Latin dictionaries (e.g. Cooper 1573, Holyoke 1606) and by Gouldman (1669) as "starch made of wheat bran, or cuckow-pit roots". Ewen & Prime's (1975) translation, "frumenty", appears to have been an unlucky guess.

[789] "Wheat with a whitish docked ear, with reddish grains."

[790] "Wheat with a docked ear, with both the ear and the grains reddish. The first kind of winter wheat lacking awns, πυρος for the Ancients, *Robus* for Columella, *of C[aspar] B[auhin]*." For Lucius Junius Columella see Chapter 4.

[791] "Wheat endowed with awns, with a whitish ear, with reddish grains."

[792] "Wheat endowed with awns, with the grains and the ear reddish. The first crested kind of winter wheat *of C[aspar] B[auhin] in [his] Theat[rum Botanicum]*."

[793] *Triticum turgidum* was probably introduced to Britain by the Normans and was widely cultivated in central and southern England by the late mediaeval period (Letts 2000). Red and Grey Pollard were land races of this species cultivated in the 17th and 18th centuries (Percival 1921, 1948). In his classification scheme Percival includes them under *T. turgidum* var. *dinurum* Körn. and var. *iodurum* Körn. respectively, two of the most widely cultivated varieties of this species in Europe. He confirms that flour from *T. turgidum* is unsuitable for bread-making.

[794] "Wheat protected by surrounding awns, with hairy husks [and] with a whitish ear." The passive participle *circumvallatus, -a, -um* suggests that the ears are surrounded by protective palisades or ramparts.

gruel[795] prepared for farmers from dehusked wheat & milk. Just as it is a common feature with all kinds of wheat that each grain that exists in the axils of the ears is covered with three husks, so also it is characteristic of this & its relatives that they have two thicker or outer ones unequal[796] both in length and in breadth; what is more the outer one, which is smaller, does not embrace the inner one directly as in the other kinds but only at the side. Not infrequently the ears of this kind, as also of the next, either are formed without awns or shed them before maturity; hence they are called by our farmers *Red Pollard, and gray Pollard*, that is Triticum tonsum rubrum & album.[797]

601 *Triticum turgidum* L.

Triticum aristis munitum, & spicâ & granis rufescentibus, glumis præ cæteris aristatis lævibus & splendentibus.[798] *Red eared bearded Wheat.*

N. *Theophr[astus]* explains *in De causis, book 4, chapter 10,* why Wheat is more favourable in food for man than for the rest of the animals although they are stronger, many of which die when sated with it.[799]

2. Not only the grains of wheat are sought for food by partridges but also the leaves & the stalk, which they eat in such quantity in early spring that it has been frequently noticed by many people that their flesh at that time even tastes quite strongly of the herbage of wheat.[800]

3. Bread made from wheat or any other kind of grain-plant that has been kept for too long not only will reduce its mould or mustiness but will become like new if it it is either placed by the fire or baked again in the oven, as will be confirmed for anyone putting it to the test.

4. Every grain-plant such as wheat, barley, &c. sends forth a root from the stout part of the grain and a leaf from the thin part; but the Bean and the rest of the pulses send forth the root & the stem from the same part, where also the junction with the pods or 'lobes' is situated, where the beginning of germination reveals itself clearly,[801] as we have shown above. Moreover in grain-plants the root comes out first and then the stem. See *Theophr[astus'] Hist[oria], book 8, chapter 7.*[802]

---

[795] *ad conficiendam alicam sive pulticulam.* For the meaning of *alica* see Ray's 'Interpretation'.

[796] *inæquales inter se*, literally "unequal among themselves".

[797] Literally "Shaven Wheat red & white".

[798] "Wheat fortified with awns, with both the ear and the grains reddish, with the husks smoother & more shining than in the remaining awned [varieties]." The passive participle *munitus, -a, -um* is somewhat less expressive than *circumvallatus, -a, -um*.

[799] The reference is to IV. 9. 1 of Theophrastus' *De causis plantarum* in the Loeb edition and Ray's text is clearly adapted from Gaza's (1529, part 2, p. 206; see footnote 21) translation, but his *ualidioribus* ("stronger") has been misread as *calidioribus* ("warmer", "hotter"). It seems that Neolithic humans in the Near East had to develop a tolerance to wheat, but the digestive systems of many individuals are still intolerant to gluten (Greco 1995; Simoons 1982). Theophrastus was probably not aware of gluten intolerance in humans, but it seems to have been first observed by the first-century AD Greek physician Aretaeus of Cappadocia though he did not correctly diagnose its cause (Adams 1856, pp. 350–351).

[800] *… eorum etiam carnes tunc tritici herbam non obscurè sapere*, literally "… of them even the flesh [plural] then of the herbage of wheat not obscurely to taste" (accusative and infinitive, equivalent to an English clause beginning with "that").

[801] *haud obscurè apparet*, literally "not obscurely appears". See also Ray's note on *Vicia faba* [193].

[802] VIII. II. 1–2 of Theophrastus' *Historia plantarum* in the Loeb edition; this is an accurate précis of what Theophrastus wrote, though it is unclear why the pods or 'lobes' (*lobi*, plural of the Greek word

This was the opinion of the Ancients & of those who were philosophising confusedly.[803] For, if we examine the matter more attentively, as that Most Illustrious Man *Master Tho[mas] Brown M.D.* has observed *in his golden little book recently published in English which is entitled The Garden of Cyrus*,[804] we shall see that in Rye & Wheat the sprouts of the root & the leaves not only break out but also appear at the same time and what is more that in Barley & Oats they arise at the same time from the same point or corcule on the seed,[805] even though the roots emerge first from the lowest part and the leaf, after creeping under the thicker husk, comes out from the top later. So it is entirely in vain & rash for anyone to hope that one part of a plant is produced from one part of the seed and another from another, since the germination of both, the root and the blade alike,[806] begins at the same point and origin at the same time, and those who think that Oats and Barley germinate from both ends are too vulgarly mistaken.

602 unidentifiable crucifer[807]

Turritis *Lob. Tab. Ger.* vulgatior *J.B.Park.* Brassica sylvestris VIII, sive sylv[estris] foliis integris et hispidis *C.B.* Vaccaria *Tab.* Tower Mustard. *Found where flax did grow about Cambridge* Ger. *We have not yet met with it.*

603 *Tussilago farfara* L.

Tussilago *Matth. Fuch. Cord in Diosc. Ger. J.B.Park.* vulgaris *C.B.* Chamæleuce Plinii *Jo. Bod[æ]. in Theophr. hist.*[808] Farfara *Ad. Lob.* Farfarella *Cord. in Diosc.* Ungula caballina *Trag. Brunf.* Bechion *Dod. gal. Cord. hist.* Colts-foot, Foals-foot. In moist places. It is called Son before father[809] because in February and

---

λοβος used by Theophrastus, meaning the lobe of an ear and secondarily a leguminous pod) are plural. The phrase "as we have shown above" is not in Theophrastus' text, but he does begin the section by stating that some plants produce their root and leaf from the same point and others from either end of the seed.

[803] *Hæc fuit sententia Veterum & crassè Philosophantium.*

[804] "... prout observavit Clarissimus Vir *D. Tho. Brown M.D. in aureo libello Angl. nuper edito, qui inscribitur Cyri hortus.*" Dr (later Sir) Thomas Browne (1605–1682), "Norfolk's leading physician" (ODNB) and a notable author, wrote (1658, pp. 130–131): "In vain we expect the production of plants from different parts of the seed, from the same *corculum* or little original proceed both germinations; ... . The seminall nebbe hath a defined and single place, and not extended unto both extremes. And therefore many too vulgarly conceive that Barley and Oats grow at both ends; For they arise from one *punctilio* or generative nebbe, and the Speare sliding under the husk, first appeareth nigh the toppe. But in Wheat and Rye being bare the sprouts are seen together. If Barley unhulled would grow, both would appear at once."

[805] *ex eodem seminali puncto sive corculo*, literally "from the same seminal point or little heart".

[806] *utriusque tum radicis tum herbæ germinatio*, literally "of each, both of the root and of the herbage, the germination".

[807] Gerarde's *Turritis* is usually interpreted as *Turritis glabra* L. The only confirmed records in Cambridgeshire are from sandy habitats at Gamlingay (Crompton 2001). Gerarde's record was not accepted as the first vice-county record by Perring *et al.* (1964) and was queried by Crompton (2001); we agree that it must be treated as very doubtful and suspect that the true identity of Gerarde's plant will never be known.

[808] See Pliny's *Natural history*, XXIV. 135 and XXVI. 30 in the Loeb edition, in both of which the use of its smoke to treat an inveterate cough is recommended and in the second of which it is also called *bechion* (the Greek βηχιον) and the statement about water-finders made in Ray's first note occurs. Bodaeus (1644; see footnote 23) discusses the names for this plant on p. 877.1.

[809] *Filius ante patrem*: see also footnote 437.

March, when the leaves have not yet come up, it flowers unexpectedly and puts forth very fleeting flowers that scarcely persist with their stalks more than two days.[810] *Schrod. pharmac.*

N. 1. Water-finders believe that, where it grows wild, there are waters underneath and they treat this as a sign according to *Pliny in book 26, chapter 6*,[808] something which I have several times observed to be true.

2. A sort of whitish down is said to grow on the root of this plant which, if it is first cleaned off sections of the roots and then wrapped up in small linen cloths and boiled a little in lye with some soda added[811] and finally dried in the sun, is the best tinder of all for obtaining fire struck from a flint. In fact the fire is so greedy that it is kindled immediately as soon as the steel is struck. *Jo. Bod[æ]. in Theophr. hist. book 7, chapter 12, page 877.*[812]

604 *Typha latifolia* L.

Typha *Ger. Matth. Fuch. Lob. Tab.* palustris *Ruel. Cæs.* palustris major *J.B. C.B.* Τυφη [Typhe] of Dioscorides and Theophrastus *of the same authors*. Typha palustris maxima *Park*. Calamus aquaticus Plinii *Trag*. *Cats-tail, Reed-mace. In a pond near a house at the hithermost end of Teversham moor, and in diverse other waters.*

V

605 *Valeriana officinalis* L.

Valeriana sylvestris major *Ger.* sylvestris *Ad. Lob. Dod. Cam.* sylvestris magna aquatica *J.B.* sylvestris major *C.B.* Phu with the likeness of Elaphoboscum[813] for Dioscorides *of the same author*. Phu Germanicum *Fuch*. Cats' herb for the Germans,[814] because cats take an eager delight in the the scent of its root, by which their eyes are strengthened. *Great wild Valerian. In many places near Cambridge: as in the closes about Madingley on the right hand as you go thither, and about the wood, &c*

606 *Valeriana dioica* L.

Valeriana sylvestris minor *Park. Lob. Ger.* palustris minor *C.B.* minor pratensis vel aquatica *J.B.* exigua vel 4 *Trag*. Phu parvum *Dod. gal.* minus *Gesn. hort. Tab.*

---

[810] *ultra biduum ferè cum caulibus non perstantes*, literally "beyond a two-day period almost with stems not persisting".

[811] *et in lixivio parùm decocta, addito salis nitri aliquanto*. "Salt of natron" is commonly called "soda"; the OED defines it thus: "An alkaline substance obtained originally from the ashes of certain marine or other salt-impregnated plants, esp. species of *Salsola*, and now manufactured artificially from common salt, or occurring in a mineral state as a deposit, esp. in certain lakes, or in solution in the water of such lakes (natron); used largely in commerce, esp. in the manufacture of glass and soap; soda-ash; sodium carbonate (Na_2CO_3)." See also footnote 306.

[812] The final sentence of Ray's note is: *Quippe adeò ignis capax est, ut primo statim chalybis concussu accendatur.* Literally this means: "In fact so capable is the fire that immediately at the first strike of steel it is kindled." The whole note is taken from Bodaeus (1644, p. 877.1, 2; see footnote 23), but he wrote *rapax* rather than *capax*, giving the meaning in the text above, and *statim* before *primo*, which is a more natural word order.

[813] i.e. parsnip.

[814] *Herba cattorum Germanis*. The usual German name is *Baldrian*, but a Spanish folk-name is *hierba de los gatos*.

MAIN CATALOGUE

minimum *Matth. Cast. Lugd. Small wild Valerian. In moist meadows, and on the moors.*[815]

Veratrum nigrum *Cord. see* Helleborus niger hortensis flore viridi. [*Helleborus viridis* 276]

Verbascum mas & vulgare, *see* Tapsus barbatus. [*Verbascum thapsus* 564]

607 *Verbascum nigrum* L.
Verbascum nigrum *Dod. Lob. ico. Tab. Ger.* alterum nigrum *Cord.* nigrum flore ex luteo purpurascente *C.B.* nigrum salvifolium luteo flore *Lob.* nigrum latifolium *Thal.* nigrum flore parvo luteo, apicibus purpureis *J.B.* Blattaria Plinii, sive Verbascum nigrum. *Eystet. Common black Mullein.* In many places towards Linton, and about Gogmagog hills, as by the lanes sides, and in the closes about Abington, Shelford, &c.

N. In Etruria the common people of the country areas use the flowers of Mullein to clot milk. *Brassavol. in comment. de Syrup. apud Cam. in hort.*[816]

Verbasculum odoratum, *see* Primula veris.[817] [*Primula veris* 467]

608 *Verbena officinalis* L.
Verbena vulgaris *J.B. Cæs.* mas surrecta & vulgaris *Park.* communis cæruleo flore *C.B.* communis & sacra recta *Ad. Lob.* Hierobotane mas *Brunf.* Herba sacra *Ang. Tur.* Hierobotane or herba sacra of Dioscorides, and of the same author Peristereon supinum for some, & Pliny's Verbenaca fœmina *Cæs. Common Vervain.* By footpaths.

Veronica fœmina, *see* Elatine.[818] [*Kickxia spuria* 176]

609 *Veronica officinalis* L.
Veronica mas supina & vulgatissima *C.B.* mas *Fuch. Lugd.* mas serpens *Dod.* mas vulgaris supina *Clus. hist. Park.* vulgatior folio rotundiore *J.B.* vera & major *Ger. Ad. Lob. ico. Cam.* Auricula muris 3 *Cæs. Speedwell or Fluellin.* In the old parke near Linton on the ant-hilles, and elsewhere.

610 *Veronica serpyllifolia* L.
Veronica pratensis *Dod.* pratensis minor *Park.* VI, sive pratensis serpillifolia *C.B.* minor serpillifolia *Ad. Lob.* Betonica Pauli *Gesn. hort. Tur.* Pauli quorundam *Cam.* Veronica fœmina quibusdam, aliis Betonica Pauli serpillifolia *J.B.* Teucrium alterum *Trag. Little Fluellin, or smooth Fluellin, or small meadow Fluellin or Speedwell.* In meadows & pastures, as about Madingley, &c.

611 *Veronica spicata* L.
Veronica spicata recta minor *J.B.* V, sive spicata minor *C.B.* Ver[onica] assurgens sive spicata *Ger.*, from which species Ver[onica] recta minima *of the same author* perhaps does not differ. Veronica mas erecta *Park. Upright male Speedwell or Fluellin.* In severall closes on Newmarket heath, as in a close near the beacon on

---

[815] The male plant of this dioecious species is referred to here. See also *Valeriana sylvestris sive palustris minor altera* [B101] in *Addenda* to the 1685 appendix, where the female plant is described.

[816] i.e. Brasavola in his treatise about syrup, quoted by Camerarius in his *Hortus medicus et philosophicus*. Antonio Musa Brasavola (or Antonius Musa Brassavolus) of Ferrara (1500–1555) published several medical treatises, including one entitled *Examen omnium syruporum, quorum publicus usus est* at Lyon in 1540 from which this statement was presumably taken by Camerarius (1588).

[817] Ray cites *Verbasculum odoratum* as a synonym of *Primula veris major* [467].

[818] Ray cites *Veronica fœmina* as a synonym of *Elatine folio subrotundo* [176].

the left hand of the way from Cambridge to Newmarket in great plenty. It flowers in July & August.

612 *Viburnum lantana* L.
Viburnum *Park. Matth. Gesn. hort. Cast. Cam. Lugd.* vulgo *C.B.* Spiræa Theophrasti Dalechampio *Lugd.* Lantana sive Viburnum *Ger.* Lantana *Gesn. hort.* because the branches are flexible.[819] Lantana vulgò, aliis Viburnum *J.B.* Viurna vulgi Gallorum & Ruellii *Lob. Cast.* Rhus coriariorum Theophrasti & Plinii *Cæs. Pliant-mealy-tree* Park. *Wayfaring-tree* Ger. *Cotten-tree* P.B. Not infrequent in hedges.

Vicia *Tab. see* Lathyrus sylvestris. [*Lathyrus pratensis* 326[820]]

Vicia multiflora sylvestris, *see* Aracus. [*Vicia cracca* 55]

613 *Vicia hirsuta* (L.) Gray
Vicia parva sive Cracca minor cum multis siliquis hirsutis *J.B.*[821] segetum cum siliquis plurimis hirsutis *C.B.* sylvestris sive Cracca minima *Ger. emac. Small wild Tare or Tine-tare.* The pods of this species are not glabrous, as Johnson observed in his revision of Gerarde, but, as C. B[auhin] stated rightly, hairy.

614 *Vicia tetrasperma* (L.) Schreb.
Viciæ sive Craccæ minimæ species cum siliquis glabris *J.B.* Aracus sive Cracca III, seu Vicia segetum singularibus siliquis glabris *C.B. Tine-tare with smooth single cods. In the corn as you goe from Hoginton to Huntington roade; and in the woods at St George Hatley.* The colour of the flowers in our plant differs both from that which is described by J.B. & from that which is described by C.B., just as they themselves also differ between themselves about this matter, for it is of a washed-out blue. J. Bauhin states that it is endowed with numerous pods & like him we have frequently seen it thus.

615 *Vicia sativa* L. subsp. *sativa*[822]
Vicia *Ger.* vulgaris sativa *Park. J.B.* sativa vulgaris semine nigro *C.B.* Orobus sativus & Vicia major & I *Trag.* Ervum *Brunf.* Βιϰιον [Bicion] vicia about which Varro writes, Ἀραϰον [Aracon] for the Athenians[823] *Lob. in Ad. Common Vetch or Tare.* It is sown in fields.

N. For three consecutive months bees collect a large amount of honey from the black spots on the upper leaves of this plant and (which is very remarkable) they do not touch the flowers on the same expedition; so what Aristotle says in *book 9, chapter 40, of his Hist[oria] animal[ium]* is true. It is the custom for bees not to seek out several kinds of flowers on one and the same expedition but for each to

---

[819] *quòd lenti sunt rami*, in the belief that *Lantana* is derived from the adjective *lentus, -a, -um*.

[820] *Vicia* Tab. is not given as a synonym of *Lathyrus luteus sylvestris dumetorum* J.B. [*Lathyrus pratensis* 326] in the *Catalogus*, but Johnson (1633, p. 1231) gives *Vicia* of *Tabernamontanus* as a synonym of his *Lathyrus sylvestris flo. luteo*, which is cited by Ray as a synonym of this species with the same English name, "Tare everlasting".

[821] Ray accidentally included this species twice in the *Catalogus*. See his comment on *Vicia parva* in *Emendanda* of the 1663 appendix.

[822] We have assumed that subsp. *sativa* was the only plant commonly grown in Britain, as argued by Sell & Murrell (2009). If Stace (1997, 2010) is correct in stating that subsp. *segetalis* is sometimes cultivated for fodder, that subspecies should perhaps be included here with subsp. *sativa* rather than in Ray's entry for native *Vicia sativa* [56].

[823] *Atticis*, i.e. for the people of Attica, of which Athens was the capital city.

seek out its own kind.[824] *Butler, in his treatise about bees,*[825] *chapter 37, note 49. What is more beans too have leaves marked with similar spots that are very pleasing to bees. The same source, note 48.*[826]

616 Vicia sepium L.
Vicia maxima dumetorum *Ger. emac. Park. C.B.* sepium perennis *J.B. Bush-vetch. In a thicket on the east side of Linton.* It flowers in the month of May.

Vinca pervinca, *see* Clematis daphnoides. [*Vinca minor* 138]

Viola aquatica, *see* Millefolium aquaticum. [*Hottonia palustris* 365]

617 *Viola odorata* L. var. *odorata*
Viola nigra seu purpurea *Ger.* Martia purpurea *J.B.* Martia rotundifolia I, sive purpurea flore simplici odoro *C.B,* præcox purpurea *Lob. Purple March Violet.*

N. 1. Those flowers whose leaves are odourless give off the most pleasing scent of all. *Verul[am], Hist[oria] nat[uralis], cent[ury] 6, exp[eriment] 589, & Theophr. de causis plant., book 6, chapter 13.*[827]

2. An emulsion of the seeds of this plant provokes very much vomiting and retirement to the privy. *Schrod. Pharm.*

618 *Viola odorata* var. *dumetorum* (Jord.) Rouy & Foucaud and/or var. ***imberbis*** (Leight.) Hensl.[828]
Viola Martia alba odora *C.B.* flore albo *Ger. Common white Violet.*

N. Coloured flowers generally give off a more pleasing scent than scented white ones of the same species, if you exclude the flowers of plants that are

---

[824] *sed singulis singula*, literally "but for single ones [i.e. each one] single ones [i.e. one each]". In his Latin translation of Aristotle's *Historia animalium* Theodorus Gaza (see footnote 21) moved book 9 of the manuscripts to follow book 6, but the Loeb edition presents the chapters in their earlier order, so Ray's book 9 is book VIII there. Aristotle does not mention extrafloral nectaries (see footnote 826 below), but he does say (in VIII. XL: 624b in the Loeb edition) that on each flight a bee "does not go on to flowers different in form; it goes for example from violet to violet, and does not touch any other before it has flown back to the hive" and (in VIII. XL: 625b) that bees are "appointed to each of the tasks, for example some are flower-gatherers, others are water-gatherers, and others smooth and level the combs".

[825] *The Feminine Monarchie* (Butler 1634): see footnote 40.

[826] The black spots on the leaf-like stipules of *Vicia* spp. are extrafloral nectaries. Free (1962) confirmed the observation reported here: 86% of the honeybees he observed visited either the extrafloral nectaries or the flowers of field beans (*Vicia faba* L.) and only 14% visited both during the period he was able to keep them under observation.

[827] For the work of Baron Verulam (Sir Francis Bacon) see footnote 104. In *De causis plantarum* (VI. 11. 4 in the Loeb edition) Theophrastus gives the violet and the rose as examples of plants with scentless leaves but scented flowers.

[828] Ray continued to treat this and the preceding, purple-flowered plant separately in his later publications. Initially (1670, 1677) he did not comment on the differences between them. In *Historia plantarum* (1688a, pp. 1049–1050) he noted that they differed only in flower colour, not in the other differences which had been suggested by J. Bauhin. In his *Synopsis* (1690, 1696) he said that they did not seem to him to be specifically distinct and suggested that perhaps the paler flower colour resulted from the dryness or the sterility of the soil, a similar suggestion to that made for *Centaurea scabiosa* [298, 299, q.v.]. The white-flowered variants of *Viola odorata* are now known to be genetically distinct and are divided by modern botanists into two varieties, both of which occur in Cambridgeshire (Leslie, Pannell & Walters 1982; Walters 1946). There is no way of knowing whether Ray saw one or both of these taxa.

succulent such as the lily & the white orchid & the like. *Verulam, Hist[oria] nat[uralis], cent[ury] 6, exp[eriment] 507.*[827]

619 *Viola reichenbachiana* Jord. ex Boreau and/or ***V. riviniana*** Rchb.[829]
Viola canina sylvestris *Ger. Trag.* Martia inodora sylvestris *C.B.* canina cærulea inodora sylvestris serotina *Lob. ico.* cærulea Martia inodora sylvatica, in cacumine semen ferens *J.B. Dogs Violet, late Violet.* By hedges & in thickets.

Viola pentagonia,[830] *see* Speculum Veneris. [*Legousia hybrida* 556]

620 *Viola arvensis* Murray[831]
Viola tricolor *Dod. Clus. Ger.* tricolor major & vulgaris *Park.* tricolor hortensis repens *C.B.* tricolor sive Trinitatis flos *J.B.* An Viola bicolor *C.B?* Jacea sive Flos Trinitatis *Matth. Hearts-ease, or Pansies.* In sandy places & among standing corn, *as upon the hill of Health.* This plant varies considerably in the colour & size of the flower; the kind that grows here generally has a small whitish flower streaked with blue lines.

621 *Clematis vitalba* L.
Viorna *Ger.* vulgi *Lob. Eyst.* Clematitis sylvestris latifolia *C.B.* Clematis sylvestris latifolia, sive Viorna *Park.* Clem[atis] latifolia dentata *J.B.* Vitis sylvestris *Trag. Lugd.* sylvestris caustica *Gesn. hort.* Atragene Theophrasti *Ang. Clus.* Uva Taminia of Celsus & the Latins, Vitis sylvestris of Dioscorides, & Athragena of Theophrastus *Dalechamp. in book 6 of Aegineta, chapter 77.* Clematis laciniata flore albo erecto, sive Vitalba Dodonæi *hort. Bat. Great White Climer, Travellers joy.* Very frequent in hedges around *Hinton, Teversham* and elsewhere.

Virga sanguinea, *see* Cornus fœmina. [*Cornus sanguinea* 151]

Virga pastoris, *see* Dipsacus minor. [*Dipsacus pilosus* 169]

622 *Viscum album* L.
Viscum *Ger.* vulgare *Park.* arborum I, sive baccis albis *C.B.* Viscus Quercûs & cæterarum arborum[832] *J.B.* Ἰξος [Ixos] of Aristotle & Ἰξια [Ixia] of Theophrastus *J.B. C.B.* Ἰξος [Ixos] of Dioscorides *C.B.* Aeneas' golden branch in the work of

---

[829] This identification is based on the habitats listed by Ray. These two species of wooded habitats were not separated by botanists until the mid 19th century. (*V. reichenbachiana* was described in 1857.) Ray may well have also seen the later-flowering *V. canina* L. in Cambridgeshire on heaths and in fens, but it was not until his *Catalogus plantarum Angliae* (1670, p. 317) that he wrote of *Viola canina sylvestris*: *Habetur & in palustribus frequens* ("It is also frequently found in marshes"), adding: *nisi fortè ea sit distincta species* ("unless perhaps that may be a distinct species").
[830] Ray does not actually cite this name as a synonym of *Speculum Veneris minus* [*Legousia hybrida* 556].
[831] Although Babington (1860) listed this synonym under *Viola tricolor* L., he commented that the Cambridgeshire plant was the segregate *V. arvensis*. Ray's description confirms that his plant was the latter, which has smaller and paler flowers than *V. tricolor* and is now accepted as a distinct species. His comment in *Emendanda* of the 1685 appendix also makes it clear that the plant that he knew in Cambridgeshire had bicoloured, not tricoloured, flowers. There is, therefore, no reason to accept Ewen & Prime's (1975) view that Ray's plant is *V. tricolor* as currently defined.
[832] "Mistletoe of Oak & of the rest of trees".

Virgil[833] *J.B.* Misseltoe. *Upon apple-trees chiefly: we have sometime found it upon White-thorne.*[834]

N. We suggest that it should be examined by diligent investigators of nature whether the seed of this shrub (when it is mature and ripe) sometimes produces a plant of its own kind. Aristotle, Pliny and the entire assembly of the Ancients maintain an affirmative position, but Julius Scaliger, J. Bauhin and most of the more recent authors a negative one. See *Scal. Subtil. exerc. 168*,[835] & *J. Bauhin, volume 1, part 2, book 7, page 91.* But just as it seems to us fairly certain that mistletoe not infrequently grows without a seed, especially on that part of a branch that faces the ground, so it is scarcely credible that nature has created a perfect seed in a whole species that is useless for generating it & infertile.

Vitis alba & nigra, *see* Bryonia alba & nigra. [*Bryonia dioica* 91 and *Tamus communis* 92]

623 *Filipendula ulmaria* (L.) Maxim.

Ulmaria *Gesn. Clus. Lugd. Cam. Tab. J.B.* vulgaris *Park*. Regina prati *Dod. Ger.* Barba capri *Ad. Lob.* capri floribus compactis *C.B.* Medesusium *Cord. hist. Meadsweet.* In moist meadows generally.

624 *Ulmus minor* Mill.

Ulmus folio glabro *Ger. emac. Park. The smooth-leaved or Wych Elm.*

625 *Ulmus glabra* Huds.

Ulmus folio latissimo scabro *Ger. emac.* folio latiore *Park*. Ulmus *J.B.* Ulmus Theophrasti *Park. Jo. Bodœ. in Theophr. hist.*[836] *C.B.* Ulmus I *Dod*. 2 sive lata *Trag*. I in latitudinem magìs fusa *Cæs*. nostras, Italica & sylvestris Plinii *Park. Jo. Bodœ.* vernacula Columellæ *Park. Dod. C.B. Jo. Bodœ.* campestris Plinio *Jo. Bodœ. C.B.* Ulmus I, sive campestris & Theophrasti *C.B. The broadest-leaved Elm or Wych-hasell. By the road side within half a mile of Bartlow and elsewhere.*

626 *Ulmus procera* Salisb.

Ulmus vulgatissimus folio lato scabro *Ger. emac.* vulgaris *Park*. Ulmi genus alterum quod in altitudinem ingentem adolescit *Cæs*. Ulmus I sive procera *Trag*. II, sive montana *C.B.* 2ª *Dod*. Gallica Columellæ *Park. Dod. Jo. Bodœ.* Atinia

---

[833] *Aureus ramus Æneæ apud Virgilium*. The reference is to Virgil's *Aeneid*, 6: 136–148 and 185–211, where Aeneas, guided by the Sybil of Cumae and by his mother, Venus, finds and plucks from a holm oak (line 209: *ilice*) the golden bough, sacred to Proserpina, which will enable him to visit Hades. In lines 205–209 Virgil compares this to mistletoe, thus suggesting that he did not regard it as actually that plant.

[834] *Crataegus* spp.

[835] *Exotericarum exercitationum liber quintus decimus, De subtilitate: ad Hieronymum Cardanum* by Julius Caesar Scaliger (Giulio Cesare della Scala or Scaligero, 1484–1558), a leading classical scholar of his period, was first published in Paris in 1557. *Exercitatio* [literally "Exercise" or "Practice"] *CLXVIII* is on pp. 230–232 and includes the statement that Mistletoe is produced from trees in the same way as horns are from animals' bones (*Sed quasi cornua ex ossibus animalium: sic ex arboribus educi Viscum.*). There were several subsequent editions. For Hieronymus Cardanus or Girolamo Cardano see footnote 5.

[836] i.e. πτελεα [ptelea], one of two kinds of elm described by Theophrastus in his *Historia plantarum* (III. XIV. 1 in the Loeb edition); see also footnote 838.

& Gallica[837] of Pliny *for the same authors*. Montiulmus or Ulmus montosa[838] of Theophrastus *Park. Dod. Jo. Bodæ*. The lofty elm *of Virgil*.[839] The two preceding species have spreading branches but this one more erect branches. *The common Elm.*

N. 1. In all species of Elm one half of the leaf projects further (understand this as meaning at the middle vein) than the other.

2. In the month of May vesicles erupt on the leaves of this tree that are full of a strongly vulnerary fluid.[840] *Cam. hort.*, about which see also *J. Bauhin, Hist[oria] plant[arum], volume 1, part 2, book 8, page 144, & Sennertus in Paralipom[ena]*[841] *in book 1, pract. chapter 19*. Henry from Heer in his *Observationes medici*[842] tells two memorable stories about this juice, the former of which it will not be inappropriate to transcribe here. Twenty-nine years ago, I travelled far over sea and land with a certain Lithuanian who professed alchemy[843] and various arts, serious, ludicrous and even magical, of which however in my presence he was never able to provide any sample although he tried earnestly. This man had gathered, from all the physicians, vagabonds,[844] old women, sorcerers and alchemists in the whole of Asia & Europe over which he had wandered, secrets, as he used to call them, either by entreaty or by payment or very many by barter. Among other things he presented a French cock to me, holding it in one hand, and ordered me to replace in the orbit in its skull the eye which had been frequently punctured with a very sharp scalpel and squeezed between my fingers; he then poured on it two drops of a liquid and shut it away, with its head bound up, in a chest, from which he removed it after eight hours, with each eye so alike that neither I who had cut it nor any of the bystanders could distinguish the cut eye from the sound one.[845] That juice, as he confessed to me, was the one that is found in the blisters on elms in the month of May and which, unless it is collected

---

[837] i.e. the elm of the town in Latium near Rome still called Atina and of Gaul.

[838] i.e. Mountain Elm, ὀρειπτελεα [oreiptelea], the first kind of elm described by Theophrastus in his *Historia plantarum* (III. XIV. 1 in the Loeb edition); see also footnote 836.

[839] *Ulmus aeria Virgilii*. The reference is to the poet's *Eclogues*, 1: 58: *nec gemere aëria cessabit turtur ab ulmo* ("nor shall the turtle-dove cease to coo from the lofty elm").

[840] *plenæ humore admodum vulnerario*: i.e. the fluid is very useful in healing wounds, not, as translated by Ewen & Prime (1975) "an injurious moisture". The vesicles are presumably one of the numerous types of gall that are found on elm leaves.

[841] Daniel Sennert's (1572–1637) medical treatise *Paralipomena* was published posthumously in Lyon in 1643.

[842] *Observationes medicae oppido rarae* by Henri de Heer (1570–1636), called by Ray "Henricus ab Heers", was published in 1630; in 1645 the book was republished posthumously, with another work, *Spadacrene*, at Leipzig and Leiden.

[843] *Ante annos 29 multum maris & terræ emensus sum cum quodam Lithuano, chemiam ... profitente*, literally "Before 29 years much of sea & land I measured out with a certain Lithuanian professing ... alchemy". These are the words of Henry from Heer himself. *Chemia* is the Greek χημεια (a variant of χυμεια), which, at least in classical times, meant "*the art of alloying metals, alchemy*" (Liddell, Scott & Jones 1940) rather than what today would be called chemistry. It therefore seems appropriate also to translate *ab omnibus ... chemicis* later in the passage as "from all the ... alchemists" rather than "from all the ... chemists". See also Ray's note 2 on *Sambucus* [519].

[844] *agyrtis*, latinised ablative plural from the Greek ἀγυρτης [agyrtes], "prop. *collector*, esp. *begging priest of* Cybele" but hence "*vagabond*" (Liddell, Scott & Jones 1940).

[845] This story may not be quite as unlikely as it sounds, if the eye was not actually detached from its orbit, as the aqueous and vitreous humours can be replaced very rapidly if they are lost when the eyes of birds are punctured (T.R. Birkhead, *pers. comm.*).

then, is transformed into some sort of red worm-like granules. The other story is about the daughter of a nobleman whose eye, which had been punctured with a needle, he healed completely with this liquid within the space of one night even after all the aqueous fluid and part of the vitreous fluid had flowed away; however it remained deprived of light & vision, but in such a condition that even after twenty years no one could distinguish the blinded eye from the seeing one even if he was examining them carefully; on the contrary she lived for many years with a husband who was unaware of her loss.

3. The Male Lime described by [Mathias] de L'Obel, Gerarde & Parkinson[846] does not seem to differ from some sort of species of elm, as *J.B.* rightly affirms *in volume 1, part 2, page 141*. Compare *Park[inson] & Gerard[e]*.

4. From the inclination of tall trees growing in the open air[847] it is known from which part of the sky the stronger & more prevalent winds in any particular region usually blow. Thus trees growing near sea-shores are bent towards the land & in the midlands of our country England the tops of the trees usually face east because those regions are especially troubled by Westerly winds[848] assaulting them often & violently.

N. in passing. Trees crowded together grow up without knots and tall while those wide apart grow up more knotted and more contorted; for it happens that one lot is in shady places and the other in sunny ones. *Marant[a], Meth[odus] cogn[oscendorum] simp[licium], book 2, chapter 11*. Compare also *Theophr. de causis, book 2, chapter 11*.[849]

Volubilis, *see* Convolvulus.[850]

Urtica Herculea *Tab.*[851] *see* Galeopsis. [*Stachys sylvatica* 209]

Urtica aculeata foliis serratis, *see* Cannabis spuria. [*Galeopsis tetrahit sensu lato* 105]

Urtica non mordax & iners, *see* Lamium & Galeopsis.[852]

627 *Urtica urens* L.
Urtica minor *Ger.* minor acrior *Lob.* minor annua *J.B.* urens minor *C.B.* urens minima *Dod. Small stinging Nettle.*

628 *Urtica dioica* L.
Urtica vulgaris urens *Trag.* major vulgaris *J.B.* major vulgaris & media sylvestris *Park.* urens *Ger.* urens I & II, sive urens maxima et altera urens *C.B.* fœmina et communis *Dod. Common Nettle.*

---

[846] *Tilia mas à Lob⁰, Ger⁰ & Park⁰. descripta.*

[847] *libero aere*, literally "with free air", which perhaps implies a location where the wind blows freely.

[848] *ventis Occidentalib⁹*: the superscript 9 represents the Latin ending *-us*: see Chapter 7.

[849] For Bartolomeo Maranta see footnote 34. Theophrastus made the same point in *De causis plantarum* (II. 9. 1 in the Loeb edition), though it seems that Maranta copied what he had written in his *Historia plantarum* (I. VIII. 2 in the Loeb edition).

[850] Ray cites *Volubilis major*, *Volubilis minor* and *Volubilis nigra* as synonyms of *Convolvulus major* [*Calystegia sepium* 148], *Convolvulus minor arvensis* [*Convolvulus arvensis* 149] and *Convolvulus niger* [*Fallopia convolvulus* 150] respectively.

[851] *Urtica Heraclea* Lugd. Trag. is actually the name Ray cites under *Galeopsis legitima Dioscoridis* [*Stachys sylvatica* 209].

[852] See *Lamium album* [*Lamium album* 317], *Lamium luteum* [*Lamiastrum galeobdolon* 318], *Lamium rubrum* [*Lamium purpureum* 319] and *Galeopsis legitima Dioscoridis* [*Stachys sylvatica* 209] as well as *Alsine hederula altera* [*Lamium amplexicaule* 25].

N. 1. More than once[853] we have seen, not without pleasure and admiration, (that which *Aldrovandus* reports from *B. Porta in book 14, chapter 1, of his Ornithologia*[854]) a capon caring like a mother for a hen's chicks,[855] which it calls together with a clucking sound in the manner of hens, feeds, keeps warm and cares for; and it does not cease from this work but rears in turn chicks hatched[856] by very many hens until it has given some leisure to them all for some time, whether adults or those removed.[857] It is induced to fulfil this service by rubbing its belly with the leaves of nettles, after the feathers have first been plucked off, which is done towards evening and by placing the chicks under it one or two nights after the wounds inflicted by the nettles until it begins to protect them. Is this because the chicks soothe with their heads the itching provoked by the sting of the nettles? And by their strong complaining do they persuade one unhappy being to succour other unhappy beings?

2. The black Nettle caterpillar of Moufet[858] is transformed into a rough, angular chrysalis brightly covered over many parts of it as it were with gold, from which emerges a butterfly that is very common with us, painted with colours, principally black, russet & yellow, to which are added two white spots situated at the very ends of the wings & more that are blue beautifully lighting up the fringed margin of the wings.

629 *Tilletia tritici* (Bjerk.) R. Wolff and/or *Ustilago segetum* Roussel *sensu lato*[859]
Ustilago J.B. Ἐρυσινη [Erysine] *of Theophrastus, book 8, chapter 10.*[860] Rubigo *of Pliny. C.B. in Pin[ax] & Ger[arde]* make three species of it, namely of rye, barley

---

[853] *Semel atque iterum*, literally "Once and again".

[854] Ulysses Aldrovandus or Ulisse Aldrovandi (1522–1605) published the 12 books of his *Ornithologia* at Bologna in 1599–1603. The source of this observation reported there is Ioannes Baptista Porta or Giambattista della Porta's (1535?–1615) *Magia naturalis*, first published in four books in Naples in 1558; an enlarged edition in 20 books was published in 1589. An English translation, *Natural magick*, was published in London in 1658, in which the relevant passage in Chapter 26 of book 4 is as follows: "But a Cock or Capon will perform what the Hen should. Do but show him the Chicken, and stroke him gently on the back, and give him meat out of your hands often, that he may become tame. Then pull the Feathers off of his breast, and rub him with Nettles. For in a few hours, not to say days, he will take care of the Chickens so well and give them their meat, that no Hens did ever do it as he will." However it seems that Ray knew this observation by Porta only through Aldrovandus. See also footnote 217.

[855] *capum pullos gallinaceos matris more curantem*, literally "a capon tending poultry [adjective] young in the manner of a mother".

[856] *exclusos pullos*, which probably means this rather than that the chicks are rejected (excluded) by their mothers.

[857] i.e. presumably chicks taken from their mothers.

[858] *Eruca nigra Urticaria Moufeti*, the larva of the Small Tortoiseshell butterfly, *Aglais urticae* (Linnaeus). For Moufet's (1634) book see footnote 76. The butterfly itself is the twelfth of Moufet's larger diurnal species, being described and illustrated on p. 101 of book 1, chapter 14.

[859] *Tilletia tritici* ("bunt") used to infect wheat and, less frequently, rye, filling the grains with spores. Infection occurs at the seedling stage and the disease is now controlled effectively by treating cereal seeds with fungicides as well as by resistant varieties. The name *Ustilago segetum* ("smut") is used here in the sense of Mordue & Ainsworth (1984). It infects a range of cereals including barley, oats and wheat and has been split by Vánky (1994) into a number of more host-specific species including *U. avenae* (Pers.) Rostr., *U. hordei* (Pers.) Lagerh. and *U. tritici* (Pers.) Rostr. The spores mature at the flowering stage, replacing the tissues within the glumes ("covered smut") or the glumes and their contents ("loose smut").

[860] *Historia plantarum*, VIII. x. 1 and 2 in the Loeb edition, where the name is given as ἐρυσιβη [erysibe].

and oats, to which *Park[inson]* adds the smut of wheat. *Burnt, blasted or smutted Corn; Wheat, Oats, Rie and Barley.*

N. Rust,[861] even though it is a common defect of grain-plants, nevertheless infects wheat more because its grain is covered with several coats & it produces the leaf closer to the ear, from which it sheds its fruit; for putrified moisture that causes rust collects in it. Add the fact that it produces an erect and denser ear, for on this account the moisture is less able to flow away; for this reason it is better for the ear to bend down a little. *J.B., volume 2, page 404. Theophr. de caus. plant., book 3, chapter 27.*[862]

Vulvaria, *see* Atriplex olida. [*Chenopodium vulvaria* 69]

Uvularia, *see* Trachelium.[863]

Vulneraria rustica *Gesn. see* Anthyllis leguminosa. [*Anthyllis vulneraria* 49]

X

630 Iris foetidissima L.

XYris *Matth. Lob. Lugd. Ger.* Xyris sive Spatula fœtida *Park.* Xyris, sive Iris agria Theophrasti *Dod. gal. Ad. Lob. ico.* Spatula fœtida *Trag. Fuch. Dod. Tab.* Spatula fœtida, plerisque Xyris *J.B.* Gladiolus fœtidus *C.B.* Hyacinthus poetarum *Jo. Bod[æ]. in Theophr. hist.* & Vaccinium Latinorum *for the same author. Stinking Gladwin or Gladdon. At Teversham and in many other places.*

FINIS.

---

[861] *Ærugo*, originally the verdigris that forms on copper, bronze and brass.

[862] Theophrastus discusses rust in *De causis plantarum* (III. 22. 1–2 in the Loeb edition), emphasising the effect of continuing wetness; he mentions two contrasting strains of κριθη [crithe, "barley"] but σιτος [sitos, "wheat", though sometimes generally "grain"] only as most susceptible to rust at the full moon.

[863] Ray cites *Uvularia major* and *Uvularia exigua* as synonyms of *Trachelium majus* [*Campanula trachelium* 572] and *Trachelium minus* [*Campanula glomerata* 573] respectively.

INDEX PLANTARUM AGRI CANTABRIGIENSIS,

In quo
Nomina Anglica Latinis
præponuntur ordine
alphabetico:

In gratiam tyronum.

CANTABRIGIÆ:
Excudebat *Joann.* Field, celeberrimæ
Academiæ Typographus.
Ann. Dom. 1660.

AN INDEX
OF PLANTS
OF THE COUNTY
OF CAMBRIDGE,

In which
the English Names
are placed before the Latin
in alphabetical order:

For the benefit of beginners.

CAMBRIDGE:
Printed by *John Field,* Printer
to the most renowned University.
A. D. 1660.

A

*A*Dders-tongue, Ophioglossum.
Agrimony, Agrimonia *Offic.* Eupatorium *Græ-corum.*
Dutch Agr. Eupatorium cannabinum mas, Eupat. Avicennæ.
Water Hemp-Agr. with an undivided leafe, Eupat. aquaticum folio integro, Cannabina aquatica folio non diviso.
Water Hemp-Agr. with a yellow flower, Eupat. cannabinum fœmina, Cannabina aquat. folio tripartitò diviso.
Alder-tree, Alnus.
Alehoof, id est, *Ground Ivie,* Hedera terrestris.
All-good, id est, *English Mercury,* Bonus Henricus.
Clowns All-heal, Sideritis Anglica strumosa radice, Panax coloni.
Allisanders, Hipposelinum, Olusatrum.
Wild Angelica, Angelica sylvestris.
Wood-Anemony, Anemone nemorum, Ranunculus nemorosus.
Apple-tree, Malus.
Archangell, Lamium, Urtica iners.
Red Arch. Lam. rubrum.
White Arch. Lam. album.
Yellow Arch. Lam. luteum.
Great Arrowhead, Sagittaria. sive Sagitta major.
Small Arrowh. Sagittar. sive Sagitta minor.
Arsmart, Persicaria urens, Hydropiper.
Small creeping Arsm. Pers. pusilla repens.
Spotted or dead Arsm. Pers. mitis maculosa.
Ash-tree, Fraxinus.
Ashweed, id est, *Goutwort,* Herba Gerardi.

Aspen

INDEX OF ENGLISH NAMES

A

A*Dders-tongue*, Ophioglossum. [*Ophioglossum vulgatum* 395]
Agrimony, Agrimonia *Offic[inarum]*, Eupatorium *of the Greeks*.[1] [*Agrimonia eupatoria* 16]
Dutch Agr[imony], Eupatorium cannabinum mas,[2] Eupat[orium] *Avicennæ*. [*Eupatorium cannabinum* 191]
Water Hemp-Agr[imony] with an undivided leafe, Eupat[orium] aquaticum folio integro, Cannabina aquatica folio non diviso. [*Bidens cernua* 190]
Water Hemp-Agr[imony] with a yellow flower, Eupat[orium] cannabinum fœmina, Cannabina aquat[ica] folio tripartitò diviso. [*Bidens tripartita* 189]
Alder-tree, Alnus. [*Alnus glutinosa* 20]
Alehoof, that is, *Ground Ivie*, Hedera terrestris. [*Glechoma hederacea* 272]
All-good, that is, *English Mercury*, Bonus Henricus. [*Chenopodium bonus-henricus* 90]
Clowns All-heal, Sideritis Anglica strumosa radice, Panax coloni. [*Stachys palustris* 540]
Allisanders, Hipposelinum, Olusatrum. [*Smyrnium olusatrum* 285]
Wild Angelica, Angelica sylvestris. [*Angelica sylvestris* 45]
Wood-Anemony, Anemone nemorum, Ranunculus nemorosus. [*Anemone nemorosa* 44]
Apple-tree, Malus.[3]
Archangell, Lamium, Urtica iners.[4]
Red Arch[angell], Lam[ium] rubrum. [*Lamium purpureum* 319]
White Arch[angell], Lam[ium] album. [*Lamium album* 317]
Yellow Arch[angell], Lam[ium] luteum. [*Lamiastrum galeobdolon* 318]
Great Arrowhead, Sagittaria, or Sagitta major. [*Sagittaria sagittifolia* 506]
Small Arrowh[ead], Sagittar[ia] or Sagitta minor. [*Sagittaria sagittifolia* 507]
Arsmart, Persicaria urens, Hydropiper. [*Persicaria hydropiper* 435]
Small creeping Arsm[art], Pers[icaria] pusilla repens. [*Persicaria minor* 434]
Spotted or dead Arsm[art], Pers[icaria] mitis maculosa. [*Persicaria maculosa* 433]
Ash-tree, Fraxinus. [*Fraxinus excelsior* 203]
Ashweed, that is, *Goutwort*, Herba Gerardi. [*Aegopodium podagraria* 46]

---

[1] Ray's original entry reads "*Agrimony*, Agrimonia *Offic*. Eupatorium *Græcorum*." In expanding abbreviated words such as *Offic*. we have replaced the full stop with a comma to bring the punctuation into line with the unabbreviated entries.

[2] In this and numerous subsequent cases Ray does not cite the exact name under which the species appears in the main alphabetical catalogue. This plant, for example, is listed there as *Eupatorium cannabinum* Park. C.B., with synonyms that include *cannab[inum] mas* Ger. emac. and *Eup[atorium] Avicennæ creditum* Ang. Gesn. hort. Cam. The differences are not always as slight as this, and Ray uses, for example, *Herba Gerardi* rather than his chosen name *Angelica sylvestris minor* in three entries in this index and *Pes cati* rather than *Gnaphalium montanum album* in another. We have not noted such discrepancies individually if Ray's meaning is clear. Users of this index should also note that not all Ray's English names are included in it and that not all the names in the index appear in the main catalogue.

[3] It is not clear whether this refers to *Malus sylvestris* [356], listed by Ray in the main catalogue as "The Crab-tree, or Wilding", or the cultivated apple, *Malus pumila* Mill., which Ray does not list even though he refers to orchards as a habitat of *Cicutaria vulgaris* [*Anthriscus sylvestris* 135]. See also the entry for "Wild Pear-tree" under P below.

[4] Ray lists the three species of "Archangell" or *Lamium* immediately after this entry; *Urtica iners* means "Deadnettle".

Aspen-tree, Populus Libyca. [*Populus tremula* 457]
Avens, Caryophyllata, Geum. [*Geum urbanum* 117]

B

B*Arberry-bush*, Berberis, Oxyacantha. [*Berberis vulgaris* 413]
Battle-door Barley, that is, *Sprit Barley*. [*Hordeum distichon* 290]
Beare Barl[ey], Hordeum polystichum. [*Hordeum vulgare* 291]
Common Barl[ey], Hord[eum] distichum. [*Hordeum distichon* 289]
Sprit Barl[ey], Hord[eum] distichum minus. [*Hordeum distichon* 290]
Wall Barl[ey], that is, *Rie grasse*, Hord[eum] spurium. [*Hordeum murinum* 288]
Stone-Basill, Acinos Anglicum *Clusii*. [*Clinopodium acinos* 9]
Wild Basill, Clinopodium majus. [*Clinopodium vulgare* 139]
Garden Beans, Faba major. [*Vicia faba* var. *faba* 193]
Horse Beans, Faba minor. [*Vicia faba* L. var. *equina* and/or var. *minor* 194]
Bearsfoot, see *Black Hellebore*. [*Helleborus viridis* 276]
Round-leaved Bellflower, Campanula rotundifolia. [*Campanula rotundifolia* 103]
Herb Benett, that is, *Avens*, Caryophyllata, Geum. [*Geum urbanum* 117]
Wood-Betony, Betonica. [*Betonica officinalis* 84]
Water-Betony, Bet[onica] aquatica, Scrophularia aquatica major. [*Scrophularia auriculata* 83]
Great Bindweed, Convolvulus major. [*Calystegia sepium* 148]
Small Bindw[eed], Conv[olvulus] minor, Helxine cissampelos. [*Convolvulus arvensis* 149]
Black Bindw[eed], Conv[olvulus] niger, Helx[ine] cissamp[elos] altera Atriplicis effigie *Lob*. [*Fallopia convolvulus* 150]
Birch-tree, Betula. [*Betula pubescens* 85]
The lesser Birds-foot, Ornithopodium minus. [*Ornithopus perpusillus* 411]
Birds-nest, that is, *Wild Carrot*, Daucus vulgaris. [*Daucus carota* subsp. *carota* 166]
Bistort or Snakeweed, Bistorta. [*Persicaria bistorta* 87]
Bittersweet, Solanum lignosum, Dulcamara. [*Solanum dulcamara* 547]
Black-thorn, that is, *Sloe-tree*, Prunus sylvestris. [*Prunus spinosa* 471]
Bladder-nut-tree, Staphylodendron, Nux vesicaria. [*Staphylea pinnata* 561]
Black-berry-bush. that is, *Bramble*. [*Rubus fruticosus* agg. 505]
Blew-bottle, Cyanus, Baptisecula. [*Centaurea cyanus* 162]
Lesser wild white Blite, Blitum album sylvestre minus. [*Chenopodium polyspermum* 88]
Small wild red Blite, Blit[um] rubrum sylv[estre] minus. [*Chenopodium polyspermum* 89]
Brake, that is, *Fern*, Filix. [*Pteridium aquilinum* 200]
Bramble-bush, Rubus vulgaris. [*Rubus fruticosus* agg. 505]
Small Bramble, Rubus minor fructu cæruleo, Chamærubus. [*Rubus caesius* 504]
Branke, that is, *Buckwheat*, Fegopyron. [*Fagopyrum esculentum* 195]
Chickweed-Breakstone, that is, *Pearlwort*, Saxifraga Anglica alsinefolia. [*Asperula cynanchica* 523]
Briar-bush, that is, *Dogs rose*, Cynosbatos. [*Rosa canina* 502]
Brooklime, Anagallis aquatica, Becabunga. [*Veronica* spp.[5]]

---

[5] This is a composite entry covering five species of *Anagallis aquatica* with "Brooklime" included in their English names [35–38, 40].

INDEX OF ENGLISH NAMES

Long-leaved Brookl[ime], Anag[allis] aquat[ica] longifolia, Berula major Tab. [*Veronica anagallis-aquatica sensu lato*[6] 37, 38]
Narrow-leaved Brookl[ime], Anag[allis] aquat[ica] angustifolia. [*Veronica scutellata* 40]
Broom, Genista. [*Cytisus scoparius* 213]
Broom-rape, Rapum Genistæ, Limodoron. [*Orobanche elatior* and *O. rapum-genistae* 412]
White Bryony, Bryonia alba, Vitis alba. [*Bryonia dioica* 91]
Black Bry[ony], Br[yonia] nigra, Vit[is] nigra, Sigillum B[eatæ] Mariæ. [*Tamus communis* 92]
Buck-beans, that is, *Marsh-Trefoile*, Trifolium palustre. [*Menyanthes trifoliata* 592]
Buck-thorn, Rhamnus Catharticus, Spina infectoria. [*Rhamnus cathartica* 498]
Buck wheat or Beech wheat, Fegopyron, Tragopyron, Erysimum Cereale. [*Fagopyrum esculentum* 195]
Bugle, Bugula, Consolida media. [*Ajuga reptans* 95]
Small wild Buglosse, Buglossa sylvestris minor. [*Anchusa arvensis* 93]
Small wild Buglosse, that is, *German Madwort*. [*Asperugo procumbens* 34]
Vipers Buglosse, Echium vulgare. [*Echium vulgare* 171]
Bull-bind, that is, *Travellers joy*. [*Clematis vitalba* 621]
Great Burnet, Pimpinella major. [*Sanguisorba officinalis* 438]
Small Burnet, Pimp[inella] minor or vulgaris. [*Poterium sanguisorba* subsp. *sanguisorba* 439]
Burnt corn, Ustilago. [*Tilletia tritici* and/or *Ustilago segetum* 629]
Burre-dock, Bardana major, Lappa major, Personata. [*Arctium lappa* and/or *A. minus* 79]
Burre-flag, that is, *Burre-reed*, see in *Reed*. [*Sparganium erectum* 553]
Butter-burre, Petasites. [*Petasites hybridus* 436]
Butter-cups, that is, *Common creeping Crowfoot*. [*Ranunculus repens* 487]
Butterwort, Pinguicula, Sanicula Eboracensis. [*Pinguicula vulgaris* 443]

C

COmmon Calamint, Calamintha vulgaris. [*Clinopodium ascendens* 100]
Water Calamint, Calamintha aquatica. [*Mentha arvensis* 99]
Water Caltrops, Tribulus aquaticus minor Quercûs floribus. [*Potamogeton crispus* 576]
Small water Caltr[ops], Trib[ulus] aquat[icus] min[or] Muscatellæ floribus. [*Groenlandia densa* 575]
Calves-snout, see *Snapdragon*. [*Chaenorhinum minus* 50]
Cammock, that is, *Restharrow*, Anonis. [*Ononis spinosa* 47]
Dogs Camomile, Cotula non fœtida, Chamæmelum inodorum. [*Anthemis arvensis* 156]
Wild white Campions, Lychnis sylvestris flore albo. [*Silene latifolia* 344]
Canterbury Bells, see *Throatwort*. [*Campanula trachelium* 572 and *C. glomerata* 573]
Carrot, Pastinaca sativa tenuifolia. [*Daucus carota* subsp. *sativus* 423–424]
Wild Carrot or Birds nest, Daucus Offic[inarum]. [*Daucus carota* subsp. *carota* 166]

---

[6] Ray does not actually use the name *Anagallis aquatica longifolia*, but *Berula major* Tab. is included in the synonymy of *Anagallis aquatica III, sive major folio oblongo*, "The greater long-leaved Brooklime" [37]. *Anagallis aquatica IV* is called "The lesser long-leaved Brooklime" [38].

Caseweed, that is, *Shepherds-purse*, Bursa pastoris. [*Capsella bursa-pastoris* 97]
Cats-foot, that is, *Mountain Cudweed*, Pes cati. [*Antennaria dioica* 233]
Catmint, that is, *Neppe*, Nepeta, Mentha cattaria. [*Nepeta cataria* 363]
Cats-tail, Typha. [*Typha latifolia* 604]
Spanish Catchflie, Sesamoides Salamanticum, Muscipula flore muscoso. [*Silene otites* 539]
Celandine, Chelidonium majus. [*Chelidonium majus* 128]
Purple Centory, Centaurium minus purpureum. [*Centaurium erythraea* 123]
Yellow Centory, Cent[aurium] luteum. [*Blackstonia perfoliata* 122]
Charlock, Rapistrum arvorum. [*Sinapis arvensis* 490]
Charl[ock] with a rough joynted cod, Rap[istrum] lut[eum] siliquâ hirsutâ articulatâ. [*Sinapis alba* 492]
Charl[ock] with a smooth joynted cod, Rap[istrum] lut[eum] siliquâ glabrâ articulatâ. [*Raphanus raphanistrum* subsp. *raphanistrum* 491]
Cheese-rening, that is, *Ladies Bedstraw*, Gallium. [*Galium verum* 211]
Wild Cheir, that is, *Wall-flower*, Leucoium luteum, Keiri Offic[inarum]. [*Erysimum cheiri* 330]
Small Hemlock-Chervill, Myrrhis sylv[estris] seminibus asperis, Myrrhis Æquicolorum nova Col. [*Anthriscus caucalis* 384]
Wild Chervill, Cerifolium sylvestre. [*Chaerophyllum temulum* 124]
Fine Chickweed, Alsine minima. [*Arenaria serpyllifolia sensu lato* 30]
Germander Chickw[eed], Als[ine] foliis Trissaginis. [*Veronica agrestis* 22]
Great Chickw[eed], Als[ine] major. [*Stellaria media* 28]
Ivy-Chickw[eed] or Henbit, Als[ine] hederacea. [*Veronica hederifolia* 24]
Long-leaved water Chickw[eed], Als[ine] longifolia uliginosis proveniens locis. [*Stellaria alsine* 27]
Great Marsh Chickw[eed], Als[ine] aquatica major. [*Myosoton aquaticum* 21]
Middle Chickw[eed], Als[ine] media. [*Stellaria media* 29]
Mouse-eare-Chickw[eed], Als[ine] hirsuta myosotis. [*Cerastium fontanum* 26]
Narrow-leaved Chick[weed], Als[ine] tenuifolia J.B. [*Minuartia hybrida* 32]
Speedwell Chickw[eed], Als[ine] foliis Veronicæ. [*Veronica arvensis* 23]
Starre-headed water Chickw[eed], that is, *water Starre-wort*. [*Callitriche* spp. 562]
Wild Cicely, Cicutaria vulgaris, Myrrhis sylv[estris] *Fuch*. [*Anthriscus sylvestris* 135]
Common Cinquefoile, Pentaphyllum vulgare. [*Potentilla reptans* 429]
Marsh Cinquefoile, Pentaph[yllum] rubrum palustre. [*Comarum palustre* 428]
Dwarfe Cistus, Chamæcistus vulgaris, Helianthemum. [*Helianthemum nummularium* 125]
Wild Clary, Horminum sylvestre, Oculus Christi *Offic[inarum]*. [*Salvia verbenaca* 292]
Claver-grasse, Trifolium majus flo[re] purpureo. [*Trifolium medium* 591]
Great wild Climer, that is, *Travellers joy*. [*Clematis vitalba* 621]
Clivers or Cleaver, that is, *Goose-grasse*, Aparine. [*Galium aparine* 51]
Cocks-combe, that is, *Yellow Rattle*, Pedicularis lutea. [*Rhinanthus minor* 426]
Cockshead, Onobrychis. Caput gallinaceum. [*Onobrychis viciifolia* 394]
Cockle, Pseudomelanthium, Nigellastrum, Githago, Lychnis segetum. [*Agrostemma githago* 472]
Coleseed, that is, *wild Navew*. [*Brassica rapa* and *B. napus* 385]
Colts-foot, Tussilago, Bechion, Farfara. [*Tussilago farfara* 603]
Columbine, Aquilegia. [*Aquilegia vulgaris* 54]
Corn-Sallet, that is, *Lambs Lettuce*, Lactuca agnina. [*Valerianella locusta* 311]

INDEX OF ENGLISH NAMES

Female Cornell, that is, *Dog-tree*, Cornus fœmina. [*Cornus sanguinea* 151]
Cotten-tree, that is, *Wayfaring tree*, Viburnum. [*Viburnum lantana* 612]
Common Cow-wheat, Melampyrum vulgare. [*Melampyrum pratense* 360]
Crested Cow-wheat, Melamp[yrum] cristatum J.B. [*Melampyrum cristatum* 359]
Eyebright Cow-wheat, Cratæogonon Euphrosynes facie. [*Odontites vernus* 158]
Cow-slips, Primula veris major, Paralysis. [*Primula veris* 467]
Bloudy Cranes bill, Geranium hæmatodes. [*Geranium sanguineum* 222]
Crowfoot-Cranes-bill, Ger[anium] batrachoides. [*Geranium pratense* 218]
Dovesfoot Cranes-bill, Ger[anium] columbinum. [*Geranium molle* and/or *G. pusillum* 219]
Unsavoury field Cranes-bill, Ger[anium] moschatum inodorum, Ger[anium] cicutæ folio inodorum. [*Erodium cicutarium* 217]
Cresse, Nasturtium.[7]
Bank Cresse,[8] Irio, Erysimum. [*Sisymbrium officinale* 186]
Dock-Cresse, that is, *Nipplewort*, Lampsana. [*Lapsana communis* 320]
Swines Cresse, Coronopus Ruellii. [*Lepidium coronopus* 153]
Water-Cresse, Nast[urtium] aquaticum, Sium Catevæ erucæfolium. [*Nasturtium officinale sensu lato* 387]
Winter-Cresse, Barbarea, Nasturtium hybernum. [*Barbarea vulgaris* 78]
Crossewort, Cruciata. [*Cruciata laevipes* 160]
Common creeping Crowfoot or Butter-cups, Ranunculus pratensis repens. [*Ranunculus repens* 487]
Corn-Crowf[oot], Ran[unculus] arvorum. [*Ranunculus arvensis* 480]
Marsh Crowf[oot] or round-leaved Spearewort, Ran[unculus] palustris rotundifolius. [*Ranunculus sceleratus* 478]
Round rooted or bulbous Crowf[oot], Ran[unculus] bulbosus. [*Ranunculus bulbosus* 481]
Water Crowf[oot],[9] Ran[unculus] aquaticus Hepaticæ facie. [*Ranunculus aquatilis* 479]
Ivy-leaved water Crowf[oot], Ran[unculus] aquat[icus] hederaceus luteus. [*Ranunculus hederaceus* 485]
Musk wood Crowf[oot], Ran[unculus] nemorosus Muschatella dictus, Radix cava minima viridi flore. [*Adoxa moschatellina* 477]
Single white wood-Crowf[oot] or wood Anemony, Anemone nemorum, Ran[unculus] nemorosus. [*Anemone nemorosa* 44]
Sweet wood-Crowfoot, Ran[unculus] auricomus, Ran[unculus] nemorosus dulcis. [*Ranunculus auricomus* 486]
Cuckow-flower, that is, *Ladies smock*, Cardamine. [*Cardamine pratensis* 106]
Cuckow-flower, that is, *Meadow-pinke*, Armerius pratensis. [*Silene flos-cuculi* 61]
Cuckow-pint, that is, *Wake-Robin*, Arum. [*Arum maculatum* 63]
Common Cudweed, Gnaphalium vulgare. [*Gnaphalium uliginosum* 231]
Lesser Cudw[eed], Gnaph[alium] minus, Herba impia. [*Filago vulgaris* 230]
Mountain Cudw[eed], Gnaph[alium] montanum, Pes cati. [*Antennaria dioica* 233]

---

[7] The various cresses are listed below; only "Swines Cresse", "Water-Cresse" and "Winter-Cresse" have *Nasturtium* synonyms.
[8] Ray does not list "Bank Cresse" as an English name for this species in the main catalogue, but it is repeated after "Hedge Mustard" under M in this index.
[9] Ray also applies the name "Water Fennell, Crowfoot, or Millefoile" to *Millefolium aquaticum ranunculi flore & capitulo* [*Ranunculus trichophyllus* 366].

Small Cudw[eed],[10] Gnaph[alium] minimum. [*Filago minima* 232]
Cumfrey, or Comfry, Consolida major. [*Symphytum officinale* 147]
Black Currans, Ribes nigrum. [*Ribes nigrum* 499]

D.

ENglish Daffodill, Pseudonarcissus Anglicus, Bulbocodium vulgatius. [*Narcissus pseudonarcissus* 386]
Dandelion, Dens leonis, Taraxacon *Offic[inarum]*. [*Taraxacum* spp. 167]
Daisie, Bellis. [*Leucanthemum vulgare* 81 and *Bellis perennis* 82]
Great Wild Daisie, Bellis major, Buphthalmum *Lon*. [*Leucanthemum vulgare* 81]
Danewort, Ebulus, Chamæacte, Sambucus humilis. [*Sambucus ebulus* 170]
Darnell, Lolium.[11]
Dew-berrie, that is, *little Bramble*. Rubus minor fructu cæruleo, Chamæbatos. [*Rubus caesius* 504]
Diers weed, that is, *Wild Woad*, Luteola. [*Reseda luteola* 343]
Diers weed, that is, *Green weed*. Genista tinctoria. [*Genista tinctoria* 215]
Divells bit, Morsus Diaboli, Succisa. [*Succisa pratensis* 374]
Round-leaved Dock, Lapathum sylv[estre] folio minùs acuto. [*Rumex obtusifolius* 322]
Sharp-pointed Dock, Lap[athum] acutum, Oxylapathum. [*Rumex* spp. 321]
Great water Dock, Lap[athum] aquaticum maximum, Hydrolapathum majus. [*Rumex hydrolapathum* 323]
Lesser water Dock, Lap[athum] aquat[icum] or Hydrol[apathum] minus. [*Rumex conglomeratus* 324[12]]
Dodder, Cuscuta, Cassutha. [*Cuscuta europaea* 161]
Dogberrie-tree, Cornus fœmina, Virga sanguinea. [*Cornus sanguinea* 151]
Doves-foot with deeply cut leaves, Geranium columbinum magìs dissectis foliis. [*Geranium dissectum* 220]
Shining or stone Doves-foot, Ger[anium] saxatile or lucidum. [*Geranium lucidum* 224]
Dranke, see *Small wild Oates*. [*Bromus hordeaceus* 197]
Dropwort, Filipendula, Oenanthe. [*Filipendula vulgaris* 198]
Water Dropwort, Oeanthe[13] or Filipendula aquatica. [*Oenanthe fistulosa* 393]
Duckes meat, Lens palustris. [*Lemna minor* 329]
Ivy-leaved Duckes meat, Lenticula aquatica trisulca. [*Lemna trisulca* 273]
Dwale, that is, *Deadly Nightshade*, Solanum lethale. [*Atropa belladonna* 548]

E.

EArthnut, Bulbocastanum. [*Conopodium majus* 96]
Elecampane, Helenium, Enula campana. [*Inula helenium* 177]
Common Elder, Sambucus. [*Sambucus nigra* 519]
Dwarfe Elder, that is, *Danewort*, Ebulus. [*Sambucus ebulus* 170]

---

[10] Ray calls this "The least Cudweed" in the main catalogue.
[11] Ray gives "Darnell" as the English name for his *Lolium album* [*Lolium temulentum* 340] and "Red Darnell-grasse" as the name for his *Lolium rubrum* [*L. perenne* 341].
[12] See footnote 400 in the main catalogue.
[13] A misprint for *Oenanthe*.

INDEX OF ENGLISH NAMES

Water Elder, Sambucus aquatica. [*Viburnum opulus* 518]
Common Elme, Ulmus vulgatissimus folio lato, scabro. [*Ulmus procera* 626]
Wych Elme, Ulmus folio glabro. [*Ulmus minor* 624]
Broad-leaved Elme, Ulmus folio latissimo scabro. [*Ulmus glabra* 625]
Eye bright, Euphrasia, Euphrosyne. [*Euphrasia* spp. 192]

F.

FEverfew, Matricaria, Parthenium. [*Tanacetum parthenium* 421]
Fellwort, see, *Gentian*. [*Gentianella amarella* 216]
Star-like water Fennell, Stellaria aquatica. [*Callitriche* spp. 562]
Water Fennell, see *Water Millefoile*.[14]
Male Ferne, Filix mas. [*Dryopteris filix-mas* 199]
Female Ferne, Filix fœmina. [*Pteridium aquilinum* 200]
Figwort, Scrophularia. [*Scrophularia nodosa* 529]
Medick Fitchling, Onobrychis. [*Onobrychis viciifolia* 394]
Fitches, see, *Vetches*.[15]
Five-leaved grasse, that is, *Cinquefoile*. [*Potentilla reptans* 429]
Flaxe, Linum.[16]
Wild blew Flaxe, Lin[um] sylvestre cæruleum. [*Linum perenne* 337]
Great Fleabane, Conyza major, Baccharis Monspeliensium. [*Inula conyzae* 141]
Dwarfe Fleabane, Con[yza] minor, Pulicaria. [*Pulicaria vulgaris* 143]
Blew-flowered Fleab[ane], Con[yza] cærulea acris. [*Erigeron acris* 144]
Great jagged Fleab[ane], Con[yza] helenitis foliis laciniatis. [*Tephroseris palustris* 145]
Marsh Fleab[ane], Con[yza] palustris. [*Senecio paludosus* 146]
Middle Fleab[ane], Con[yza] media. [*Pulicaria dysenterica* 142]
Flixeweed, Sophia chirurgorum, Seriphium Germanicum. [*Descurainia sophia* 552]
Yellow water Flower de luce, Acorus palustris, Pseudoiris, Iris lutea palustris. [*Iris pseudacorus* 10]
Female Fluellin, Elatine folio subrotundo. [*Kickxia spuria* 176]
Sharp-pointed Fluell[in], El[atine] folio acuminato. [*Kickxia elatine* 175]
Upright male Fluell[in], Veronica spicata recta. [*Veronica spicata* 611]
Male creeping Fluell[in], Ver[onica] mas supina. [*Veronica officinalis* 609]
Small meadow Fluell[in], Ver[onica] pratensis. [*Veronica serpyllifolia* 610]
Frogs-lettuce, that is, *Water Caltrops*, Tribulus aquaticus. [*Potamogeton crispus* 576[17]]
Frog-bit, Nymphæa alba minima, Morsus ranæ. [*Hydrocharis morsus-ranae* 375]
Double flowered Frog-bit, Morsus ranæ flore pleno odoratissimo. [*Hydrocharis morsus-ranae* 376]
Fumitory, Fumaria. [*Fumaria officinalis* 204]

---

[14] This presumably refers either to "Water Fennell, Crowfoot, or Millefoile" [*Ranunculus trichophyllus* 366] or perhaps to "Fennell-leaved water Millefoile" [*Potamogeton pectinatus* 370], unless it is a generic entry referring to all "Water Millefoiles".

[15] The name "Fitches" does not appear elsewhere in this index or in the main catalogue, though Ray has "Medick-Fitchling" in the preceding entry. Under "Vetch", he has a further cross-reference to "Tare" as well as the names of various vetches and vetchlings.

[16] In addition to the "Wild blew Flaxe", listed below, Ray gives "Manured Flax" as the English name for *Linum sativum* [*Linum usitatissimum* 336] and "Dwarf wild Flax" for *Linum sylvestre catharticum* [*Linum catharticum* 338].

[17] See also *Groenlandia densa* [575], called by Ray "Small Frogs Lettuce or water Caltrops".

Furze-bush, Genista spinosa. [*Ulex europaeus* 212]
Needle-Furze or petty Whin, Genistella aculeata. [*Genista anglica* 214]

G

Row Garlick, Allium sylvestre tenuifolium. [*Allium vineale* 19]
Gatter tree, that is, *Dog-berry*, Cornus fœmina. [*Cornus sanguinea* 151]
Gaule, that is, *Dutch Myrtle*, Elæagnus Cordi. [*Myrica gale* 173]
Dwarfe autumnall Gentian, Gentianella fugax autumnalis. [*Gentianella amarella* 216]
Herbe Gerard, that is, *Ashweed*, Herba Gerardi. [*Aegopodium podagraria* 46]
Water Germander, Scordium. [*Teucrium scordium* 528]
Wild Germ[ander], Cham[ædrys] sylvestris, Teucrium pratense. [*Veronica chamaedrys* 126]
Gill go by ground, that is, *Ground-Ivy*. [*Glechoma hederacea* 272]
Stinking Gladdon or Gladwin, Xyris, Spatula fœtida. [*Iris foetidissima* 630]
Water Gladiole, Gladiolus palustris, Juncus floridus. [*Butomus umbellatus* 225]
Yellow Goats beard, Tragopogon luteum. [*Tragopogon pratensis* 574]
Go to bed at noon, The same. [*Tragopogon pratensis* 574]
Goldilocks, see *Maiden-hair, and wood Crowfoot*.[18]
Goose-grasse, or Clivers, Aparine. [*Galium aparine* 51]
Great Goose-grasse, that is, *German Madwort*, Aparine major Plinii, Alysson Germanicum echioides. [*Asperugo procumbens* 34]
Goose-foot, Atriplex sylvestris latifolia, Pes anserinus. [*Chenopodium murale* 72]
Gorsse, that is, *Furze*, Genista spinosa. [*Ulex europaeus* 212]
Goutwort, or Ashweed, Herba Gerardi, Podagraria, Angelica sylvestris minor. [*Aegopodium podagraria* 46]
Cats tail-Grasse, Gramen typhinum. [*Phleum pratense* 267 and *P. bertolonii* 268]
Cyperus-Grasse, Gr[amen] cyperoides.[19]
Fox-tail Grasse, Gr[amen] alopecuroides.[20]
Bastard Fox-tail Grasse, Gr[amen] alopecurinum.[20]
Flote-Grasse, Gr[amen] fluviatile. [*Glyceria fluitans* 238]
Meadow-Grasse, Gr[amen] pratense paniculatum.[21]
Quaking Grasse, Gr[amen] tremulum, Phalaris pratensis. [*Briza media* 266]
Rie-Grasse, Gr[amen] secalinum, Hordeum spurium. [*Hordeum murinum* 288]
Rush Grasse, Gr[amen] junceum.[22]

---

[18] For "Maiden-hair" see *Adianthum aureum majus* [*Polytrichum commune* 12], *Adianthum aureum minus* [unidentifiable mosses 13] and *Trichomanes* [577] in the main catalogue; the name "Goldilocks" is also given as an alternative to "Sweet wood-Crowfoot" [*Ranunculus auricomus* 486] but is clearly unsuitable for "Single white Wood Crowfoot" [*Anemone nemorosa* 44].

[19] Ray lists eight species [243–250] with chosen names or (in one case) a synonym starting with *Gramen cyperoides* and with English names that are variants of "Cyperus-grasse".

[20] Ray gives the English name "Small Fox-tail grass" for *Gramen alopecuroides minus* [*Alopecurus pratensis* 234], "The lesser bastard Fox-tail-grass" for *Gramen Alopecurinum minus* [*Alopecurus myosuroides* 235] and "Small rough-eared bastard Fox-tail-grass" for *Gramen alopecuroides minus spicâ asperâ brevi* [*Cynosurus cristatus* 236].

[21] Ray lists three species [259–261] with chosen names starting with *Gramen pratense* and English names that are variants of "Meadow-grasse".

[22] Ray gives English names based on "Rush-grasse" for three species with names beginning with *Gramen junceum* [253–255], although he calls *Gramen junceum* itself [256] "Toad-grasse", a name that he does not list in this index.

INDEX OF ENGLISH NAMES

Sea Spike-Grasse, Gr[amen] marinum spicatum, Gr[amen] triglochin. [*Triglochin palustris* 257]
Water Grasse with crooked cods, Potamogeito affine Gr[amen] aquaticum. [*Zannichellia palustris* 466]
Grasse of Parnassus, Gramen Parnassi. [*Parnassia palustris* 264]
Greenweed, that is, *Woodwaxen*, Genista tinctoria. [*Genista tinctoria* 215]
Gromill or Gromwell, Lithospermum. [*Lithospermum officinale* 339]
Bastard Gromw[ell], Anchusa degener facie milii Solis, Lithospermum arvense radice rubente. [*Lithospermum arvense* 43]
Ground-pine, Chamæpitys, Iva arthritica. [*Ajuga chamaepitys* 127]
Groundsell or herb Simpson, Senecio, Erigeron, Carduncellus. [*Senecio vulgaris* 535]
Cotten Groundsell, Senecio hirsutus viscidus major odoratus. [*Senecio ?sylvaticus* 534]

H

HArebells, that is, *English Hyacinth*. [*Hyacinthoides non-scripta* 293]
Hares-lettuce, that is, *Sowthistle*, Sonchus lævis. [*Sonchus oleraceus* 550]
Hares-foot, Lagopus. [*Trifolium arvense* 316]
Dutch Haver-grasse, that is, *Bearded Wild-Oats*. [*Avena fatua* 14]
Harts-tongue, Lingua cervina, Phyllitis.[23] [*Asplenium scolopendrium* A29]
Hawkweed with bitten roots, Hieracium minus præmorsâ radice. [*Scorzoneroides autumnalis* 281]
Dandelion Hawkw[eed], Hier[acium] caule aphyllo hirsutum. [*Leontodon hispidus* 278]
Clusius his least Hawkw[eed], Hier[acium] minimum *Clus.* Hyoseris mascula *Ger.* [*Arnoseris minima* 280]
Long-rooted Hawkw[eed], Hier[acium] longiùs radicatum. [*Hypochaeris radicata* 279]
Rough Hawkw[eed] with a large flower, Hier[acium] asperum flore majore. [*Picris hieracioides* 284]
Smooth Succory-Hawkw[eed], Hier[acium] luteum glabrum, Cichoreum pratense luteum. [*Crepis capillaris* 282]
Rough Succory-Hawkw[eed], Hier[acium] minus Cichorei folio hirsutum. [*Crepis foetida* 283]
Hawthorn-tree, Oxyacanthus, Spina appendix. [*Crataegus laevigata* and *C. monogyna* 414]
Hearts-ease, that is, *Pansies*, Viola tricolor. [*Viola arvensis* 620]
Common Heath, Erica vulgaris. [*Calluna vulgaris* 184]
Narrow-leaved bastard Hellebore, Helleborine angustifolia palustris.[24] [*Epipactis palustris* 274]
Bastard Hellebore with whitish green flowers, Helleborine latifolia. [*Epipactis helleborine* 275]

---

[23] Ray does not include *Phyllitis* in the main catalogue, although it is mentioned in passing in a note under *Filix fœmina* [200] and listed for Cherry Hinton in 'Index of places'. A detailed record from Cherry Hinton is given in the 1663 appendix.

[24] In the main catalogue Ray only tentatively links this synonym to his chosen name *Helleborine angustifolia*, to which he gives the English name "Narrow-leaved wild white Hellebore". The next species is there called "Wild white Hellebore with whitish green flowers".

Wild black Hellebore, Helleborus niger hortensis flore viridi, Veratrum nigrum *Cord*. [*Helleborus viridis* 276]
Common Hemlock, Cicuta. [*Conium maculatum* 132]
The lesser Heml[ock], Cicutaria fatua, Petroselini vitium. [*Aethusa cynapium* 133]
Marsh Heml[ock], Cicutaria palustris. [*Oenanthe aquatica* and/or *O. fluviatilis* 134]
Hemp, Cannabis. [*Cannabis sativa* 104]
Wild Hemp, Cannabis spuria. [*Galeopsis tetrahit sensu lato* 105]
Henbane, Hyoscyamus. [*Hyoscyamus niger* 294]
Great Henbit, Alsine hederulæ folio major, Morsus gallinæ, &c. [*Lamium amplexicaule* 25]
Small Henbit, Als[ine] hederacea, Morsus gallinæ minor. [*Veronica hederifolia* 24]
High-taper, that is, *Mullein*, Tapsus barbatus. [*Verbascum thapsus* 564]
Holly, Agrifolium, Aquifolium. [*Ilex aquifolium* 15]
Dwarf Holy-rose, see *Cistus*. [*Helianthemum nummularium* 125]
Honeysuckle, see *Woodbind* and *Trefoil*. [*Lonicera periclymenum* 432 and *Trifolium pratense* 590]
Hop, Lupulus salictarius. [*Humulus lupulus* 342]
White Horehound, Marrubium album, Prassium. [*Marrubium vulgare* 357]
Black or stinking Horeh[ound], Marrub[ium] nigrum, Ballote. [*Ballota nigra* 77]
Hornbeam-tree, Carpinus, Betulus, Ostrys. [*Carpinus betulus* 86]
Horse-beech, that is, *Hornbeam*. [*Carpinus betulus* 86]
Corn Horsetail, Equisetum segetale, Hippuris minor. [*Equisetum arvense* 178]
Female Horset[ail], Equis[etum] palustre brevioribus setis polyspermon, Polygonum fœmina *Dod*. [*Hippuris vulgaris* 181]
Great Marsh Horset[ail], Equis[etum] palustre longioribus setis. [182 *Equisetum* sp.]
Naked Horset[ail], Equis[etum] nudum. [*Equisetum fluviatile* 180]
Stinking water Horset[ail], Equis[etum] fœtidum sub aquis repens. [*Chara* spp. 179]
Great Houndstongue, Cynoglossum majus vulgare. [*Cynoglossum officinale* 164]
Hedge Hyssop, that is, *Hooded willowherb*. [*Scutellaria galericulata* 345]
Small hedge Hyssop, Gratiola angustifolia, Hyssopoides. [*Lythrum hyssopifolia* 269]

I

Jack by the hedge, Alliaria. [*Alliaria petiolata* 17]
Jews-eare, Fungus Sambucinus.[25]
Narrow-leaved Ironwort, Ladanum segetum, Sideritis arvensis angustifolia rubra. [*Galeopsis angustifolia* 315]
Juniper, Juniperus. [*Juniperus communis* 310]
Ivray, that is, *Darnell*, Lolium. [*Lolium temulentum* 340]
Climing or berried Ivy, Hedera arborea. [*Hedera helix* 270]
Barren or creeping Ivy, Hed[era] helix, or provoluta sterilis. [*Hedera helix* 271]
Ground Ivy, Hed[era] terrestris, Chamæcissus. [*Glechoma hederacea* 272]
Water Ivy, Lenticula aquatica trisulca. [*Lemna trisulca* 273]

---

[25] This fungus, *Auricularia auricula-judae*, is mentioned in the main catalogue in a long note after the entry for *Sambucus* [*Sambucus nigra* 519], in which the falsity (*vanitas*) of the Doctrine of Signatures is discussed.

INDEX OF ENGLISH NAMES

K

K*Napweed*, Jacea nigra vulgaris. [*Centaurea nigra sensu lato* 297]
Great Knapweed, Jacea segetum major. [*Centaurea scabiosa* 298 and 299]
Knawell, that is, *Germane Knotgrasse*. [*Scleranthus annuus* 453]
Common Knotgrasse, Polygonum vulgare. [*Polygonum aviculare sensu lato* 454]
Germane Knotgrasse, or Knawell, Polyg[onum] Germanicum. [*Scleranthus annuus* 453]

L

W*hite Ladies-bedstraw*, Gallium album. [*Galium palustre* 210]
Yellow Ladies-bedstraw, Gallium luteum. [*Galium verum* 211]
Ladies-finger, that is, *Kidney-vetch*, Anthyllis leguminosa. [*Anthyllis vulneraria* 49]
Ladies-smock, Cardamine, Nasturtium pratense majus. [*Cardamine pratensis* 106]
Ladies-traces, Triorchis. [*Spiranthes spiralis* 409]
Lambes-tongue, that is, *Hoary Plantain*. [*Plantago media* 448]
Lang de bœuf, Buglossum luteum. [*Helminthotheca echioides* 94]
Dwarfe Lawrell, that is, *Spurge-Lawrell*, Laureola. [*Daphne laureola* 327]
Cut-leaved wild Lettuce. Lactuca sylvestris foliis dissectis. [*Lactuca serriola* forma *serriola* 312]
Endive-leaved wild Lettuce, Lactuca sylvestris odore opii minùs gravi. [313 *Lactuca serriola* forma *integrifolia* 313]
Small wild Lettuce, Lact[uca] sylv[estris] minima not yet described. [*Lactuca saligna* 314]
White water-Lilly, Nymphæa alba, Nenuphar album. [*Nymphaea alba* 390]
Great yellow water-Lilly, Nymphæa lutea major, Nenuphar luteum. [*Nuphar lutea* 391]
Small yellow water-Lilly with a fringed flower, Nymph[aea] lutea minor flore fimbriato. [*Nymphoides peltata* 392]
Ling, that is, *Heath*, Erica. [*Calluna vulgaris* 184]
Live-long, that is, *Orpine*, Telephium. [*Sedum telephium* 566]
Liverwort, Lichen, Hepatica vulgaris. [*Marchantia polymorpha* and/or *Lunularia cruciata* 331]
Liquorice, Glycyrrhiza. [*Glycyrrhiza glabra* 229]
Wild Liquorice, Glaux vulgaris, Glycyrrhiza sylvestris. [*Astragalus glycyphyllos* 228]
Loose-strife, see *Willowherbe*.[26]
Lousewort, that is, *Red-rattle*, Pedicularis rubra. [*Pedicularis sylvatica* 427]
Lowry, that is, *Spurge-Lawrell*. [*Daphne laureola* 327]
Cow-Lungwort, see, *Mullein*, Tapsus barbatus. [*Verbascum thapsus* 564]
Tree Lungwort, Lichen arborum. [unidentifiable foliose lichens 332]

M

G*Reat bastard Madder*, Mollugo montana or Gallium. [*Galium album* 373]
Little field Madder, Rubeola arvensis. [*Sherardia arvensis* 503]
Germane Madwort, Alysson German[icum] echioides, Aparine major Plinii. [*Asperugo procumbens* 34]

---

[26] Five of Ray's species of *Lysimachia* [346, 348–349, 352–353] have English names containing "Loose-strife" and "Willowherb" as alternatives.

English Maidenhair, Trichomanes, Polytrichum *Offic[inarum]*. [*Asplenium trichomanes* 577]
Great golden Maidenhair, Adianthum aureum majus, Polytrichum aureum majus. [*Polytrichum commune* 12]
Small golden Maidenhair, Adian[thum] aur[eum] minus, Pol[ytrichum] aur[eum] minus. [unidentifiable mosses 13]
Common Mallow, Malva vulgaris. [*Malva sylvestris* 355]
Dwarfe Mallow, Malva sylvestris minor. [*Malva neglecta* 354]
Marsh Mallow, Althæa, Ibiscus, Bismalva.[27] [*Althaea officinalis* 33]
Common Maple, Acer minus. [*Acer campestre* 6]
Great Maple, Acer majus, Sycomorus Ruellii. [*Acer pseudoplatanus* 5]
Corn Marigold, Chrysanthemum segetum. [*Glebionis segetum* 130]
Marsh Marigold, Caltha palustris. [*Caltha palustris* 101]
Wild Marjerome, Origanum vulgare, Majorana sylvestris. [*Origanum vulgare* 410]
Black Matfellon, that is, *Common Knapweed*, Jacea nigra vulgaris. [*Centaurea nigra sensu lato* 297]
Great Matfellon, that is, *Great Knapweed*, Jacea segetum major. [*Centaurea scabiosa* 298 and 299]
Small Matweed, Spartum parvum. [*Nardus stricta* 555]
Mayweed or Maithes, Cotula fœtida. [*Anthemis cotula* 155]
Meadow-sweet, Ulmaria, Regina prati. [*Filipendula ulmaria* 623]
Melilote, Melilotus. [*Melilotus altissimus* 361]
Dogs Mercury, Mercurialis sylvestris, Cynocrambe. [*Mercurialis perennis* 163]
English Mercury, Bonus Henricus, Tota bona. [*Chenopodium bonus-henricus* 90]
Milkewort, Polygala. Flos ambarvalis *Dod*. [*Polygala vulgaris* 451]
Yellow Milkewort with knotted cods, Polygalon Cortusi. [*Hippocrepis comosa* 452]
Millefoile, that is, *Yarrow*, Millefolium. [*Achillea millefolium* 371]
Purple flowered Millef[oile], Millef[olium] flore purpureo. [*Achillea millefolium* 372]
Water-Millef[oile] or Violet, Millef[olium] aquaticum dictum Viola aquatica. [*Hottonia palustris* 365]
Feathered water-Millef[oile], Millef[olium] aquat[icum] pennatum, spicatum. [*Myriophyllum spicatum* 368]
Horse-taile water-Millef[oile], Millef[olium] aquat[icum] equisetifolium. [*Ceratophyllum demersum* 183]
Fennel-leaved water-Millef[oile], Millef[olium] tenuifolium. [*Potamogeton pectinatus* 370]
Hooded water-Millef[oile], Millef[olium] palustre galericulatum. [*Utricularia vulgaris* 369]
Small water-Millef[oile], Millef[olium] aquaticum minus. [*Myriophyllum verticillatum* 367]
Star-like water-Millef[oile], Stellaria aquatica. [*Callitriche* spp. 562]
Mill-mountain, Linum catharticum. [*Linum catharticum* 338]
Water-Mint, Sisymbrium, Mentha aquatica. [*Mentha aquatica* 362]
Small purple flowered Moneywort, Nummularia minor flore purpurascente. [*Anagallis tenella* 389]
Yellow Moneywort, Nummularia vulgaris. [*Lysimachia nummularia* 388]
Common earth-Mosse, Muscus terrestris vulgaris. [unidentifiable mosses 380]

---

[27] The name *Bismalva* does not appear either as a synonym of this species or elsewhere in the main catalogue.

INDEX OF ENGLISH NAMES

Cup or Challice-Mosse, Muscus pyxoides. [*Cladonia* spp. 379]
Tree-Mosse, Muscus arboreus, Lichen arborum. [*Usnea* spp. or other fruticose lichens 377]
Mouse-eare, Pilosella, Auricula muris, Myosotis. [*Cerastium fontanum* 26, *C. arvense* 76, *Myosotis arvensis* 381 and *Pilosella officinarum* 437]
Mouse-taile, Cauda muris, Myosuros. [*Myosurus minimus* 383]
Mugweed, that is, *Crossewort*, Cruciata. [*Cruciata laevipes* 160]
Mugwort, Artemisia. [*Artemisia vulgaris* 62]
Mullein, Tapsus barbatus, Verbascum album vulgare. [*Verbascum thapsus* 564]
Black Mullein, Verbascum nigrum. [*Verbascum nigrum* 607]
Hedge Mustard, or Bank-Cresse, Irio, Erysimum. [*Sisymbrium officinale* 186]
Mustard, Sinapi.[28]
Mushrome,[29] Fungus.
Honey-combe Mushrome, Fungus favaginosus. [*Morchella* sp. 205]
Dusty Mushrome, Fungus pulverulentus. [unidentifiable puff-ball fungi 207]
Dutch Myrtle, Elæagnus Cordi, Myrtus Brabantica. [*Myrica gale* 173]

N.

Wild Navew, Napus sylvestris. [*Brassica rapa* and *B. napus* 385]
Neesewort, that is, *white Hellebore*.[30]
Neppe or Catmint, Nepeta, Mentha cattaria. [*Nepeta cataria* 363]
Common Nettle, Urtica urens. [*Urtica dioica* 628]
Dead Nettle, see *Archangel*. [*Lamium album* 317 and *L. purpureum* 319]
Hedge Nettle, Galeopsis legitima Dioscoridis. [*Stachys sylvatica* 209]
The lesser stinging Nettle, Urtica minor. [*Urtica urens* 627]
Nightshade, Solanum vulgare. [*Solanum nigrum* 546]
Deadly Nightshade, Solanum lethale, Bella donna. [*Atropa belladonna* 548]
Enchanters Nightshade, Circæa Lutetiana. [*Circaea lutetiana* 136]
Woody Nightshade, that is, *Bittersweet*, Solanum lignosum, Dulcamara. [*Solanum dulcamara* 547]
Nipplewort or Dock-Cresse, Lampsana. [*Lapsana communis* 320]
Hasell-Nut, Avellana sylvestris, Corylus. [*Corylus avellana* 154]
Wall-Nut, Juglans, Nux regia. [*Juglans regia* 303]

O

Oneberry, that is, *Herb Paris*, Herba Paris, Solanum tetraphyllon. [*Paris quadrifolia* 277]
Oak-tree, Quercus. [*Quercus robur* 476]
Oats, Avena. [*Avena sativa* 74]
Black Oats, Avena nigra. [*Avena sativa* 75]

---

[28] Variants of the Latin name *Sinapi* and the English name "Mustard" appear in the main catalogue in the entries for *Erysimum Dioscordis*, "Hedge Mustard" [*Sisymbrium officinale* 186], *Rapistrum arvorum*, "wild Mustard" [*Sinapis arvensis* 490] and *Sinapi 2 sive vulgare*, "Ordinary Mustard" [*Brassica nigra* 541].

[29] The two species with "Mushrome" in their English names are listed immediately below.

[30] The name "Neesewort" is not given in the main catalogue for either of Ray's "Wild white Hellebores", *Helleborine angustifolia* [*Epipactis palustris* 274] and *Helleborine latifolia montana* [*Epipactis helleborine* 275].

Wild Oats, Bromos sterilis, Festuca, Ægilops. [*Anisantha sterilis* 196]
Small wild Oats, Festuca altera *Dod*. [*Bromus hordeaceus* 197]
Bearded wild Oats, Ægilops aristis recurvis, Avena pilosa. [*Avena fatua* 14]
Butterflie Orchies, see *Satyrion*. [*Platanthera chlorantha* 407]
Dwarfe Orchies of Zealand, Orchis lilifolius minor sabuletorum Zelandiæ. [*Liparis loeselii* 397]
Fly-Orchies, Orchis myodes. [*Ophrys insectifera* 400]
Great female handed Orchies, see *Satyrion royall*. [*Dactylorhiza fuchsii* 402]
Great male handed Orchies, that is, *male Satyrion royall*. [*Dactylorhiza incarnata* and/or *D. praetermissa* 401]
Handed Orchies with a green flower, by some Frog-Orchies, Orchis palmata flore viridi. [*Coeloglossum viride* 404]
Purple pyramidall Orchies, Orch[is] purpurea spicâ congestâ pyramidali. [*Anacamptis pyramidalis* 406]
Red handed Orchies, Or[chis] palmata minor flore rubello. [*Gymnadenia conopsea sensu lato* 403]
Little purple-flowered Orchies or Souldiers cullions, Or[chis] militaris Pannonica. [*Neottia ustulata* 405]
Wasp or Humble-bee Orchies, Orchis sphegodes or fucum referens. [*Ophrys apifera* 408]
Other species of orchids see in *Satyrion & Stones*.
Goosefoot-Orrache, see *Goosefoot*. [*Chenopodium murale* 72]
Orpine or Live-long, Telephium, Crassula, Fabaria. [*Sedum telephium* 566]
The other Goosefoot Orrache; Atriplex sylvestris latifolia altera. [*Chenopodium rubrum* 73]
Narrow-leaved Orrache, Atr[iplex] angusto oblongo folio. [*Atriplex patula* 68]
Stinking Orrache, Atr[iplex] olida, Garosmum, Vulvaria. [*Chenopodium vulvaria* 69]
Common wild Orrache, Atr[iplex] sylvestris. [*Chenopodium album* 70]
The lesser wild white Orrache, Atr[iplex] sylv[estris] angustifolia altera. [*Atriplex prostrata* 71]
Osiar, Salix folio longissimo. [*Salix viminalis* 515]
Ox-eye, that is, Great Daisie, Bellis major. [*Leucanthemum vulgare* 81]
Ox-slips or great Cow-slips, Primula pratensis inodora lutea. [*Primula elatior* 468]

P

P*Aigles, that is, Cowslips*, Primula veris major, Paralysis Offic[inarum]. [*Primula veris* 467]
Pansies or Hearts ease, Viola tricolor, Herba Trinitatis. [*Viola arvensis* 620]
Herb Paris or Oneberry, Herba Paris, Solanum tetraphyllon. [*Paris quadrifolia* 277]
Bastard stone-Parsley, Sison, Amomum Germanicum. [*Sison amomum* 542]
Fine-leaved bastard Parsley, Caucalis tenuifolia flosculis subrubentibus. [*Caucalis platycarpos* 121]
Fools Parsley, that is, lesser Hemlock, Cicutaria fatua. [*Aethusa cynapium* 133]
Hedge Parsley, Caucalis minor flosculis rubentibus. [*Torilis japonica* 119]
Knotted or Hedgehog-Parsley, Cauc[alis] nodosa echinato semine. [*Torilis nodosa* 120]
Purple-flowered bastard Parsley, Lappula Canaria latifolia or Caucalis *J.B.* [*Turgenia latifolia* 118]
Parsley-piert or Breakstone, Perchpier Anglorum. [*Aphanes arvensis sensu lato* 430]

INDEX OF ENGLISH NAMES

Parsenep, Pastinaca sativa. [*Pastinaca sativa* subsp. *sativa* 422]
Cow-Parsenep, Sphondylium. [*Heracleum sphondylium* 560]
Great water-Parsenep, Sium majus latifolium. [*Sium latifolium* 543]
Middle water-Parsenep, Sium majus angustifolium *Ger*. [*Berula erecta* 544]
Small or common water-Parsenep, Sium umbellatum repens. [*Apium nodiflorum* 545]
Wild Parsenep, Elaphoboscum, Pastinaca latifolia sylvestris. [*Pastinaca sativa* subsp. *sylvestris* 174]
Pasque-flower, Pulsatilla. [*Pulsatilla vulgaris* 475]
Pearlwort, Saxifraga Anglica Occidentalium. [*Asperula cynanchica* 523]
Wild Pear-tree, Pirus sylvestris.[31]
Everlasting Pease, Lathyrus major perennis. [*Lathyrus sylvestris* 325]
Common gray Pease, Pisum flore roseo, semine fusco. [*Pisum sativum* 444]
White Pease, Pis[um] parvum flore & semine albis. [*Pisum sativum* 444]
Maple Pease, Pis[um] flore roseo, semine variegato. [*Pisum sativum* 444]
Pellitory of the wall, Parietaria, Helxine. [*Parietaria judaica* 418]
Bastard Pellitory, that is, *Sneezwort*, Ptarmica. [*Achillea ptarmica* 473]
Penny royall, Pulegium. [*Mentha pulegium* 474]
Pennywort or Navelwort, Cotyledon palustris. [*Hydrocotyle vulgaris* 157]
Periwincle, Clematis daphnoides, Vinca pervinca. [*Vinca minor* 138]
Pestilent-wort, that is, *Butter-burre*, Petasites. [*Petasites hybridus* 436]
Pilewort, Chelidonium minus. [*Ficaria verna* 129]
Male Pimpernell, Anagallis mas. [*Anagallis arvensis* subsp. *arvensis* 42]
Female Pimpernell, Anagallis fœmina. [*Anagallis arvensis* 41]
Yellow Pimpernell, Anagallis lutea nemorum.[32] [*Lysimachia nemorum*]
Round-leaved water Pimpernell, Anagallis aquatica rotundifolia non crenata, that is, the third of [Matthias] de L'Obel. [*Samolus valerandi* 39]
Meadow Pink, Armerius pratensis, Flos cuculi. [*Silene flos-cuculi* 61]
Besome Plantain, Plantago Sarathrodes, Pl[antago] paniculâ sparsâ. [*Plantago major* 449]
Great Plantain, Plant[ago] latifolia vulgaris. [*Plantago major* 447]
Hoary Plantain or Lambs tongue, Plant[ago] incana. [*Plantago media* 448]
Ribwort Plantain, Plant[ago] quinquenervia. [*Plantago lanceolata* 450]
Great water Plantain, Plant[ago] aquatica major. [*Alisma plantago-aquatica* 445]
Small water Plantain, Plant[ago] aquatica minor. [*Baldellia ranunculoides* 446]
Pliant mealy tree, that is, *Wayfaring tree*, Viburnum. [*Viburnum lantana* 612]
Plowmans Spikenard, Baccharis Monspeliensium, Conyza major vulgaris. [*Inula conyzae* 141]
Polypody, Polypodium. [*Polypodium vulgare sensu lato* 455]
Broad-leaved Pondweed, Potamogeiton latifolium. [*Potamogeton natans* 460]

---

[31] The wild pear, *Pyrus pyraster* (L.) Burgsd., is not mentioned in the main catalogue, nor does it appear in either of the appendices. It is, however, marked C as occurring in Cambridgeshire by Ray in *Catalogus plantarum Angliae* (1670, p. 256). See also "Harts-tongue" and "Yellow Pimpernell", which are likewise absent from the main catalogue but are listed in this index and, in the case of the former, in 'Index of places' too.

[32] Neither the scientific name nor the English name appears in the main catalogue or in the appendices. *Anagallis lutea nemorum* is C. Bauhin's (1623) name for *Lysimachia nemorum* L. (Linnaeus 1753, p. 148), a plant which was recorded from Cambridgeshire by S. Corbyn in 1657 in a manuscript list (Druce 1912) but is not annotated as occurring in the county in Ray's *Catalogus plantarum Angliae* (1670, 1677).

Long-leaved Pondw[eed], Potamog[eiton] longis acutis foliis. [*Potamogeton lucens* 462]
Narrow-leaved Pondw[eed], Potamog[eiton] angustifolium. [*Persicaria amphibia* 461]
Perfoliate Pondw[eed], Potamog[eiton] perfoliatum. [*Potamogeton perfoliatus* 463]
Small branched Pondw[eed] with a flat stalk, Potamog[eiton] ramosum caule compresso. [*Potamogeton compressus* 464]
Small grasse-leaved Pondw[eed], Potam[ogeiton] pusillum gramineo folio. [*Potamogeton berchtoldii* and/or *P. pusillus* 465]
Black Poplar, Populus nigra. [*Populus nigra* 458]
Trembling poplar or Aspen-tree, Pop[ulus] Libyca or tremula. [*Populus tremula* 457]
White Poplar, Pop[ulus] alba, Λευϰη [Leuce]. [*Populus alba* and/or *P. alba* × *P. tremula* 456]
Long rough-headed bastard Poppy, Argemone capitulo longiore. [*Papaver argemone* 58]
Round-headed bastard Poppy, Arg[emone] capit[ulo] rotundiore. [*Papaver hybridum* 59]
Red Poppy, that is, *Corn Rose*, Papaver Rhœas. [*Papaver rhoeas* 416]
Spatling Poppy, Pap[aver] Spumeum, Behen album. [*Silene vulgaris* 80]
Violet-coloured horned Poppy, Pap[aver] corniculatum flore violaceo. [*Papaver bivalve* 415]
Wild Poppy, Pap[aver] spontaneum sylvestre. [*Papaver somniferum* 417]
Prickmadam, see *Stone-crop*. [*Sedum rupestre* 531 and *Sedum album* 532]
Prickwood, that is, *Spindle tree*, Euonymus. [*Euonymus europaeus* 188]
Primrose, Primula veris. [*Primula vulgaris* 469]
Privet or Primprint, Ligustrum. [*Ligustrum vulgare* 333]
Pudding-grasse, that is, *Pennyroyall*, Pulegium. [*Mentha pulegium* 474]
Puff-balls, see *Dusty Mushrome*. [unidentifiable puff-ball fungi 207]

R

H*Orse Radish*, Rhaphanus rusticanus. [*Armoracia rusticana* 489]
Water-Radish, Rhaph[anus] aquaticus. [*Rorippa amphibia* 488]
Ragwort, Jacobæa. [*Senecio jacobaea* 300]
Broad-leaved Hungarian Ragwort, Jacobæa latifolia Pannonica. [*Senecio aquaticus* 301]
Mountain Ragwort, Jac[obæa] montana lanuginosa, angustifolia, non laciniata. [*Tephroseris integrifolia* 302]
Rampions with Scabious heads, Rapunculus Scabiosæ capitulo, Scabiosa ovina. [*Jasione montana* 496]
Ramsons, Allium ursinum. [*Allium ursinum* 18]
Rape, that is, *Wild Turnep*, Rapum sylvestre. [*Brassica napus* 495]
Red Rattle or Lousewort, Pedicularis, Alectorolophos. [*Pedicularis sylvatica* 427]
Yellow Rattle or Cockscombe, Pedic[ularis] lutea, Crista galli. [*Rhinanthus minor* 426]
Ray, that is, *Darnell*, Lolium album. [*Lolium temulentum* 340]
Reed, Arundo vallatoria. [*Phragmites australis* 64]
Reed-grasse, Calamogrostis.[33]

---

[33] The name *Calamogrostis* does not appear in the main catalogue. This entry presumably refers to the only two "Reed-grasses" listed there, *Gramen aquaticum majus* [*Glyceria maxima* 239] and

INDEX OF ENGLISH NAMES

Branched Burre-Reed, Sparganium ramosum. [*Sparganium erectum* 553]
Great Burre-Reed, Sparg[anium] non ramosum. [*Sparganium emersum* 554]
Restharrow without prickles, Anonis non spinosa. [*Ononis repens* 48]
Prickly Restharrow, Anonis or Ononis spinosa. [*Ononis spinosa* 47]
Ribwort, that is, *Ribwort-Plantain*. [*Plantago lanceolata* 450]
Rie, Secale. [*Secale cereale* 530]
Wild Rie or Rie-grasse, Hordeum spurium. [*Hordeum murinum* 288]
Hairy Riverweed, Conferva *Plinii*. [unidentifiable filamentous algae 140]
Base wild Rocket or Italian Rocket,[34] Reseda Italica *Ger.* vulgaris *Park*. [*Reseda lutea* 497]
Water-Rocket, Eruca aquatica. [*Rorippa sylvestris* 185]
Rosa Solis or Sun dew, Ros Solis. [*Drosera intermedia* 500 and *D. rotundifolia* 501]
Dogs Rose, or Briar-bush, Rosa canina, Cynosbatos. [*Rosa canina* 502]
Red Rot, that is, *Rosa Solis*, Ros Solis. [*Drosera intermedia*[35] 500]
White Rot, that is, *Marsh Pennywort*, Cotyledon palustris. [*Hydrocotyle vulgaris* 157]
Meadow-Rue or Rubarbe, Thalictrum. [*Thalictrum flavum* 567]
Goates Rue, Galega, Ruta capraria. [*Galega officinalis* 208]
The lesser meadow-Rue. Thalictrum minus. [*Thalictrum minus* 568]
Wall-Rue or white Maiden-hair, Adianthum album, Ruta muraria, Salvia vitæ. [*Asplenium ruta-muraria* 11]
Aglet-headed Rush, Juncus aquaticus minor capitulis equiseti. [*Eleocharis palustris* 305]
Bull-Rush. Juncus aquat[icus] maximus. [*Schoenoplectus lacustris* 306]
Club-Rush, that is, *Aglet-headed Rush*. [*Eleocharis palustris* 305]
Flowring Rush, that is, *water Gladiole*, Junc[us] floridus, Gladiolus palustris *Cord*. [*Butomus umbellatus* 225]
Hard or sharp Rush, Juncus acutus, Oxyschœnos. [*Juncus inflexus* 304]
Common soft Rush, Juncus lævis vulgatior paniculâ sparsâ. [*Juncus effusus* 307]
Marsh soft Rush with a round blackish head, Juncus sylvaticus Tabernæmontano *J.B.* [*Schoenus nigricans* 309]
Round-headed soft Rush, Juncus lævis paniculâ non sparsâ. [*Juncus conglomeratus* 308]

S

SAffron, Crocus. [*Crocus sativus* 159]
Saint Johns wort, Hypericum, Perforata. [*Hypericum perforatum* 296]
Small upright Saint Johns wort, Hypericum minus erectum. [*Hypericum pulchrum* 295]
Saint Peters-wort, Ascyron. [*Hypericum elodes* 66]
Round-leaved hoary Saint Peters-wort, Asc[yron] supinum villosum palustre. [*Hypericum tetrapterum* 65]

Gramen arundinaceum acerosâ glumâ nostras [*Phalaris arundinacea* 240]. Parkinson uses the name *Calamogrostis* for the terrestrial "reede grasses" (1640, pp. 1180–1183) but he also says (p. 1273) that it is as fitly applied to these aquatic species.

[34] *Barbarea* [*Barbarea vulgaris* 78] is called "Winter-cresses or Rocket" in the main catalogue but "Rocket" is not indexed here for that species.

[35] Ray mentions "Red rot" in a note in the main catalogue under *Rorella sive Ros solis foliis oblongis* [*Drosera intermedia*] but the note is probably meant to apply to *Ros solis folio rotundo* [*D. rotundifolia* 501] as well.

Salfern, that is, *Bastard Gromwell*, Anchusa degener facie Milii solis. [*Lithospermum arvense* 43]
Sallow, Salix latifolia. [*Salix cinerea* 513]
Sanicle, Sanicula, Diapensia. [*Sanicula europaea* 520]
Yorkeshire Sanicle, that is, *Butterwort*, Pinguicula. [*Pinguicula vulgaris* 443]
Butterfly-Satyrion, Orchis bifolia calcare oblongo, *see* in *Orchies*. [*Platanthera chlorantha* 407]
Fly-Satyrion, *see* in *Orchies*. [*Ophrys insectifera* 400]
Female Satyrion royall, Orchis palmata fœmina, Palma Christi fœmina. [*Dactylorhiza fuchsii* 402]
Humble-bee Satyrion, *see* in *Orchies*. [*Ophrys apifera* 408]
Male Satyrion royall, Orchis palmata mas, Palma Christi mas. [*Dactylorhiza incarnata* and/or *D. praetermissa* 401]
Saw-wort, Serratula. [*Serratula tinctoria* 538]
Sawce-alone, that is *Jack by the hedge*, Alliaria. [*Alliaria petiolata* 17]
Great Burnet-Saxifrage, Pimpinella Saxifraga hircina major. [*Pimpinella major* 441]
Little round-leaved Burnet-Saxifr[age], Pimp[inella] Saxifr[aga] major altera, Saxifr[aga] hircina minor foliis Sanguisorbæ. [*Pimpinella saxifraga* subsp. *saxifraga* var. *saxifraga* 440]
Small Burnet Saxifr[age], Pimp[inella] Saxifr[aga] minor. [*Pimpinella saxifraga* subsp. *saxifraga* var. *dissecta* 442]
Meadow or green Saxifr[age], Saxifr[aga] pratensis, Seseli prat[ense] nostras *Park*. [*Silaum silaus* 524]
Small grassie Saxifr[age], Saxifr[aga] graminea pusilla. [*Sagina procumbens* 525]
Small water Saxifr[age], Alsine palustris foliis tenuissimis, Saxifr[aga] palustris Anglica. [*Sagina nodosa* 31]
White Saxifr[age], Saxifr[aga] alba. [*Saxifraga granulata* 522]
Common Scabious, Scabiosa major vulgaris. [*Knautia arvensis* 526]
Hairy sheepes Scab[ious], that is, *Rampions with Scabious heads*. [*Jasione montana* 496]
The lesser field Scabious, Scabiosa minor or Columbaria. [*Scabiosa columbaria* 527]
Mouse-ear-Scorpion-grasse, Myosotis scorpioides arvensis. [*Myosotis arvensis* 381]
Water-Scorpion-grasse, Myos[otis] scorp[ioides] palustris. [*Myosotis scorpioides* 382]
Seggrum, that is, *Ragwort*, Jacobæa vulgaris. [*Senecio jacobaea* 300]
Selfe-heale, Prunella. [*Prunella vulgaris* 470]
Water-Sengreen, that is, *Water Souldier*, Stratiotes, Militaris aizoides. [*Stratiotes aloides* 364]
Shepherds-needle or Venus comb, Scandix, Pecten Veneris. [*Scandix pecten-veneris* 425]
Shepherds-purse, Bursa pastoris. [*Capsella bursa-pastoris* 97]
Lesser Shepherds-purse, Bur[sa] past[oris] minor. [*Teesdalia nudicaulis* 98]
Shepherds rod or staffe, Dipsacus minor, Virga pastoris. [*Dipsacus pilosus* 169]
Silverweed, that is, *wild Tansy*, Argentina. [*Potentilla anserina* 60]
Simpson herb, that is, *Groundsell*, Senecio. [*Senecio vulgaris* 535]
Sloe-tree or Black Thorn, Prunus sylv[estris], Acacia German[ica]. [*Prunus spinosa* 471]
Smallage, Apium palustre, Paludapium, Eleoselinum. [*Apium graveolens* 53]
Snakeweed or Bistort, Bistorta. [*Persicaria bistorta* 87]

INDEX OF ENGLISH NAMES

The least Snapdragon ar[36] *Calves snout*, Antirrhinum minimum. [*Chaenorhinum minus* 50]
Sneezwort, Ptarmica. [*Achillea ptarmica* 473]
Sopewort, Saponaria, Struthium. [*Saponaria officinalis* 521]
Common Sorrell, Acetosa vulgaris. [*Rumex acetosa* 7]
Sheeps Sorrell, Acetosa arvensis lanceolata. [*Rumex acetosella* 8]
Wood Sorrell, Acetosella, Oxys, Trifolium acetosum, Lujula. [*Oxalis acetosella* 578]
Field Sothernwood, Abrotanum campestre. [*Artemisia campestris* 1]
Unsavoury Sothernwood, Abrot[anum] inodorum. [*Artemisia campestris* 2]
Fresh water Souldier, Militaris aizoides. [*Stratiotes aloides* 364]
Sowbane, that is, *Goosefoot*, Pes anserinus. [*Chenopodium murale* 72]
Great Sowthistle, Sonchus repens, Hieracium majus. [*Sonchus arvensis* 551]
Prickly Sowthistle, Son[chus] asper. [*Sonchus asper* 549]
Unprickly Sowthistle, Son[chus] lævis. [*Sonchus oleraceus* 550]
Great Spearwort, Ranunculus flammeus major. [*Ranunculus lingua* 482]
Indented Spearwort, Ran[unculus] flam[meus] serratus. [*Ranunculus flammula* 484]
Lesser Spearwort, Ran[unculus] flam[meus] minor. [*Ranunculus flammula* 483]
Round leaved Spearw[ort], that is, *Marsh Crowfoot*. [*Ranunculus sceleratus* 478]
Speedwell or Fluellin,[37] Veronica mas supina. [*Veronica officinalis* 609]
Small meadow Speedw[ell] or Fluel[lin], Ver[onica] pratensis. [*Veronica serpyllifolia* 610]
Lesser upright Speedw[ell] or Fluel[lin], Ver[onica] spicata recta minor. [*Veronica spicata* 611]
Plowmans Spikenard, see *Plowman*. [*Inula conyzae* 141]
Spindle-tree or Prickwood, Euonymus Theophrasti. [*Euonymus europaeus* 188]
Spurry, Saginæ spergula. [*Spergula arvensis* 557]
Chickweed Spurry, Spergula alsineformis. [*Spergularia rubra* 558]
Sea-Spurry, Sperg[ula] marina. [*Spergularia marina* and/or *S. media* 559]
Dwarf Spurge, Esula exigua, Tithymalus leptophyllos. [*Euphorbia exigua* 187]
Sun-Spurge Tithym[alus] helioscopius. [*Euphorbia helioscopia* and *E. peplus* 569]
Wood Spurge. Tithym[alus] characias amygdaloides. [*Euphorbia amygdaloides* 570]
Spurge-Lawrell, Laureola. [*Daphne laureola* 327]
Squinancy-berries, that is, *Black Currans*. [*Ribes nigrum* 499]
Sea Starwort, Tripolium. [*Aster tripolium* 594]
Water Starwort, Stellaria aquatica. [*Callitriche* spp. 562]
Stichwort, Gramen leucanthemum, Holosteum *Ruel*. [*Stellaria holostea* 286]
Lesser Stitchw[ort], Holosteum minus. [*Stellaria graminea* 287]
Ordinary Stonecrop or Prickmadam, Sedum minus hæmatoides. [*Sedum rupestre* 531]
White-flowered Stonecrop or Prickmadam, Sedum minus *Officin[arum]*, Vermicularis flore albo. [*Sedum album* 532]
Stonecrop or Wall-Pepper, Sedum minimum acre, Illecebra *Officin[arum]*. [*Sedum acre* 533]
Strangle-tare or wild Vetch, Aracus or Cracca major. [*Vicia sativa* subsp. *nigra* and/or subsp. *segetalis* 56]
Strawberry, Fragaria. [*Fragaria vesca* 201]
Barren Strawberry, Fragaria minimè vesca. [*Potentilla sterilis* 202]

---

[36] An obvious misprint for *or*.
[37] The name "Female Fluellin, or Round-leaved female Fluellin, or Speedwell" is also given for *Elatine folio subrotundo* [*Kickxia spuria* 176].

Succory, Cichoreum. [*Cichorium intybus* 131]
Little Sunflower or Dwarf Cistus, Helianthemum, Chamæcistus vulgaris flore luteo. [*Helianthemum nummularium* 125]
Long-leaved Sundew or Rosa solis, Ros Solis folio oblongo. [*Drosera intermedia* 500]
Round-leaved Sundew or, &c. R[os] Sol[is] fol[io] rotundo. [*Drosera rotundifolia* 501]
Sycomore tree, that is, *Great Maple*, Acer maius, Sycomorus Ruellii. [*Acer pseudoplatanus* 5]

T

Tansy, Tanacetum. [*Tanacetum vulgare* 563]
Wild Tansy, Argentina. [*Potentilla anserina* 60]
Tare everlasting, Lathyrus sylv[estris] flore luteo. [*Lathyrus pratensis* 326]
Birds Tares, see *Tufted Vetches*. [*Vicia cracca* 55]
Common Tare, Vicia vulgaris sativa. [*Vicia sativa* subsp. *sativa* 615]
Dioscorides his Milk Tare, Glaux Dioscoridis. [*Astragalus danicus* 227]
Tine Tare, Cracca minima, Vicia segetum cum siliquis plurimis hirsutis. [*Vicia hirsuta* 57 and 613]
Tine Tare with smooth cods, Viciæ or Craccæ minimæ species cum siliquis glabris. [*Vicia tetrasperma* 614]
Wild Teasell, Dipsacus sylvestris, Labrum Veneris. [*Dipsacus fullonum* 168]
Tentwort, that is, *Wall-Rue*, Adianthum album. [*Asplenium ruta-muraria* 11]
Tetterwort, see *Dock-cresses*. [*Lapsana communis* 320]
Dwarf Carline Thistle, Carlina acaulis Septentrionalium, Chamæleon exiguus, Carduus acaulis. [*Cirsium acaule* 107]
Wild Carline Thistle, Carlina sylvestris. [*Carlina vulgaris* 116]
Cotten Thistle, Acanthium, Spina alba sylv[estris] *Fuch*. [*Onopordum acanthium* 4]
Thistle with a bending head, Carduus nutans. [*Carduus nutans* 111]
Ladies Thistle or Milk Thistle, Carduus lacteus, Card[uus] Mariæ, Leucographis, Silybum. [*Silybum marianum* 108]
Marsh Thistle, Carduus palustris. [*Cirsium palustre* 114]
Single-headed or Melancholy Thistle, Cirsium Anglicum. [*Cirsium dissectum* 137]
Sow-Thistle, see in the letter S.
Spear-Thistle, Carduus lanceatus. [*Cirsium vulgare* 109 and 110]
Starre Thistle, Card[uus] stellatus, Calcitrapa. [*Centaurea calcitrapa* 112]
Way-Thistle, Card[uus] viarum vulgatissimus. [*Cirsium arvense* 113]
Woolly-headed Thistle, Card[uus] eriocephalus, Card[uus] capite tomentoso, Corona fratrum. [*Cirsium eriophorum* 115]
Thorow-wax, Perfoliata. [*Bupleurum rotundifolium* 431]
Common purging Thorn, see *Buck-thorn*. [*Rhamnus cathartica* 498]
Black Thorn, see *Sloe-tree*. [*Prunus spinosa* 471]
White-Thorn, that is, *Hawthorn-tree*, Oxyacanthus. [*Crataegus laevigata* and *C. monogyna* 414]
Great Throatwort, Trachelium majus. [*Campanula trachelium* 572]
Little Throatwort, or Canterbury bells, Trachel[ium] minus. [*Campanula glomerata* 573]
Toadflax, Linaria. [*Linaria vulgaris* 335]
Bastard Toadflax, Linar[ia] adulterina. [*Thesium humifusum* 334]
Tormentill, Tormentilla, Heptaphyllum. [*Potentilla erecta* 571]

INDEX OF ENGLISH NAMES

Travellers joy, Clematis sylvestris, Viorna vulgi. [*Clematis vitalba* 621]
Birds-foot-Trefoile, Trifolium corniculatum primum *Dod*. Lotus pentaphyllos siliquis rectis glabra minor. [*Lotus corniculatus* 579]
The greater Birdsfoot-Trefoile Trifolium corniculatum tertium *Dod*. Loti corniculatæ major species. [*Lotus pedunculatus* 580]
Heart-Trefoile or Claver, Trif[olium] cochleatum folio cordato maculato, Medica Arabica *Cam*. [*Medicago arabica* 581]
The smallest hedgehog Trefoile, Trif[olium] echinatum arvense. [*Medicago minima* 582]
Hop-Trefoil, Trif[olium] luteum lupinum. [*Trifolium campestre* 584]
The lesser Hop-Trefoile, Trif[olium] lupulinum alterum minus. [*Trifolium dubium* 585]
Marsh Trefoil, or Buckbeans, Trif[olium] palustre. [*Menyanthes trifoliata* 592]
Knotted Trefoile,[38] Trifol[ium] nodiflorum. [*Trifolium striatum* 593]
White-flowered meadow-Trefoile. Trif[olium] pratense album. [*Trifolium repens* 588]
The great rough yellowish or white Meadow Trefoile, Trif[olium] prat[ense] hirsutum majus flore albo-sulphureo. [*Trifolium ochroleucon* 589]
Yellow Medick Trefoile with flat wreathed cods, Trif[olium] sylv[estre] luteum siliquâ cornutâ, Medica frutescens flavo flore *Clus*. [*Medicago sativa* subsp. *falcata* 587]
Common purple Trefoile or Honey-suckle, Trif[olium] pratense purpureum. [*Trifolium pratense* 590]
Great purple Trefoile or Claver-grasse, Trif[olium] majus flore purpureo. [*Trifolium medium* 591]
Strawberry Trefoil, Trif[olium] fragiferum. [*Trifolium fragiferum* 583]
Little yellow Trefoile, or Melilote Trefoile, Trif[olium] luteum minimum. [*Medicago lupulina* 586]
True-love or Oneberry, Herba Paris. [*Paris quadrifolia* 277]
Tunhoof, that is, *Ground Ivy*, Chamæcissus, Hedera terrestris. [*Glechoma hederacea* 272]
Long Turnep, Rapa or rapum radice oblonga. [*Brassica rapa* 494]
Round Turnep, Rap[a or rapum] radice rotunda. [*Brassica rapa* 493]
Wild Turnep or Navew, Rapum sylvestre. [*Brassica napus* 495]
Twayblade, Bifolium. [*Neottia ovata* 396]
Herb Twopence, that is *Moneywort*. Nummularia. [*Lysimachia nummularia* 388]
Wild Tyme or Mother Tyme, Serpyllum vulgare. [*Thymus polytrichus* 537]
Hoary wild Tyme, Serpyllum hirsutum. [*Thymus polytrichus* 536]

V

G*Reat wild Valerian*, Valeriana sylv[estris] major. [*Valeriana officinalis* 605]
Small wild Valerian, Val[eriana] sylv[estris] minor. [*Valeriana dioica* 606]
Venus-bason, that is, *wild Teasell*, Labrum Veneris, Dipsacus sylvestris. [*Dipsacus fullonum* 168]
Venus combe that is, *Shepherds needle*, Scandix, Pecten Veneris. [*Scandix pecten-veneris* 425]
The lesser Venus looking-glasse, Speculum Veneris minus. [*Legousia hybrida* 556]

---

[38] This name must refer to the plant that Ray describes in detail in the main catalogue and tentatively ascribes to *Trifolium nodiflorum* of *Phytologia britannica* (How 1650), i.e. *Trifolium striatum* [593], though he does not give any English name there.

Vervain, Verbena. [*Verbena officinalis* 608]
Vetch, see *Tare*.
Bush-Vetch, Vicia maxima dumetorum. [*Vicia sepium* 616]
Kidney Vetch or Ladies finger, Anthyllis leguminosa. [*Anthyllis vulneraria* 49]
Tufted Vetches or Birds Tares, Vicia sylvestris multiflora, Aracus *Tab*. Galega sylv[estris] *Dod*. [*Vicia cracca* 55]
Yellow Vetchling, Aphaca. [*Lathyrus aphaca* 52]
Corn Violet, that is, *Venus looking-glasse*. [*Legousia hybrida* 556]
Purple Violet, Viola Martia purpurea. [*Viola odorata* L. var. *odorata* 617]
Water Violet, see, *water Millefoile*. [*Hottonia palustris* 365]
White Violet, Viola Martia flo[re] albo. [*Viola odorata* var. *dumetorum* and/or var. *imberbis* 618]
Wild Violet or Dogs Violet, Viola canina. [*Viola reichenbachiana* and/or *V. riviniana* 619]

W.

Wake-Robin or Cuckow-pint, Arum. [*Arum maculatum* 63]
Wall-flower, Leucoium luteum, Keiri. [*Erysimum cheiri* 330]
Wall-Pepper or Stonecrop, Illecebra, Sedum minimum acre. [*Sedum acre* 533]
Wallwort, that is, *Danewort*, Ebulus. [*Sambucus ebulus* 170]
Way-benett, see, *Rie-grasse*. [*Hordeum murinum* 288]
Way-bread, see, *Great Plantain*. [*Plantago major* 447]
Wayfaring tree, Viburnum. [*Viburnum lantana* 612]
Gray Wheat, see Triticum cinericeum,[39] &c. [*Triticum turgidum* 600]
Red Wheat, Trit[icum] spicâ rufâ, Robus. [*Triticum aestivum* 597]
White Wheat, Trit[icum] spicâ albâ, Siligo. [*Triticum aestivum* 595]
Whins or Furze, Genista spinosa, Scorpius. [*Ulex europaeus* 212]
Petty Whin,[40] Genista or Genistella aculeata. [*Genista anglica* 214]
Whitlow or Whitlow-grasse, Paronychia. [*Erophila verna sensu lato* 419]
Rue-Whitlow-grasse, Par[onychia] rutaceis foliis, Sedum tridactylites tectorum. [*Saxifraga tridactylites* 420]
Wilding or Crab-tree, Malus sylvestris. [*Malus sylvestris* 356]
Wild Williams, that is, *Meadow-pink*, Armerius pratensis. [*Silene flos-cuculi* 61]
Crack-Willow, Salix folio lato splendente fragilis. [*Salix fragilis* 510]
Common red Willow, Salix folio utrinque glauco, viminibus rubris. [*Salix alba* 508]
Rose Willow, Salix rosea *Ger*.[41]
Round-eared shining Willow,[42] Salix fol[io] splendente, auriculato, flexilis. [*Salix triandra* 511]
Sweet Willow, that is, *Gaule*, Elæagnus *Cordi*, Myrtus Brabantica. [*Myrica gale* 173]
Common white Willow, Sal[ix] fol[io] utrinque glauco, viminibus albidioribus. [*Salix alba* 509]
Great codded Willowherbe or Loosestrife, Lysimachia siliquosa hirsuta magno flore. [*Epilobium hirsutum* 352]

---

[39] Ray does not cite this name in the main catalogue, but it is a direct translation of his English name Gray Wheat.
[40] Ray also gives "petty Whinn" as an English name for *Anonis sive Ononis* [*Ononis spinosa* 47].
[41] This variant is discussed at the end of *Salix* in the main catalogue.
[42] From here until the end of the index and in places earlier in the list the capital W is formed from two capital Vs.

INDEX OF ENGLISH NAMES

Lesser hairy codded Willowh[erbe], Lysim[achia] siliq[uosa] hirs[uta] parvo flore. [*Epilobium parviflorum* 353]

The least smooth codded Willowh[erbe], Lys[imachia] siliq[uosa] glabra minor angustifolia. [*Epilobium palustre* 351]

The lesser smooth codded Willowh[erbe], Lys[imachia] siliq[uosa] glabra minor. [*Epilobium tetragonum* 350]

Greater wild codded Willowh[erbe], Lys[imachia] siliq[uosa] glabra major, that is, campestris *Ger.* [*Epilobium montanum* 349]

Hooded Willowh[erbe], Lys[imachia] galericulata, Gratiola palustris latifolia. [*Scutellaria galericulata* 345]

Three-leaved purple Willowh[erbe] or Loosestrife, Lys[imachia] purp[urea] trifolia caule hexagono. [*Lythrum salicaria* 348]

Purple spiked Willowh[erbe], Lys[imachia] purp[urea] spicata. [*Lythrum salicaria* 347]

Yellow Willowh[erbe], Lys[imachia] lutea. [*Lysimachia vulgaris* 346]

Woad, Isatis, Glastum. [*Isatis tinctoria* 226]

Wild Woad or Diersweed, Luteola. [*Reseda luteola* 343]

Wood-bind, Caprifolium, Periclymenium, Matrisylva. [*Lonicera periclymenum* 432]

Woodroof, Asperula or Aspergula, Hepatica stellaris. [*Galium odoratum* 67]

Woodwaxe or Diersweed, Genistella tinctoria. [*Genista tinctoria* 215]

Treacle-Wormseed, Camelina, Myagrum siliquâ longâ. [*Erysimum cheiranthoides* 102]

Wormwood, Absinthium. [*Artemisia absinthium* 3]

Y.

YArrow or Millefoile, Millefolium, Achillea. [*Achillea millefolium* 371]
Yellow-weed, that is, *Wild Woad*, Luteola. [*Reseda luteola* 343]
Yew tree, Fraxinus.[43] [*Taxus baccata* 565]

---

[43] This must be a slip for *Taxus*.

INDEX LOCORUM ALIQUOT CUM *PLANTIS* quæ inibi crescunt non passim obviis.

CHESTERTON,
In the lanes and closes.

Anonis non spinosa, Anthyllis leguminosa, Armerius pratensis, Cynosorchis purpurea spicâ congestâ pyramidali, Geranium lucidum saxatile, Orchis morio fœmina, Orch. palmata fœmina, Orch. palmata flore viridi, Orch. palmata mas, Or. palmata rubella cum longis calcaribus, Pimpinella vulgaris, Polygala flore cæruleo, Populus alba, Populus Libyca, Radix cava minima viridi flore, Reseda Italica, Trifolium nodiflorum, Trifolium lupulinum majus & minus.

DITTON,
In the lanes, and in Eigwin and other closes.

Allium ursinum, Bifolium, Bryonia alba, Bry. nigra, Geranium batrachoides, Helleborus niger flore viridi, Nepeta, Ophioglossum, Ranunculus auricomus, Virga pastoris.

INDEX
OF SOME
PLACES
WITH THE PLANTS
which grow in them
that are not generally met with.

CHESTERTON,
In the lanes and closes.

ANonis non spinosa, Anthyllis leguminosa, Armerius pratensis, Cynosorchis purpurea spicâ congestâ pyramidali, Geranium lucidum saxatile, Orchis morio fœmina, Orch[is] palmata fœmina, Orch[is] palmata flore viridi, Orch[is] palmata mas, Or[chis] palmata rubella cum longis calcaribus, Pimpinella vulgaris, Polygala with a blue flower, Populus alba, Populus Libyca, Radix cava minima viridi flore, Reseda Italica, Trifolium nodiflorum, Trifolium lupulinum majus & minus.[1]

[*Ononis repens* 48, *Anthyllis vulneraria* 49, *Silene flos-cuculi* 61, *Anacamptis pyramidalis* 406, *Geranium lucidum* 224, *Anacamptis morio* 399, *Dactylorhiza fuchsii* 402, *Coeloglossum viride* 404, *Dactylorhiza incarnata* and/or *D. praetermissa* 401, *Gymnadenia conopsea* 403, *Poterium sanguisorba* subsp. *sanguisorba* 439, *Polygala vulgaris* 451, *Populus alba* and/or *P. alba* × *P. tremula* 456, *P. tremula* 457, *Adoxa moschatellina* 477, *Reseda lutea* 497, *Trifolium striatum* 593, *Trifolium campestre* 584 and *T. dubium* 585.
Other species listed from Chesterton in the main catalogue but not included in the above list include *Linum catharticum* 338, *Humulus lupulus* 342 and *Berberis vulgaris* 413.]

DITTON,
In the lanes, and in Bigwin and other closes.

Allium ursinum, Bifolium, Bryonia alba, Bry[onia] nigra, Geranium batrachoides, Helleborus niger flore viridi, Nepeta, Ophioglossum, Ranunculus auricomus, Virga pastoris.

[*Allium ursinum* 18, *Neottia ovata* 396, *Bryonia dioica* 91, *Tamus communis* 92, *Geranium pratense* 218, *Helleborus viridis* 276, *Nepeta cataria* 363, *Ophioglossum vulgatum* 395, *Ranunculus auricomus* 486, *Dipsacus pilosus* 169.]

---

[1] These names do not match any in the main alphabetical list of the main catalogue, but are presumably the plants with the chosen names *Trifolium luteum lupinum* [584] and *Trifolium lupinum alterum minus* [585].

GAMLINGAY,

Acetosa arvensis lanceolata, Alsine longifolia uliginosis proveniens locis, Ascyrum supinum villosum palustre, Bulbocastanum, Bursa pastoris minor, Coronopus vulgaris, Filix mas & fœmina, Genista, Genistella aculeata, Gnaphalium minimum, Holosteum minus, Hyoseris mascula, Hypericum pulchrum *Trag.*, *Knawell* Germanorum, Lactuca agnina, Ornithopodium minus, Pentaphyllum rubrum palustre, Pimpinella saxifraga hircina major, Rapum genistæ, Rapunculus scabiosæ capitulo, Ros solis rotundifolius, Spartum parvum *Lob.*, Saginæ spergula, Spergula purpurea, Trifolium lupulinum.

[*Rumex acetosella* 8, *Stellaria alsine* 27, *Hypericum elodes* 66, *Conopodium majus* 96, *Teesdalia nudicaulis* 98, *Plantago coronopus* 152, *Dryopteris filix-mas* 199 and *Pteridium aquilinum* 200, *Cytisus scoparius* 213, *Genista anglica* 214, *Filago minima* 232, *Stellaria graminea* 287, *Arnoseris minima* 280, *Hypericum pulchrum* 295, *Scleranthus annuus* 453, *Valerianella locusta* 311, *Ornithopus perpusillus* 411, *Comarum palustre* 428, *Pimpinella major* 441, *Orobanche rapum-genistae*[2] 412, *Jasione montana* 496, *Drosera rotundifolia* 501, *Nardus stricta* 555, *Spergula arvensis* 557, *Spergularia rubra* 558, *Trifolium campestre* 584.
Other species listed from Gamlingay in the main catalogue but not included in the above list include *Ulex europaeus* 212, *Luzula multiflora* 252 and *Lotus pedunculatus* 580.]

GOGMAGOG HILLS.

Acinos Anglicum *Clusii*, Anthyllis leguminosa, Campanula minor rotundifolia, Erica vulgaris, Euphrasia, Filipendula, Gentianella fugax Autumnalis, Glaux Dioscoridis, Helianthemum, Jacobæa montana lanuginosa non laciniata, Linaria, Linaria adulterina, Linum catharticum, Linum sylvestre cæruleum, Onobrychis, Orchis militaris Pannonica, Orchis sphegodes, Polygala, Pimpinella vulgaris, Pulsatilla, Reseda Italica, Saxifraga Anglica Occidentalium *Lob.*, Serpyllum hirsutum, Trachelium minus.

[*Clinopodium acinos* 9, *Anthyllis vulneraria* 49, *Campanula rotundifolia* 103, *Calluna vulgaris* 184, *Euphrasia* spp. 192, *Filipendula vulgaris* 198, *Gentianella amarella* 216, *Astragalus danicus* 227, *Helianthemum nummularium* 125, *Tephroseris integrifolia* 302, *Linaria vulgaris* 335, *Thesium humifusum* 334, *Linum catharticum* 338, *Linum perenne* 337, *Onobrychis viciifolia* 394, *Neottia ustulata* 405, *Ophrys apifera* 408, *Polygala vulgaris* 451, *Poterium sanguisorba* subsp. *sanguisorba* 439, *Pulsatilla vulgaris* 475, *Reseda lutea* 497, *Asperula cynanchica* 523, *Thymus polytrichus* 536, *Campanula glomerata* 573.
Other species listed from Gogmagog Hills in the main catalogue but not included in the above list include *Chaenorhinum minus* 50, *Carlina vulgaris* 116, puff-ball fungi 207, *Hippocrepis comosa* 452 and *Vebascum nigrum* 607.]

HILL of HEALTH.

Alsine hederulæ folio, Alsine Veronicæ folio, Auricula muris pulchro flore, Centaurium minus purpureum, Echium vulgare, Fungus pulverulentus, Geranium

---

[2] Although Ray did not distinguish the two big broomrapes *Orobanche rapum-genistae* and *O. elatior*, the Gamlingay plant grew "at the roots of the broom" and was clearly the former.

cicutæ folio inodorum, Geranium columbinum,[3] Lagopus vulgaris, Marrubium album & nigrum, Muscus pyxioides, Paronychia vulgaris & rutaceis foliis, *Perchepier Anglorum*, Saxifraga alba, Scabiosa minor, Tussilago *in the ditches fast by*.

[*Veronica hederifolia* 24, *Veronica arvensis* 23, *Cerastium arvense* 76, *Centaurium erythraea* 123, *Echium vulgare* 171, puff-ball fungi 207, *Erodium cicutarium* 217, *Geranium molle* and/or *G. pusillum* 219, *Trifolium arvense* 316, *Marrubium vulgare* 357 and *Ballota nigra* 77, *Cladonia* spp. 379, *Erophila verna sensu lato* 419, *Saxifraga tridactylites* 420, *Aphanes arvensis sensu lato* 430, *Saxifraga granulata* 522, *Scabiosa columbaria* 527, *Tussilago farfara* 603.
Also listed from the Hill of Health in the main catalogue but not included in the above list is *Viola arvensis* 620.]

HINTON or *CHERRY-HINTON;*
In the lanes and closes.

Adianthum album *on the church*. Amomum Germanicum, Antirrhinum minimum, Circæa Lutetiana, Cirsium Anglicum, Clinopodium vulgare,[4] Geranium batrachoides, Glaux vulgaris, Hipposelinum, Linum catharticum, Linum sylvestre cæruleum, Lysimachia lutea, Majorana sylvestris,[5] Orchis myodes, Orchis palmata rubella cum longis calcaribus, Orchis purpurea spicâ pyramidali, Orchis sphegodes, Orobanche, Phyllitis,[6] Saponaria, Trifolium majus albo-sulphureum.

[*Asplenium ruta-muraria* 11, *Sison amomum* 542, *Chaenorhinum minus* 50, *Circaea lutetiana* 136, *Cirsium dissectum* 137, *Clinopodium vulgare* 139, *Geranium pratense* 218, *Astragalus glycyphyllos* 228, *Smyrnium olusatrum* 285, *Linum catharticum* 338, *Linum perenne* 337, *Lysimachia vulgaris* 346, *Origanum vulgare*[7] 410, *Ophrys insectifera* 400, *Gymnadenia conopsea* 403, *Anacamptis pyramidalis* 406, *Ophrys apifera* 408, *Orobanche elatior*[8] 412, *Asplenium scolopendrium* A29, *Saponaria officinalis* 521, *Trifolium ochroleucon* 589.
Other species listed from Cherry Hinton in the main catalogue but not included in the above list include *Centaurium erythraea* 123 (perhaps on Hinton Moor, as Hinton is

---

[3] In the main catalogue Ray also gives this locality for "Geranium malacoides sive columbinum minimu[m] *Park. C.B. The least Doves-foot*." [221], which we also regard as *Geranium molle* and/or *G. pusillum*: see footnote 136 in *Emendanda* of the 1685 appendix.

[4] This is not a name that Ray uses elsewhere but it must presumably be *Clinopodium majus* of the main catalogue [*Clinopodium vulgare* 139].

[5] In the main catalogue Ray gives this name as a synonym for *Origanum vulgare* [410], which he describes as growing in five locations including "a close called Bowyers at Cherry-Hinton, not far from the Church", but in *Emendanda* of the 1663 and 1685 appendices he explains that the plant at this site was in fact *Clinopodium vulgare* (see footnote 4 above), which he mistook "when it was young".

[6] Ray does not include *Phyllitis* in the main catalogue, although it is included as "Harts-tongue" in the index of English names. Despite the record from Cherry Hinton above, Ray listed *Phyllitis* in a letter to Peter Courthope written in July 1661 as one of "about twenty-six species that are not in our Catalogue" which he had discovered "this year" (Gunther 1934; Thompson 1974). A detailed record from Cherry Hinton is given in the 1663 appendix.

[7] See, however, footnote 5 above.

[8] Although Ray did not distinguish the two big broomrapes *Orobanche rapum-genistae* and *O. elatior*, the Cherry Hinton plant must have been *O. elatior*, which is a chalkland species.

linked there with Teversham), *Vinca minor* 138, *Sambucus ebulus* 170 and *Clematis vitalba* 621.]

On the moor and in the water there.

Anagallis aquatica tertia *Lob.*, Centaurium luteum perfoliatum, Cotyledon palustris, Cyperus longus inodorus, Glaux Dioscoridis, two kinds of Gramen cyperoides, various kinds of Gramen junceum, Gramen palustre echinatum, Gramen Parnassi, Gratiola minor, Juncus sylvaticus *Tab.*, Nummularia minor flore purpurascente, Orchis lilifolius minor sabuletorum Zelandiæ, Pinguicula, Polygala, Polygalon Cortusi, Polytrichum aureum majus, Potamogeiton latifol[ium], Ros solis longifolius, Saxifraga palustris Anglica, Tormentilla.

[*Samolus valerandi* 39, *Blackstonia perfoliata* 122, *Hydrocotyle vulgaris* 157, *Cladium mariscus* 165, *Astragalus danicus* 227, two *Carex* spp.,[9] various *Juncus* spp.,[10] *Carex lepidocarpa* 263, *Parnassia palustris* 264, *Lythrum hyssopifolia* 269, *Schoenus nigricans* 309, *Anagallis tenella* 389, *Liparis loeselii* 397, *Pinguicula vulgaris* 443, *Polygala vulgaris* 451, *Hippocrepis comosa* 452, *Polytrichum commune* 12, *Potamogeton natans* 460, *Drosera intermedia* 500, *Sagina nodosa* 31, *Potentilla erecta* 571.
Other species listed from Hinton Moor in the main catalogue but not included in the above list include *Erigeron acris* 144, *Triglochin palustris* 257, *Eriophorum angustifolium* 265, *Epipactis palustris* 274 and *Potamogeton berchtoldii* and/or *P. pusillus* 465.]

KINGSTON WOOD,
and in the closes and corn fast by.

Betonica, Blitum album sylvestre minus, Bugula, Caucalis tenuifolia floribus albis, Gramen leucanthemum, Helenium, Helleborine latifolia, Herba Paris, Lappula canaria latifolia, Lathyrus major perennis, Ligustrum, Lithospermum, Nummularia, Orchis morio mas foliis maculatis, Primula pratensis inodora lutea, Trachelium majus, Viburnum, Viola canina.

[*Betonica officinalis* 84, *Chenopodium polysermum* 88, *Ajuga reptans* 95, *Caucalis platycarpos* 121, *Stellaria holostea* 286, *Inula helenium* 177, *Epipactis helleborine* 275, *Paris quadrifolia* 277, *Turgenia latifolia* 118, *Lathyrus sylvestris* 325, *Ligustrum vulgare* 333, *Lithospermum officinale* 339, *Lysimachia nummularia* 388, *Orchis mascula* 398, *Primula elatior* 468, *Campanula trachelium* 572, *Viburnum lantana* 612, *Viola reichenbachiana* and/or *V. riviniana* 619.
Other species listed from Kingston Wood in the main catalogue but not included in the above list include *Anemone nemorosa* 44, *Euonymus europaeus* 188, *Hyacinthoides non-scripta* 293, *Lamiastrum galeobdolon* 318, *Melampyrum cristatum* 359 and *Scrophularia nodosa* 529.]

---

[9] Perhaps *Carex riparia* [243], *Carex* sp. [244], *C. otrubae* [247] and/or *C. disticha* [249] in the wetter places.
[10] Probably *Juncus articulatus* [253], *J. acutiflorus* and/or *J. subnodulosus* [255] and *J. bufonius* [256].

INDEX OF PLACES

MADINGLEY;
In the wood.

Anemone nemorum, Betonica, Bifolium, Carduus capite tomentoso, Cruciata, Cynocrambe, Euonymus Theophrasti, Gramen leucanthemum, Gramen cyperoides sylvarum tenuiùs spicatum, Hyacinthus Anglicus, Lamium luteum, Lathyrus major perennis, Lathyrus sylvestris flore luteo, Ligustrum, Melampyrum cristatum, Orchis bifolia calcare oblongo, Orchis morio mas foliis maculatis, Quercus, Rhamnus catharticus, Sambucus aquatica, Sanicula, Scrophularia, Serratula, Valeriana sylvestris major, Viburnum, Ulmaria.

[*Anemone nemorosa* 44, *Betonica officinalis* 84, *Neottia ovata* 396, *Cirsium eriophorum* 115, *Cruciata laevipes* 160, *Mercurialis perennis* 163, *Euonymus europaeus* 188, *Stellaria holostea* 286, *Carex sylvatica* 250, *Hyacinthoides non-scripta* 293, *Lamiastrum galeobdolon* 318, *Lathyrus sylvestris* 325, *L. pratensis* 326, *Ligustrum vulgare* 333, *Melampyrum cristatum* 359, *Platanthera chlorantha* 407, *Orchis mascula* 398, *Quercus robur* 476, *Rhamnus cathartica* 498, *Viburnum opulus* 518, *Sanicula europaea* 520, *Scrophularia nodosa* 529, *Serratula tinctoria* 538, *Valeriana officinalis* 605, *Viburnum lantana* 612, *Filipendula ulmaria* 623.
Other species listed from Madingley Wood in the main catalogue but not included in the above list include *Primula elatior* 468 and *Campanula trachelium* 572.]

In the lanes and closes about the Town.

Besides the forenamed, Ebulus, Enula Campana, Genista tinctoria, Geranium sanguineum,[11] Polypodium *on the church-porch*, Veronica pratensis.

[*Sambucus ebulus* 170, *Inula helenium* 177, *Genista tinctoria* 215, *Geranium sanguineum* 222, *Polypodium vulgare sensu lato* 455, *Veronica serpyllifolia* 610.]

NEWMARKET;
About the town.

Aparine major Plinii, Coronopus vulgaris, Medica echinata minima,[12] Sesamoides magnum Salamanticum, Thalictrum minus.

[*Asperugo procumbens* 34, *Plantago coronopus* 152, *Medicago minima* 582, *Silene otites* 539, *Thalictrum minus* 568.]

---

[11] This species is not recorded from Madingley in the main catalogue. The record was overlooked or disregarded by Babington (1860) and Perring *et al.* (1964) and listed as a "casual/garden escape" by Crompton (2001). If this is not an identification or compilation error, the plant must presumably have been a garden escape.

[12] This grew with the two following species at a site which Ray later admitted "may be in Suffolk" (see the entries in the main catalogue [582, 539 and 568] and especially footnote 730 there).

On the heath.

Abrotanum campestre,[13] Abr[otanum] inodorum,[13] Acetosa ovilla, Genista spinosa, Geranium sanguineum, Veronica recta minor, Veronica mas repens.[14]

[*Artemisia campestris* 1, *A. campestris* 2, *Rumex acetosella* 8, *Ulex europaeus* 212, *Geranium sanguineum* 222, *Veronica spicata* 611, *V. officinalis* 609.
Other species listed from Newmarket Heath in the main catalogue but not included in the above list include *Clinopodium acinos* 9, *Chaenorhinum minus* 50, *Cerastium arvense* 76, *Carlina vulgaris* 116, *Helianthemum nummularium* 125, *Calluna vulgaris* 184, puff-ball fungi 207, *Astragalus danicus* 227, *Antennaria dioica* 233, *Tephroseris integrifolia* 302, *Thesium humifusum* 334, *Neottia ustulata* 405, *Hippocrepis comosa* 452, *Asperula cynanchica* 523, *Sagina procumbens* 525, and *Campanula glomerata* 573.]

On the bank of the great Ditch called
Devils ditch.

Besides all mentioned to grow on Gogmagog hills, Auricula muris flore specioso, Geranium sanguineum, Orchis myodes, Orchis palmata rubella cum longis calcaribus, Orch[is] purpurea spicâ congestâ pyramidali, & other species of orchids.

[*Cerastium arvense* 76, *Geranium sanguineum* 222, *Ophrys insectifera* 400, *Gymnadenia conopsea* 403, *Anacamptis pyramidalis* 406, etc.
The only species listed from the Devil's Ditch in the main *Catalogus* but not included above or under the Gogmagog Hills is *Hippocrepis comosa* 452.]

About STRETHAM ferry, in
the river and ditches and on the banks.

Conyza palustris serratifolia, Millefolium aquaticum pennatum spicatum, Myriophyllum tenuifolium *Ger. emac.* Nymphæa aquatica minor flore fimbriato, Stratiotes.
[*Senecio paludosus* 146, *Myriophyllum spicatum* 368, *Potamogeton pectinatus* 370, *Nymphoides peltata* 392, *Stratiotes aloides* 364.]

TEVERSHAM,
About the closes and on the moor.

Besides many mentioned in Cherry-Hinton, which also grow there, Anagallis aquatica quarta *Lob.*, Aquilegia sylvestris, Gramen triglochin, Juncus floridus, Laureola, Lithospermum, Nymphæa alba major, Ranunculus flammeus major, Triorchis, Xyris.

---

[13] See footnote 2 in the main catalogue.
[14] This is not a name that Ray cites elsewhere; it is presumably *Veronica mas supina & vulgatissima* and is perhaps a slip for the synonym *Veronica mas serpens* [*Veronica officinalis* 609].

[*Veronica scutellata* 40, *Aquilegia vulgaris* 54, *Triglochin palustris* 257, *Butomus umbellatus* 225, *Daphne laureola* 327, *Lithospermum officinale* 339, *Nymphaea alba* 390, *Ranunculus lingua* 482, *Spiranthes spiralis* 409, *Iris foetidissima* 630.
Species listed in the main text as growing at both Hinton and Teversham include *Sagina nodosa* 31, *Samolus valerandi* 39, *Centaurium erythraea* 123, *Hydrocotyle vulgaris* 157, *Parnassia palustris* 264, *Eriophorum angustifolium* 265, *Epipactis palustris* 274, *Schoenus nigricans* 309, *Linum catharticum* 338, *Anagallis tenella* 389, *Liparis loeselii* 397, *Ophrys insectifera* 400 and *Clematis vitalba* 621. Species listed for Teversham but not included in the above list include *Galium palustre* 210, *Epilobium palustre* 351, *Populus alba* and/or *P. alba* × *P. tremula* 456, *P. tremula* 457, *Rhamnus cathartica* 498 and *Typha latifolia* 604.]

TRIPLOW Heath,

Alsine tenuifolia *J.B.*, Chamæpitys.

[*Minuartia hybrida* 32, *Ajuga chamaepitys* 127.]

In one small close at CHERRY-HINTON, (which we call the Chalk-pit close, because it is near the place where they now burn lime, and likewise full of chalk pits) besides many vulgar plants which we took no notice of, we remember to have seen the following.

Absinthium vulgare, Anthyllis leguminosa, Anonis non spinosa, Artemisia, Bellis major, Bifolium, Bryonia alba, Bryonia nigra, Centaurium minus luteum,[15] Cerefolium sylvestre, Clinopodium,[16] Cruciata, Daucus vulgaris, Elaphoboscum, Galeopsis vera, Glaux vulgaris, Gramen tremulum minus, various species of Hieracium, Hypericum, Jacea segetum major, Ligustrum, Linaria, Linum catharticum, Linum sylvestre cæruleum, Luteola, Melilotus vulgaris, Morsus diaboli, Onobrychis, Orchis purpurea spicâ pyramidali, the same with a white flower, Orchis palmata fœmina, Orchis palmata rubella cum longis calcaribus, Orobanche, Paralysis, Pimpinella sanguisorba,[17] Polygala vulgaris, Polygalon *Cortusi*, Primula veris, Reseda Italica, Saxifraga Anglica alsinefolia, Scabiosa vulgaris, Scabiosa minor, Serpyllum, Trachelium minus, Tragopogon, Trifolium lupulinum, Trifol[ium] majus albo-sulphureum, Verbena, Vicia sylvestris multiflora.

[*Artemisia absinthium* 3, *Anthyllis vulneraria* 49, *Ononis repens* 48, *Artemisia vulgaris* 62, *Leucanthemum vulgare* 81, *Neottia ovata* 396, *Bryonia dioica* 91, *Tamus communis* 92, *Blackstonia perfoliata* 122, *Chaerophyllum temulum* 124, *Clinopodium vulgare* 139, *Cruciata laevipes* 160, *Daucus carota* subsp. *carota* 166, *Pastinaca sativa* subsp. *sylvestris* 174, *Stachys sylvatica* 209, *Astragalus glycyphyllos* 228, *Briza media* 266, composites with yellow, wholly ligulate flower-heads (of the tribe Lactuceae),[18] *Hypericum perforatum* 296, *Centaurea scabiosa* 298, *Ligustrum vulgare* 333, *Linaria vulgaris* 335, *Linum catharticum* 338, *Linum perenne* 337, *Reseda luteola* 343, *Melilotus altissimus* 361, *Succisa pratensis* 374, *Onobrychis viciifolia* 394, *Anacamptis pyramidalis* 406, *Dactylorhiza fuchsii* 402, *Gymnadenia conopsea* 403, *Orobanche elatior*[19] 412, *Primula veris* 467, *Poterium sanguisorba* subsp. *sanguisorba* 439, *Polygala vulgaris* 451, *Hippocrepis comosa* 452, *Primula vulgaris* 469, *Reseda lutea* 497, *Asperula cynanchica* 523, *Knautia arvensis* 526, *Scabiosa columbaria* 527, *Thymus polytrichus* 537, *Campanula glomerata* 573, *Tragopogon pratensis* 574, *Trifolium campestre* 584, *Trifolium ochroleucon* 589, *Verbena officinalis* 608, *Vicia cracca* 55.]

---

[15] This is not a name that Ray cites elsewhere but it must be the plant with the chosen name *Centaurium luteum* and the synonym *Centaurium parvum luteum* [*Blackstonia perfoliata* 122]; the next species in the main catalogue is *Centaurium minus* with the synonym *Centaurium minus flore purpureo & albo*.

[16] Presumably *Clinopodium majus* of the main catalogue [*Clinopodium vulgare* 139].

[17] Presumably *Pimpinella vulgaris, sive minor* of the main catalogue [*Poterium sanguisorba* subsp. *sanguisorba* 439], which has a synonym *Pimpinella sanguisorba minor hirsuta*.

[18] Some at least of *Leontodon hispidus* [278], *Hypochaeris radicata* [279], *Scorzoneroides autumnalis* [281], *Crepis capillaris* [282], *Picris hieracioides* [284] and, if species with *Hieracium* names as synonyms are included, *Helminthotheca echioides* [94], *Taraxacum* spp. [167] and *Sonchus arvensis* [551]. Ewen & Prime (1975) misleadingly render this as "*Hieracium* spp.", even though Ray recorded no species now included in this genus in the *Catalogus*.

[19] See footnote 8.

In one small close at CHERRY-HIN-TON, (*which we call the Chalk-pit close, because it is near the place where they now burn lime, and likewise full of chalk pits) besides many vulgar plants which we took no notice of, we remember to have seen the following.*

Absinthium vulgare, Anthyllis leguminosa, Anonis non spinosa, Artemisia, Bellis major, Bifolium, Bryonia alba, Bryonia nigra, Centaurium minus luteum, Cerefolium sylvestre, Clinopodium, Cruciata, Daucus vulgaris, Elaphoboscum, Galeopsis vera, Glaux vulgaris, Gramen tremulum minus, Hieracii variæ species, Hypericum, Jacea segetum major, Ligustrum, Linaria, Linum catharticum, Linum sylvestre cæruleum, Luteola, Melilotus vulgaris, Morsus diaboli, Onobrychis, Orchis purpurea spicâ pyramidali, eadem cum flore albo, Orchis palmata fœmina, Orchis palmata rubella cum longis calcaribus, Orobanche, Paralysis, Pimpinella sanguisorba, Polygala vulgaris, Polygalon *Cortusi,* Primula veris, Reseda Italica, Saxifraga Anglica alsinefolia, Scabiosa vulgaris Scabiosa minor, Serpyllum, Trachelium minus, Tragopogon, Trifolium lupulinum, Trifol. majus albo-sulphureum, Verbena, Vicia sylvestris multiflora.

NOMINUM
quorundam Ratio,
SIVE
ETYMOLOGIA.

A

Abrotănum, Ἀβρότανον, ut Nicandri testatur interpres, indè nomen sibi usurpavit, quòd conspectu tenerum, molle & delicatum appareat, διὰ τὸ πρὸς ὄψιν ἁβρὸν κỳ ἁπαλὸν φαίνεϑς: aut quòd gravem acrémque & difficilem halitum prospiret, ἄλλως τε κỳ διὰ τὸ δύσπνυν ἐῖ. Casp. Hofmannus Abrotonum scribit per o, ex autoritate Plinii, Theophrasti & aliorum, ut sit ab ἁβρὸς & τίνυς, id est, molle & delicatum robur. Cæterùm legitur etiam Abrotanus apud *Lucret.*

Abrotaníque graves & tristia Centaurea.

Absinthium, Ἀψίνϑιον, plurimi dictum putant; quasi ἀπίνϑιον ab α priv. & πίνω, quòd propter amarorem insignem nemo illud libenter bibat. Alii nomen hoc inditum esse huic plantæ existimant ἀπὸ τῦ ἅπτεϑς, quòd est tangere & manibus tractare, per antiphrasin, quòd nullum animal ob amaritudinem eximiam eam attingat. Cognomentum Pontici assumpsit à Ponto, Asiæ minoris provincia mari adjacente, ubi pecora pinguescunt illo, & ob id sine felle reperiuntur. *Theophr. Plin.*

Acanthium, Ἀκάνϑιον, ab ἄκανϑος vel ἄκανϑα spina.

Ace‐, Σφένδαμνος, neutrius est generis. Acer vel aceris

The Meaning of certain
NAMES,
OR
ETYMOLOGY.

A

Abrotănum, Ἀβροτανον [Abrotanon], as Nicander's interpreter bears witness; he brought the name into use from it because it appears tender, soft & delicate, *owing to its seeming delicate [ἁβρον, habron] and soft to the sight*, or because it exhales a heavy, penetrating & difficult breath, *above all owing to its being unfit to breathe*.[1] Casp[ar] Hofmann writes Abrotonum with o on the authority of Pliny, Theophrastus & others, as it were from ἁβρος [habros] & τονος [tonos], that is a soft & delicate force [robur].[2] On the other hand it is also written as Abrotanus in the works of *Lucret[ius]*:
    And heavily-scented Southernwoods & foul Centauries.[3]

Absinthium, Ἀψινθιον [Apsinthion], as most people think it is pronounced, as if it was ἀπινθιον [apinthion] from α priv[ative][4] & πινω [pino, "I drink"], since nobody would willingly drink it on account of its extraordinary bitterness. Others think that this name was given to this plant *from its being touched*,[5] because it is, by antiphrasis,[6] something to touch & handle with the hands, since no animal touches it on account of its exceptional bitterness. It acquired the name of Ponticum from Pontus, a province of Asia Minor bordering the sea, where livestock grow fat on it & are found to be without any bitterness on that account. *Theophr[astus]. Plin[y].*

---

[1] Throughout this translation of Ray's 'Etymology' a sanserif typeface is used to represent passages originally in Greek. Ray himself frequently provides a translation in Latin, which is given here in English, often in the same or similar words to the translation from Greek. Nicander of Colophon was a Greek poet, physician and grammarian of the second century BC, whose two surviving hexameter poems, *Alexipharmaca* and *Theriaca*, are principally about poisons and their antidotes. Ray's first sentence is taken directly from Bauhin & Cherler (1651, vol. 3(1), p. 179), whose source appears to be *Definitionum medicarum libri XXIIII* (1564, p. 1) by Jean de Gorris (Johannes Gorraeus, 1505–1577). Nicander's interpreter is the unnamed author of Σχολια εἰς Θηριακα ("Commentaries on *Theriaca*") referred to as *Interpretatio innominati authoris in Theriaca* in the extended title of an edition of Nicander's works published in 1523 (Anon. 1523); in this the two explanations are given in almost identical Greek words to Ray's on p. 30 *verso* in a commentary on ἀβροτανοιο [abrotanoeo] in line 66 of *Theriaca*.
[2] Caspar Hofmann (1572–1648), Professor of Medicine in Altdorf, Switzerland, published his *De medicamentis officinalibus* at Paris in 1646. For Pliny the Elder (Gaius Plinius Secundus) and Theophrastus see Chapter 4. All three are frequently cited in the 'Etymology'.
[3] *Abrotanique graves & tristia Centauria*. The source is Lucretius' *De rerum natura*, 4: 125; the Loeb Classical Library edition has the final word as *centaurea*. Titus Lucretius Carus was a Roman poet of the first century AD.
[4] i.e. an alpha at the beginning of a word signifying a negative meaning (as, for example, in "atheist" or "asymmetrical").
[5] from the Greek verb ἁπτειν [haptein, "to touch"].
[6] This is the use of a word or phrase to mean the opposite of its usual or literal sense.

Acanthium, Ἀκανθιον [Acanthion], from ἀκανθος [acanthos] or ἀκανθα [acantha], a thorn.

Acer, Σφενδαμνος [Sphendamnos], is of neuter gender. Acer or aceris (as Solinus calls it in his Memorabilia[7]) was called thus by some (the reason for the name having been sought after for quite a long time) because it exercises the shrewd talents of craftsmen[8] on account of the excellence & elegance of that timber. *J.B.*

Acetōsa is named from vinegar [acetum] on account of its sap; moreover it is called Ὀξαλις [Oxalis] by the Greeks from its acid juice & *Sorrell*, as if *Sowrell*, by the English from its acid (English *sowre*) sap. Acetosella is a diminutive from Acetosa.

Achillēa, Ἀχιλλεως & Ἀχιλλειος σιδηριτις [Achilleos & Achilleios sideritis],[9] was named after Achilles, its first discoverer. *Plin[y]*, book 25, chapter 5: *Achilles too, the disciple of Chiron, found a plant with which he might heal wounds and which for that reason is called Achillea; with this plant he is said to have cured Telephus.*[10]

Acĭnos, Ἀκινος, because, in the opinion of Dioscorides, *it stops*, that is *it arrests, the bowels and menstruation when drunk*, it arrests the bowels and the menses.[11] [Mathias] de L'Obel,[12] by writing υ argues that it is ἀκυνος [acynos], as it were from α priv[ative] & κυω [cyo], I am pregnant, because it signifies a sterile and barren plant, for Theophr[astus] & Pliny do not recognise any flower or seed in it, perhaps because it produces flowers rather obscurely & rather late, so that it is frequently destroyed by the arrival of cold weather.

Aconitum, Ἀκονιτον [Aconiton], said by Pliny to be from ἀκονη [acone, "whetstone"], because it grows on bare sharp rocks [in nudis cautibus], but by Theophrastus from Ἀκοναι [Aconae], a village of the Maryandini.[13] Ovid supports the former derivation when he says:

---

[7] For Gaius Julius Solinus and his geographical work see Chapter 4.

[8] *artificum acria ingenia*, with the improbable implication that the tree's name is derived from the adjective *acer* (with neuter plural *acria* as here and meaning "sharp", "acrid", "keen", "shrewd").

[9] "Achilles'& Achillean ironwort".

[10] The quotation is from Pliny's *Natural history*, XXV. 42 in the Loeb edition. According to legend, on their way to attack Troy the Greeks landed by mistake at Mysia, where Achilles wounded the king, Telephus, with his famous ashen spear (see footnote 290 in the main catalogue). The wound would not heal, so Telephus consulted the Delphic oracle and was told that "what wounded shall heal". According to one version of the story he convinced Achilles to heal his wound in return for showing the Greeks the way to Troy; Achilles had been tutored in herbalism by the centaur Chiron, so he knew of the virtues of *Achillea*. However, according to another version he refused, claiming that he had no such knowledge; the cunning Odysseus then argued that the oracle was referring not to Achilles but to the spear, from which Achilles scraped some verdigris, which duly cured Telephus, and from the scrapings the herb *Achillea* sprang up. See also the entry on *Telephium* below.

[11] The first Greek word is ἀκινει [acinei], a hypothetical verb compounded from α privative (see footnote 4 above) and κινειν [cinein, "to move"] on the basis of ἀκινητος [acinetos, "motionless"]. For Pedanius Dioscorides, an influential ancient writer on the medical uses of plants frequently cited in the 'Etymology', see Chapter 4.

[12] For Mathias de L'Obel (Lobelius) see *Lob.* in Ray's 'Explanation' and Chapter 4. He is frequently cited in the 'Etymology', as in the main catalogue, sometimes just by the abbreviation *Lob.*

[13] *Enquiry into plants*, IX. XVI. 4 in the Loeb edition, where the people are called Μαριανδυνοι [Mariandynoe]. The text continues with the statement that the plant specially likes rocky places, so Pliny says in his *Natural history* (XXVII. 10 in the Loeb edition) that the bare rocks where it grows are called *aconae*. The alternative explanation that follows after the quotation from Ovid is given at the end of the same section of Pliny's *Natural history*.

ETYMOLOGY

Which countrymen call aconites because they grow vigorously from a hard rock.[14] Others derive the name from rocks indeed but ascribe a different reason to it, *namely because* this plant *has the same power in causing death as whetstones have in grinding the sharp edge of an iron implement and its rapid action is immediately felt when it is applied. Plin[y].*

Acorus, Ἄκορος [Acoros], is so called because it heals *the pupils*,[15] that is the pupils or the misty pupil of the eyes.

Adiantum, Ἀδίαντον [Adianton], because *it may not be wetted*,[16] i.e. it does not get wet in waters because of its dryness, evidently of its surface, which is resistant to water & regularly repels it, as happens on dust & feathers & other dry things on which water does not rest but either runs off or forms round droplets.[17] *Plin[y]*, book 22, chapter 21:[18] *Another wonder relates to Adianthum: it rejects waters and so great a dissociation is observed that it is as if dry when sprinkled or immersed; whence also its name given by the Greeks.* However nothing like this is observed in Adianthum today, whence others say that it is not called Ἀδίαντον for the reason that it never gets completely wet, as if, when submerged in water for a longer space of time, it would not be wetted just as much as other plants, but because the drops of a shower falling from the sky do not settle on its leaves. Rather we agree with those who are of the opinion that, *instead of the verb to wet*, it is named from the verb to wither[19] & α priv[ative], for it does not wither immediately like all other herbs.

Ægilops Αἰγίλωψ means a goat's eye and it is so called because it cures lachrymal fistulas,[20] i.e. defects & abscesses of the eyes. *The herb with the same name cures lachrymal fistulas.*[20] *Plin[y], book 25, chapter 13.*[21] In fact abscesses of the eyes between the nostrils & the greater corner of the eye are called ægilopes because that defect is almost particular & continuous for goats, from αἴξ [aix], a she-goat, & ὤψ [ops], an eye, or from ἀγχι [anchi], near, & ὤψ [ops], an eye, because they arise near the eye.

---

[14] *Quæ quia nascuntur dura vivacia caute*
Agrestes aconita vocant.
The source is Ovid's *Metamorphoses*, 7: 418–419. For Ovid (Publius Ovidius Naso) see Chapter 4.

[15] ταῖς κόραις, dative plural of κόρη [core] with the definite article. The Greek word is explained by two different Latin ones, first *pupilla* in the plural (a word, like the Greek one, of which the primary meaning is "girl") and then *acies* in the singular (of which the primary meaning is "sharp edge", as in the quotation from Pliny in the previous entry, whence "sharpness of vision" and, concretely, "pupil of the eye").

[16] from α privative and the verb διαίνειν [diaenein], meaning "to wet", "to moisten", as also in the phrase near the end of the entry, παρα το διαινειν ("instead of to wet").

[17] *aut elabitur, aut in guttulas rotundatur*, literally "either slides away or into droplets is rounded".

[18] Pliny's *Natural history*, XXII. 62 in the Loeb edition.

[19] *marcescere*, which is puzzling because Ray gives no equivalent Greek word; the obvious equivalent is μαραίνειν [marainein] but this bears no resemblance to Ἀδίαντον [Adianton]. Ferns generally wither more, not less, quickly than other herbs, but *Asplenium ruta-muraria* [11; see footnote 282 in the main catalogue] is more robust than most ferns; possibly Ray intended *reliquæ herbæ* to mean "the rest of the ferns" but this seems rather unlikely. The 'unwettable' plant of Theophrastus (*Enquiry into plants*, VII. XIV. 1 in the Loeb edition) and Pliny was probably Maidenhair Fern, *Adiantum capillus-veneris* L., rather than *Asplenium ruta-muraria*.

[20] *Ægilopas* (accusative plural of *Ægilopa*).

[21] Pliny's *Natural history*, XXV. 146 in the Loeb edition.

Agrifolium is a barbarous word; monks have Acrifolium,[22] but Aquifolium, Ἄγρια [Agria], is read in the works of Plin[y], so called from the prickly leaves, for the Ancients said aquus instead of acus.[23] *Gassend[us]*.[24]

Agrimonia, Ἐυπατωριον [Eupatorion]: We, with the Lexicographers, think this plant is so called because it is abundant in the fields.[25] Some think this word is derived from ἀργεμώνη [argemone] because it heals the ἀργεμα [argema] or ἀργεμώνη [argemone], that is a defect of the eyes, but without credible reason.[26]

Ἀιρα [Aera], that is White Darnel,[27] as [Mathias] de L'Obel conjectures, *from its being raised*, to be raised up, to be lifted up, to be seized upwards;[28] for taken internally both in bread and in drink it causes a dark giddiness and intoxication akin to dementia, whence it is called in France *Yuray*, i.e. drunken or inebriating. *J.B.*

Alchimilla, which is so called; it had its name from the fact that it was honoured by the commendations of the Alchemists.

Alectorolŏphus, Ἀλεκτορολοφος [Alectorolophos], Cock's Crest, from the likeness of the leaves.

Alleluia, Ὀξυς [Oxys], i.e. Trifolium acetosum,[29] is called Juliola by the Calabrians; the barbarians[30] have ridiculously corrupted it to Lujula & Alleluia. *Scaliger in Theophr[astus], de causis plant[arum]*.[31]

Alliaria, because it smells of allium[32] when crushed with the fingers. Thankfully the name is not found in botanical works.[33]

Allium, Σκοροδον [Scorodon], either perhaps from *springing up*,[34] because it grows by springing up, or from the strong smell exhaling from it, & then it ought to be written as alium. Either derivation is inappropriate. It is commonly believed that a magnet smeared with garlic does not attract iron, but Gilbertus[35] asserts that that is a fiction & a lie.

---

[22] i.e. "Sharp-leaf".

[23] *aquum* [accusative] *pro acu* [ablative after the preposition *pro*] *dixerunt*: *acus* means "needle".

[24] See *P. Gassend.* in Ray's 'Explanation' and Chapter 4.

[25] *in agris*, as if the plant was named from the noun *ager* (plural *agri*).

[26] See the entry for *Argemone* below.

[27] *Lolium album* [340], i.e. Darnel, *Lolium temulentum* L.

[28] παρα το αἱρεως [para to aereos] from the Greek verb αἱρειν [aerein, "to raise", "to lift up"], explained by the passive infinitives of three Latin verbs, *efferi*, *attolli* and *sursum rapi*.

[29] "Vinegary Trefoil", i.e. Wood-sorrel, *Oxalis acetosella* [578].

[30] See footnote 170 in the main catalogue.

[31] Julius Caesar Scaliger (Giulio Cesare della Scala or Scaligero, 1484–1558) published his *Commentarii et animadversiones* on Theophrastus' *De causis plantarum* at Lyon and Geneva in 1566.

[32] i.e. Garlic, *Allium sativum* L.

[33] *Grætè nomen non invenit apud botanicos*. Literally "Thankfully, the name does not find in the works of botanists." The spelling of the first word, normally *gratè*, is odd and the verb should be passive, *invenitur*.

[34] from the Greek deponent verb ἅλλομαι [hallomae, "I leap", "I spring up"].

[35] William Gilbert (*c.* 1544–1603), born in Colchester, was a leading London physician with close ties with the city's maritime community. He rejected the accepted doctrines of Aristotelian natural philosophy and Galenic medicine. His major work, *De magnete* ("On the lodestone"), published in 1600, not only summarised the results of numerous experiments on magnetism and for the first time explained the action of the nautical magnetic compass but also set out a new philosophy (ODNB). Ray's reference is to the start of the book (p. 2) where he surveys the fables propagated by earlier authors. For an English translation see Mottelay (1990).

ETYMOLOGY

Alnus, Κληθρα [Clethra], because it is nourished by a river.³⁶ *C.B.* What if we were to say that rather the tree itself nourishes herbage? *It feeds* (says *Pliny*) *sown crops & it has been observed that under the shelter of an Alder lusher herbage grows.*³⁷ *J.B.*

Alsine is so called because it especially likes *groves*,³⁸ i.e. groves & shady places.

Alyssum, Ἀλυσσον [Alysson], from α priv[ative] & λυσσα [lyssa], madness, because it is helpful against insanity or against the bites of a rabid dog.

Amellus,³⁹ the flower of the herb Amella, which grows by the river Amella in France.

Anacampsĕros, Ἀνακαμψερως, *from the returning*⁴⁰ *of love,* so called from returning love, as Pliny declares *in book 23, chapter 17,* since, including it among magic herbs, he asserts that love returns at its touch, even that renounced with hatred.⁴¹

Anagallis, Ἀναγαλλις, perhaps received its name from its hyacinthine colour, for γαλλις [gallis] is the hyacinth.⁴²

Anchūsa, Ἀγχουσα, either *from* ἀγχειν [anchein, "to strangle"], because it possesses a suffocating power, or from the florid colour of the root if indeed ἀγχουζειν [anchuzein] was to daub with purple.⁴³

Androsæmum: this plant was named from ἀνδρος αἱμα [andros haema, "a man's blood"] for the reason that the bruised leaves stain as it were with blood. It is called by our countrymen by its Italian name *Tutisan*, as if curing everything.⁴⁴

Anemōne, Ἀνεμωνη, *from the wind,*⁴⁵ because the flower opens only when the wind is blowing or rather because it exults in flowering in open places & those exposed to wind.

Angelica seems to barbarians & vagabonds⁴⁶ to be called after angels because of its remarkable qualities; for the same people it is called the root of the Holy Spirit for the same reason.

Anōnis, Ἀνωνις, seems to be so called from not gratifying because it affords no use, since it is inimical to ploughmen. Ononis, ὀνωνις: some derive this from *ass & I*

---

³⁶ *quòd amne alatur*, as if the tree was named from the verb *alere* ("to nourish").
³⁷ Pliny, in a series of brief comments about the shade of different trees (*Natural history*, XVII. 90 in the Loeb edition), says no more than that the shade of the alder is thick (*pinguis*) but that it feeds sown crops (*pascens sata*). The roots of alders growing near the soil surface bear nodules which support nitrogen-fixing bacteria in the actinomycetous genus *Frankia* in an "actinorhizal symbiosis" (Huss-Danell 1997). The extra nitrogen is available to plants other than the host tree and it could conceivably produce visible effects on species growing in soils with low nitrogen levels.
³⁸ ἀλση [alse], plural of ἀλσος [alsos].
³⁹ See footnote 55 in the main catalogue.
⁴⁰ from the Greek verb ἀνακαμπτειν [anacamptein, "to return"] with an aorist infinitive ἀνακαμψαι [anacampsae].
⁴¹ Pliny's *Natural history*, XXIV. 167 in the Loeb edition. Pliny makes "love" plural, *amores*.
⁴² We can find no evidence to support this statement.
⁴³ We have not been able to trace this Greek verb.
⁴⁴ *quasi omnia sanans*. The modern English name for *Hypericum androsaemum* L. is "Tutsan". One of the current Italian names is *tutta-sana*.
⁴⁵ i.e. supposedly from the Greek word for "wind", ἀνεμος [anemos], whence one of the English names of *Anemone nemorosa* [44], "Windflower". A more probable source of the name is Naaman, the Phoenician name of the Sumerian god of food and vegetation, Tammuz, also identified with Adonis (derived from the Semitic word for "Lord"), from whose blood the scarlet flowers of *Anemone coronaria* L. and similar species were believed to have sprung. *Adonis* is now the name of a related genus.
⁴⁶ See footnotes 170 and 844 in the main catalogue.

gratify,[47] as if you might say an ass's delight. For some of the Greeks assert that asses roll themselves around on it & rub their backs with pleasure on the prickles of this herb. *C.B.pin.*

Anserina, *see* Argentina.

Anthyllis, Ἀνθυλλις.

Antirrhinum, for Dioscorides Ἀντιρρινον [Antirrhinon][48] & Ἀναρρινον [Anarrhinon], because it has a fruit like the nostrils of a bull-calf. It is also called Os leonis by Caesalpinus,[49] who thinks that this plant is the same as the one that Columella[50] understands by those words, *And the savage Mouth of a fierce gaping Lion*,[51] since its flowers imitate the open jaws of a lion's mouth.

Aparine, Ἀπαρινη. Some call it Philanthropon from its roughness that clings onto clothes. *Plin[y], book 27, chapter 5.*[52]

Aphaca, Ἀφακη [Aphace].

Apium, Σελινον [Selinon], appears to some to be derived from bees[53] because bees delight in it. Among the Greeks the victors' garlands were not infrequently made from smallage, especially at the Isthmian Games.[54] *But its prizes wild olive, apples, smallage, stone pine.*[55]

Aquilegia, i.e. Aquilina, on account of the curved tips of the flowers as if they are eagles' claws; it is called by the English *columbines* because the recurved spurs of the flowers recall, both in shape and in colour, the neck of doves with the head & beak.[56] It is not certain whether it would have been known to the Ancient Greeks or Latins.

Aracus or Arachus, Ἀρακος [Aracos] & Ἀραχος [Arachos].

Argemōne is so called because it cleanses *ulcers*[57] & cloudiness of the eyes.

Argentīna is from the silvery lustre of the leaves; the Germans call it 𝕲𝖊𝖓𝖘𝖊𝖗𝖎𝖈𝖍, i.e. Anserina, because geese take delight in this in their food.[58] It has not been related by anyone by what name the Greeks & Latins call this herb. Certainly more recent

---

[47] ὀνος [onos] & ὀνημι [onemi], the latter correctly ὀνινημι [oninemi].
[48] Literally "Instead of a nose".
[49] The name means "Lion's mouth". For Andreas Caesalpinus Aretinus see *Cæs.* in Ray's 'Explanation' and Chapter 4.
[50] For the Roman author Lucius Junius Moderatus Columella see Chapter 4.
[51] *Et hiantis sæva Leonis*
Ora feri
[52] Pliny's *Natural history*, XXVII. 32 in the Loeb edition. Φιλανθρωπον [Philanthropon] means "Human-loving", just as children today call this plant, *Galium aparine* [51], "Loveman" and "Sweethearts" because it clings to clothes.
[53] *ab apibus*, from *apis*, meaning "bee".
[54] Smallage is the name used by Ray for celery, *Apium graveolens* [53]. The Isthmian Games, held by the Ancient Greeks at Corinth every two years, one and three years after the Olympic Games, were named after the Isthmus of Corinth.
[55] Ἀθλα δε της κοτινος, μηλα, σελινα, πιτυς. This is a dactylic pentameter, but we have not traced its source. Κοτινος [Cotinos] is Oleaster, the wild tree from which the cultivated olive, *Olea europaea* L., is derived; μηλα [mela] are apples, the fruit of *Malus pumila* Mill. (*M. domestica* Borkh.); σελινα [selina] is the plural of σελινον [selinon]; πιτυς [pitys] is Stone or Umbrella Pine, *Pinus pinea* L.
[56] *Aquilegia* and *Aquilina* are derived from *aquila*, meaning "eagle", while *columba* is Latin for "dove".
[57] ἀργεμα [argema]. See also the entry for *Agrimonia* above.
[58] *Argentum* is the Latin for "silver"; *Gänserich* [sic] means "gander" in German; *anser* is the Latin for "goose".

authors call it Potentilla, from the exceptional powers with which it acts, and some Wild Agrimony from the similarity of its appearance.[59] *Trag.* in the works of *J.B.*

Armerii flores were so called from the French name *armoires* & *armoiries*, on which they represent these plants.[60] *Armoiries* are clan insignia depicted on shields, as they call them.

Artemisia was so called from Artemisia, Queen of Caria, who adopted this plant for herself, though previously it was called Parthenis because the Virgin Goddess had given her name to it.[61] Others derive the name from Ἄρτεμις [Artemis], i.e. Diana, because it cures women's ills. The following plants found by heroes and gods and dignified by their names are recounted by Pliny *in book 25*:[62] Mercurialis from Mercury [38]; Gentiana from Gentius, King of the Illyrians [71]; Lysimachia from Lysimachus [72]; Artemisia, presently mentioned, from Artemisia, wife of King Mausolus [73]; Clymenos from King Clymenos [70]. Mithridation, named by Cratevas, is attributed to Mithridates [62].[63] Eupatorium is named, as some would have it, after Eupator, which was Mithridates' surname [65]; Telephium after Telephus, King of Mysia [42–44];[64] Teucrium after Teucer, King of Troas [45]; Melampodion, a kind of hellebore, after Melampus [47]; Helenium after Helen, wife of Menelaus, King of Sparta [12]; Achillea after Achilles [42]; Panax Asclepium after Æsculapius [30]; Panax Heraclion after Hercules [32]; Panax Chironium after the Centaur Chiron [32] There is also another kind of Panax found by the same Chiron called Centaurion, but also Pharnaceon after King Pharnaces in a controversy over its discovery.[65] King Juba called Euphorbium by the name of his physician & his brother Musa.[66] Country folk[67] & monks of a later age have labelled very many plants with the names of Saints & of Gods whom they invoke, such as are Barbarea minor, called Herb of St Albert; a certain species of Geranium named as St Rupert's;[68] the Lepidium called Dentellaria as St Anthony's; Nasturtium

---

[59] i.e. *Potentilla* is "the powerful little one", related to *potentia* ("power"). *Agrimonia sylvestris* ("Wild Agrimony") is *Agrimonia eupatoria* [16].

[60] *Armoires* are normally cupboards or wardrobes, while *armoiries* means a coat of arms; heraldic flowers include roses, *fleurs-de-lis* (irises) and lilies.

[61] Artemisia II was the sister, wife and successor of King Mausolus, whose monument built by her was one of the seven wonders of the ancient world; she died two years after him in 350 BC. *Parthenis* is derived from the Greek παρθενος [parthenos], meaning "virgin". The Virgin Goddess here is the Greek Artemis (the Roman Diana).

[62] The section numbers of book XXV of Pliny's *Natural history* in the Loeb edition are given in square brackets after each example; some of these passages are also referred to under the appropriate entries elsewhere in the 'Etymology'.

[63] Cratevas was a Greek herbalist, frequently quoted by Pliny and Dioscorides and mentioned by Galen, who may have been physician to King Mithridates VI Eupator of Pontus: see the entry for *Eupatorium* below and the second footnote to it. Pliny gives the name of the herb as *Mithridatia*.

[64] Pliny here gives several alternative versions of the healing of Telephus but does not actually name the herb as *Telephium* or *Telephion*. See also the entry for *Achillea* above, the footnote to it and the entries for *Fabaria* and *Telephium* below.

[65] See the entry on *Centaurium* below and the footnote to it referring to Pliny's *Natural history*, XXV. 33 in the Loeb edition.

[66] For Juba and Musa see Chapter 4.

[67] *Barbari*, literally "Barbarians", but see footnote 170 in the main catalogue.

[68] *Geranium robertianum* [223], usually called Herb Robert, for which Ray gives *Herba Ruperti* as a synonym in his main catalogue. Bauhin (1591) gives as synonyms *Geranium Ropertianum, Ropertiana*

hybernum as St Barbara's Herb or Barbarea; Jacobæa as St James's Herb;[69] Costus hortorum as St Mary's;[70] Eupatorium cannabinum as St Kunigunde's; Hypericum vulgare as St John's; Ascyron as St Peter's;[71] Acetosa semine villoso as the Infant St John's; Alcea vulgaris minor as St Simeon's; Circæa Lutetiana as St Stephen's; & finally (God willing[72]) Nicotiana as of the Holy Cross and Cyanus as Flower of St Zacharias.

Ascyrum, Ἀσκυρον [Ascyron]: the Greeks seem to have imposed a name on this herb from its opposite, for σκυρος [scyros] signifies roughness to them.[73] *J.B.*

Asperula obtained its name from the rather rough & hairy leaves: it is also called Aspergula.[74]

Astragalus, Ἀστραγαλος [Astragalos]: the *dice-like* shape of the seeds gave to this plant the name of Astragalus; ἀστραγαλος certainly means knuckle-bone & the game of dice, the shape & use of which the seeds of this plant recall; certainly on the upper side they have a rounded shape and on the lower side a flat one or a concave one like those of bone-sockets, &c.[75] *Column[a] in the works of J.B.*

Atriplex, from the Greek Ἀτραφαξις [Atraphaxis], seems to be a made-up name, for some people from the colour black [ab atro colore], because it makes people bluish.

Avellana nux is so called from Abellæ or Avellæ, a city of Campania or, as today, of Apulia, *on the authority of Serv[ius]. The Avellan nut is from the tree which is called hazel.*[76] *Macrob[ius], Saturn[alia], book 2, chapter 14.*[77]

---

and *herba D. Ruperti*. The saint is St Rupert (or Ruprecht) of Salzburg (died *c.* 718), sometimes known as Hrodbert or Robert and "venerated as an apostle of Bavaria and Austria" (Watkins 2002). Uniquely in this list Ray prefixes his name here and in the entry for *Geranium* below with *D.* (for *Divus*), not *S.* (for *Sanctus*): see the similar treatment of St Stephen under *Circæa Lutetiana* [136] and of St John the Baptist in the note about *Scrophularia major* [529] in the main catalogue. The attribution of the plant's name to a saint seems to be a case of 'sanitisation', as German folklore associates it with the house goblin Knecht Ruprecht and English folklore with Robin Goodfellow: see Grigson (1955, 1974) and Oswald (1992).

[69] Ragwort, *Senecio jacobaea* [300], comes into flower around 25 July, the feast day of St James the Great; *Jacobus* is the Latin for James.

[70] i.e. Costmary, *Tanacetum balsamita* L.

[71] Perforate St John's-wort, *Hypericum perforatum* [296], comes into flower around St John the Baptist's Day or Midsummer Day, 24 June, while Square-stemmed St John's-wort, *Hypericum tetrapterum* [65], comes into flower a little later; St Peter's Day is on 29 June.

[72] *Si Diis placet*, literally "If it pleases the Gods" – perhaps a surprising phrase for a Christian of Ray's period; it is not quite clear whether it relates to the end of so long a list or specifically to the following entry or two but probably the former.

[73] as in the name of the island Scyros, so called from its ruggedness.

[74] *Asper* means "rough".

[75] Both ἀστραγαλος and the Latin *talus* (used by Ray here) mean "knuckle-bone" and hence – because originally games of dice were played with animal knuckle-bones – "die"; for this reason Greek and Roman dice had pips on only four of the six faces. *Acetabulum* primarily means "vinegar vessel" but has secondary meanings in anatomy, zoology and botany. Here the meaning seems to be "bone-socket"; see also the entries for *Cotyledon* and *Polypodium* below.

[76] *Nux Avellana ex arbore est quæ dicitur corylus*. The town was usually called Abella in Roman times and is the modern Avella.

[77] Servius Maurus Honoratus was a grammarian and Ambrosius Theodosius Macrobius a critic, both of the late fourth century AD. For the latter see also the entry for *Scrophularia* below and Chapter 4.

ETYMOLOGY

B

B Allōte, Βαλλωτη *or black horehound.*[78]

Baptisecŭla, i.e. Cyanus, seems to Hermolaus[79] to be a word derived from Latin, undoubtedly from batuo or bato, an ancient word that means to strike & to beat, whence the English *beate*, & secula, namely from striking & obstructing the reapers' sickles, which we now call falces; secula is from cutting [à secando] and from it comes the English *Sickle. Varro. In Campania these (sickles) are seculæ, from cutting.*[80]

Barbarea is named after St Barbara, just as very many other plants are named among country folk[81] after other Saints, some of which we have recounted above.[82]

Bardana: this word in the apothecaries' shops has been taken from the Spanish language, since this plant is called Lappa major & Personata by the Latins.

Becabunga, that is Anagallis aquatica; the German word is 𝔅𝔢𝔠𝔨𝔭𝔲𝔫𝔤𝔢𝔫, and this passed into the apothecaries' shops from the German language.

Bedeguar is an Arabic word; it is accepted by the apothecaries' shops as the 'sponge' of dog-rose, but others maintain that the Bedeguar of the Arabs is a kind of thistle.[83]

Behen likewise is an Arabic word. Behen album of the apothecaries' shops is not that about which the Mauritanians have written. *J.B.*

Bella donna, i.e. Solanum lethale, Στρυχνος μανικος [Strychnos manicos]:[84] it is called thus by the Italians & Venetians either by Antiphrasis,[85] since its berries are scarcely beautiful, or rather because from its juice or from water distilled from it women make a concoction with which they smear their face & by the strength of the cooling make it pale from being ruddy.

Bellis is from the beauty & pretty colour [bello colore] of the flowers. Bellis does not seem to have been described by the Greeks.

Berberis, Ὀξυακανθος [Oxyacanthos], Spina acida: the word is Arabic, used by Averroës[86] & the apothecaries' shops. Oxyacanthus & Oxyacantha are different: Dioscorides' Oxyacantha is Pyracantha in the opinion of Cordus or Spina alba, but Theophrastus' Oxyacanthus & Galen's Oxyacantha is Berberis.[87]

---

[78] μελαν πρασιον [melan prasion].

[79] Hermolaus Barbarus (Ermolao Barbaro, 1453/4–1493) was an Italian Renaissance scholar; see also *Hermol. in corol.* in Ray's 'Explanation' and Chapter 4. The plant is Ray's *Cyanus*, i.e. Cornflower or Bluebottle, *Centaurea cyanus* [162].

[80] *Hæ (falces) in Campania seculæ à secando*. The standard Latin word for "sickle" is *falx*, with plural *falces*. The quotation is from Varro's *De lingua latina*, 5: 137. For Marcus Terentius Varro see Chapter 4.

[81] *apud Barbaros*, literally "among Barbarians", but see footnote 170 in the main catalogue.

[82] See the entry for *Artemisia* above.

[83] See Ray's note on the entry for *Rosa sylvestris* [502] and the entry for Acanthium, i.e. *Onopordum acanthium* [4], both in the main catalogue.

[84] This is Dioscorides' name for *Atropa belladonna* [548], meaning "maddening nightshade".

[85] See footnote 6 above. See also footnote 746 in the main catalogue.

[86] Ibn Rushd, known in European literature as Averroës, was an Andalusian Muslim polymath of the 12th century.

[87] For Valerius Cordus see *Cord. & Cord. hist.* in Ray's 'Explanation' and both for him and for Galen, the influential Greek doctor and philosopher of the second century AD, see Chapter 4.

Berula is latinised from the French name *Birle* or *Berle*:[88] Anagallis aquatica longifolia.

Betonica, for others Vetonica. Bettonica by reason of its origin, for it is agreed that the Vettones or Vectones were a people of Spain. *C. Hofman[n]*. The Vettones found in Spain the plant which is called Vettonica in Gaul. *Plin[y], book 25, chapter 8*.[89] Betonica of Paulus, namely of Ægineta.[90]

Betula or Betulla, Σημυδα [Semyda], seems to some to be so called because it bubbles out with bitumen: Pliny[91] relates that people in Gaul refined bitumen from it. *Constant[inus]* wants Betula to be named from beating,[92] because boys are thrashed with its rods, and not from bitumen.

Blitum, Βλιττον [Blitton], i.e. βλητον [bleton] to be thrown away, either on account of its uselessness & worthlessness or because of its unsalted or insipid taste. Hence comic poets have taken the opportunity of joking about men of no substance, whom they call blitei[93] & blitomammæ. Festus derives it from βλαξ [blax], a fish of no use.[94]

Bonus Henricus[95] from some useful faculty through which it acts, whence also a certain pernicious plant is bad Henry for Dodoens[96] in the French edition. *J.B.*

[Bor]ago seems to be formed from corago with one letter altered. Moreover this herb is called corago by the Lucanians in the work of Apuleius, because it helps conditions of the heart as much as is possible.[97] Hence that well-known little verse:
Says Borage to this,
I always bring bliss.[98]

---

[88] The latter name now applies to *Berula erecta* [544] and related umbellifers; the word is of Celtic origin, probably meaning water-cress, but *Berula* was in fact used in Latin as synonymous with *Cardamine* by Marcellus Empiricus, a medical writer of Gaul who flourished at the end of the fourth century AD; his only extant work is his *De medicamentis*. Its transfer to *Veronica* spp. [35–38], albeit aquatic ones, is mysterious.

[89] Pliny's *Natural history*, XXV. 84 in the Loeb edition.

[90] See also the entry for *Veronica* below. For Paulus Aegineta see under Aegineta in Chapter 4.

[91] Pliny's *Natural history*, XVI. 75 in the Loeb edition, where he says *Gallica haec arbor* ("This tree [is] Gallic"), *terribilis magistratuum virgis* ("a cause of terror because of the magistrates' rods") and *bitumen ex ea Galli excoquunt* ("the Gauls refine bitumen from it").

[92] *à batuendo*, from the verb *batuere*, "to beat". Robert Constantin (Robertus Constantinus, who died in 1605) was one of the commentators whose annotations were included in Johannes Bodaeus' (1644) great edition of Theophrastus' *Historia plantarum* (see *Jo. Bodæ*. in Ray's 'Explanation'); this comment is in column 2 of page 215 of that work.

[93] This is an anteclassical Latin word used by the comic playwright Plautus, meaning "insipid", "silly" or "useless", here in the plural.

[94] Βλαξ was used in Ancient Greek to mean a dolt. For Sextus Pompeius Festus see footnote 309 below.

[95] See the equivalent entry in the main catalogue and the second footnote to it.

[96] For Dodoens (Rembertus Dodonaeus) see *Dod.* in Ray's 'Explanation' and Chapter 4. Dodoens' *Histoire des plantes* (1557) has been published in a modern facsimile (Opsomer 1978) but we cannot find "*malus Henricus*" in this or in Lyte's translations of it (1578, 1619). It is generally thought to be a German name, *Böser Heinrich*, applied to *Mercurialis perennis* [163] (Grigson 1955; Oswald 1992).

[97] *Corago* is presumably derived from *cor*, meaning "heart". The Lucanians were a people of Lower Italy. The author is the Apuleius Platonicus discussed under *Apul.* in Ray's 'Explanation'; see also Chapter 4. The name is given as "*corrago*" under *Herba Bovis Lingua* in Howald & Sigerist's edition (1927), but the work survived as numerous manuscripts.

[98] *Dicit Borago, gaudia semper ago* (a rhyming dactylic pentameter). Elyot (1539, p. 26) says of "Bourage": "Comforteth the harte, and maketh one merye, eaten rawe before meales, or layde in wyne

ETYMOLOGY

Moreover the flowers of Borage are among the number of the four that are cordial.[99]

Branca ursina, for the Italians *Branqua*, for the French *Branch*, that is bear's arm[100] because the leaves bring to mind the front foot of a bear.

Britannica of Pliny *Caes[ar]*, that is Persicaria mitis *Gesn[er]*, that is Cochlearia Batavica. The Frisians showed this plant to the Romans. I am curious to know (says Pliny) the cause of the name, unless perhaps they named it thus, living alongside the ocean of Britain as if it were adjoining, for it is certain that it was not called after it because it grew there in quantity, Britain being still independent at the time.[101]

Bryonia, Βρυωνια, from βρυω [bryo] I shoot up, I sprout forth, because it shoots up & sows itself & becomes aggressively rank, or from βριαω [briao] I raise up, I exalt, for the reason that it raises itself up by climbing into neighbouring shrubs. It seems to be called Melothron from the preparation of skins, for the Greeks use the name μηλωθρα [melothra] for *dyes & ornaments*,[102] that is dyeings with colours & preparations, perhaps because they were resembling the skins from the fruit of Bryonia alba. *J.B.*

Buglossum, Βουγλωσσον [Buglosson], from the likeness of the leaves, which portray an ox's tongue both in shape and in roughness.[103]

Bugula is a word of French origin, *Bugle*.[104]

---

that is drunke: Also mollifieth the bealy, and prepareth the stoole." Gerarde (1597, p. 654) says: "*Pliny* calleth it *Euphrosinum* [Greek Εὐφροσυνον, "Cheerful"], bicause it maketh a man merrie and ioyfull: which thing also the old verse concerning Borage doth testifie: *Ego Borago gaudia semper ago*. In English. I Borage bring alwaies courage."

[99] *quin & flores Boraginis sunt è numero quatuor cordialium*. The OED says of the adjective "cordial": "Of medicines, food, or beverages: Stimulating, 'comforting', or invigorating the heart; restorative, reviving, cheering." It also says that Borage "was formerly much esteemed as a cordial, and is still largely used in making *cool tankard*, claret cup, etc.". Indeed "Cool tankard" became an alternative name for Borage (OED) and its leaves and flowers are still used in drinks based on Pimm's Cup. The four cordial flowers of Galenic medicine (see Chapter 3) were borage, [viper's] bugloss (*Echium vulgare* [171]), roses and violets (e.g. Quincy 1718, p. 66). In his *Healths improvement* (1655, p. 117; see footnote 4 in the main catalogue) Thomas Moufet writes: "*Borrage, Bugloss* [again *Echium vulgare*] and *Langdebeif* [Bristly Oxtongue, *Helminthotheca echioides* [94], very surprisingly to a modern botanist], are of so great a temperature in all qualities, that they are not only commended for special Cordials being steeped in Wine, or made into Conserves; but also their flowers, herbs and roots are esteemed restorative, nourishing weak bodies sufficiently, and strengthening the parts of nourishment more then meanly, being sodden in broths, cullises, or gellies." Perhaps there was no single standard list of cordial flowers.

[100] *brachium ursinum*: the original application was to *Acanthus mollis* L., which has the corrupted English name "Bear's-breech" (see Stearn 1996), but *Heracleum sphondylium* [560] is intended here. *Branc-ursine* in French is now applied to *Heracleum sphondylium* but also to *Cirsium oleraceum* (L.) Scop. Today *Acanthus mollis* is *brancalupo* in Italian as well as *brancorsino*, the wolf having joined the bear.

[101] The reference is to Pliny's *Natural history*, XXV. 21 in the Loeb edition. *Cochlearia Batavica* and *Britannica* are synonyms of *Cochlearia officinalis* L., a predominantly coastal species which is not known from Cambridgeshire.

[102] βαμματα [bammata] and καλλωπισματα [callopismata], both neuter plurals, as is μηλωθρα [melothra].

[103] The Greek name, explained by the Latin *lingua bovis*, means "ox-tongue", being derived from βους [bus] and γλωσσα [glossa].

[104] The OED suggests that the French *bugle* was derived from the Late Latin *bugula* rather than the other way round, but in any event *bugillo*, used by the physician Marcellus Empiricus (see footnote 88 above) around AD 400, seems to denote the same plant.

Bulbocastanum, Βολβοκαστανιον [Bolbocastanion], received its name from the bulbous shape of the root and from the sap imitating chestnuts.[105]

Bunias, Βουνιας, from βουνος [bounos], a hill, on account of the swollen shape of the root.

Butŏmos, Βουτομος, from Βους [Bus, "ox"] and τεμνω [temno, "I cut"], because it is a herb suitable for the fodder of oxen. *Ruellius*[106] in the works of *J.B*.

C

CAlamintha, as if a good & useful mint,[107] for with its vapour (as Aristotle has it) it regularly puts snakes to flight whether it is set on fire or spread around. For C. Hofmann καλος [calos] is an ambiguous word [medium vocabulum] (like σεμνος [semnos, "solemn" or "majestic" but also "pompous"] &c.) so that Calamintha is a Mint that is rough, hoary & with an unpleasant smell. Consequently it is called *calamint as a euphemism*.

Calcitrapa is so called because its top is protected as if by caltrops,[108] which the common people of the French call *calcatrepæ*, & is remarkably armed with the points of spines. *Ruel[lius]*, *book 3, chapter 84*.

Campanula is from the shape of the flowers resembling a bell. Moreover campana ["bell"] is so called from Campania, a region of Italy where the bronze is very highly praised. Paulinus, Bishop of Nola and the equal of Jerome,[109] is reputed to have first introduced the use of a Bell into his Church & hence perhaps, from the name of Nola, bells were regarded[110] as being from this Nola in Campania. Campana, essentially,[111] is a modern name for a bell.

Cannăbis is thought by some to take its name from canna ["reed"] because it is sought after above all on account of its reed or culm. Moreover for the Greeks καναβοι [canaboe] are little fountains and impure springs,[112] from which places, since cannabis abounds there both naturally & planted, it seems to *[Mathias] de l'Obel* that it obtained that as its name.

---

[105] Βολβος [Bolbos] is the Greek word for "bulb" and chestnuts are καστανα [castana]; καστανιον [castanion] is a little chestnut.

[106] See *Ruel.* in Ray's 'Explanation'.

[107] This is because the Greek adjective καλος [calos] means "beautiful" or "good" (particularly in the sense of useful for some purpose).

[108] *velut muricibus*: the primary meaning of *murex* is the shell of a marine mollusc from which purple dye was made (see footnote 374 below), but here and in the entry on *Rumex* below it seems to mean a caltrop.

[109] St Paulinus (*c*. 354–431) was a Roman senator who became a monk and eventually Bishop of Nola, a town in Campania. He corresponded about monasticism with the major Doctor of the Church St Jerome (or Hieronymus, *c*. 347–420).

[110] *& hinc fottè Nolæ vocabulo censebantur tintinnabula*: *fottè* here is clearly a misprint for *fortè*; the meaning of the next two words is somewhat unclear; *tintinnabulum* is the classical Latin word for a bell.

[111] *substant.*, presumably an abbreviation for *substantialiter*.

[112] The meanings given for καναβος (the singular of καναβοι) by Liddell, Scott & Jones (1940) are "*a wooden framework* round which artists moulded wax or clay" and "*a rough drawing of the human frame*", but they mention that it was wrongly explained as "*cistern*" by Joannes Philoponus in his 6th-century commentary on Aristotle's *De generatione animalium*.

ETYMOLOGY

Capnos, that is Fumaria, seems to be named from the fact that it provides clarity and watering to eyes smeared with it, as if it was smoke. *Plin[y], book 25, chapter 13*.[113]

Caprifolium, because the leaves are tendrilled in their outermost curves.[114] *Lob. in Adv.*

Cardamīne was called thus by some for the reason that it exhibits the taste of cardamum, that is Nasturtium.[115] But καρδαμον [cardamon] is as it were καρηδαμον [caredamon], no doubt because *it tames a head*, it tests & tames a head with its fierce heat, or it is *from the heart* because it makes people prudent or because it is very effective for what is called cardiac Syncope.[116]

Carduus, *Martin[i]*.[117] As there are many kinds of thistles, he first took this as the name for what fullers use in carding wool. Carere is an old word for ξαινειν [xaenein, "to card"], whence both carduus & carmino ["I card"] & also the English **to card wooll**.[118] Carduus lacteus & Mariæ is called Leucographis from the fact that its leaves are variegated with white & as it were milky drawings & marks.[119]

Carlina, as if Carolina, because it is believed that this herb was revealed to Charlemagne[120] by an Angel, as if it was a most certain remedy for driving the plague out of his army.

---

[113] Pliny's *Natural history*, XXV. 155–156 in the Loeb edition, where he says of his first species (which he calls *capnos trunca*, probably *Fumaria officinalis* 204) *suco caliginem discutit, itaque in medicamenta oculorum additur* ("with its sap it disperses mist before the eyes, and so it is added to remedies for the eyes"). Pliny explains the name Capnos (Καπνος in Greek), which means "smoke", in his account of his second species, *capnos fruticosa*, described as a bushy species similar both in name and properties to *capnos trunca*. The Latin name *Fumaria* is derived from *fumus*, also meaning "smoke". Both names more probably relate to "the very slender and sparse branches" of *Fumaria* species, which Pliny also describes. See also footnote 205 below.

[114] *quòd folia sunt extremis flexibus capreolata*: a strange phrase especially since honeysuckles climb by the stems twisting rather than by tendrils, as correctly described in the entry for *Periclymenum* below.

[115] ... *gustu cardamum, id est, Nasturtium repræsentat*, literally "... by taste it manifests cress, that is Cress". Both *cardamum* (from the Greek καρδαμον) and *nasturtium* were quite probably Garden Cress, *Lepidium sativum* L. See also the entry for *Nasturtium* below.

[116] The two Greek phrases καρα δαμα [cara dama] and καρδια [cardia] include words suggesting *caredamon* or *cardamon*; the Latin word used here for "prudent" is *cordatus, -a, -um*; Syncope cardiaca is a loss of consciousness due to a shortage of oxygen in the brain.

[117] Matthias Martini (1572–1630) published an enormous etymological Latin dictionary (folio with 4138 half-page columns), *Lexicon philologicum, praecipue etymologicum*, at Bremen in 1623. Ray does not refer to this author in earlier parts of his work, but his dictionary is a major source in the 'Etymology', though not always acknowledged. An enlarged version was published at Frankfurt am Main in 1655. Ray may only have consulted this second edition, but the only firm evidence that he used it is in the suggested explanation of *Meum*, for which there is no entry in the 1623 edition; nevertheless references in later footnotes here are to the earlier version because the later one is not paginated.

[118] The plant is Fuller's Teasel, *Dipsacus sativus* (L.) Honck., closely related to *D. fullonum* [168]. Carere is the infinitive of the rare third-conjugation verb caro ("I card"), a verb not to be confused with careo ("I lack") which also has *carere* as its infinitive. Rather than giving the English phrase Martini (1623, column 385) provides the German and French equivalents: "Nos **kartendistel** vocamus, à *carduo*. unde Gallis *carder*, ut *carder la laine* pectere, carminare lanam. vide & *Dipsacus*."

[119] The Latin names mean "Milky or Mary's Thistle" and *Leucographis* (Greek Λευκογραφις) means "white writing". The white markings on the leaves of *Carduus lacteus*, i.e. *Silybum marianum* [108], were thought of as the milk of the Virgin Mary, hence *Mariæ*.

[120] Carolus Magnus or Charlemagne, Carolingian king of the Franks, was crowned as Emperor of the Romans (i.e. the first Holy Roman Emperor) by Pope Leo III on 25 December 800.

Carpinus: *see* under Ostrys.

Caryophyllus from some sort of relationship in the scent.[121] Others think it is so called from the similarity of the flower. Moreover there are Caryophylli that we call in English 𝕮𝖑𝖔𝖚𝖊𝖘: the word means *nut's leaf*, καρυου φυλλον [caryu phyllon].

Caryophyllata is so called from from the scent of the root, because when dug up in spring-time it recalls the fragrant clove with its scent; it is commonly called Herba benedicta & Sanamunda from the remarkable potency & efficacy with which it acts.[122]

Caucalis for [Mathias] de L'Obel is from the shape of the seeds, which resemble oblong & hemispherical little vessels, or from the clatter which these seeds emit as they are shaken by the wind when they are fully dried after the harvest has been completed; καυκαλιον [caucalion] is a kind of vessel, so called from the sound that it emits while water goes into it in a compact mass, with air escaping, and contrariwise.[123]

Centaurium, Κενταυρειον [Centaureion] & κενταυριον [centaurion], from the Centaur Chiron, Achilles' tutor, a Physician by profession on the evidence of Pliny, who makes Centaurium the fourth kind of all-heal. *Plin[y], book 25*.[124]

Cercis, Κερκις of Theophr[astus],[125] that is Populus tremula. It may be that[126] κερκις [cercis] has crept in instead of κεκρη [cecre] or κεκρις [cecris] or of κρεκις [crecis] from κρεκειν [crecein],[127] from the rustling & trembling movement of the leaves shaken by the wind, something which Pliny often attributes to the Poplar in general & which we have observed above all in this kind. *J.B.*

Chamædrys is so called because it has leaves like the oak in their cleavage, like oak in their cutting.[128]

Chondrilla, Χονδριλλη [Chondrille], *from* χονδρος [chondros, "grain"] & ιλλειν [illein, "to pack together"], which means to keep together, as if containing grains, which Dioscorides says are like mastic.

Chamæmēlum, Χαμαιμηλον [Chamaemelon], ἀνθεμις [anthemis], because it recalls the smell of an apple, for example of a quince.[129]

---

[121] i.e. to cloves from the Clove-tree: see footnote 339 below.

[122] This is Wood Avens or Herb Bennet, *Geum urbanum* [117]. *Herba benedicta* means "blessed herb" and *Sanamunda* perhaps means something like "world's-health".

[123] Liddell, Scott & Jones (1940) define the late Greek word βαυκαλιον [baucalion] or καυκαλιον as "*a narrow-necked vessel, that gurgles when water is poured in* or *out*".

[124] Pliny's *Natural history*, XXV. 33 in the Loeb edition, where he is listing the various kinds of *panaces* (Greek πανακες, meaning "all-healing" as in "panacea"). This plant is traditionally supposed to have been a species of *Centaurea* rather than a *Centaurium*, though that too is listed by Pliny as a kind of *centaurium*. See also the entries on *Achillea* and *Artemisia* above and on *Panax* below.

[125] The reference is to Theophrastus' *Enquiry into plants*, III. XIV. 2 in the Loeb edition rather than to I. XI. 2, which seems to relate to Judas-tree, *Cercis siliquastrum* L.

[126] *Fieri potest ut* … , literally "It is able to happen that …".

[127] Κερκις primarily refers to the rod used to drive home the threads of the weft on a loom and κρεκειν describes the action of striking with this rod; the sound generated is referred to in the plays of Sophocles. We have not traced the other three Greek nouns here.

[128] The Greek for an oak is δρυς [drys] and χαμαι [chamae] is an adverb and prefix meaning "on the ground". See also the entry for *Trisago* below.

[129] See footnote 123 in the main catalogue. The Greek name is derived from χαμαι again and μηλον [melon], meaning "apple". The quince was regarded as a kind of apple: the Latin here is actually *putà Cydonii*, meaning "for example of a Cydonian [one]"; Cydonia is the modern Chania in Crete.

ETYMOLOGY

Chelidonium, Χελιδονιον [Chelidonion], from χελιδων [chelidon], a swallow. *Swallows have shown that Celandine is very salubrious for the sight by healing the damaged eyes of their chicks with it; but for some it is named thus because it flowers with the swallows' arrival & begins to wither at their departure. Plin[y], book 8, chapter 27.*[130]

Chrysanthemum, Χρυσανθεμον [Chrysanthemon], on account of the yellow flowers shining with a golden colour.

Cichorium, Κιχωριον [Cichorion] & Κικορειον [Cicoreion] or κιχοριον [chichorion], perhaps from κιω [cio] I go & χωρος [choros, "place", "land"]. *Martin[i], to whom go if you will.*[131] But Cichorium is a foreign word for the Greeks, says Marcellus Virgilius.[132]

Cicūta, on the evidence of *Servius*[133] *in his Eclog[ues], 2*, is an internode in wheat-like or culm-bearing plants;[134] since this plant also has such internodes, it seems to be named thence. The Greek is κωνειον [coneion] perhaps *from the verb* κωναν [conan, "to spin round"], which is περιδινειν [peridinein], that is to twist round, explains the Great Etymology.[135] *Of making foolish* Galen, for it adds stupidity to the cooled brain.[136] *C. Hofman[n]*.

Circæa is believed to be named after that famous witch Circe & hence it is called in English 𝔈𝔫𝔠𝔥𝔞𝔫𝔱𝔢𝔯𝔰 𝔑𝔦𝔤𝔥𝔱𝔰𝔥𝔞𝔡𝔢.[137]

---

[130] Pliny's *Natural history*, VIII. 98 in the Loeb edition, in a passage listing *Chelidonia* [sic] among examples of cures that other species have revealed to mankind. Only the first half of the sentence is found in this passage; a longer version is to be found in XXV. 89–90, where, as part of another passage on the same theme, two species of *Chelidonia* are described (usually interpreted as *Chelidonium majus* [128] and *Ficaria verna* [129], q.v. in the main catalogue) and the statement about the flowering and withering is appended without any suggestion that this is the reason for the name.

[131] *quem adi sis*: see footnote 359 in the main catalogue. The etymological suggestion is from Martini (1623, column 453).

[132] Marcellus Virgilius Adrianus (1464–1521), often known simply as Marcellus Virgilius, published a commentary on Discorides in Florence in 1518 and a second edition in Cologne in 1529; he was a professor at the University of Florence and from 1498 until his death he was Secretary of the Republic, where Machiavelli worked under him (Cosenza 1962; Riddle 1980). The relevant phrase is buried in a long discussion about the application of various Greek and Latin names to wild and cultivated chicories on p. 267 of the 1529 edition in Marcellus' commentary on Dioscorides' Book II, Chapter CXXI, Περι Σερεως (*De Intybis*), but Ray almost certainly took it from Bauhin & Cherler (1651, vol. 2, p. 1008), where it is prominently attributed to Marcellus. See also footnotes 319, 387 and 480 below.

[133] *teste Servio*, literally "with Servius witness"; Servius Maurus Honoratus was a Latin grammarian of the late fourth century AD with the contemporary reputation of being the most learned man of his generation in Italy. His commentaries on the works of Virgil (including his *Eclogues*) were printed in 1471 by Bernado Cennini in Florence, the first book to be published there.

[134] *internodium in triticeis sive culmiferis*.

[135] Ἐτυμολογικον Μεγα or *Etymologicon Magnum* is the traditional title of a Greek lexical encyclopedia compiled in Constantinople around AD 1150. It was first printed in Venice in 1499 and there were later editions in 1549 and 1594; the latter, edited by Sylbergius, is cited in our bibliography. An explanation of the Greek name similar to this one but without the verb περιδινειν is given, followed by two alternative ones, in column 551 of this edition.

[136] The Greek verb is μαραινειν [maraenein], meaning "to make foolish". Presumably "cooled" is a reference to the well-known symptom of hemlock poisoning, a feeling of coldness, as related by Plato about the death of Socrates. For Galen see Chapter 4.

[137] In the story told in Homer's *Odyssey*, 10: 133–399, Circe was thwarted in her plan to turn Odysseus into a pig. According to some people she would have used this herb, Enchanter's Nightshade, *Circaea*

Cirsium because it settles the pains of κιρσοι [cirsoe], that is of varicose veins.

Clematis is named from κλημα [clema, "vine-twig"], because it is twiggy & clambers through trees & shrubs.

Clinopodium, Κλινοποδιον [Clinopodion], because it has flowers *as if like the feet of a couch*[138] or because it presents the appearance of the feet of a couch in the round circuit of its flowers. *Plin[y]*.[139] *With flowers arranged at intervals in the manner of horehound.*

Conferva was named from soldering together. *I know that a pruner who fell from a tall tree with nearly all his bones broken and was treated with it by having his whole body enveloped in it, with sprinklings of its water as often as it might become dry and with infrequent unbindings other than for changing the herb when it was losing its potency, recovered with scarcely believable speed.* Plin[y].[140]

Conyza, Κονυζα: could it be from κονις [conis] a nit?[141] or rather from κωνωψ [conops] a gnat, for it drives away gnats & kills fleas, whence it has also been called Pulicaria.[142]

Consolida, Συμφυτον [Symphyton], because it cements & consolidates wounds. *So outstanding is it for curing wounds that it cements even pieces of meat when added to them while they are being cooked, whence also the Greeks have given it its name; it also heals broken bones.* Plin[y], book 27, chapter 6.[143]

Convolvulus, Σμιλαξ [Smilax], is so called because it climbs by interweaving itself [convolvendo se] into neighbouring plants.

Coronŏpus, Κορωνοπους,[144] a crow's foot, from the shape of the leaves, which are so arranged into a circle that they bring it to mind somewhat.

Corylus, Καρυα Ποντικα [Carya Pontica] & λεπτοκαρυα [leptocarya], is from καρυα [carya] nut tree or καρυον [caryon] nut bush.[145] *Martin[i], Lex[icon]*. Moreover the tree Corylus is *from the word* κορυλος [korylos], on the authority of Priscianus

---

lutetiana [136], but according to others Bittersweet or Woody Nightshade, *Solanum dulcamara* [547]; the two differing traditions explain the application of "nightshade" to an unrelated plant. See also footnote 24 in the main catalogue.

[138] as if derived from κλινη [cline, "couch"] and ποδες [podes, "feet"], plural of πους [pus].

[139] Pliny's *Natural history*, XXIV. 137 in the Loeb edition. The nearest match to the final sentence quoted here that we have found in Pliny's work is in XXV. 43, where he discusses the identity of Achillea (q.v. above) and says that some say that it has heads like those of horehound (*capitulis marrubii*).

[140] Pliny's *Natural history*, XXVII. 69 in the Loeb edition. The explanation of the name, from the verb *confer[r]uminare* ("to solder together"), is from earlier in the same passage.

[141] Both the Greek κονις and the Latin *lens* can mean "nit" (a louse's egg), though the former also means "dust" and the latter "lentil".

[142] The Latin for "flea" is *pulex* (with the stem *pulic-*). The plant described seems to be the annual *Pulicaria vulgaris* [143] rather than the well-known perennial, *P. dysenterica* [142].

[143] Pliny's *Natural history*, XXVII. 42 in the Loeb edition, where he uses the Greek name given here but calls it *alum* rather than *consolida* in Latin. See also the entry for *Symphytum* below, where the etymology of the Greek name is explained in the related footnote.

[144] This word is derived from the Greek κορωνη [corone], which was used for various corvids, and πους [pus], meaning "foot".

[145] Καρυα is a walnut tree, but the first two Greek names relate to Hazel, *Corylus avellana* [154]: see Ray's second note on this species in the main catalogue. Καρυον is the Greek word for the nut rather than the bush: Ray has miscopied Martini's (1623, column 515) *nux fructus* as *nux frutex*.

ETYMOLOGY

& Curtius quoting from him,[146] although that word (which seems to denote some place) has not yet become fully known to us.[147] *J.B.*

Cotylēdon, Κοτυληδων, received its name from the shape of the leaf like a hip-bone socket,[148] since κοτυλη [cotyle] is a hip-bone socket, the hollow in which the head of the femur turns.

Crespinus, that is Berberis, is an Italian word; but Hermolaus is the author of the idea that Crespinus is so called as being acris spinus,[149] as many like to think.

Crocus, Κροκος [Crocos] or κροκη [croce], means a thread and a tuft of hair, of which kind are the stamens or tufts of hair of Crocus, which are called κροκιδες [crocides].[150]

Cruciāta is named as a new word from the positioning of the leaves in the form of a cross.

Cyănus received its name from the deep blue or azure colour of the flower. Κυανος [Cyanos] for the Greeks.[151]

Cypērus, Κυπειρος [Cypeiros] or κυπερις [cyperis], from the shape of the root, which resembles a small box or very small vessel.[152]

D

Daucus, Δαυκος [Daucos], because it has a fierce & fiery power or a hot taste, from δαιω [daeo] I burn. *Martin[i], Lex[icon]*.

Dens leonis[153] was so called from the resemblance of the leaves. A modern name.

Dipsăcus, Διψακος [Dipsachos], that is Sitibundus,[154] finds a name from its opposite since it collects dew or rainwater in the hollow curve of the axils, which it makes use of as it were to banish the ravages of thirst. The rainwater collecting in the axils of the leaves is recommended for getting rid of warts and hence perhaps it obtained the name of Venus' Basin [Labrum Veneris].

Dodecatheon from the twelve gods, namely of the greater clans. *The highest reputation is for the herb which they call Dodecatheon, honouring the majesty of all the Gods*. Plin[y].[155] Ennius has encompassed the names of the twelve gods whom they describe as of the greater clans in this Distich:

---

[146] Priscianus Caesariensis was a grammarian who lived around AD 500, but we have not traced Curtius, who cannot be the Roman historian of the first century AD.

[147] The word is not known to Liddell, Scott & Jones (1940) either, though Lewis & Short (1890) give it under their entry for *corylus* or *corulus*.

[148] *à figura folii acetabulo simili*: see also footnotes 75 above and 388 below.

[149] i.e. "sharp thorn-tree". For Hermolaus Barbarus see footnote 79 above and Chapter 4.

[150] Κροκη [Croce] does indeed mean "woof", or more generally "thread", and κροκις [crocis], a variant of κροκυς [crocys] with plural κροκιδες [crocides], means the nap on woollen cloth or else a piece of wool (Liddell, Scott & Jones 1940).

[151] Κυανος was originally a dark blue enamel used in the Homeric age to adorn armour, later *lapis lazuli* and Cornflower or Bluebottle, *Centaurea cyanus* [162]. The two Latin adjectives used here are *cyaneus, -a, -um* (derived from Greek) and *cæruleus, -a, -um*.

[152] Possibly the reference is to κυπελλον [cypellon], which means "beaker" or "goblet".

[153] "Lion's-tooth": this is the origin of the English "dandelion" (first recorded in 1513) through the French *dent de lion* (OED).

[154] i.e. "thirsty", just as διψα [dipsa] means "thirst".

[155] Pliny's *Natural history*, XXV. 28 in the Loeb edition, where he says that this herb is second only in its reputation to Homer's *moly* (μωλυ in Greek): see footnote 24 in the main catalogue. Δωδεκαθεον

Juno, Vesta, Minerva, Ceres, Diana, Venus, Mars,
Mercurius, Jovis, Neptunus, Vulcanus, Apollo.[156]
Six, as you see, are male and the same number female. Herodotus[157] asserts that the Egyptians first used the names of the twelve gods and that the Greeks took them from them. Diodorus[158] says that the most ancient of the gods ruled for more than 1200 years. Perhaps these may have been the antediluvian patriarchs.[159]

Dulcamāra, or Amara dulcis,[160] because, if anyone chews a twig or its bark, certainly at first it seems unpleasant to the taste, but, the longer it is chewed, the more remarkably its sweetness reveals itself.

E

EBulus, Χαμαιακτη [Chamaeacte]: Martini thinks it is named as if it was Ebusus, because it grows commonly in the island of Ebusus or because, as Pliny relates, snakes, which are so frequent in the aforesaid island that it is called for that reason Ophiusa by the Greeks, are put to flight by the smoke of Danewort.[161]

Echium, Ἔχιον [Echion]: they want it called thus because it is efficaceous against the bites of vipers, others because the opening flower-heads are like a viper,[162] but Dioscorides says because it has the seeds *like the head of a viper*,[163] that is similar to a viper's head.

Elæagnus, from *olive and chaste-tree*;[164] that is the name was composed from olive & chaste-tree. Ἄγνος [Agnos] means a shrub which for the Latins is vitex, because, on

---

[Dodecatheon] is compounded from the Greek words for "twelve" and "god". The *majores gentes* ("greater clans") consisted of the patrician families of ancient Rome.

[156] *Jovis* is usually the genitive of *Juppiter* (derived from *Jovis pater* – "Father Jove" – and spelled "Jupiter" in English), the king of the Roman gods, but the early Roman poet Quintus Ennius (*c.* 239–*c.* 169 BC) used it for the nominative also; however these lines may be wrongly attributed to him because they were unknown to Castagna (1996), who compiled a concordance of his works. They are a brave attempt at two dactylic hexameters, but there is a false quantity in the second syllable of *Jovis*, which, ending in a consonant, cannot be short before *Neptunus*, which begins with another consonant.

[157] Herodotus' *Histories* (Ἱστορίες, originally meaning "Enquiries"), written in the second half of the fifth century BC in the Ionic dialect of Ancient Greek, are one of the seminal works of western literature and are indeed the origin of the modern meaning of "history".

[158] Diodorus Siculus, a Greek author of the first century BC, compiled a history of the world, *Bibliotheca historica*, in 40 books, 25 of which survive, the first six being partially mythological accounts of Egypt, Mesopotamia, Greece and other countries up to the fall of Troy.

[159] In modern perception the reluctance of writers of Ray's time to treat classical mythology as incompatible with prevalent Judaeo-Christian beliefs is puzzling.

[160] i.e. "Bitter sweet", the origin of the English name of *Solanum dulcamara* [547].

[161] The entry is derived from Martini (1623, columns 631–632). Pliny's *Natural history* (XXV. 119 in the Loeb edition) contains only the simple statement *Ebuli fumo fugantur serpentes*. Ὀφιουσσα [Ophiussa] is the feminine form of a Greek adjective (to agree with νησος [nesos, "island"]) meaning "abounding in snakes"; the name was used for several islands, but the Latin name identifies this one as the modern Ibiza. Elsewhere Pliny (XXIV. 163) describes *ophiusa* as a halucinogenic plant reported by Democritus from Ethiopia. It is clear from other passages that Pliny calls Danewort [170] *Ebulum* (neuter).

[162] *quòd florescat capitibus viperæ similibus*, literally "because it begins to flower with heads like a viper".

[163] The Greek word is ἔχις [echis], so ἔχιον [echion] literally means "little viper".

[164] The Greek names of these two trees, *Olea europaea* L. and *Vitex agnus-castus* L., are ἔλαια [elaea] and ἄγνος [agnos] or λυγος [lygos] (both of which are used by Pliny in the passage referred to in the

the evidence of Pliny, during the Thesmophoria matrons guarding their chastity used it for bedding. It is called vitex from weaving or binding, because, as the same Pliny says, it is not much different in use from willow.[165]

Elaphoboscum, Ἐλαφοβοσκον [Elaphoboscon], that is food of a stag, for the tradition is that with this food stags withstand snakes, on the evidence of Dioscorides & Pliny.[166]

Enŭla or Inula, Ἐλενιον [Helenion], seems to be a word corrupted from Helenium.

Equisētum, Ἱππουρις [Hippuris], Cauda equina, because the leaves correspond to horse-hairs.[167]

Erica Ἐρικη [Erice] or Ἐρεικη [Ereice] from ἐρεικω [ereico] I break, because it acts powerfully in breaking up bladder and kidney stones. *Martin[i]*.[168]

Erigĕron, Ἠριγερων, Senecio *[in] Lat[in]*. The Greeks have given it this name because it becomes hoary in spring. *Plin[y]*, book 25, chapter 13,[169] from ἐαρ [eär], spring, & γερων [geron], old man.

Erūca,[170] Ἐυζωμον [Euzomon],[171] is [so] called from the fact that it erodes things, for, if you have tasted it, it will pucker your mouth & tongue quite sharply, but for others it is from burning,[172] as if it was Uruca with the u turned into e, because it is of fiery potency & when taken in food incites the heat of Passion,[173] whence it is said by the poets to be exciting of lust.

   And no less I prescribe the avoidance of Lust-exciting Rockets. Ovid
   Rocket excites reluctant husbands to Passion. Columel[la]
   And Rocket renewing reluctant Passion. Mart[ial][174]

---

next footnote). The combined name, Ἐλαιαγνος [Elaeagnos], was used by Theophrastus (*Enquiry into plants*, IV. x. 2 in the Loeb edition) for what was probably a sallow, perhaps *Salix caprea* [514], but was adopted by Valerius Cordus (see footnote 87 above) for the superficially willow-like Sweet Gale, *Myrica gale* [173]; Linnaeus used it for Oleaster or Russian Olive, *Elaeagnus angustifolia* L., native to western and central Asia and first cultivated in Europe in 1736 in Germany.

[165] See Pliny's *Natural history*, XXIV. 59 in the Loeb edition, where the statement at the end of this entry opens a long account of *Vitex* that also includes the use made of it by the Athenian matrons. In the Middle Ages monks and nuns ate the seeds to suppress sexual desire. The Thesmophoria were a three-day women's festival held in Athens to celebrate the goddess Demeter, the "Law-giver" (Θεσμοφορος [Thesmophoros]). Pliny correctly relates the plant name *vitex* to the Latin verb *viere* ("to twist together", "to weave"), though less certainly to *vincire* ("to bind"): see also footnote 89 in Ray's 'Interpretation'.

[166] This plant name is compounded from ἐλαφος [elaphos, "stag"] and βοσκη [bosce, "food"]. See Pliny's *Natural history*, XXII. 79 in the Loeb edition, where he says that it is like Giant Fennel, *Ferula communis* L. (*ferulaceum*), but with the leaves of Alexanders, *Smyrnium olusatrum* [285] (*foliis olusatri*). In XXV. 92 he includes it among plants that other species have revealed to mankind, but it is the hinds (*cervae*) that are reputed to have done this.

[167] The first name means "horse's-hair" and the next two "horse's-tail". In Ray's time *Equisetum* and *Hippuris* were synonyms for the same genus, which was believed to include what are now recognised as unrelated plants [178–183] (Oswald & Preston 1998).

[168] *Calculi* are literally "pebbles". Ray's source is Martini (1623, columns 697–698).

[169] Pliny's *Natural history*, XXV. 167 in the Loeb edition. See also the entry for *Senecio* below.

[170] i.e. Garden Rocket, *Eruca vesicaria* subsp. *sativa* (Mill.) Thell.

[171] The Greek name means "making good broth", from the adverb and adverbial prefix εὐ [eu, "well"] and ζωμος [zomos, "broth"].

[172] *ab urendo*, suggesting a relationship between *Uruca* and the verb *urere* ("to burn").

[173] *Veneris incendium*, literally "the fire of Venus".

[174] *Nec minùs Erucas jubeo vitare Salaces.*

Eruca is also the name of an insect,[175] which seems to be so called because it gnaws the leaves of herbs.

Erysĭmum, Ἐρυσιμον [Erysimon], is so called for some *from the verb* ἐρεικειν [ereicein, "to rend"] from the multiple division of the leaves, for others as if it was ἐριτιμον [eritimon], which means precious, and for others *from the verb* ἐρυειν [eryein, "to drag"] because, on account of its heat, it has been endowed with the capability of dragging. *C.B.*

Esŭla is a word of the apothecaries' shops, Arabic in origin, as they say. It is a kind of Tithymalus.[176]

Eupatorium, Ἐυπατωριον [Eupatorion] or ἡπατιριον [hepatirion][177] because it heals the liver. There are those who would like this plant to take its name from its finder, Mithridates, with the surname Eupator. *Plin[y].*[178]

Euonymus, Ἐυωνυμος [Euonymos], that is of good name, by antiphrasis says J. Bauhin.[179] Daléchamps is of the opinion that this is called Tetragonium by Theophrastus because the branches appear quadrangular because of the running of four spurred red lines standing out like a ridge extended directly along the green bark, which is peculiar to this tree, & because the pod is quadrangular in fruit.[180] It is also called Fusanus because from it spindles that are famous from women's yarn are regularly made among the Germans & Gauls.[181]

Excitat ad Venerem tardos Eruca maritos.
Et Venerem revocans Eruca morantem.
The first quotation is from line 799 of Ovid's *Remedia amoris*, but for the fourth word the original has *aptum* ("[it is] appropriate") rather than *jubeo* ("I command", "I prescribe"). For Ovid (Publius Ovidius Naso) and Lucius Junius Moderatus Columella see Chapter 4. Marcus Valerius Martialis (died AD 102) wrote numerous epigrams which are strikingly frank and often bawdy. The last line seems not to appear in his poems, though he does mention the supposed aphrodisiac power of rocket in epigram 75 of book 3

[175] i.e. a caterpillar.
[176] Τιθυμαλος [Tithymalos] was the Greek name for the spurges, as used by Dioscorides and others and still favoured in Ray's time rather than *Euphorbia*.
[177] supposedly derived from ἡπαρ [hepar, "liver"] with the stem ἡπατ- [hepat-] (as in "hepatitis").
[178] *Natural history*, XXV. 65 in the Loeb edition, part of a series of accounts of plants reputedly discovered by kings, where, before a brief description of the plant, Pliny says merely that *Eupatoria* [*sic*] has "royal authority" (*regiam auctoritatem*). Eupator (literally "well-fathered", i.e. "born of a noble father") was the surname of Mithridates VI, King of Pontus (reigned *c.* 120–63 BC) and a formidable enemy of Rome. He was famously expert in poisons, so the naming of this plant after him is more credible than that it has been so improbably corrupted as the initial phrase suggests.
[179] Again the adverb and prefix εὐ [eu] means "well" (as also in the preceding and succeeding entries); ὀνυμα [onyma] is a variant of ὀνομα [onoma], meaning "name". For "antiphrasis" see footnote 6 above.
[180] Theophrastus mentions τετραγωνια [tetragonia] only twice in his *Enquiry into plants* (III. IV. 2 and 6 in the Loeb edition) without any details about it; the word means "four-cornered" and hence "square". Daléchamps' opinion appears in his *Historia generalis plantarum* (1586–1587, p. 272), the work described under *Lugd.* in Ray's 'Explanation'. However, Ray's wording is exactly the same as that in Bauhin & Cherler (1650, vol. 1(2), p. 201) except for discrepancies in abbreviations and punctuation, so it seems that he took this information from this secondary source. See also footnotes 182 and 320 below.
[181] *Dicitur & Fusanus, quòd fusos ex ea muliebri netu celebres apud Germanos & Gallos confici solent*. Literally "It is called also *Fusanus* because spindles from it, from womanly yarn famous, among the Germans & Gauls to be constructed are accustomed." There is a grammatical error here: *fusos* (accusative plural) should be *fusi* ("spindles" in the nominative).

ETYMOLOGY

Euphrasia is called Ἐυφροσυνη [Euphrosyne] because it pleases & delights the eyes, a word which those ignorant of the Greek language have corrupted and turned into Euphrasia, says Daléchamps, whom Jean Bauhin refutes because nothing about this little herb is found in the works of the ancient Greeks or Latins.[182]

F

FAba, Κυαμος [Cyamos]: it was called haba by the Falisci, a people of Etruria, whence the name seems to be taken. Martini derives it from παω [pao] I feed myself, I graze, as if it was Paba.[183] It is called Boona in Latin by Dodoens,[184] *who, supported by a Germanism and having abused an Alemannic idiom, in order that he might seem more learned to others, invented this new name. J.B.* In the whole of Insubria & Liguria fresh beans are called by the word used by those selling them, Baianæ, a name that Apitius[185] also used; thence the English *Beane*. Moreover they are called κυαμοι [cyamoe] by the Greeks, because they are *terrific at conceiving and responsible for conceiving*, & Plutarch gives this as the reason why Pythagoras has commanded abstention from them,[186] no doubt because they implant in people's bodies an impure spirit and humour and for that reason incite to Passion. Cicero assigns a slightly different reason *in his book De divinat[ione], 1.*[187] *Plato gives this instruction for getting off to sleep, that there should be nothing that may bring uncertainty and disquiet to the mind. For which reason also is thought to be the prohibition for Pythagoreans on eating beans.* Plutarch, *in his book About the education of boys*, interprets that Pythagorean mystery or symbol *of abstaining from beans* differently, undoubtedly that they must abstain from public office, *that they must not engage in government; for the elections through which they used to set a limit on the magistrates were of old made with beans*, *because in former times the votes for creating magistrates were registered by beans instead of pebbles*. Agellius in book 4, chapter 11,[188] explains this precept quite otherwise with these words: *The musician Aristoxenus, a most diligent man of true letters, a disciple of the philosopher Aristotle, says in a book that he left about Pythagoras that Pythagoras used no pulse more often than beans, since that food*

---

[182] Ἐυφροσυνη is an abstract noun meaning "merriment", also the name of one of the Graces, derived from ἐυ [eu] (see footnote 171 above) and the stem of φρην [phren, "heart" or "mind"]; ἐυφρασια [euphrasia] was in fact a synonym for it. Daléchamps' opinion is again given in his *Historia generalis plantarum* (1586–1587, p. 1166). It is reproduced in the same words by Bauhin & Cherler (1651, vol. 3(2), p. 432), who use the phrase "inquit Dalechampsius" ("Daléchamps says") which also appears in Ray's Latin text; this suggests that he has again taken his information from the later work.

[183] Martini's Greek verb παω [pao] seems to be a hypothetical one based on the stem found in the deponent verb πατουμαι [patumae, "I eat"] and the Latin *pasco* ("I feed"). Much of this long entry, and particularly the part about the Pythagoreans (though excluding the reference to Plutarch), derives from his *Lexicon philologicum* (1623, columns 750–752).

[184] For Dodoens (Rembertus Dodonaeus) see *Dod.* in Ray's 'Explanation'.

[185] Correctly Apicius: see footnote 771 in the main catalogue.

[186] The literal meaning of δεινος [deinos], like that of "terrific", is "fear-inducing", as in "dinosaur". The Greek verb used here is κυειν [cyein], meaning "to conceive". For Plutarch (Mestrius Plutarchus) see Chapter 4 and for Pythagoras and the Pythagoreans see footnote 439 in the main catalogue.

[187] For Marcus Tullius Cicero see Chapter 4. See also Ray's second note on *Pulegium* [474] in the main catalogue.

[188] *Agellius* seems to be an error for *A. Gellius*. Martini (1623, column 752) has *Sed Gellius lib. c. 11. ex Aristoxeno refert, ...* . For Aulus Gellius, author of *Noctes Atticae* (the work referred to here), see footnote 75 in Ray's 'Interpretation' and Chapter 4.

both gently purges & relaxes the bowels.[189] Moreover it seems that the cause of the error about the bean not being eaten was because in a poem of Empedocles, who followed the teachings of Pythagoras, this verse is found:
   Wretched, all-wretched ones, keep your hands away from beans.[190]
For most people have believed that κυαμος [cyamos] generally means a pulse, but those who have thought more diligently and carefully about the poems of Empedocles say that κυαμοι in this passage mean testicles and that these are called κυαμοι figuratively & symbolically in the manner of Pythagoras because they are terrific at conceiving and responsible for conceiving & may offer the power of human begetting; and for that reason men have wished to derive the meaning in that verse of Empedocles not from eating a bean but from an excess of affairs of Passion.[191] Gesner[192] asserts that by beans in this precept Pythagoras understood eggs, from which he wished there to be abstinence for no other reason than from the flesh of any animals, thinking that it would be an equal crime for a man to sin against a bird or a bird's egg. Moreover he called an egg Cyamos because it was so to speak a κυησις [cyesis, the embryo of an animal] & enclosed its conception within itself.

Fabaria or Faba inversa, Τελεφιον [Telephion], because the bottom of the root abounds in innumerable tubers, which reach the size of beans.[193]

Farfara is perhaps a word imported into Italy by the Phoenicians. *Martin[i]*. Hermolaus conjectures that it is perhaps named from Farfarus, a river of Damascus.[194] But Tussilago is thought to be called Farfara & Farfarella from the likeness that it has with the leaves of poplar, for, as Cordus[195] notes, the ancient Latins called both poplars Farfarus. Others on the contrary are of the opinion that poplar is so called because it has leaves like Tussilago.[196]

Fasēlus is from the likeness of the pods & fruits, which resemble a small boat, a skiff or a cutter.[197]

---

[189] Aristoxenus of Tarentum (fourth century BC) learned music from his father, the learned musician Spintharus (otherwise Mnesias), and finally became a pupil of Aristotle, whom he appears to have rivaled in the variety of his studies; only fragments of his reputed 453 books, including a life of Pythagoras, survive. For Aristotle see Chapter 4.

[190] Δειλοι πανδειλοι κυαμων απο χειρας εχεσθε. Empedocles, who wrote this dactylic hexameter, was an important pre-Socratic philosopher of the fifth century BC from Acragas or Agrigentum, a Greek city in Sicily, is best known for his cosmogenic theory of the four classical elements; he was influenced by the Pythagoreans and supported their doctrine of reincarnation.

[191] *sed à rei Veneriæ proluvio*, literally "but from an excess of the Venereal thing/matter".

[192] For Conrad Gesner (Gesnerus) see *Gesn*. in Ray's 'Explanation' and Chapter 4.

[193] *Faba inversa* means "upside-down bean". For *Telephion* see footnote 64 above.

[194] The initial statement is from Martini (1623, column 779). For Hermolaus Barbarus see footnote 79 above. For the river see II Kings, 5: 12, where Naaman, "captain of the host of the king of Syria" (verse 1), angry at Elisha's command that he should wash in the River Jordan to cure his leprosy, says: "Are not Abana and Pharpar, rivers of Damascus, better than all the waters of Israel?" Pharpar is Φαρφαρ [Pharphar] in the Septuagint version. This seems among the more unlikely of Ray's explanations of plant names, but coincidentally there was a tributary of the Tiber called Farfarus, possibly from its poplars.

[195] For Valerius Cordus see *Cord. & Cord. hist.* in Ray's 'Explanation' and Chapter 4.

[196] See also Ray's synonyms under the accounts of *Tussilago farfara* [603] and *Populus alba* and/or *P. alba × tremula* [456] in the main catalogue.

[197] … *qui lintrem lembúmve naviculam exprimunt*. The Greek φασηλος [phaselos], latinised as *phaselus* or *faselus*, means either a bean or a skiff.

ETYMOLOGY

Favagello is an Italian word, so called, Martini thinks, because it produces somewhat fat leaves like beans.[198]

Fegopyron or Fagotriticum[199] is so called because its seed somewhat resembles beech-mast[200] in both its triangular shape & its dirty colouring, although much smaller in size. *J.B.*

Ficaria & Scrophularia mean Chelidonium minus[201] because it cures piles & because its roots recall those piles by their similarity; hence it is also called herba hæmorhoidum.[202]

Filipendula, *see* Oenanthe.

Filix, Πτερις [Pteris], from the verb I split, because it has finely split & divided leaves.[203] *It has leaves pinnate on the sides, whence the Greeks have given it its name* (i.e. πτερις). Plin[y], *book 27, chapter 9.*[204]

Fumaria: the most notable thing about the herb is that smoke at one time gave it its name, since it has tear-provoking power against the eyes of man such as smoke has. *J.B. It gives clarity and watering to eyes smeared with it as if it was smoke, whence it received the name* καπνος *[capnos].* Plin[y], *book 25, chapter 13.*[205]

Fungus, Μυκης [Myces], as some people wish, is a variant of funus[206] because by its poison it brings those who eat it to their funeral, or more correctly it is named *from the bringing forth*[207] *out of the earth,* because it arises of its own accord from the ground, whence also φυμα [phyma] a tubercle. The Greeks named Mycetæ from the hilt of a sword or from the highest point closing the scabbard,[208] from the corruption of which Mycenae took its name.

Fragaria (Κομαρον [Comaron] *Apul.*) is so called by the Latins from fragro ["I am fragrant"] because this fruit has an excellent smell.

Fraxinus, Μελια [Melia], because it grows in rough places, or from frago flecto, because its branches are easily broken. *Martin[i].*[209]

---

[198] See footnote 192 in the main catalogue. We have been unable to trace this reference to Martini.
[199] These names mean "Oak-wheat" and "Beech-wheat" in Greek and Latin respectively; they refer to *Fagopyrum esculentum* [195]. The English name "buckwheat" probably has the same meaning as the latter (OED).
[200] *glandem faginam*, literally "a beechen acorn" (accusative).
[201] i.e. Lesser Celandine, now called *Ficaria verna* [129]. The Latin word *ficus*, meaning "fig", can also mean "pile".
[202] "herb of haemorrhoids" or pilewort.
[203] The verb is *findo* with a passive past participle *fissa* (agreeing with *folia*).
[204] Pliny's *Natural history*, XXVII. 79 in the Loeb edition; πτερις is related to πτερον [pteron], meaning "feather".
[205] Pliny's *Natural history*, XXV. 156 in the Loeb edition, with reference to what he calls *capnos fruticosa*. Both the Latin *fumus*, from which *Fumaria* is derived, and the Greek καπνος [capnos] mean "smoke". See also footnote 113 above.
[206] *Funus*, with the stem *funer-*, means "funeral" or, by extension, "death" or "ruin".
[207] The Greek verb is φυειν [phyein], meaning "to bring forth" or "to beget".
[208] *vel summo vertice vaginam occludente*, the meaning of which is not obvious.
[209] Martini (1623, column 939) attributes the first theory (which derives the tree's name from *fragosus, -a, -um*, meaning "rough" or "uneven") to Nicolaus Perottus, Bishop of Siponto in Apulia, who published a commentary on the Latin language, *Cornucopia*, at Venice in 1513; this is cited in Martini's catalogue of principal sources at the beginning of his book. Both verbs in the second suggestion are in the first person singular, cited ungrammatically after *à* ("from"), like *fragro* in the previous entry. Martini's (1623, column 940) text clarifies this theory: "Sed Latina vox est; ut puto à *frago*

G

Galega is considered by Fracastoro to be as it were spoken in French; however Galega is an Italian word, as Dodoens, Mattioli and others witness.[210] What name prevailed among the ancients is not settled.

Galeopsis, Γαλιοψις [Galiopsis] or γαληοψις [galeopsis]: those who write Galiopsis with i derive it from the helmeted shape of the flowers because its flowers present an appearance as it were of a helmet. But since it was unusual for the Greeks, and especially the ancient Greeks, to combine Latin names with Greek ones, γαληοψις is rather to be written with η from the shape of the flowers to some extent portraying the face & open jaws of a weasel, as Pliny too bears witness.[211]

Gallium, Γαλλιον [Gallion], *from the mixing in milk instead of beestings*,[212] because it is said that, when infused in milk instead of beestings, it coagulates & thickens it, as Diosc[orides] and Galen bear witness & from experience of the present time.

Garosmum, because it smells of γαρον [garon] or fish-sauce. *J.B.* So Atriplex olida[213] is called thus.

Genista is thought to be named from a knee [à genu], either just because a knee is flexible or because it heals painful knees. *Lob. Martin[i], Lex[icon].*[214]

Gentianella is a diminutive from Gentiana, which was named after Gentius, king of the Illyrians, who first discovered its powers in war.[215]

---

flecto, lento, de quo in *Frango*." ("But it is a Latin word, as I think from *frago* I bend, I make flexible, about which [see] under *Frango*.") *Frago* is not a recognised Latin verb, but Martini states in his entry for *Frango* ("I break") in column 936: "*Frago* prius fuit, unde præteritum *fregi*, tanquam α in η verso, augmento temporali." ("*Frago* was the first [form], whence the past [tense] *fregi*, as if by α [being] changed into η with a temporal augment.") *Fregi* is simply the irregular perfect tense of *frango*. Martini's ingenious but improbable theory is based on the fact that Greek verbs beginning with alpha change this to eta (a temporal augment) in their past tenses.

[210] Girolamo Fracastoro or Hieronymus Fracastorius (1478–1553) was an eminent Italian physician and scholar and a professor at the University of Padua; he discovered and named syphilis (newly arrived from the New World), was elected physician of the Council of Trent in 1545 and suggested that epidemic diseases were spread by tiny particles (spores). For Dodoens (Rembertus Dodonæus) and Mattioli (Petrus Andreas Matthiolus) see *Dod.* and *Matth.* in Ray's 'Explanation' and Chapter 4.

[211] See Pliny's *Natural history*, XXVII. 81 in the Loeb edition, where he confirms the spelling without however giving any reason for it: γαλεη [galeë], usually contracted to γαλη [gale], is the Greek name for a weasel or a polecat (while in Latin *galea* means "helmet") and ὀψις [opsis] means "appearance". *Galeopsis legitima Dioscoridis* (i.e. "the genuine *Galeopsis* of Dioscorides") is Ray's chosen name for *Stachys sylvatica* [209], which is the species that best fits Pliny's description. Pliny treats *galeobdolon* as a synonym: βδολος [bdolos] means "stench" (for which the polecat was notorious) and so fits *Stachys sylvatica* well. Today both names are used for other labiates [e.g. 105, 315, 318].

[212] The Greek for "milk" is γαλα [gala] and πυετια [pyetia] is "beestings", the first milk secreted by a cow after giving birth, sometimes used in cheese-making.

[213] See the entry for this plant, Stinking Goosefoot, *Chenopodium vulvaria* [69], and the related footnote in the main catalogue.

[214] Martini (1623, column 1015) attributes this theory to "Trag. in hist. stirp." (i.e. Bock 1552).

[215] See Pliny's *Natural history*, XXV. 71 in the Loeb edition, again part of a series of accounts of plants reputedly discovered by kings, though he does not mention war. Gentius, the last Ardiaean king of Illyria (modern Dalmatia and Albania), was defeated by the Romans in 168 BC. The plant named after him was probably the alpine Great Yellow Gentian, *Gentiana lutea* L.

ETYMOLOGY

Geranium, Γεϱανιον [Geranion],[216] from the likeness of a crane's head which is at the top of this plant's flower-head with a most exact resemblance. There are various species of this plant, of which one is called Geran[ium] sanguineum[217] both from the blood-red colour of the root within & without and from its wonderful potency in staunching blood. *J.B.* A second is Ger[anium] Rupertianum or D[ivi] Ruperti or Robertianum, from the red stem & red leaves, as it seems to *Tabernaemontanus*.[218]

Glaux, Γλαυξ, as if Galaux from γαλα [gala], milk: certainly it is cooked in a broth with barley meal, salt & oil to restore the abundance of milk that has dried up. *Dioscor[ides]*. Pliny too seems to assent to this derivation when he writes *in book 27, chapter 9,* that Glaux was called in former times Eugalacton.[219]

Gnaphalium, Γναφαλιον [Gnaphalion], because they use its leaves *instead of flock*, that is for stuffing, or rather because its leaves are covered with felt.[220] *Lob. in Adv. The Greeks use the word* γναφειν [gnaphein] *for to pull out or pluck off wool or cloth, whence Gnaphalion, which the Romans named Centunculus & Tomentaria*.[221] It is called Herba impia by Pliny.[222] They have called it impious because the children (the later little heads) exceed the parent, that is the earlier one. Others have thought rather that it was called thus because no animal touches it. In English it is called *Cudweed* because it helps animals chewing the cud & recovers food from the rumen. In Northumbria *Chafe-weed* because it is efficacious for chafings of the skin.[223]

---

[216] from γεϱανος [geranos], meaning "crane".

[217] Bloody Crane's-bill, *Geranium sanguineum* [222].

[218] This is Herb Robert, *G. robertianum* [223]. Once again Ray prefixes the name Rupert with *D.* (for *Divus*): see footnote 68 above. Tabernaemontanus is the Latin title used by Jakob Theodor (*c.* 1525–*c.* 1590) after his home town of Bergzabern in the Palatinate region of Germany: see *Tab.* in Ray's 'Explanation' and Chapter 4.

[219] See Pliny's *Natural history*, XXVII. 82 in the Loeb edition, where he says that it is cooked in broth made from the finest wheat flour (*in sorbitione similaginis*). *Eugalacton* is compounded from εὐ [eu, "well"] and γαλα [gala, "milk" again] with the stem γαλακτ- [galact-] (as in the English word "galactic", which originally related to the Milky Way). Pliny provides a fairly convincing description of Sea Milkwort, *Glaux maritima* L., though *Glaux* has generally been regarded as *Lepidium coronopus* [153] or by some as *Astragalus* spp. [227, 228].

[220] The usual Greek word for flock used for stuffing cushions and pillows is κνεφαλλον [cnephallon], of which γναφαλον [gnaphalon], used here, is a variant. The Latin word used here for both "stuffing" and "felt" is *tomentum*, now a common botanical term in the latter sense; see also the entry for this word in Ray's 'Interpretation'.

[221] Again the usual Greek word for "to card wool" or "to dress or full cloth" is κναπτειν [cnaptein], of which γναφειν [gnaphein] is a variant. Pliny uses *Centunculus* for a different plant and does not mention *Tomentaria*.

[222] See Pliny's *Natural history*, XXIV. 173 in the Loeb edition, where he describes a plant like rosemary, but hoary, and continues with a passage similar to what follows; however he does not equate this plant with his *Gnaphalium* (mentioned in XXVII. 88 as being used for flock). See also the entry for *Gnaphalium minus* [*Filago vulgaris* 230] and footnote 437 in the main catalogue.

[223] The final phrase of the penultimate sentence is *rumen amissum revocat*, literally "it recalls a lost/dropped rumen", but presumably the reference is to the cud in the rumen (a ruminant animal's first stomach). The last four words are *quoniam ad intertrigines valet*. The original source of the two English names is Turner (1551, folio I.i. *verso*; Chapman & Tweddle 1995, p. 124), who says: "Centunculus called in Yorke shyre Cudweede, and in Northumberlande Chafwede because it is thought to be good for chafynge of any mans fleshe wyth goynge or rydynge." Ray clearly took his information from Bauhin & Cherler (1651, vol. 3(1), p. 159), who give a slightly abbreviated Latin translation of the relevant passage, citing its origin and including several phrases that Ray uses; they include both *In Eboracensi agro* and *In Northumbria*, but Ray mentions only the latter, Turner's native county.

Gramen, Ἄγρωστις [Agrostis]: the Grammarians talk nonsense about the origin of this name; while some derive it from gradior, others from germen, others from γραω [grao] & some from ager, ἀγρωστις is indeed no doubt from ἀγρος [agros] since it grows generally.[224] Some grass is called of Parnassus[225] because it grows on Mount Parnassus. *Diosc[orides], book 4, chapter 32.*

H

HEdera, Κισσος [Cissos] or Κιττος [Cittos], seems to be named by the Latins from eating [ab edendo], and indeed it is a churlish lover of trees &, fastened like a flatterer onto the fissured bark, it demolishes older enclosure-walls &, having embraced trees with its rooted growths, it tortures & devours them, as Jean Bauhin rhetorically announces. In fact Pompeius is of the opinion that it is called hedera because it sticks on or rather because it seeks heights.[226] Moreover they think that Κισσος is from Cissus, whom the Greeks tell was Father Liber's favourite boy[227] & was turned into the shrub bearing his name when, while leaping with the God as his attendant, he was dashed to the ground and died.

Hellebŏrus, Ἑλλεβορος [Helleboros], is also written without aspiration,[228] *as compared with the seizing in the food,*[229] because it destroys when eaten;[230] it is called veratrum because it turns the mind.[231] *Etymol[ogicum].*[232]

Hedypnois;[233] Dens Leonis is called thus from the sweetness of the scent that is exhaled from the flowers.

Hedysarum, Ἡδυσαρον [Hedysaron], for the Greeks perhaps by antiphrasis since its seed is very bitter.[234]

Helxīne, Ἑλξινη, that is Parietaria, from *the verb to drag,*[235] because it attracts the clothing of passers-by, since it has seeds in the bur-like heads that adhere tenaciously to clothing. It is called Parietaria & muralium, since it loves to grow on walls,

---

[224] The five words mean respectively "I step", "bud" or "sprout", "I gnaw", "field", and "field" or "the country".

[225] i.e. Grass of Parnassus, *Parnassia palustris* [264].

[226] *quòd hæret, vel potiùs quòd edita petat.* The reference, which is taken from Bauhin & Cherler (1651), vol. 2, p. 111, is probably to the second-century Sextus Pompeius Festus, who abridged a now lost work by Marcus Verrius Flaccus, *De verborum significatione*; the surviving part of this abridgement was published in various places in 1560, 1576, 1584 and 1593.

[227] *pusionem Liberi patris.* Liber was an old Italian god later identified with the Greek Bacchus.

[228] i.e. as Ἐλλεβορος [Elleboros].

[229] The relevant Greek words are ἑλειν [helein, "to seize"] and βορα [bora, "food"].

[230] *quòd esu perimat,* literally "because by the eating it destroys".

[231] This suggests a relationship with the verb *verto* ("I turn").

[232] This seems to be another reference to *Etymologicon Magnum* (1594; see footnote 135 above), where the phrase in Greek appears in column 331 with the words in a different order. The explanation of *Veratrum* is given in column 4004 of Martini (1623) under *Veratrum*, attributed to Nicolaus Perottus (see footnote 209 above), but he also suggests an alternative explanation (see entry for *Veratrum* below).

[233] Ἡδυπνοις is derived from ἡδυς [hedys, "sweet"] and πνοη [pnoë, "blowing", "exhalation"].

[234] This suggests that ἡδυσαρον is derived from ἡδυς [hedys, "sweet"] and ἀρον [aron, "arum"]. For "antiphrasis" see footnote 6 above.

[235] The Greek verb is ἑλκειν [helcein]. From the name and synonyms the plant must be Pellitory-of-the-wall, *Parietaria judaica* [418], though the description of the seeds and some other details suggest Cleavers, *Galium aparine* [51].

ETYMOLOGY

Perdicium from Partridges, which are accustomed to feed on it, and Urceolaria & Vitrearia because it is effective in cleansing small pitchers & glass vessels.[236]

Herba Gerardi is called 𝕲𝖊𝖗𝖆𝖊𝖗𝖙𝖘 𝖈𝖗𝖚𝖞𝖙[237] by the Brabantians[238] & Belgians.

Helenium, Ἑλένιον [Helenion]: some want it to be named thus from Helen's tears, from which the Poets tell that it sprang, others because from it a remedy against snakes was first found by Helen, which she showed to her husband Menelaus when he was voyaging to Egypt. Others say it is from the island of Helene, where it abounds far the best.[239] *J.B.*

Hieracium, Ἱεράκιον [Hieracion]: it is called Hawkweed from the fact that hawks, by scraping it and wetting their eyes with its juice, dispel poor vision when they have become aware of it. *Plin[y], book 20, chapter 7*,[240] or perhaps on account of the fact that its pappi are useful to hawks for provoking vomit.

Holosteum, Ὁλόστεον [Holosteon], is a herb without any hardness; quite the opposite, or it is so called by the Greeks by antiphrasis, just as one says sweet gall. *Plin[y], book 27, chapter 10.*[241]

Holcus, *as compared with the dragging*,[242] for, tied around the head or around the arm, this plant draws awns out of the body; for that reason some people call it Aristis. *Plin[y], book 27, chapter 10.*[243]

Hormīnum, Ὅρμινον [Horminon]: [Mathias] de L'Obel derives it from ὁρμᾶν [horman, "to urge on"] because *it urges anyone on to intercourse*, it urges on to & inflames Passion. *But drunk with wine this seems to urge on to intercourse. Diosc[orides]*.

---

[236] *Parietaria* is derived from *paries* ("wall of a house"), *muralium* from *murus* ("wall", originally of a city), *Perdicium* from *perdix* ("partridge"), *Urceolaria* from *urceolus* ("small pitcher") and *Vitrearia* from *vitreus* ("made of glass"). In the main catalogue the synonyms equivalent to the last two are *Urceolaris* and *Vitriola*.

[237] This too means "Gerard's herb": presumably the suggestion is that the Latin name is a translation of the Brabantic one. *Aegopodium podagraria* [46] was called Herb Gerard because it was dedicated to the 10th-century St Gerard of Brogne, who was formerly invoked to cure gout, against which the herb was chiefly employed.

[238] The Duchy of Brabant was a historical region in the Low Countries consisting of the three modern Belgian provinces of Brabant, the Brussels area and the present-day Dutch province of North Brabant.

[239] In his *Natural history* (XXI. 59 in the Loeb edition) Pliny writes that *Helenium* is said to have sprung from Helen's tears and therefore is highly praised on the island of Helene.

[240] Pliny's *Natural history*, XX. 60 in the Loeb edition. Ἱεράκιον [Hieracion, "Hawkweed"] is derived from ἱέραξ [hierax], meaning "hawk", for which the Latin name is *accipiter*. Pliny used the Greek name for the plant, but Ray here uses the equivalent Latin word *Accipitrina*.

[241] Pliny's *Natural history*, XXVII. 91 in the Loeb edition. Ὁλόστεον means "All-bone", being derived from ὅλος [holos, "whole"] and ὀστέον [osteon, "bone"]. For "antiphrasis" see footnote 6 above; the example given here is *fel dulce*.

[242] The verb ἕλκειν [helcein, "to drag", "to draw"] is indeed the origin of the Greek word ὅλκος [holcos], originally meaning an agent of dragging, "hauling-engine", and then the result of drawing a plough, "furrow", before being applied to this grass.

[243] Pliny's *Natural history*, XXVII. 90 in the Loeb edition. We have treated *aristas* (accusative plural) as awns (as defined in Ray's 'Interpretation') rather than as ears of corn (as in the Loeb translation) and presumably these have become lodged under the skin, but it is not obvious how the grass extracts them. Pliny has apparently given a Greek form, *Aristida*, as the accusative of *Aristis* as if he thought the name was Greek, but it is not. The grass seems not to be a member of the modern genus *Holcus* but Ray's *Hordeum spurium*, now called (as also by Pliny in XXII. 135) *Hordeum murinum* [288], which see in the main catalogue, where it can be seen that Anguillara (1561) reported Pliny's name as *Aristida*.

Hyacinthus: the poets tell that the flower sprang from the blood of the boy Hyacinthus. *Ovid, Metamorph[oses]*.[244] Nannius thinks it is so called as being ἰα Κινθιου [ia Cinthiu], that is the Violet of the Cinthian or of Apollo.[245]

Hyoscyămus, 'Υοσκυαμος [Hyoscyamos], that is Faba fuilla,[246] because wild boars are weakened & brought down by feeding on it, and that with instant danger of death unless they have immediately washed themselves within & without with plenty of water. *Aelian[us]*,[247] book 1.

Hyssōpus, 'Υσσωπος [Hyssopos]: the Grammarians inappropriately want it to be named from *being rained on* for *being poured upon the face*,[248] because it may be poured over the face. Actually its use was for besprinkling the people with lustral water in temples. We think it is named from the Hebrew **'ēzōv**, since the LXX always translate ezob as ὑσσωπος.[249] I am not unaware that some people think that they themselves translated it thus, deceived by the similarity of the names, since ezob does not signify the Hyssop of the Greeks but Rosemary,[250] but in my case the authority of the Seventy Elders, who could scarcely have been deceived over so common a plant, is worth more than these people's conjectures. And this should not be seen as extraordinary

---

[244] See the story in book 10: 205–219.

[245] Apollo was called the Cynthian because he and his sister Artemis were supposedly born on Cynthos, a mountain on the island of Delos; the correct spelling in both Greek and Latin is with y rather than i. Petrus Nannius (1500–1557) published a commentary on Virgil's bucolic poems at Basel in 1559, on pp. 122–123 of which this statement can be found, in comments on Virgil's reference in his *Eclogues*, 3: 106 to flowers *inscripti nomina regum* ("inscribed with the names of kings"): the petals of the original hyacinth (probably not the plant so called today) were supposedly marked with letters, variously interpreted as AIAI ("Alas!"), Y (for Hyacinthos) or AI (for Aias, i.e. Ajax); both Hyacinthos and Ajax were the sons of kings. Ray seems to have taken this explanation from Martini (1623, column 1164), who wrote: "Sed Nannius vult esse ἰα κινθιου *violas Apollinis*." ("But Nannius wants [Hyacinth] to be ia cynthiu, *violets of Apollo*.") Ray's singular *Viola* should be plural as in Martini's text.

[246] i.e. "Swine's bean", which is also the meaning of the Greek name.

[247] For Claudius Aelianus see footnote 439 in the main catalogue.

[248] Neuter of the present passive participles of the verbs "to rain" and "to pour", ὑομενον [hyomenon] and χεομενον [cheomenon], and the accusative singular of ὠψ [ops, "face"], ὠπα [opa].

[249] Throughout this entry words in bold type are romanisations from Ray's Hebrew or Chaldean. In his original text many of these words after this first one are followed by his own transliterations in italics, but these are not reproduced here and we have adopted a more exact system of romanisation (including the marking of long vowels, the use of a superscript e or a to represent an indistinct vowel like that in "the" or "about", ' for 'aleph (a glottal stop) and ' for 'ayin, and the addition of dots under t and s to represent ṭet and ṣin); there are problems here because Ray's Hebrew and Chaldean words are represented by consonants only and, where he supplies vowels in his transliterations, they often do not match those in the actual words. Ray's romanisations of words from other Semitic languages are retained unchanged in italics. In addition, transliterations of Greek words and English translations are added in square brackets. LXX is the standard abbeviation for the Septuagint, the Greek version of the Old Testament supposedly translated from the Hebrew by 70 (or 72) scholars, or, as here, for the scholars themselves. *'Ēzōv* was not Hyssop, *Hyssopus officinalis* L., but probably *Origanum syriacum* L.; it holds liquids well and was used by the Israelites for purging and cleansing, for example for sprinkling the lintel and door-posts with the blood of the Passover lamb (Exodus 12: 22; Walker 1958). Unlike most of the explanations of words that follow, the etymology here is correct, ὑσσωπος being a loan-word from Hebrew or another Semitic language; a few other similarities are also explainable as loan-words in either direction (see later footnotes). *Kar'kōv* ["border"] should be *kar'kōm* ["saffron"].

[250] 'Υσσωπος is mentioned by Dioscorides but not by Theophrastus; if it is not *'ēzōv*, its identity is uncertain. Interestingly, Rosemary, *Rosmarinus officinalis* L., is used today in some churches for ritual sprinkling of the congregation.

ETYMOLOGY

when very many words, both Greek and Latin, have become current from Hebrew sources; such are from **'ella'** ἀλλα [alla, "but"]; from **'awwāh** aveo ["I desire"]; from **'ozen** οὐς [us, "ear"]; from **'elleh** ille ["that man"]; from **'āthown** asina ["she-ass"]; from *anbub*, Punic for a kind of reed pipe, *anbubai*, one who makes music with it, whence that phrase in Horace *Ambubaiarum collegia* &c.;[251] from *esua* in Punic asso ["I roast", "I broil"], which is derived from the Hebrew **'ēsh**; from **'ᵃnuw** νω [no, "we two"] and nos ["we"]; from *Rab[binical] besas* in Punic basio ["kiss"]; from *baccara* in Punic vacca ["cow"]; from *borg* in Punic πυργος [pyrgos, "tower"]; from **bāmāh** βωμος [bomos, "altar"]; from *Buq* in Syriac buccina ["trumpet"]; from *barra* in Punic foris ["door"]; from **bā'** βαινω [baeno, "I go"] & formerly βαω [bao]; from *Gorab* in Punic κοραξ [corax, "raven"]; from *Gelida* in Syriac Gelu ["frost"]; from **gibbēn** gibbus ["hump"]; from **chemer** merum ["unmixed wine"]; from *huth* in Punic ἰχθυς [ichthys, "fish"]; from *harag* in Punic ὀργη [orge, "anger"]; from *hesba* in Syriac herba ["plant", "herb"]; from *zavag* in Punic ζευγω & ζυγος [zeugo & zygos, "I yoke" & "yoke"];[252] from **zāra'** sero ["I sow"]; from *tal* in Punic tellus ["earth"], & from **'erets** or Chaldean **'ar'ā'** ἐρα [era, "earth"]; from *talla* in Punic tollo ["I raise"]; from **yayin** οἰνος [oenos, "wine"]; from **kᵉraz** in Chaldean κηρυσσω [cerysso, "I announce"] & from *keroz* κηρυξ [ceryx, "herald"]; from **qeren** κερας [ceras] & cornu [both "horn"]; from **karᵉkōv** crocus ["crocus"]; from **pethiy** fatuus ["foolish"]; from **lāvān** albus ["white"] & in the fut[ure] Hiph[il][253] *albin*; from **mar** amarus ["bitter"], whence perhaps mare ["sea"]; from *Farasa* in Syriac φαρσος [pharsos, "portion"]; from *Mard* in Punic morbus ["disease"]; from *Sarab* in Punic Syrupus ["syrup"]; from **ṣāvar** Spero ["I hope"]; from **lāqaṭ** legit ["he chose"]; from *Quedra* in Punic χυτρα [chytra, "pipkin"]; from *Radd* in Punic redeo ["I return"]; from **peleg** πελαγος [pelagos, "sea"]; from **pāgar** piger ["sluggish"]; from *Sul* in Syriac scala ["stair"]; from *ferot* in Syriac fructus ["fruit"]; from **qāṭal** κτεινω [cteino, "I kill"] & κτειλω [cteilo, a verb we cannot trace]; from *Luz* in Punic νυξ [nyx, "night"]; from *Lachia* λαχνη [lachne, "down", "fur"]; from **lappīyd** lampas ["torch"]; from **sādīyn** σινδων [sindon, "fine linen cloth"]; from **tuppīyn** tympana, τυμπανα ["kettle-drums"];[254] from *Seif* in Punic ξιφος [xiphos, "sword"]; from **shēsh** Sex [six]; from *Sarab* in Punic Sorbeo ["I swallow"]; from **hēylēyl**[255] ὀλολυζω [ololyzo, "I cry aloud"]; from **har** ὀρος [oros, "mountain"]; from **rā'āh** ὁρω [horo] video [both "I see"]; from *Fistaq* in Punic πιστακιον [pistacion, "pistachio nut"]; from *haref* ἁρπω[256] [harpo, "I snatch"]; from **pittāh** πειθω [peitho, "I persuade"]; from **tōwr** turtur ["turtle-dove"]; from **karpas** carbasus ["fine flax", "fine linen"],[257] from **shānāh** ἐνος [enos, "last year's"]; from **mezeg** μισγω [misgo,

---

[251] See Horace's *Satires*, 1, 2, 1. *Ambubaiae* were "a class of Syrian girls in Rome, who supported themselves by their music and immorality", but the Syriac or Punic word from which it is indeed derived supports Ray's version spelled with *n* rather than *m* (Lewis & Short 1890). For Horace (Quintus Horatius Flaccus) see Chapter 4.

[252] The first Greek word appears to be an error, as the verb has an irregular form for the first person singular, ζευγνυμι [zeugnymi]. The two related stems ζευγ- and ζυγ- are exemplified in the English words "zeugma" and "zygote".

[253] a tense and mood of a Hebrew verb.

[254] The similarity in the words is in fact less than suggested here, since the correct Hebrew plural is *tuppīym*, but it does mean the same as the Latin. *Tympana* were used in the worship of Cybele and Bacchus.

[255] This is part of the Hebrew verb found in *halᵉlūwyāh*, "Hallelujah!" ("Praise God!").

[256] The correct form of the verb is ἁρπαζω [harpazo].

[257] These words really are related, both of them and also the Greek καρπασος [carpasos] being derived from the Sanscrit *karpâsa*, though this means "cotton".

"I mix"]; from **kᵉthōneh** κιτων [citon, "tunic"];²⁵⁸ from *Faza in Punic* φυζα [phyza, "headlong flight"]; from *vai* οὐαι [uae, "woe!"]; from **pūrpa'** πορπη [porpe] fibula [both "buckle", "clasp"]; from **būwts** βυσσος²⁵⁹ [byssos, "flax", "linen"]; from **b*s*ys**²⁶⁰ βασις [basis, "step"]; from *thaur in Arabic*, which is from *Hebrew* **shōwr**, ταυρος [tauros, "bull"]; from **shākhan** σκηνη [scene, "tent"]; from *taraq*, which is from *Hebrew* **dārak**, τρεχω [trecho, "I run"]; from *kala*, which is from **qōwl** *in Hebrew*, καλεω [caleo, "I call"], whence the old Latin word calo ["I call"] & the English to call; from **m*sh*k** in Syriac μεσκος [mescos]²⁶¹ pellis ["hide", "leather"]; from **pilegesh** πολλακις;²⁶² from **gāmāl** καμηλος²⁶³ [camelos, "camel"]; from **'ēden**²⁶⁴ ἡδονη [hedone, "pleasure"]; from **'ōren** ornus ["Manna Ash"];²⁶⁵ so that I may omit very many other entirely extraneous examples,²⁶⁶ such as are ἀρραβων [arrhabon, "pledge", "earnest"], παραδεισος [paradeisos, "park", "paradise"], ἀγγαρευω [angareuo, "I press into sevice"], σατραπης [satrapes, "satrap"], παρασαγδα [parasagda: meaning not traced], πελεκυς [pelecus, "axe"], πεπερι [peperi, "pepper"], χομμι [chommi: meaning not traced] and very many names of plants & precious stones & musical instruments, such as Jaspis ["jaspar", from ιασπις in Greek] from **yoshpēh**, σαπφειρος [sappheiros, "sapphire"] from **sappīyr**, &c., which it would be easy to enumerate if this were not to deviate too far from our theme. Moreover the Reader is to be warned that we have selected words with a Punic origin from the alphabet of Canini,²⁶⁷ who on the analogy of other names makes it seem probable²⁶⁸ that both all those mentioned and very many others are derived from it.

Hypericum, Ὑπερικον [Hypericon], is from Ὑπεριων [Hyperion],²⁶⁹ for it warms & dries.

---

²⁵⁸ The Greek word is probably of oriental origin and so related to the Hebrew one.

²⁵⁹ This is another loan-word from Hebrew or another Semitic language.

²⁶⁰ We have been unable to trace a Hebrew word made up of the consonants b s y s.

²⁶¹ We are uncertain about the vowels in the Syriac word; μεσκος is a very rare word for a fleece.

²⁶² Πολλακις [pollakis] means "many times"; Ray must have intended παλλακις [pallakis], which means "concubine" and from which *pilegesh* is a loan-word. This seems to be confirmed by an entry in Canini (1624, p. 5): see footnote 267 below.

²⁶³ In this case it is the Greek name of the animal that is a loan-word. *Gimel* is the third letter of Semitic alphabets, originally written as Λ like a camel's hump and, from the Phoenician alphabet, the origin of the Greek letter *gamma* (Γ).

²⁶⁴ The Hebrew word means "delight" and was thus used for the Biblical Garden of Eden.

²⁶⁵ *Ornus* is Manna Ash, *Fraxinus ornus* L., but *'ōren* is probably Stone or Umbrella Pine, *Pinus pinea* L.

²⁶⁶ Most of the examples that follow are loan-words in Greek. Two of them we cannot trace, but of these παρασαγδα is possibly an error for παρασαγγης [parasanges, *farsang*, a Persian unit of length equivalent to 30 stades], which Canini (1624, p. 53) compares to the Punic *Farsach*.

²⁶⁷ Angelo Canini or Angelus Caninius (1521–1557), one of the greatest linguists of the 16th century, published an Aramaic grammar in 1554, but Ray is here referring to his etymological treatise on the Greek language, Ἑλληνισμος [Hellenismos], originally published in Paris in 1555 and frequently reprinted including in London in 1624; the first part of this is headed "ALPHABETVM" and has sections headed by each letter of the Greek alphabet. Ray has extracted many of his examples from a longer passage setting out Canini's theories about the ways in which words in one language or dialect are modified in others, including numerous examples of supposed relationships between Punic and Greek words (the former transliterated into the Roman alphabet); however Ray has also introduced Hebrew and Chaldean words in the Hebrew alphabet.

²⁶⁸ Clearly *probale* in Ray's text is a misprint for *probabile*.

²⁶⁹ Hyperion was the Homeric sun-god, one of the twelve Titan gods of Ancient Greece, which were supposedly later supplanted by the Olympian gods.

ETYMOLOGY

I

Jacea is said by the Italians to be from ιov [ion, "violet"], because the flowers are like the March violet.[270] Jacea nigra[271] is so called for its difference from the three-coloured viola, which is also called Jacea.[272]

Irio, Ἐρυσιμον [Erysimon]: the Latins named irio from rushing in [ab irruendo] because it rushes into one's tasting with fiery force & burning relish. *J.B.*

Iris, Ἴρις, for Dioscorides received its name from its likeness to the celestial bow,[273] which to some extent it reproduces with the varied colour of its flower.

Juglans, καρυα βασιλικα [carya basilica]: *Servius* wrote: According to the opinion of some people that nut is thought to be named from helping and an acorn.[274] But, more impressively, in his book about the meaning of words Bassus[275] reports this: *The tree was called Juglans, as if Jovis glans* ["Jupiter's acorn"] *because that kind of tree has nuts that have a sweeter flavour than the acorn has; those ancients, who thought it was excellent and like an acorn and the tree itself worthy of God, called this fruit Jovis glans, which, with letters erased, is now called Juglans.* Moreover Cloatius Verus in his book of tracts from the Greeks relates this matter thus: *Juglans: di has been omitted, as if it was dijuglans, that is* Διος βαλανος [Dios balanos, "acorn of Zeus"]. *Macrob[ius], Saturn[alia], [book] 2, chapter 14.*[276]

Juncus, Σχοινος [Schoenos], is from joining [à jungendo] because, since it is flexible, things can be joined with it.

Juniperus, Ἄρκευθος [Arceuthos], is from junis & pario, because it produces younger & new fruits while the old ones are ripening, as one that produces with its fruit still clinging to its tree up to the third year.[277]

L

Lactūca, Θριδαξ [Thridax], is from the abundance of milk with which it is swollen[278] & because it increases milk for women.

---

[270] *viola Martia*: see [*Viola*] *Martia purpurea* under *Viola nigra seu purpurea*, i.e. *Viola odorata* [617], in the main catalogue.

[271] i.e. *Centaurea nigra sensu lato* [297]. See also *Jacea nigra* under *Morsus Diaboli*, i.e. *Succisa pratensis* [374], in the main catalogue.

[272] ... *ad differentiam tricoloris violæ, quæ etiam Jacea nuncupatur.* See *Jacea sive Flos Trinitatis* under *Viola tricolor* in the main catalogue, an aggregate species represented in Cambridgeshire by *Viola arvensis* [620]. The use of the same word in names for two such different groups of plants is strange.

[273] i.e. a rainbow, for which the Greek name is ἰρις [iris].

[274] *à juvando & à glande*, from the verb *juvo* ("I help") and the noun *glans* ("acorn") respectively. For Servius Maurus Honoratus see footnote 77 above.

[275] Cassianus Bassus (6th or 7th century AD) compiled *Geoponica*, a collection of agricultural literature revised in about 950 by an unknown writer (see Chapter 4).

[276] Cloatius Verus was a grammarian who probably lived at the end of the first century AD. Ambrosius Theodosius Macrobius was a critic who lived at the end of the fourth century AD (see Chapter 4).

[277] *Junis* is presumably a hypothetical shorter form of *juvenis* ("young"), based on the comparative form *junior* (as later in the sentence), while *pario* means "I produce". For a modern study that shows that the fruiting of *Juniperus communis* [310] is just as Ray describes see Ward (2010).

[278] *à copia lactis quo turget*.

Lagōpus, Λαγωπους,[279] is called thus because its heads, which finally disintegrate into a kind of scurf, recall a hare's foot.

Ladănum segetum: *This herb is called Ladanum of standing corn by Pliny from the similarity of the leaves to Ladanum & from its habitat, but by others narrow-leaved Tetrahit*. Hist. Lugd.[280]

Lamium, Γαλιοψις [Galiopsis], is perhaps from the shape of the flowers, as if it was a cowl representing some demented demon.[281]

Lampsāna, Λαμψανη [Lampsane],[282] is otherwise written Lapsana, from λαπτω [lapto, "I lap up"], as some wish, because it is eaten as if by lapping,[283] or rather *as compared with the lapping*,[284] whence λαπαζειν [lapazein, "to empty"] to evacuate, because it relaxes, cleanses and evacuates the bowels.

Lapăthum, Λαπαθον [Lapathon], *is from the emptying*,[285] to evacuate, to empty, because a decoction of its leaves relaxes the bowels. It is Hippolapathum to the Greeks, as if it was a large Lapathum, for they are accustomed to put Bu & Hippo in front of big and large things, borrowing the size from the ox & horse, animals of outstanding size.[286] *J.B.*

Lappa may be so called *from the lapping*, that is to lap, because it sticks to the clothes of passers-by.[287]

Laureola is a diminutive from Laurus.[288] For Isidore Laurus was from the word laus, for the heads of victors were crowned with laudes;[289] afterwards, with the letter d taken away & r substituted, it was called Laurus, as in audiculæ, which are now auriculæ, & medidies, which is now called meridies.[290] For others Laurus is from

---

[279] This is correctly derived from λαγως [lagos, "hare"] and πους [pus, "foot"].

[280] Pliny twice (XXVI. 47 and 115 in the Loeb edition of his *Natural history*) refers to this *Ladanum, quod in segetibus nascitur* ("which grows in standing corn"), as different from the gum of the same name derived from *Cistus* species. The source of the quotation is Daléchamps (1586–1587, p. 443), a work referred to in Ray's 'Explanation' as *Lugd*. from the place of publication, Lyons (Lugdunum), but, as with other references to this work, Ray may have taken the explanation from Bauhin & Cherler (1651, vol. 3(2), p. 855), where it appears with the same wording as in his *Catalogus*. The plant is Ray's *Ladanum segetum*, i.e. Red Hemp-nettle, *Galeopsis angustifolia* [315].

[281] *velut cucullo larvatam lamiam quandam repræsentante*. The Greek Λαμια [Lamia] was "*a fabulous monster said to feed on man's flesh*, a bugbear to frighten children with" (Liddell, Scott & Jones 1940). In classical Latin and still in Ray's time *larva* meant an evil spirit or demon (or sometimes an actor's mask, as in the entry for *Personata* below) and so *larvatus, -a, -um* means "possessed" or "demented" (OLD).

[282] i.e. Nipplewort, *Lapsana communis* [320], but see also the entry for *Papillaris* (listed as a synonym in the main catalogue) below.

[283] *quasi lambendo*.

[284] λαψαι [lapsae], the aorist infinitive of λαπτω.

[285] The Greek verb is again λαπαζειν [lapazein].

[286] The Greek Ἱππολαπαθον [Hippolapathon] incorporates the stem ἱππο- of ἱππος [hippos, "horse"]; the other stem mentioned by Bauhin, βου- [bu], is of βους [bus, "ox"]. Similarly we have English names such as "Horse Radish" and "Cow Parsley".

[287] The same Greek and Latin verbs are cited as under *Lampsana* above but with a different interpretation.

[288] See *Laureola*, i.e. *Daphne laureola* [327], in the main catalogue. The similarity of the leaves of this species to those of Laurel or Bay-tree, *Laurus nobilis* L., is responsible for the use of *Daphne*, originally the Greek name for the latter, as the generic name of an unrelated genus.

[289] For Isidore, Bishop of Seville, see Chapter 4. *Laus*, with plural *laudes*, means "praise".

[290] *Auriculæ*, a diminutive of *auris* ("ear"), are ear-lobes and *meridies* is midday, but the word for "to hear" is *audio* and that for "middle" is *medius, -a, -um*.

ETYMOLOGY

Lavo ["I wash"]; it is thought to mean purgo ["I purge"], for it acts with extraordinary power to purge the blood.

Lens, Φακος [Phacos], because it makes people slow or patient. *I find* (says Pliny *in book 17, chapter 12*[291]) *in earlier authors' works that equanimity is engendered in those eating it.*[292] Lenticula palustris or lens palustris is so called because its leaf is in the shape of a Lentil, on the authority of Dioscorides.[293]

Ligustrum seems to be so called from tying,[294] whence also the true Ligustrum is considered by the poets to be a kind of convolvulus.[295]

Lichen, Λειχην [Leichen], is so called because it cures λειχηνες [leichenes], that is scabby eruptions on the skin.[296]

Limodōron, Λιμοδωρον for Theophrastus, is because it weakens neighbouring plants & does not permit them to grow to maturity by drawing their nourishment to itself. *It gives hunger*[297] it gives hunger. It is called Orobanche because *it strangles vetches,*[298] that is it strangles & tortures vetch & other pulses. Because of the shape of its root & because it frequently grows on the roots of Genista vulgaris the rabble of Herbalists call it Broomrape.[299]

Limonium, Λειμωνιον [Leimonion], is from λειμων [leimon] meadow. *Limonium obtained its name from the Greeks because it rejoices in irrigated meadows & similar places.* Cordus[300] in the works of J.B.

Lupulus is called Salictarius[301] because it steals into and climbs up willows & all plantations by twining around them. [Mathias] de L'Obel thinks it is called Salictarius because the poles which it climbs are for the most part made from willow.

Lolium, Ἀιρα [Aera],[302] is perhaps *from the destroying of a standing crop,*[303] because it destroys standing corn, or Lolium is as if it was δολιον [dolion, "treacherous"] just as lachryma is from dachryma. *Martin[i].*[304]

---

[291] Pliny's *Natural history*, XVIII. 123 in the Loeb edition; book 17 is evidently an error for book 18.

[292] The suggestion is that *lens* (with the stem *lent-*) is related to *lentus* ("slow").

[293] See the accounts of *Lens* [328] and *Lens palustris* [329] in the main catalogue.

[294] *videtur dici à ligando*.

[295] See the entry for *Convolvulus* above and that for *Convolvulus major*, i.e. *Calystegia sepium* [148], in the main catalogue.

[296] The Greek word λειχην has two meanings; the Latin word is *impetigo* (with plural *impetigines*).

[297] The Greek deponent verb δωρουμαι [doroumae, "I give"] is derived from δωρον [doron, "gift", "present"] and λιμος [limos] means "hunger".

[298] Ὀροβος [Orobos] is a kind of vetch grown as a crop and ἀγχειν [anchein] means "to strangle".

[299] *Herbariorum turba Rapum Genistæ appellat*: *turba* (literally "crowd") is here used derogatively. In this entry, as in his account of *Orobanche* [412] in the main catalogue (q.v.), Ray failed to realise that there are distinct species parasitising different plants. The English generic name for these, broomrape, means the same as *Rapum Genistæ* and originates from *Orobanche rapum-genistae* Thuill., which chiefly parasitises Broom, *Cytisus scoparius* [213], previously called *Genista*. *Limodorum* is now the generic name for an unrelated European (but not British) parasitic orchid, *L. abortivum* (L.) Sw.

[300] For Valerius Cordus see *Cord.* & *Cord. hist.* in Ray's 'Explanation' and Chapter 4.

[301] This word is derived from *salix*, with stem *salic-*, meaning a willow.

[302] i.e. Darnel, *Lolium temulentum* [340].

[303] The Greek noun is λαιον [laeon] and the verb ὀλειν [olein], the future infinitive of the irregular ὀλλυμι [ollymi].

[304] This suggestion comes from Martini (1623, column 1661). *Lacrima* (correctly spelled thus), meaning a tear, was in earlier Latin *dacrima*, related to the Greek δακρυ [dacry] or δακρυον [dacryon] with the same meaning; δακρυμα [dacryma] means something that is wept for.

Lychnis is from λυχνος [lychnos, "lamp"].

Lysimachia, Λυσιμαχια, took its name from King Lysimachus,[305] its discoverer, on the evidence of Pliny, or from breaking up a fight, for it is *to loose the fight*,[306] for it has the power of reconciling even the fierce dispositions of beasts & mutually hostile combats. *J.B. Lysimachus too discovered a plant that still keeps the name coming from him. Its power is so great that, if placed on the yoke of quarrelsome beasts, it restrains their ferocity.* Plin[y], *book 25, chapter 7.*[307]

Lotus seems to some learned men to be a word with an Egyptian origin, Λωτος [Lotos].[308]

M

Malva, Μαλαχη [Malache]: It is Malva for the Latins, as if Molva, because it relaxes the bowels, as Festus says, according to that familiar verse in Schola Salern[itana]:
  The ancients called it Malva because it relaxes the bowels.[309]
For the Greeks also it was Μαλαχη for the same reason, *from the softening*,[310] because it means to soften.

Marrubium, Πρασιον [Prasion]: Marrubium is a Latin word & not of Greek origin, for the second r written without aspiration is to be read everywhere.[311]

Matricaria, Παρθενιον [Parthenion], is so called because it confers on the matrix & the uterus[312] and cures women's ills.

Melampyrum, Μελαμπυρον [Melampyron], is from μελας [melas, "black"] and πυρος [pyros, "wheat"], for melampyrum is produced from an alteration of wheat so that it may become black from quince-yellow, on the evidence of Galen.[313] It is commonly

---

[305] Lysimachus (360–282/1 BC) was a close companion of Alexander the Great who ruled Thrace after Alexander's death in 323 and became king of Thrace, Asia Minor and Macedonia in 306 or 305.

[306] The Greek verb λυειν [lyein] means "to loose" and μαχη [mache] means "fight" or "battle". Compare the English name "Loosestrife" for *Lysimachia vulgaris* [346] and *Lythrum salicaria* [347, 348].

[307] Pliny's *Natural history*, XXV. 72 in the Loeb edition, again in one of a series of accounts of plants reputedly discovered by kings.

[308] Λωτος [Lotos] was used by the Ancient Greeks for five distinct plants or groups of plants, of which the Egyptian lotus was three different species of water-lilies.

[309] *Dixerunt Malvam veteres quia molliat alvum.*
This is a dactylic hexameter. Sextus Pompeius Festus was a grammarian of the mid second century AD who wrote *De verborum significatione* ("About the meaning of words") or *De interpretatione linguae latinae* ("About the interpretation of the Latin language"). *Schola Salernitana* is a mediaeval work of uncertain authorship containing metrical precepts for maintaining good health.

[310] The Greek verb is μαλασσειν [malassein].

[311] Greek words containing double rho were written in Ray's time with the first rho bearing a smooth breathing and the second a rough one (indicating an aspirate), hence transliterations containing rrh: see, for example, Ἀντιρρινον in the entry for *Antirrhinum* above. See also the entry for *Prassium* and the related footnote below.

[312] *quòd matrici & utero conferat*: the phrase is obscure because both *matrix* and *uterus* mean "womb" and there is no word indicating what it is that is conferred. Παρθενιον is derived from παρθενος [parthenos, "maiden", "virgin"], like the Parthenon (named after the virgin goddess Pallas Athene).

[313] *Ex tritici enim mutatione nascitur melampyrum, ut ex melino nigrum fiat, teste Galeno.* This rather obscure statement, taken from Bauhin & Cherler (1651), vol. 2, p. 439, may reflect a belief that crops could be transformed into weeds; for example Theophrastus says in his *Historia plantarum*

ETYMOLOGY

called Cow-wheat [Triticum vaccinium] from its seed resembling Wheat & because excellent fodder is made from it for fattening oxen & cows. *[Charles] de l'Écluse writes that in the western part of Frisia or Flanders (where he says it is very abundant) it spoils the bread & makes it darker and what is more those who eat it are generally assailed by a feeling of heaviness in the head not unlike if they had fed on darnel.*[314] Jo. B., who asserts that he had frequently seen bread coloured by it but had noticed no unpleasant taste in it & that no harm had ever come to anyone from its consumption.

Melilōtus, Μελιλωτος [Melilotos], is from μελι [meli, "honey"] & λωτος [lotos],[315] as if it was Honey Trefoil [Lotus mellea], perhaps from the colour of the flowers or from the scent of honey that it emits when dried.

Mentha, Ἡδυοσμος [Hedyosmos], is from the ancient Greek word μινθη [minthe], which the Grammarians want to be named *as compared with the curtailing,*[316] because it checks the menstruation of women. *The fables of the poets relate that Mintha, Pluto's concubine, was transformed into garden Mint by Proserpina (as it is always the way of wives to hate concubines*[317]*) as an eternal monument of a human being in an eternal plant.*[318] *The Romans, loving this word more because in it the memory of a girl badly treated because of love might be preserved, made the Greek Minthe their own Mentha with just one or two letters changed. J.B. or Marcellus in his works.*[319] For Daléchamps μινθη is so named by the Greeks *from human faeces,*[320] from the stink of dung, by ἀντιφρασις [antiphrasis][321] because it is not in the least fetid.

(II. IV. 1 in the Loeb edition) that πυρος [pyros, "wheat"] can change into αἰρα [aera, "darnel", *Lolium temulentum* 340], though he later (VIII. VIII. 3) says that, if this is not so, at least "darnel loves chiefly to appear among wheat, as does the Pontic *melampyros*". The plant described is Field Cow-wheat, *Melampyrum arvense* L., an attractive cornfield weed not recorded from Cambridgeshire in the main catalogue or the appendices. There may be some play on the similarity of the words *melinum* (μηλινον, "quince-yellow", the colour of the flowers of *M. arvense*) and *melan* (μελαν, "black", the colour of its seeds). For Galen see Chapter 4.

[314] *non secus ac si lolio victitâssent*. This almost certainly refers to *Lolium temulentum* [340] rather than *Bromus hordeaceus* [197], for which Ray gives *Lolium* as a synonym, especially since the entry for *Lolium* above gives *Aera* as a synonym.

[315] Here the Greek kind of λωτος, a leguminous plant, is meant: see the entry for *Lotus* above and the footnote to it.

[316] The Greek verb is μινυθειν [minythein].

[317] *ut solent pellices semper odisse uxores*, literally "as are accustomed concubines always to hate wives"; since both *pellices* and *uxores* could be either nominative or accusative, the Latin could be read to make either wives or concubines the subject of the sentence, but the version given above seems more likely.

[318] *æterno in æterna planta hominis monumento*. This is obscure: although *homo* (literally "man") can refer to women as well as men it is not a word one would expect to describe Mintha. The story is very briefly referred to by Ovid in his *Metamorphoses*, 10: 728–730. For Ovid (Publius Ovidius Naso) see Chapter 4.

[319] Bauhin's source is p. 356 of Marcellus Virgilius' (1529) commentary on Dioscorides' Book III, Chapter XXXVI, Περι Ἡδυοσμου ἡμερου (*De Menta satiua*), but Ray clearly took his text from Bauhin & Cherler (1651, vol. 3(2), p. 213), since he follows their word order and spelling of *Mentha*. For Marcellus, see footnote 132 above.

[320] The Greek noun is μινθος [minthos]. The information about Daléchamps is taken from Bauhin & Cherler (1651, vol. 3(2), p. 213); Ray follows their wording almost exactly. See also footnotes 180 and 182.

[321] See footnote 6 above.

Mercuriālis, the discovery of Mercury. *Plin[y], book 25, chapter 5.*[322] *Linozostis* or *Parthenium* was discovered by Mercury, so many among the Greeks call it *Hermupoa* but among us everyone calls it *Mercurialis*. Ἑρμου βοτανιον [Hermu botanion] *for Dioscorides, book 4, chapter 191.*[323]

Muscus, Βρυον [Bryon]: the Grammarians derive it from μοσχος [moschos, "calf"], that is ἁπαλος [hapalos, "tender"], τρυφερος [trypheros, "delicate"]. *Eustath[ius]*[324] declares calves to be tender and delicate.

Meum, Μειον [Meion] & Μηον [Meon]: is it from μειον [meion], that is less, because of its small & minute leaflets? *Martin[i].*[325]

Milium Solis, a barbarous name, might perhaps be called more correctly Milium Soler in imitation of the Mauritanians on the evidence of Mattioli, [Mathias] de L'Obel & C. Bauhin, since (as Serapio writes) it grows in quantity in the Soler mountains.[326] *J.B. C.B.* In English it ought to be called *Graymill*, that is Milium griseum ["grey Millet"], not, as it commonly becomes, *Grummell. Tur.* For the Greeks Λιθοσπερμον [Lithospermon],[327] Lithospermum, was so called because of the hardness & shininess of the seed.

Myosōtis is from μυος [myos] & ὠτις [otis],[328] ear-lobe of a mouse, because it has leaves similar to the ear-lobes of a mouse and also pilose on the upper side, whence it is also called Pilosella.[329]

N

NApus, Βουνιας [Bunias], is either from Rapum, because it is like Rapum in its leaves & seed, or from the Greek ναπυ [napy, "mustard"], i.e. Sinapi.[330]

Narcissus, Ναρκισσος [Narcissos]. The poets relate the origin of the flower Narcissus to the youth Narcissus. *Ovid, Metam[orphoses].*

---

[322] Pliny's *Natural history*, XXV. 38 in the Loeb edition, in a series of accounts of plants reputedly discovered by gods and heroes.

[323] The three alternative names given are all Greek: Λινοζωστις [Linozostis] suggests "flax-girdled", Παρθενιον [Parthenion] is connected with maidens or the maiden goddess Pallas Athene, and Ερμουποα [Hermupoa] means Hermes'-grass; Hermes was the Greek name for the Roman god Mercury. The final plant name means "Hermes' little grass/herb".

[324] This is presumably Eustathius of Thessalonica, a bishop and great scholar of the twelfth century who wrote commentaries on Homer and other works.

[325] As mentioned in footnote 117 above, this suggested explanation derives from the expanded (1655) edition of Martini's *Lexicon philologicum*.

[326] For Mattioli (Petrus Andreas Matthiolus) and de L'Obel (Mathias Lobelius) see *Matth.* and *Lob.* in Ray's 'Explanation' and Chapter 4. The ninth-century Joannes Serapio (Yuhanna Ibn Serapion), sometimes called Serapion the Moor, wrote about herbal medicines; his work was published at Lyon in 1525.

[327] This is derived from λιθος [lithos, "stone"] and σπερμα [sperma, "seed"]. The plant is Ray's *Lithospermum* or *Milium solis*, i.e. Common Gromwell, *Lithospermum officinale* [339].

[328] The first word is the genitive of μυς [mys, "mouse"] and the second is derived from οὐς [us, "ear"], which has the stem ὠτ- [ot-].

[329] The species referred to here is evidently not a forget-me-not but Mouse-ear Hawkweed, *Pilosella officinarum* [437]; although Ray does not include *Myosotis* among its synonyms in the main catalogue, he does give *Auricula muris minor* (i.e. Lesser Mouse-ear) as one of them.

[330] See also the entry for *Sinapi* below: σινηπι [sinepi] is the Greek name, but the Attic form of it is ναπυ [napy].

ETYMOLOGY

His body was nowhere; instead of his body a yellow flower
They find, in the middle of surrounding white leaves.[331]

On the other hand Pliny asserts that Narcissus is called after numbness,[332] not a fabled boy, as also Plutarch,[333] who relates it as dulling the nerves & creating a torpid heaviness. *Jo. B.*

Nasturtium, Καρδαμον [Cardamon],[334] is from nasus ["nose"] & tero ["I rub"], no doubt because it torments the nostrils & with the odour & pungency of its seeds provokes sneezing.

Nepĕta is from nepa Scorpius [both "scorpion"], because it acts against these & other poisonous small animals. *Martin[i]. The common people call it cattaria, a name perhaps unknown to the ancients, on the basis of a very pleasing sight, for scarcely will a cat have seen it and smelled at it when it will embrace & kiss it and, while frolicking, now approaching it, now withdrawing itself with both feet, it will rub against it for a long while & very thoroughly, and finally, while playing with amazing movements, it avidly licks the whole plant, &c.* J.B.: see in *Catal[ogus]*.[335]

Nummularia is because it has round leaves like a coin [nummus].

Nymphæa, Νυμφαια. *It seems to have been named Nymphaea because of its liking a watery place,*[336] *because it likes watery places.* Dioscor[ides]. Pliny, in book 25, chapter 7, derives it from *a Nymph who died from jealousy towards Hercules, whence, he says, some call it Heraclion, others ropalon from the root like a club, and so that those who have drunk it are deprived of intercourse and begetting for twelve days.*[337]

Nenuphar is named thus in the apothecaries' shops because the Arabs have made Nufar & Nenufar out of the Greek Νυμφαια [Nymphaea].[338]

O

OCymum, 'Ωκιμον [Ocimon] & ὠκυμον [ocymon], is named from its speed of coming up. *Theophr[astus], Diosc[orides], Plin[y]. Fuchs relates that there are*

---

[331] *Nusquam corpus erat, croceum pro corpore florem*
Inveniunt, foliis medium cingentibus albis.
These two dactylic hexameters are from Ovid's *Metamorphoses*, 3: 509–510. The "white leaves" are the outer perianth segments of Pheasant's eye Daffodil, *Narcissus poeticus* L., which according to the myth sprang up when the beautiful but vain youth Narcissus died after pining away when he saw his reflection in a forest pool and fell in love with it.

[332] *à narce*, from the Greek word ναρκη [narce]: see Pliny's *Natural history*, XXI. 128 in the Loeb edition.

[333] For Plutarch (Mestrius Plutarchus) see Chapter 4.

[334] See the entry for *Cardamine* above and the first footnote to it.

[335] Ray's text follows fairly closely that of Martini (1623, column 2203), where he attributes the observation to "Lobel. in Advers." (see *Ad. & Ad. Lob.* in Ray's 'Explanation'). However Ray omits the end of the passage, perhaps because of his own observations recorded in his second note on *Mentha cattaria*, i.e. *Nepeta cataria* [363], in the main catalogue. Martini's text continues (in our translation): "but more avidly if it has been transplanted into gardens from the fields, for then it is more tender and does not have as strong a smell."

[336] presumably like a water-nymph.

[337] Pliny's *Natural history*, XXV. 75 in the Loeb edition. The Greek word ῥοπαλον [rhopalon] means "club" and was used in particular of Heracles' (in Latin Hercules') famous club.

[338] See the entries for *Nymphæa alba* [390] and *Nymphæa lutea*, i.e. *Nuphar lutea* [391], in the main catalogue.

those who by writing z derive ozymum more expressly from ὄζω *[ozo], since the whole herb exhales a very agreeable odour. From the same odour, because it is worthy of a kingly & royal house, what was usually called Ocymum garyophyllatum is today called Basilicum.*[339] J.B. For Varro and other authors of the Latins on country matters it is not the name of some kind of herb but of a fodder, namely one made from cereals & pulses, as Varro asserts *in book 2, chapter 31. That kind of fodder,* he says, *are green crops before they bear pods.*[340] Pliny asserts the same thing *in book 18, chapter 16.*[341]

Odontītis, from ὀδους [odus], tooth, because a decoction of its stalks in sour wine, as much as the hand can hold, cures toothache so long as it is kept in the mouth. *Plin[y], book 17, chapter 12.*[342]

Oenanthe, Οινανθη,[343] is *the flower of the vine*, because it is not unlike the vine in scent & to some extent in colour, or rather because it flowers with the vine & in its flower a scent is perceived such as is found in a maturing vine. It is also called Filipendula because its roots hang like tonsils from long & slender fibres or threads;[344] for the same reason it is called *Dropwort* in English, because the bulbules or tubercles of its roots are like drops[345] hanging from long fibres or threads.

Onobrychis, Ὀνοβρυχις, is from ὀνος [onos] ass & βρυχις [brychis], as if an ass's vine-sprout;[346] others derive it from βρυχω [brycho] I gnash my teeth and say it takes its name from the braying of an ass.

Ophioscorōdon, Ὀφιοσκοροδον,[347] is so called because its head with the topmost part of the stem twists itself like a snake.

---

[339] Pliny (see next but one footnote) correctly states that Varro derived *ocinum* [*sic*] from the Greek adverb ὠκεως [oceos, "quickly"], hence "from its speed of coming up". Ὄζω means "I smell", i.e. "I have a scent". The word used here for "kingly" is *basilicus, -a, -um* (here *basilicâ*), from the Greek βασιλικος, -η, -ον, hence the plant name *Basilicum*. The adjective *garyophyllatus, -a, -um* (more correctly *caryophyllatus, -a, -um*) means "smelling of cloves", from the Greek καρυοφυλλον [caryophyllon], literally "nut-leaf", Clove-tree, *Syzygium aromaticum* (L.) Merr. & L.M. Perry (*Caryophyllus aromaticus* L.).

[340] Varro's actual words, in his *De re rustica*, book 2: 31. 4, are: *Id est ex fabali segete viride sectum, antequam genat siliquas*. ("That [i.e. ocimum] is cut green from a bean crop before it produces pods.") For Marcus Terentius Varro see Chapter 4.

[341] i.e. in his *Natural history*, XVIII. 143 in the Loeb edition, where he suggests that the growing of this fodder was a thing of the past, citing Cato, Mamilius Sura and Varro.

[342] Pliny's *Natural history*, XXVII. 108 in the Loeb edition, where he recommends the cure but does not explain the derivation of the name; book 17 is evidently an error for book 27. The stem of ὀδους is ὀδοντ- [odont-] (as in "odontology").

[343] The name is derived from οινος [oenos, "wine"] and ἀνθος [anthos, "flower"]. The species described here is Dropwort, *Filipendula vulgaris* [198], rather than any species of water-dropwort, *Oenanthe* [e.g. 134 or 393], but, doubtless because of the root tubers in both and perhaps from a superficial resemblance between them, both *Filipendula* and *Oenanthe* have been used for each of these, as Ray's synonyms in the main catalogue show, just as "dropwort" has in English.

[344] The suggestion is that the name is derived from *filum* ("thread") and *pendere* ("to hang"), from which the adjective is *pendulus, -a, -um* ("hanging down").

[345] *velut iguttæ*, clearly a misprint for *veluti guttæ*.

[346] The source of the explanation seems to be Martini (1623, columns 2411–2412). Βρυχις is known to Liddell, Scott & Jones (1940) as a synonym for κλημα ("*twig* or *branch*, esp. *vine-twig*") only on the authority of Hesychius, a lexicographer probably of the fifth century AD, who is cited by Martini. It is remarkable that he knew this word and παιπαλλω [paepallo] (see footnote 390 below), both of which were unknown to Liddell & Scott (1890).

[347] The name is derived from ὀφις [ophis, "snake"] and σκοροδον [scorodon, "garlic"].

ETYMOLOGY

Opŭlus, because it brings help to the vine,[348] for it is most particularly suitable for training vines. *Colum[ella]*,[349] *book 5*.

Orchis, Ὄρχις, that is testicle. *All herbs that have two or three bulbous roots like testicles smelling somewhat of human semen are regarded as Orchids*. Spigel[ius], *Isagoge[s]*.[350]

Origanum, Ὀρίγανον [Origanon], *from the rejoicing in the mountains*,[351] because, for the authors Dioscorides & Pliny, it rejoices in mountains.

Orobanche, *see* in Limodoron.

Osyris or Osiris (for it is written [in Greek] with both ι & υ) is an appellation from the Egyptian God, the husband of Isis, because it is efficacious against all poisons, on the authority of Bock, who asserts that it is the Osirites or Cynocephalia of Pliny.[352]

Oxălis, *see* in Acetosa.

P

PAnax, Πάναξ, as if πασιν ἀκος [pasin acos],[353] *bringing a cure to all. In fact Panax by its very name promises remedies for all diseases*. Plin[y], *book 25, chapter 4*.[354] Clown's All-heal is the name given to this plant by our Gerarde because he was first taught its extraordinary ability to heal wounds by a certain rustic mower who had wounded his leg with a scythe while mowing. See the remarkable & almost incredible story in his work *on page 1005* or *in book 2, chapter 390*.[355]

Papāver, Μηκων [Mecon], is also read as Μηκων;[356] it is so called because it is endowed with papa or pappa, that is young children's food. *Martin[i]*.[357] For the

---

[348] *quia viti fert opem*.

[349] For the Roman author Lucius Junius Moderatus Columella see Chapter 4.

[350] *Isagoges in rem herbarium* by Adrianus Spigelius (Adriaan van der Spiegel) was published in Padua in 1606; a second edition was published posthumously in Leiden in 1633.

[351] The suggested derivation is from the Greek ὄρος [oros, "mountain"] and the deponent verb γάννυμαι [gannymae, "I am glad"].

[352] For Jerome Bock (Hieronymus Tragus) see *Trag.* in Ray's 'Explanation' and Chapter 4. The reference is to Pliny's *Natural history*, XXX. 18 in the Loeb edition, part of an account of the fraudulence of magicians. *Cynocephalia* (Greek Κυνοκεφαλια) means "Dog-headed", perhaps with reference to another Egyptian god, Anubis.

[353] The dative plural of the Greek word for "all", which has neuter singular παν [pan], so "for all", and the word for "cure".

[354] Pliny's *Natural history*, XXV. 30 in the Loeb edition, where he in fact calls the plant *Panaces*.

[355] The reference is actually to Thomas Johnson's (1633, pp. 1004–1006) revised version of Gerarde's *Herball*, though the story comes from the original edition of 1597. The account of the plant, *Stachys palustris* [540] (q.v. in the main catalogue), begins with "Clownes All-heale, or the Husbandmans Wound-wort" (in honour of the "rustic mower"), for which the Latin name chosen by Gerarde and used here by Ray is *Panax Coloni*, literally "All-heal of the Farmer". "Clown" is used in its original sense of "countryman, rustic or peasant" (OED). Gerarde's story (Johnson 1633, p. 1005) is dramatic and is indeed difficult to believe in its entirety: "it chanced that a poore man in mowing of Peason did cut his leg with a sithe, wherein he made a wound to the bones"; he "tied a great quantity" of the herb "vnto the wound with a piece of his shirt, which presently stanched the bleeding, and ceased the paine"; he went on working, but healing "was fully performed in seuen dayes, that would haue required forty dayes with balsam it selfe".

[356] the second time accented on the second syllable.

[357] *Papa* or *pappa* is onomatopoeic, being "*the word with which infants call for food*" (Lewis & Short 1890). This could be simply a comparison between a baby's milk and the milky latex of Opium

Greeks it is μηκων *as compared with not waiting at table*,[358] with not waiting at table, because those eating it were not able to perform their duties. By these it is otherwise called Rhœas, from ῥέω [rheo] I flow, because, as some want it, fluid flows from it. Gaza[359] translates it as fluid, because the flower drops immediately after it has once unfolded. Papaver spumeum is so called on account of the foam like spittle which is frequently found on its branchlets.[360]

Paronychia, because it cures the abscesses at the roots of the nails called paronychias ["whitlows"].[361]

Pastināca, Σταφυλινος [Staphylinos], seems to be so called from feeding,[362] since it feeds & fattens the body abundantly.

Paralysis is so called because it is applied by those skilled in medicine to curing the disease of paralysis. It is Primula veris because it flowers at the beginning of spring.[363]

Papillaris, because it is efficacious in curing sore nipples [papillæ], that is Lampsana, & thence the English *Nipplewort*.[364]

Pedicularis, that is Crista Galli ["Cock's-crest"]: [Mathias] de L'Obel writes that this herb is called thus because it has poisonous powers, is troublesome for meadows & is an enemy to lice [pediculis hostis]. Contrariwise Dodoens[365] names it Pedicularis from its effect, because it produces a host of lice on a flock of sheep or on beasts of

---

Poppy, *Papaver somniferum* [417], but Martini's (1623, columns 2481–2482) text suggests that it reflects the former practice of calming a baby, during teething or when fevered or otherwise fractious, by feeding it this plant (cf. Allen & Hatfield 2004), since he quotes Meursius: "*Papaver, quia papæ indebatur adversus insomniam.*" ("*Papaver*, because it is put into a baby's pap to combat sleeplessness.") (The verb should be *indabatur* but is uncorrected in the 1655 edition.) Johannes van Meurs or Meursius (1579–1639), Professor of Greek History at Leyden, was a prolific Dutch scholar whose collected works (*Opera omnia*) occupied 12 volumes when they were reprinted in the 18th century; this statement is not to be found, at least under Μηκων [Mecon, "Poppy"], in his *Glossarium Graecobarbarum* (1614), cited by Martini in his catalogue of principal sources at the beginning of his book. Martini then quotes a longer passage from Tragus (Bock) "de *Papavere candido*", describing *inter alia* how the seeds can be boiled "*in lacte vel pulte*" ("in milk or pap") "*inquieti & insomniis*" ("for restlessness and sleeplessness") in infants.

[358] το μη κονειν [to me conein], literally "the not to wait at table"; the existence of the simple verb is presumed from the compound one, διακονειν [diaconein], related to διακονος [diaconos, the origin of "deacon"], but this is a misunderstanding of the origin of the compound verb.

[359] See Gaza (1529). Theodorus Gaza (*c*. 1400–*c*. 1475) was a Greek humanist who translated the works of Aristotle and Theophrastus on natural science into Latin. His translation of the latter's *Historia plantarum* or *Enquiry into plants*, published in Treviso in 1483 and Paris in 1529, was included in Johannes Bodaeus' great edition published in Amsterdam in 1644 (see *Jo. Bodæ.* in Ray's 'Explanation').

[360] This is Ray's *Behen album*, i.e. Bladder Campion, *Silene vulgaris* [80], called "Spatling Poppy" in the main catalogue. See Ray's account of "the frothy or spittle-like spume (in English called *Woodseare* & *Cuckoo-spittle*)" in his note on *Papaver spumeum* there.

[361] Παρονυχια [Paronychia] is derived from the elided prepositional prefix παρα [para, "beside"] and ὀνυξ [onyx, "nail"] with stem ὀνυχ- [onych-]. The plant is Whitlowgrass, *Erophila verna sensu lato* [419]. The name *Paronychia* is now used for another non-British genus in Caryophyllaceae.

[362] *à pascendo*.

[363] *Primula veris quia primo vere floreat*. Literally "Firstling of spring because in first spring it flowers." The name is now applied to Cowslip [467] but in Ray's time it applied equally to Primrose, *Primula vulgaris* [469], and Oxlip, *P. elatior* [468].

[364] See also the entry for *Lampsana* above and the related footnote.

[365] For Dodoens (Rembertus Dodonaeus) see *Dod.* in Ray's 'Explanation' and Chapter 4.

burden that graze in the meadows where this plant grows. To me the leaves of this herb seem extremely like a louse's back & thence perhaps comes the name.

Perforāta, Ὑπερικον [Hypericon], is so called because the side of the leaves turned towards the ground transmits both our vision and the rays of the Sun through almost innumerable pores as if pierced with a needle; and likewise it is called Demons' Flight [Fuga dæmonum], because very many people are of such an opinion that they assert that it puts Demons to flight not otherwise than it resists poison as some sort of remedy.[366]

Periclymĕnum, Περικλυμενον [Periclymenon], as if it was περικυλιομενον [pericyliomenon],[367] because it winds itself around poles & climbs.

Persicaria, Ὑδροπεπερι [Hydropeperi], was no doubt named thus because it has leaves resembling the Peach tree.[368] It obtains the name Hydropiper from its habitat & its taste, for it grows in waters or around waters & is very hot & biting like pepper.[369]

Personāta, because it had been the custom for its very large leaves to be stretched over the face like an actor's mask or face-covering.[370]

Petasītes, from the breadth of the leaves, which recall a πετασος [petasos, "broad-brimmed felt hat"] or cap.

Phalăris, Φαλαρις, pratensis[371] is so called for the reason that it has heads like a Coot[372]; moreover it was called Phalaris after the bird of the same name, which feeds on it.

Phœnix, that is Lolium rubrum,[373] is so called from the colour Tyrian purple.[374]

Phu, Φου, is from the interjection φευ [pheu, "alas"] of someone who detests it, for its root smells of something cat-like, but not without a pleasing scent of nard.[375] C. Hofman[n].

---

[366] *non aliter atque veneno alexiterium aliquod resistat*. *Alexiterium* is an adaptation of the Greek ἀλεξητηριον [alexeterion], the neuter of an adjective which, with φαρμακον [pharmacon] understood, means "remedy" or "protection", derived from the verb ἀλεξω [alexo, "I ward off"]. See also the entry for *Hypericum*, i.e. Common St John's-wort, *Hypericum perforatum* [296], and the related footnote in the main catalogue.

[367] This is the neuter of the passive present participle of περικυλιειν [pericyliein, "to wind round"].

[368] *mali Persicæ æmula*, literally "of the Persian apple-tree rivals".

[369] *Hydropiper* is a latinisation of the Greek name given earlier in the entry, which is derived from ὑδωρ [hydor, "water"], with stem ὑδρ- [hydr-], and πεπερι [peperi, "pepper"]. The plant is Ray's *Persicaria urens*, i.e. Water-pepper, *Persicaria hydropiper* [455].

[370] *quòd folia ejus prægrandia vultui larvæ aut personæ vice obtendi solita essent*. Both *larva* and *persona* were used in classical times to mean an actor's mask; see also footnote 281 above. The plant is Ray's *Bardana major*, i.e. *Arctium lappa* and/or *A. minus* [79].

[371] For Ray this is Quaking Grass, *Briza media* [266], not a species of the modern genus *Phalaris*, e.g. Canary Grass, *P. canariensis* L., or Reed-grass, *P. arundinacea* [240]: see the entry for *Gramen tremulum* in the main catalogue. However the text here does rather suggest *P. arundinacea*.

[372] *Phalaris*, the Greek φαλαρις, a Coot, *Fulica atra* Linnaeus, so called because of its bald white head, from φαλος [phalos], the peak of a Homeric helmet.

[373] i.e. Perennial Rye-grass, *Lolium perenne* [341].

[374] *à colore Phœniceo*. The Greek adjective φοινικεος, -η, -ον [phoeniceos, -e, -on] is derived from φοινιξ [phoenix], the colour Tyrian purple, which was derived from a marine mollusc, *murex*, and reputedly discovered by the Phoenicians of Tyre and Sidon (Stearn 1992, pp. 230–232 and 450).

[375] In the main catalogue Ray cites synonyms beginning with *Phu* under both *Valeriana sylvestris major* [*Valeriana officinalis* 605] and *Valeriana sylvestris minor* [*V. dioica* 606]. Nard (*Nardus*) is Spikenard, *Nardostachys grandiflora* DC. See also the entry for *Valeriana*.

Pimpinella is a modern name, perhaps from Bipennella, since it has leaves arranged on the midrib in two rows like a feather,[376] or rather in the form of a two-edged or double-sided axe. One kind of this is called sanguisorba because it checks & absorbs flows of blood.[377]

Pisum, Πισον [Pison], is from πτισσειν [ptissein] to crush in a mortar: it was the custom for the ancients to crush pulses. For others it is from Pisa, which is between [Mounts] Ossa & Olympus, on which it grows very abundantly. *P. Gassend.* However I do not remember having read anywhere about another Pisa than that which was located by the River Alpheus in Elis in the Peloponnese. Finally for others it is *from the falling*,[378] because it falls on the ground.

Plantago is from the breadth of the leaves, resembling the sole of a foot.[379] Ἀρνογλωσσον [Arnoglosson][380] is because of the likeness to a lamb's tongue.

Platanaria, that is Sparganium ramosum, because it produces prickly globules like the globules of a Plane-tree.[381] *Jo. B.*

Polytrichum, Πολυτριχον [Polytrichon], is so called from the branchlets, which are slender like hairs, & also because it causes hair to grow.[382] Adianthum, Ceterach, Phyllitis, &c.[383] are called Herbæ capillares ["hair-bearing herbs"] because they have things like hairs[384] spread along the length of the midrib, whence, from the signature,[385] they are believed to cause hair to grow.

Polemonium or Polemonia: others call it Philetæria from the rivalry of the kings about its discovery, says *Plin[y]*.[386] *For in times past kings contended to be called the discoverers of some extraordinary powers of plants & of the benefit to the human race coming from them.* Marcell[us].[387]

Polypodium, Πολυποδιον [Polypodion], is so called because there are small swellings on its root like those suckers that are seen on the arms of Octopuses.[388] These small

---

[376] *binis ordinibus pennatim sive plumatim ad mediam costam digesta.*

[377] *quòd sanguineos fluxus sistat & sorbeat.*

[378] πεσειν [pesein], the aorist infinitive of the verb πιπτειν [piptein, "to fall"].

[379] *plantam pedis æmulantium.*

[380] This is derived from ἀρνος [arnos, "lamb"] and γλωσσα [glossa, "tongue"].

[381] The plant is Ray's *Sparganium ramosum*, i.e. Branched Bur-reed, *Sparganium erectum* [553], which has spiny fruits somewhat like those of Oriental Plane, *Platanus orientalis* L.

[382] The Greek name is compounded from the adjective πολυς [polys, "many"] and the noun θριξ [thrix, "hair"] with stem τριχ- [trich-].

[383] See the two entries for *Adianthum aureum*, i.e. *Polytrichum commune* and other mosses [12, 13], and footnote 282 in the main catalogue.

[384] *quia habent quosdam quasi pilos*, literally "because they have certain so to speak hairs".

[385] For Ray's views on the Doctrine of Signatures see Chapter 3 and footnote 707 in the main catalogue.

[386] Pliny's *Natural history*, XXV. 64 in the Loeb edition, again in the series of accounts of plants reputedly discovered by kings; in this case two kings, Polemon of Pontus and Philetaerus of Cappadocia, claimed to have first found the plant. Φιλεταιρος [Philetaeros] means "Loving to comrades" and φιλεταιρια [philetaeria] is the related abstract noun meaning "love of comrades" or "friendship".

[387] This is another reference to the edition of Dioscorides by Marcellus Virgilius, but, as with the story about *Mentha* (see above), Ray's text is taken from Bauhin & Cherler (1651, vol. 3(2), p. 356), who condensed a long passage on p. 465 of Marcellus' work (1529) commenting on Dioscorides' Book IIII, Chapter IX, Περι Πολεμωνιου (*De Polemonia*). For Marcellus, see footnote 132 above.

[388] *in Polyporum cirris*. Aristotle's name for an octopus was πολυπους [polypus], a word compounded from the adjective πολυς [polys, "many"] again and the noun πους [pus, "foot"] with stem ποδ- [pod-].

swellings or suckers are the vestiges of leaves falling off annually, as Caesalpinus[389] says.

Popŭlus is perhaps from πιπαλλω [pipallo],[390] because it has trembling leaves, which applies especially to the Libyan poplar.[391] *Martin[i]*.

Portulāca, Ἀνδραχνη [Andrachne] & Ἀδραχνη [Adrachne]: Turnebus[392] thought that Portulaca should be called also Portulata, because it resembles little gates [portulæ] in its leaves; but examples from ancient writings are lacking.

Potentilla was named Argentina ["of silver"] from the remarkable power [potentia] with which it acts in breaking up a bladder or kidney stone. *Lob*.

Prassium seems, says Pena,[393] to be a leek-green colour,[394] or certainly its fetid odour to have given Marrubium its name, both that black fetid one that is called Ballote and the white & pleasantly scented one, in the works of *J.B., book 3, chapter 317*.[395]

Primula veris is so called because it flowers in early spring.[396]

Prunella, commonly Brunella, received its name from its effect, for the reason that it cures conditions of the throat or tongue or rather that evil of the military camp, in German 𝔇𝔦𝔢 𝔟𝔯𝔲𝔶𝔫𝔢, that is quinsy [angina]. *C.B*. But for Casp[ar] Hofmann it is named thus because its use is very great in the Hungarian fever[397] which they call 𝔡𝔦𝔢 𝔟𝔯𝔞𝔲𝔫𝔢.

Prunus, Κοκκυμηλεα [Coccymelea],[398] is perhaps from πυρην [pyren, "stone (of a fruit)"] because it has a large kernel. Goropius introduces something new for us from his Cimbrian victuals.[399] We (he says) call them *Pruinnen*, as if it was

---

Πολυποδιον [Polypodion] is the diminutive form of the word. The word used for "sucker", *acetabulum*, is the same as that used elsewhere for "bone-socket": see the entries for *Astragalus* and *Cotyledon* above and related footnotes. See also Pliny's *Natural history* (IX. 85 and XXVI. 58 in the Loeb edition).

[389] For Andreas Caesalpinus Aretinus see *Cæs*. in Ray's 'Explanation' and Chapter 4.

[390] The verb cited by Martini (1623, column 2794) is in fact παιπαλλω [paepallo], not πιπαλλω. This verb is known to Liddell, Scott & Jones (1940) as a synonym for σειω ("I shake") only on the authority of Hesychius (see footnote 346 above) and is unlikely to have been familiar to Ray, so πιπαλλω is probably his copying error rather than a misprint.

[391] i.e. Aspen, *Populus tremula* [457].

[392] Adrianus Turnebus or Adrien Turnèbe (1512–1565) was a French classical scholar and the tutor of Joseph Scaliger (see footnote 67 in Ray's 'Interpretation'). Ray's wording is that of Bauhin & Cherler (1651, vol. 3(2), p. 678) but the original source is Turnèbe's *Adversaria* (1564, p. 159).

[393] For Petrus Pena see *Ad. & Ad. Lob*. in Ray's 'Explanation' and Chapter 4.

[394] *prasinus viror*, literally "a leek-green greenness"; the first word is from the Greek adjective πρασινος [prasinos], which is derived from πρασον [prason, "leek"].

[395] *Prassium* is from the Greek Πρασιον [Prasion] and *Ballote* from Βαλλωτη [Ballote]. In Ray's synonomy in the main catalogue *Prassium* is *Marrubium album*, i.e. White Horehound, *Marrubium vulgare* [357], the "white & pleasantly scented" species, but *Marrubium* is used also for the fetid Black Horehound, *Ballota nigra* [77], while *Ballote* is also *Ballota nigra* but is used as well for Henbit, *Lamium amplexicaule* [25].

[396] *quòd primo vere floreat*, literally "because in the first spring it flowers". *Primula* is derived from *primus, -a, -um*, meaning "first", and *veris* is the genitive singular of *ver*, meaning "spring".

[397] *Morbus Hungaricus* (the Hungarian disease) in Ray's time was typhus; *causus* was adopted in post-classical medical Latin from the Greek καυσος [causos], defined by Liddell, Scott & Jones (1940) as "*causus*, i.e. *bilious remittent fever* (the endemic fever of the Levant)"; presumably *causus* came to be used more generally for fevers, so *in causo Hungarico* here effectively means "in treating typhus".

[398] This literally means "cuckoo-apple-tree".

[399] *Goropius è penu suo Cimbrico nobis aliquid novi importat*. Ray's comments are sarcastic. Jan Gerartsen (1518–1572), a Dutch physician and linguist from Gorp in the municipality of Hilvarenbeek,

𝔓ur-ruinnen, that is purging mulberries. Consequently the Latins ought to tolerate the name of Pruna, which is agreeable to our own language.[400] *J.B.* However he talks remarkable nonsense.[401] It is more probable that the Latin *Prunus* is made from the Greek προυνος [prunos] or προυνη [prune], a word that is to be read in the works of Theophrastus &, as is probable, means Prunus.[402]

Ptarmĭca has its name from the thing [ex re], for it provokes πταρμος [ptarmos], i.e. sneezing.

Pulegium, Βληχων [Blechon], because, when burnt, its fresh flower kills fleas with the smell.[403] Βληχων, as if it was the bleating herb,[404] because, when tasted by a flock of sheep when it is flowering, it provokes bleating from them with its acrid taste.

Pulsatilla, because the tremulous pappi of the seeds are tossed hither and thither by the lightest breath, whence it is also called herb of the wind [herba venti]. The name Pulsatilla is Italian, since this plant is commonly called thus in that country.

R

Ranunculus, Βατραχιον [Batrachion], either because like frogs [ranarum more] it delights in moist paths and shady margins or because frogs are frequently found among its clumps. It may be mentioned here in passing that it is advantageous for gardeners to foster a great army of frogs in their gardens; to be sure, leaving the vegetables untouched [vegetabilibus intactis], they search for food from the small creatures that injure plants, e.g. the smaller land snails, woodlice [aselli] & various kinds of insects, as we have frequently seen by dissecting their stomachs. Ranunculus flammeus, which is also Flammula, is so called because when chewed in the mouth it contracts & as it were sets on fire [veluti inflammat] the tongue & palate with its fiery & biting heat.

---

adopted the latinised name Johannes Goropius Becanus, despite his theory that after Noah's flood the Cimmerians of Scythia migrated to northern Europe, avoiding the confusion of languages at the Tower of Babel and becoming the Cimbrians, and thus that his native dialect Brabantic (which he called Cimbrian) was the original language of Adam and Eve (Considine 2008, pp. 142–143). See also Ray's fifth note on *Linum sativum* [336] in the main catalogue, the entry for *Herba Gerardi* above and the two footnotes to it.

[400] *Latini itaque Prunorum nomen nostrati linguæ acceptum ferre debent.*

[401] *Verùm illè egregiè ineptit.* Literally "However that man singularly talks foolishly." The adverb *egregiè* normally suggests admiration and *ille* often implies a famous person, but this phrase is probably again sarcastic. Ray is perhaps echoing the words of the great scholar Joseph Scaliger (see footnote 67 in Ray's 'Interpretation'), whose verdict on Goropius' theory is often quoted as: "Never have I read greater nonsense." Similarly Gibson's (1695) edition of Camden's *Britannia* says: "And as for Goropius's conjecture, that the *Angli* are derived from an *angle*, i.e. a *Fishing-rod*, or *Fishing-hook*, because (as he adds) *they hook all to them*, and are, as we commonly say, *good anglers*; this does not deserve so much to be credited, as laughed at."

[402] The two Greek words are usually spelled προυμνον [prumnon] (as a neuter noun) and προυμνη [prumne]. See Theophrastus' *Enquiry into plants*, IX. 1. 2 (in the Loeb edition), where, however, no clue is given to the identity of the tree other than that its sap is like tears – δακρυωδης [dacryodes].

[403] *...pulices necet odore.*

[404] *quasi balans herba*: like the Latin *balare*, the Greek deponent verb βληχασθαι [blechasthae] means "to bleat" and the noun βληχη [bleche] "bleating"; all four words are onomatopoeic. See also the entry for *Pulegium*, i.e. *Mentha pulegium* [474], in the main catalogue.

ETYMOLOGY

Raphănus vel Rhaphanus, Ῥαφανος [Rhaphanos], received the name for itself *from the appearance of a root*,[405] on the authority of Athenaeus,[406] that is because it appears readily, for it springs up on the third day after sowing. The name of Radix, as Varro maintains,[407] came to Latium from Greece, since the ancient Greeks called what they now call Rhaphanus radix.[408] *J.B.*

Rapum, Γογγυλη [Gongyle], Γογγυλις [Gongylis], likewise Ῥαφυς [Rhaphys] & ῥαπυς [rhapys]: Rapa was so named on the evidence of Varro[409] because it rushes out of the ground, i.e. ruapa.[410] Others think the name of Rapum was given to it because it is generally snatched up [rapiatur]. However it can easily be seen that that name came from the Greeks, who on the authority of Athenaeus[411] call it Rhaphys & Rhapys. *Jean Bauhin.*

Resēda was named from assuaging pain, for it disperses swellings & inflammations, as Pliny writes *in book 28, chapter 12*. Those who are treating them with it add these words: *Rocket, rock away diseases; do you know, do you know who here drove away the chicks? May the roots have neither head nor feet!* They say these words three times and spit the same number of times. *The same author in the same place.*[412]

Rogga is a Belgian name 𝕽𝖔𝖌𝖌𝖍𝖊, that is rye.[413]

Rosa, Ῥοδον [Rhodon], a Latin name from the Greek; moreover the Greek name is from ῥεω [rheo] & ὀζω [ozo], because it emits *a big flow of scent.*[414] *Plutarch, Sympos[iaca], 3.*[415] Why Dog-rose is so called can be discovered from Pliny, who, *in book 25, chapter 2*, recommends it against the bite of a rabid dog and tells a memorable story about a certain soldier in Lacetania who, having been bitten by a rabid dog and already beginning to dread waters, by the use of the root of this rose (which his mother had seen in her sleep and been told to send a message to her son that he should drink), was restored to his previous health; & afterwards unexpectedly so was anyone who tried a similar remedy.[416]

---

[405] This seems to be the meaning of this Greek phrase, which, to explain ῥαφανος, includes ῥα [rha, a root of a plant reputedly of the genus *Rheum*] and an invented verbal noun, φαινις [phaenis] with genitive φαινεως [phaeneos], derived from φαινομαι [phaenomae, "I appear"].

[406] For Athenaeus see Chapter 4.

[407] *ut autor est Varro*, literally "as the authority is Varro". Ray took this from Bauhin & Cherler (1651, vol. 2, p. 848) but the original source is Varro's *De lingua latina*, 5: 103.

[408] *Radix*, which usually meant simply "root", was sometimes used for "radish"; it is not however Greek, so perhaps the equivalent Greek word ῥιζα [rhiza] is intended here.

[409] The reference is to Varro's *De lingua latina*, 5: 108, although Ray's source may be Bauhin & Cherler (1651, vol. 2, p. 839).

[410] *... quòd ex terra rueret, q. ruapa.*

[411] For Athenaeus see Chapter 4.

[412] See Pliny's *Natural history*, XXVII. 131 in the Loeb edition; book 28 is evidently an error for book 27. The implication is that *Reseda* [*Reseda lutea* 497] is connected with the verb *sedare*, meaning "to assuage", "to soothe". The spell includes a pun on *reseda* as both the plant name in the vocative and the imperative of the verb *resedare*, which our translation attempts to replicate. The Latin original is: *Reseda morbos reseda, scisne, scisne quis hîc pullos egerit? Radices nec caput, nec pedes habeant.*

[413] *secale*, i.e. Rye, *Secale cereale* [530].

[414] The two Greek verbs mean "I flow" and "I smell", i.e. "I have a scent"; the equivalent nouns, ῥευμα [rheuma] and ὀδωδη [odode], follow.

[415] For Plutarch (Mestrius Plutarchus) see Chapter 4.

[416] Pliny's *Natural history*, XXV. 17 in the Loeb edition, where he calls the wild rose *cynorrhodon* (Greek for "dog-rose") and says that the soldier was in the praetorian guard serving in Lacetania, *Hispaniae proxima pars* ("the nearest part of Spain", i.e. the nearest to Rome).

Rubus, Βατος [Batos], either because its thickets are red [rubeant] or, as the Grammarians wish, because its blackberries when crushed give out a blood-red juice.

Rumex, Λαπαθον [Lapathon]:[417] properly the Romans called Rumex Oxalis,[418] perhaps because the sap from it is regularly sucked out by those who are thirsty rather as from a breast, which the Ancients called ruma, for the sourness of the leaves is drunk in by anyone like milk by lambs that are suckling from their mothers, unless anyone may think that the name was given from its likeness to a weapon [telum], because it is recurved in the manner of a foot like a Hunting-spear [Sparum], of which Lucilius makes mention:

> At one time Hunting-spears are carried, at another Javelins, and next a spear with a throwing-strap,

as cited by Nonius.[419] However this verse is also set forth by Festus, but he has caltrops instead of javelins.[420] *J.B.*

Ruta, Πηγανον [Peganon], is without doubt of Greek origin, if indeed ῥυτη [rhyte] is to be read in the works of Athenaeus.[421]. *C. Hofman[n].* Iolaus also, that most ancient writer of the Greeks & one whose name was heard so much even in times past, writes in his Peloponnesians that what the whole of Greece called πηγανον [peganon] was called ῥυτιον [rhytion] in the Peloponnese.[422] *J.B. They say that πηγανον is so called from its potency, for it congeals*[423] *the sperm with its dryness on account of its heat and it is wholly inimical to conception.* They say it is called πηγανον from its potency because with its dryness & heat it thickens & coagulates sperm; hence it is inimical to those bearing a womb. *Plutarch*[424] in the works of *J.B.*

---

[417] These names are applied more widely in the main catalogue than they seem to be here, where the description of their sap indicates that Sorrel, *Rumex acetosa* [7], and Sheep's Sorrel, *R. acetosella* [8], are meant. See also the entry for *Acetosa* above.

[418] In fact this is the Greek name for species of sorrel, ὀξαλις, derived from ὀξος [oxos, "vinegar"]. Today *Oxalis* is the name of an unrelated genus including Wood-sorrel, *O. acetosella* [578], which shares with the sorrels [7, 8] the sour but refreshing taste of oxalic acid.

[419] *Tum Spara, tum Rumices portantur, tragula porro.* This dactylic hexameter is a fragment of uncertain origin from the works of Caius Ennius Lucilius, a satirist of the late second century BC, quoted by Nonius Marcellus, a grammarian of the late third century AD (*citante Nonio*, literally "with Nonius citing"). *Rumex*, defined in OLD as "A kind of javelin or hunting-spear", seems to have been more or less a synonym of *sparum*, while *tragula* is defined there as "A spear fitted with a throwing-strap, used as a hunting and military weapon".

[420] *... sed pro rumices habet murices. Murex* is the shellfish from which Tyrian purple was obtained, but the word was also used for a sharp rock or stone of a similar shape to a *murex* shell and for a caltrop (OLD), which seems the most probable meaning here; see also footnotes 108 and 374 above. Martini (1623, column 3209), before quoting Lucilius' line, writes: *Rumex teli genus, simile spari Gallici, cujus meminit Lucilius:* ("*Rumex* is a kind of weapon similar to the Gallic *sparum*, of which Lucilius makes mention:") He adds "aliter *murices*" ("otherwise *murices*") and says that in a military context the Greeks call these τριβολα [tribola]. Liddell, Scott & Jones (1940) give one of the meanings of τριβολος [tribolos, *sic*] as "caltrop": see the entry on *Tribulus* below. For Sextus Pompeius Festus see footnote 309 above.

[421] For Athenaeus see Chapter 4.

[422] Liddell, Scott & Jones (1940) state that both ῥυτη [rhyte] and ῥυτον [rhyton, *sic*] mean the same as πηγανον, at least the former being a Peloponnesian word. In Greek mythology Iolaus was a nephew of Heracles (Hercules in Latin) who helped him in some of his labours; we can trace no tradition that this hero or anyone else with the same name was an author.

[423] The Greek verb is πηγνυμι [pegnymi, "I make solid", "I freeze"].

[424] For Plutarch (Mestrius Plutarchus) see Chapter 4.

ETYMOLOGY

S

SAgitta is so called because its leaves very closely resemble the point of a winged Arrow.[425]

Salix for the Latins, from springing up, Greek Ἰτεα [Itea] *from the verb to be about to go*, because it quickly grows to a height & so to speak springs up.[426] Willow is called ὠλεσικαρπος [olesicarpos] in Homer,[427] perhaps because it loses its fruit or seed very quickly[428] before it wholly experiences maturity, as envisaged by Theophr[astus], Plin[y] and Democritus? But perhaps because it resists the power of Venus and the leaves induce sterility and impede conception, as Dioscorides & Plin[y] suggest.[429] And this is more probable, for not all Willows are fruit-losing, since both Bock[430] & J.B. have observed seeds on some of them & assert that Willows are produced from seeds. And we indeed have seen very minute pappose seeds of Willows wrapped in down but have not yet investigated whether or not they are fertile.

Salicastrum, that is Dulcamara, because it grows in willow plantations. *Plin[y]*.[431]

Salvia, Ἐλελισφακος [Elelisphacos],[432] is from saving,[433] since it is salutary for many things, especially for fecundity.

Sambucus, Ἀκτη [Acte], is named from the musical instrument the Sambucus,[434] which some call pectis or magadis;[435] others prefer it to be named after an author whose name was Sambyx. Perhaps the Sambucus was first made from the branches of this tree.

---

[425] *quòd ejus folia alatæ Sagittæ cuspidem proximè referant.*

[426] The suggestion is that the Latin name is derived from *salire* ("to leap", "to spring up") and the Greek name from the irregular verb ἰεναι [ienae], which acts as a future tense for the verb "to go". The true etymology of ἰτεα is explained in footnote 89 of Ray's 'Interpretation'.

[427] The epithet occurs only once, in *Odyssey*, 10, 510, in a description of the groves of Persephone at the approach to Hades, meaning "fruit-losing", to translate which Pliny (see below) coined the Latin word *frugiperdus, -a, -um*.

[428] *ocyssimè*, a superlative adverb perhaps thought by Ray to be derived from the Greek word for "quick", ὠκυς [ocys]; but there is a normal Latin version *ocissimè*, as used, for example, in the Loeb text.

[429] See Pliny's *Natural history*, XVI. 110 in the Loeb edition, where he mentions Homer's reference and then says that a later age has interpreted its meaning in accordance with its own wickedness (*scelere suo*) because it is well known that willow seed makes women sterile.

[430] For Jerome Bock (Hieronymus Tragus) see *Trag.* in Ray's 'Explanation' and Chapter 4.

[431] See Pliny's *Natural history*, XXIII. 20 in the Loeb edition and the entry for *Solanum lignosum sive Dulcamara*, i.e. *Solanum dulcamara* [547], in the main catalogue.

[432] This probably means "whorled sage"; ελελισφακιά [elelisfakiá] is still one of the names for a wild *Salvia* bush in Modern Greek.

[433] *à salvando.*

[434] In fact the instrument was *sambuca*, from the Greek σαμβυκη [sambyce], originally a Chaldean word meaning "*a triangular musical instrument with four strings*" (Liddell, Scott & Jones 1940). It was among the instruments played as a sign that everyone should "fall down and worship the golden image that Nebuchadnezzar the king hath set up", leading to the disobedience of Shadrach, Meshach and Abednego and their being cast into the burning fiery furnace (Daniel, 3). The King James Bible calls it a sackbut, apparently because of the similarity of the word to the Chaldean one, but this is actually not a string instrument but an early trombone.

[435] These are both Greek names of instruments which differed little if at all: πηκτις was a "*stringed instrument* used by the Lydians" and μαγαδις "an instrument with twenty strings arranged in octaves", also Lydian according to some (Liddell, Scott & Jones 1940).

Sanicula seems to have obtained its name from healing wounds,[436] for it is a modern name.

Saponaria is from its potency in cleaning, by which it washes & cleanses garments like soap, & Struthium can perhaps be named from the word στρουθιζειν [struthizein],[437] that is to cleanse woollen garments & make them white. Saponaria also is a recently formed name.

Satyrium, Σατυριον [Satyrion], is named from the Satyrs on account of the heat of lust because it is an incentive to Passion more than all other bulbs. Those which have a single bulb are called Satyria and those which have twin bulbs orchids, on the authority of Dioscorides. Spigelius says in his *Isagoges*:[438] Properly Satyrium is a plant which has three roots like the roots of orchids, one that is raised erect like a penis[439] and two others lower down growing from the same point, hanging down like testacles. However C.B^9.[440] & [Mathias] de L'Obel follow Dioscorides.

Scabiosa is a modern name, perhaps ψωρα [psora] of Aetius:[441] some people derive the name from the scabrous appearance of the herb,[442] for it is hairy, but others, & better, from scabies, because it alleviates scabies; & the word Psora seems to claim this too.

Scirpus, σχοινος [schoenos], is from I plait, I bind, I intertwine,[443] for it is written also as sirpus without c.

Scordium, Σχορδιον [Scordion], is *from* σχοροδον [scorodon, "garlic"], because it assails the nostrils with the smell & pungency of garlic.

Scrophularia received its name from its knotted root & from its effect, because it is believed to be useful for scrophulae.[444] *C.B*. There are some[445] who want Scrofulæ or strumæ & chœrades,[446] tumours of the hardened glands in the neck, armpits & groins, in English *Kings evil*,[447] to be so called because they are common in pigs or because they are numerous in progeny & prolific like breeding Sows [Scrofæ]. Now Scrofa is a pig that is reared for its offspring. Scropha was also a personal

---

[436] *à sanandis vulneribus*.

[437] It seems that this verb is derived from the name of the plant rather than the other way round.

[438] *Isagoges in rem herbarium* by Adrianus Spigelius (Adriaan van der Spiegel) was published in Padua in 1606; a second edition was published posthumously in Leiden in 1633.

[439] *sursum erigitur virilis membri formâ*.

[440] As explained in Chapter 6, the superscript 9 here signifies the nominative masculine ending *-us*, making it clear that *Bauhinus*, together with *Lobelius* but unlike *Dioscoridem*, is part of the subject of the sentence.

[441] Ψωρα [Psora] is a skin disease, *scabies* in Latin. For Aëtius Amidenus see Chapter 4.

[442] *à scabro herbæ habitu*.

[443] *à sirpo, ligo, implico*: the first of these verbs is also written as *scirpo* and is used specifically of plaiting rushes; it is much more likely to have been derived from the noun than to have been the origin of it.

[444] Despite the spelling with ph this is not a Greek word and the version Scrofulæ used later is correct in Latin.

[445] *Sunt a. ...* : presumably *a*. stands for *alii*, meaning "some".

[446] *Scrofulae* are defined as "*a swelling of the glands of the neck, scrofula*" and *struma* as "*a scrofulous tumour, struma*" by Lewis & Short (1890); χοιραδες [choerades] are defined by Liddell, Scott & Jones (1940) as "*scrofulous swellings in the glands of the neck*". This last word is derived from χοιρος [choeros, "young pig", "porker"], seeming to confirm that *scrofula* is, for whatever reason, derived from *scrofa*.

[447] Scrofula was so called because it was believed to be cured by the King's touch.

name which, being by some chance given to a certain Roman, stuck to his family. Macrob[ius],[448] *Sat[urnalia]*, book 1, chapter 7: *In fact Tremellius was surnamed Scrofa through such an event. He was at his country-house with his family and children when his slaves steal and consume a sow when it has strayed from a neighbour. The neighbour summons his guards and surrounds everything so that nothing can be carried off,*[449] *and he calls upon the master of the house to restore the animal to him. Tremellius, who has found out about the affair from his bailiff, places the sow's corpse under the coverings on which his wife used to lie and allows the neighbour to make a search. When the bedroom is reached he expresses the words of an oath that there is no Sow in his country-house except her, he says, who is lying on the coverings, and he shows the couch. That very witty oath gave Tremellius the surname of Scropha.*

Secāle is from cutting,[450] because in Autumn, when other foods fail, this may be supplied to beasts of burden & oxen. See *Columel[la]*,[451] De farragine.

Selīnum, Σελινον [Selinon]: Eustath[ius][452] derives this from ἑλος [helos, "marsh"], as if it was ἑλινον [helinon],[453] because it rejoices in marshes. [Mathias] de L'Obel contends that Selinum is so called from its effect as if it was lunaticum.[454]

Senecio, Ἡριγερων [Erigeron], as if a vernal old man,[455] because it becomes hoary in spring, says Pliny,[456] or because it puts forth flowers continuously & through the whole summer, which suddenly vanish into pappi & display down. *Casp[ar] Hofman[n]*.

Serapias, a kind of orchid, is so called on the authority of Dioscorides because the root is useful for many purposes. There are those who would like it to have been named thus from Serapis, a God of the Alexandrines, on account of the shameless lasciviousness with which that God was worshipped at Canopus,[457] where he had a temple excelling both in worship and in sanctity, as Strabo relates in his *Geogr[aphica]*, *[book] 17*,[458] for the flowers of Serapiades recall fertile & lascivious insects.

Serpyllum, Ἑρπυλλος [Herpyllos]: A Latin word made from Greek by the rough breathing being changed into s, as happens in very many others, e.g. ἑξ [hex] sex ["six"], ἑπτα [hepta] septem ["seven"], ἑδος [hedos] sedes ["seat"], ἱδρως [hidros] sudor ["sweat"], ἑκυρος [hecyros] socer ["father-in-law"], ἁλς [hals] sal ["salt"],

---

[448] Ambrosius Theodosius Macrobius was a critic of the late fourth century AD (see Chapter 4); see also the entry for *Avellana nux* above.
[449] This seems a classic case of closing the stable door after the horse has bolted!
[450] *à secando*.
[451] For the Roman author Lucius Junius Moderatus Columella see Chapter 4.
[452] For Eustathius see footnote 324 above.
[453] There are a number of related words in Greek and Latin beginning with an aspirate in the former and s in the latter, as described by Ray in the later entry on *Serpyllum*.
[454] The moon, the supposed cause of madness, is σεληνη [selene] in Greek and *luna* in Latin.
[455] *quasi vernus senex*, explaining the Greek name compounded from ἐαρ [eär, "spring"], often contracted to ἠρ [er], and γερων [geron, "old man"]. See also the entry for *Erigeron* above.
[456] Pliny's *Natural history*, XXV. 167 in the Loeb edition.
[457] Canopus or Canobus was an island-town in Lower Egypt at the western mouth of the Nile which the Roman poet Juvenal also says was notorious (*famosus*) for its luxury (*Satires*, 15: 46). It was reputedly named Κανωβος [Canobos] after Menelaus' pilot who died there.
[458] Strabo (*c*. 63 BC–AD 24) was a Greek historian, geographer and philosopher most famous for his 17-volume work *Geographica*, which presented a descriptive history of the people and places known in his time; he describes Canobus in chapter 17. 1. 17 of this work.

ἄλλομαι [hallomae] salio ["I leap"], ὑπο [hypo] sub ["under"], ὑπερ [hyper] super ["above"], ὕλη [hyle] sylva ["wood"], ἠμι [hemi] semi ["half"], ἑρπω [herpo] serpo ["I creep"], ὅλκος [holcos] sulcus ["furrow"], ἑ [he] se ["self"], ὑς [hus] sus ["pig"], ὑαδες [hyades] suculæ ["piglets", hence the constellation the Hyades] *in Gellius*.[459]

Serratula is from the leaves, which are finely incised all round, & it has its name from serrati.[460] *Jo. B.*

Sertula campana, that is Melilotus officinarum. It is called Sertula because in former times garlands were regularly made from it and Campana because the most praised kind is found in Campania in Italy. *Plin[y], book 21, chapter 9.*[461]

Sideritis, Σιδηριτις, *from iron*, on account of its strength in gluing wounds together. *Ironwort has that strength that, when bound on even a recent wound of a gladiator, it shuts off the blood.* Plin[y], book 26, chapter 13.[462]

Sināpi, Σιναπι [Sinepi], *because it harms*[463] *the eyes with its smell*, because it harms the eyes with its smell, or as if it was οινον ναπυ [sinon napy], for ναπυ [napy] & σιναπι [sinepi] are the same for the Athenians;[464] but we think rather that this name is foreign, as also is χομμι [chommi] & so are πεπερι [peperi][465] & some others. But what is called *Moustarde* by the French seems to be made from it, because it is usual for Sinapi to be ground with must for sauces.[466] Camerarius[467] likes to call it Mustarda, as if it was heating the must [mustum ardens], because its seed seizes the nostrils & brain with a fiery force, evoking tears, with no preceding smell of its very pungent taste. *J.B.*

Sium, Σιον [Sion] is *from the shaking*,[468] from agitating, or because it has the power of driving out bladder and kidney stones or because *it is shaken*, that is it is agitated,

---

[459] All of these are genuinely related words except perhaps ὑαδες and *suculae*, the latter of which may have been adopted as a name for the constellation on the model of the Greek name (which some think was related to rain rather than piglets). After Homer ἑ [he] was used only in the compound words ἑαυτον [heauton] and ἑαυτην [heauten], meaning "himself" and "herself". All but two of the examples are listed by Canini (1624, p. 60), but, unlike in the case of the list in the entry for *Viscus* below, Ray does not mention this. For Aulus Gellius, author of *Noctes Atticae*, see footnote 75 in Ray's 'Interpretation' and Chapter 4; the punctuation in both Canini's and Ray's texts suggests that only the last example is taken from his work.

[460] *Serrati* were "*silver coins notched on the edge*" (Lewis & Short 1890).

[461] Pliny's *Natural history*, XXI. 53 in the Loeb edition. *Sertula* is literally "little garland", being derived from one of two verbs *serere*, this one meaning "to plait", "to interweave"; the neuter plural of its passive past participle, *serta*, means "wreaths of flowers" or "garlands". See also the entry for *Melilotus vulgaris*, i.e. *Melilotus altissimus* [361], and Ray's first note on it in the main catalogue.

[462] Pliny's *Natural history*, XXVI. 135 in the Loeb edition, where the text has "so great strength" (*tantam vim*) rather than "that strength" (*eam vim*). *Sideritis* is derived from the Greek σιδηρος [sideros, "iron"], as explained in the introductory phrase, hence the English plant name "Ironwort" – spelled "Ironwoort" by Johnson (1633) – used for various plants in Ray's time though not by Ray himself.

[463] The Greek deponent verb is σινομαι [sinomae].

[464] *Atticis enim idem est* ναπυ [napy] & σιναπι [sinepi]. The first word means the Athenians, the principal inhabitants of Attica. There is no such word as οινον and ναπυ is in fact the original Attic form of σιναπι, σινηπι and similar words (Liddell, Scott & Jones 1940).

[465] See near the end of the long list of supposedly related words in the entry on *Hyssopus* above.

[466] *quòd cum musto Sinapi ad intinctus conteri soleat*. Must (*mustum*) is unfermented wine and the sauces (*intinctus*) seem to be a kind of French mustard.

[467] For Joachimus Camerarius see *Cam. & Cam. hort.* in Ray's 'Explanation' and Chapter 4.

[468] The Greek verb is σειειν [seiein].

by the waters in which it grows flowing past, whence it is also called laver by the Latins because it is washed in flowing waters.[469]

Σμίλαξ [Smilax] is for some from σμαω [smao, "I cleanse"] I shave, whence σμιλα [smila] a lancet; but the Poets tell that a maiden of the same name was changed into this shape on account of her love for the youth Crocos. *Ovid, book 4 of Metam[orphoses]*:
And Crocos turned into little flowers together with Smilax.[470]

Solānum, Στρυχνος [Strychnos], as if it was Sulanum because it is good for pigs [suibus] against burning fevers, on the authority of Bock. *Martin[i]*:[471] very ridiculously as it seems to us.

Sonchus, Σογχος [Sonchos], seems to some to be corrupted from συμφος [symphos][472] loose, evidently because it has a loose & hollow stem, but to others to be *from the pouring safe and sound*,[473] because it pours out from itself a salubrious sap very useful for burning gripings of the stomach & bilious obstructions of the liver.[474]

Sparganium, Σπαργανιον [Sparganion] is named *from the swaddling-clothes*[475] because nurses used to use its leaves, which are broad & pliant like wickerwork, as swaddling-bands. *C.B.*

Spatŭla fœtida, Ξυρις [Xyris], was so named both because of the shape of the leaf recalling a spatula and because it emits an unpleasant smell; for the same reason they also called it ξυρις[476] as if it was a shaving knife or razor, for its leaf is rather like a two-edged razor with a sharper edge & a finer and more rigid point than Gladiolus or Iris. Its leaves smell as if recooked, as the Italians say, whence it is called Recoctaria in certain parts of Italy.[477]

Spergula is a modern name created (as is conjectured by Dodoens) from the Brabantian *Spuery*.[478]

Sphondylium, Σφονδυλιον [Sphondylion], was perhaps so named because of the unpleasantness of its smell, from a sphondyle, a kind of insect with an unpleasant

---

[469] *quòd aquis fluentibus lavetur*.

[470] *Et Crocon in parvos versum cum Smilace flores*.
This line, a dactylic hexameter, is from *Metamorphoses*, 4: 283. Ovid uses the Greek form of the youth's name.

[471] For Jerome Bock (Hieronymus Tragus) see *Trag*. in Ray's 'Explanation' and Chapter 4. Martini (1623, column 3671) provides this information, adding that the German name is **serukraut**.

[472] We have been unable to trace this word, explained by the Latin *rarus*.

[473] ἀπο του σωον χεειν [apo tu soön cheein], one of the more far-fetched explanations in the 'Etymology'.

[474] *quia salubrem succum de se fundit, ad æstuosas stomachi lancinationes, & biliosas hepatis obstructiones utilissimam*. Here *utilissimam* should be *utilissimum* to agree with *succum* ("sap").

[475] These are called σπαργανα [spargana].

[476] This name is related to ξυρον [xyron, "razor"].

[477] The plant is *Xyris* in the main catalogue, i.e. Gladdon, *Iris foetidissima* [630], also known popularly as "roast beef plant".

[478] For Dodoens (Rembertus Dodonaeus) see *Dod*. in Ray's 'Explanation' and Chapter 4; for Brabant and Brabantian the entry on *Herba Gerardi* and the related footnote above.

smell. *Pena*.[479] Marcellus[480] thinks Sphondylium is named from the shape of the vertebrae of the spine and of a phalangium of the same name,[481] since these two are similar to one another in their roundness & both have a hollow passage,[482] in which they are also like the whorl of women's spindles.

Spiræa Theophrasti, that is Viburnum, because it is flexible & weavable, from σπειρα [speira, "something wound round", "cord"]. *Martin[i]*.[483]

Stachys, Σταχυς, seems to be so called from the fact that it bears at the tops of its heads flowers as if in a spike but in distinct whorls.[484]

Staphylodendron, that is arbor vitis:[485] I cannot find the reason for this name.

Stellaria aquatica is so called because its leaves at the top recall the rays of stars.[486]

Succisa, a modern name, is because its roots are bitten off; for the same reason others superstitiously call it Devil's Bit,[487] as if he, grudging men this root, cuts it off from below with his teeth[488] and gnaws it all round. *Jo. B.* We think the root of this plant appears to have been cut or bitten off for the reason that after the space of a year the central & directly descending fibre of the root deteriorates after numerous fibres have grown out of it to the sides. Almost the same thing happens in the plantain according to [Mathias] de L'Obel & in one or two species of Hawkweed.[489] Likewise also in every kind of orchid the testicle of the previous year becomes flabby & deteriorates & each year a new one grows on it.

Symphytum, Συμφυτον [Symphyton], from συν [syn] and φυω [phyo], because it heals wounds & makes them coalesce. *So outstanding is it for curing wounds that*

---

[479] This insect (probably a beetle) and its smell (presumably a defence mechanism) are mentioned by Aristophanes in line 1077 of his comedy *Peace*. Pliny says in his *Natural history* (XXVII. 143 in the Loeb edition) that this is the only one of external animals (*animalium exterorum*) that attacks the roots of the useful plants that he has described and that it attacks them all. (The word "external" presumably excludes grubs within the roots.) For Petrus Pena see *Ad. & Ad. Lob.* in Ray's 'Explanation' and Chapter 4.

[480] This is another reference to Marcellus Virgilius' edition of Dioscorides; Ray's text is greatly condensed from a long passage on pp. 395–396 of the 1529 edition commenting on Dioscorides' Book III, Chapter LXXXI, Περι Σφονδυλιου (*De Spondylio*), much of it quoted by Bauhin & Cherler (1651, vol. 3(2), p. 159), from whom Ray almost certainly obtained the information. For Marcellus, see footnote 132 above.

[481] *Spondylus* is from the Greek σπονδυλος [spondylos], or σφονδυλος [sphondylos] in Attic Greek, meaning "vertebra". *Phalangium* is apparently the Greek φαλαγγιον [phalangion], a venomous spider.

[482] *perviam concavitatem*, literally "a passable concavity".

[483] This suggestion is made by Martini (1623, column 3723), who describes the plant as *flexilis & vitilis*; the last word normally means "interwoven" but presumably here means that the plant can be used for this purpose. The identification with *Viburnum* is not made by Martini.

[484] The original meaning of σταχυς is an ear of corn (Latin *spica*).

[485] Σταφυλοδενδρον is derived from σταφυλη [staphyle, "bunch of grapes"] and δενδρον [dendron, "tree"] and is translated into Latin by Ray as "tree vine".

[486] *stellarum radios referant*.

[487] *alii ... Morsum Diaboli superstitiosè vocant*.

[488] *dentibus eam succidat*: *succisa* (as in the next sentence) is the feminine form of the passive past participle of this verb.

[489] *Idem ferè accidit in plantagine secundum Lobelium, & in una & altera* [literally "one & the second"] *Hieracii specie*. One of the "Hawkweeds" referred to must be *Scorzoneroides autumnalis* [281], which Ray calls "Hawkweed with bitten roots, Yellow Devils-bit" in the main catalogue.

ETYMOLOGY

it cements even pieces of meat when added to them while they are being cooked, whence also the Greeks have given it its name. Plin[y], *book 27, chapter 6*.[490]

T

Tanacētum, called thus by more recent authors, perhaps from Tanacum, by which name Pliny relates that the herb Parthenium was formerly called.[491] *C.B.*

Tapsus or rather Taxus, because it resists poisons, and bearded from its dense down.[492] *Lob*.

Taraxacon is an Arabic name taken from Avicenna & Serapio & distributed into the apothecaries' shops; but in fact it owes its origin to the Greek word τρωξιμον [troximon].[493]

Taxus, Μιλος [Milos] of Theophr[astus], Σμιλος [Smilos] of Nicand[er],[494] Σμιλαξ [Smilax] of Dioscor[ides], *C.B.*, perhaps because *bows*[495] were made from it? There are those who say that poisons in which arrows are dipped, which we now call "toxic", were from this originally called "taxic". See *Plin[y], book 16, chapter 10*.[496]

Telephium, Τελεφιον [Telephion], is so called because it may help the kind of malignant & incurable ulcers through which Telephus King of Mysia lost his strength when wounded by Achilles (which for that reason are also called Telephia).[497]

Tertianaria, that is Lysimachia galericulata, because it is believed to cure tertian fever.[498]

Teucrium, for that plant because Teucer is said to have noticed its use or to have discovered it. *Plin[y], book 25, chapter 5*.[499]

Thalictrum, Θαλικτρον [Thalictron] & Θαλιητρον [Thalietron], was perhaps so called from θαλλω [thallo] I flourish, for the reason that while it is putting out sprouts it flourishes elegantly.

---

[490] Pliny's *Natural history*, XXVII. 42 in the Loeb edition, as quoted also in the entry for *Consolida* above. Συμφυτον [Symphyton] is derived from the Greek verb συμφυειν [symphyein], meaning "to grow together" (including of wounds) or "to knit" (of bones), compounded from the preposition and prefix συν [syn, "with", "together"] and the verb φυειν [phyein, "to grow"]; it is given here in the first person singular ("I grow").

[491] We can find no evidence for this statement in Pliny's *Natural history*.

[492] ... *quia toxiacis adversetur, barbatus à densa lanugine.* The second word must be a misprint for *toxicis*. *Tapsus barbatus* is Ray's chosen name for Great Mullein or Aaron's Rod, *Verbascum thapsus* [564].

[493] For Avicenna (Ibn Sīnā) see Chapter 4 and for Serapio see footnote 326 above. Τρωξιμον is the neuter of a verbal adjective from τρωγειν [trogein], meaning "to munch" (of herbivorous animals) and hence "to eat raw vegetables or fruit" (of human beings), becoming the ordinary word for eating in Modern Greek. Thus τρωξιμον means "munchable".

[494] For Nicander see footnote 1 above.

[495] τοξα [toxa], which becomes, later in the passage, the explanation for the use of the word "toxic" just as the Latin name of the tree explains "taxic".

[496] Pliny's *Natural history*, XVI. 51 in the Loeb edition.

[497] For a different tradition about the identity of the plant that healed Telephus see the entry on *Achillea* and the related footnote.

[498] See the entry on *Lysimachia galericulata*, i.e. Skullcap, *Scutellaria galericulata* [345], in the main catalogue and the related footnote about tertian malaria.

[499] Pliny's *Natural history*, XXV. 45 in the Loeb edition, again in the series of accounts of plants reputedly discovered by gods and heroes. Teucer was the son of King Telamon of Salamis, a nephew of King Priam of Troy, half-brother of Ajax and a cousin of Hector and Paris.

Tithymālus is from τιτθος [titthos] nipple & μαλος [malos] soft, because it pours out a soft milk not unlike some breast.[500] [Mathias] de L'Obel is the authority that this word ought to be written with a diphthong as Tithoimalus because it may be interpreted not unskilfully as a deadly or poisonous breast or teat of an udder.[501] *J.B.* One species of this is called characias, that is of a stockade, so called because it is suitable for fortifying stockades and barriers[502] or because it grows mainly in these places.

Tormentilla, Ἑπταφυλλον [Heptaphyllon, "Seven-leaved"], because a powder made from its root put into cavities in teeth with a particle of alum & pyrethrum[503] settles the torment [tormentum] or torture of the same. *Lob.* Others are of the opinion that it was named thus by more recent authors from curing dysenteric gripes [tormina].

Trachelium: the neck is called τραχηλος [trachelos] by the Greeks, and what is more the whole spine from the crown of the head right to the lowest bone,[504] *from the roughness* from the roughness,[505] which is the reason for the name & can suit this plant, unless it is to be called Trachelium because of the several properties for which it is recommended in conditions of the neck, whence it is also called cervicaria.[506] *J.B.*

Tragopōgon, Τραγοπωγων, received its name from the rather large grey pappus that is produced at the top of its head & which resembles a he-goat's beard. *Theophr[astus]* in the works of *J.B.* It seems to us to be so called because the flower when not yet expanded strongly recalls a he-goat's beard by its shape.

Tribŭlus, Τριβολος [Tribolos], is from τρεις βολαι [treis bolae, "three throws/strikes"], a herb armed with three-fold points;[507] for τριβολοι [triboloe, "caltrops"] are little iron machines, cunning devices made in a triangular or square shape which, on whichever part they lie, project one or more dangerous points.

---

[500] Authorities have disagreed in the past about the meaning of this word, some making it a synonym for μαλακος [malacos], meaning "soft", but others suggesting that it meant "white" or "woolly" (Liddell & Scott 1890).

[501] Τιθυμαλος [Tithymalos] is the Greek name for *Euphorbia* species: perhaps de L'Obel was trying to make something of its ending *malus* ("bad"), but why turning y into oi helps the origin of the word is unclear.

[502] *... characias dicitur, id est, vallaris, sic dictus quia vallis septisque muniendis aptus sit, ...* Χαρακιας [Characias], derived from χαραξ [charax, "*pointed stake*", "*pale*"], means "*of or fit for a stake, pale or palisade*" (Liddell, Scott & Jones 1940), explained by Ray as *vallaris*. Ray uses *characias* for Wood Spurge, *Euphorbia amygdaloides* [570]. This plant would not be much use for the purpose, but *E. tirucalli* L. is commonly used for defensive hedges in East Africa, just as Prickly Pear, *Opuntia ficus-indica* (L.) Mill., is in the Mediterranean region today, and perhaps a Mediterranean species, Tree Spurge, *E. dendroides* L., was originally meant. Today the name is the specific epithet of another species, Large Mediterranean Spurge, *E. characias* L. See also footnote 85 in the main catalogue.

[503] Πυρεθρον [Pyrethron] in Greek, probably an extract from Spanish Chamomile, *Anacyclus pyrethrum* (L.) Link (*Anthemis pyrethrum* L.).

[504] *tota spina à vertice usq; ad os sacrum.*

[505] The Greek noun is τραχυτης [trachytes] and the Latin *asperitas*.

[506] *Cervicaria* is derived from *cervix* ("neck"), with stem *cervic-*.

[507] Ray has the name as a synonym for *Carduus stellatus*, i.e. Red Star-thistle, *Centaurea calcitrapa* [112], but it probably originally referred to Maltese Cross, *Tribulus terrestris* L., a weed of southern Europe with a fearsome fruit. See also footnote 420 above.

ETYMOLOGY

Trit́icum is πυρος [pyros, "wheat"] that has been threshed from the ears & has been cleaned by the threshing. *Varro*.[508] Πυρος, i.e. ἀπυρος [apyros] for ἀπορος [aporos], *from the verb* ἀπειρειν. *Etymol[ogicum]*.[509]

Four kinds of bread [panis] were made from wheat among the ancients.
1. Siligineus panis, σιλιγνιτης [silignites],[510] from winter wheat [ex siligine], that is from the flower of the flour or the purest, softest & finest part, very thoroughly cleansed of all bran.
2. Similagineus, σεμιδαλιτης [semidalites], from fine wheat flour,[511] that is indeed from a fine part of the flour but less pure than winter wheat in as much as not all the bran has been sieved out but only the coarser part.
3. Confusaneus, αὐτοπυρος [autopyros] & συγκομιστος [syncomistos],[512] from flour containing in it every part of the corn and from which nothing, either of the flower or of the bran, has been removed.
4. Cibarius, gregarius, sordidus, ῥυπαρος [rhyparos],[513] composed of meal or bran-like flour from which the softest & purest flour has been sieved. From *Dodoens*.[514]

Trisāgo, Χαμαιδρυς [Chamaedrys],[515] i.e. drysago, from δρυς [drys], as *-ago* is a termination signalling similarity, for it has leaves like Oak, whence it is also called Chamædrys. *Mart[inius]*.

Tussilāgo, Βηχιον [Bechion],[516] from the most certain benefit that it offers to those short of breath & afflicted with a cough [tussi].

Typha, Τυφη [Typhe], is either from τιφος [tiphos] marsh, because it grows in marshes, or from the turbinate[517] shape of the flowers, because it expresses a whirlwind [turbinem] of Typhon[518] spinning himself around or the menacing stature of a giant. *C.B.*

---

[508] ... *quòd ex spicis tritum, & trituratione expurgatum sit*. *Tritus, -a, -um* is the passive past participle of the verb *tero* ("I rub"). The reference is to Varro's *De lingua latina*, 5: 106.

[509] This seems to be a third reference to *Etymologicon Magnum* (1594; see footnotes 135 and 232 above), where two alternative explanations of the Greek name are given in column 697, one with some of the same Greek words. We find it impossible to make any sense of this one, as only the first two of the last three Greek words are known to Liddell, Scott & Jones (1940) and they are derived from different roots, the first meaning "without fire" and the second "impassable", "unmanageable" or "without resources"; the verb πειρειν means "to pierce" but there is no verb ἀπειρειν.

[510] The adjective *siligineus* is derived from *siligo*, a kind of very white wheat known as "winter wheat": see footnote 786 in the main catalogue. Galen (see Chapter 4) gives the Greek adjective, derived from σιλιγνις [silignis], adding that it is not good Greek.

[511] *ex simila seu similagine*. The Greek adjective, derived from σεμιδαλις [semidalis], also describes bread made from fine wheaten flour.

[512] The Latin name *confusaneus* means "mingled"; both Greek adjectives are used of unbolted wheaten flour, the first perhaps meaning literally "of simple wheat" and the second "of the harvest", from συγκομιζειν [syncomizein, "to gather in"].

[513] The three Latin names mean respectively "relating to food", from *cibus* ("food"), and hence, from the food given to slaves, "ordinary"; "belonging to the common herd", from *grex* ("herd"); and "dirty" and hence "mean". The Greek ῥυπαρος has the same meaning as the last.

[514] For Dodoens (Rembertus Dodonaeus) see *Dod*. in Ray's 'Explanation' and Chapter 4.

[515] This is a compound of χαμαι [chamae, "on the ground"] and δρυς [drys, "oak"]. See also the entry for *Chamædrys* above.

[516] This word is related to βηξ [bex, "cough"], the stem of which is βηχ- [bech-].

[517] i.e. shaped like a *turbo*, usually a spinning-top rather than a whirlwind in this context, but the latter seems more appropriate for Typhon (from whose name "typhoon" is derived).

[518] Typhon or Typhoeus was a mythical giant who was struck with lightning by Jupiter and buried under Mount Etna.

V

Valeriāna is either from some man called Valerius[519] or rather from valour or valiantness, because it is efficacious with many powers. Whether it was known to the Ancients or not is uncertain. Apothecaries' shops use it in place of Phu.[520]

Verbascum, Φλομος [Phlomos], i.e. Barbascum, because its leaves are hairy, for it is called *Barbasso* by the Italians. Pliny derives it from verbena. Φλομος *from the word flame*, from a flame, because they use its leaves & stems in lamps for wicks.[521]

Verbēna, Ἱεροβοτανη [Hierobotane], for the Grammarians as if it was herbena, because it means any herb used in sacred rites. It is called Hierobotane or Sacred Herb because, on the authority of Dioscorides, when suspended or attached in purificatory sacrifices, its use is auspicious:[522] it was called Verbena, on the authority of Pliny, because it was employed for sweeping the table of Jupiter or for making preparations for a sacrifice. See *Plin[y], book 22, chapter 2, & book 25, chapter 9*.[523] What is more, moreover, this herb was used in sacred rites by the heathens and so they were accustomed to use its name to describe any herb used in sacred rites, as the Critics & Grammarians unanimously bear witness with Servius.[524]

Veronica, as if it was φερονικη [pheronice],[525] because it triumphs among newly discovered plants. *C. Hofman[n]*. For it is a modern name &, as we think, made from Bettonica.[526]

Viburnum is from binding or tying,[527] for among the ancients Viburnum is not the name of any particular shrub; by more recent authors it is allotted to this tree as its own name because its branches are flexible,[528] whence it is also called Lantana.

Vicia for some is from binding, for others from the vine[529] because, like the vine, it has

---

[519] *vel à Valerio nescio quo*, literally "either from a Valerius I don't know which".

[520] See also the entry for *Phu* above.

[521] The suggestion is that *Verbascum* is derived from *barba*, meaning "beard". The Italian name of the plant is now *tasso barbasso*. Pliny (*Natural history*, XXV. 108 in the Loeb edition) describes a plant like mullein (*similis verbasco herba*), called *blattaria*, which, he says, is often taken for the sacred plant called by the Romans *verbenaca*, but he does not relate the two words etymologically. The Greek word for "flame" is φλοξ [phlox, now the name of a quite different genus of plants] with genitive singular (as written in this entry) φλογος [phlogos], but, despite its apparent relevance, the word is most unlikely to be related to φλομος.

[522] *in lustrationibus suspensa alligatáve, felicem habeat usum*, literally "in lustrations suspended or tied on, it may have a happy use".

[523] Pliny's *Natural history*, XXII. 5 and XXV. 105 in the Loeb edition; in the former he describes how lowly plants were used in holy ceremonies and ambassadorial visits to other states and in the latter he uses the name *hiera botane* (the Greek ἱερα βοτανη, meaning "holy grass/herb") and again mentions the use by ambassadors, adding the sweeping of Jupiter's table and the cleansing and purifying of homes.

[524] For Servius Maurus Honoratus see footnote 133 above.

[525] The suggestion is that the name is compounded from φερω [phero, "I bear"] and νικη [nice, "victory"].

[526] This plant name is usually spelled *Betonica*, but see the entry for it above; see also *Betonica*, i.e. *Betonica officinalis* [84], in the main catalogue and note that Ray gives *Betonica Pauli* as a synonym for *Veronica serpyllifolia* [610].

[527] *à viendo seu ligando*, a phrase taken from Martini (1623, column 4033).

[528] *quia rami ejus lenti sunt*, in the belief that *Lantana* is derived from the adjective *lentus, -a, -um*. See also the entry for *Viburnum*, i.e. *Viburnum lantana* [612], in the main catalogue.

[529] *aliis à vinciendo, aliis à vitis*.

tendrils; rather it is from βικιον [bicion] because this means vetch among the inhabitants of Asia.[530]

Viola seems to be so called by making a diminutive from the Greek ἰον [ion, "violet"] with the smooth breathing converted into v.[531]

Viscus is from the Greek Ἰξος [Ixos] of the same meaning, with the smooth breathing likewise changed into v, which it also may be in very many other Latin words derived from Greek, e.g. οινος [oenos] vinum ["wine"], ἐμω [emo] vomo ["I vomit"], ἰταλος [italos] vitulus ["bull-calf"], εἰδω [eido] video ["I see"], ἐαρ [eär] ἠρ [er] ver ["spring"], οὐλω [ulo] valeo ["I am well"], οἰκος [oecos, "house"] perhaps vicus ["row of houses", "quarter"], ἰν [in] ἰνος [inos] vena ["vein"], ἰς [is] vis ["strength"], &c., which can be read in the work of Canini.[532]

Urtica is so called from burning.[533] Ἀκαλυφη [Acalyphe] *as compared with not having a good touch*[534] because it has an unpleasant touch and arouses an itch. *Athen[aeus]* in the works of *C.B.* It is also called κνιδη [cnide, "nettle"] because it plucks and pricks with its noxious hairs & irritates the feelings with its bite, from the word κνιζειν [cnizein], which means to prick & to pluck.[535]

Uvularia, that is Trachelium, because it relates to the uvula, that is to the Columella.[536]

Verātrum: Martini thinks it is named as if it was verè atrum, for it has a black root, whence also it is called μελανορριζον [melanorrhizon] by the Greeks.[537]

X

Xyris, *see* in Spatula fœtida.

---

[530] Galen (see Chapter 4) uses this word for *Vicia sativa* [56] (Liddell, Scott & Jones 1940).

[531] The derivation is correct, but the absence of v from the Greek word is because it originally began with the extinct letter known as digamma which was pronounced as w; the English "violet" is thus a double diminutive.

[532] For Angelo Canini see footnote 267 above. The derivation of *Viola* and all the examples in this entry except οἰκος are given by Canini (1624, p. 66), whose etymology is generally sound, though ἰν ἰνος (the second word being the genitive singular) should be ἰς ἰνος [is inos, "sinew", "tendon"]. Possibly Ray added "perhaps" to the derivation of οἰκος because it was his own suggestion, but it too is correct.

[533] *ab urendo*.

[534] The far-fetched suggestion seems to be that the Greek name is compounded from alpha (α) privative (see footnote 4 above), the adjective καλος [calos, "beautiful", "good"] and some derivative of the verb ἁπτειν [haptein, "to touch"], some parts of which have a phi (φ) in them.

[535] The derivation of the name is correct; the verb is also used especially in the sense of "to tease" or "to nettle". For Athenaeus see Chapter 4.

[536] *Columella* means a small post or column and was thus applied in post-classical medical Latin to the human uvula because of the shape of this organ. See also the entry for *Trachelium* above.

[537] Martini's (1623, column 4004) explanation means "truly black" and the Greek name is compounded from μελας [melas, "black"], with stem μελαν- [melan-], and ῥιζα [rhiza, "root"]. See also the entry for *Helleborus* above.

Terminorum quorundam & Vocum generaliorum interpretatio & explicatio brevis.

A

ACinus, non pro uvæ vinaceo apud idoneos autores, sed pro uvæ fructu toto accipitur, qui ex succo, & parte quæ velut caro est, & ex vinaceis, & ex ambiente cute sive folliculo constat. *The berry of a grape.* Vide *Columell.* lib. 11. c. 2. ubi hæc verba habet, *Cùm expresseris vinacea, quæ acinis celantur, eáque jam infuscata & propemodum nigra fuerint.* Extenditur & hoc nomen latiùs ad fructus minutiores arborum fruticúmve spissiùs crescentes & coactiùs, ut hederæ, &c. & sic à bacca distinguitur, quæ rariùs crescit, ut in olea, lauro, &c. *Jun.* A medicis recentioribus pro gigarto seu vinaceo accipitur, cùm dicunt *uvæ exacinatæ,* &c.

Acus eris, purgamentum frumenti, palea. *The chaffe.* Porrò *acus* Plinio dicitur, cùm per se spica pinsitur: *palea* verò, ubi cum stipula teritur in area.

Adonidis horti vocantur οἱ μετέωροι κ̀ ἄωροι κ̀ μὴ ἐῤῥιζωμένοι, quales sunt qui testis & fictilibus è fenestris ad solem exponi solent.

Ala caulis, est angulus quem cum caule vel ramus, vel folium, vel pediculus folii constituit; qui semper ferè acutus est, vel rectus saltem, quia pediculi sursum nituntur. *Jung.*

Alabastri sunt folia illa viridia herbacea quæ flores sepiunt. *Junius* exponit *alabastrum* rosæ globum nondum explicitum & patentem totum, sed conniventem.

Alica

Interpretation &
Brief explanation of certain terms
& more general words.

A

A Cinus, in the works of sound authors, is not understood as the pip [*vinaceum*] of the grape but as the whole fruit[1] of the grape, which consists of the juice [*succus*] & the part that is as it were the flesh [*caro*], of the pips & of the surrounding skin or husk [*folliculus*]. *The berry of a grape.* See *Columell[a]*,[2] book 2, chapter 2, where he has these words: *When you have pressed out the pips which are concealed in the fruit of the grapes and they have now been made dark & almost black.* This name is also extended more widely to the smaller fruits of trees and shrubs growing rather densely & crowdedly, such as of ivy [*hedera*], &c., & thus it is distinguished from a berry [*bacca*], which grows rather widely spaced, as in olive [*olea*], bay [*laurus*], &c. *Jun[ius]*.[3] By more recent physicians it is understood as a stone [*gigartum*[4]] or pip, when they speak of depipped grapes [*uvæ exacinatæ*], &c.

Acus -eris,[5] the refuse of a grain-plant [*frumentum*], chaff [*palea*]. *The chaffe.* Moreover *acus* is the word used by Pliny when the ear of corn [*spica*] is crushed by itself, but *palea* when it is threshed with the stalk [*stipula*] on a threshing-floor [*area*].

Adonidis horti:[6] οἱ μετεωροι και αωροι και μη ἐρριζωμενοι[7] are called thus, such as are those that are regularly exposed to the sun out of windows on bricks & earthen vessels.

Ala, of the stem, is the angle which either a branch or a leaf or a leaf-stalk [*pediculus folii*] makes with the stem, which is almost always acute, or at most a right-angle, because the stalks strive upwards. *Jung[ius]*.[8]

---

[1] The Latin versions of the following common terms will not be given: plant [*planta*, plural *plantæ*], tree [*arbor*, plural *arbores*], shrub [*frutex*, plural *frutices*], subshrub [*suffrutex*, plural *suffrutices*], herb [*herba*, plural *herbæ*], root [*radix*, plural *radices*], stem [*caulis*, plural *caules*], branch [*ramus*, plural *rami*], leaf [*folium*, plural *folia*], flower [*flos*, plural *flores*], fruit [*fructus*, plural *fructus* with a long second *u*], seed [*semen*, plural *semina*].
[2] Lucius Junius Moderatus Columella from Cadiz (flourished AD 50) wrote *De re rustica* (12 volumes), a comprehensive description of Roman agriculture; see also Chapter 4.
[3] Ray's references to *Jun.* relate to the *Nomenclator* of Adrianus or Hadrianus Junius whereas references to *Jung.* relate to the unpublished manuscript of Joachim Jung's *Isagoge phytoscopica* (see Chapter 3).
[4] From the Greek γιγαρτον [gigarton]: see the entry for *Vinaceum* below.
[5] i.e. *aceris* in the genitive singular.
[6] "Gardens of Adonis". These were 'gardens' of wheat, barley, lettuces, fennel and other quickly germinating plants grown in baskets and shallow pots on the roofs of the houses in ancient Athens during the Adonia, the women's festival to mourn the death of Adonis.
[7] "The raised and pendulous and not firmly rooted".
[8] See footnote 3 above.

Alabastri[9] are those green herbaceous leaves that surround the flowers. *Junius* explains *alabaster* as the ball [*globus*] of a rose that is not yet unfurled & wholly open but still closed as in sleep.

Alica or *Halica*[10] is sometimes taken to mean a particular kind of grain-plant [*frumentum*] (see *Pliny*, book 18, chapter 7, & *Cels[us]*,[11] book 3, chapter 6), but more frequently as an artificial product which may be made from Spelt [*Zea*[12]], which is called *Semen*, to which chalk [*creta*] has been mixed in, which has gone into its body and taken into it its colour & softness. This special & genuine alica is called a Roman thing by Pliny and it was not found before the times of Pompey the Great, in whose work see the method of preparing it *in book 18, chapter 11*. But *Alica*, taken broadly, differs from the Chondrus [Χονδρος] of the Greeks as a genus differs from a species, since Chondrus is cleansed, not by chalk but by gypsum & sand, like the other kind of alica mentioned by Pliny, called counterfeit, prepared in Africa from degenerating Spelt.

Amylum, that which may be prepared χωρις μυλου [choris mylu], without a millstone, for the method of preparing which see the work of C. Bauhin *in his Theat[rum]*. In fact Wheat [*Triticum*] or Winter Wheat [*Siligo*] used frequently to be steeped daily in water poured on it & then poured away until it became so soft that its pap [*crema*] could be squeezed out, and it was then dried in the sun & they called this *amylum*.

Anthera [evidently neuter plural here] of the Apothecaries' shops [*Offic.*] are the summits or little extremities [*cacuminula*] in the middle of flowers pressing upon the stamens [*stamina*], i.e. ἀνθος ῥοδιου [anthos rhodiu], that is the flower of the rose. *The chives*.

Apices [plural], the same as *Anthera*.

Arbor is that which arises from the root with a single stock [*stipes*] which is called the trunk [*caudex*] & later is divided into many larger branches, which the Greeks call ἀκρεμονας [accusative plural of ἀκρεμων, acremon] & ὀζους [accusative plural of ὀζος, ozos], then into numerous twigs [*ramusculi*], which the same people call κλαδους [accusative plural of κλαδος, clados] and the Latins *surculos* [accusative plural of *surculus*], and it is not possible for it to be easily destroyed. *Spigel[ius in] Isag[oges]*,[13] A tree.

Arilli [plural] of the Apothecaries' shops [*Offic.*]: Pips [*Vinacea*] or stones [*gigarta*[14]], i.e. somewhat dry [*ariduli*], because they are dry & provide no juice [*succus*]. *Grape stones*.

Arista, the slender and long needle sticking out from the glume [*gluma*], i.e. dry [*arida*]. *The beard or awne*.

Asparagus, the first sprout [*germen*] of herbs which is for eating. *Jun[ius]*. Or the shoot [*turio*] of any vegetable before it is unfolded into leaves: from ἀπειρος

---

[9] In ancient times *alabaster* (with plural *alabastri*, from the Greek ἀλαβαστρος) was a perfume box tapering to a point at the top; hence Pliny used the word for a rosebud.

[10] Spelt or the grits prepared from it: apparently originally the feminine of an adjective meaning "nourishing", perhaps with *farina* (= "meal") understood; so there seems no justification for the spelling *halica*.

[11] Aurelius Cornelius Celsus was a Roman physician of the first century AD. He wrote several works, of which only one remains entire, his treatise *De medicina* in eight books, one of the first medical works to be printed (in 1478).

[12] i.e. *Triticum spelta* L.

[13] For Adrianus Spigelius' work see footnote 31 in the main catalogue.

[14] Plural from the Greek γιγαρτον [gigarton]: see the entry for *Vinaceum* below.

INTERPRETATION

[apeiros].[15] *The tender sprouts or shoots of any herbe from the ground*. Hence, by Antonomasia,[16] a certain herb is so called which produces large & edible sprouts [*asparagi*].

B

B *Accæ* [plural]: small round fruits that grow sparsely on trees & shrubs and even herbs are called thus. *Berries*. & they are distinguished thus from *acini*, which are berries hanging down in a cluster.

Balaustium, the flower of the wild Pomegranate-tree.[17] *The flower of the wild Pomegranate tree*.

Bulbus, a spherical root clothed in its own membranes and sometimes stuck together with several sections [*capitula*], with one small sack [*folliculus*] surrounding them all, sometimes separated into separate sections as in cultivated garlic [*allium sativum*]. *Calep[ino]*.[18]

C

C*Alyx*, the small sack [*folliculus*] or wrapper [*involucrum*] of a flower, especially of a rose, before it splits open: sometimes the flower of the rose itself half closed & not yet expanded. But now it is most frequently accepted as the small sack [*folliculus*] in which first the flower and then the seed of herbs & the fruit of trees is covered.

Capillamenta [plural] are those very fine filaments arising in the middle of flowers, generally provided with anthers [*apices*].

Capitulum, whence plants are called capitate [*capitatæ*, feminine plural of *capitatus*], is that which consists of several flowers & stamens closely compacted into a globular or circular shape or that of a disc, as in cornflower, scabious, knapweed, thistles, &c.[19] *Jung[ius]*.

---

[15] Literally "inexperienced", a most unlikely derivation; the word is more probably related to the Greek verbs σπαργαω [spargao] and σφριγαω [sphrigao] meaning "I swell", "I am full to bursting".

[16] A figure of speech in which an epithet is used instead of a true name (Greek ἀντονομασια).

[17] *sylvestris mali Punicæ flos*, literally "the flower of the wild Punic (i.e. Carthaginian) apple-tree"; the pomegranate, *Punica granatum* L., reached Rome from the area around Carthage in North Africa. "Wild" trees must have escaped from cultivation, since the native source of this tree, cultivated since the Bronze Age, is unknown. Compare the later entry for *Cytinus*.

[18] The Augustinian monk Ambrogio Calepino or Ambrosius Calepinus (c. 1435–c. 1510) published a multilingual dictionary in Latin, expanded versions of which continued to be produced long after his death. The actual words on p. 214 of vol. 1 of the 1609 edition (which Ray owned: see Chapter 5) are *Generaliter verò apud idoneos autores bulbi sunt radices rotundæ, membranis suis tunicatæ, siue pluribus capitulis coagmententur, vno folliculo omnes ambiente, siue quadantenus distinctæ separe[n]tur: cuiusmodi sunt croci, narcissij, colocasiæ, scillæ, & arundinu[m]*. ("Generally, in fact, in the works of sound authors bulbs are spherical roots clothed in their own membranes, or they may be stuck together from several sections with one small sack surrounding them all, or they may be separated, [being] to some extent distinct; of which kind are [the bulbs] of crocus, narcissus, taro, scilla & reeds."). Of these plants only *Narcissus* and *Scilla* – as well as Garlic (*Allium sativum* L.), which Ray mentions here and Calepino refers to elsewhere in his account of *Bulbus* – have true bulbs in the modern sense, the rest having corms, tubers or rhizomes, all of stem rather than leaf tissue.

[19] *ut in cyano, scabiosa, jacea, carduis, &c*.

Capreolus for Varr[o], *clavicula* for Cic[ero] and Colum[ella], *viticula & crinis* for Plin[y], *annulus vitis* for Caelius Aurelianus.[20] A little stalk of a vine [*coliculus viteus*] twisted like a curl of hair [*cincinnus*] by which it takes hold by creeping.[21] *Jun[ius]*. Spigelius extends this appellation more widely when he writes in his *Isag[oges]*, book 1, chapter 7: *Capreoli arise from the branches and extremities of plants, especially of climbing ones, which are twisted little stalks [cauliculi] with which Vine, Pea, Bryony & other pulses*[22] *entwine themselves and entangle themselves with other plants or trunks [stipites]*. In English *Claspers or tendrells*.

Caro in Botany denotes that softer substance of succulent fruits which we otherwise call pulp [*pulpa*], such as the pulp or flesh [*caro*] of Cassia, the flesh of Plums [*Pruni*], &c. *Schrod[er]*.

Catulus in the works of more recent Botanists means the same as catkin [*julus aut nucamentum*], such as the catuli of Willow or Walnut, that is Juli.[23]

Caudex is the trunk of a tree with its foliage [*frondes*].[24] *Jun[ius]. The stemme or body of a tree.*

Caulis is that which arises singly above the ground, whence the leaves and foliage [*frondes*] or the twigs [*ramusculi*] spring forth. *Joachim Jung of Lübeck* defines a stem thus in *Isagoge Phytoscopica*, not yet published, communicated to us by that Illustrious Man, Master *Samuel Hartlib*:[25] *The stem [Caulis] is the upper part of a herb, so extended in height that the front parts do not differ from the back ones or the right-hand ones from the left-hand ones: in trees & shrubs it is called the trunk [caudex], in cereals the culm [culmus]. The stalke of any herbe.*

Cicus & ciccum, the fine membrane dividing the grains [*acini*] inside a Pomegranate:[26] hence the proverbial saying *I wouldn't give a bit of pith*;[27] that is I care not at all, likewise it's mine.[28]

Clavicula, *A Clasper or tendrell*. See Capreolus.

Comæ [plural] are the summits of plants or the foliage [*frondes*] of trees.

Conus in Botany means a fruit composed of a compacted mass [*strues*] of woody scales [*squammæ*] emulating the shape of a cone, that is ending in a point [*mucro*] from a broad & circular base, such as is the cone of Pine, Fir, &c.[29]

---

[20] Ray seems to have combed the classical Latin authors here! Marcus Terentius Varro (died 27 BC) wrote a treatise on farming, *De re rustica*; Marcus Tullius Cicero (died 43 BC) is best known for his works of oratory; for Lucius Junius Moderatus Columella see footnote 2 above; Caius Plinius Secundus ('Pliny the Elder': died AD 79) wrote his *Naturalis historia* in 37 books and is the principal source of our knowledge about Roman plant names; Caelius Aurelianus (flourished about AD 420) was a physician. See also Chapter 4. For Adrianus Spigelius' work see footnote 31 in the main catalogue.

[21] *quo locum serpendo capit*, literally "by which it takes a place by creeping".

[22] *Vitis, Pisum, Bryonia & alia legumina*.

[23] *ut catuli Salicis aut Juglandis, id est, Juli*.

[24] Compare the later entry for *Truncus*.

[25] Ray's actual text is "*Joachimus Jungius Lubecensis in* Isagoge Phytoscopica nondum editâ, nobis a Cl. Viro D. *Samuele Hartlib* communicatâ, caulem ita definit, …". See also footnote 3 above.

[26] *in malo Punico*, literally "in a Carthaginian apple"; see footnote 17 above.

[27] *ciccum non interduim*, equivalent to our "I wouldn't give a peppercorn" or "I don't care a fig". The phrase comes from Plautus' *Rudens*, 2, 7, 22; elsewhere he uses *floccum* ["a bit of wool"] *non interduim*.

[28] *hoc est, nihil curo, perinde mihi est* [literally "it is to me"].

[29] *qualis est Pini, Abietis, &c.*

INTERPRETATION

Cor or *corculum* of a seed is a small portion [*portiuncula*] of a seed from which both the root and the shoot [*germen*] arise.

Cortex, the outer as it were crust [*crusta*] of a tree: for the inner tunic [*tunica*], adhering to the wood [*lignum*] itself, is called *liber*, i.e. covering the rind [*corium*] Isid[*ore*].[30]

Corymbus signifies in general whatever is the top, *Corymbi* the extremities [*cacumina*], the summits: for the first notion of the corymb [του κορυμβου] is the knot of hair [*tutulus*] or the cone [*meta*] of the head-dress of hair of the ancient Athenians.[31] In particular, corymbs are grapes or racemes of ivy [*racemi hederæ*] crowded with berries [*acinis stipati*] & rounded into an orb: likewise the rather broad extremities of florets [*flosculi*] & seeds such as are seen in Helichrysum, Tansy [*Tanacetum*], Sweet Maudlin [*Ageratum*],[32] the female Southernwood [*Abrotanum*], &c. Indeed Pliny calls the flowers or, as he himself says, the fruits of Chrysocoma corymbs bedecked with the brightness of gold. Jean Bauhin names as corymbiferous those plants that produce a flower that is dense & crowded, being of a kind that is massed into a head of very minute stamens, & under this title he embraces Tagetes [*Flos Africanus*], Marsh Marigold [*Caltha*], Corn Marigold [*Chrysanthemum*], Helenium, Sunflower [*Flos solis*], Daisy [*Bellis*], Chamomile [*Chamæmelum*], Feverfew [*Parthenium*], Ox-eye [*Buphthalmum*], Tansy [*Tanacetum*], Sweet Maudlin [*Ageratum*],[32] Sneezewort [*Ptarmica*], Yarrow [*Millefolium*], Costmary [*Costus hortorum*], Helichrysum, Stœchas citrina,[33] Cudweed [*Gnaphalium*], Mouse-ear [*Pilosella*], Wormwood [*Absinthium*], Mugwort [*Artemisia*], Southernwood [*Abrotanum*], Rue [*Ruta*] & some other plants.[34]

Crenæ [plural] are incisions in the extreme margin of a leaf, whence it is called crenate; such as are seen in Betony [*Betonica*], Lady's-mantle [*Alchimilla*], &c.

Culmus, the stalk of a grain-plant [*frumenti calamus*] from the root to the ear [*spica*]. The blade of any corn.

Cyma -æ, & *Cyma -atis*,[35] i.e. κυημα [cyema, a variant of κυμα, cyma], that is a young foetus that is produced at first germination; it is the more delicate and tender little stalk [*cauliculus*] of the stems themselves, & it is used especially about Cabbage [*Brassica*]. However, in general, *Cymæ* are also the more tender & delicate little stalks of scapes [*scapi*], which they outstrip from the first cutting.[36]

Cytinus, the flower of the cultivated Pomegranate-tree:[37] but this is also called balaustium in the apothecaries' shops, for whom the appellation of cytinus is unknown.

---

[30] For Isidore of Seville (*c*. 560–636), see Chapter 4.
[31] In fact in *Iliad*, 9: 241 Homer used a neuter plural form of the word, κορυμβα [corymba], for the high stern of a ship.
[32] i.e. *Achillea ageratum* L., an aromatic plant of southern Europe with a sweet smell and bitter taste introduced into English gardens in 1570.
[33] i.e. *Helichrysum stoechas* (L.) Moench, Eternal Flower, Goldilocks or Shrubby Everlasting.
[34] & *alia nonuulla* (a misprint for *nonnulla*), literally "& others not none".
[35] i.e. either *cymæ* or *cymatis* (as a result of the word's Greek origin) in the genitive singular.
[36] Presumably this is a reference to meadow grasses after cutting for hay.
[37] *flos mali Punicæ sativæ*, literally "the flower of the sown Carthaginian apple-tree"; see footnote 17 above.

D

Actylus, the fruit of the Palm tree. *A Date*. From its similarity to a finger.[38] *palmula, caryotis*.

E

EChinus, the rough & spiny covering of a chestnut [*castanea*]; & in general whatever is crowded with a manifold mass of spines [*aculei*], whether it may be a covering [*tegmen*] or a head [*caput*] or an extremity [*cacumen*].

F

Fibræ [plural] are 1. as it were stamens [*stamina*] interwoven with some things, e.g. with trees, leaves, &c. [meaning obscure]. 2. They are the most minute & extreme rootlets [*radiculæ*].

Flos, flosculus. *The flower*. It is defined by *Jungius* as *the more delicate part of a plant, remarkable for its colour or shape or both, united to the rudiment of the fruit*.

Folium, as defined *by Jungius*,[39] *is that which extends from its point of attachment*[40] *in height or in length & breadth in such a way that the boundaries of the third dimension differ among themselves, that is the inner surface of the leaf from the outer*. *A leafe*.

Folliculus is the fine sheath [*theca*] which contains the grain [*granum*] or seed enveloped in it. *A cod or huske*.

Frit for Varro[41] is that which is smaller than the grain [*granum*] at the top of a mature ear of corn [*spica*].

Fructus [plural, with the second *u* long] are those larger and succulent productions of plants, receptacles of their seeds, such as apples, pears, plums, &c., derived from the verb fruor,[42] but the name of fruit is extended more widely to all the produce of the fields & is defined thus by *Jungius*: *Fructus means the annual part of a plant, connected to the flower & succeeding it, which, when it has matured, that is has come to its perfection, of its own accord breaks off from the plant &, with the ground or other suitable nurse excepted*,[43] *becomes the beginning of a new plant*. In English *Fruit*.

Fruges [plural]: Whatever of the produce of the earth results in food. Consequently those people are wrong who distinguish pulses [*legumina*] from the fruits of the earth [*fruges*].[44]

Frumenta or *Frumentacea* [plurals]: Whatever plants consist of a jointed culm [*geniculatus culmus*] & reedy leaves [*arundinacea folia*] and produce, in ears [*spicæ*] or tassels [*jubæ*], seed suitable for bread-making or pottage are called thus by

---

[38] In Greek δακτυλος [dactylos].
[39] *definiente Jungio*, literally "with Jungius defining".
[40] *à sede cui adhæret*, literally "from the seat/spot to which it adheres".
[41] For Marcus Terentius Varro see footnote 20 above.
[42] *ut poma, pyra, pruna, &c. à fruendo* [ablative singular of the gerund of *fruor*, meaning "I enjoy"].
[43] *terrâ aliáve commodâ nutrice exceptâ*, the meaning presumably being that a fruit does not mature into a new plant without this assistance.
[44] See the following entry.

INTERPRETATION

Botanists, such as Wheat, Barley, Common Millet, Italian Millet, &c.[45] For there are two principal kinds of fruges, that is frumentacea & legumina. On the other hand those little grains of figs [*ficuum grana*] are called frumenta *by Pliny*.

Frutex is that which sends up several stocks [*stipites*] from the root, each of which is afterwards divided into many branches & does not arise to the true size of a tree. *A Shrub*.

Fundus of a plant, πυθμην [pythmen],[46] is the common boundary at which the root & the upper part are united. *Jung[ius]*.

G

GEmma, in Botany, is the same as *oculus*, in English *A budde*. That which swells in trees when they begin to be conspicuous.

Geniculum is a sort of knot [*nodus*] in the stem of herbs. *A joynt or knot*.

Glans means that which consists of a crusted shell [*cortex crustosus*] containing only a single seed, of which the posterior part that adheres to the tree is covered by the calyx [*calix*], with the upper part naked. *Spigel[ius in] Isag[oges]*,[47] book 1, chapter 47.

Gluma, utriculus, the husk of a grain or the wrapper of seeds in fruits of the earth,[48] derived from the verb glubo,[49] because the grain is husked by that wrapper,[50] *The husk or chaffe of any grain*.

Granum properly means the seed of fruits of the earth [*fruges*]; hence, more generally, whatever is minute and has some likeness to the grain [*granum*] of a grain-plant [*frumentum*].

H

HErba is a kind of plant more tender and softer than a tree, a shrub or a subshrub, which puts forth leaves from its root before it produces a stem. *An herbe*. In medical practice it is used for the most part only for leaves.[51]

I

INternodium is the interval between two joints [*genicula*]. *Jun[ius]*. such as is observed in culms [*culmi*] and reeds [*arundines*]. *The space between two joynts*.

Juba, in Botany, is a loosely flowing tassel of a reedy plant [*arundinacea coma*], as there is in common millet [*milium*].

---

[45] *ut Triticum, Hordeum, Milium, Panicum*, &c. The penultimate plant is *Panicum miliaceum* L., also known as Broomcorn Millet. The last was named by Linnaeus (1753) *Panicum italicum* but is now called *Setaria italica* (L.) P. Beauv.; it is also known as Foxtail Millet.

[46] This must be the word intended, though it looks like πυθμιω [pythmio] in the text.

[47] For Adrianus Spigelius' work see footnote 31 in the main catalogue.

[48] *folliculus grani seu involucrum seminum in frugibus*. For the last phrase see the entries on *Fruges* and on *Frumenta* or *Frumentacea* above.

[49] *à glubendo* [ablative singular of the gerund of *glubo*, meaning "I bark", "I peel", "I husk"]. This is another improbable derivation.

[50] *quòd eo folliculo deglubiter granum*.

[51] *pro foliis solis*, literally "for leaves alone".

Julus, that which hangs down in Hazel, Walnut, &c.[52] like a very long worm & seems to be nothing other than a chain of threadlike flowers.[53] *A catkin, and in Willowes a palme.*

L

Achryma[54] is the fluid flowing out of plants both naturally (whether it may congeal, e.g. into oil, resin or gum,[55] or not) and when pouring forth after an incision is made. *Schrod[er]*. Spigelius[56] distinguishes lachryma from juice [*succus*] in this way, that liquid which is pressed out or extracted from a bruised herb should be called succus but that which flows either naturally or by incision lachryma.

Lacinia is the portion of a leaf enclosed between two fissures [*fissuræ*] (a fissure is a deeper cleft [*scissura*]). *Jung[ius]*.

Lanugo, in Botany, is the soft wool [*lana*] of leaves or fruits, as of quince-trees [*cydonii*], &c. *Jun[ius]*. or, as Schroder defines it, somewhat delicate fibres [*filamenta*] growing like hairs or wool on the outermost skin, e.g. in Mullein [*Verbascum*]. *Downe.*

Legumen, legumentum, whatever of the crops of the earth grows in pods [*siliquæ*], such as Bean, Pea, Chickpea and Lentils;[57] derived from the verb lego according to Varro,[58] because it may not be cut but is picked by hand by plucking. *Pulse.*

Liber, the inner tunic of a tree adhering to the wood [*lignum*]. *Jun[ius]*. Whence *liber* means as well that which is written in, out of whatever material it may be made, because, before the use of papyrus or parchment, volumes were composed from the inner bark [*libri*] of trees. *The innermost barke of a tree.*

Loculamentum is a certain separate & distinct chamber [*camera*] in seed-capsules, as can be seen in Poppy-heads.[59] *Jun[ius]*.

Locustæ, in Botany, are single husks of oats [*avenaceæ glumæ*] with their seeds hanging down on long threads [*fila*], so called because of some similarity to locusts. This name is even extended to the panicles [*paniculæ*] of certain Grasses [*Gramina*].

M

Malicorium, the rind [*cortex*] of a Pomegranate.[60]

Matrix of a tree is the same as its medulla or Heart [*Cor*].

---

[52] *in Corylo, Juglande, &c.*

[53] *florum stamineorum series.* The adjective *stamineus* might here mean "consisting of stamens" but it is uncertain whether Ray intended this.

[54] In classical Latin *lacrima*, meaning "a tear". The spelling here suggests a Greek origin, which is false.

[55] *in oleum, resinam, gummi.* The last word is an indeclinable noun adapted from a word already foreign in Greek, κομμι [commi].

[56] For Adrianus Spigelius see footnote 31 in the main catalogue.

[57] *ut Faba, Pisum, Cicer, Lentes.*

[58] *à legendo, teste Varrone*, literally "from picking, with Varro [as] witness". For Marcus Terentius Varro see footnote 20 above.

[59] *in seminum vasculis, uti videre est in capitulis Papaverum*, literally "in the capsules of seeds, as is to see in the little heads of Poppies".

[60] *Mali Punici cortex*, literally "the rind of a Carthaginian Apple". See footnote 17 above.

INTERPRETATION

Medulla, in Botany, refers to the softer and superior middle part of anything, which they call also *cor* or *matrix*.

Merges, a bundle of ears of corn [*spicæ*], namely as much as can be reaped in one sweep with a reaper's tool called *mergus*.[61]

Mutica ear of corn [*spica*], one which lacks awns [*aristæ*], as if maimed [*mutila*].

N

NErvus, in Botany, the very long and continuous filament that creeps through the leaves, as in a plantain [*plantago*], which from it is called five-nerved [*quinquenervia*, feminine of *quinquenervius*] or seven-nerved [*septinervia*].

Nucamentum, the same as *Julus*, that is something like a panicle [*panicula*] hanging down from the branches of nut-trees [*nuces*]. *A catkin or palm.*

Nucleus, for the ancients[62] *nuculeus*, that little thing which lies hidden under the shell of a nut[63], and more generally any seed included and hidden in nuts [*nuces*] or fruits. *The Kernell.*

Nux of agriculture means for authors every kind of Fruit [*Pomum*] that is covered by a tough rind [*cortex*] on the outside but inside contains what is for eating. *Macrob[ius]*.[64] These, to be sure, have considered a Fruit [*Pomum*] not to be the same as an apple [*malum*], but that the name *Pomum* includes both nuts [*nuces*] and apples [*mala*].

O

OCulus, Gemma, a swelling [*tuberculum*] on the twigs [*surculi*] of trees from which they sprout, *A bud for inoculation*. Schroder explains *oculi* as the forerunners of flowers and as unfolding into flowers.

Olus, any edible herb, the name being derived from *olla*[65] because it may come into food. *Jun[ius]*. *Holus* for Julius Scaliger is from ὅλος [holos],[66] with a clear indication of the frugality of ancient times which considered that nothing in the way of food was lacking for sustenance so long as a vegetable [*olus*] was not lacking; for this is ὅλον [holon] just as bread [*panis*] also is πᾶν [pan]. Jos[eph] Scalig[er] in *a note on Varro* derives *olus* from oil [*ab oleo*],[67] because, when taken into a garden from the wild, they become cultivated. *Pot-herbs*.

---

[61] *Merges, manipulus spicarum, quantus scil[icet] semel instrumento messorio, mergo dicto, meti potest. Merges* is a rare word meaning "sheaf", apparently found only in Virgil's *Georgics*, 2: 517. The literal meaning of *semel* is "once". We have been unable to trace *mergus* in the sense indicated by this supposed derivation of *merges*.

[62] *veteribus*, i.e. for the Romans. In fact the shortened form *nucleus* already occurred in ancient times.

[63] *sub putamine nucis* (genitive singular of *nux*, here meaning the nut itself, rather than the nut-tree as in the preceding entry).

[64] Ambrosius Theodosius Macrobius (who perhaps flourished AD 410) wrote *Saturnalia*, a dialogue in seven books covering a multitude of subjects. His definition of nuts appears at the start of Book 3, Chapter 19 (Davis 1969, p. 248).

[65] i.e. a pot or jar.

[66] "whole", as later the neuter singular form ὅλον [holon] and similarly πᾶν [pan], meaning "everything". The form *holus* is in fact more correct than *olus*, but there is no substance in these supposed derivations or in the one that follows.

[67] This is followed by *cresco* ("I grow", "I increase"), but the meaning here is obscure. Julius Caesar Scaliger (Giulio Cesare della Scala or Scaligero, 1484–1558) and his son Joseph Justus Scaliger (1540–1609) were leading classical scholars of their period. The latter, described in such extravagant

Ossiculum, the inner bone [*os*, i.e. stone] of fruits covered with the flesh [*caro*], within which is the kernel [*nucleus*], as in cherries [*cerasa*] and plums [*pruna*]. *Jun[ius]*. *The stone of a Plum, or Cherry.*

P

PAlmula, Dactylus, caryotis.[68] *A Date.*

Panicula, a woolly tassel [*lanosa coma*] from which the seed hangs down, as in Common Millet [*Milium*], Rush [*Juncus*], Grass [*Gramen*], &c. or, as *Jungius* defines it, *A loosely diffuse spike* [*spica*]*, especially with hanging stalks* [*petioli*]*, as in common millet* [*milium*].

Pappus, deciduous & wind-blown down [*lanugo*], such as is seen in Thistles [*Cardui*], Fleabane [*Erigeron*], &c. *Jun[ius]*; hence the pappus-flowered class [*classis pappi-flora*] of herbs. *Thistle-down, and the like.*

Pecten of a tree, the tract of lines which extends in length in timber [*materia*] or wood [*lignum*], of a kind in which teeth are observed in the grain [*pecten*]. *The grain of the wood.*

Pediculus is that from which a leaf or a flower or a fruit hangs. *The stalk.*

Perianthium is that which covers a flower and is also called calyx.

Pericarpium, any sheath [*theca*] or wrapper [*involucrum*] of a seed, whether that may be membranous, as in the bladders [*vesicæ*] of Elm [*Ulmus*], Bladdernut [*Staphylo-dendron*[69]], &c. or hard, as in Chestnut [*Castanea*] or fleshy, as in many other plants.

Petiolus, the same as pediculus. *A stalk.*

Pomum is a general name within which every fruit of trees suitable for eating is included, whether it may be covered with a soft skin [*corium*] or a hard one, namely a nut [*nux*] and an apple [*malum*]. We, when we call certain trees *pomiferæ*, interpret the name of *pomum* more strictly as a fruit whose flesh [*caro*] is edible and which is not enclosed in a hard shell [*putamen*] and does not contain a small bone [*ossiculum*, i.e. a stone] in it. *An Apple or Pear*, &c.

Pulpa is the flesh [*caro*] or succulent contents [*substantia*] of fruits enclosed between the skin [*cortex*] and the kernel [*nucleus*], e.g. of apples [*poma*], cherries [*cerasa*[70]] and similar fruits. Primarily it signifies the fleshy part in the body of an animal.

R

RAcemus, a whole bunch of grapes [*uva*] consisting of many berries [*acini*]. *A cluster of Grapes*. It is extended to other fruits also that are composed of numerous berries like a bunch of grapes.[71]

---

terms as "Phoenix of Europe" and "bottomless pit of erudition", published an annotated edition of the works of Marcus Terentius Varro in 1573 (reprinted in 1581). For the derivation of *olus* see Scaliger (1581, p. 173). For Varro see footnote 20 above.

[68] From the Greek καρυωτις, a kind of date shaped like a walnut.

[69] i.e. *Staphylea pinnata* L. This has inflated 2–3-celled seed-capsules that can well be described as bladders, but the fruits of *Ulmus*, though membranous, are winged achenes or samaras; see later entry in this glossary for the alternative form *samera*.

[70] The text has *cerasornm*, a misprint for the genitive plural *cerasorum*.

[71] ... *quæ* [an error for *qui*] *uvæ in modum ex pluribus acinis sunt conglobati*, literally "... which in the manner of a grape-bunch from more berries are conglobated/gathered into a ball".

INTERPRETATION

Radix, as defined by *Jungius*,[72] is the lower part of a plant, which is concealed within a more solid medium [*corpus*] that offers a foundation [*sedes*] to the Plant (whether that may be the ground, as generally, or a rock [*saxum*] or limestone [*calx*] or sand [*arena*] or wood [*lignum*] or whatever else) and which is designed to absorb its nourishment. *The root*.

Ramus is that into which a stem is divided and is like a plant's arm. *A bough*.

S

SAmera is the Seed of an Elm [*Ulmus*]. *Colum[ella]*,[73] book 1, chapter 6.

Sarmentum, a very long rod [*virga*], especially of a luxuriant vine [*vitis*] but also of any tree or shrub, derived from the verb sarpo,[74] because such rods are cut off mostly while trees are being pruned, that is cleansed. *A long trayling branch*.

Scapus, an upright stem like a column. *Jun[ius]*. For Scapus is the very body of a column between the base & the architrave, whence the English *Shaft*. Or: *Scapus is the part of a branched stem that extends from the bottom to the top as it were in one line, to which the branches adhere on both sides*. Jung[ius].

Scobs, that which is worn away from wood [*lignum*] by a saw [*ferra*]. *Sawdust*. Filings are also called thus, or that sort of flour [*farina*] that is rubbed off with a file by scraping.

Semen, in Apothecaries' shops [*Officinæ*], denotes the more minute fruits of plants, from which a similar one can grow. *Schrod[er]*.

Servia, a garland [*sertum*] made from flowers or a roll [*pastillus*] put together from flowers. *A Garland*. *Jun[ius]*.

Siliqua, a husk [*folliculus*] in which the grains [*grana*] of pulses [*legumina*] & similar plants are enclosed. *A cod or husk*.

Spadix, a branch of a Palm [*Palma*] broken off together with the fruits;[75] thence a certain colour is called spadiceus, i.e. *Bay, or Chesnut colour*.

Spica is that which the culm [*culmus*] brought forth, and it contains the grain [*granum*], the husk [*gluma*] & the awn [*arista*]. *An ear of corn*. However this name is extended more widely: so Spica is rightly defined by Jungius as that which is densely composed of flowers or stamens, so that an erect but oblong or very acute cone may result from it, as in Purple Loosestrife [*Lysimachia purpurea*[76]], Mullein [*Verbascum*], Plantain [*Plantago*], Weld [*Luteola*], Mignoncttc[77] [*Reseda*], Bistort [*Bistorta*], Rye [*Secale*]. Barley [*Hordeum*], &c.

---

[72] *definiente Jungio*, literally "with Jungius defining".
[73] For Lucius Junius Moderatus Columella see footnote 2 above.
[74] *à sarpendo* [ablative singular of the gerund of *sarpo*, meaning "I prune"].
[75] This definition is taken from *Noctes Atticae* of Aulus Gellius, a Roman author of the second century AD, who wrote (2, 26, 10) that *spadix* (i.e. σπαδιξ) was the name that the Dorians (i.e. Greeks) gave to such a branch; he also (2, 26, 9) explained the colour. His work is a collection of discussions of law, antiquities and sundry other subjects in 20 books, chiefly valuable as a storehouse of quotations from lost works.
[76] This species, which often grows in the same wet habitats as Yellow Loosestrife, *Lysimachia vulgaris* [346], is not closely related to it; its modern name is *Lythrum salicaria* [347].
[77] This is not a name used by Ray, who called *Reseda lutea* [497] "Italian Rocket" or "Common base wild Rocket".

Servia, sertum è floribus, sive pastillus è floribus compactus. *A Garland*. *Jun*.
Siliqua, folliculus cui leguminum & similium grana includuntur. *A cod or husk*.
Spadix, ramus Palmæ unà cum fructibus revulsus; inde color quidam spadiceus denominatur, viz. *Bay, or Chesnut colour*.
Spica, est quod culmus extulit, continétque granum, glumam, & aristam. *An ear of corn*. Verùm hoc nomen latiùs extenditur: unde rectè à Jungio definitur Spica, quod ex floribus vel seminibus spissè compositum est, ità ut conus erectus, sed oblongus, sive admodum acutus, inde evadat, ut in Lysimachia purpurea, Verbasco, Plantagine, Luteola, Reseda, Bistorta, Secali, Hordeo, &c.
Stamina idem sunt quod capillamenta, quæ sic describit Spigelius *Isag. l. 1. c. 6. Sunt autem stamina in floris medio, quod umbilicus appellatur, partes oblongæ, tenues veluti capillamenta, interdum paulò crassiores, paululum prominentes in summitate, apicibus frequenter præditæ, quæ stylum partem similiter oblongam, sed paulò crassiorem, atq; è centro umbilici emergentem, ambiunt.*
Stirps, Planta firmitate prædita. *Jun*.
Stolones sunt cauliculi inutiles è radicibus aut caudicis lateribus succrescentes. *Suckers from the root*.
Stylus, in Phytologia, est pars floris medium ejus occupans, & rudimento fructûs aut seminis cohærens. Dicitur stylus quia in longitudinem tenuem plerumque extenditur. *Jungius*.
Suffrutex est quod unicum stipitem illico ex radice emittit, post bina illa folia quæ cunctis ferè plantis statim è seminis pulpa prodeunt. Ità Spigelius Scaligerum secutus. Quid si ità definiamus? Planta humilior, perennis, à radice non foliata, & ab imo statim ramosa; quales sunt Satureia, Salvia, Ruta, Lavendula, &c.

Surcu-

INTERPRETATION

Stamina [plural] are the same as capillamenta, which Spigelius describes thus in *Isag[oges]*,[78] book 1, chapter 6. Moreover stamens [*stamina*] are oblong parts in the middle of the flower, which is called the navel [*umbilicus*], slender like capillamenta or sometimes a little thicker, projecting a little at the top and frequently provided with anthers [*apices*], which surround the style [*stylus*], a part that is similarly oblong but a little thicker and emerging from the centre of the navel.

Stirps, a Plant endowed with stability. *Jun[ius]*.

Stolones [plural] are useless little stems [*cauliculi*] growing up from the roots or the sides of the trunk [*caudex*]. *Suckers from the root*.

Stylus, in Botany, is the part of a flower occupying its centre & united to the rudiment of the fruit or seed. It is called the style [*stylus*] because it is generally long and slender.[79] *Jungius*.

Suffrutex is that which sends forth a single stock [*stipes*] from the root immediately after those twin leaves that in nearly all plants spring up at once from the flesh [*pulpa*] of the seed. Thus Spigelius, following Scaliger. What if we might define it thus? A more lowly perennial plant, not leafy from the root & branched at once from the base; such are Savory [*Satureia*], Sage [*Salvia*], Rue [*Ruta*], Lavender [*Lavendula*], &c.

Surculus is that which arises singly in branches, tender & slender; it is also a young twig [*ramulus*] that is split off for planting. *A cyon or graff, or a smal twig of a tree*.

T

TAlea, a cutting [*segmentum*] of a tree or a scion [*surculus*] cut through at each end so that it may be planted in the ground.

Termes, a branch broken off from a tree with the fruit. Properly (says Porphyr[ius][80]) termes is a branch of the olive [*oliva*] just as palmes is a branch of the vine [*vitis*]; Gellius,[81] however, calls a branch of the Palm [*Palma*] broken off with the fruit termes. *Calep[ino]*.[82]

Thyrsus, 1. It is that which arises in plants like a wand [*virgula*] or to the straightness of a dart [*telum*] & is sometimes confused with a spike [*spica*] in the works of Botanists. 2. A small branch [*hastula*] clothed with leaves that Bacchantes[83] used to carry.

Tomentum, short wool cut with shears [*forpices*] or derived from smoothing by a fuller from rough or refurbished garments, with which any kind of cushions are stuffed. *Jun[ius]*. Hence it is transferred to plants which are provided with a soft & woolly tassel that is prominent in their heads [*capitula*], such as Woolly-headed Thistle [*Carduus tomentosus*], Cotton-grass [*Gramen tomentosum*],[84] &c.

---

[78] For Adrianus Spigelius' work see footnote 31 in the main catalogue.

[79] *quia in longitudinem tenuem plerumque extenditur*, literally "because into slender length generally it is extended".

[80] Presumably Porphyrius of Tyre, a prolific Neoplatonic philosopher of the third century AD.

[81] For Aulus Gellius see footnote 75 above.

[82] For Ambrogio Calepino see footnote 18 above.

[83] Female drunken revellers following the god of wine, Bacchus, and brandishing what are usually called *thyrsi*.

[84] i.e. *Cirsium eriophorum* [115] and *Eriophorum angustifolium* [265].

Topiarium is a work of ornamental gardening [*topia*, plural], which obtained its name from the cords [*funiculi*] by which clipped shrubs or small trees [*arbusculæ*] are made like representations[85] of animals or things with splendid variety; tied in many places on this side and on that, they are separated, twisted and bent to represent subjects that are imitations of things & animals.

Truncus, the trunk [*caudex*] of a tree bereft of branches & leaves.[86]

Tubera [plural] in their more precise meaning concerning soils are used of[87] the swollen round roots of certain plants.

Turiones [plural], the somewhat tender extremities [*cacumina*] of trees, shrubs or herbs, as if they were called teneriones, or, as Schroder says, tenerates of the very summits of trees which grow year by year.

V

VErticillatæ plantæ are so called because the small flowers [*flosculi*] surround the stem joint by joint [*articulatim*] like spindle-whorls [*verticilli*], with leaves interspersed. Moreover whorls [*verticilli*] are certain round instruments that are employed on spindles so that they may turn more easily: in English we call them Wherles.[88]

Vimen, any pliant and flexible cutting [*virgultum*] suitable for binding things together, derived from the verb vieo,[89] or as it were a binding [*vincimen*]. *A bending wythe or twig.*

Vinaceum, the grain [*granum*] of a grape-berry [*acinus*], γιγαρτον [gigarton[90]]. *A grape stone.* See Acinus. It is also sometimes understood as the distorted remains of the pressed grape [*uva*].

Virgultum, cremium, a mass of sprouting shrubs; likewise a sprig [*germen*] the use of which is for burning.[91] *Brushwood, or small slender twigs.*

Vitilia, withies [*vimina*], properly those to which vines [*vites*] are bound.

Umbella is the name for[92] the summit [*fastigium*] curved into a circle with several longish stalks [*pediculi*], such as is seen in Fennel [*Fœniculum*], Anise [*Anisum*],[93]

---

[85] *historias*, literally "stories".

[86] Compare the earlier entry for *Caudex*.

[87] *dicuntur* (+ ablative), literally "are said by".

[88] OED defines "whorl" (earlier also "whirl") as "a small fly-wheel fixed on the spindle of a spinning-wheel to maintain or regulate the speed". Like Ray, both Turner (1551) and Lyte's edition of Dodoens (1578) compare botanical whorls to spindle-whorls, but the first example found by OED of the direct use of the English word with this meaning, which relates to rosemary having "Wharles or small slender leaves set at distances about the stalk", is in a work of 1688.

[89] *à viendo* [ablative singular of the gerund of *vieo*, meaning "I twist together", "I weave"]. This time Ray's etymology is correct: related words describing this activity and a common material for it, willow, but also chaste-tree and grape-vine, are found in many Indo-European languages including Greek (ἰτεα [itea, originally pronounced as witea], meaning "willow"), Latin (*vitex*, meaning "chaste-tree", and *vitis*, "vine"), German (*Weide*, meaning "willow") and English ("withe", "withy" and perhaps "wire"). See also the entry for *Elæagnus* in Ray's 'Etymology'.

[90] See the entries for *Acinus* and *Arilli* above.

[91] *ad cremandum*, explaining the alternative name *cremium*, a word usually found in the plural, *cremia* ("dry firewood").

[92] *vocatur*, literally "is called".

[93] i.e. *Pimpinella anisum* L.

INTERPRETATION

&c., from its likeness to a parasol or bonnet[94] with which women protect their face from the Sun. Otherwise from J. Bauhin: Umbella is as it were a prop[95] of the flower or seed, divided into several longish stalks, which arise from the same summit [*fastigium*] and immediately radiate widely,[96] and each bears a flower curved into a circle. Finally otherwise from Jungius: Umbella is the extremity of a stem arranged through many subdivisions into the shape of an inverted cone, as in Smallage [*Apium*],[97] Anise [*Anisum*],[93] Fennel [*Fœniculum*] & the remaining umbellifers.

Umbilicus, 1. that which in apples [*poma*] either projects or is concealed by the adjoining part of the stalk [*pediculus*], 2. the middle part of a flower from which the style [*stylus*] & the stamens [*stamina*] emerge.

Unguis, the whitish lowest part of a rose petal.[98]

Volva, the middle of an Apple [*Pomum*] in which the seeds are hidden as if in a purse [*loculus*] according to Scribonius.[99]

Uruncus,[100] that which, at the bottom of an ear of corn [*spica*] by the stalk [*culmus*] of the straw [*stramentum*], is smaller than a grain [*granum*].

Utriculi [plural] are the wrappers [*involucra*] & coverings [*operimenta*] of seeds in fruits of the earth [*fruges*], which, when threshed & shaken out in a riddle [*cribrum*], are called glumes [*glumæ*] or husks [*acus*].[101] It also has other meanings, for which see in the works of the Lexicographers.

---

[94] *à similitudine umbellæ seu galeri*.
[95] *veluti pedamentum*; but perhaps here "as it were a pediment (or gable-end)", this being also the primary meaning of *fastigium*.
[96] *in latius*, the syntax of which is unclear but the meaning apparent.
[97] i.e. Celery, *Apium graveolens* [53].
[98] *folii rosæ*. literally "of the leaf of a rose". *Unguis* literally means "nail".
[99] Scribonius Largus was a Roman physician of the middle of the first century AD who wrote *De compositione medicamentorum*, where this definition is to be found (104, *fin*.).
[100] The only occurrence of this term that has been traced in classical literature is in Varro's *De re rustica* (see footnote 20 above), 1, 48, 3, in the neuter form *urruncum*, supposedly derived from the Greek word οὐρα [ura], meaning "tail".
[101] See the entries for *Acus* and for *Gluma*, *utriculus*. The first of these entries declines *acus* as a third-declension neuter noun (the plural of which would be *acera*), as in the works of Pliny, Cato and Varro, but here Ray makes it a fourth-declension feminine noun (with plural *acus* with a long *u*), as does Columella (2, 10, 40); the primary meaning of this latter form is "needle".

Capita seu divisiones plantarum usitatiores.

Planta perfecta & propriè dicta communi sententia in *arborem, fruticem, suffruticem, & herbam* dividitur.

Arbores & frutices dividuntur in *semper virentes*, & quæ *foliis sunt deciduis*.

Arbores sunt vel,

1. *Pomiferæ*, ut Malus in genere sic dicta, Malus Arantia, Ma. Limonia, M. Citria, M. Punica, M. Cydonia, Pyrus, Ficus.

2. *Pruniferæ*, Per Pruniferas intelligo quarum fructus osseam nucem singularem obtinent, carne contectam esculenta: tales sunt Prunus, Malus Persica, Malus præcox seu Armeniaca, Cornus, Cerasus, Palma, Olea.

3. *Nuciferæ*, ut Juglans, Corylus, Castanea, Nux moschata, Pistacia.

4. *Bacciferæ*, ut Laurus, Morus, Juniperus, Buxus, Myrtus, Sambucus, & alia plurima.

5. *Glandiferæ*, ut Quercus, Robur, Ilex, Suber, Fagus, &c.

6. *Coniferæ*, ut Abies, Pinus, Cupressus, Larix, Cedrus, &c.

7. *Siliquosæ*, ut Laburnum, Arbor Judæ, Cassia.

8. Reliquæ his classibus non comprehensæ, quarum nonnullæ julos tantum proferunt, ut Betulus, Salix &c. aliæ in folliculis membranaceis semen inclusum habent, ut Fraxinus, Acer, Ulmus, Tilia, &c. nonnullæ fortè sunt infœcundæ, sub titulo *sylvestrium* censeri possunt.

Frutices sunt vel,

1. *Spinosi*, ut Berberis, Paliurus, Rhamnus, Grossularia, &c.

The more customary headings or divisions of plants.[1]

PLants, perfect[2] & properly so called, are by common opinion divided into *trees, shrubs, subshrubs & herbs*.[3]

Trees & shrubs are divided into *evergreens* [*semper virentes*] & those that have *deciduous leaves*.

Trees are alternatively:[4]

1. *Pome-bearing* [*Pomiferæ*],[5] such as Apple [*Malus*], so called in general, Orange [*Malus Arantia*], Lemon [*Mal[us] Limonia*], Citron [*M[alus] Citria*], Pomegranate [*M[alus] Punica*], Quince [*M[alus] Cydonia*], Pear [*Pyrus*], Fig [*Ficus*].

2. *Plum-bearing* [*Pruniferæ*]: By Pruniferæ I understand those whose fruit contains a single stony nut [*ossea nux*] covered in edible flesh [*caro esculenta*]; such are Plum [*Prunus*], Peach [*Malus Persica*], Apricot [*Malus præcox seu Armeniaca*], Dogwood [*Cornus*], Cherry [*Cerasus*], Palm [*Palma*], Olive [*Olea*].

3. *Nut-bearing* [*Nuciferæ*], such as Walnut [*Juglans*], Hazel [*Corylus*], Chestnut [*Castanea*], Nutmeg [*Nux moschata*], Pistachio [*Pistacia*].

4. *Berry-bearing* [*Bacciferæ*],[6] such as Laurel [*Laurus*], Mulberry [*Morus*], Juniper [*Juniperus*], Box [*Buxus*], Myrtle [*Myrtus*], Elder [*Sambucus*] & very many others.

5. *Acorn-bearing* [*Glandiferæ*],[7] such as Oaks [*Quercus*],[8] Common Oak [*Robur*], Holm Oak [*Ilex*], Cork Oak [*Suber*], Beech [*Fagus*], &c.

6. *Coniferous* [*Coniferæ*],[9] such as Fir [*Abies*], Pine [*Pinus*], Cypress [*Cupressus*], Larch [*Larix*], Cedar [*Cedrus*], &c.

7. *Podded* [*Siliquosæ*],[10] such as Laburnum, Judas-tree [*Arbor Judæ*],[11] Senna [*Cassia*].[12]

8. The remainder not included in these classes, of which some bear only catkins [*juli*] such as Hornbeam [*Betulus*], Willow [*Salix*], &c., others have their seed

---

[1] *Capita seu divisiones plantarum usitatiores*. For the chosen plants we have tried, as far as possible, to give the English names which Ray himself used or with which he would have been familiar, but we have used modern spelling for them (e.g. "Woodruff" rather than "Woodroof"). In this translation, once again the Latin versions of the terms listed in footnote 1 in 'Interpretation' will not be given.

[2] The meaning of this term is uncertain, but Ray's examples exclude fungi, lichens and perhaps mosses.

[3] The Latin here is in fact in the singular throughout.

[4] *vel*, literally "either", but there are no further corresponding conjunctions.

[5] i.e. bearing fruits without stones: see the entry for *Pomum* in Ray's 'Interpretation'.

[6] See the entry for *Baccæ* in Ray's 'Interpretation'.

[7] i.e. bearing fruits with a crusted shell containing only a single seed: see the entry for *Glans* in Ray's 'Interpretation'.

[8] We have interpreted *Quercus* here as plural, despite the absence of a circumflex to indicate a long *u*, since the three following examples are all species of *Quercus*; compare *Campanulæ* in division 10 of *Herbæ*.

[9] i.e. bearing fruits composed of a compacted mass of woody scales: see the entry for *Conus* in Ray's 'Interpretation'.

[10] i.e. bearing pods containing the seeds of pulses and similar plants: see the entry for *Siliqua* in Ray's 'Interpretation'.

[11] i.e. *Cercis siliquastrum* L., a tree covered with beautiful purple flowers in spring. There is a long-standing myth that Judas Iscariot hanged himself from this tree, but it has been suggested that the name is corrupted from the French *arbre de Judée*, meaning tree of Judea and referring to a region where it occurs wild; if this is so, it must have happened before Ray's time.

[12] i.e. *Cassia acutifolia* Delile, an important medicinal plant grown for its purgative leaves and pods and first cultivated in England in about 1640.

enclosed in membranous sacks [*folliculi membranacei*], such as Ash [*Fraxinus*], Maple [*Acer*], Elm [*Ulmus*], Lime [*Tilia*], &c., and some are perhaps infertile, can be classified under the title of woodland trees [*sylvestres*].

Shrubs are alternatively:

1. *Thorny* [*Spinosi*], such as Barberry [*Berberis*], Christ's-thorn [*Paliurus*],[13] Buckthorn [*Rhamnus*], Gooseberry [*Grossularia*], &c.

2. *Not thorny* [*Non spinosi*], such as Broom [*Genista*], Black Alder [*Frangula*],[14] Jasmine [*Jasminum*], Privet [*Ligustrum*], Chaste-tree [*Vitex*].[15]

Shrubs can also be divided otherwise into *flowering* [*floriferi*], *fruiting* [*fructiferi*], *climbing* [*scandentes*], &c.

Of subshrubs there is one family and that not numerous. Moreover they are nearly all scented, such as Hyssop [*Hyssopus*], Lavender [*Lavendula*], Savory [*Satureia*], Sage [*Salvia*], Poley [*Polium*],[16] Southernwood [*Abrotanum*], French Lavender [*Stœchas*],[17] &c.

Herbs are so numerous that it is certainly difficult & even next to impossible to divide them into fixed classes that include all of them so that no plant may belong to several different classes or later have to be retracted. We merely propose the more customary headings.

Herbs are alternatively:

1. *Bulbous* [*Bulbosæ*],[18] such as Lily [*Lilium*], Narcissus, Tulip [*Tulipa*], Hyacinth [*Hyacinthus*], Crocus, Garlic [*Allium*], &c.

2. *Tuberous* [*Tuberosæ*],[19] such as Figwort [*Scrophularia*], Potato [*Battata*],[20] Asphodel [*Asphodelus*], &c.

3. *Umbelliferous* [*Umbelliferæ*],[21] such as Fennel [*Fœniculum*], Anise [*Anethum*],[22] Parsnip [*Pastinaca*], &c.

4. *Whorled* [*Verticillatæ*],[23] such as White Horehound [*Marrubium*],[24] Motherwort [*Cardiaca*],[25] some Mints [*Menthæ quædam*], Water-Horehound [*Marrubium aquaticum*],[26] &c.

---

[13] i.e. *Paliurus aculeatus* Lam. (*P. spina-christi* Mill.), also called "Jerusalem thorn" and "Crown-of-thorns", traditionally believed to be the shrub from which came the thorns with which Christ was crowned before his crucifixion.

[14] i.e. *Frangula alnus* [B4], now known as "Alder Buckthorn".

[15] i.e. *Vitex agnus-castus* L., a Mediterranean shrub, the stems of which have long been used in basket-making and the seeds of which were supposed to suppress sexual desire (hence "Chaste-tree" and "Monk's Pepper" as popular names).

[16] Most probably *Teucrium polium* L., a Mediterranean subshrub long valued for its medical properties.

[17] i.e. *Lavandula stoechas* L. (*Stoechas officinarum* Mill.), another Mediterranean subshrub, yielding an essential oil and having aromatic leaves and flowers that are used in pot-pourri, as an insect repellent, etc.

[18] See the entry for *Bulbus* in Ray's 'Interpretation' and the latter part of the related footnote; here all the plants listed have true bulbs in the modern sense except *Crocus*, which has a corm.

[19] See the entry for *Tubera* in Ray's 'Interpretation'.

[20] i.e. *Ipomoea batatas* (L.) Lam., now called "Sweet Potato".

[21] See the entry for *Umbella* in Ray's 'Interpretation'.

[22] i.e. *Pimpinella anisum* L.

[23] See the entry for *Verticillatæ plantæ* in Ray's 'Interpretation', the only one of these divisions listed there as such.

[24] i.e. *Marrubium vulgare* [357].

[25] i.e. *Leonurus cardiaca* [B16].

[26] i.e. *Lycopus europaeus* [358].

HEADINGS OR DIVISIONS OF PLANTS

5. *Spiked* [*Spicatæ*],[27] such as Loosestrife [*Lysimachia*], Speedwell [*Veronica*], &c., which can almost all be referred to verticillatæ.

6. *Climbing* [*Scandentes*], which either intertwine themselves with other plants, such as those more specifically called Hops [*Lupuli*] & Bindweeds [*Convolvuli*], or attach themselves to them with tendrils [*claviculi*]. They are also either *pome-bearing* [*pomiferæ*], such as Melon [*Melo*], Pompion [*Pepo*], Cucumber [*Cucumis*], Gourd [*Cucurbita*], Citrull [*Citrullus*], &c., or *leguminous* [*leguminosæ*], such as Pease [*Pisum*], Vetch [*Ervum*], Lentil [*Lens*], Pease Everlasting [*Lathyrus*], &c.[28]

7. *Corymb-bearing* [*Corymbiferæ*],[29] such as Sweet Maudlin [*Ageratum*],[30] Tansy [*Tanacetum*], &c. or, as J. Bauhin says, Daisy [*Bellis*], Marigold [*Calendula*], &c.[31] Others take the name of *corymb* in other ways. For Jungius *Corymbus* is the extremity of the stem so subdivided and laden with flowers or fruits that a spherical shape comes about, as in Climbing or Berried Ivy [*Hedera arborea*],[32] Marsh or Water Elder [*Sambucus aquatica*],[33] Onion [*Cepa*], Leek [*Porrum*], &c.

8. *Pappus-bearing* [*Pappiferæ*], of which the flower is transformed into a pappus, such as Dandelion [*Dens leonis*], Groundsel [*Senecio*], Goat's-beard [*Tragopogon*], &c.

9. *Capitate* [*Capitatæ*],[34] such as Knapweed [*Jacea*], Scabious [*Scabiosa*], Silver Knapweed [*Stæbe*],[35] Saw-wort [*Serratula*], &c.

10. *Bell-shaped* [*Campaniformes*], that is those of which the small flowers [*flosculi*], which are rounded like bells, end in a point [*mucro*] or a top-shaped cone, such as Bell-flowers [*Campanulæ*], Foxglove [*Digitalis*], Rampion [*Rapunculus*], Coventry-bells [*Viola mariana*].[36] Most of these are also climbing [*scandentes*].

11. *Garland-making* [*Coronariæ*], which are sought after for their flower and serve for making up garlands[37] or crowns, such as Sweet-William [*Armerius*], Pink [*Caryophyllus*], &c.

12. *Round-leaved* [*Rotundifoliæ*], such as Asarabacca [*Asarum*], Sowbread [*Cyclamen*], &c.

---

[27] See the latter part of the entry for *Spica* in Ray's 'Interpretation'. In his two examples here Ray is probably referring to *Lysimachia purpurea*, Purple Loosestrife, now *Lythrum salicaria* [347], and to *Veronica officinalis* [609], which he regards as the common species of *Veronica* and calls simply "Speedwell or Fluellin" in the main catalogue.

[28] See footnote 5 above for *Pomiferæ* and the entry for *Legumen* in Ray's 'Interpretation'. Pompions were species of *Cucurbita* including *C. pepo* L., now called pumpkins or vegetable marrows, while Ray's *Cucurbita* was probably a gourd; Citrull is most likely to have been Water Melon, *Citrullus vulgaris* Schrad. *Pisum* is Garden Pea, *Pisum sativum* [444], *Lens* is *Lens culinaris* Medik. (*Ervum lens* L.) and *Ervum* is a kind of vetch, supposedly *Vicia ervilia* (L.) Willd. (*Ervum ervilia* L.), Bitter Vetch or Bastard Lentil; the last plant is presumably *Lathyrus sylvestris* [325].

[29] See the entry for *Corymbus* in Ray's 'Interpretation'.

[30] i.e. *Achillea ageratum* L., an aromatic plant of southern Europe with a sweet smell and bitter taste introduced into English gardens in 1570.

[31] Interestingly, one of the two plants mentioned here, *Calendula*, does not appear in Ray's 'Interpretation' in the long list of 22 plants regarded as corymbiferous by J. Bauhin.

[32] i.e. climbing and fertile *Hedera helix* [270].

[33] i.e. *Viburnum opulus* [518], now called "Guelder Rose".

[34] See the entry for *Capitulum* in Ray's 'Interpretation'.

[35] i.e. *Centaurea stoebe* L., a European species grown by Gerarde (1597) in his garden and no doubt by others too; it has become a noxious weed in the United States.

[36] The last two plants are *Campanula rapunculus* L., now called "Rampion Bellflower", and *C. medium* L., now called "Canterbury-bells".

[37] *serviis*, which must be a misprint for *sertis*, dative of *serta*, the neuter plural of the past participle passive of *sero* ("I entwine"), so meaning "wreaths", "garlands".

13. *Vein-leaved* [*Nervifoliæ*],[38] such as Plantain [*Plantago*], Wild White Hellebore [*Helleborine*],[39] &c.

14. *Star-like* [*Stellatæ*], the leaves of which surround the stem like a radiating star, such as Woodruff [*Asperula*], Bedstraw [*Gallium*], Madder [*Rubia*], Goosegrass [*Aparine*], &c.

15. *Cereals* [*Cerealia*], *fruits of the earth* [*fruges*]; they include *grain-plants* [*frumentacea*] and *Pulses* [*Legumina*].[40]

16. *Succulents* [*Succulentæ*], such as Stonecrop [*Sedum*], Orpine or Livelong [*Telephium*], &c.

17. *Grass-leaved* [*Graminifoliæ*], such as Grain-plants [*Frumenta*], Grasses [*Gramina*], Reeds [*Arundines*], Rushes [*Junci*], &c.

19. *Vegetables* [*Oleraceæ*],[41] such as Succory or Endive [*Endivia*], Spinach [*Spinachia*], Beet [*Beta*], Cabbage [*Brassica*], &c.

20. *Aquatics* [*Aquaticæ*], and they are of two kinds,[42] for they grow around waters, like Arsmart [*Persicaria*], Brooklime [*Anagallis aquat[ica]*] &c.,[43] or they always float in the waters themselves, like Water-Sengreen or Freshwater Soldier [*Stratiotes*],[44] White Water-lily [*Nymphæa*], Water-Milfoil [*Millefolium aquaticum*],[45] &c.

21. *Marine* [*Marinæ*], such as Grass-wrack [*Alga*], Sea-wrack [*Fucus*], Sea Navelwort [*Androsace*],[46] &c.

22. *Rock-dwelling* [*Saxatiles*], to which belong maidenhairs [*capillares*], such as Fern [*Filix*], Maidenhair [*Adianthum*], Hart's-tongue [*Phyllitis*], Spleenwort [*Asplenium*], &c.[47]

Finally in general a Herb is alternatively:

1. Remaining [*Restibilis*], named from remaining in the ground,[48] of which the root is perennial, that is it lives for several years, although the stems and leaves perish every year.

2. Perennial [*Perennis*], of which the stem or stock [*stipes*] or even the leaves last for some years.

3. Annual [*Annua*], which does not survive beyond the year, whether it may be sown in spring or in Autumn.

Plants might be divided in various other ways as well, e.g. with regard to their roots, stems, flowers, seeds, leaves, &c., which it is not part of our plan to pursue further for the present.

FINIS.

---

[38] See the entry for *Nervus* in Ray's 'Interpretation'.
[39] i.e. orchids of the genera *Epipactis* and *Cephalanthera*.
[40] See the entries for *Fruges*, *Frumenta* or *Frumentacea* and *Legumen*, *legumentum* in Ray's 'Interpretation'.
[41] There is no division 18 in Ray's list. See the entry for *Olus* in Ray's 'Interpretation'.
[42] *eæque in duplici sunt differentia*, literally "and they are in double difference".
[43] i.e. *Persicaria* spp. [433–435] and *Veronica* spp. [35–38].
[44] i.e. *Stratiotes aloides* [364], now called "Water-soldier".
[45] i.e. one or more of *Hottonia palustris* [365], *Ranunculus trichophyllus* [366] and *Myriophyllum* spp. [367, 368].
[46] The organism named ἀνδροσακες [androsaces] by Dioscorides was unrelated to the species of the Linnaeus' genus *Androsace*, being probably a marine zoophyte rather than a plant; see, for example, Johnson (1633, p. 531).
[47] See footnote 282 in the main catalogue.
[48] *à restando in terrâ*.

(103)

3. Annua, quæ non ultra annum durat, sive vere, sive Autumno seratur.

Possent adhuc plantæ variis modis aliter dividi, v. g. respectu radicum, caulium, florum, seminum, foliorum &c. quæ ulteriùs persequi impræsentiarum nostri non est instituti.

FINIS.

[CAMBRIDGE]

APPENDIX

Ad
Catalogum Plantarum
Circa
CANTABRIGIAM
nascentium:

Continens
ADDENDA
Et
EMENDANDA.
[By John RAY]

CANTABRIGIÆ:
Excudebat *Joan.* Field, celeberrimæ
Academiæ Typographus.

Impensis Gulielmi Morden, *Bibliopolæ.*
Ann. Dom. 1663.

APPENDIX

To

the Catalogue of Plants

Growing around

CAMBRIDGE:

Containing

ADDENDA

And

EMENDANDA.

CAMBRIDGE:

Printed by *John Field*, Printer
to the most renowned University.

At the expense of William Morden, *Bookseller.*
A. D. **1663**.

Lectori S.

Ex *quo* Catalogum plantarum circa Cantabrigiam nascentium *in lucem emisi, alias aliquam multas stirpes in agris oppido ei vicinis sponte provenientes animadverti; quas quoniam jam dilectissimæ nutrici* almæ Academiæ *non sponte supremum vale dixerim, & in transmarinas regiones mox discessurus sum, in unum fasciculum collectas Phytologiæ studiosis seorsim exhibendas censui. Seorsim inquam, nam quoad* Cantabrigiæ vixi, *cùm nova eis quotannis accessio fieret, penes me diu servâsse, nec antea publici juris fecisse decreveram, quàm Catalogum generalem, quem meditabar, plantarum omnium quæ in* Anglia *& insulis adjacentibus sponte oriuntur penitus absolvere & unà evulgare potuissem. Huic scilicet studio sedulus incubui, & plerásque omnes etiam remotissimas hujus regni provincias in hunc præcipuè finem obivi & perlustravi. Verùm quoniam opus illud ut debitam perfectionem acquirat, plurium adhuc annorum observationibus indiget, plurimæque subinde emergunt difficul-*

¶ 2   *tates*

Greetings to the Reader.[1]

*S*ince I published[2] Catalogue of plants growing around Cambridge, *I have noticed a considerable number of other plants growing wild in the fields near that town.* Because I have now unwillingly said my last farewell to my most beloved nurse the bountiful University *& shall soon be departing overseas,*[3] *I have resolved to collect these into one pamphlet and publish them separately for students of Botany. I say 'separately' because, as long as I lived* in Cambridge, *since new additions were being made to them every year, I had decided that it was in my power to keep them for a while and not to make them public property until I had been able fully to complete & publish in one piece a general Catalogue, which I was planning, of all the plants that grow wild in* England *& the adjacent islands. In fact I have devoted myself diligently to this study & I have visited & wandered over almost all of even the remotest parts of this kingdom principally for this purpose. But, since that work still requires several years' observations to achieve its due perfection and since numerous difficulties are continually emerging for which I do not know whether the whole of what is left of my life would suffice to overcome and rectify all of them, having changed my mind,*[4] *I have decided that* the Catalogue[5] *should be separated from the Appendix & that neither publication of the former should be hastened nor that of the latter held back. And now a brief word about the Addenda.*[6] *Now, since I have detected even quite a number of errors in* the Catalogue of Cambridge plants, *even though they are neither so serious that I could not have suppressed & concealed them without damage to my good reputation nor so obvious & manifest that they could easily be detected by anyone and brought as a charge against me, yet, lest my sense of shame should cause young students to be deceived*[7] *I have of my own accord*[8] *succeeded in frankly bringing them into the open & in serving truth and the benefit of others rather than my own good name.*

---

[1] *Lectori S.* (a common abbreviation for *Salutem*).
[2] *in lucem emisi*, literally "into the light I sent out".
[3] *in transmarinas regiones*, literally "into overseas regions".
[4] *mutatâ sententiâ*, literally "with changed opinion".
[5] i.e. Ray's *Catalogus plantarum Angliae* (1670, 1677), from which much material was drawn for the later appendix of 1685.
[6] *Atque hæc breviter de Addendis*. Literally "And these things briefly about the Addenda."
[7] *nè tamen pudor meus studiosæ juvetuti fraudi foret*, literally "lest, however, my sense of shame to studious youth for fraud should be".
[8] *à me ipse*, an error for *à me ipso*, literally "by me myself", corrected in the 1685 appendix.

Addenda.

Alcea vulgaris *J.B. Dod. Ad.Lob. Cluf.hift.* vulgaris major *C.B.*vulgaris sive Malva verbenacea *Park.* Malva verbenacea *Ger.Vervain mallow. In Kingston wood.*

Alsine aquatica surrectior *J.B.* forté. Plantula Portulacæ aquaticæ affines. *Small upright water Chickweed, called by some Blinkes. On the boggy grounds about Gamlingay. This is very frequent in many other places of England, in watery grounds, and bogs, especially where the ground is sandy.*

Alsine hirsuta altera viscosa *C. B.* Spuria 4 *Dod. The broader-leaved Mouse-ear-Chickweed.* Hæc omnino distincta est ab Alsine arvensi hirsuta magno flore *C.B.* Nimiru*m florem* habet minorem, evanidum; *folia* breviora, latiora, pallidiora; caules magis erectos quàm Alsine myosotis hirsuta *Ad.* Quinetiam tota planta [caules, folia, flores] in siccioribus viscosa est. *This flowers before the other.*

Alsine fontana credita flosculorum foliolis non divisis. An Alsine fontana *Ger?* aquatica media *C.B? Fountain-chickweed.* Ramulos habet tenues, infirmos, humi procumbentes; folia pallidè virentia, acutiora quàm Alsine media; Flosculorum foliola non ut in illa ad unguem usque fissa, sed integra & in acutum desinentia. *Found in many places but especially about Balsham. There is no reason why this should be called Fountain-chickweed: we never found it growing in the water.* Floret circa initium Aprilis & deinceps.

Aparine semine læviore. *Cleavers or Goose-grass with smoother seed.* Hæc (quod sciam) nondum descripta est. Inter segetes passim.

Auricula leporis minima *J.B* An Bupleurum minimum *Col. Park?* An Bupleurum angustissimo folio *C.B? The smallest Hares-ear. By the way-side as you ride to S. Neotes beyond*

Addenda.

A1 *Malva moschata* L.
Alcea vulgaris *J.B. Dod. Ad. Lob. Clus. hist.* vulgaris major *C.B.* vulgaris sive Malva verbenacea *Park.* Malva verbenacea *Ger. Vervain mallow. In Kingston wood.*

A2 *Montia fontana* L.
Alsine aquatica surrectior *J.B.* perhaps. A small plant similar[9] to Portulaca aquatica. *Small upright water Chickweed, called by some Blinkes. On the boggy grounds about Gamlingay. This is very frequent in many other places of England, in watery grounds, and bogs, especially where the ground is sandy.*

A3 *Cerastium glomeratum* Thuill.
Alsine hirsuta altera viscosa *C.B.* Spuria 4 *Dod. The broader leaved Mouse-ear-Chickweed.* This is absolutely distinct from Alsine arvensis hirsuta magno flore *C.B.*[10] Certainly it has a smaller, short-lived *flower*, shorter, broader, paler *leaves* and more erect stems than Alsine myosotis hirsuta *Ad.*[11] What is more in drier places the whole plant [stems, leaves, flowers] is sticky. *This flowers before the other.*

A4 *Moehringia trinervia* (L.) Clairv.
Alsine fontana credita flosculorum foliolis non divisis. An Alsine fontana *Ger?* aquatica media *C.B? Fountain-chickweed.*[12] It has slender, weak little branches, spreading on the ground, pale green leaves, more acute than those of Alsine media,[13] and petals[14] not, as in that species, divided all the way to the base[15] but entire & ending in a point. *Found in many places but especially about Balsham. There is no reason why this should be called Fountain-chickweed: we never found it growing in the water.* It flowers around the beginning of April & subsequently.

A5 *Galium tricornutum* Dandy
Aparine semine læviore. *Cleavers or Goose-grass with smoother seed.* This (as far as I know) has not yet been described. Among standing corn generally.[16]

---

[9] *affines*, a misprint for *affinis*, literally "neighbouring", "allied", "related". See the footnote on *Portulaca sylvestris* [459] in the main catalogue and the entry for *Alsine rotundifolia sive Portulaca aquatica* [B8] in *Addenda* of the 1685 appendix; *Montia fontana* is in fact superficially similar but not related to *Lythrum portula*.

[10] Ray later (1690, p. 146; 1696, p. 208) listed this in the synonymy of *Alsine hirsuta myosotis* Ad., which Linnaeus (1759, p. 101) equated with his own *Cerastium vulgatum*, now treated as a *nomen ambiguum* but here almost certainly *C. fontanum* Baumg. (see below).

[11] i.e. *Cerastium fontanum* [26], though this is listed in the order *Alsine hirsuta myosotis* in the main catalogue.

[12] Ray abandons all four of these names in *Addenda* of the 1685 appendix (q.v. under *Alsine Plantaginis folio*).

[13] i.e. *Stellaria media* [29].

[14] *Flosculorum foliola*, literally "Leaflets of the little flowers".

[15] *ad unguem usque fissa*, literally "to the nail all the way split": *unguis* was used in classical Latin for a tip or extremity in plants, but Ray defines it in his 'Interpretation of terms' as *pars ima folii rosæ albida* ("the whitish lowest part of the leaf [presumably here the petal] of a rose").

[16] Ray (1670, p. 25) refined this to "Inter segetes præsertim solo cretaceo, *as about the Gogmagog-hills and other places of Cambridge-shire*" and there is an equivalent comment in English in the 1685

A6 *Bupleurum tenuissimum* L.
Auricula leporis minima *J.B.* An Bupleurum minimum *Col. Park?* An Bupleurum angustissimo folio *C.B?* The smallest Hares-ear. *By the way-side as you ride to St Neotes beyond Elles-ly. This I have found in sundry other places in England, viz. in the road to Stilton a little beyond Huntington. At Maldon in Essex, about and in the yard where they use to build vessels, near Fullbridge; and upon the bank by the channel side there.* Compare with J. Bauhin's description.

C

A7 *Clinopodium calamintha* (L.) Stace
Calamintha flore minore, odore pulegii *J.B.* odore pulegii *Ger.* altera odore pulegii foliis maculosis *Park.* pulegii odore sive Nepeta *C.B. Field Calamint.* I had this plant in the Catalogue as Calamintha vulgaris;[17] however I think that a Calamintha other than only vulgaris grows wild around Chesterton & elsewhere.

A8 *Carduus crispus* L.
Cardui polyacanthi prima species *Ger. emac.* Aculeosa Gazæ *Ad.* Polyacantha *Tab.* in Johnson's opinion.[18] *Thistle upon Thistle.*

A9 *Carduus tenuiflorus* Curtis[19]
Cardui polyacanthi secunda species *Ger. emac.* Sylvestris 3 *Dod.* Polyacantha *Lob.* Carduus spinosissimus capitulis minoribus *P.B. Welted Thistle with small flowers.* Look for descriptions of these two plants among the names in chapter 490 of book 2 of *Ger. emac.*[20] *The first of these grows commonly on ditch sides and among bushes. The second plentifully on the banks of the river Cam toward the isle of Ely and elsewhere.*

---

appendix. He cited no synonyms in his later works except, very tentatively, *Aparine lævis* Park. (see footnote 22 in *Addenda* of the 1685 appendix).

[17] i.e. *Clinopodium ascendens* [100].

[18] The reference is to p. 1176 of Johnson (1633), who says that his first figure on p. 1173 (labelled "*Polyacanthos*. Thistle vpon Thistle.") is of "*Polyacantha Theophrasti* of *Tabern.*".

[19] Linnaeus (1759, p. 106) identified this species as his *Carduus acanthoides* and the preceding one as his *C. crispus*. Babington (1860) reversed these identifications, commenting that he had "not seen or heard" of *C. acanthoides* "being found here". *C. acanthoides* is not a native British plant but there have been a few subsequent Cambridgeshire records, although here as elsewhere the situation is obscured by nomenclatural and taxonomic confusion (see Sell & Murrell 2006). Johnson's description of his first thistle (quoted in the next footnote) fits both *C. acanthoides* and *C. crispus* and we assume that it is *C. crispus*, which is frequent in Britain. His second thistle clearly differs from the first in more characters than those that separate *C. crispus* and *C. acanthoides* and it is in fact a good match for *C. tenuifolius*. It is not clear why Clarke (1900) does not accept this Johnson record as the first British record of *C. tenuifolius*, as the first record that he cites is Johnson's (1634) *Carduus spinosissimus capitulis minoribus sive Polyacantha* Lob., presumably the same plant. *C. tenuiflorus* now has a submaritime distribution and the first Cambridgeshire record has hitherto been regarded as made by Skrimshire on the river side at Outwell Bridge, *c.* 1800 (Perring *et al.* 1964; Crompton 2001). *C. tenuifolius* may have extended further inland along the rivers of Cambridgeshire in the past. There is evidence for a contraction in range in the London area, where it was described as frequent by Ray's contemporary Buddle and was abundant in Holland Park and Shepherd's Bush in the 19th century, but declined so greatly that it was not seen anywhere in Middlesex between 1929 and 1976 (Kent 1975, 2000).

[20] Again the reference is to p. 1176 of Johnson (1633), where, in the section headed *The Names*, he seeks to "amend, and giue as much light as [he] can, to the obscurities of our Author [Gerarde] and some others". He says: "Of this Thistle I obserue three kindes [though there is only one figure: see also below]: the first is a Thistle some two cubits and a halfe high, with many slender stalkes

APPENDIX (1663)

A10 *Geum rivale* L. × *G. urbanum* L.
Caryophyllata vulgaris flore majore C.B. Common Aveus[21] with a large flower. This was found in the fields somewhere about the Town, and brought into our gardens: I do not remember the place.[22]

A11 *Lathyrus nissolia* L.
Catanance leguminosa quorundam J.B. Ervum sylvestre Ger. Dod. Sylvestre sive Catanance Park. Lathyrus V, sive sylvestris minor C.B. *Crimson grass Vetch. Found by Mr Dent about Hadnam in the isle of Ely.*

A12 *Veronica montana* L.
Chamædrys spuria foliis pediculis oblongis insidentibus. *Wild Germander with leaves standing on long footstalks. This plant (as I am informed from Mr Goodyer[23]) is figured and described by* Fabius Columna *p.* 288. *under the title of* Alysson montanum.

D

A13 unidentifiable fern[24]
DRyopteris Tragi Ger. emac. Filix querna C.B. pumila saxatilis prima Clus. pumila saxatilis prima Clusii Park. ramosa minor J.B. *The lesser branched or female Fern. Observed by Mr Dent on Kings-Colledge walls.*

and branches exceeding prickly, hauing commonly fiue prickly welts running alongst the stalks: the leaues on the vpper sides as also the stalkes are of a reasonable fresh greene colour, but the vnderside of the leafe is somwhat whitish: the heads consist of sundry hairy greene threds which looke like prickles, but they are weake, and not prickly: the floure is of the bignesse, and of the like colour and shape as the common Knapweed, yet somwhat brighter: it grows on ditch sides, and floures in Iuly. ... The second of these I take to be that which *Lobel* hath figured for *Polyacantha*, and *Dodonæus* for *Carduus syl.* 3. (which figure we here giue you [a puzzling statement since earlier he has said that the single figure portrays the first thistle]) and in the *Hist. Lugd. pag.* 1473. it is both figured and described by the name of *Polyacanthos Theophrasti*. In the figure there is little difference: in the things themselues this; the stalkes of this are as high as those of the last, but slenderer, with fewer and straighter branches, and commonly edged with foure large welts, which haue fewer, yet longer prickles than those of the former: the leaues and stalkes of this are of a grayish or whitish colour: the heads are longish, but much smaller than those of the former, and they seldom open or spred abroad their floures, but onely shew the tops of diuers reddish threds of a feint colour. This growes as frequently as the former, and commonly in the same places." Πολυακανθος [Polyacanthos] is mentioned only once in Theophrastus' *Enquiry into plants*, in a list of thistles and thistle-like plants (VI. IV. 3 in the Loeb Classical Library edition), so speculation about which species he meant is fruitless.

[21] A misprint for Avens.
[22] Ray reported this rather differently in 1670 (p. 59): "*We found this in our Gardens at Cambridge, which we had brought out of the neighbour-fields, but do not particularly remember the place.*"
[23] The Hampshire botanist John Goodyer (c. 1592–1664) was one of the most able field botanists of his generation, but the period of his greatest activity was short (1616–1621). He sent much information to Johnson when he was revising Gerarde's *Herball* (1633), but he "was seemingly undone by an overriding bookishness" (ODNB, vol. 22, pp. 831–832) and, in particular, devoted his energies to a translation of Dioscorides. His reference is to Colonna (1616).
[24] The name refers to *Gymnocarpium robertianum* (Hoffm.) Newman, "a deciduous fern of cracks, fissures and scree in limestone rock" (Preston, Pearman & Dines 2002), which is unlikely to have been growing on walls in Cambridge in Ray's time, though it is now occasionally established as a garden escape on walls and culverts and has occurred on a wall near the Botanic Garden in Cambridge in recent years (Crompton & Preston 2000). Juvenile ferns growing on walls can be difficult to identify as they lack the characteristic morphology of the mature adults, so it is likely that Dent

F

A14 *Hippocrepis comosa* L.

FErrum equinum Germanicum siliquis in summitate *C.B.* equinum comosum *Park*. Ornithopodio affinis vel potiùs Soleæ aut Ferro equino herba[25] *J.B.* Bush-headed Hors-shoe-vetch. *On Gogmagog hills, and on the drier part of Hinton moor and on Newmarket heath, &c.*

G

A15 *Gnaphalium sylvaticum* L.

GNaphalium Anglicum *Ger.* Anglicum vulgare majus *Park.* majus angusto oblongo folio alterum *C.B.* rectum *J.B.* Great English Cudweed. *In the pastures about Gamlingay.*

A16 *Carex pendula* Huds.[26]

Gramen cyperoides spicâ pendulâ longiore *Park.* cyperoides spicâ pendulâ longiore & angustiore *C.B. J.B.* Pseudocyperus *Ger.* Bastard Cyperus-grass with long pendulous heads. *In the great ditch at the end of the little Thicket adjoyning to Teversham moor, and in other great ditches.*

A17 *Carex remota* L.

Gramen cyperoides angustifolium paniculâ multiplici. Narrow-leaved Cyperus-grass with many heads. *On the banks of many ditches. I cannot find any figure or description of this.*[27] The panicles or heads grow out of the axils of the leaves[28] one above another on the stalk, every head having a long leaf under it.

H

A18 *Helleborus foetidus* L.

HElleboraster maximus *Ger.* maximus sive Consiligo *Park.* Helleborus niger fœtidus *C.B.* Enneaphyllon of Pliny, *book* 27, *chapter* 9. Great bastard Hellebore or Bears-foot. *At Cherry-hinton, in the hedges of a close near to the moor.*

---

mistook a juvenile plant of a commoner species for *Gymnocarpium robertianum*. Ray (1670, pp. 95–96) says "*we found it in the shady Lanes near Tintern-Abbey in Monmouthshire, and in divers places of Westmorland, plentifully*", but he does not mention this record.

[25] Literally "A herb related to Small Bird's-foot or rather to Shoe or Iron for a horse". Bauhin & Cherler (1651, vol. 2, p. 348) describe this species immediately after one called *Solea equina* with the synonym *Ferrum equinum*; both these names mean 'horseshoe'.

[26] Although the identification of this species seems clear, Ray's subsequent treatment of it is puzzling. In *Catalogus plantarum Angliae* (1670, p. 147; 1677, p. 144) he does not indicate that it is recorded from Cambridgeshire and in both editions, and even in *Historia plantarum* (1688a, p. 1294), the only locality for the species cited is Notley in Essex. By contrast, in the English catalogue Ray annotates with C *Gramen cyperoides spicâ pendulâ breviore* C.B. [*Carex pseudocyperus* L.], indicating that it was a Cambridgeshire plant. It seems unlikely that he decided that he had previously confused these two species, as *C. pseudocyperus* is well illustrated by Parkinson (1640, p. 1266) and Bauhin & Cherler (1651, vol. 2, p. 496), unlike *C. pendula*, which they do not illustrate, and Ray would surely have recognised it if he had seen it in Cambridgeshire.

[27] Ray did not cite an earlier name in his subsequent works (e.g. 1686, p. 1295; 1696, p. 267).

[28] *è foliorum sinubus*, literally "out of the folds/bosoms of the leaves".

APPENDIX (1663)

A19 *Marchantia polymorpha* L., female plants[29]
Hepatica stellata *Ger*. Lichen sive Hepatica minor stellaris *Park*. Lichen petræus stellatus *C.B*. *Star headed Liverwort*. *On the north side of Peterhouse chappel abundantly, observed by Mr Dent.*

A20 *Marchantia polymorpha* L., male plants
Hepatica umbellata *Ger*. Lichen sive Hepatica minor umbellata *Park*. Lichen petræus umbellatus *C.B*. *Small ground Liverwort with round heads*. *On the moors.*

A21 *Hypochaeris maculata* L.
Hieracium montanum caule aphyllo non ramoso flore pallidiore.[30] An Hieracium montanum hirsutum minus *C.B*? *Mountain Hawk-weed with a pale flower*. *On Gogmagog hills and Newmarket heath. The flower of this plant doth much resemble the flower of Mouse-ear. The leaf (as far as I remember) is not laciniated, but in some waved a little.*

A22 *Hypericum hirsutum* L.
Hypericum majus sive Androsæmum Matthioli *Park*. Hypericum Androsæmum dictum *J.B*. Androsæmum Hypericoides *Ger*. alterum foliis Hyperici, quod aliquibus Hypericoides[31] *C.B*. *Tutsan St Johns wort*. *In thickets & hedges generally.*[32]

L

A23 *Blechnum spicant* (L.) Roth
Lonchitis aspera *Dod*. *Ger*. minor *C.B*. aspera minor *Park*. *Matth*. *Cam*. *Lugd*. altera folio Polypodii *J.B*. *Rough Spleen-wort*. *About Gamlingay.*

M

A24 *Thuidium tamariscinum* (Hedw.) Schimp.
Muscus filicinus *Ger*. *J.B*. *Park*. filicinus major *C.B*. *Fern-moss*. *In Kingston and Eversden woods, and the woods about Balsham.*

A25 *Fontinalis antipyretica* Hedw.[33]
Muscus triangularis aquaticus. *Triangular water-moss*. *This moss is very similar to* [Muscus] terrestris denticulatus *of* [Mathias] de L'Obel. *In the river beyond Stretham ferry. I found it since in a well in Carnarvan-shire in Wales, called Gods*

---

[29] The plant reported as *Lichen* in the main catalogue [331] was almost certainly *Marchantia polymorpha* and/or *Lunularia cruciata*. *Hepatica stellata* and the next entry *Hepatica umbellata* can be identified as *Marchantia polymorpha* with greater certainty, as *stellata* describes the deeply divided receptacle of female plants and *umbellata* the less deeply divided receptacle of male plants. Both sexes were present in Old Court, Peterhouse, less than 30 metres from Dent's locality, in 2009.

[30] "Mountain Hawkweed with a leafless, unbranched stem [and] a palish flower."

[31] "The second [*Androsæmum*] with leaves of a St John's-wort, which for some [authors is] *Hypericoides*".

[32] Ray listed this in a letter to Peter Courthope (July 1661) as one of the species he had discovered in 1661, commenting: "This we use to confound with Hypericum vulgare [*Hypericum perforatum* 296]; it growes commonly almost in every hedge about this towne." (Gunther 1934; Thompson 1974).

[33] This is the one moss in the *Catalogus* and its appendices that can be identified with absolute certainty. Ray saw the Caernarvonshire plant on 30 May 1662, when "we passed over from Aberdaren [Aberdaron] to Pulhely [Pwllheli], and by the Way saw a Well which the *Welch* call *Funnan Deuw*, i.e.

445

well, but always immersed in the water. I took it once to be the Fontalis minor lucens *J.B. but he doth not mention the leaves of that to be so disposed all along the stalk as to make it appear triangular.*

O

A26 *Ophrys sphegodes* Mill.
ORchis sive Testiculus Sphegodes hirsuto flore *J.B.* An Orchis fucum referens colore rubiginoso *C.B.?* What our authors Gerarde & Parkinson may mean[34] in their account of the Orchids I do not sufficiently understand; however I think that Parkinson understood this orchid to be his Sphegodes altera, but he describes it wrongly (namely with purplish wings when those of this species are green tinged with yellow[35]). Perhaps for Gerarde Testiculus vulpinus major Sphegodes.[36] *The green-winged Humble-bee Orchis.* It flowers around the end of April & the beginning of May, together with Orchis morio fœmina.[37] *In an old gravel-pit near Shelford by the foot way from Trumpington to the church*[38] *we found hundreds of them.*

A27 *Herminium monorchis* (L.) R.Br.
Orchis pusilla odorata *Park.* parva autumnalis lutea *J.B.* odorata moschata sive Monorchis *C.B. The sweet yellow musk Orchis. In the chalk-pit close at Cherry-hinton.*

P

A28 *Euphorbia peplus* L.
PEplus sive Esula rotunda *J.B. C.B. Ger.* Esula rotunda sive Peplos *Park. Petty Spurge.* Along roads & on the margins of gardens.[39]

A29 *Asplenium scolopendrium* L.
Phyllitis *Ger. Dod. ut.* Phyllitis sive Lingua cervina vulgi *J.B.* Lingua cervina officinarum *C.B.* Phyllitis sive Lingua cervina vulgaris *Park. Harts-tongue. On the walls of Cherry-hinton church and elswhere.*

A30 *Arabidopsis thaliana* (L.) Heynh.
Pilosella siliquata major & minor *Thal.* Bursa pastoria sive Pilosella siliquosa *J.B.* Bursæ pastoris similis siliquosa major & minor[40] *C.B.* Paronychia major &

GOD'S Well. They fancy it works strange Cures. We did not observe any notable Taste in it, neither would it tincture silver, as we were told it would." (Derham 1760, p. 234). The only well of this name in Caernarvonshire listed by Jones (1954) is Ffynnon Dduw near Llanbedrog village. J. Bauhin's *Fontalis minor lucens* is *Fontinalis squamosa* Hedw., a calcifuge which is widespread in northern and western Britain. It was not mentioned in the first and second editions of Ray's *Synopsis* (1690, 1696) and its inclusion as a synonym of a species reported from the Thames in London on p. 79 of the 1724 edition must be an error. The true plant was eventually discovered in Britain by Dillenius on his trip to North Wales in 1726 and published in *Historia muscorum* (1741, pp. 258–259).

[34] *sibi velint*, literally "for themselves may wish".
[35] *cùm hujus ex luteo virescant*, literally "when of this out of yellow they grow green".
[36] Literally "Greater testicle of a fox [Fox-stones], Drone-like". See footnotes 480, 496 and 497 in the main catalogue.
[37] i.e. *Anacamptis morio* [399].
[38] Ray (1670, p. 228) described the locality as "*an old Gravel-pit in the open field near great Shelford*".
[39] See *Tithymalus helioscopius* below in *Emendanda*.
[40] Literally "Greater & lesser podded [plant] like Shepherd's-purse".

Paronychia altera minor *Park*. Jean Bauhin thinks that actually these two do not differ in species but only in height & size. *Codded Mous-ear. In some closes on the north side of Gamlingay.*

A31 *Limosella aquatica* L.

Plantaginella palustris *C.B.* Plantago aquatica minima *Clus.* aquatica minima Clusii *Park. The smallest water Plantain.* Along roads where waters have flooded during the winter. This is either the same as Plantago aquatica minima of [Charles] de l'Écluse or very close to it: it does not accord with its description in all respects.

A32 *Hieracium* sp.[41]

Pulmonaria Gallica sive aurea latifolia *Ger.* Hieracium murorum folio pilosissimo *C.B.* Hieracium murorum Bauhini quod est Pulmonaria Gallorum Lobelii *Park*. Pilosella major quibusdam, aliis Pulmonaria flore luteo *J.B. French or golden Lungwort. About Gamlingay.*

Q. R.

A33 *Ranunculus parviflorus* L.

Ranunculus hirsutus arvensis flore minimo. *Field Crowfoot with a small flower.* It has not been discovered by me whether this plant has already been published & described before by others and the authors whom I might consult are not to hand.[42] The *leaves* are hairy and pale green; the flowers are yellow and the smallest of all that I have observed in the Ranunculi. The plant itself, as far as I remember, is not erect[43] but covers the ground as it were with a dense sward.

A34 *Ranunculus sardous* Crantz

Ranunculus rectus foliis pallidioribus hirsutus *J.B.,* in whose work there is a full & exact description. *Upright pale-leaved sweet Crowfoot.* In moist & muddy places, especially where waters have flooded during the winter.

A35 *Ranunculus acris* L.

Ranunculus rectus, non repens flore simplici luteo *J.B.* Ranunculus surrectis cauliculis *Ger.* pratensis erectus acris an dulcis *C.B?* pratensis erectus acris vulgaris an dulcis *Park? Upright Field Crowfoot.* I am uncertain whether this is biting or mild,[44] but if I remember rightly that which I happened to taste was mild. Let Jean Bauhin's description be attended to. In meadows.

---

[41] Linnaeus (1753, pp 802–803; 1759, p. 105) identified C. Bauhin's *Hieracium murorum folio pilosissimo* as his own *H. murorum*. The latter is often used as an aggregate name. Few of the apomictic microspecies of *Hieracium* that it covers occur in Cambridgeshire, but it is nevertheless impossible to identify Ray's plant more precisely.

[42] Ray discovered this species at Gamlingay in the spring of 1662 (Gunther 1934; Thompson 1974). He then left Cambridge on an extended tour of England and Wales with Willughby, returning for only a few days in late summer to settle his affairs before spending the following winter in Suffolk (Raven 1950, pp. 121–131). This perhaps explains why he had not got the relevant literature to hand when writing this entry. He did not cite an earlier description in his subsequent works (e.g. 1670, p. 260; 1686, p. 583; 1690, p. 86).

[43] *non erigitur*, literally "is not raised up".

[44] *An hic acris sit, an dulcis mihi non certò constat*, literally "Whether this may be sharp/biting or sweet/mild for me not for certain is sure". See also Ray's account of this species in *Addenda* of the 1685 appendix.

A36 *Asperula cynanchica* L.
Rubia cynonchica[45] *J.B.* VI sive cynanchica *C.B.* Synanchica Lugd. p. 1125 *Ger.* Squinancy-wort. *On Gogmagog hills and Newmarket heath.*

A37 *Ruscus aculeatus* L.
Ruscus *J.B. C.B. Park. Ad. Lob.* Ruscus sive Bruscus *Ger.* Myrtacantha, murina spina *Lob. ico.* Oxymyrsine *Ang.* Butchers broom. *About Anglesey-abbey.* The root of this plant is one of the five best known aperients;[46] the rest are those *of Celery, Asparagus, Fennel and Parsley.*[47]

S

A38 *Knautia arvensis* (L.) Coult.
Cabiosa vulgaris flore pleno. *Double flowered Scabious.* In the flower of this form the tubular petals that are on the periphery are not larger than the inner ones, as in the common Scabious,[48] but equal, just as in Succisa;[49] moreover it lacks those capitate Stamens that are seen in the common form. *In many closes about Teversham and elsewhere*[50] *together with the common sort.*

A39 *Petroselinum segetum* (L.) W.D.J. Koch
Selinum Sii foliis *Ger. Park.* Honewort. *On many ditch-banks about Cambridge.* This plant is related to that called Amomum Germanicum[51] & certainly ought to be reckoned among the parsnips, with which it accords in smell & taste & outward appearance.

A40 *Thymus* sp.
Surpylli[52] vulgaris primum genus *J.B.*[53] *The other common Mother Thyme. In many places together with the common sort.*[54] Consult J. Bauhin about Serpyllum vulgare, who notes exactly the differences between these two species.

---

[45] A misprint for *cynanchica*, corrected in the 1685 appendix.
[46] ...*è numero quinque illarum aperientium famosarum*, literally "... out of the number five of those opening renowned".
[47] *Apii, Asparagi, Fœniculi, Petroselini* (singular genitives of *Apium, Asparagus, Fœniculum* and *Petroselinum*).
[48] i.e. *Knautia arvensis* [526].
[49] i.e. *Succisa pratensis* [374].
[50] Ray (1670, p. 277) says "*In many Meadows and Closes about Teversham, Balsham, and else-where in Cambridge-shire*" and adds that it differs from *Knautia arvensis* only in non-essential ways (*accidentaliter*), not in species. This illustrates the fourth of the five or six meanings of the word "accident" in Ray's works enumerated by Cain (1996).
[51] i.e. *Sison amomum* [542].
[52] A misprint for *Serpylli*, corrected in the 1685 appendix.
[53] In *Catalogus plantarum Angliae* (1670, p. 282) and all his later works Ray calls this *Serpylli vulgaris secundum genus* J.B., suggesting that *primum* here is a slip for *secundum* or that he had changed his mind about the equivalent Bauhin taxon by 1670. The identification of the species is discussed in footnote 109 in *Addenda* of the 1685 appendix.
[54] i.e. *Thymus polytrichus* [537].

APPENDIX (1663)

T

A41 *Trifolium subterraneum* L.
Trifolium pumilum supinum flosculis longis albis nondum descriptum[55] *P.B. Dwarf Trefoil with long white flowers. At Gamlingay by the lanes-side as you go up from the Town to the Windmils.*

V

A42 *Solidago virgaurea* L.
Virga aurea *Ger. Matth. Dod. ut.* aurea vulgaris *Park.* aurea vulgaris latifolia *J.B.* aurea IV. sive angustifolia minùs serrata *C.B. Golden Rod. In some pastures about Gamlingay.*

---

[55] "Dwarf prostrate Trefoil with long white florets, not yet described." This is another species that Ray discovered at Gamlingay in the spring of 1662; he described it to Courthope in May 1662 as "a very pretty sort of trefoile, wch I intend to examine hereafter further & describe" (Gunther 1934; Thompson 1974).

Emendanda, in quibus quædam mutanda & supplenda.

A

ABrotanum campestre & Abrotanum inodorum] *Si placet expungantur ambo. I beleeve neither of them is to be found on Newmarket heath, at least that part of it which is in Cambridge-shire.*

Allium sylvestre] *Huic adde Synon:* Allium campestre junci-folium capitatum majus *C.B.*

Anagallis aquatica I,& Anagallis aquatica III *C.B.*] *Et hæc pariter expungantur. Ego sc. has differentias apud nos nunquam potui observare. D. Nid ex conjectura & præsumptione Catalogo inseruit, quoniam Botanici vulgò & præcipuè Jo. Bauhinus passim & promiscuè quatuor varietates hic recensitas crescere affirmat.*

B

BLitum album minus] *I now think the plant we found in Kingston wood to be the same with the common red Blite and that it differed in colour by reason of the shadiness of the place where it grew.*

Bursa pastoris minor] *Hæc planta non videtur specie differre à Nasturtio Alpino petræo.* Certè Jo. Bauhinus *Nasturtium petræum Tab.* hujus synonymum facit. *Verùm nihil temerè pronuncio.*

C

CArduus lanceatus angustifolius] *facessat: nec enim puto apud nos in hoc genere observabile esse ullum discrimen.*

Carduus nutans] An Carduus spinosissimus latifolius Sphærocephalus vulgaris *C. B*? Intelligo jam hunc censeri Carduum moschatum *Ger.* Nec tamen moschi odor in floribus mihi unquam perceptus, neque florum color pallidè purpureus est sed saturatior, nec tandem figura ipsa respondet.

T   Echium

Emendanda, in which certain things are to be changed & corrected.

A

ABrotanum campestre & Abrotanum inodorum] *Please expunge both.*[56] *I beleeve neither of them is to be found on Newmarket heath, at least that part of it which is in Cambridge-shire.* [*Artemisia campestris* 1 and 2]

Allium sylvestre] *Add to this as a Synon[ym]:* Allium campestre junci-folium capitatum majus *C.B.* [*Allium vineale* 19]

Anagallis aquatica I, & Anagallis aquatica III *C.B.*] *And let these equally be expunged. Actually I have never been able to observe these differences with us. Master Nid inserted them in the Catalogue from conjecture & presumption because Botanists commonly, & especially Jean Bauhin, generally & indiscriminately affirm that the four varieties enumerated here really do grow.* [*Veronica beccabunga* 35 and *V. anagallis-aquatica* 37]

B

BLitum album minus] *I now think the plant we found in Kingston wood to be the same with the common red Blite, and that it differed in colour by reason of the shadiness of the place where it grew.* [*Chenopodium polyspermum* 88]

Bursa pastoris minor] *This plant does not seem to differ in species from Nasturtium Alpinum petræum. Certainly Jean Bauhin makes Nasturtium petræum Tab. a synonym of this species. But I pronounce nothing rashly.*[57] [*Teesdalia nudicaulis* 98]

C

CArduus lanceatus angustifolius] *let it depart: for I do not think that with us any distinction is observable in this kind.* [*Cirsium vulgare* 110]

Carduus nutans] An Carduus spinosissimus latifolius Sphærocephalus vulgaris *C.B?* I understand now that this should be regarded as Carduus moschatus *Ger.* However I have never noticed the odour of musk in the flowers,[58] nor is the colour of the flowers pale purple but deeper, nor finally does the figure itself accord with it. [*Carduus nutans* 111]

E

EChium alterum] I confess that I have not yet perceived any specific difference between Echium vulgare & that which springs up on walls generally with us. Let others examine this more carefully. [*Echium vulgare* 172]

---

[56] *Si placet expungantur ambo*: literally "If it pleases let be expunged both."

[57] In later works Ray (e.g. 1690, p. 116; 1696, p. 174) made *Nasturtium petræum* his chosen name for this plant.

[58] *Nec tamen moschi odor in floribus mihi unquam perceptus ... est*, literally "Neither however the odour of musk in the flowers to me ever has been perceived".

G

GEranium malacoides sive columbinum minus] *I am not sufficiently certain whether the little plant that we know here is absolutely the same species as* Geranium columbinum minus C.B. [*Geranium molle* and/or *G. pusillum* 221[59]]

Gramen alopecuroides minus spicâ asperâ brevi] The London Botanists call this Gramen cristatum and I would not dissent. [*Cynosurus cristatus* 236]

Gramen caninum] This is the same as that which Jean Bauhin describes under his published title and so the things that I have noted here are to be retracted, for I had not myself observed the panicle before I published the Catalogue but rashly believed others. [*Elytrigia repens* 242]

Gramen hirsutum majus] I now suspect that this Grass does not differ specifically from Gramen exile hirsutum[60] but only in size and other non-essential characters by reason of the place where it grows. [*Luzula multiflora* 252]

M

MOllugo montana sive Gallium] *These two synonyms*, Rubia sylvatica altera *Gesn.* & Matrisylva 2 *Trag.*, do not belong here but to Mollugo prima of Dodoens, which has a round stem & is erect. [*Galium album* 373]

I have not yet quite discovered where to place this plant,[61] which grows generally in hedges, *namely* whether in Mollugo, where we have placed it here, or in Rubia sylvestris. I have no doubt that this is the same as Jean Bauhin described under the title of *Rubia angulosa aspera* and that the London Botanists call it *Rubia sylvestris*. It seems likely to me that the same plant is described under each of these two titles. Jean Bauhin evidently doubts whether Mollugo *of [Charles] de L'Écluse* corresponds to his Rubia angulosa or not, but in fact Mollugo *of de L'Écluse* is now the same as Mollugo prima *of Dod[oens]*, Mollugo montana *of Ger[arde]* and Mollugo vulgatior *of Park[inson]*, for the figure is everywhere the same one and both the figure and the description match this plant, but the very short descriptions of Rubia sylvestris in the works of *Ger[arde]* & *Park[inson]* are not likewise the same.

C. Bauhin wrongly assigns Mollugo *of de L'Écluse* & Mollugo montana *of Ger[arde]* to his own Mollugo prima or montana ramosa latifolia since these synonyms relate[62] to his Mollugo secunda no less than to Mollugo prima *of Dod[oens]*.

O

ORiganum vulgare] *The plant that grows in Bowyers close at Cherry-hinton is not this, but* Clinopodium vulgare.[63] *We mistook it when it was young.* [*Origanum vulgare* 410]

---

[59] See the footnote to this taxon in *Emendanda* of the 1685 appendix.

[60] i.e. *Luzula campestris* [251]. Ray later (e.g. 1690, p. 193; 1696, pp. 262–263) revised this opinion and kept the two species separate.

[61] *Nondum mihi planè compertum est, quò referenda sit hæc planta*, literally "Not yet for me wholly has it been discovered whither is to be referred this plant".

[62] *spectant*, literally "look".

[63] This was *Clinopodium vulgare* [139], but Ray's other records for *Origanum vulgare* were sound.

APPENDIX (1663)

P

POlygalon Cortusi] For this substitute Ferrum equinum Germanicum siliquis in summitate *C.B.*, the synonyms of which look for in the Addenda. [*Hippocrepis comosa* 452, A14]

R

RAnunculus hederaceus rivulorum, &c.] The flowers of this plant are white, not yellow as we were guessing, & it arises generally in watery places; accordingly the synonyms listed do not correspond to it unless perhaps it may vary in the colour of the flower overseas. [*Ranunculus hederaceus* 485]

S

SAxifraga Anglica Occidentalium] In the place of Saxifraga Anglica alsinefolia *Ger. emac.* substitute Synanchica Lugd. p. 1125 *Ger. emac.* & add the synonyms Rubia cynanchica *J.B.* VI, or cynanchica *C.B.*[64] [*Asperula cynanchica* 523]

Saxifraga graminea pusilla] *I finde upon comparing that this is the same plant described by Johnson upon Gerard under the name of* Saxifraga Anglicana alsinefolia *or Pearlwort.*[65] [*Sagina procumbens* 525]

T

TIthymalus helioscopius] The second kind of this plant, which we said has a non-crenate leaf & is smaller, is the same as Peplus or Esula rotunda Herbariorum, which see in the preceding appendix.[66] [*Euphorbia helioscopia* 569]

V

VIcia parva sive Cracca minor cum siliquis plurimis hirsutis] let it be deleted, for it is included also above under Aracus.[67] [*Vicia hirsuta* 613]

FINIS.

---

[64] See also footnote 716 in the main catalogue.
[65] The reference is to Johnson (1633, pp. 567–568), where there is a long discussion with a full description; Johnson's English name is, however, not "Pearlwort" but "Chick-weed Breake-stone".
[66] i.e. *Peplus sive Esula rotunda* J.B. C.B. Ger., *Euphorbia peplus* [A28], in *Addenda* above.
[67] *Bis* [literally "twice"] *enim ponitur, superiùs sub Araco*; i.e. as *Aracus sive Cracca minor* Park. [57] as well as, like here, as *Vicia parva sive Cracca minor cum siliquis plurimis hirsutis* J.B. [613].

APPENDIX
AD
Catalogum Plantarum
Circa
CANTABRIGIAM
NASCENTIUM:
Continens
Addenda & Emendanda.

Editio secunda, aucta Plantis sexaginta.

CANTABRIGIÆ,

Ex officinâ *Joh. Hayes*, Celeberrimæ
Academiæ Typographi. 1685.

APPENDIX TO the Catalogue of Plants GROWING Around *CAMBRIDGE*:

Containing

Addenda & Emendanda.

Second edition, increased by sixty Plants.

CAMBRIDGE,

From the workshop of *John Hayes*, Printer to the Most Renowned University. 1685.

Lectori S.

EX quo Catalogum Plantarum circa *Cantabrigiam* nascentium in lucem emisi, alias aliquammultas stirpes in agris oppido ei vicinis sponte provenientes animadverti; quas, quoniam jam dilectissimæ Nutrici almæ Academiæ non sponte supremum vale dixerim, & in transmarinas regiones mox discessurus sum, in unum fasciculum collectas Phytologiæ studiosis seorsim exhibendas censui. Seorsim, inquam, nam quoad *Cantabrigiæ* vixi, cum nova eis quotannis accessio fieret, penes me diu servasse, nec antea publici juris fecisse decreveram, quam Catalogum generalem, quem meditabar, Plantarum omnium, quæ in *Anglia* & Insulis adjacentibus sponte oriuntur, penitus absolvere, & una evulgare potuissem. Huic scil. studio sedulus incubui, & plerasque omnes etiam remotissimas hujus regni provincias in hunc præcipue finem obivi & perlustravi. Verum quo-

Greetings to the Reader.[1]

Since I published[2] Catalogue of Plants growing around *Cambridge*, I have noticed a considerable number of other plants growing wild in the fields near that town. Because I have now unwillingly said my last farewell to my most beloved Nurse the bountiful University & shall soon be departing overseas,[3] I have resolved to collect these into one pamphlet and publish them separately for students of Botany. I say 'separately' because, as long as I lived *in Cambridge*, since new additions were being made to them every year, I had decided that it was in my power to keep them for a while and not to make them public property until I had been able fully to complete & publish in one piece a general Catalogue, which I was planning, of all the plants that grow wild in *England* & the adjacent Islands. In fact I have devoted myself diligently to this study & I have visited & wandered over almost all of even the remotest parts of this kingdom principally for this purpose. But, since that work still requires several years' observations to achieve its due perfection and since numerous difficulties may emerge[4] from time to time for which I do not know whether the whole of what is left of my life would suffice to overcome and rectify all of them, having changed my mind,[5] I have decided that the Catalogue[6] should be separated from the Appendix & that neither publication of the former should be hastened nor that of the latter held back. And now a brief word about the addenda.[7] Now, since I have detected even quite a number of errors in the Catalogue of *Cambridge* Plants, even though they are neither so serious that I could not have suppressed & concealed them without damage to my good reputation nor so obvious & manifest that they could easily be detected by anyone and brought as a charge against me, yet, lest my sense of shame should cause young students to be deceived[8] I have of my own accord[9] succeeded in frankly bringing them into the open & in serving truth and the benefit of others rather than my own good name.

---

[1] *Lectori S.* (a common abbreviation for *Salutem*). The preface of the 1663 appendix is repeated here before a new one, but with changes in the italicisation, capitalisation and punctuation and two other minor alterations mentioned in the succeeding footnotes.
[2] *in lucem emisi*, literally "into the light I sent out".
[3] *in transmarinas regiones*, literally "into overseas regions".
[4] *emergant*, subjunctive here instead of the indicative in the original version.
[5] *mutatâ sententiâ*, literally "with changed opinion".
[6] i.e. Ray's *Catalogus plantarum Angliae* (1670, 1677), records in which contributed substantially to the 1685 appendix.
[7] *Atq; hæc breviter de addendis*. Literally "And these things briefly about the addenda."
[8] *nè tamen pudor meus studiosæ juvetuti fraudi foret*, literally "lest, however, my sense of shame to studious youth for fraud should be".
[9] *à me ipso*, literally "by me myself", correcting the grammatical error in the original version.

Lectori Benevolo.

Bibliopolæ Cantabrigienses multoties conquesti sunt Appendicem ad Catalogum Plantarum circa Cantabrigiam nascentium à Phytologiæ studiosis sæpius fuisse expetitam, verum omnia ejus exemplaria jamdudum adeo distracta esse ut nulla omnino superessent, ex eo Catalogum quodammodo mancum videri; sua idcirco interesse ut Appendix ista denuo excuderetur. Quo vero perfectior prodiret, petierunt à D. Dent, ut, siquas ipse in hoc agro Plantas observasset in Catal. Cantabrig. non numeratas, conferret Appendici adjiciendas. Is quadraginta fere penes se habuit, quibus additæ aliorum quorundam symbolæ sexaginta effecerunt. Quas tibi, Lector, exhibemus, atque ex illis Appendicem plusquam duplo auctiorem. Omnium autem nomina, synonyma, descriptiones, quæve ad descriptionem aliquo modo facerent ex Doctissimi Raii Cata-

APPENDIX (1685)

To the Kind Reader.

He Cambridge Booksellers have many times complained that the Appendix to the Catalogue of Plants growing around Cambridge has frequently been sought after by students of Botany but that all copies of it have been sold so long ago that absolutely none survive, that as a result the Catalogue seems in some way defective and that therefore it is in their interest that that Appendix should be printed again. But, so that it might come out in a more complete form,[10] they requested Master Dent *that, if he himself had observed any Plants in this county[11] not listed in* the Cambridge Catalogue, *he should collate them for addition to the Appendix. He had nearly forty in his possession, added to which the contributions of certain others made sixty. We present these to you, Reader, and from them an Appendix enlarged more than two-fold. Moreover we have transcribed the names, synonyms and descriptions of them all, or anything else that might in some way help in their description,[12] from* the Catalogue of the Plants of England *of the Most Learned* Mr Ray; *only you should except one or two which do not occur therein, as well as a very small number of short notes to distinguish Plants, interspersed as occasion demanded.[13] We have also added certain things to the Emendanda, but none except those that* Master Ray *himself thought should be emended (as was evident to us from* the Catalogue of the Plants of England*). Moreover in these we have used entirely his exact words, for we neither knew nor could wish for anyone else more accurate or careful whom we might follow.[14]*

Farewell.

---

[10] *Quo vero perfectior prodiret*, literally "But so that more perfect it might come forth".
[11] *in hoc agro*, literally "in this field", but from mediaeval times *ager* was often applied to a county; compare footnote 15 in Ray's 'Preface'.
[12] *quæve ad descriptionem aliquo modo facerent*, literally "or what things to the description in some way might produce [i.e. contribute]".
[13] *ubi res exigebat*, literally "where the matter was requiring".
[14] *In his autem ipsissimis plane illius verbis usi sumus, nec enim accuratioris diligentiæ alterum, quem sequeremur, aut novimus aut exoptare potuimus*. Literally "Moreover in these [the *Emendanda*] entirely the words of that man their very selves we have used, for neither the other of more accurate diligence, whom we might follow, did we either know or were able to wish for."

ADDENDA.

A.

ADianthum nigrum vulgare *Park.* nigr. Officinarum *J. B.* foliis longioribus pulverulentis, pediculo nigro *C. B.* Onopteris mas. *Ger.* Dryopteris nigra *Dod. Black Maidenhair. On Ditton Church walls, chiefly on the North side, and on Hildersham Church.*

Alcea vulgaris *J. B. Dod. Ad. Lob. Cluſ. hiſt.* vulgaris major *C. B.* vulgaris ſive Malva verbenacea *Park.* Malva verbenacea *Ger. Vervain Mallow. In Kingston wood.*

Alchimilla *Ger.* vulgaris *C. B.* major vulgaris *Park.* Pes Leonis ſive Alchimilla *J. B. Ladies Mantle. Abundantly in ſome paſtures about Balſham and Gamlingay.*

Alnus nigra baccifera *J. B. C. B.* nigra ſive Frangula *Ger.* Frangula ſive Alnus nigra baccifera *Park. The Black Alder. In Gamlingay Park and thereabouts.*

Alſine hirſuta altera viſcoſa *C. B.* Spuria 4. *Dod. The broader leav'd Mouſe-ear'd-Chickweed.* Hæc omnino diſtincta eſt ab Alſine arvenſi hirſuta magno flore *C. B.* Nimirum florem habet minorem, evanidum; folia breviora, latiora pallidiora; caules magis erectos quam Alſine myoſotis hirſuta *Ad.* Quinetiam tota planta (Caules, folia, flores) in ſiccioribus viſcoſa eſt. Floret circa finem Aprilis & initio Maii & poſtea brevi marceſcit.

Alſine paluſtris Portulacæ aquaticæ ſimilis. Alſine aquaticæ ſurrectiori *J. B.* ſimilis, ſi non eadem. An Alſine paluſtris minor ſerpyllifolia *Ger. Park. Small water Chickweed by ſome call'd Blinks. In the*

ADDENDA.

A.

B1 ***Asplenium adiantum-nigrum*** L.

A Dianthum nigrum vulgare *Park.* nigr[um] Officinarum *J.B.* foliis longioribus pulverulentis, pediculo nigro *C.B.* Onopteris mas. *Ger.* Dryopteris nigra *Dod*. Black Maidenhair. *On Ditton Church walls, chiefly on the North side, and on Hildersham Church.*

B2(A1) ***Malva moschata*** L.

Alcea vulgaris *J.B. Dod. Ad. Lob. Clus. hist.* vulgaris major *C.B.* vulgaris sive Malva verbenacea *Park.* Malva verbenacea *Ger.* Vervain Mallow. *In Kingston wood.*

B3 ***Alchemilla filicaulis*** subsp. ***vestita*** (Buser) M.E. Bradshaw[15]

Alchimilla *Ger.* vulgaris *C.B.* major vulgaris *Park.* Pes Leonis sive Alchimilla *J.B.* Ladies Mantle. *Abundantly in some pastures about Balsham and Gamlingay.*

B4 ***Frangula alnus*** Mill.

Alnus nigra baccifera *J.B. C.B.* nigra sive Frangula *Ger.* Frangula sive Alnus nigra baccifera *Park.* The Black Alder. *In Gamlingay Park and thereabouts.*

B5(A3) ***Cerastium glomeratum*** Thuill.

Alsine hirsuta altera viscosa *C.B.* Spuria 4. *Dod.* The broader leav'd Mouse-ear'd-Chickweed. This is absolutely distinct from Alsine arvensis hirsuta magno flore *C.B.* Certainly it has a smaller short-lived flower, shorter, broader, paler leaves and more erect stems than Alsine myosotis hirsuta *Ad*.[16] What is more in drier places the whole plant (Stems, leaves, flowers) is sticky. It flowers around the end of April & the beginning of May and shortly afterwards it begins to wither.[17]

B6(A2) ***Montia fontana*** L.

Alsine palustris Portulacæ aquaticæ similis. Similar to if not the same as Alsine aquatica surrectior *J.B*.[18] An Alsine palustris minor serpyllifolia *Ger. Park.* Small water Chickweed by some call'd Blinks. *In the boggy grounds about Gamlingay.* The flower of this plant is small, five petalled and white and the seed vessel is tripartite, containing three black grains, and dehisces at maturity into 3 keels. The plant is of its own kind & far different from Alsine.

---

[15] The current taxonomic treatment of the *Alchemilla vulgaris* aggregate, which recognises numerous, very similar apomictic agamospecies, was not developed until the 20th century. Only *A. filicaulis* subsp. *vestita* has been recorded in Cambridgeshire. A small population survived in a pasture at Balsham until 1953 and was confirmed as this segregate, which was also recorded in Balsham churchyard as recently as 1990 (Walters 1954; Crompton 2001).

[16] i.e. *Cerastium fontanum* [26]; see also footnotes 10 and 11 in *Addenda* of the 1663 appendix.

[17] The information on flowering time, which is more detailed than that in the 1663 appendix, comes from Ray's *Catalogus plantarum Angliae* (1670, pp. 16–17; 1677, p. 16).

[18] This was the name that Ray chose for this plant in the 1663 appendix. The new name and the description that follows were taken from Ray's *Catalogus plantarum Angliae* (1670, pp. 17–18; 1677, p. 17).

B7(A4) *Moehringia trinervia* (L.) Clairv.
Alsine Plantaginis folio *J.B. Plantain-leav'd Chickweed*.[19] It has slender, weak little branches spreading on the ground and pale green leaves, more acute than Alsine media;[20] the small flower consists of five undivided petals. It flowers in spring. By hedges & in marshy thickets especially. *Especially about Balsham.*

B8 *Lythrum portula* (L.) D.A. Webb
Alsine rotundifolia sive Portulaca aquatica *Ger. emac.* aquat[ica] minor folio oblongo sive Portulaca aquatica *Park.* palustris minor folio oblongo *C.B. Water Purslane. In the wet grounds nigh Gamlingay Park.*

B9(A5) *Galium tricornutum* Dandy
Aparine semine læviore *Goose-grass with smoother seed.* Among standing corn, especially on chalky soil *as about Gogmagog Hills.* The seeds of this are much smoother than those of [Aparine] vulgaris[21] but not absolutely glabrous; accordingly it seems to be different from Aparine lævis *Park.*[22]

B10 *Aristolochia clematitis* L.
Aristolochia Clematitis *Park.* Clematitis vulgaris. *J.B.* Clematitis recta *C.B. Running-rooted Birthwort. In several Hedges at Wittlesford.*

B11 *Narthecium ossifragum* (L.) Huds.
Asphodelus Lancastriæ *Ger.* luteus acorifolius palustris Anglicus Lobelii *J.B.* Pseudoasphodelus primus *Clus.* major sive palustris *Park.* palustris Anglicus *C.B. Lancashire Asphodil, or Bastard Asphodil. In the first Bog nigh the Park at Gamlingay.*

B12(A6) *Bupleurum tenuissimum* L.
Auricula leporis minima *J.B.* An Bupleurum minimum *Park?* An Bupleurum angustissimo folio *C.B? The least Hares-ear. By the way-side as you go to S. Neots beyond Ellesly.* It flowers & produces its seed in the months of August & September. See a figure and a description of it, both accurate, in the work of *J.B., volume 3, part 2, page 201.*[23]

B.

B13 *Arctium minus* (Hill) Bernh.
Bardana major altera *Ger.* major lanuginosis capitulis *Park.* Personata altera cum capitulis villosis *J.B.* Lappa major montana capitulis tomentosis sive

---

[19] These are both new names not found in *Addenda* of the 1663 appendix (q.v. under *Alsine fontana credita flosculorum foliolis non divisis*) but later adopted by Ray (1670, p. 18; 1677, p. 17), although the English name appears as *Plantain-leafed Chickweed* in 1670 and *Plantain leaved Chickweed* in 1677.

[20] i.e. *Stellaria media* [29].

[21] i.e. *Galium aparine* [51].

[22] Parkinson (1640, p. 567) describes his plant as smaller than *Aparine vulgaris* and with smooth rather than rough seeds; he says it was "brought me out of *Spaine*". Ray never expressed a definite conclusion about its identity in works published during his lifetime (e.g. 1670, p. 25; 1690, p. 74), and it was presumably J.J. Dillenius who was responsible for removing the qualifications that Ray had expressed about Parkinson's name as a synonym of *Aparine semine læviore*, when he edited the third edition of *Synopsis* (1724, p. 225) after Ray's death.

[23] These remarks are more detailed than those in the 1663 appendix and are based on Ray's *Catalogus plantarum Angliae* (1670, pp. 36–37; 1677, p. 34).

APPENDIX (1685)

Arctium *C.B. Woolly-headed Burdock. In Kingston wood.* It does not differ from the common burdock in any way except in the burs,[24] the spines of which are interwoven with a greyish down like spiders' webs.[25]

C.

B14(A7) *Clinopodium calamintha* (L.) Stace
Calamintha flore minore odore pulegii *J.B.* odore pulegii *Ger.* altera odore pulegii, foliis maculosis *Park.* pulegii odore sive Nepeta *C.B. Field Calamint.* I had this plant in the Catalogue as Calamintha vulgaris;[26] however I think that a Calamintha other than only vulgaris grows wild around Chesterton & elsewhere.

B15 *Cardamine pratensis* L.
Cardamine altera flore pleno *Ger. emac. Double flower'd Ladies-smock.*[27] *In the Closes on the North-side of Chesterton, and in a little ditch on the left hand about mid-way to Gogmagog-hills.*

B16 *Leonurus cardiaca* L.
Cardiaca *J.B. Park. Ger.* Marrubium nigrum I, seu Marr[ubium] Cardiaca dictum, forte primum Theophrasti *C.B. Motherwort on the bank of a close on the backside of Trumpington looking towards Shelford.*

B17(A8) *Carduus crispus* L.
Carduus polyacanthos primus *Ger. emac.* Aculeosa Gazæ *Ad.* Polyacantha *Tab. in* Johnson's *opinion. Thistle upon Thistle. On the banks of ditches & among thornbushes.* The flower of this thistle is like the flower of Carduus nutans but considerably smaller[28] & sometimes nods too but only slightly.

B18(A9) *Carduus tenuiflorus* Curtis
Carduus polyacanthos secundus *Ger. emac.* sylvestris 3. *Dod.* Polyacantha *Lob.* Carduus spinosissimus capitulis minoribus *P.B. Welted Thistle with small flowers.* Look for the descriptions of these thistles in the work of *Ger. emac., chapter*

---

[24] *Non in alia re à vulgari differt quam echinis*, literally "Not in another thing from the common [one] it differs than by the 'hedgehogs'". The common one is *Arctium lappa* and/or *A. minus* [79].

[25] This description is taken from *Catalogus plantarum Angliae* (1670, pp. 38–39; 1677, p. 36) and is presumably based on plants reported therein from Craven in Yorkshire. In *Historia plantarum* (1686, p. 332) Ray said that, very recently, Samuel Dale of Braintree had shown it to him elsewhere in England, growing with *Bardana major* [79]. He later concluded (in part as a result of cultivating seed sent to him by Samuel Doody) that the montane taxon from Yorkshire was a different species from the less tomentose lowland plant (Ray 1696, p. 88). In the classification put forward by Sell & Murrell (2006), which is based on a completely different interpretation of the breeding biology of *Arctium* from that of Stace (1997, 2010), these tomentose variants are treated as *A. lappa* subsp. *pubens* (Bab.) P.D. Sell, an extremely variable taxon which tends to form distinct local populations.

[26] i.e. *Clinopodium ascendens* [100].

[27] Ray (1677) commented in the unnumbered preliminary pages of the second edition of *Catalogus plantarum Angliae* that he did not regard plants distinguished by a double [*pleno*, literally "full"] flower or a multiple flower as distinct species but that this variety should not have been omitted without a mention. There is no reference to a *flore pleno* variant in *Historia plantarum* (1686, p. 814). In *Synopsis* (1690, p. 114; 1696, p. 171) Ray says that "*Mr. Lawson observed it* flore pleno *in* Little Strickland *Pastures in* Westmorland" but does not mention Dent's Cambridgeshire record.

[28] *sed multis numeris minor*, literally "but by many numbers smaller". The entire sentence is based on *Catalogus plantarum Angliae* (1670, p. 57; 1677, p. 53).

490 of book 2, where they are distinguished from the other species accurately enough.[29]

B19 *Carum carvi* L.

Carum sive Careum *Ger.* Carum vulgare *Park.* Caros *J.B.* Cuminum pratense, Carui Officinarum *C.B. Carawaies. In Christs College piece, in the Pastures beyond, and in several places nigh Sturbridge very plentifully.*

B20(A10) *Geum rivale* L. × *G. urbanum* L.

Caryophyllata flore majore *C.B. This was found in the fields somewhere about the Town, and brought into our Gardens; I do not remember the place.*[30]

B21 *Dianthus deltoides* L.

Caryophyllus Virgineus *Ger. emac.* perhaps. Caryophylli montani purpurei[31] *Ger.* The picture corresponds very well to our plant but the description not likewise, as also not the description of [Caryophyllus] Virgineus in all respects and much less the Picture. In our plant the flowers are flesh-pink marked with purple spots around the base; the leaves (except those that grow on the flowering stem) recall the leaves of pinks very little but resemble more closely Auricula muris flore pulchro albo *J.B.*[32] both in appearance and in colour. Betonica corononararia[33] sive Caryophyllea repens rubra *J.B.* Caryophyllus sylvestris humilis flore unico *C.B.*, but he wrongly adds *with a single flower* if he means our plant, which produces two or three or even more flowers on individual stems. *Maiden Pinks. On a little hill or brow whereon Furze grows, beyond Hildersham near Juniper hill.*[34]

B22(A11) *Lathyrus nissolia* L.

Catanance leguminosa quorundam *J.B.* Ervum sylvestre *Ger. Dod.* sylvestre sive Catanance *Park.* Lathyrus V. sive sylvestris minor *C.B. Crimson Grass Vetch. Found by Mr. Dent about Hadnam in the Isle of Ely.*

B23 *Torilis arvensis* (Huds.) Link

Caucalis segetum minor Anthrisco hispido similis. *Small Corn Parsly.* This differs from Caucalis minor flosculis rubentibus[35] in some ways. In the first place that plant is taller, not infrequently reaching in fact a height of two cubits, while this is short, scarcely ever exceeding a foot or certainly one cubit. Secondly, that plant is more erect & has longer internodes, while this is tortuous & diffuse with more frequent knots. Thirdly, the stem of that plant is hairier near the ground than this one's. Fourthly, the flowers of that plant are white or even pale reddish and consist of five petals of almost equal size & contain stamens tipped with purple

---

[29] This remark was first published by Ray (1670, p. 57; 1677, p. 54) and the reference is to p. 1176 of Johnson (1633): see footnotes 18 and 19 in *Addenda* of the 1663 appendix.

[30] See the footnote 22 in *Addenda* of the 1663 appendix.

[31] This plural name could refer to two or more of the plants described and illustrated by Johnson (1633, pp. 592–596) as *Caryophyllus montanus purpureus*, *Caryophyllus montanus Clusij* and *Caryophyllus montanus humilis latifolius*, but none of these seem to be British species. Neither this name nor the two following sentences appear in Ray's *Catalogus plantarum Angliae* (1670, 1677) and they may therefore be Dent's own work; the Bauhin synonyms and the subsequent comment are from Ray (1670, pp. 59–60; 1677, p. 57).

[32] i.e. *Cerastium arvense* [76].

[33] This is a misprint for *coronaria*.

[34] This record was first published in *Catalogus plantarum Angliae* (1670, pp. 59–60); in *Synopsis* Ray (1690, p. 138) attributes it to Mr Dent.

[35] i.e. *Torilis japonica* [119].

heads, while this one's flowers, verging towards yellow, have the two outer petals clearly broader than the inner ones; moreover the stamens are crowned with white tips. Fifthly, the seeds of this plant are much bigger, with their tips green at the top, while that one's are far more fragrant and more aromatic, of a darker colour with purple tips. Finally, with us you will not easily find that one away from hedges & thorn-bushes, while this one never grows in hedges but for the most part among standing corn.[36] Generally.

B24 *Prunus avium* (L.) L.
Cerasus sylvestris fructu rubro *J.B. Common Wild Cherry. In a Thicket nigh Elsworth wood.*

B25 *Prunus cerasus* L.
Chamæcerasus *Ger. Dwarfe Cherry-tree. In some closes of Teversham going from the Church towards Gains.*

B26(A12) *Veronica montana* L.
Chamædrys spuria foliis pediculis oblongis insidentibus. Alysson Dioscoridis montanum *Col.* An Chamædry spuriæ affinis rotundifolia scutellata *C.B?*[37] *Wild Germander with leaves standing on long foot-stalks. In moist woods & by hedges on moist ground, but rather rarely.*

B27 *Teucrium* sp.[38]
Chamædrys vulgo vera existimata *J.B.* vulgaris *Park. Germander. In a little Island nigh the road by Qui water on the right hand from Cambridge.*

B28 *Convolvulus arvensis* L.
Convolvulus minimus. *The least Bindweed.*[39] This plant has narrow leaves, cuspidate almost in the manner of Sagittaria,[40] much smaller than the leaves of Convolvulus minor vulgaris;[41] it also has smaller flowers; and so it seems to differ from that as a species, to be sure because, when they grow side by side in the same soil, they still show the above-mentioned differences.[42] *In the Corn between Harleston and little Eversden.*

---

[36] This description is based on the account by Ray (1670, pp. 61–62; 1677, pp. 58–59).
[37] Literally "Is it C. B[auhin]'s round-leaved [plant] bearing little dishes, related to false Ground-oak?"
[38] *Chamædrys vulgo vera existimata* J.B. is *Teucrium chamaedrys* L., which is a very uncommon alien plant in Britain, usually found on walls, rocks and dry banks (Preston, Pearman & Dines 2002). Ray (1686, p. 527; 1690, p. 78) described it as frequent in Germany, France and Italy; in England he said he had found it a number of times on the margins of fields fairly remote from any building but nevertheless he did not dare to assert that it was a wild plant. It is possible that Dent's record is of a similar occurrence, but we agree with Babington (1860) that "there is every reason to consider that his record is a mistake". The habitat suggests that he may have seen *Teucrium scordium* L., another species with pinkish-purple flowers which was reported by Ray in the main catalogue as *Scordium* [538].
[39] Ray (1677, preliminary pages) omitted this species from the second edition of *Catalogus plantarum Angliae*, commenting that, although it might be distinct from the plant we now call *Convolvulus arvensis*, he had not seen it for himself and would not dare to determine it for certain. However, he included it as a full species, *Convolvulus arvensis minimus* D. Dent, in *Historia plantarum* (1686, p. 725), with the same description as that in this appendix, and retained it in *Fasciculus* (1688b, p. 5, as *Convolvulus minimus* Cat. Cant. App.) and *Synopsis* (1690, p. 102; 1696, p. 157).
[40] i.e. *Sagittaria sagittifolia* [506, 507].
[41] i.e. *Convolvulus arvensis* [149].
[42] ... *supradictas tamen differentias sortiantur.* Literally "... the above-mentioned, however, differences they divide." Sell & Murrell (2009) suggest that *C. arvensis* is phenotypically plastic, with the size of its leaves and flowers depending on the habitat in which it grows.

D.

B29(A13) unidentifiable fern[43]

DRyopteris Tragi *Ger. emac*. Filix saxatilis sive ramosa nigris maculis notata *C.B.* ramosa minor *J.B.*[44] pumila saxatilis prima Clusii *Park*. *The lesser branched or Female Fern. Observ'd by Mr. Dent on Kings College walls.*

E.

B30 *Erica tetralix* L.

ERica pumila altera Belgarum *Lob*. Brabantica folio Coridis hirsuto quaterno *J.B.* pumila Belgarum Lobelii, Scoparia nostras *Park*. folio Coridis 13. *Clus*. folio Coridis 7, sive ex rubro nigricans Scoparia *C.B.* major flore purpureo *Ger. emac*. *Low Dutch Heath or Broom Heath. Upon all the boggy grounds about Gamlingay.*

F.

B31(A14) *Hippocrepis comosa* L.

FErrum equinum Germanicum siliquis in summitate *C.B.* equinum comosum *Park*. Ornithopodio affinis vel potius soleæ aut Ferro equino herba[45] *J.B. Bush-headed Horse-shooe-Vetch. On Gogmagog hills, and the dryer part of Hinton Moor and on New-Market Heath, &c.*

B32 *Osmunda regalis* L.[46]

Filix florida sive Osmunda Regalis *Ger*. Osmunda regalis sive filix florida *Park*. Filix ramosa non dentata florida *C.B.* Filix floribus insignis *J.B. Water Fern or flowring Fern or Osmund Royal. Within Gamlingay Park, and without by the Pales in the corner next to Sandy.*

B33 *Athyrium filix-femina* (L.) Roth

Filix mas non ramosa pinnulis angustis raris profunde dentatis *Ger. emac*. mollis sive glabra vulgari mari non ramosæ accedens[47] *J.B. Male Fern with thin set deeply indented leaves. Nigh the* Filix florida.[48]

---

[43] See the footnote on this taxon in *Addenda* of the 1663 appendix.

[44] From here until the end of letter H, and frequently in the subsequent text, the B in C.B. and J.B. is in roman rather than italic font. The capital B in some other italicised text is also in roman type, e.g. "B*utchers* B*room*" in the entry for *Ruscus* [B87] and "B*oth in the* B*og* …" in that for *Salix humilis repens rotundifolia* [B89], suggesting that the printer was short of italic capital B type. We have kept to the usual convention of italicising these letters.

[45] Literally "A herb related to Small Birdsfoot or rather to a shoe or Iron for a horse". See also footnote 25 in *Addenda* of the 1663 appendix.

[46] This record was doubted by Babington (1860) on the grounds that the species was not seen there subsequently and that the locality was not included by Ray in his national Floras. However, it is one of the most distinctive of ferns and one which does occur in lowland bogs such as Roydon Common in Norfolk (Petch & Swann 1962) and Askham Bog near York (Fitter & Smith 1979, p. 65). By 1670 Ray (p. 113) already knew it from "boggy places in most Countreys I have passed through, as in Middlesex, Suffolk, Cornwal, York-shire, and Westmorland". We therefore accept the record, as did Perring *et al*. (1964), though it was queried by Crompton & Whitehouse (1983).

[47] "Soft or glabrous [fern] coming close to the common unbranched male [fern]", i.e. to *Dryopteris filix-mas* [199].

[48] See the previous entry (*Filix florida sive Osmunda Regalis* Ger.).

APPENDIX (1685)

B34 *Oreopteris limbosperma* (All.) Holub
Filix palustris seu aquatica. An Filix minor non Ramosa *J.B?* Both the description and the habitat are appropriate. *Marsh or Water Male Fern.* It is similar to Filix mas communis[49] but much smaller, paler & yellowish; the pinnules also are very little toothed around their edges. *Found together with the two last mention'd within the Park and nigh the Park Pales.*

B35 *Fumaria* sp.[50]
Fumaria major scandens flore pallidiore. *The greater ramping Fumitory. In the Corn upon the borders of Gogmagog hills towards Hinton Moor.* This is larger in all its parts than the common fumitory.[51] The flower is flesh-pink with the middle part deep red. It seizes neighbouring stems with the stalks of its leaves in almost the same way as Nasturtium Indicum[52] (for it does not have claspers[53] or tendrils) sprawls on the ground elsewhere.

G.

B36(A15) *Gnaphalium sylvaticum* L.
Gnaphalium Anglicum *Ger.* Anglicum vulgare majus *Park.* majus angusto oblongo folio alterum *C.B.* rectum *J.B. Great English Cudweed. In the pastures about Gamlingay, and in the Park.*

B37(A17) *Carex remota* L.
Gramen cyperoides angustifolium paniculis plurimis sessilibus in foliorum alis. *Narrow leav'd Cyperus-grass with small panicles. On the banks of many ditches.* There is a long leaf under each panicle.

B38(A16) *Carex pendula* Huds.
Gramen cyperoides spica pendula longiore *Park.* [cyperoides] sp[ica] pend[ula] longiore & angustiore *C.B.J.B.* Pseudocyperus Ger. *Bastard Cyperus grass with long pendulous heads. In the great ditch at the end of the little Thicket adjoyning to Teversham Moor, and in other great Ditches.*

B39 *Juncus bulbosus* L.
Gramen junceum capsulis triangulis minimum. Graminis juncei varietas minor. *Ger. emac.,* where it is wrongly given as a variety of Gramen bufonium.[54] *The least*

---

[49] i.e. *Dryopteris filix-mas* [199]. This sentence comes from Ray (1670, p. 114; 1677, p. 108).
[50] The plant described is clearly one of the taller fumitories, all of which have predominantly western distributions in Britain. If the Cambridgeshire plant was indeed one of these, the most likely species is *Fumaria muralis* Sond. ex W.D.J. Koch, which is recorded from time to time in Cambridgeshire, especially in the Fens, but it may not be an established member of the flora and there are no confirmed records from the Gogs, where fumitories have long been studied. However, the description comes from *Catalogus plantarum Angliae* (1670, p. 122; 1677, p. 117) and is almost certainly based on material from outside Cambridgeshire. Babington (1860) listed Ray's chosen name under *F. capreolata* L. but suggested that some of the Cambridgeshire records of this species (including a later one from the Gogs) might have been based on a rampant form of *F. officinalis* [204].
[51] i.e. *Fumaria officinalis* [204].
[52] i.e. Nasturtium, *Tropaeolum majus* L.
[53] See *Clavicula* in Ray's 'Interpretation of terms'.
[54] i.e. *Juncus bufonius* [256]. The reference is to Johnson (1633, p. 4), where he says: "There is a varietie of this to be found in bogs, with the seeds bigger, and the leaues and whole plant lesser." He had found this "on the bogges vpon Hampstead heath". Clarke (1900) suggests that Johnson's plant was

Rush-grass with triangular seed vessels. In some wet places of the Heath between Gamlingay & Potton.

B40 *Rhynchospora alba* (L.) Vahl
Gramen junceum leucanthemum *Ger. emac.*, book 1, chapter 22, number 7.[55] An Gramen junceum polysttachion[56] *C.B?* *Boggy Rush-grass with white chaffy heads. On a boggy ground nigh Gamlingay Park.*

H.

B41(A18) *Helleborus foetidus* L.
Helleboraster maximus *Ger.* maximus sive Consiligo *Park.* Helleborus niger fœtidus *C.B.* Enneaphyllon *of Pliny*, book 27, chapter 9. *Great Bastard Hellebore or Bears-foot. In the second Close on the left hand as you go from Hinton Church to Fulbourn: also in the hedges of a close near Hinton Moor.*

B42 unidentifiable orchid
Helleborine minor flore albo *Park*. *Narrow-leav'd Bastard white Hellebore with white flowers.*[57] *On Teversham Moor nigh Qui water.*

B43(A19) *Marchantia polymorpha* L., female plants[58]
Hepatica stellata *Ger.* Lichen sive Hepatica minor stellaris *Park*. Lichen petræus stellatus *C.B.* Star-headed Liverwort. *Not infrequent in shady gravelly places.*

B44(A20) *Marchantia polymorpha* L., male plants
Hepatica umbellata *Ger.* Lichen sive Hepatica minor umbellata *Park*. Lichen petræus umbellatus *C.B. Small Ground Liverwort with round heads. On the Moors.*

---

probably *J. bulbosus* and this seems very likely, although, as Ray suggests, it is not closely related to *J. bufonius*. Ray's (1670, p. 150) record of *Gramen junceum capsulis triangulis minimum*, usually cited as the first British record of *J. bulbosus*, was from "boggy places by rills of water on Hampsted-heath, and else-where" and Kent (1975) reported that the species was "still present in the bog on the West Heath" three centuries later. Edgington (2007, pp. 288–289) has suggested that Johnson's plant was in fact *Juncus ranarius* Songeon & E.P. Perrier (*J. ambiguus* Guss.), but this is implausible as the differences between this species and *J. bufonius* are very slight and *J. ranarius* is a plant of seasonally flooded, brackish habitats.

[55] The reference is to Johnson (1633, p. 30), where the seventh, unillustrated species in his new chapter entitled *Of diuers other Grasses* is described as "a pretty rushie grasse some foote or better in height" with "a slender stalke being some 2 or 3 inches high, and at the top … as in a little vmble a prety white chaffie floure; and at, or nigh to the top of the maine stalke … three or foure such floures clustering together vpon little short and slender foot stalkes". He says that this "may very fitly be called *Gramen junceum leucanthemum*: White floured rush-grasse".

[56] This is a misprint for Bauhin's *polystachion*.

[57] Ray (1677, preliminary pages) omitted this species from the second edition of *Catalogus plantarum Angliae*. He did not give his reasons, but it seems likely that he did not believe the record. *Helleborine minor flore albo* Park. is the woodland orchid *Cephalanthera damasonium* (Mill.) Druce and, as Babington (1860) remarked, the locality is a very unlikely one for this species. It was first recorded from Britain by Ray (1670, p. 339) from woods in Oxfordshire and Surrey. In Cambridgeshire it is an apparently natural colonist of planted woodland of a type that would not have been present in the 17th century; it was first found by J.S. Henslow & L. Jenyns in 1825 in a belt of woodland at Bottisham (Crompton 2004). Dent's plant may have been, as Babington (1860) implies, *Epipactis palustris*, reported from Teversham Moor by Ray in the main catalogue [274].

[58] See the footnote on this species in *Addenda* of the 1663 appendix.

APPENDIX (1685)

B45 *Hieracium sabaudum* L. forma **sabaudum**

Hieracium fruticosum latifolium hirsutum *Park. C.B.* Hieracii Sabaudi varietas altera *J.B. Bushy Hawkweed with rough broad leaves.* This plant is distinct specifically from Pulmonaria Gallorum,[59] since it grows taller, not infrequently exceeding two cubits; the stem also is more robust & rigid, clothed with very many leaves.[60]

B46 *Hieracium sabaudum* forma **bladonii** (Pugsley) P.D. Sell

Hieracium fruticosum latifolium glabrum *Park.* frut[icosum] latif[olium] foliis dentatis glabrum *C.B.* Hieracii Sabaudi varietas I, Erinus quibusdam *Matth.* dicta[61] *J.B. The smaller broad-leav'd bushy Hawkweed.* There is a clear difference between this & the preceding hawkweed, for that has leaves & stems pubescent with a soft down while this is covered with a short & scarcely noticeable down; next, that one flowers later and this one earlier. It is badly described in *Parkinson.*[62]

B47 *Hieracium umbellatum* L.[63]

Hieracium fruticosum angustifolium majus *C.B. Park.* intybaceum *Ger.* rectum rigidum, quibusdam Sabaudum *J.B. Narrow leav'd bushy Hawkweed. These three last in Gamlingay Park, and on the sides of the great Road beyond the Park.*

B48(A21) *Hypochaeris maculata* L.

Hieracium montanum caule aphyllo non ramoso flore pallidiore. An Hieracium montanum hirsutum minus *C.B. Mountain Hawkweed with a pale flower. On Gogmagog hills, and New Market heath.* The flower of this species is pale like that of Auricula muris.[64] The leaves, as far as I remember, are not jagged but in some cases a little wavy-edged. It is not unlike Hieracium Narbonense rotundifolium caule aphyllo[65] *J.B.*

---

[59] This species is treated in both the 1663 and the 1685 appendices as *Pulmonaria Gallica sive aurea latifolia* Ger. [A32, B81]; for a brief discussion about its identity see footnote 41 in *Addenda* of the former.

[60] This description is taken from Ray (1670, p. 165; 1677, p. 160), where the species is not reported from Cambridgeshire.

[61] "The first variety of *Hieracium Sabaudum*, called by Mattioli *Erinus* according to some [authors]."

[62] These comments come from Ray (1677, p. 160) and, like the description of the preceding taxon, are not based on the Cambridgeshire plant. It is very likely that both the plants reported by Dent were variants of *Hieracium sabaudum*. This species varies greatly in hairiness and the two extremes have been given the names cited above. "When the two forms grow together the hairiness makes them look very different." (Sell & Murrell 2006). However, whereas forma *sabaudum* still occurs at Gamlingay, there do not appear to be any confirmed records of forma *bladonii*. *H. vagum* Jord. can be confused with this form, but it is not confirmed from Gamlingay either (Crompton 2003).

[63] Babington's (1860) identification of Ray's plant as this species is supported by Perring *et al.* (1964).

[64] i.e. *Pilosella officinarum* [437] in this case, as *Cerastium fontanum* [26] and *Myosotis arvensis* [381] are clearly not intended.

[65] "Round-leaved Hawkweed of Narbonne with a leafless stem", a plant described and illustrated by Bauhin & Cherler (1651, vol. 2, p. 1037) from rocks at Montpellier and treated under the same name by Ray (1686, p. 247). The comparison was first made by Ray (1670, p. 167). In *Flora Galloprovincialis* Gerard (1761) lists Bauhin & Cherler's plant in the synonymy of the species which later became *Taraxacum officinale* Wigg. (in the broadest sense), but the leaves of the plant illustrated by Bauhin & Cherler are quite unlike those of any dandelion and the plant depicted does not appear to resemble any British species.

B49 *Moenchia erecta* (L.) P. Gaertn., B. Mey. & Scherb.
Holosteum minimum tetrapetalon sive Alsine tetrapetalos Caryophylloides.[66] *The least Stitchwort*. Its root is white, equipped with fibres. The numerous little stems are an inch and a half high,[67] slender, weak, round in section, reddish and separated by frequent joints. Paired opposite leaves are situated at each joint, almost encircling the stem, scarcely half an inch long and gradually narrowing from the rather broad base to an acute point. The flowers at the tops of the branchlets are white, four-petalled from a four-sepalled calyx, by which character it is distinguished sufficiently from all the remaining species of this genus. It flowers in spring in rather barren gravelly places[68] *as nigh that gate of Gamlingay Park which is next the Town*.

B50(A22) *Hypericum hirsutum* L.
Hypericum majus sive Androsæmum Matth[ioli] *Park*. Hyp[ericum] Androsæmum dictum *J.B.* Androsæmum Hypericoides *Ger.* alterum foliis Hyperici, quod aliquibus Hypericoides[69] *C.B. Tutsan S. Johns wort*. In thickets & hedges generally. The leaves in this species are hairy and larger than those of Hypericum vulgare,[70] the flowers of a paler yellow colour, and the stems that bear the flowers rise up into the shape of a pyramid while the flowers of Hyp[ericum] vulg[are] rather more recall an umbel.[71]

B51 *Hypericum humifusum* L.
Hypericum minus supinum *Park*. minimum supinum *J.B.* minus supinum vel supinum glabrum. *C.B.* supinum glabrum *Ger. Small creeping S. Johns wort*. *On the first hill next Hildersham towards Juniper hill.*

J.

B52 *Senecio erucifolius* L.
JAcobæa Senecionis folio incana perennis.[72] An Jacobæa vulgaris minor *Park?* if indeed his description fits this better than Jacobæa latifolia palustris sive aquatica.[73] *Perenniall Ragwort with Groundsell leaves*. On the banks of hedges & in thickets generally.

B53 *Isolepis setacea* (L.) R.Br.
Juncellus omnium minimus, Chamæschœnos *Lob. Ad.* Juncus humilis *J.B.* An Juncus lævis nonus sive Juncellus inutilis *C.B.?* who is of the opinion that Juncellus omnium minimus *Ad.* should be referred to it. *The least Rush*. It never

---

[66] "The least four-petalled Stitchwort or four-petalled *Alsine* like a pink."
[67] *semi-palmates*, literally "of half a palm" (in the plural), the width of the palm of the hand being regarded as three inches.
[68] The description up to this point comes from Ray (1670, p. 168; 1677, p. 163); only in the 1677 edition is the plant marked as occurring in Cambridgeshire.
[69] "The second [*Androsæmum*] with leaves of a St John's-wort, which for some [authors is] Hypericoides."
[70] i.e. *Hypericum perforatum* [296].
[71] This description does not appear to be taken from Ray's *Catalogus plantarum Angliae* (1670, 1677).
[72] "Hoary perennial St James'-wort with the leaf of Groundsel."
[73] i.e. *Senecio aquaticus* [301], which is included in the main catalogue without a synonym from Parkinson.

APPENDIX (1685)

attains a height of three inches,[74] but it produces a panicle of almost the same shape as Juncus aquaticus maximus.[75] *In Gamlingay Park, and many other watry places thereabouts.*

B54 *Juncus squarrosus* L.

Juncus acutus Cambro-Britannicus *Park. Moss Rush, Goose Corn. Upon all the boggy and heathy grounds about Gamlingay.* It has rather large reddish fibrous roots.[76] The numerous leaves are not erect but generally spread over the ground like a star, short, hard, sharply pointed and not precisely round in section but channelled in the upper part. From the middle of the leaves there arises a stem which is nine inches or a foot high, firm, filled with pith,[77] encircled by only one leaf close to the ground but otherwise naked, bearing at its top a rushy panicle composed of several parts which emerge singly from the axil of the leaf that is based on the stem at their base. The meagre flower of this hairy grass is six-petalled with the same number of fringed yellow stamens.[78] There follow fairly large, roundish capsules containing the seed.

L.

B55 *Rumex crispus* L.

Lapathum folio acuto crispo *C.B.* acutum crispum *J.B.* acutum minus *Park.* acuti varietas folio crispo *Ger. Sharp pointed Dock with curled leaves.*[79] Frequent in ploughed fields.

B56 *Rumex maritimus* L.

Lapathum folio acuto flore aureo *C.B.* anthoxanthon *J.B. Golden Dock.*[79] *In a Fen two mile from Ely call'd the North-Fen.* The seeds of this plant are small & paler than those of the rest of its congeners.

B57 *Rumex pulcher* L.

Lapathum pulchrum Bononiense sinuatum *J.B. Fiddle-Dock.*[79] *In Trinity and Caius College courts among the stones.*

B58 *Lepidium latifolium* L.

Lepidium latifolium *C.B.* Lep[idium] Pauli *J.B.* Piperitis sive Lepidium vulgare *Park.* Raphanus sylvestris Officinarum, Lep[idium] Æginetæ Lobelio *Ger.*

---

[74] *Palmarem altitudinem nunquam assequitur*, literally "A palm's height never it attains": see footnote 67. This phrase is from Ray's description in *Catalogus plantarum Angliae* (1670, p. 181; 1677, p. 175), but the second half of the sentence is different there as Ray adds "paniculam tamen profert, unde nulla ratio est cur inutilis dicatur" ("but it produces a panicle, whence there is no reason why it should be called useless").

[75] i.e. *Schoenoplectus lacustris* [306].

[76] *Fibris majusculis radicatur rubentibus*, literally "With somewhat greater reddish fibres it takes root". The entire description is from *Catalogus plantarum Angliae* (Ray 1670, p. 179; 1677, pp. 172–173); it was not until the second edition that Ray marked the species with a C, indicating that it was known from Cambridgeshire.

[77] ... *caulis dodrantalis aut pedalis, firmus, medulla farctus*, ...

[78] The words "of ... hairy grass" (*graminis hirsuti*) and "with ... fringed stamens" (*staminulis fimbriatis*) do not fit *Juncus squarrosus*, but nevertheless there seems to be no doubt that this is the species described here.

[79] See also Ray's notes on the species of *Lapathum* occurring wild in England after the entry for *Lapathum aquaticum sive Hydrolapathum minus* in *Emendanda* below.

Dittander or Pepperwort. In a little Close on the right hand of Maids Causey in the way to Barnwell.

B59(A23) ***Blechnum spicant*** (L.) Roth
Lonchitis aspera *Dod. Ger.* minor *C.B.* aspera minor *Park. Matth. Cam. Lugd.* altera folio Polypodii *J.B. Rough Spleenwort. About Gamlingay.*

B60 ***Botrychium lunaria*** (L.) Sw.
Lunaria minor *Ger. Park.* botrytis *J.B.* racemosa minor vel vulgaris *C.B. Small Moonwort. Upon the same hill where* Hypericum supinum glabrum[80] *grows, As also in an old Gravell Pit on the left hand of the road to Balsham about half a Mile from the Town.*

B61 ***Silene gallica*** L.
Lychnis sylvestris annua angustifolia flore rubente. An Lychnis segetum parva viscosa *P.B.? Among the Corn near the Devils ditch on Newmarket heath.*[81]

M.

B62 ***Mentha* ? *spicata*** L. × ***suaveolens*** Ehrh.[82]
MEntastrum folio rugoso rotundiore spontaneum flore spicato odore gravi *J.B.* Mentha sylvestris I. sive rotundiore folio *C.B.* an Mentastrum *Ger.* & mentastrum hortense sive Mentha sylvestris *Park?* for the figures in no way match it. *Horsemint or round leav'd Wild Mint. In the Chalk-pit close at Cherry Hinton.*

B63 ***Mentha spicata*** L.[83]
Mentastrum spicatum flore longiore candicante *J.B.* Mentha sylv[estris] folio longiore *C.B.* Mentastrum *Ger.* hortense sive Mentha sylvestris *Park.* as far as the Picture goes. *In a bushy close near Exning.*

B64 ***Mercurialis annua*** L.
Mercurialis mas & fœmina *J.B.* vulgaris mas & fœmina *Park.* Mercurial[is] mas & Merc[urialis] fœm[ina] *Ger.* Mercurialis I, i.e. testiculata sive mas Dioscoridis & Plinii *C.B.* likewise Mercurialis II, i.e. spicata sive fœmina Diosc[oridis] & Plinii of the same author. *French Mercury.* It occurs very frequently among vegetables & on footpaths.[84]

---

[80] i.e. *Hypericum humifusum* [B51].
[81] This record was first published by Ray in *Catalogus plantarum Angliae* (1670, p. 202), where it is attributed to "*Mr. Dent Apothecary in Cambridge*".
[82] This was identified by Babington (1860) as *Mentha rotundifolia*, but in his day this covered the species now known as *M. suaveolens* Ehrh. and its hybrids with *M. longifolia* (*M.* × *rotundifolia* (L.) Huds.) and *M. spicata* (*M.* × *villosa* Huds.). As Crompton (2004) points out, any allocation of the historic records of "*M. rotundifolia*" from Cambridgeshire to these taxa must be conjectural, but we follow her in tentatively assigning this record to *M.* × *villosa* as there are virtually no modern records for either of the others.
[83] Babington (1860) and Perring *et al.* (1964) identified this as *Mentha longifolia* (L.) Huds. (*Mentha sylvestris* L.), but these names are now believed to have been misapplied in Britain to hairy variants of *M. spicata*.
[84] See footnotes 99 and 157 in the main catalogue.

APPENDIX (1685)

B65 *Utricularia minor* L.
Millefolium palustre galericulatum minus flore minore. *The smaller hooded water Milfoil. On Teversham Moor.* The flowers in this species are much paler than in the greater one.[85]

B66 *Galium saxatile* L.
Mollugo montana minor Gallio albo similis. *Small Mountain Bastard Madder.* It grows both in marshes and in moorish places[86] & heaths; it is distinguished from Gallium album[87] by its smallness & hairlessness. *On the Heath beyond Gamlingay.*

B67(A24) *Thuidium tamariscinum* (Hedw.) Schimp.
Muscus filicinus *Ger. J.B. Park.* filicinus major *C.B. Fern Moss. In Kingston and Eversden woods, and the Woods about Balsham.*

B68(A25) *Fontinalis antipyretica* Hedw.
Muscus triangularis aquaticus. *Triangular Water-Moss.* This floats in waters and has larger leaves than Muscus terrest[ris] vulg[aris] & minus ramosus, though otherwise very similar. Perhaps Fontalis minor lucens *J.B.*, who in fact does not mention about the leaves of his Fontalis that they are so disposed along the stem that it appears triangular.[88]

N.

B69 *Narcissus poeticus* L. × *N. tazetta* L.
NArcissus medio-luteus *Ger. Dod.* pallidus circulo luteo *C.B. Primrose-Pearles or the Common white Daffodill.*[89] *In Barnwell Abby, and other places in this County.*

---

[85] i.e. *Utricularia vulgaris* [369]. *Millefolium palustre galericulatum minus flore minore* was an addition to the second edition of the *Catalogus plantarum Angliae* and the name alone appears in the alphabetical list there (Ray 1677, p. 200). Even the name is not included in *Historia plantarum*, although under *Millefolium palustre galericulatum* Ray (1688a, p. 1322) notes: "Duas species observavit D. Dent, majorem et minorem." ("*Master Dent* has observed two species, a greater and a lesser.") The species is not treated in full until the first edition of *Synopsis* (1690, p. 208), where the first record cited is Dent's from Teversham Moor. These are the first British records of *Utricularia minor* (Clarke 1900).

[86] See footnote 154 in the main catalogue.

[87] i.e. *Galium palustre* [210]. The entire sentence is based on Ray (1670, p. 212); the comment about hairlessness presumably refers to the absence of small prickles on the stem.

[88] See also Ray's similar comment on this moss in English in *Addenda* of the 1663 appendix [A25] and the accompanying footnote.

[89] This is one of Dent's records that were omitted by Ray (1677, preliminary pages) from the second edition of *Catalogus plantarum Angliae*. He commented that he too had seen it several times on the margins of fields but suspected that it owed its origin either to remnants of gardens or to rejects from them. He later admitted it into his Floras (e.g. 1688b, p. 17; 1690, p. 166). This daffodil was described by Gerarde (1597, p. 109) as "that sort of *Narcissus* or Primrose peerelesse, that is most common in our countrey gardens, generally knowne euery where", and its subsequent history is described by Bowles (1934, as *Narcissus biflorus* Curtis). The valid binomial for this hybrid is *Narcissus* × *medio-luteus* Mill.

O.

B70(A26) *Ophrys sphegodes* Mill.
ORchis sive testiculus sphegodes hirsuto flore *J.B.* Serapias IX. sive fucum referens colore rubiginoso *C.B.* This I suspect to be what the Herbalists of our country call Spider Orchid.[90] *The green-wing'd Humble-bee Orchis. In an old Gravel-pit near Shelford by the foot way from Trumpington to the Church plentifully.* It flowers around the end of April & the beginning of May.

B71(A27) *Herminium monorchis* (L.) R.Br.
Orchis pusilla odorata *Park.* parva Autumnalis lutea *J.B.* odorata moschata sive Monorchis *C.B.* The sweet Yellow Musk Orchis. *In the chalk-pit Close at Cherry Hinton, and in some pits not far thence about Gogmagog Hills.*

B72 *Ornithogalum pyrenaicum* L.
Ornithogalum angustifolium majus floribus ex albo virescentibus *C.B.* Asphodelus bulbosus *Ger.* Asphod[elus] bulbosus Dodonæi sive Ornithogalum spicatum flore virente *J.B. Spiked Star of Bethlehem with a greenish flower.*[91] *In a bushy Close in little Eversden near the Church.*

P.

B73 *Potentilla argentea* L.
PEntaphyllum erectum foliis profunde sectis, subtus argenteis, flore luteo[92] *J.B.* rectum minus *Park.* Quinquefolium erectum I sive erectum folio argenteo *C.B.* Quinquef[olium] Tormentillæ facie *Ger. Tormentill Cinquefoil. In many places about Linton and Gamlingay.*

B74(A28) *Euphorbia peplus* L.
Peplus sive Esula rotunda *J.B. C.B. Ger.* Esula rotunda sive Peplos *Park. Petty Spurge. In more fertile fields & Gardens among vegetables.*[93]

B75(A29) *Asplenium scolopendrium* L.
Phyllitis *Ger.* Phyll[itis] sive Lingua Cervina vulgi *J.B.* Phyll[itis] seu Lingua Cervina vulgaris *Park.* Phyllitis I Lingua Cervina Officinarum *C.B. Hartstongue Was found on Hinton Church.*

B76(A30) *Arabidopsis thaliana* (L.) Heynh.
Pilosella siliquata major & minor. *Thal.* Bursa Pastoria sive Pilosella siliquosa *J.B.* Bursæ pastoriæ similis siliquosa major & minor[94] *C.B.* Paronychia major & Paronychia altera minor *Park.* For they do not differ in species but only in size. *Codded Mousear. In some Closes on the North side of Gamlingay.*

---

[90] Hanc esse suspicor quam Herbarii nostrates Orchin Arachnitidem appellant.
[91] Ray (1677, preliminary pages) omitted this species from the second edition of *Catalogus plantarum Angliae*, commenting that he too had sometimes observed it in orchards but that he believed it to be of the same origin as *Narcissus medio-luteus* (see footnote 89). This must throw considerable doubt on the traditional view that this species is native in a small area of Bedfordshire and adjacent counties of eastern England, an outlier of its main British range (see Preston, Pearman & Dines 2002).
[92] "Erect Five-leaf with leaves deeply cut [and] silver beneath, with a yellow flower."
[93] See *Tithymalus helioscopius* below in *Emendanda*.
[94] Literally "Greater & lesser podded [plant] like Shepherd's-purse".

APPENDIX (1685)

B77(A31) *Limosella aquatica* L.
Plantaginella palustris *C.B.* Plantago aquatica minima *Clus.* aquat[ica] minima Clusii *Park.* *The smallest Water-Plantain.* Along roads where waters have flooded during the Winter. This is either the same as Plantago aquatica minima of [Charles] de L'Écluse or very close to it: it does not accord with its description in all respects.

B78 *Plantago major* L.
Plantago rosea glabra non spicata. *Rose Plantain with a flat Flower.* In Jesus College close, and in the Fields in other places. It seems to differ from the common garden plant in some respects.[95]

B79 *Prunus domestica* subsp. *insititia* var. *syriaca* (Borkh.) Koehne[96]
Prunus sylvestris fructu majore albo. *The white Bullace.*

B80 *Prunus domestica* subsp. *insititia* var. *nigra* Asch. & Graebn.[96]
Prunus sylvestris fructu majore nigro. *The Black Bullace.* Both grow very plentifully in many Hedges of this County.

B81(A32) *Hieracium* sp.[97]
Pulmonaria Gallica sive aurea latifolia *Ger.* Hieracium murorum folio pilosissimo *C.B.* Hieracium murorum Bauhini, quod est Pulmonaria Gallorum Lobelii *Park.* Pilosella major quibusdam, aliis Pulmonaria flore luteo *J.B.* *French or Golden Lungwort.* About Gamlingay.

R.

B82 *Ranunculus penicillatus* subsp. *pseudofluitans* (Syme) S.D. Webster[98]
Ranunculo sive Polyanthemo aquatili albo affine Millefolium Maratriphyllum fluitans[99] *J.B.* Millefolium Maratriphyllum Ranunculi flore *Park.* Millefolium aquaticum 7. sive Foliis fæniculi[100] Ranunculi flore & Capitulo *C.B.* *Fennell leav'd Water-Crowfoot.* In the river between Cambridge and Chesterton.

B83(A33) *Ranunculus parviflorus* L.
Ranunculus hirsutus annuus flore minimo. *Field Crowfoot with a small flower.* The leaves are hairy and pale green; the flowers are yellow and the smallest of all that

---

[95] Parkinson (1640, p. 495) described *Plantago latifolia rosea multiformis* ("The broad leafed Plantaine of much variety") as "although a strange Plantane, yet naturally found in our owne country, and ... cherished in gardens". This was another plant omitted by Ray (1677, preliminary pages) from the second edition of *Catalogus plantarum Angliae*, in this case because he thought that it had arisen from seed at some time thrown out of gardens. See also footnotes 551 and 552 in the main catalogue about another teratological variant of *Plantago major*.

[96] The varietal names of both bullaces are based on Sell's (1991) account of the cherries and plums of Cambridgeshire. Both were omitted by Ray from the second edition of *Catalogus plantarum Angliae*, although he conceded that, as Dent had asserted that both were not infrequent in hedges, it was possible that they should be accepted as wild.

[97] See the footnote about this taxon in *Addenda* of the 1663 appendix.

[98] For the identity of the large Batrachian *Ranunculus* in the River Cam, see Murrell & Sell (1990). They treat *R. pseudofluitans* (Syme) Newbould ex Baker & Foggitt as a species distinct from *R. penicillatus*.

[99] Literally "Floating Fennel-leaved Milfoil related to water Crowfoot/Buttercup or Many-flower".

[100] This is wrongly spelled as Bauhin's original (1623, p. 141) has *fæniculi*.

I have observed in the Ranunculi. The plant itself, as far as I remember, is not erect[101] but covers the ground as it were with a dense sward. In fields tilled every year & on the banks of ditches, but rather rarely.

B84(A35) *Ranunculus acris* L.
Ranunculus pratensis erectus acris *C.B.* pratensis erectus acris vulgaris. *Park.* surrectis cauliculis *Ger.* rectus non repens flore simplici luteo *J.B. Upright Meadow Crowfoot*. Common in meadows. The leaves of this species resemble the leaves of Monk's-hood or Aconite.[102] Sometimes it is found to be mild in taste.

B85(A34) *Ranunculus sardous* Crantz
Ranunculus rectus foliis pallidioribus hirsutus *J.B.*, in whose work there is a full & exact description. *Upright pale-leav'd Crowfoot*. In moist & muddy places where waters have flooded during the winter.

B86(A36) *Asperula cynanchica* L.
Rubia cynanchica *J.B.* VI sive Cynanchica *C.B.* Synanchica *Lugd. p.* 1125. *Ger. Squinancywort*. On Gogmagog hills, and Newmarket heath.

B87(A37) *Ruscus aculeatus* L.
Ruscus *J.B. C.B. Park. Ad. Lob.* Ruscus sive Bruscus. *Ger.* Myrtacantha murina spina *Lob. Icon.* Oxymyrsine *Anguill. Butchers Broom*. About Anglesey Abby.

S.

B88 *Salix repens* L.
SAlix humilis repens *Ger.* pumila angustifolia II. sive pumila brevi angustóque folio incano *C.B.* angustifolia Lobelii *J.B.* angustifolia repens *Park. The Narrow-leav'd creeping Dwarf-Willow*.

B89 *Salix repens* L.
Salix humilis repens rotundifolia. *The round-leav'd creeping Dwarf-Willow*.[103] Both in the Bog near Gamlingay, where the Asphodelus Lanc[astriæ][104] grows.

B90(A38) *Knautia arvensis* (L.) Coult.
Scabiosa vulgaris flore pleno. *Double Flowerd Scabious*. In the flower of this form the tubular petals that are on the periphery are not larger than the inner ones, as in the common Scabious,[105] but equal, just as in Succisa;[106] moreover it lacks those capitate stamens that are seen in the common form. *In many Closes about Teversham, Balsham, and elsewhere together with the Common sort.*

---

[101] *non erigitur*, literally "is not raised up".
[102] *Folia hujus Napelli aut Aconiti foliis accedunt*. The comparison, which comes from Ray (1670, p. 260; 1677, p. 250), is with *Aconitum napellus* L., which is not listed in the Cambridge catalogue and was not discovered until much later as a naturalised British species but was presumably known to Ray in gardens; Gerarde (1597, p. 823) says it "is vniuersally knowne in our London gardens, and else where". Ray has now resolved the uncertainties that he had about *Ranunculus rectus* in the 1663 appendix (q.v.).
[103] Ray (1688a, p. 1423) treated this as a separate species but did not describe it, saying simply that it grew in marshes principally in montane regions. However, we have not traced it even as a synonym in *Synopsis* (1690, 1696). It was presumably one of the variants of the variable *S. repens*.
[104] i.e. *Narthecium ossifragum* [B11].
[105] i.e. *Knautia arvensis* [526].
[106] i.e. *Succisa pratensis* [374].

APPENDIX (1685)

B91(A39) *Petroselinum segetum* (L.) W.D.J. Koch
  Selinum Sii foliis *Ger.* segetale *Park. Honewort.* On a bank on the left hand in the road to Madingley, as soon as you are out of Cambridge, and on many other ditch-banks about Cambridge.

B92(A40) *Thymus* sp.
  Serpylli vulgaris secundum genus *J.B.*[107] *Mother of Time with large flowers.* It is often found with the common sort.[108] Consult J. Bauhin, who notes exactly the differences between these two species.[109]

B93 *Stachys arvensis* (L.) L.
  Sideritis humilis lato obtuso folio *Ger. emac.* Alsines Trissaginis folio *C.B. Park.*, who presents it again, as if new & found by him, under the title *of Sideritis hederulæ folio.*[110] An Sideritis folio rotundiore *J.B.? Petty All-heal. In a close call'd Gains nigh Teversham.*

B94 *Sorbus aucuparia* L.
  Sorbus sylvestris foliis domesticæ similis *C.B.* sylvestris sive Fraxinus Bubula *Ger.* Aucuparia *J.B.* Ornus sive Fraxinus sylvestris *Park. Quicken-Tree. In Gamlingay Park near* Filix florida.[111]

---

[107] In the 1663 appendix Ray called this J. Bauhin's first kind of thyme, *Serpylli vulgaris primum genus* (see footnote 53 in *Addenda* of that appendix).

[108] i.e. *Thymus polytrichus* [537].

[109] The relevant part of the text of Bauhin & Cherler (1651) on *Serpyllum vulgare* (vol. 3(2), p. 269) can be translated thus: "There are two kinds of it. *In the first* the flower is smaller in the middle part, with the apex sometimes bifid in the middle: with the colour for the most part purplish: the leaves, as of the following, with a smell somewhat approaching Bastard-balm. *In the second* the flower is twice as large, five apices in the middle, the colour gentler & less purplish: it does not differ from the preceding in its leaves, except in smell, which seems to us sweeter. Thyme sometimes degenerates into tomentose, whitish heads, which are in the place of the flowers. We have besides some kind, similar to the [two] preceding in leaves & stems, but very hairy." Ray retains both species in his later works but provides little additional information on them. His description of Bauhin's second species in *Historia plantarum* (1686, p. 521) is based on Bauhin & Cherler's account and the only possible modification to his views is in *Synopsis* (1690, p. 78) where he says that the second species grows with the first and seems to be merely a variety of it. All three British species of *Thymus* occur in Cambridgeshire, the common *T. polytrichus* A. Kern. ex Borbás, which is morphologically variable and gynodioecious, the female plants often having smaller flowers than the hermaphrodites (Pigott 1955), and two rarer species, *T. serpyllum* L. (a small plant similar to *T. polytrichus*) and *T. pulegioides* L. (which often has larger inflorescences and is strongly aromatic). Ray's comment on the exactness of Bauhin & Cherler's description suggests that he had little difficulty in identifying Bauhin's second kind with a Cambridgeshire plant, but we are unable to do so. (*T. polytrichus* and *T. pulegioides* are extremely variable in Europe and one would not necessarily expect descriptions of Central European material to match British plants.) Ray's English name and the widespread national range implied in *Synopsis* suggest the possibility that his plant may simply have been the hermaphrodite *T. polytrichus*, though the flowers of these plants are not twice as large as those of female ones. For notes on the hairy variants described at the end of Bauhin's account, see footnote 728 in the main catalogue.

[110] Literally "Ironwort with the leaf of a little ivy"; Parkinson (1640, p. 587) says the leaves are "like to Alehoofe [*Glechoma hederacea* 272], or rather unto the red Archangell leaves [*Lamium purpureum* 319]" although *hederula* suggests a comparison with Ivy-leaved Speedwell, *Veronica hederifolia* [24], and/or Henbit, *Lamium amplexicaule* [25].

[111] i.e. *Osmunda regalis* [B32].

B95 *Sorbus torminalis* (L.) Cranz
Sorbus torminalis *Ger.* torminalis sive vulgaris *Park.* torminalis & Cratægus Theophrasti[112] *J.B.* Mespilus 7 seu Apis folio sylvestris non spinosa sive Sorbus torminalis *C.B. Common Service or Sorbetree. At Gransden.*

B96 *Heracleum sphondylium* forma *angustisectum* Gremli
and/or forma *stenophyllum* (Gaudin) P.D. Sell[113]
Sphondylium majus aliud laciniatis foliis. *Park.* An Sphondylium hirsutum foliis angustioribus *C.B.? Jagged Cow-Parsnep. In Kingston wood.*

T.

B97 *Euphorbia platyphyllos* L.[114]
Tithymalus segetum longifolius. *Long leav'd Corn Spurge.* It scarcely reaches the height of nine inches.[115] The oblong leaves end in a point. The whole plant is of a yellowish-green colour apart from the part of the stem closest to the ground, which is reddish. An Esula minor sive Pityusa *Ger. emac.?* The picture is not unlike it except in the leaves, which end more obtusely than those of our spurge.[116] *Among the Corn beyond Kingston wood in the way to Gamlingay in several places.*

B98 *Lotus tenuis* Waldst. & Kit. ex Willd.
Trifolium corniculatum minus angustioribus foliis fruticosius. *The lesser bushy narrow-leav'd Birds-foot Trefoil.* Among standing corn & in moister places. An Lotus pentaphyllos frutescens tenuissimis glabris foliis *C.B.?*

B99(A41) *Trifolium subterraneum* L.
Trifolium humilum supinum flosculis longis albis nondum descriptum[117] *P.B. Dwarf Trefoil with long white flowers. At Gamlingay by the way side as you go up from*

---

[112] The reference is to Theophrastus' *Enquiry into plants* (III. xv. 6 in the Loeb Classical Library edition), where there is quite a full description of this tree, κραταιγος [crataegos], called by some κραταιγων [crataegon], which he compares with μεσπλη [mespile], i.e. Medlar, *Mespilus germanica* L., of which he says it may be a wild form. As he also says that the ripe fruits become yellowish and then black, this cannot in fact be *Sorbus torminalis*. It is usually identified as *Crataegus heldreichii* Boiss., but that has red fruits, so it is possibly *C. pentagyna* Willd.

[113] See Sell & Murrell (2009), who recognise these two variants of *Heracleum sphondylium* with unusually narrow leaf segments rather than the one variant usually recognised by British botanists as "var. *angustifolium*". In *Historia plantarum* Ray (1686, p. 408) treated this plant in a separate entry but said that it seemed to be a variety of the plant that we now call *Heracleum sphondylium* [560] rather than a distinct species; he commented that in many cases nature plays tricks in this way.

[114] Ray (1670, p. 299) knew *Euphorbia platyphyllos* as *Tithymalus platyphyllos Fuchsii* J.B. He treated *Tithymalus segetum longifolius* as a separate species in *Historia plantarum* (1686, p. 868), where he added a record from Dale, "inter segetes circa *Comberton* cis Kingstoniam" ["among standing corn around *Comberton* on the near side of Kingston"]. He suspected that it was identical to one of two other species described in *Historia plantarum* but said that he had not affirmed this confidently as he had not yet seen the plant. He retained it as a separate species in *Fasciculus* (1688b, pp. 20–21) and the first two editions of *Synopsis* (1690, pp. 124–125; 1696, p. 183), but in the posthumous 1724 edition (p. 312) there is added at the end of the account "(Idem cum priori.)" [The same as the preceding, i.e. *Tithymalus platyphyllos Fuchsii* J.B.].

[115] Altitudinem dodrantalem vix assequitur.

[116] The reference is to figure 15 on p. 502 of Johnson (1633).

[117] "Dwarf prostrate Trefoil with long white florets, not yet described." Here *humilum* is a misprint for *pumilum*.

the Town to the Windmill. Morison, in his *Historia Plantarum Oxonienis*, *p*. 138, describes this plant thus under the title of Trifolium album tricoccum subterraneum Gastonium reticulatum.[118] At the beginning of summer this plant bears long white widely spaced flowers, of which the involucres follow after each as it falls, formed like a net and containing included in their midst[119] for the most part three seeds each with their three distinct little pods. When the flower dies away[120] the rudiments of the pods immediately descend into the ground and hide themselves in the same place at a depth of several inches, especially in sandy meadows or fields; therein also they are perfected, & we have called it by that name of subterraneum & from the three seeds (for the most part, as said above, unless weather & soil[121] thwart them) we have named it tricoccum; next we have added Regium[122] or Gastonium, since it was first detected (like several other plants described here) by his Royal Highness the Duke of Orléans, paternal uncle of King Louis XIV.[123]

V.

B100 *Vaccinium oxycoccos* L.

Vaccinia palustria *Ger*. Vaccinium palustre *Park*. Oxycoccos sive vaccinia palustria *J.B*. Vitis Idea palustris *C.B*. *Marsh Whortle-berries, Mosse-berries or Moor-berries*. *In the boggs about Gamlingay plentifully.*

B101 *Valeriana dioica* L.

Valeriana sylvestris sive palustris minor altera. *The other small Wild or Marsh Valerian*. With [Valeriana] sylvestris minor *Park*.[124] generally. This is somewhat larger than [Valeriana] minor in all its parts except the flowers; but the flowers, by which character it is sufficiently distinguished from that plant, are a fifth of the size[125] and more densely compressed than those of that plant.

B102(A42) *Solidago virgaurea* L.

Virga aurea *Ger*. aurea vulgaris *Park*. aurea vulgaris latifolia *J.B*. aurea 4. sive angustifolia minus serrata *C.B*. *Golden Rod*. *In the Park and some Pastures about Gamlingay.*

---

[118] "Gaston's white, three-seeded, subterranean, reticulate Trefoil/Clover." The description that follows is also taken from Morison. Robert Morison (1620–1683) was a Scottish botanist and physician whose royalism forced him to flee to France, where he became physician to Gaston Jean-Baptiste de France, Duc d'Orléans (1608–1660). After the restoration Charles II made him royal physician, he was elected Professor of Botany at Oxford in 1669 and he published part 2 of his lavish *Plantarum Historia Universalis Oxoniensis* in 1680. (Part 1 on trees and shrubs was never issued.)

[119] *in ipsorum gremio*, literally "in the lap/bosom of them themselves".

[120] *Evanescente flore*, literally "With the flower dying away".

[121] *cœlum & solum*.

[122] i.e. "Royal".

[123] *à Celsitudine ejus Regia Aurelianensi Duce Ludovici XIV patruo*. The Roman name of Orléans was Aurelianum and Ludovicus is the Latin for Louis.

[124] i.e. the male plant of this dioecious species [606]. The description here is of the female plant and is taken from Ray (1670, p. 310; 1677, p. 299). Few modern Floras make the distinctions between them as clear.

[125] *quintuplo minores*, literally "by five times smaller".

ADDENDA.

Valeriana sylvestris sive palustris minor altera. *The other small Wild or Marsh Valerian.* Cum sylvestri minore *Park.* passim. Hæc minore *Park.* omnibus suis partibus, exceptis floribus aliquanto major est. Flores autem, qua nota ab illa satis distinguitur quintuplo minores & densius stipati sunt quam illius.

Virga aurea *Ger.* aurea vulgaris *Park.* aurea vulgaris latifolia *J. B.* aurea 4. sive angustifolia minus serrata *C. B. Golden Rod. In the Park and some Pastures about Gamlingay.*

EMENDANDA.

A.

Abrotanum campestre & Abrotanum inodorum] *Si placet expungantur ambo.* I believe neither of them is to be found on Newmarket heath, at least that part of it which is in Cambridgeshire.

Allium sylvestre] *Huic adde synon.* Allium sylvestre juncifolium capitatum majus *C. B.*

Alsine major *Dod*] *Deleatur cum synonymis. Non enim specie differt à sequente* media *ejusdem. Quæ ratione loci aliorúmque accidentium magnitudine insigniter variat unde Botanici eam in plures species diremerunt.*

Anagallis aquat. I. & Anagallis aquat. III. *C. B.*] *Et hæc pariter expungantur. Ego sc. has differentias apud nos nunquam potui observare. D. Nid ex conjectura & præsumptione Catalogo inseruit, quoniam Botanici vulgo & præcipue J. Bauhinus passim & promiscue* 4 *varietates hic recensitas crescere affirmat.*

Aracus

APPENDIX (1685)

EMENDANDA.

A.

A Brotanum campestre & Abrotanum inodorum] *Please may both be expunged.*[126] *I believe neither of them is to be found on Newmarket heath, at least that part of it which is in Cambridgeshire.* [*Artemisia campestris* 1 and 2]

Allium sylvestre] *Add to this as a synon[ym]* Allium sylvestre juncifolium capitatum majus *C.B.* [*Allium vineale* 19]

Alsine major *Dod*] *Let it be deleted together with the synonyms, for it does not differ in species from the following plant,* media *of the same author. This plant varies remarkably in size by reason of the locality and other extraneous circumstances, whence Botanists have divided it into several species.* [*Stellaria media* 28]

Anagallis aquat[ica] I. & Anagallis aquat[ica] III. *C.B.*] *And let these equally be expunged. Actually I have never been able to observe these differences with us. Master Nid inserted them in the Catalogue from conjecture & presumption because Botanists commonly, & especially J. Bauhin, generally & indiscriminately affirm that the 4 varieties enumerated here really do grow.* [*Veronica beccabunga* 35 and *V. anagallis-aquatica* 37]

Aracus sive Cracca minor] *Let it be deleted, for it is entered a second time, lower down under Vicia.*[127] [*Vicia hirsuta* 57]

Atriplex sylvestris altera *C.B.*] *Delete* sylv[estris] vulgatior sinuata minor. *Park. Authors are so brief & obscure in describing this plant that I do not fully know what they mean.*[128] *So we shall try to describe more accurately that plant which we want to be understood by this title. It has a white root, which is simple, woody and also hairy with little fibres, and a stem that is indeed erect but weak, throwing out oblong procumbent branches close to the ground, even longer than the stem itself, in opposite pairs.*[129] *The leaves, especially the lower ones, are triangular, narrowing from a broad base to a sharp point & to some extent like the point of a spear and incidentally wavy or toothed; those which are at the top of the stem are longer, narrower & generally smooth at the margins. The triangular seed-vessels are made up of two connivent leaflets.* [*Atriplex prostrata* 71] *It will be convenient for the more accurate identification of Atriplex species to add here what* Cat[alogus] Plant[arum] Angl[iæ][130] *also notes about the other common Atriplex, namely* sylvestris *J.B. This plant, it says, is correctly described by J. Bauhin but by no means by others. It is easily distinguished from the remaining like-named plants on this account, that, when the seed-vessel is closed, it displays to observers a star of five rays; moreover it contains a single small, black, shiny, flattened seed, whence it seems to be referable to Blitum rather than to Atriplex.* [*Chenopodium album* 70]

---

[126] *Si placet expungantur ambo.* Literally "If it pleases let be expunged both."

[127] i.e. as *Vicia parva sive Cracca minor cum siliquis plurimis hirsutis* J.B. [613] as well as, like here, as *Aracus sive Cracca minor* Park. [57].

[128] *quid sibi velint,* literally "what for themselves they may want". This sentence and the following description are based on Ray (1670, pp. 34–35; 1677, p. 32).

[129] *ex adverso binos,* literally "opposite [each other] two by two" (accusative).

[130] Ray (1670, p. 34; 1677, p. 32).

B.

BLitum album minus] *I now think the plant we found in Kingston wood to be the same with the common red Blite,*[131] *and that it differed in colour by reason of the shadyness of the place where it grew.* [*Chenopodium polyspermum* 88]

Bursa Pastoris minor] *This plant does not seem to differ in species from Nasturtium Alpinum petræum. Certainly* J. Bauhin *makes Nasturtium petræum Tab. a synonym of this, but I pronounce nothing rashly.*[132] [*Teesdalia nudicaulis* 98]

C.

CArduus lanceatus angustifolius] Let it depart, for I do not think that with us any distinction is observable in this kind. [*Cirsium vulgare* 110]

Carduus nutans] *An* Carduus spinosissimus latifolius sphærocephalus vulgaris *C.B.?* I understand now that this should be regarded as Carduus moschatus *Ger.*; however I have never noticed the odour of musk in the flowers,[133] nor is the colour of the flowers pale purple but deeper, nor finally does the figure itself accord with it. [*Carduus nutans* 111]

E.

EChium alterum] I confess that I have not yet perceived any specific difference between Echium vulgare & that which springs up on walls generally with us. Let others examine this more carefully. [*Echium vulgare* 172]

Equisetum palustre ramosum aquis immersum seu Millefolium aquat[icum] equisetifolium. The description of Millefolium aquaticum cornutum *J.B.* agrees with this in all respects, even though the figure, which is C. Bauhin's, does not correspond, as neither does his description in his Prodromus.[134] [*Ceratophyllum demersum* 183]

G.

GEranium malacoides sive columbinum minus[135]] *I am not sufficiently certain whether the little plant that we know here is absolutely the same species as* Geranium columbinum minus *C.B. Master Ray omits this plant in his Cat[alogus] Plant[arum] Angl[iæ] and indeed it does not seem to differ from Ger[arde's] columb[inum] vulg[are]*[136] *otherwise than in a difference in the leaf.* [*Geranium molle* and/or *pusillum* 221]

---

[131] *Blitum rubrum minus* I.B. C.B. Ger. emac., i.e. *Chenopodium polyspermum* [89].

[132] In later works Ray (e.g. 1690, p. 116; 1696, p. 174) made *Nasturtium petræum* his chosen name for this plant.

[133] *nec tamen moschi odor in floribus mihi unquam perceptus ... est*, literally "neither however the odour of musk in the flowers to me ever has been perceived".

[134] This entry is based on a more detailed account in Ray (1670, p. 210; 1677, p. 200).

[135] The last word should be *minimum*.

[136] i.e. *Geranium molle* [219]. As we implied in the main catalogue (footnotes 317 and 318), Babington's (1860) interpretation of Ray's *Geranium columbinum* [219] as *Geranium molle* L. and *Geranium malacoides sive columbinum minimu[m]* [221] as *G. pusillum* L. is not entirely convincing. Ray (1670, pp. 131–132) suggested that *Geranium malacoides sive columbinum minimum* was not specifically distinct from the species above it in *Catalogus plantarum Angliae*. (The species

APPENDIX (1685)

Genista[137] aculeata *Park.*] *Please expunge*[138] *the synonyms* aculeata foliosa *J.B.* & spinosa III sive minor Germanica *C.B., in place of which should follow* Genista minor aspalathoides seu Genista spinosa Anglica *C.B.* & Genistella minor aspalathoides *J.B.*[139] [*Genista anglica* 214]

Gentianella fugax Autumnalis minor *Ger. emac.*] *Let all the synonyms of both Bauhins be deleted, in the place of which put* Gentianellæ species quibusdam, an Cordo Pneumonanthe aut Gentiana fugax altera Clusii[140] *J.B., to whose description, which I have no doubt is of our plant, give your attention, having disregarded the others.*[141] [*Gentianella amarella* 216]

Gramen alopecuroides minus spica aspera brevi] *The London Botanists call this Gramen cristatum and I would not dissent.* [*Cynosurus cristatus* 236]

Gramen caninum] This is the same as that which *J.B.* describes under his published title and so the things that I have noted here are to be retracted, for I had not myself observed the panicle before I published the Catalogue but rashly believed others. [*Elytrigia repens* 242]

Gramen hirsutum majus] I now suspect that this Grass does not differ specifically from Gramen exile hirsutum[142] but only in size and other non-essential characters by reason of the place where it grows. [*Luzula multiflora* 252]

H.

HElleborine angustifolia *Tab.*] Delete this with all the synonyms and substitute in their place: Helleborine palustris nostras. An Helleborine angustifolia palustris *C.B.?* With us it occurs nowhere other than in marshes & moist meadows. The description of de L'Écluse's third Helleborine (which is [Helleborine] angustifolia sive palustris *C.B.,* Damasonium flore herbaceo intus nonnihil candicante[143] *J.B.*) does not match this plant of ours at all, for he says that its leaves are wider than are those of Helleborine *Dod.*, while those of this one of ours are narrower, and that the colour of the flowers is grass-green, which in this species are white or somewhat purplish.[144] [*Epipactis palustris* 274]

immediately above was the one currently known as *G. columbinum* L., but he may have intended to refer to *G. molle* further up the page.) In the second edition (1677) he omitted *Geranium malacoides sive columbinum minimum* but inserted after his *Geranium columbinum* [our *G. molle*] a new species, *Geranium columbinum majus flore minore cærulescente* (p. 130), followed by a convincing description of *G. pusillum* L. This is not marked C as occurring in Cambridgeshire. Ray never cited *Geranium malacoides sive columbinum minimum* in the synonymy of the later species, but it was included as a synonym in the 1724 edition of his *Synopsis*, presumably added by Dillenius. We have therefore concluded that Ray did not reliably distinguish these two species until 1677 and that the 1660 records must be treated as doubtful.

[137] Parkinson's name is correctly cited as *Genistella* in the main catalogue.

[138] *Expungas*, literally "May you expunge".

[139] The reasons for these changes are discussed by Ray in *Catalogus plantarum Angliae* (1670, p. 128; 1677, p. 127).

[140] Literally "A species of *Gentianella* for some [authors], perhaps *Pneumonanthe* for Cordus or de l'Écluse's second short-lived Gentian".

[141] Ray (1670, pp. 129–130; 1677, pp. 128–129) discusses the reasons for these changes at some length.

[142] i.e. *Luzula campestris* [251]. Ray later (e.g. 1690, p. 193; 1696, pp. 262–263) revised this opinion and kept the two species separate.

[143] Literally "*Damasonium* with a grass-green flower somewhat whitish within".

[144] This entry is based on Ray (1670, pp. 161–162; 1677, p. 157).

L.

Actuca sylvestris 2. *Ger. emac.*] *Delete all the synonyms. In their place add* Lactucæ sylvestris sive Endiviæ multis dictæ folio laciniato dorso spinoso varietas *J.B. This plant does not differ in any way from Lactuca sylv[estris] costa spinosa C.B.*[145] *other than in its leaves, which in it are not cut even on the stem. We now retract what we propounded about this plant in the Catalogue & we yield to Johnson's opinion.*[146] [*Lactuca serriola* L. forma *integrifolia* 313]

Lactuca sylv[estris] laciniata minima, not yet described] *This plant is presented in Cat[alogus] Plant[arum] Angl[iæ]*[147] *under the title* Chondrilla viscosa humilis *C.B. Park. Ger. emac. Dwarf Gum-succory.* [*Lactuca saligna* 314]

Lapathum acutum *Ger. emac.* & Lapathum sylv[estre] folio minus acuto *of the same author. Cat[alogus] Angl[iæ] does not have them as distinct.*[148] [*Rumex* spp. 321 and *Rumex obtusifolius* 322]

Lapathum aquaticum sive Hydrolapathum minus *Ger. emac.*] Let it be removed with the synonyms.[149] Replace by Lapathum acutum minimum[150] *Ger. Park. J.B.*, minimum *C.B. Small sharp-pointed Dock.* In moister places. It will offer much to the clarification of a Study of Docks[151] to add here what *Master Ray* notes[152] about the differences between the Docks of our country.

The species of Lapathum occurring wild in England are: 1. Hydrolapathum maximum.[153] 2. Lap[athum] anthoxanthon *J. B.*[154] 3. Lapathum folio acuto & non-nunquam obtusiore plano.[155] The leaves of this dock are broader than those of the following, paler green, flat & not curled at the margins; those leaflets that make up the seed-vessel have as it were some teeth on the margins. Moreover the seed is paler than that of Oxylapathum crispum. 4. Lapathum folio acuto crispo *C. B.*[156] The leaves of this dock are narrower than those of the preceding one, longer, deeper green and wavy at the margins. The seed grows so compactly that it completely covers up & conceals the stems, of the same size as the seed of the preceding species but redder. Moreover its seed-vessels are similar except that they are level at the margins and do not have any toothing.[157] 5. Lapathum acutum minimum,[158] the flowers

---

[145] i.e. *Lactuca serriola* forma *serriola* [312].

[146] See the entry on this species and footnote 394 in the main catalogue.

[147] Ray (1670, p. 71; 1677, p. 69).

[148] Ray (1670, p. 187; 1677, p. 180). For the identification of *Lapathum acutum* [321], see the footnote to that taxon in the main catalogue.

[149] *Rumex palustris* [324] should be removed from the list of species recorded from Cambridgeshire by Ray: see footnote 400 in the main catalogue.

[150] *Succedat Lapathum acutum minimum*, literally "Let smallest sharp Dock replace". This is *Rumex conglomeratus*: see the fuller description below and footnote 158.

[151] *Multum conferet ad dilucidandam Historiam Lapathorum.*

[152] Ray (1670, pp. 188–189; 1677, p. 181).

[153] i.e. *Rumex hydrolapathum* [323].

[154] i.e. *Rumex maritimus* [B56].

[155] i.e. *Rumex obtusifolius* [322].

[156] i.e. *Rumex crispus* [B55].

[157] *nisi quod ad margines plana sint, nec ullis denticellis prædita*, literally "except that at the margins they are flat, nor with any little teeth endowed".

[158] The description of the leaflets in the inflorescence allows this plant to be identified as *Rumex conglomeratus* Murray. For the reasons for rejecting Babington's (1860) identification of *Lapathum acutum* [321] as *Rumex conglomeratus*, see the footnote to that taxon in the main catalogue.

of which grow at intervals in whorls, with almost every whorl with a leaflet contiguous to it like a little tongue; the seed is less than half the size of those preceding most closely, than which it is also smaller in all its parts although it not infrequently rises to an equal height. 6. Lapathum pulchrum Bononiense sinuatum.[159] The Study of Docks in the works of Botanists suffers from great confusion & obscurity and so we have considered it worth the trouble[160] to distinguish them more accurately.

Linum sylvestre radice perenni flore cœruleo] Neither the names nor the descriptions of the Authors suit well the plant that we know.[161] [*Linum perenne* 337]

M.

MOllugo montana sive Gallium] *Master Ray* presents this plant in his Cat[alogus] Angl[iæ] under the title of Rubia angulosa aspera *J.B.*, sylvestris lævis *C.B.*, an sylvestris *Ger.?* Read this also from him about it:[162] This plant, unless I am much mistaken, is entered by [Mathias] de L'Obel in his *Obs[ervationes]* as Mollugo vulgatior Herbariorum, Gallium album quorundam, and is Mollugo prima *Dod.*, montana *Ger.*, vulgatior *Park.*, Mollugo montana angustifolia vel Gallium album latifolium *C.B.*, as will be obvious to anyone examining the figures; and so they have made two plants out of one, as often elsewhere. [*Galium album* 373]

O.

ORiganum vulgare] *The Plant that grows in Bowyers close at Cherry Hinton is not this, but* Clinopodium vulgare;[163] *we mistook it when it was young. It is found in other places.* [*Origanum vulgare* 410]

P.

POlygalon Cortusi] *For this substitute* Ferrum equinum Germanicum siliquis in summitate *C.B.*, *the synonyms of which look for in the Addenda.* [*Hippocrepis comosa* 452, A14]

Potamogeiton longis acutis foliis *Ger. emac.*] Delete the synonyms, but add Fontalis lucens major *J.B.*, An Potamogeiton foliis angustis splendentibus *C.B.?* Long leav'd pellucid Pondweed. This species is generally submerged in the water. The leaves are scarcely serrated; accordingly we wrongly denominated it as Lapathum fluitans longo serrato folio *J.B.* in the Cambridge Catalogue.[164] [*Potamogeton lucens* 462]

---

[159] i.e. *Rumex pulcher* [B57].

[160] *ideoque operæ pretium duximus*, literally "and so a reward for the work/trouble we have calculated/reckoned", this being one of the many senses in which the verb *duco* ("I lead") is used.

[161] This entry is taken from Ray (1670, p. 197; 1677, pp. 188–189), where there is a detailed account of the differences.

[162] *De qua hæc etiam ab illo accipe*, literally "About which these [things] also from that [man] receive".

[163] This was *Clinopodium vulgare* [139], but Ray's other records for *Origanum vulgare* were sound.

[164] This entry is based on Ray (1670, p. 251; 1677, p. 241).

R.

Ranunculus hederaceus rivulorum, &c.] The flowers of this plant are white, not yellow as we were guessing, & it arises generally in watery places; accordingly the synonyms listed do not correspond to it unless perhaps it varies in the colour of the flower overseas. [*Ranunculus hederaceus* 485]

Rapistum luteum siliqua hirsuta articulata] Add Sinapi siliqua hirsuta semina[165] albo vel russo *J.B.*, Sinapi album *Ger.*, under which titles it appears in *Cat[alogus] Angl[iæ]*.[166] [*Raphanus raphanistrum* L. subsp. *raphanistrum* 491]

Rapum sylvestre *Ger.*] *Master Ray* in his *Cat[alogus] Angl[iæ]*[167] thinks that the plant which[168] is sown for the production of oil in the Isle *of Ely* is Napus sylvestris,[169] but that in fact [Mathias] de L'Obel, since he had heard the plant from which they produce oil called *Rape-seed* by Cambridgeshire people, had perhaps imagined to himself that some Rapum sylvestre was provided[170] of which this might be the seed. [*Brassica napus* 495]

S.

Salix folio utrinque glauco viminibus rubris & Salix folio utrinque glauco viminibus albidioribus.] I do not regard these varieties as distinct species.[171] For I have observed that in the same species some willow trees produce only flowers (for I name those catkins[172] in which there is no seed thus) but others seeds, the latter, if you wish, female and the former male; but in fact whether those that produce flowers or fruit or seeds once may do that consistently or forthwith may change & produce flowers & seeds alternately I confess I have not yet discovered.[173] *Cat[alogus] Angl[iæ]*.[174] [*Salix alba* 508 and 509]

Salix folio longo non auriculato vimine rubro] *Master Ray* omits this in *Cat[alogus] Angl[iæ]*,[175] as he certainly does not think that it differs from the other species with yellowish withies.[176] [*Salix* sp. 517]

Saxifraga Anglica Occidentalium] Delete this with the synonyms and substitute Rubia cynanchice,[177] which see in the Addenda. [*Asperula cynanchica* B86(A36), 523]

Saxifraga graminea pusilla] *I find upon comparing, that this is the same plant described by Johnson upon Gerard under the name of* Saxifraga Anglicana

---

[165] This must be a misprint for *semine* ("with a seed").
[166] Ray (1670, pp. 284–285; 1677, p. 274).
[167] Ray (1670, p. 219; 1677, p. 210).
[168] The Latin *qua* must be a misprint for *quæ*.
[169] i.e. *Brassica rapa* and *B. napus* [385], but in fact only the latter: see footnote 470 in the main catalogue.
[170] *dari*, literally "to be given".
[171] *Hasce varietates pro distinctis speciebus non habeo*. Literally "These varieties for distinct species I do not have."
[172] See footnote 674 in the main catalogue: here the word that Ray uses is *nucamentum*.
[173] *mihi nondum compertum esse fateor*, literally "for me not yet to have been discovered I confess".
[174] Ray (1670, p. 271; 1677, pp. 260–261).
[175] Ray (1670, p. 274; 1677, p. 263).
[176] i.e. *Salix folio longo, & folio & vimine subluteo, non auriculata* [516].
[177] This must be a misprint for *cynanchica*. See also footnote 716 in the main catalogue.

APPENDIX (1685)

alsinefolia *or Pearlwort*.[178] An Caryophyllus minimus muscosus nostras *Park.?* [*Sagina procumbens* 525]

T.

Tithymalus helioscopius] The second kind of this plant, which we said has a non-crenate leaf & is smaller, is the same as Peplus sive Esula rotunda Herbariorum, which see in the preceding Appendix.[179] [*Euphorbia helioscopia* 569]

Trifolium majus flore purpureo sive 4. *Ger.*] It is not what with us is taken for Trifolium majus purpureum 3 of de L'Écluse as he formerly believed, but either not specifically distinct from [Trifolium] purpureum pratense vulgare[180] or not yet described. For Gerarde & Parkinson have Trif[olium] majus purp[ureum] 3 of de L'Écluse as the name for our *Claver*,[181] in which they are mistaken. *Cat[alogus] Angl[iæ]* presents this Trifolium under the name of purpureum majus vel sativum. [*Trifolium medium* 591]

V.

Viola tricolor *Dod.*] *Let it depart with the synonyms. The plant which is designated here is* Viola bicolor arvensis C.B. Jacea bicolor frugum & hortorum vitium Trago J.B. Viola tricolor 5, *chapter* 313 *of book* 2 *of Ger[arde]*.[182] An viola tricolor sylvestris *Park?* Small wild Hearts-ease or Pansies. [*Viola arvensis* 620]

FINIS.

---

[178] The reference is to Johnson (1633, pp. 567–568), where there is a long discussion with a full description; Johnson's English name is, however, not "Pearlwort" but "Chick-weed Breake-stone".
[179] i.e. *Peplus sive Esula rotunda* J.B. C B. Ger., *Euphorbia peplus* [B74(A28)], in *Addenda* above.
[180] i.e. *Trifolium pratense* [590]. This entire entry is taken from Ray's *Catalogus plantarum Angliae* (1670, pp. 304–305; 1677, p. 292).
[181] See footnote 774 in the main catalogue.
[182] The meaning of the first synonym is literally "Two-coloured *Jacea*, a bane of fruits & gardens, for Tragus". *Jacea* is a name for wild pansy that is still used in homeopathic circles. For Hieronymus Tragus (Jerome Bock) see *Trag.* in Ray's 'Explanation'; he seems to have had a low opinion of this pretty weed! The reference in the second synonym is to Johnson (1633, pp. 853–854), who describes his fifth "Hearts-ease or Paunsie" (unillustrated and individually unnamed) thus: "There is found in sundry places of England a wilde kinde hereof, bringing floures of a faint yellow colour, yet hauing a deeper yellow spot in the lowest leafe [i.e. petal] with foure or fiue blackish purple lines, wherein it differeth from the other wilde kinde: … ." Ray (1670, pp. 317–318; 1677, pp. 305–306) separated *Viola tricolor* Ger. from *Viola bicolor arvensis* C.B. and annotated the latter C, thus making it clear that the plant that he knew in Cambridgeshire had bicoloured, not tricoloured, flowers.

GAZETTEER

The localities mentioned in the *Catalogus* and its appendices are listed below. Ray's spelling is given in bold type with the modern spelling in square brackets. We do not include variants of place-names which differ from the listed names solely in minor details of spelling, punctuation and capitalisation. The O.S. grid reference for each locality is given and normally indicates the approximate centre of the site (for villages, it is the square including the parish church); the plants mentioned by Ray from the locality did not necessarily grow at this grid reference. If Ray lists a plant as present between two villages, both are listed in the gazetteer. However, we have not listed localities outside the county if they are included by Ray only as the destination of roads (e.g. Bedford and Sandy) or the sites elsewhere in Britain that are mentioned in the appendices (e.g. Gods Well, Caernarvonshire, and Maldon, Essex).

A1 and A2 indicate that a site is mentioned in the *Catalogus* (1660) and in the 1663 (A1) and/or 1685 (A2) appendices; *A1 and *A2 indicate that the locality is not mentioned in the *Catalogus* but only in the appendix or appendices listed. Sites included in Ray's 'Index of places' are marked with an obelus (†). The Watsonian vice-counties are given for sites outside Cambridgeshire (v.c. 29). Species are listed by their modern names with a reference in square brackets to our entry numbers in the main catalogue or the 1663 and 1685 appendices (the latter prefaced by A and B respectively).

Map 1 shows the main sites mentioned in the *Catalogus* and its appendices. It shows that the *Catalogus* was based on records from most areas of the county except Whittlesey and Thorney in the north-west. The mapped sites are very unlikely to represent a comprehensive list of the places where Ray botanised. Numerous sites are mentioned in the *Catalogus* only once, and there must presumably have been many other places where Ray saw nothing worth a special mention. The few localities in Fenland presumably reflect the distance from Cambridge, the sparsely populated terrain and perhaps the problems of travelling in this area. There is no indication in Ogilby's atlas (1675, plate 43) that the road from Cambridge to Ely through Stretham Ferry posed any problems, but between Ely and Soham he notes (plate 73) "fenny Ground on both Sides and the way bad & dirty" and near Soham "a very Dirty way".

Abington [Abington Pigotts] TL3044.
Anglesey-abbey TL5262, *A1, *A2. The Priory later known as Anglesey Abbey was dissolved in 1536 and granted to Sir John Hynde of Madingley three years later, but he probably dismantled parts of it to build Madingley Hall (q.v.) and neglected the rest. However, the surviving buildings were modified for domestic use in the

---

Map 1. Localities mentioned in Ray's *Catalogus* and its appendices. The solid county boundary is that of vice-county 29, which is defined as the boundary of the county in 1852 and can be taken as that of the historic county; changes made between 1852 and the local government reorganisation of 1974 are indicated by dotted lines. The Ordnance Survey grid squares are marked along the borders of the map at 10 km intervals. The map shows all the main sites listed in our gazetteer except for the places outside Cambridgeshire (Potton, St Neots) from which plants were not recorded. Circles indicate places which appear only in the 1685 appendix. The following abbreviations are used: Ang, Anglesey Abbey; Bar, Barton; Beacon, Newmarket Heath beacon; CAMB, Cambridge; Che, Chesterton; ChH, Cherry Hinton; Chi, Childerley; Com, Comberton; Cot, Coton; DeDi, Devil's Ditch; Dit, Ditton; Gra, Grantchester; Harl, Harlton; Hau, Hauxton; LiEv, Little Eversden; LiSh, Little Shelford; Mil, Milton; Tev, Teversham; Tru, Trumpinton.

GAZETTEER

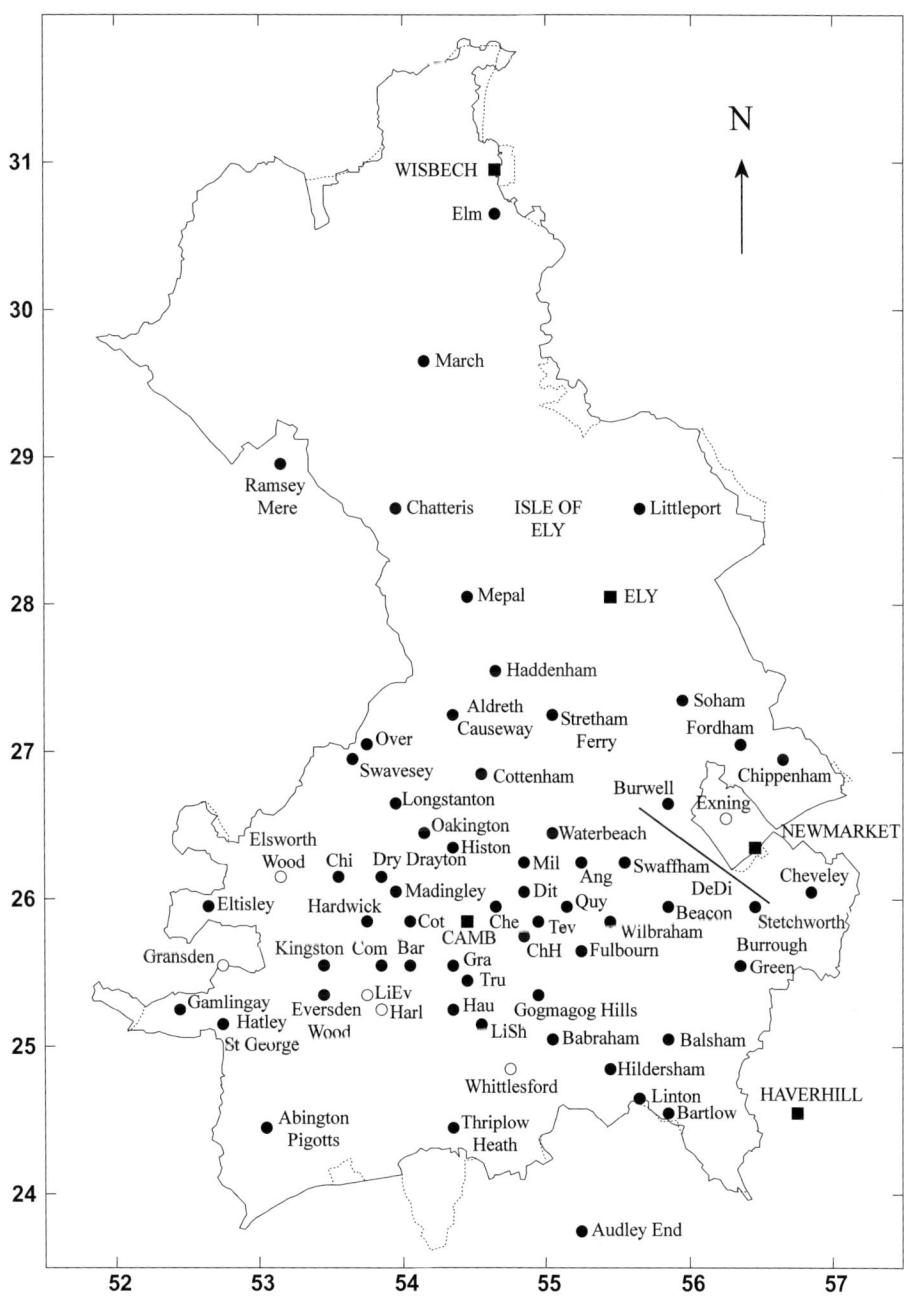

GAZETTEER

early 17th century and are the core of the house now owned by the National Trust (RCHM 1972).

Audley end, North Essex (v.c. 19), TL5237; **the great house**, also in v.c. 19, is at TL5238. The house at Audley End is "one of the great Jacobean 'prodigy' houses and it retains much of its original character" (Drury 2010). It was built by Thomas Howard, first Earl of Suffolk, between 1605 and 1614 and is now owned by English Heritage.

Audrey causey [Aldreth Causeway] TL47; the causeway crosses the Cam at TL4372.

Babram [Babraham] TL5050.

Balsham TL5850, *A1, *A2.

Bartlow TL5845.

Barton TL4055.

Bigwin closes, see Ditton.

Burrough-green [Burrough Green] **church** TL635554.

Burwell TL5866.

Cambridge TL4458, A1, A2.

~~**Barnwell** TL4658, A2; **Barnwell Abby** [Barnwell Priory] TL462589, A2. Barnwell Priory was built on a lavish scale; it was dissolved in 1538 and by the end of the century it was being used as a source of stone for the college buildings of Cambridge (RCHM 1959). Richard Relhan's drawings of those remnants which survived in the early 19th century are reproduced by Taylor (1999) and one fragment can still be seen on Priory Road.

~~**brook Stoure, rivulet Stoure** TL4759. Ray refers to this stream as the site for three aquatic species [346, 367, 369], all found by an islet in the river above the Paper Mills (q.v.). Later Cambridgeshire botanists also refer to the Stour near the Paper Mills; it must be the stream currently called Coldham's Brook (cf. Babington 1860, p. xxiv).

~~**Caius College** [Gonville and Caius College] TL4458, *A2.

~~**Castle, castle hill**, TL4459.

~~**Christs Colledge** [Christ's College] TL4558; **Christs College piece** [Christ's Pieces], TL4558, *A2.

~~**Clare-hall** [Clare College] TL4458.

~~**Common called Coldhams** [Coldham's Common] TL4758; **Pesthouses**, c. TL4658. The "*new Pesthouses*" were, according to Ray's description under *Cirsium acaule* [107] and *Nepeta cataria* [363], on Coldham's Common and approached along a lane from Barnwell. In July 1655 the Cambridge corporation paid £30 towards the building of pesthouses (Cooper 1845, p. 464), and these were perhaps those referred to by Ray. In the plague outbreaks of 1665 and 1666 (the last to afflict the town) 40 acres of Coldham's Common were enclosed for pesthouses, as described by Gray (1925), but the common had clearly been used for this purpose in earlier years.

~~**Garret-hostle-bridge** [Garrett Hostel Bridge] TL4458.

~~†**Hill of health** c. TL4359. According to Babington (1860, p. xxviii), "After this name had been in use for about two centuries it totally disappeared upon the enclosure of St Giles's Parish. It is the moderately elevated ground lying to the south of the Huntingdon road just outside of the town of Cambridge. It is now partly planted and partly occupied by a gentleman's house." Ray's habitat descriptions show that it was at least partly cultivated as arable land in his time. We have not found the site marked on any map. The species recorded (see 'Index of places') are described by Ray as plants of dry, barren and sandy soils. This suggests that

490

the site must have been on the 'Head Gravel and Observatory Gravels', shown on modern geological maps as extending north-west from the eponymous Observatory to Howes and just reaching the north side of Huntingdon Road at Girton College.

~~**Huntingdon rode** (sometimes given as **Huntington rodc**) runs north-west from Cambridge, TL4459, towards Huntingdon; **Howes bowling-green** *c*. TL4360. Howes was a settlement near the city boundary (and the current National Institute of Agricultural Botany) on Huntingdon Road. It is represented as place-names both north and south of the road on historic maps (e.g. Howes Close on the south side, How House on the north) and, although it has long been engulfed by the city, by current street names on the north side (Howes Close, Howes Place).

~~**Iesus Colledge** [Jesus College] TL4558, A2; **the gate which opens out of the rode into Garlick fair** TL4558. Garlick Fair was originally held in the Nuns' close on the south side of the college; the gate led from here into Jesus Lane. The college Audit Books record the expenditure of 5s. 10d. for the repair of the highway "from Garlick fair gate to ye Col: gate" in 1654–1655 (Willis & Clark 1886, vol. 2, p. 115).

~~**Kings-Colledge** [King's College] TL4458, *A1, *A2.

~~**London rode** [Trumpington Road, Cambridge Road] runs south from Cambridge, TL4557, through Trumpington, Hauxton and Harston to Royston.

~~**Magdalen Colledge** [Magdalene College] TL4459.

~~**Maids Causey** [Maid's Causeway], TL4558, *A2.

~~**Newnham** TL4457; **Newnham mills** TL4457.

~~**Paper mills** TL472594. The Paper Mills were on Coldham's Brook about 0.75 km south of the point where it enters the River Cam, close to the point where it is now crossed by Newmarket Road (see RCHM 1959, p. 386). The "paper myll" was first recorded in 1557 (Cooper 1843, p. 473). According to Fuller (1662, p. 149), "there are *Mills*, nigh *Sturbridge-fair,* where *Paper* was made in the memory of our Fathers", and the site was probably used for milling flour in Ray's time. Ray's reference to *Angelica sylvestris* [45] growing "*By the rivers side above the mills*" probably refers to the Cam above Newnham Mill and, on the other side of the river, Bishop's Mill and King's Mill (which shared a single building).

~~**Pesthouses**, see Common called Coldhams.

~~**Peterhouse** TL4457; **Peterhouse chappel** TL448579, *A1; **Peter-house Tennis-court** TL448578. The chapel was consecrated in 1632. Loggan's plan of Cambridge (1688) shows 'The Tennis Courts' on the west side of the college property, on the inner side of the wall separating it from Coe Fen.

~~**St Johns Colledge** [St John's College] TL4458.

~~**Spittle-house end** TL4557. This is shown on Loggan's map of Cambridge (1688) as that part of Trumpington Street south of Peterhouse. The name was derived from the Hospital of St Anthony and St Eligius, founded in 1361 at a site which was then outside the town (Gray 1925; Taylor 1999).

~~**Sturbridge faire place** [Stourbridge Common] TL4759, A2; **the old chappell** TL471594. Stourbridge chapel dates from the 12th century and was associated with a hospital for lepers; it was later acquired by the Monastery of Ely and after the dissolution leased to the Corporation of Cambridge, but by the 17th century it was in private hands (RCHM 1959, p. 298). It probably survived because of its use in connection with the Stourbridge Fair, which was one of the great fairs of mediaeval England and maintained its importance until the 18th century (Roach 1959; Taylor 1999).

GAZETTEER

~~**Trinity Colledge** [Trinity College] TL4458, A2; **the Fellows garden** TL4458. For the history of the walks and gardens of the college see Willis & Clark (1886, vol. 2, pp. 635–650), who reproduce (pp. 460–461) Loggan's print of the college in 1688.

~~**Trinity-hall** [Trinity Hall] TL4458.

Chatteresse [Chatteris] TL3986.

†**Cherry-hinton, Hinton** [Cherry Hinton] TL4857, A1; **a close called Bowyers** TL488567, A1, A2; †**chalk-pit close** TL4856, A1, A2; **church** TL489571, A1, A2; **Hinton moor** TL4756, A1, A2; **springs near the chalk-pits** TL4856. Bowyers is marked as the name of a field (eight acres) on the 1804 enclosure award map of Cherry Hinton (Cambridgeshire Archives Q/RCc13). Hinton Moor was an extensive wetland, described by Babington (1860, p. xxviii) as "once an exceedingly wet fen. It is now completely drained and cultivated. The footpath leading from Cambridge to the church at Cherry Hinton and the carriage-road to the same village, cross it. It once extended from the Hills Road, near Red Cross [TL470550], nearly to the Stone Bridge in Coldham's Lane [presumably TL476579]."

†**Chesterton** TL4659, A1, A2; **church** TL462596; **great brick house** [Chesterton Hall] TL459597. Chesterton Hall, built in the early 17th century, survives, marooned on a small island surrounded by roads.

Cheveley TL6860; **Sr John Cotton's park** TL6761. Sir John Cotton (1615–1689) inherited the Cheveley estate in 1620 and was created a Baronet in 1641. The estate had been bought by William Cotton in 1450 but Sir John was the last of the family to own it, selling it in 1671. He is one of the Cottons commemorated by fine memorials in Landwade church. The park initially developed around the castle but was extended in the 16th century; Sir John occupied a house which had been built in the south-west corner of the park by his father. The estate was extended and the park greatly modified after it was bought by the Duke of Somerset in 1732. See Wareham & Wright (2002) and Cokayne (1902).

Chidley is mentioned as a site for *Geranium sanguineum* [222], "*in a wood adjoyning to the highway betwixt Stitchworth and Chidley*". It is presumably Cheveley (q.v.), as Babington (1860) suggested, although Ray also refers to Cheveley by its current name. However, Reaney (1943) gives no such variant of Cheveley.

Childerley TL3561.

Chippenham TL6669; **Sir Francis Russells house** TL6669. Sir Francis Russell (*c*. 1616–1664) of Chippenham Hall, second Baronet, M.P. for Cambridge from 1641 until the Restoration, was a friend of Oliver Cromwell and a colonel in the Eastern Association during the Civil War. His eldest daughter Elizabeth married Cromwell's son Henry in 1653 and after the Restoration his son Sir John Russell became the second husband of Cromwell's daughter Frances (Holmes 1974; Fraser 1997). The house was newly built in Ray's time but it was remodelled in the 1690s and much of it was demolished in the 1790s (Wareham & Wright 2002).

Comberton TL3855.

Coton TL4058; **Whitwell** TL4058.

Cottenham TL4568.

†**Devils ditch** [Devil's Ditch, Devil's Dyke], A2. A linear earthwork 7½ miles long which runs from Woodditton, TL6558, north-east through TL66 to Reach, TL5666. Its original purpose is unknown but it might have been intended as a defensive structure; it is almost certainly post-Roman in date, built at some time between the fifth and seventh centuries (RCHM 1972; Malim 1997). For an account of its flora, see Leslie (2011).

GAZETTEER

†Ditton [Fen Ditton] TL4860; **Bigwin closes** TL4861; **church** TL483603, *A2. The Biggin or Biggin Abbey is at the north end of Fen Ditton parish just east of the River Cam. It was never an abbey, but it was the property of the Bishops of Ely from the 12th century and the current building dates from the late 14th century (RCHM 1972). A copy of the 1807 enclosure map (Cambridge University Library MS.Plans.5^a) shows a series of small hedged fields west and south-west of the Abbey on the east side of the river and these might perhaps be the closes mentioned by Ray. At least one of the plants that Ray recorded here, *Helleborus viridis* [276], is clearly a garden escape.

dry Draiton [Dry Drayton] TL3861.
Ellesly [Eltisley] TL2659, *A1, *A2.
Elme [Elm] TF4606.
Elsworth wood TL3161, *A2. An ancient wood, described briefly by Howlett & Moule (1979).
Ely TL5480; **Cathedrall church** TL541802; **North-Fen** TL58, *A2. Northfen was mentioned in the Ely Coucher Book of 1251 as one of a number of wastes, areas of marshland between Ely and other townships; the Bishops of Ely had inclosures here for peat-diggings (Pugh 1953). It was mapped in 1666 on Theophilus Byrd's "A map or a description of part of the Common belonging to the Cittie of Eley devided in to dollvers by the consent and appointment of the inhabitants and owners thereof, commonly called Westmore; Burnt fen, Little Shell, Padnall Ruff North Fen Byall Fen; and Byall Fen Wash."[1] Unfortunately Byrd's map, which shows North Fen alongside a watercourse and divided into seven plots, lacks a scale and any landmarks, and the whereabouts and extent of this fen in 1660 is uncertain. Unlike West Fen, North Fen does not appear on several later maps on which one might expect to see it named. A North Fen is shown on modern Ordnance Survey maps north-east of Little Downham, at TL5483, and a "Plan of the Turnpike Road in the North District of Ely" (1857)[2] shows what appears to be a north-eastern corner of North Fen west of the present Pyper's Hill Farm (then Piper's Hill) at TL5684, currently the site of a farm reservoir. This suggests that North Fen may have occupied the low-lying land between Ely and Littleport between these points, an area which is almost three miles north of Ely cathedral and thus more or less compatible with the locality for *Rumex maritimus* in the 1685 appendix, "*In a Fen two mile from Ely call'd the North-Fen*". This area is shown as Wood Fen on the first Ordnance Survey maps and much of it is named Lower Woodhouse Fen on modern maps. Although it now lies in the parishes of Downham and Littleport, it was formerly fen or marsh common to the parishes of Ely St Mary and Holy Trinity (Pugh 1953). More work is needed on the interpretation of Byrd's map and the location of North Fen.
Eversden wood TL3453, A1, A2. An ancient wood, described by Rackham (2003).
Exning, West Suffolk (v.c. 26), TL6265, *A2.
Fordham TL6370.
Fulborn [Fulbourn] TL5256, A2.

---

[1] Cambridge University Library MS Plans 120. Despite the punctuation, Padnall Ruff, North Fen and Byall Fen appear to be separate sites; 'North Fen' is written in red ink and is therefore distinguished from the names on either side. There are six individual maps on the sheet which appear to correspond with the six places listed after Westmore, although one of the names (probably Padnall Ruff) is virtually obliterated by a hole in the vellum and one of the maps (presumably, by elimination, Byall Fen) is unnamed.

[2] Cambridge University Library MS Plans 132.

GAZETTEER

†**Gamlingay** TL2452, A1, A2; **church** TL241523; **Sir Roger Burgoynes park** TL2251 (as **Gamlingay Park** in A2); **Windmills** TL2351, A1, A2. The estate atlas of Gamlingay produced by Thomas Langdon for Merton College, Oxford, in March 1602 provides a detailed representation of the village as it was known to Ray (see Map 2).[3] "Greate Heathe" is mapped to the south-west of the village, in the area shown as Gamlingay Great Heath on modern maps (but no longer heathland); the Bedford–Cambridge road crosses the northern part of the heath. The park, which in 1602 was owned by Sir John Burgoyne (and is shown as "The Graundge Io. Burgan" on Map 2) but by the 1650s had been inherited by Sir Roger, borders the heath on its north-east side. It is shown as a fenced park, as one would expect from the description of the locality for *Osmunda regalis* [B32] in the 1685 appendix (*"Within Gamlingay Park, and without by the Pales in the corner next to Sandy"*).[4] The area of turf-digging described under *Hypericum elodes* [66] as *"In a boggy ground beyond Gamlingay, near Sir Roger Burgoynes park, where they dig turfs"* and under *Drosera rotundifolia* [501] as *"Near Gamlingay in a moorish place where they digge turfes, not far from the high-way leading to Bedford"* would therefore appear to be the Great Heath, as more obviously is *"a heath beyond Gamlingay by S*[r] *Roger Burgoynes park"*, Ray's site for *Genista anglica* [214]. Sir Roger Burgoyne (1618–1677), second Baronet, was M.P. for Bedfordshire from 1641 until 1648, when, with other moderate members, he was forcibly excluded by the army in Pride's Purge. His letters to his life-long friend and fellow moderate M.P. Sir Ralph Verney Bart (1613–1696) of Claydon Hall, Buckinghamshire, survive in the remarkably rich Verney archive, and his portrait, still at Claydon, is reproduced by Tinniswood (2007).

Ray also refers to *"the pastures and Broom fields about Gamlingay"* and Langdon's map shows "Greate Broom Close" and "Broome Close" east of the village (and thus not on Map 2), as well as "Broome Woode", the current White Wood (Rackham 1992), on the north-west side of the Great Heath. Ray's locality near the windmills, described under *Teesdalia nudicaulis* [98] as *"a sandy lay near the windmills beyond Gamlingay toward Sandy"*, is the current Mill Hill. Langdon's map shows a standing windmill on a small triangular plot on the west side of "The way from Potton to Gamlingay", just south of the current Mill Bridge, and a similar plot on the east of the road which has no windmill but is labelled as "mill hill" on his detailed map.[5] An area of strip-cultivated fields south of the windmill is called "The Sandes".

---

[3] There is one map of the parish, and a further 14 maps cover the area in more detail; the former is particularly useful in interpreting Ray's localities and is reproduced by Bendall (2000). The original maps are in the muniments of Merton College (MCR 6.17), but we have also consulted photostats at Shire Hall, Cambridge (Cambridgeshire Archives TR274/p1). For further details, see Bendall (1992).
[4] The small sketch of "Gamlinghay Park" in Ogilby's map of the Bedford–Cambridge road (1675, plate 80) also shows it as fenced.
[5] There were two mills in the 19th century; they were demolished in about 1890 and in 1975 (Bruce & Sharpe 2002).

Map 2. Langdon's map of Gamingay. Although Thomas Langdon's estate atlas of Gamlingay dates from 1602, it provides a vivid illustration of the sites mentioned by Ray nearly 60 years later (see text). The south-western part of the general map of the parish is shown here. The scale is in perches; Langdon used the standard perch of 16.5 feet, so that 200 perches equals 0.63 miles or 1 km. Reproduced by kind permission of The Warden and Fellows of Merton College Oxford from a coloured original (Merton MCR 6.17).

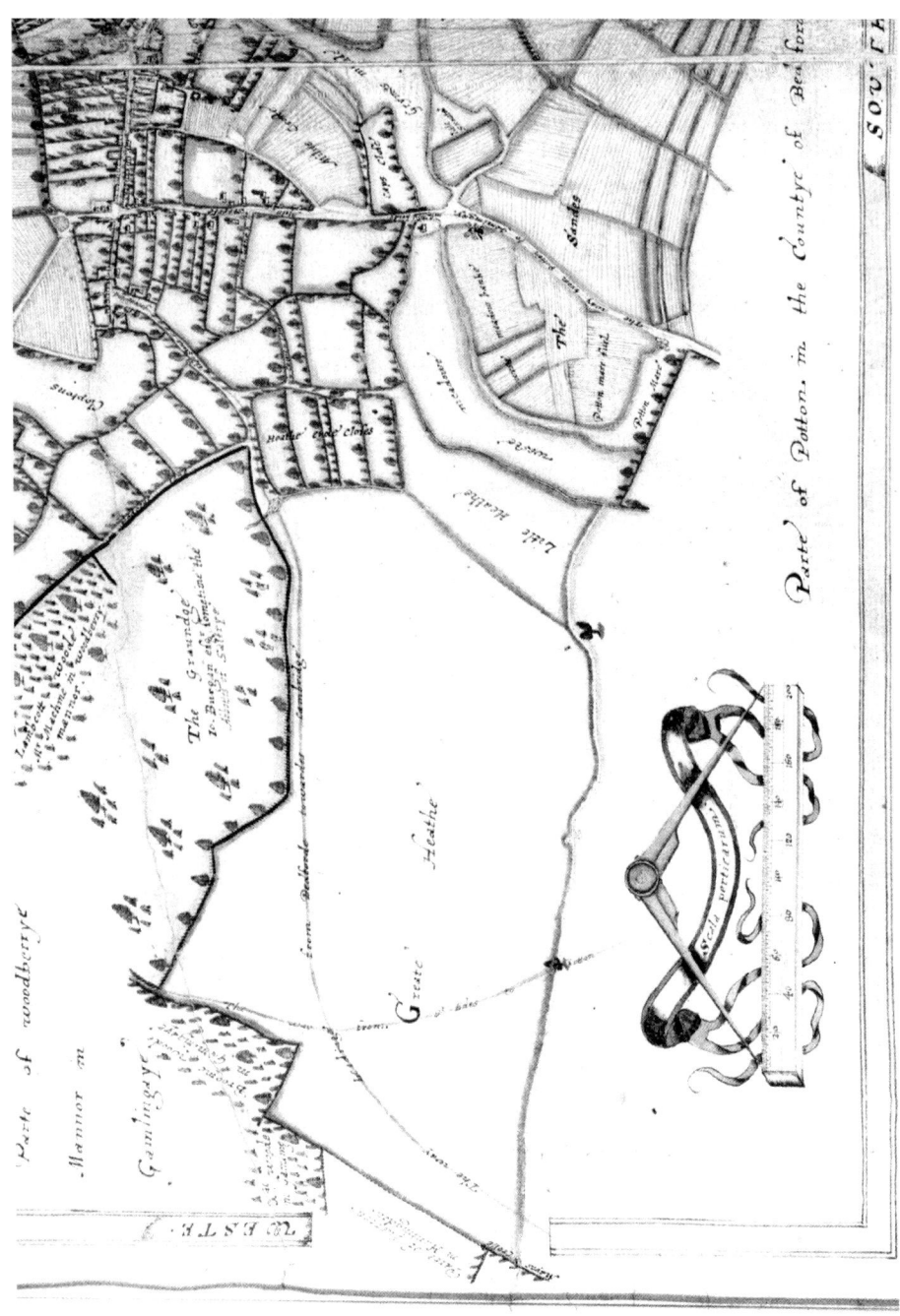

GAZETTEER

†**Gogmagog hills** TL4953, A1, A2. Ray's record of *Spiranthes spiralis* [409] "*on the skirts of Teversham moor towards Gogmagog hills*" suggests that the site was more extensive in Ray's time than it is now.

Gransden, *A2, could refer to either Great Gransden TL2755 (Huntingdonshire, v.c. 31) or Little Gransden TL2755 (Cambridgeshire, v.c. 29). *Sorbus torminalis* [B95], the species reported from here, was recorded in the 20th century from Waresley Wood, Little Gransden, Cambridgeshire as well as in the Huntingdonshire part of the contiguous Gransden and Waresley Woods (Crompton 2001).

Grantcester [Grantchester] TL4355; **Grantcester meadow** TL4356. "*Grantcester meadow*" is mentioned by Ray under *Ophioglossum vulgatum* [395] and, as "*Grantcester meadows*", under *Petasites hybridus* [436]. The names of the open fields and meadows of Grantchester are well documented, especially in a terrier and associated map of 1666, a pre-enclosure map of 1795 and the enclosure map of 1802 (King's College Archive Centre GRA/277, 892, 894, 896; see also Elrington 1973). We can find nothing corresponding to Ray's name in any of these sources. Before enclosure Grantchester was dominated by strip-cultivated open fields; its meadowland lay along the west side of the River Cam between the village and Cambridge. Until 1912 the parish extended to Newnham Croft, but the area close to Cambridge was then transferred to the town. The name Grantchester Meadows is currently used for the grassland between Grantchester and Cambridge, but it does not appear on Ordnance Survey maps (except as a street name in Newnham); it is tempting to suggest that it had the same informal meaning in Ray's time as it does now.

Hadnam [Haddenham] TL4675, *A1, *A2.

Hardwick TL3758.

Harleston [Harlton] TL3852, *A2.

Haverill [Haverhill], West Suffolk (v.c. 26), TL6745; **close behinde the Bell Inn**, **Haverill** TL6745. Ray mentions Haverhill twice, once simply as the destination of the road over the Gogmagog hills, a Cambridgeshire site. The Bell Inn is, however, on the High Street in Haverhill and the close mentioned by Ray as a site for *Ophrys apifera* [408] is therefore in West Suffolk.

Hawkston [Hauxton] TL4352.

Hildersham TL5448, A2; **church** TL545488, *A2; **Juniper-hill** ?TL5647, A2. Ray reports *Juniperus communis* [310] from Juniper-hill itself. This may be Rivey Hill, Linton, where juniper was known in the 19th century (see Crompton 2001). The "*little hill or brow whereon Furze grows, beyond Hildersham near Juniper hill*", from which *Dianthus deltoides* [B21] is reported, is almost certainly a reference to the site now known as Hildersham Furze Hill, TL5548 (see Trist 1988 for a description of its flora).

Hinton, see Cherry-hinton.

Histon TL4363; **old Chappell** TL4364. In the Middle Ages Histon consisted of two parishes, St Andrew's and St Etheldreda's, with their land intermingled. The "*ruines of an old Chappell*" mentioned by Ray under *Tanacetum vulgare* [563] are presumably those of St Etheldreda's church, which was 200 m west-north-west of the current church, St Andrew's. It was partly demolished around 1599 and not replaced, although the chancel was still standing in 1728. The parishes were merged in the 17th century and St Etheldreda's churchyard was taken into the grounds of Abbey Farm (Wright & Lewis 1989).

Hoginton [Oakington] TL4164.

Huntington rode, see under Cambridge.

Isle of Ely TL57 & TL58, A1.

GAZETTEER

†Kingston TL3455; **Kingston wood** TL3254, A1, A2. For a detailed account of the plant records from Kingston Wood, an ancient wood, see Reynolds (2003). The current byway along the north side of Kingston Wood was the main road to Gamlingay in Ray's time.

Linton TL5646, A2; **old parke near Linton** TL5447. The 'old parke' would, paradoxically, appear to be the 'New Park', Little Linton, which was by the Cambridge–Horseheath road at the western edge of the parish (Wright 1978). In the 1650s it was owned by the Parys family, who were papists and staunch Royalists and consequently heavily fined; their estate was mortgaged by 1659. From the 1560s they had usually lived in Norfolk, and even in Linton they transferred their main residence from the manor house at Little Linton to Catley Park in the early 17th century.

Little Eversden TL3753, *A2; **church**, TL374532, *A2.

Littleport TL5686.

Little Shelford church-yard TL453516.

London rode, see under Cambridge.

Long-Stanton [Longstanton] TL3966.

†Madingley TL3960, A2; **church** TL395603; **the great house** [Madingley Hall] TL392604; **Madingley wood**, TL4059. Madingley Hall was built by the Henrician lawyer Sir John Hynde between 1543 and 1547 and completed by his son Sir Francis Hynde before his death in 1597 (Salter 1970). In the 1650s the owners were Jane Hynde, the sole heir of the Hynde family, and her husband Sir John Cotton of Landwade (see Cheveley), although the Royalist Sir John spent much of the Interregnum abroad. For a detailed account of the plant records from Madingley Wood, an ancient wood which has been much modified since the 17th century, see Rackham & Coombe (1996).

Marsh [March] TL4196.

Mepole [Mepal] TL4480.

Milton TL4862.

†Newmarket, Cambridgeshire (v.c. 29) & West Suffolk (v.c. 26), TL6463; **church by the Kings house** [All Saints' church], (v.c. 29), TL644632; **nearest windmill on the Northside of Newmarket town**, (v.c. 26), *c.* TL640638; **Newmarket heath** (v.cc. 29 & 26), TL56, 66, A1, A2; **beacon, Newmarket heath** (v.c. 29), TL581598. Newmarket straddles the boundary between the old counties of Cambridgeshire and Suffolk. As the name implies, it is a relatively recent settlement (first recorded *c*. 1220), and it grew up alongside the Icknield Way, which was both a busy highway and the county boundary (May 1982). The new town developed in the parish of Exning (Suffolk), spreading south of the road into Woodditton (Cambridgeshire), and the vice-county boundary still runs along the High Street. There were two mediaeval churches in Newmarket, but Ray's locality "*church by the Kings house*" identifies his site for *Asperugo procumbens* [34] as All Saints' church south of the High Street in Cambridgeshire rather than the parish church of St Mary's in Suffolk. The royal residence was built by James I and had a frontage along the south side of the High Street. The building known to Ray was designed by Inigo Jones but it was sold in 1650 and fell into disrepair in the Interregnum; Charles II built a replacement residence further east on the High Street. The whereabouts of the old palace was established by May (1984) and Ray's locality provides additional evidence to support his view that it lay between All Saints' church and the High Street. Both the church and "Old Kings-yard" are shown on Chapman's map of Newmarket (1787). The current All Saints' church is a Victorian building but it replaced an earlier church on the same site.

GAZETTEER

Ray's locality for *Silene otites* [539] *"Near the gravell pits as you go to the nearest windmill on the Northside of Newmarket town"* is clearly from its position on the north side of the town in West Suffolk (v.c. 26), as Ray (1695) later suspected (see footnote 730 in the main catalogue, which names the other species recorded here). The Newmarket mill was on the road from Newmarket to Exning (still called Mill Hill at the Newmarket end); Chapman's map shows a single mill on the east side of the road at approximately TL 640638. May's (1982) account of Newmarket does not suggest that there was more than one mill here, but Ogilby's much less detailed map (1675, plate 46) shows two mills.

The name Newmarket Heath was formerly applied to a much more extensive tract of land than that shown on current maps, as illustrated by Ogilby (1675, plate 46) and described by Coombe (1987). This is also apparent from Ray's references to the beacon on Newmarket Heath; this was on a tumulus (now ploughed out) near the start of the Beacon Course, near Four Mile Stable Farm (see the account of *Veronica spicata* in Crompton 2001). Coombe also discusses the historical ecology and some of the current vegetation on the heath. The county boundary crosses the eastern part of Newmarket Heath, as Ray appreciated (see the first entry in Emendanda in the 1663 appendix).

Over church TL372707.

Potton, (Bedfordshire, v.c. 30), TL2249, *A2. The locality for *Juncus bulbosus* [B39] in the 1685 appendix, *"In some wet places of the Heath between Gamlingay & Potton"*, is presumably in Cambridgeshire.

Quoy church [Stow-cum-Quy, Quy church] TL515599; **Qui water** [Quy Water] TL5059, *A2.

Ramsey mear [Ramsey Mere], Huntingdonshire (v.c. 31), TL3189. Ray's record of *"Portulaca sylvestris"* [459] *"About Ramsey mear in the foot pathes by the rape mills"* was made by George Bowle or Bowles and published by How (1650). Ray says he has not visited the place himself. Although the site is in Huntingdonshire, Ray may have been misled by published maps into thinking that it was in Cambridgeshire. The maps of Cambridgeshire published by Camden (1607, pp. 354–355) and Speed (1611, between pp. 37 and 38) show Whittlesey Mere, Ug Mere and Ramsey Mere in the Isle of Ely rather than in Huntingdonshire, and Speed lists the site in "An Alphabeticall Table of all the Townes, Riuers, and places mentioned in Cambridge-shire". However, his map of Huntingdonshire in the same work (between pp. 57 and 58) provides what appears to be a more accurate map of the meres, and on this map they appear on the Huntingdonshire side of the county boundary and are listed in the alphabetical table of places mentioned in this county. Both Nidd and Ray owned abridged versions of Speed's atlas: the catalogue of Ray's library (Feisenberger 1975) identifies his copy as *England, Wales, Scotland and Ireland described and abridged* (Speed '1627')[6] and Nidd's copy, listed simply as "Speeds little maps" (Leedham-Greene 1986), was presumably an edition of the same work. This pocket-size volume had similar maps to those in the folio edition. The discrepancy in the Speed volumes arose because Speed (like Camden) followed Saxton's map of Cambridgeshire but redrew the Huntingdonshire map, perhaps on the basis of information received from the antiquary Sir Robert Cotton, a native of the county. An accurate manuscript map of Fenland survives in the Cotton

---

[6] There were in fact three distinct editions of the abridged atlas with the engraved title-page dated 1627; Skelton (1970) suggests that they were published with the folio editions of 1627, 1632 and 1648.

GAZETTEER

collection in the British Library (Lynam 1936; Skelton 1970). Other published maps resemble Speed's Cambridgeshire map in showing Ramsey Mere in the Isle of Ely rather than Huntingdonshire, including Jansson's *Comitatis Cantabrigiensis* (1646) and Blaeu's *Regiones inundatae* (1648). Ramsey Mere survived long enough to appear on the first Ordnance Survey maps but it was drained later in the 19th century.

River Cam the branches of the river flow into the county from Hertfordshire and Essex and north through Cambridge to the junction with the Great Ouse at TL5374; most of Ray's observations were probably made in the vicinity of Cambridge, TL45 and TL46. A1.

St George Hatley [Hatley St George] TL2751. Ray refers to "*a wood at St George Hatley*" [360, 580] and "*woods about S^t. George Hatley*" [441, 614]. The large ancient woods near Hatley St George are Buff Wood and Hayley Wood in Cambridgeshire (v.c. 29) and Cockayne Hatley Wood in Bedfordshire (v.c. 30). Littlehound Wood (v.c. 29) was destroyed around 1650 (Rackham 1975).

St Neotes, [St Neots], Huntingdonshire (v.c. 31), TL1860, *A1, *A2. Although the town is in Huntingdonshire, Ray's locality for *Bupleurum tenuissimum* [A6] was "*By the way-side as you ride to St Neotes beyond Elles-ly*" and was presumably in Cambridgeshire.

Shelford presumably covers both Great Shelford TL4551 and Little Shelford TL4551. A1, A2. "Shelford" in the site for *Ophrys sphegodes* [A26, B70] reported in the 1663 appendix as "*an old gravel-pit near Shelford by the foot way from Trumpington to the church*" refers to Great Shelford. See also Little Shelford.

Soham TL5973.

Stitchworth [Stetchworth] TL6459.

†Stretham ferry TL5072, A1. The ferry crossed the River Great Ouse.

Swafham presumably covers both Swaffham Bulbeck TL5562 and Swaffham Prior TL5663.

Suacy [Swavesey] **church** TL362693.

†Teversham TL4958, A1, A2; **church** TL496585 *A2; **Gains** TL5057, *A2; **Teversham moor** TL5058, A1, A2. Gains was on the south side of the parish; Dengaines or Engaines manor house was on the site of the current Manor Farm (see Wareham & Wright 2002).

†Triplow heath [Thriplow Heath] TL4344. In 1647 the army camped on Thriplow Heath, at a time when it was in dispute with Parliament (Wright 1982).

Trumpington TL4454, A1, A2; **Trumpington moor** TL45. The whereabouts of Trumpington moor are uncertain, but Baker's map of Cambridge (1830) shows as Trumpington Fen the fields along the east side of the River Cam from Trumpington to Cambridge. The only low-lying areas of the parish are these, which extend southwards beyond the area mapped by Baker to the edge of the parish at Hauxton Mill. The rivulet which comes from Trumpington, mentioned by Ray under *Hippuris vulgaris* [181], is presumably Vicar's Brook and its associated watercourses, which had been greatly modified around 1610 in the construction of Hobson's Conduit (Bushell 1938).

Water-Beach load [Waterbeach Lode] TL5064. The lode connected the River Cam to a small dock on the south-east side of the village (Ravensdale 1974).

Whitwell, see Coton.

Wilborham [Wilbraham] TL55; **Wilborham church**, TL55. Great Wilbraham church is at TL548577 and Little Wilbraham church at TL545585. Ray's "*old gravell-pit in the corn field near Wilborham church*", mentioned as a site for *Medicago minima*

[582], was probably near Little Wilbraham church, as that is built over Fourth Terrace Deposits rather than Lower Chalk.
Wisbich [Wisbech] TF4609.
Wittlesford [Whittlesford] TL4748, *A2.

VOCABULARY OF EPITHETS USED TO DESCRIBE PLANT SPECIES

This vocabulary is intended to provide explanations of the epithets used in Ray's chosen Latin names and in the synonyms from earlier literature in his main catalogue and in its two appendices. Our translations relate to the words as used in these lists and do not attempt to cover the full range of meanings that some of them have. The masculine, feminine and neuter nominative singular forms of adjectives are listed (e.g. *albus, -a, -um* and *aquatilis, -is, -e*) except where all three forms are identical, when the stem is added (e.g. *fugax, fugac-* and *repens, -ent-*, which has, for example, ablative singular *repente*). In the case of nouns the nominative and genitive singular are normally given (e.g. *arbor, -oris* and *gluma, -æ*), but nouns (especially proper names) that occur only in the genitive, dative and / or ablative are usually listed without their nominative singular. Phrases are sometimes listed alone (e.g. *capsulis triangulis*) but usually under the nominative singular of their first word (e.g. *acutiore folio* under *acutior, -ior, -ius* and *floribus albis* under *flos, floris*) or under a preposition (e.g. *cum paniculis nigris* under *cum*); some more complex phrases that are translated and explained in the footnotes are not repeated here. Synonymous nouns (e.g. *Bunias* for *Brassica rapa* and *B. napus* [385]) are not included, but semi-popular phrase-names (e.g. *Bonus Henricus*, *Herba Benedicta* and *Lingua bovis*) and most phrases that compare a species with another (e.g. *cotyledonis folio* and *Trifolii genus medicæ simile*) are. The datives and / or genitives of proper nouns (e.g. *Galeni, Galeno, Græcis, Græcorum* and *Lancastriæ*) are included where they form part of a polynomial phrase-name. Ray's capitalisation and accentuation have been retained as far as possible, but he is not always consistent.

acaulis/os, -is/os, -e/on stemless (Greek ἀκαυλος, -ος, -ον)
accedens, -ent- (after a dative or *ad* + accusative) resembling
acer/acris, acris, acre sharp, bitter, acrid
acerosâ glumâ with a chaffy glume
acetosus, -a, -um acidic, sour-tasting
Achilleos (Greek genitive) of Achilles (the Homeric hero: Greek Ἀχιλλευς, Ἀχιλλεως)
acidus, -a, -um sharp, sour-tasting
acorifolius, -a, -um with leaves [like those] of sweet flag (*Acorus calamus* L.)
acrior, -ior, -ius sharper, stinging more
aculeatus, -a, -um prickly
acutior, -ior, -ius more acute, sharper; *acutiore folio* with a more acute / sharper leaf
acutus, -a, -um acute, sharp, with acute leaves or perianth-segments; *acuti varietas, -atis* a variety of the acute[-leaved dock]
adulterinus, -a, -um bastard, false, counterfeit
Æginetæ (genitive singular) of Aegineta (Paul of Aegina)
æstivalis, -is, -e of summer
affinis, -is, -e (after a dative) related [to]
agninus, -a, -um of a lamb, of lambs
agrarius, -a, -um of fields, of cultivated land
agrestis, -is, -e of fields, of cultivated land
agrius, -a, -um wild (Greek ἀγριος, -α, -ον)
agrorum (genitive plural) of fields, of cultivated land

VOCABULARY

Aizoi facie planta, -æ a plant with the appearance of an Everlasting (e.g. *Aizoon* or *Sedum* sp.)

aizoides, -oid- like an everlasting (see previous entry)

albus, -a, -um white; *Album olus* White vegetable / potherb, *Valerianella locusta* [311]; *albis floribus* with white flowers

Alexandrinus, -a, -um Alexandrine, of Alexandria

alius, -a, -ud another, other (of more than two); *aliis, aliorum* for / of other / some [people, authors or botanists]; *aliorum colorum* of other colours; *aliorum Græcorum* of other [Ancient] Greeks

Alopecurinus, -a, -um foxtail-like, like *Alopecurus* spp.

alopecuroides, -oid- foxtail-like, like *Alopecurus* spp.; *Alopecuroidi accedens & Phalaridi* resembling a Foxtail-like [grass] & a Reed-grass (*Alopecurus* and *Phalaris* spp.)

Alpinus, -a, -um Alpine, of the Alps

alsinefolius, -a, -um chickweed-leaved (see next but one entry)

alsineformis, -is, -e chickweed-like (see next entry)

alsines folio (genitive and ablative singular) with a leaf [like that] of a chickweed (*Stellaria* sp.)

Alsines Trissaginis folio (genitive and ablative singular) with a leaf [like that] of Germander-chickweed (*Veronica agrestis* [22]), called by Ray *Alsine foliis Trissaginis*

alter, -era, -erum the other (of two), second; *alterum genus* (after a genitive) the other / second kind [of]

altissimus, -a, -um tallest, very tall

amarus, -a, -um bitter

Amygdalino folio (ablative singular) with a leaf [like that] of Almond (*Prunus dulcis* (Mill.) D.A. Webb)

amygdaloides, -oid- almond-like

an or (or indicating doubt, like a question mark)

anatus, -a, -um floating

Anemones folio rotundo (genitive and ablative singular) with the round leaf of an Anemone (*Anemone* sp.)

Anglic[an]us, -a, -um English, of England

Anglorum (genitive plural) of the English

angulosus, -a, -um strongly angled

angustifolius, -a, -um narrow-leaved

angustiore folio et spicâ (ablative singular) with a narrower leaf and spike

angustissimo, & acuto vel gramineo folio (ablative singular) with a very narrow & acute or grass-like leaf

angustissimo folio (ablative singular) with a very narrow leaf

angustus, -a, -um narrow; *angustis & longissimis foliis crispis subtus albicantibus* with narrow & very long wavy leaves, whitish beneath; *angusto lanuginosóque folio* with a narrow and downy leaf; *angusto oblongo folio* with a narrow oblong leaf

annuus, -a, -um annual

anserinus, -a, -um of a goose, of geese

anthoxanthos, -os, -on golden-flowered (Greek ἀνθοξανθος, -ος, -ον)

Anthrisco hispido affinis / similis, -is, -e like / related to Hedge Parsley (*Torilis japonica* [119])

Antiquorum (genitive plural) of the Ancients (i.e. the Greeks and / or Romans)

aphyllos/us, -os/a, -on/um leafless (Greek ἀφυλλος, -ος, -ον)

VOCABULARY

apicibus purpureis (ablative plural) with purple tips (i.e. stamen filaments)
apii folio/is (genitive singular and ablative singular/plural) with a leaf / leaves [like that / those] of celery (*Apium graveolens* [53]), celery-leaved
apiifolius, -a -um celery-leaved (see previous entry)
Apis folio (genitive and ablative singular) with a leaf [like that] of *Apis* (perhaps a corruption of Greek ἄπιος, -ου [apios, "pear-tree"])
Apium risûs Celery of laughter (literally), *Ranunculus sceleratus* [478]
appendix, -icis an appendage
Aprili florens, -ent- flowering in April
Apuleii (genitive singular) of Apuleius (Apuleius Platonicus)
aqua, -æ water; *aquis [cœnosis] innatans, -ant-* floating in [muddy] waters; *aquis immersus, -a, -um* submerged in waters
aquaticus, -a, -um aquatic, growing in or near water; *aquaticum* (neuter) *duorum generum* aquatic of two kinds
aquatilis, -is, -e aquatic, growing in or near water
Arabicus, -a, -um Arab, Arabic, of the Arabs
Arabum (genitive plural) of the Arabs
arbor, -oris a tree; *arborum* of trees
arborescens, -ent- arborescent, becoming tree-like
arboreus, -a, -um arboreal, growing on trees
arista, -æ an awn; *aristis recurvis* with recurved awns
Arnoglossa, -æ Lamb's-tongue, *Plantago media* [448]
aromaticus, -a, -um aromatic, fragrant
arthriticus, -a, -um treating arthritis
articularis, -is, -e jointed
arundinaceus, -a, -um like a reed (*Phragmites australis* [64])
arvensis, -is, -e of ploughed fields
arvorum (genitive plural) of ploughed fields
asinus, -i an ass, a donkey
aspalathoides, -oid- like aspalathos (Greek ἀσπαλαθος, -ου, a Mediterranean broom-like shrub)
asper, aspera, asperum rough, prickly
asperior, -ior, -ius rougher, more prickly
assurgens, -ent- ascending, rising up
Aster, -eris an Aster (?*Inula crithmoides* L.), *Asteris flore luteo* with the yellow flower of an Aster
at but
Athenæi (genitive singular) of Athenaeus (of Naucratis in Egypt)
atque [etiam] and [also]
atrâ maculâ notatus, -a, -um marked with a dead-black spot
Atriplicis effigie (genitive and ablative singular) with a likeness to an Orache (*Atriplex* or *Chenopodium* sp.)
atrorubens, -ent- dark red
Atticus, -a, -um Attic, of Attica (i.e. Athenian, of Athens)
aureus, -a, -um golden
auricomus, -a, -um with golden hair, goldilocks
Auricula, -æ leporis Little ear of a hare, Hare's-ear, *Bupleurum* sp.
Auricula, -æ muris Little ear of a mouse, Mouse-ear, *Cerastium fontanum* [26] or *Pilosella officinarum* [437]; also *Antennaria dioica* [233], *Myosotis arvensis* [381] or *Veronica officinalis* [609]

VOCABULARY

auriculatus, -a, -um auriculate, eared
aut or
autumnalis, -is, -e autumnal, of autumn
avenaceus, -a, -um like oats (*Avena* spp.)
Avicennæ of Avicenna (or Ibn Sina)
avis, -is a bird; *avium pennas* (genitive and accusative plural) the feathers / wings of birds
bacca, -æ a berry; *baccis albis* with white berries; *baccis parvis purpureis* with small purple berries; *baccis rubris* with red berries
baccifer, -fera, -ferum berry-bearing, berried
Barb[ul]a, -æ hirci [little] beard of a he-goat, Goat's-beard, *Tragopogon pratensis* [574]
Barba, -æ capri Goat's-beard, *Filipendula ulmaria* [623]
barbatus, -a, -um bearded
basilicus, -a, -um royal, kingly
basis, -is a base
Batavia, -æ Batavia (the Dutch Netherlands)
Batavicus, -a, -um Batavian (see previous entry)
batrachoides, -oid- of a frog, of frogs
Bauhini (genitive singular) of Bauhinus (either Jean or Caspar Bauhin)
Belga, -æ a Belgian; *Belgarum* of the Belgians
Belgicus, -a, -um Belgian, of Belgium
Bella, -æ Donna, -æ Beautiful Lady (literally, in Italian), *Atropa belladonna* [548]
belliformis, -is, -e (correctly *bellidiformis, -is, -e*) like a daisy, *Bellis perennis* [82], describing *Silene otites* [539], presumably because of the similar rosettes
Bellonio (dative singular) for Bellonius (Pierre Belon)
benedictus, -a, -um blessed
bicolor, -color- bicoloured, two-coloured
bifolius, -a, -um bifoliate, two-leaved
bituminosus, -a, -um smelling of tar or bitumen
bombycinus, -a, -um like silk, silky
Bononiensis, -is, -e of Bologna (in northern Italy)
bonus, -a, -um good; *Bonus Henricus* Good Henry, *Chenopodium bonus-henricus* [90]
bos, bovis an ox, a bullock
botrytis, -is like a bunch of grapes (Greek βοτρυιτης, -ου)
Brabanticus, -a, -um Brabantian, of Brabantia (a province of the Low Countries)
Branca, -æ ursina, -æ Bear's-arm, *Heracleum sphondylium* [560]
brevior, -ior, -ius shorter; *brevioribus foliis* with shorter leaves
brevis, -is, -e short; *brevi angustóque folio incano* with a short and narrow hoary / greyish leaf; *brevibus pediculis* with short petioles
brevissimus, -a, -um shortest, very short
Britannicus, -a, -um British
bromoides, -oid- resembling a brome-grass (*Bromus* sp.)
bubulus, -a, -um relating to cows or oxen
bufonius, -a, -um of a toad, of toads
bulbosus, -a, -um bulbous, bulb-bearing
Bursa pastoris Shepherd's Purse, *Capsella bursa-pastoris* [97]
caballinus, -a, -um of a horse, of horses
caduco flore Phœniceo (ablative singular) with a readily falling scarlet flower

VOCABULARY

cæruleus, -a, -um blue; *cæruleo flore [majore]* with a [larger] blue flower
Cæsalpini (genitive singular) of Caesalpinus (Andrea Cesalpino)
calcar, -aris a spur; *calcare oblongo* with an oblong spur; *calcaribus oblongis* with oblong spurs
calyce luteo (ablative singular) with a yellow cup
calyculis paleaceis (ablative plural) with chaffy perianth-segments
Cambro-Britannicus, -a, -um Welsh
Camerarii (genitive singular) of Camerarius (Joachim Camerarius the Younger)
campanus, -a, -um Campanian, of Campania (a province of middle Italy)
campestris, -tris, -tre of open plains
canarius, -a, -um canine, of a dog (hence common or worthless)
candicans, -ant- shining white, becoming pure white; *candicantibus floribus* with shining white flowers
candidus, -a, -um pure white
caninus, -a, -um canine, of a dog (hence common or worthless)
cannabinus, -a, -um like hemp (*Cannabis sativa* [104])
Cantabricus, -a, -um Cantabrican, northern Spanish
capillaceus, -a, -um hair-like, very slender; *capillaceo folio* with a hair-like leaf
capillaris, -is, -e hair-like, very slender
capillatus, -a, -um hair-like
capillus, -i hair; *capillúsve* or hair
capitatus, -a, -um capitate, with flowers in a dense head
capitulum, -i a head (of flower[s] or fruit[s]); *capitulis cardui benedicti* with heads [like those] of blessed thistle (*Cnicus benedictus* L.); *capitulis equiseti* with heads [like those] of a horsetail (*Equisetum* sp.); *capitulis longis* with long heads; *capitulis minoribus* with smaller heads; *capitulis nigricantibus* with blackish heads; *capitulis Psyllii* with heads [like those] of *Plantago psyllium* L. (Greek ψυλλιον, -ου); *capitulis rotundis echinatis* with round spiny heads; *capitulis tomentosis* with tomentose heads; *capitulo breviore [hispido]* with a shorter [bristly] head; *capitulo globoso* with a spherical head; *capitulo hirsuto longiore* with a longer, hairy head; *capitulo longiore spinoso* with a longer, spiny head; *capitulo lupuli* with a head [like that] of a hop (*Humulus lupulus* [342]); *capitulo minore* with a smaller head; *capitulo purpureo* with a purple head; *capitulo rotundiore* with a rounder head; *capitulo squamoso* with a scaly head; *capitulo torulis canulato* with a head somewhat hoary with tufts; *capitulo torulo* with a head with a tuft
caprarius/capreus/caprinus, -a, -um of a goat, of goats
capsulis echinatis (ablative plural) with spiny capsules
capsulis triangulis (ablative plural) with triangular capsules
caput, capitis a head (usually of flower[s] or fruit[s]); *capite rotundo* with a round head; *capite tomentoso* with a tomentose head; *Caput gallinaceum* Head of a hen, Cock's-head, *Onobrychis viciifolia* [394]; *Caput monachi* Monk's-head, *Taraxacum* sp. [167]; *caput tentans, tent-* assailing the head
Carduus, -i fullonius, -i Thistle of fullers, Teasel, *Dipsacus fullonum* [168]
Carduus, -i fullonum Fullers' Thistle, Teasel, *Dipsacus fullonum* [168]
carens, carent- (after an ablative) lacking
caryophylli foliis (genitive singular and ablative plural) with leaves [like those] of a pink (*Dianthus* sp.)
Caryophylloides, -oid- Pink-like, like a Pink (*Dianthus* sp.)
catharticus, -a, -um cathartic, purging
cattarius, -a, -um of a cat, of cats

VOCABULARY

catus, -i a cat
cauda, -æ a tail; *Cauda equina* Horse's-tail, Horsetail, *Equisetum* spp.; *Cauda muris* Mouse's-tail, Mousetail, *Myosurus minimus* [383]
caulis, -is a stem; *caule aphyllo [non ramoso]* with a leafless [unbranched] stem; *caule compresso* with a flattened stem; *caule crispo* with a crisped / kinky stem; *caule et semine striato* with a striate / striped stem and seed; *caule folioso* with a leafy stem; *caule hexagono* with a hexagonal stem; *caule maculoso / nudo* with a spotted / bare stem; *caule purpurascente* with a purplish stem; *caule quadrangulo / triangulo* with a square / triangular stem; *caule rotundo* with a round stem; *caulibus procumbentibus* with procumbent stems
causticus, -a, -um caustic, burning, acrid
cavernosus, -a, -um full of cavities
cavus, -a, -um hollow
ceanothos, -i Theophrastus' name for a prickly plant (Greek κεανωθος, -ου), applied to *Cirsium arvense* [113]
Cerealis, -is, -e cultivated (from Ceres, the Roman goddess of agriculture)
cervinus, -a, -um of a stag, of stags
Chærophyllo nonnihil similis, -is, -e somewhat like Wild Chervil (*Chaerophyllum temulum* [124])
chamædryfolius, -a, -um, with leaves [like those] of ground-oak (*Teucrium chamædrys* L.)
chamædryodes, -od- like ground-oak (*Teucrium chamædrys* L.)
characias, -æ of or fit for a stake, pale or palisade (Greek χαραχιας, -ου)
Chironius, -a, -um of Chiron (the centaur)
chirurgus, -i a surgeon; *chirurgis*, *Chirurgorum* for / of surgeons
chrysanthemus, -a, -um golden-flowered
Chymicus, -i an Alchemist; *Chymicis* for Alchemists
Ciceri sylvestri minori affinis, -is, -e related to the smaller wild Chick-pea
cicutæ folio (genitive and ablative singular) with a leaf [like that] of hemlock (*Conium maculatum* [132])
cineracei coloris (genitive singular) of a grey colour, grey-coloured
circulo luteo (ablative singular) with a yellow circle
circumvallatus, -a, -um surrounded by a fortification, protected by something surrounding
cissampelos, -i ivy-vine (literally; Greek κισσαμπελος, -ου, supposedly a bindweed [149, 150])
clavatus, -a, -um clavate, club-shaped
Clusii, Clusio (genitive and dative singular) of / for Clusius (Charles de l'Écluse)
cochleatus, -a, -um spirally twisted (like a snail-shell)
Cœli donum Heaven's gift, *Chelidonium majus* [128]
cognatus, -a, -um (after a dative) related [to], similar [to]
collinus, -a, -um of hills
colore rubiginoso (ablative singular) with a rusty colour
columbarius/columbinus, -a, -um of a dove, of doves, dove-like
Columellæ (genitive singular) of Columella (Lucius Junius Moderatus Columella)
Columnæ (genitive singular) of Columna (Fabio Colonna)
communior, -ior, -ius commoner, more common
communis, -is, -e common
comosus, -a, -um tufted (like hair)
conifer, -fera, -ferum coniferous, cone-bearing

VOCABULARY

conyzoides, -oid- like fleabane (*Conyza, Erigeron* or *Pulicaria* sp.)
cordatus, -a, -um cordate, heart-shaped
Cordi (genitive singular) of Cordus (Valerius Cordus)
coriariorum (genitive plural) of leather-workers
corniculatus, -a, -um with small horns
cornu, -ûs a horn; *Cornu cervi[num]* Stag's-horn, *Plantago coronopus* [152] or *Lepidium coronopus* [153]
cornutus, -a, -um horned, horn-shaped
Corona, -æ fratrum brothers' Crown, *Cirsium eriophorum* [115]
coronarius, -a, -um used for garlands
coronatus, -a, -um crowned, garlanded
Cortusi (genitive singular) of Cortusus (Jacobi Antonio Cortusi)
corymbosus, -a, -um corymbose, bearing corymbs
costâ spinosâ (ablative singular) with a spiny midrib
cotyledonis folio (genitive and ablative singular) with a leaf [like that] of navelwort (*Umbilicus rupestris* (Salisb.) Dandy)
cranium, -i a skull
crassus, -a, -um thick; *crassa radice* with a thick root
Cratevæ (genitive singular) of Cratevas (physician to King Mithridates VI of Pontus)
creditus, -a, -um believed [to be]
crenatus, -a, -um crenate, scalloped
Crescentio (dative singular) for Crescentius (Pietro Crescenzi)
creticus, -a, -um Cretan, of Crete
crispus, -a, -um crisped, kinky, curled, wavy-margined
crista, -æ a crest; *Crista galli* Cock's-comb, *Rhinanthus minor* [426] or *Pedicularis sylvatica* [427]
cristatus, -a, -um crested
cuculus, -i a cuckoo
cujusdam see *quidam*
cum (+ ablative) with; *cum alis virentibus* with green wings; *cum capitulis villosis* with woolly heads; *cum cauda muris purpurascente* with a purplish tail of a mouse; *cum longis calcaribus rubellis* with long reddish spurs; *cum longissima panicula* with a very long panicle; *cum maculis nigricantibus et sine illis* with blackish spots and without them; *cum multis siliquis hirsutis* with many hairy pods; *cum paniculis nigris* with black panicles; *cum parva cauda muris* with a small tail of a mouse; *cum pericarpiis rotundis* with round fruits; *cum rubore* with redness; *cum siliquis glabris* with glabrous pods; *cum siliquis plurimis hirsutis* with very many hairy pods
cyanoides, -oid- like the blue cornflower (*Centaurea cyanus* [162])
cynanchicus, -a, -um curing or treating quinsy (*alias* squinancy) (Greek κυναγκη [cynance])
cyparissias, -æ a kind of spurge (*Euphorbia* sp.) (Greek κυπαρισσιας, -ου)
cyperoides, -oid- sedge-like, like a *Carex* sp.
Dalechampii, Dalechampio (genitive and dative singular) of / for Dalechampius (Jacques Daléchamps)
Damocratis (genitive singular) of Damocrates (Damocrates Servilius)
daphnoides, -oid- like laurel (bay tree, *Laurus nobilis* L., Greek δαφνη, -ης [daphne])
degener, degener- false, not genuine
dens, dentis a tooth; *Dens leonis* Lion's-tooth, Dandelion, *Taraxacum* sp. [167]; *Dentis leonis folio acuto* with the sharp leaf of Dandelion

VOCABULARY

dentatus, -a, -um dentate, toothed
denticulatus, -a, -um denticulate, with small teeth
dictus, -a, -um called
dilutè purpureus, -a, -um pale purple
Dioscoridi, Dioscoridis (dative and genitive singular) for / of Dioscorides (Pedanius Dioscorides)
dissectus, -a, -um dissected, divided, with cut leaves; *dissectâ paniculâ* with a divided panicle; *dissectis foliis* with dissected / cut leaves
distichus, -a, -um consisting of two rows
diversitas, -atis a diversity, a difference
Dodonæi, Dodonæo of / for Dodonaeus (Rembert Dodoens)
dolichanthus, -a, -um with a long flower-spike
domesticus, -a, -um domesticated
dorso spinoso (ablative singular) with a spiny back [to the leaf]
dulcis, -is, -e sweet, mild
dumetorum (genitive plural) of thickets, of thorn-bushes
durus, -a, -um hard
Eboracensis, -is, -e of York, of Yorkshire
echinatus, -a, -um echinate, spiny (like a hedgehog or sea-urchin); *echinato magno fructu* with a large spiny fruit; *echinato semine* with a spiny seed
echioides, -oid- like *Echium vulgare* [172], prickly
effigies, -ei likeness; *effigie* (ablative singular after a genitive) with a likeness [to]
elatior, -ior, -ius taller
endiviæ foliis (genitive singular and ablative plural) with leaves [like those] of Belgian endive, formerly called also "scariole" (OED), the French *chicorée de Bruxelles* (or locally *chicon*) and the Dutch *witloof* (*Cichorium intybus* L. var. *foliosum* Hegi) (The epithet *endiviifolius, -a, -um* has more recently been used for comparisons to the related frilly-leaved species, Endive (*C. endivia* L.), but this is clearly not the case here.)
equinus, -a, -um equine, of a horse, of horses
Equiseti facie (genitive and ablative singular) with the appearance of a Horsetail (*Equisetum* sp.)
equisetifolius, -a, -um with horsetail-like leaves, with leaves [like those] of a horsetail (*Equisetum* sp.)
erectus, -a, -um erect, upright
eriocephalus, -a, -um woolly-headed (Greek ἐριοκεφαλος, -ος, -ον)
erraticus, -a, -um wandering, invasive
erucæfolius, -a, -um with leaves [like those] of rocket (*Eruca vesicaria* (L.) Cav.)
esculentus, -a, -um edible
et and
etiam also; *etiámq;* (= *etiamque*) and also
Euboicus, -a, -um Euboean, of Euboea (a large island near the mainland of Greece)
Euphrosynes facie (genitive and ablative singular) with the appearance of Eyebright (*Euphrasia* spp. [192])
ex (+ ablative) out of, from; *ex cranio [humano]* out of a [human] skull; *ex monte Ballon* from Mount Ballon; *ex rotunditate / rotundo* from roundness, from a round [shape]; *ex rubro nigricans, -ant-* blackish tinged with red
excelsior, -ior, -ius taller
exiguus, -a, -um small
exilis, -is, -e lank, meagre

VOCABULARY

facies, -ei a face, an appearance; *facie* (ablative singular + genitive) with the appearance of; *facie milii solis* with the appearance of millet of the sun, i.e. like *Lithospermum officinale* [339]; *facie Seseli pratensis* with the appearance of meadow Seseli (apparently *Silaum silaus* [524]); *facie urticæ* with the appearance of nettle (*Urtica dioica* [628])

fæminus, -a, -um female

fætidus, -a, -um stinking, fetid

fatuus, -a, -um foolish, insipid, tasteless

favaginosus, -a, -um like a honeycomb

felinus, -a, -um feline, of a cat, of cats

Ferrum, -i equinum, -i Iron for a horse (literally), Horseshoe, *Hippocrepis comosa* [A14/B31]

fertilis, -is, -e fertile

fervidi gustûs (genitive singular) of / with a burning taste

fibrinus, -a, -um of a beaver, of beavers

filicinus, -a, -um fern-like

filius, -i a son; *Filius ante patrem* Son before father, *Epilobium hirsutum* [352] or *Tussilago farfara* [603]

fimbriatus, -a, -um fimbriate, fringed

flammeus, -a, -um flame-like, flame-coloured

Flandrorum (genitive plural) of the Flemings, of the Flemish

flavo flore (ablative singular) with a yellow flower

florens, florent- flowering

floridus, -a, -um flowery, flowering

flos, floris a flower, an inflorescence; *flore albo [erecto]* with a[n erect] white flower; *flore albo aut roseo* with a white or pink flower; *flore albo et purpurascente* with a white and purplish flower; *flore albo simplici* with a simple white flower; *flore albo-sulphureo* with a whitish sulphur-yellow flower; *flore aureo [odorato]* with a golden [scented] flower; *flore cæruleo [interdum lacteo aut dilutè purpurascente]* with a blue flower [sometimes milk-white or pale purplish]; *flore candido* with a pure white flower; *flore calcari donato* with a flower provided with a spur; *flore carneo* with a flesh-pink flower; *flore Codii i.e. campanulæ* with the flower of a Bell (from Greek κωδων [codon]) i.e. of a little bell (literally), with a bell-like flower; *flore conglomerato* with a clustered inflorescence; *flore dilutè purpurascente* with a pale purplish flower; *flore diluti ruboris* with a flower of pale redness; *flore & semine albis* with a white flower & seed; *flore ex luteo purpurascente* with a purplish flower tinged with yellow (the usual meaning, though more applicable to *Verbascum nigrum* [607] would be: with a flower becoming purple [in the centre] from yellow); *flore fimbriato* with a fringed flower; *flore flavo* with a yellow flower; *flore gallii albi* with a flower [like that] of Marsh Bedstraw (*Galium palustre* [210]); *flore globoso* with a spherical inflorescence; *flore herbaceo intus nonnihil candicante* with a grass-green flower somewhat whitish within; *flore incarnato* with a flesh-pink flower; *flore laciniato simplice* with a simple ragged flower; *flore longiore candicante* with a rather long shining white inflorescence; *flore luteo [galericulato]* with a [helmet-shaped] yellow flower; *flore luteo simplici* with a simple yellow flower; *flore magno [luteo]* with a large [yellow] flower; *flore majore [cæruleo & albo]* with a larger [blue & white] flower; *flore majore luteo splendente* with a larger shining yellow flower; *flore lupulino* with an inflorescence [like that] of a hop (*Humulus lupulus* [342]); *flore melino* with a quince-yellow (Greek μηλινος, -η, -ον) flower; *flore minimo / minore* with a very small /

smaller flower; *flore minùs pulchro* with a less beautiful flower; *flore muscoso* with a mossy flower; *flore nigræ menthæ* with the flower of black mint (*Mentha* sp.); *flore odorato incarnato* with a scented flesh-pink flower; *flore pallidiore* with a paler / rather pale flower; *flore parvo [herbido et muscoso]* with a small [grass-like and mossy] flower; *flore parvo luteo* with a small yellow flower; *flore phœniceo* with a scarlet flower; *flore pleno [odoratissimo]* with a [very sweet-smelling] double flower; *flore pulchriore, sive lupulino* with a more beautiful or hop-like inflorescence; *flore punicante* with a purplish flower; *flore purpurante/ purpurascente* with a purplish flower; *flore purpureo [cæruleóve magno]* with a [large] purple [or blue] flower; *flore roseo* with a pink flower; *flore rotundiore* with a rounder inflorescence; *flore rubente / rubicundo* with a reddish flower; *flore rubro [elegantissimo]* with a [very elegant] red flower; *flore simplici [luteo]* with a simple [yellow] flower; *flore simplici odoro* with a simple scented flower; *flore spadiceo* with a deep reddish-brown (date-coloured) flower; *flore spicato* with a spicate / spiked inflorescence; *flore subcæruleo* with a somewhat blue flower; *flore subvirente* with a somewhat green flower; *flore unico* with a single flower; *flore violaceo* with a violet-coloured flower; *flore virente / viridi* with a green flower; *flore viridante* with a greenish flower; *floribus ad latera pilosis* with flowers hairy on the sides; *floribus albis / cæruleis* with white / blue flowers; *floribus compactis* with compact inflorescences; *floribus & loculis cauliculis adhærentibus* with the flowers & pods clasping the stems (literally little stems); *floribus ex albo virescentibus* with the flowers greenish tinged with white; *floribus ex singularibus pediculis majoribus, simplicibus* with larger simple flowers growing on individual pedicels; *floribus insignis, -is, -e* remarkable for [its] flowers; *floribus luteopallescentibus* with palish yellow flowers; *floribus odoratis luteis* with scented yellow flowers; *floribus pallidis ac dilutiùs coloratis quasi albidi* with pale and more palely coloured flowers, as it were whitish; *Flos crucis* Flower of the cross, *Polygala vulgaris* [451]; *Flos cuculi* Flower of the cuckoo, Cuckoo-flower, *Cardamine pratensis* [106] or *Silene flos-cuculi* [61]; *Flos solis* Flower of the sun, *Helianthemum nummularium* [125]; *Flos Trinitatis* Flower of the Trinity, *Viola arvensis* [620] or *V. tricolor* L.

flosculus, -i floret, small flower; *flosculis albicantibus / albis* with whitish / white florets; *flosculis cauliculis adhærentibus* with the florets clasping the stems (literally little stems); *flosculis exiguis* with small florets; *flosculis luteis* with yellow florets; *flosculis longis albis* with long white florets; *flosculis pediculis oblongis insidentibus* with florets standing on oblong pedicels; *flosculis rubentibus* with reddish florets; *flosculis subrubentibus* with somewhat reddish florets; *flosculorum foliolis non divisis* with the petals (literally little leaves) of the florets undivided

fluitans, -ant- floating on the surface

fluviatilis, -is, -e of a river, of rivers

fœminus, -a, -um female

fœtens, -ent- fetid, stinking

fœtidissimus, -a, -um most fetid, very stinking

fœtidus, -a, -um fetid, stinking

foliaceus, -a, -um foliaceous, leafy

foliosus, -a, -um leafy, many-leaved

folium, -i a leaf; *foliis angustioribus* with narrower leaves; *foliis angustis splendentibus* with narrow shining leaves; *foliis angustis, subcæruleis* with narrow bluish leaves; *foliis betæ aut plantaginis* with leaves [like those] of a beet (*Beta* sp.) or a plantain (*Plantago* sp.); *foliis caryophylleis* with pink-like (*Dianthus*-like) leaves;

VOCABULARY

foliis caulem ambientibus with leaves surrounding the stem; *foliis convolvuli minoris* with leaves [like those] of field bindweed (*Convolvulus arvensis* [149]); *foliis crenatis* with crenate / scalloped leaves; *foliis crispis* with crisped / kinky / wavy-margined leaves; *foliis dentatis* with toothed leaves; *foliis domesticæ similis, -is, -e* like the domesticated [kind] in [its] leaves; *foliis dissectis* with dissected / cut leaves; *foliis erucæ* with leaves [like those] of rocket (*Eruca vesicaria* (L.) Cav.); *foliis & spica junci* with leaves & spike [like those] of a rush (*Juncus* sp.); *foliis fœniculi [latioribus]* with [broader] leaves [like those] of fennel (*Foeniculum vulgare* Mill.); *foliis fumariæ bulbosæ, de qua Cordus* with leaves [like those] of bulbose fumitory (*Corydalis solida* (L.) Clairv.), about which Cordus [has written]; *foliis glabris* with glabrous leaves; *foliis Hyperici* with leaves [like those] of St John's-wort (*Hypericum perforatum* [296]); *foliis in profundas lacinias divisis* with leaves divided into deep divisions; *foliis incisis / integris* with cut / uncut leaves; *foliis integris et hispidis* with entire / uncut and hairy leaves; *foliis laciniatis* with laciniate / jagged leaves; *foliis latioribus et crassioribus* with broader and thicker leaves; *foliis latis splendentibus* with broad shining leaves; *foliis longioribus pulverulentis* with longer powdered leaves; *foliis longiusculis* with longish / rather long leaves; *foliis maculatis/maculosis* with spotted leaves; *foliis matricariæ* with leaves [like those] of feverfew (*Tanacetum parthenium* [421]); *foliis oblongis* with oblong leaves; *foliis pediculis oblongis insidentibus* with leaves standing on oblong petioles / footstalks; *foliis pallidioribus* with paler / rather pale leaves; *foliis profundè incisis* with deeply cut leaves; *foliis profunde sectis, subtus argenteis* with the leaves deeply cut [and] silver beneath; *foliis Sanguisorbæ* with leaves [like those] of Burnet (*Sanguisorba officinalis* [438] or *Poterium sanguisorba* [439]); *foliis serratis / sinuatis* with toothed / wavy-margined leaves; *foliis subrotundis* with nearly circular leaves; *foliis tenuissimis* with very slender leaves; *foliis Trissaginis* with leaves [like those] of *Trissago* (i.e. *Teucrium chamaedrys* L.); *foliis urticæ* with leaves [like those] of nettle (*Urtica dioica* [627]); *foliis veronicæ* with leaves [like those] of a speedwell (*Veronica* sp.); *foliis vetonicæ* with leaves [like those] of betony (*Betonica officinalis* [84]); *foliis viciæ* with leaves [like those] of a vetch (*Vicia* sp.); *folio acuminato* with an acuminate leaf; *folio acuto [crispo]* with an acute [wavy-margined] leaf; *folio acuto & nonnunquam obtusiore plano* with a flat leaf [that is] acute & sometimes somewhat obtuse; *folio acuto plano* with a flat sharp / acute leaf; *folio ad myrrhidem accedente* with a leaf approaching [that of] cow parsley (*Anthriscus sylvestris* [135]); *folio Amygdalino utrinque virente aurito* with a leaf [like that] of Almond (*Prunus dulcis* (Mill.) D.A. Webb), green on both sides [and] eared; *folio anguloso* with a strongly angled leaf; *folio argenteo* with a silver leaf; *folio articulato* with a jointed leaf; *folio aspero* with a rough / prickly leaf; *folio caulem ambiente* with a leaf surrounding the stem; *folio cicutæ vel myrrhidis* with a leaf [like that] of hemlock (*Conium maculatum* [132]) or cow parsley (*Anthriscus sylvestris* [135]); *folio cordato maculato* with a heart-shaped spotted leaf; *folio Coridis [hirsuto quaterno]* with a [hairy fourfold] leaf [like that] of Coris (probably *Hypericum coris* L.); *folio crassiore* with a thicker leaf; *folio crispo* with a wavy-margined leaf; *folio cubitali* with a leaf a cubit long; *folio Cyclamini* with a leaf [like that] of a Cyclamen (*Cyclamen* sp.); *folio & vimine subluteo* with a yellowish leaf & withy; *folio ex rotundo acuminato* with a leaf [that is] acuminate from a round [shape]; *folio glabro [spisso]* with a [thick] glabrous leaf; *folio hederæ* with a leaf [like that] of ivy (*Hedera helix* [270]); *folio [hederulæ] hirsuto* with a hairy leaf [of a little ivy]; *folio integro* with an undivided leaf; *folio laciniato [dorso spinoso]* with a laciniate / jagged leaf [with a spiny back]; *folio*

VOCABULARY

lanuginoso with a downy leaf; *folio latiore* with a boader leaf; *folio lat[issim]o scabro* with a [very] broad rough leaf; *folio lato splendente* with a broad shining leaf; *folio longiore* with a longer / rather long leaf; *folio longissimo* with a very long leaf; *folio longiusculo tereti* with a longish / rather long terete / smooth leaf; *folio longo [non auriculato]* with a long [earless] leaf; *folio malvæ rotundo* with a round leaf [like that] of a mallow (*Malva* sp.); *folio maximo* with the largest / a very large leaf; *folio minùs acuto* with a less acute / sharp leaf; *folio minùs secto glauco* with a glaucous, less dissected leaf; *folio multùm laciniato* with a very laciniate / jagged leaf; *folio myricæ* with a leaf [like that] of a tamarisk (*Tamarix* sp.); *folio nigricante, levitérq; (= leviterque) hirsuto* with a blackish and lightly hairy leaf; *folio non dissecto / non diviso* with an uncut / undivided leaf; *folio non laciniato* with a non-laciniate / unragged leaf; *folio non maculato* with an unspotted leaf; *folio oblongo [serrato]* with an oblong [serrate] leaf; *folio obtuso* with an obtuse leaf; *folio olente* with a smelling leaf; *folio pilosissimo* with a very hairy leaf; *folio Polypodii* with a leaf [like that] of Polypody (*Polypodium vulgare sensu lato* [455]); *folio rotundiore* with a rounder / rather round leaf; *folio rotundo [multùm serrato]* with a round [much serrated] leaf; *folio rotundo et capillaceo* with a round and hair-like leaf; *folio rugoso rotundiore* with a rather round rugose / wrinkled leaf; *folio serrato* with a serrate leaf; *folio sinuato [candicante]* with a wavy-margined leaf [becoming pure white]; *folio sonchi* with a leaf [like that] of a sow-thistle (*Sonchus* sp.); *folio splendente auriculato* with a shining eared leaf; *folio subrotundo [non crenato]* with a nearly circular [unscalloped] leaf; *folio trifido* with a trifid / three-cleft leaf; *folio tripartitò diviso* with a leaf divided into three; *folio umbilicato* with a navel-like leaf; *folio urticæ* with a leaf [like that] of nettle (*Urtica dioica* [627]); *folio utrinque glauco* with the leaf glaucous on both sides

fontanus, -a, -um growing in or by a spring or fountain
fortè perhaps, by chance
fragifer, -fera, -ferum bearing strawberries (or heads like them)
fragilis, -is, -e fragile, breaking easily
frequens, -ent- frequent
Frisicus, -a, -um Frisian, of Friesland
fructus, -ûs a fruit; *fructu albido aut cinericeo* with a whitish or grey fruit; *fructu cæruleo / cæsio* with a blue / grey-blue fruit; *fructu echinato* with a spiny fruit; *fructu majore albo* with a larger white fruit; *fructu majore nigro* with a larger black fruit; *fructu minore serotino* with a smaller late fruit; *fructu [in umbella] nigro* with black fruit [in an umbel]; *fructu rubro* with a red fruit
frumentum, -i grain, corn; *frumenti genus* a kind of grain
frutescens, -ent- shrubby, bushy
frutex, fruticis a shrub, a bush
fruticosior, -ior, -ius more shrubby, more bushy, rather bushy
fruticosus, -a, -um shrubby, bushy
frux, frugis a fruit; *frugum* of fruits; *frugum & hortorum vitium, -i* a bane of fruits & gardens
Fuchsii, Fuchsio (genitive and dative singular) of / for Fuchsius (Leonhart Fuchs)
fuciflorus, -a, -um drone-flowered, with flowers like drones
fucum referens, -ent- recalling a drone
Fuga Dæmonum Demons' Flight, *Hypericum perforatum* [296]
fugax, fugac- fleeting, transitory
fullonius, -a, -um of a fuller, of fullers
fullonum of fullers

VOCABULARY

fumus, -i smoke; *Fumus terræ* Smoke of the earth, *Fumaria officinalis* [204]
fungosus, -a, -um fungoid, of a fungus, of fungi
fungus, -i a fungus; *Fungorum præcipuum genus* a special kind of Fungus (plural in Latin)
funis, -is a rope, a cord
furiosus, -a, -um raging
galea, -æ a helmet; *galea & alis herbidis* with grasslike helmet & wings; *galeâ & alis purpurascentibus* with purplish helmet & wings
Galeni, Galeno (genitive and dative singular) of / for Galenus (Galen of Pergamum)
galericulatus, -a, -um helmeted, helmet-shaped
Gallia, -æ Gaul (in earlier contexts), France (later)
Gallicus, -a, -um Gallic, French
gallina, -æ a hen
gallinaceus, -a, -um of a hen, of hens
Gallio albo similis, -is, -e like Marsh Bedstraw (*Galium palustre* [210])
garyophyllum olens, -ent- smelling of cloves (*Syzygium aromaticum* (L.) Merr. & L.M. Perry (*Caryophyllus aromaticus* L.))
gaudens, -ent- (after an ablative) rejoicing [in]
Gazæ (genitive singular) of Gaza (Theodoros Gaza)
geniculatus, -a, -um geniculate, with little knees, bent at the nodes
genuinus, -a, -um genuine, true
genus, generis (usually after a genitive) a kind (not necessarily at generic level)
Gerardi (genitive singular) of Gerardus (John Gerarde or St Gerard of Brogne)
Germania, -æ Germany
Germanicus, -a, -um German
Germanis, Germanorum (dative and genitive plural) for / of the Germans
Gesneri, Gesnero (genitive and dative singular) of / for Gesnerus (Conrad Gesner)
glaber, glabra, glabrum hairless, glabrous
globosus, -a, -um globose, spherical
globularis, -is, -e globular
glomeratæ rotunditatis (genitive singular) of rounded rotundity
glomerato flore (ablative singular) with a clustered flower / flower-head
glomerulis florum oblongis ... caulibus adnatis (ablative, genitive and dative plural) with oblong clusters of flowers ... adnate to the stems
gluma, -æ a glume, a husk; *glumis hirsutis* with hairy glumes / husks
glutinosus, -a, -um sticky, glutinous
Glycyrrhizatus, -a, -um sweet-rooted, i.e. like liquorice (*Glycyrrhiza glabra* [229])
gracilis, -is, -e thin, slender; *gracilibus & longis foliis* with slender & long leaves
Græcus, -i a Greek; *Græcis, Græcorum* for / of the [Ancient] Greeks
gramen, graminis grass; *gramen aquaticum* water-grass, *Zannichellia palustris* [466]; *Gramen bufonis* Grass of a toad, Toad-grass
gramineus, -a, -um grassy, grass-like, grass-leaved; *gramineo folio* with a grass-like leaf
grandis, -is, -e large, great
granum, -i grain, seed; *granis & spica rufescentibus* with the grains and the ear reddish; *granis rubentibus* with red seeds; *granis rufescentibus* with reddish grains
gratia, -æ grace; *Gratia Dei* God's Grace, *Helianthemum nummularium* [125]
grave olens, -ent- strong-smelling
griseus, -a, -um grey
gruinus, -a, -um of a crane, of cranes

VOCABULARY

hæmato[i]des, -o[i]d- bloody, blood-red (Greek αἱματωδης, -ης, -ες)
harundinaceus, -a, -um like a reed (*Phragmites australis* [64])
hederaceus, -a, -um like ivy (*Hedera helix* [270]), ivy-leaved
hederula, -æ little ivy; *hederulæ folio* with the leaf of a little ivy
helioscopius, -a, -um looking to the sun (Greek ἡλιοσκοπιος, -ος, -ον)
Hepatica, -æ alba, -æ White Liverwort, *Parnassia palustris* [264]
Hepatica, -æ stellata, -æ Star-like Liverwort, *Galium odoratum* [67]
hepaticæ facie (genitive and ablative singular) with the appearance of liverwort (probably *Hepatica nobilis* Schreb.)
Hepaticus, -i flos, floris Liverwort flower, *Parnassia palustris* [264]
Heracleus, -a, -um of Heracles, of Hercules (the mythical hero)
herba, -æ a herb; *Herba Benedicta* Blessed Herb, *Geum urbanum* [117]; *Herba Cancri* Herb of the Crab (i.e. the constellation Cancer), *Capsella bursa-pastoris* [97]; *Herba D[ivi] Stephani* St Stephen's Herb, *Circaea lutetiana* [136]; *Herba fullonum* Fullers' Herb, *Saponaria officinalis* [521]; *Herba Gerardi* St Gerard's Herb, *Aegopodium podagraria* [46]; *Herba impia* Disrespectful Herb, Herb impious, Common Cudweed, *Filago vulgaris* [230]; *Herba muscarum* Herb of flies, *Persicaria hydropiper* [435]; *Herba Paralysis* Paralysis Herb, *Primula veris* [467]; *Herba Paris* Herb Paris, *Paris quadrifolia* [277]; *herba Phœnicea* Phoenician herb, *Lolium perenne* [341]; *Herba radioli* Herb of a fern-like plant, *Polypodium vulgare sensu lato* [455]; *Herba Ruperti* Rupert's Herb, Ruprecht's Herb, Herb Robert, *Geranium robertianum* [223]; *Herba sacra* Sacred Herb (i.e. Greek Ἱεροβοτανη [Hierobotane]), *Verbena officinalis* [608]; *Herba S. Antonii* St Anthony's Herb, *Epilobium hirsutum* [352]; *Herba S. Barbaræ* St Barbara's Herb, *Barbarea vulgaris* [78]; *Herba S. Jacobi* St James's Herb, *Senecio jacobaea* [300]; *Herba S. Kunigundis* St Kunigunde's Herb, *Eupatorium cannabinum* [191]; *Herba Sardoa* Sardinian Herb, *Ranunculus sceleratus* [478] or *Pulsatilla vulgaris* [475]; *Herba stella* Star herb, *Plantago coronopus* [152]; *Herba venti* Herb of the wind, *Pulsatilla vulgaris* [475]
herbaceus, -a, -um herbaceous, grass-green
herbarius, - i a herbarist, a botanist; *herbariorum* of botanists
herbido muscoso flore (ablative singular) with a grass-like mossy flower
Herculeus, -a, -um of Hercules, of Heracles
Hermaphroditicus, -a, -um Hermaphrodite, bisexual
hibernus, -a, -um wintry, of winter
hieracio cognatus, -a, um related to a hawkweed (i.e. some yellow-flowered composite)
Hippocratis (genitive singular) of Hippocrates (physician of Cos)
hircinus, -a, -um smelling like a goat
hirsutie asper, aspera, asperum hairily rough
hirsutus, -a, -um hirsute, hairy; *hirsuto flore* with a hairy flower
Hispanicus, -a, -um Spanish, of Spain
hispidus, -a, -um hispid, bristly
holoschœnos, -i all-rush (literally) (Greek ὁλοσχοινος, -ου)
holosteus, -a, -um all-bone (jokingly applied to *Stellaria holostea* [286])
Holostio affinis, -is, -e related to All-bone (perhaps *Juncus bufonius* [256])
Homeri, Homero (genitive and dative singular) of / for Homerus (the Ancient Greek poet Homer)
hordeaceus, -a, -um like barley (*Hordeum* spp. [289–291])
hortensis, -is, -e of a garden, of gardens

VOCABULARY

humi repens/serpens, -ent- creeping on the ground
humidus, -a, -um wet, of wet places
humilior, -ior, -ius lower-growing
humilis, -is, -e low-growing
hybernus, -a, -um of winter
hyemalis, -is, e of winter
igniarius, -a, -um fire-making
imus, -a, -um lowest
in (+ ablative) in, on; *in agrorum limitibus* on the margins of fields; *in aquis fluitans, -ant-* floating in waters; *in basi auriculatus, -a, -um* auriculate at the base; *in cacumine semen ferens, -ent-* bearing seed at the top; *in foliorum alis* in the axils (literally wings) of the leaves; *in palustribus* in marshes; *in segetibus nascens, -ent-* growing in cornfields, growing in standing corn
in (+ accusative) into, in; *in latitudinem magis fusus, -a, -um* more spread out in breadth
incanus, -a, -um hoary, greyish
iners, -ert- inert, stingless
infectorius, -a, -um dyed, coloured
infernè hirsutus, -a, -um hairy underneath
ingratior, -ior, -ius more unpleasant, rather unpleasant
innatans, -ant- (after an ablative) floating [in]
inodorus, -a, -um scentless
insipidus, -a, -um tasteless
inter segetes nascens, -ent- growing among standing corn
interdum sometimes
intybaceus, -a, -um chicory-like, like chicory (*Cichorium intybus* [131])
inutilis, -is, -e useless
inversus, -a, -um turned upside down
invicem alternately
Italia, -æ Italy
Italicus, -a, -um Italian
Italorum (genitive plural) of the Italians
junceus, -a, -um rushlike, like *Juncus* spp.
juncifolius, -a, -um rush-leaved, with leaves like those of a rush (*Juncus* sp.)
juxta (+ accusative) next to; *juxta muros proveniens, -ent-* growing next to walls
Labrum, -i Veneris Venus's Basin, *Dipsacus fullonum* [168]
laciniatus, -a, -um laciniate, jagged, ragged; *laciniatis foliis* with jagged leaves
lacteus, -a, -um milk-white; *lacteis maculis notatus, -a, -um* marked with milk-white spots
Lactuca, -æ ranarum Frogs' Lettuce, *Potamogeton crispus* [576]
lævicaulis, -is, -e smooth-stemmed
lævior, -ior, -ius smoother
lævis, -is, -e smooth; *lævi folio* with a smooth leaf
lanatus, -a, -um woolly, like wool
Lancastriæ (genitive singular) of Lancastria, of Lancashire
lanceatus, -a, -um lanceolate, lance-shaped
lanceolatus, -a, -um lanceolate, lance-shaped
laniger, -gera, -gerum wool-bearing, woolly
lanuginosus, -a, -um downy; *lanuginosis capitulis* with downy heads
lanugo, lanuginis down (like wool); *lanugine* (ablative singular) with down

VOCABULARY

lappulatus, -a, -um burred
latifolius, -a, -um broad-leaved
Latinis, Latinorum (dative and genitive plural) for / of the Latins (i.e. Romans)
latior, -ior, -ius broader, wider; *latiore folio* with a broader leaf
latus, -a, -um broad, wide; *latâ siliquâ* with a broad pod; *lato [obtuso] folio* with a broad [obtuse] leaf
legitimus, -a, -um genuine, true
legumen, leguminis a pulse, a bean, a leguminous plant
leguminosus, -a, -um leguminous, pod-bearing
lenis, -is, -e smooth, soft
lenticulatus, -a, -um like duckweed (*Lemna* spp.) (?)
leoninus, -a, -um of a lion, of lions
leporinus, -a, -um of a hare, of hares, hare-like
leptophyllos, -os, -on thin-leaved (Greek λεπτοφυλλος, -ος, -ον)
lepus, leporis a hare
lethalis, -is, -e deadly, lethal
leucanthemus, -a, -um white-flowered
lignosus, -a, -um woody
lilifolius s, -a, -um lily-leaved, with leaves [like those] of a lily (*Lilium* sp.)
limonius, -a, -um of a meadow, of meadows (Greek λειμωνιος, -α, -ον)
limosus, -a, -um of muddy places
lingua, -æ a tongue; *Lingua avis* Tongue of a bird, Bird's-tongue, *Senecio paludosus* [146]; *Lingua bovis* Tongue of an ox, Oxtongue, *langue de boeuf*, *Helminthotheca echioides* [94]; *Lingua cervina* Stag's-tongue, Hart's-tongue, *Asplenium scolopendrium* [A29/B75]; *Lingua passerina* Sparrow's-tongue, *Polygonum aviculare sensu lato* [454]; *Lingua serpentina* Serpent's-tongue, Adder's-tongue, *Ophioglossum vulgatum* [395]; *Lingua vulneraria* Vulnerary Tongue, *Ophioglossum vulgatum* [395]
lini folio (genitive and ablative singular) with a leaf [like that] of flax (*Linum usitatissimum* [336])
Lobelii, Lobelio (genitive and dative singular) of / for Lobelius (de L'Obel)
loculis oblongis (ablative plural) with oblong pods
loculo oblongo (ablative singular) with an oblong pod
lolio similis, -is, -e like darnel (*Lolium temulentum* [341])
longiflorus, -a, -um long-flowered
longifolius, -a, -um long-leaved
longior, -ior, -ius longer, rather long; *longioribus & magis atris foliis, & quodammodo splendentibus* with longer & blacker & to some extent shining leaves; *longioribus setis* with longer bristles
longiùs radicatus, -a, -um longer-rooted
longus, -a, -um long; *longis acutis / angustis foliis* with long acute / narrow leaves; *longis calcaribus* with long spurs; *longo calyce* with a long cup; *longo serrato folio* with a long serrate leaf
Loto affinis, -is, -e related to Birdsfoot Trefoil (*Lotus* spp.)
lucens, -ent- shining
lucidus, -a, -um shining
Lugdunensis, -is, -e of Lyon (Lugdunum in Roman times)
Lumbricorum herba, -æ Herb of worms, *Descurainia sophia* [552]
Lupi crepitus, -ûs Wolf's-fart, a puff-ball fungus [207]
lup[ul]inus, -a, -um hop-like, of hops (*Humulus lupulus* [342])
luteo-pallidus, -a, -um pale yellow

VOCABULARY

Lutetianus, -a, -um of Paris (Lutetia in Roman times)
luteus, -a, -um yellow; *luteo flore* with a yellow flower
Lybicus, -a, -um Lybian, of Lybia
Macedonicus, -a, -um Macedonian
macrocaulos, -os, -on long-stemmed (Greek μακροκαυλος, -ος, -ον)
macula, -æ a spot; *maculis candidis vel nigris* with white or black spots
maculatus, -a, -um spotted; *maculato folio* with a spotted leaf
maculosus, -a, -um spotted; *maculoso folio* with a spotted leaf
magis more (as an adverb); *magìs sparsâ paniculâ* with a sparser panicle
magnus, -a, -um great, large; *magno & oblongo tubo* with a large oblong tube; *magno flore* with a large flower
Magonis (genitive singular) of Mago (a Carthaginian writer on agriculture)
maior/major, -or, -us greater, larger; *maiore/majore flore [purpureo]* with a larger / rather large [purple] flower; *major, -oris species, -iei* a larger species / kind
majusculis vasculis (ablative plural) with somewhat larger vessels / capsules
malacoides, -oid- mallow-like, like a *Malva* sp.
malvæ folio (genitive and ablative singular) with a leaf [like that] of a mallow (*Malva* sp.)
manicus, -a, -um manic, causing madness (Greek μανικος, -η, -ον)
marinus, -a, -um marine, growing in or by the sea
maritimus, -a, -um maritime, growing by the sea
Martius, -a, -um of March
mas, maris a male, male (as an adjective)
masculus, -a, -um male
mater, matris a mother; *mater herbarum* the mother of herbs
Matthioli (genitive singular) of Matthiolus (Pietro Andrea Mattioli)
maximus, -a, -um largest, greatest, very large; *maximo flore* with a very large flower
medioluteus, -a, -um yellow in the middle
medius, -a, -um middle-sized, intermediate; *media à nonnullis cognominata* (feminine) called by some [people] intermediate
melanocerasos, -i black-cherry-tree (Greek μελανοκερασος, -ου)
militaris, -is, -e military, of a soldier, of soldiers
milium, -i millet
millefolius, -a, -um one-thousand-leaved
minimè not at all
minimus, -a, -um smallest, least, very small; *minimus omnium* smallest of all
minor, -or, -us smaller, lesser; *minore purpureo flore* with a smaller purple flower; *minori folio* with a smaller leaf
minùs less (as an adverb); *minus serratus, -a, -um* less serrate, less saw-toothed
mitior, -ior, -ius milder
mitis, -is, -e mild
mollis, -is, -e soft
monoclonos, -os, -on single-stemmed (Greek μονοκλωνος, -ος, -ον)
monococcos, -os, -on one-berried (Greek μονοκοκκος, -ος, -ον)
mons, montis a mountain
Monspeliacus, -a, -um of Montpelier
monspeliensis, -is, -e of Montpelier; *Monspeliensium* of the Montpelier [botanists / authors]
montanus/montosus, -a, -um montane, of mountains

VOCABULARY

Montbelgardensis, -is, -e   of Montbéliard (in Franche-Comté, France, formerly Mömpelgard); *Gros* is presumably German *groß* (= "great")

mordax, -ac- / mordens, -ent-   biting (literally), stinging, acrid

morio, morionis   a fool (whence *Fools-stones* for *Orchis mascula* [398] and *Anacamptis morio* [399])

morsus, -ûs   a bite; *Morsus Diaboli*   Bite of the Devil, Devil's-bit, *Succisa pratensis* [374]; *Morsus gallinæ*   Bite of a hen, Henbit, *Veronica hederifolia* [24]; *Morsus ranæ*   Bite of a frog, Frogbit, *Hydrocharis morsus-ranae* [375]

moschatus, -a, -um   musky, musk-scented

multicaulis, -is, -e   many-stemmed

multiflorus, -a, -um   many-flowered

multiplex, -plic-   of many parts, multiple, complex; *multiplici spica*   with a multiple / complex spike

multus, -a, -um   much, many; *multis*   for many [people, authors or botanists]; *multis dictus, -a, -um*   called by many [people]; *multis punctis*   with many spots

muralis, -is, -e   of walls

murarius, -a, -um   of walls

muricatus, -a, -um   muricate, rough with short hard points (like the shell of *murex*)

murinus, -a, -um   of a mouse, of mice

murorum   (genitive plural)   of walls

mus, muris   a mouse

muscam referens, -ent-   recalling a fly

muscatellæ floribus   (genitive singular and ablative plural)   with flower[-head]s [like those] of moschatel (*Adoxa moschatellina* [477]), as explained by Johnson (1633)

muscoso flore   (ablative singular)   with a mossy flower

muscus, -i   a moss; *Musci capillaris altera species*   the second kind of Hair-like Moss; *Muscus capillúsve aureus minor*   smaller golden Moss or maidenhair; *Muscus lenticulæ similis*   Moss like a small lentil – hence a duckweed (*Lemna* sp.)

Myagro affinis, -is, -e   related to *Myagrum* (some yellow-flowered crucifer)

myodes, myod-   mouse-like (Greek μυώδης, -ης, -ες)

myosotis, -is   mouse-ear (*Cerastium*, *Myosotis* and *Pilosella* spp.)

myrtifolius, -a, -um   myrtle-leaved, with leaves [like those] of myrtle (*Myrtus communis* L.)

nascens, nascent-   springing forth, growing

Neapolitanus, -a, -um   Neapolitan, of Naples

nemorensis, -is, -e   of groves

nemorosus, -a, -um   of groves

nemorum   (genitive plural)   of groves

neotericus, -a, -um   modern; *neotericis* (dative plural)   for modern [authors]

Nicandri   (genitive singular)   of Nicander (Ancient Greek poet of Colophon)

niger, nigra, nigrum   black; *nigris maculis notatus, -a, -um*   marked with black spots

nigricans, nigricant-   blackish

nigrior, -ior, -ius   blacker, rather black

nigroluteus, -a, -um   blackish-yellow

nitente flore   (ablative singular)   with a shining flower

nodosus, -a, -um   knotted, knobby

non   not

nondum descriptus, -a, -um   not yet described

nonnihil similis, -is, -e   (after a dative)   somewhat like

VOCABULARY

nonus, -a, -um ninth
noster, nostra, nostrum our
nostras, nostrat- of our country, our own [familiar]
notatus, -a, -um marked
novus, -a, -um new
noxius, -a, -um noxious, harmful, poisonous
nudus, -a, -um naked, without branches
nuperorum (genitive plural) of modern [authors]
nutans, nutant- nodding
nux, nucis a nut, a nut-tree; *Nux vesicaria* bladder-like nut, Bladder-nut, *Staphylea pinnata* [561]
oblongior, -ior, -ius more oblong, rather oblong
oblongus, -a, -um oblong (with almost parallel sides); *oblongis siliquis* with oblong pods; *oblongo flore / folio* with an oblong flower / leaf
Occidentalium (genitive plural) of Western [authors]
Octavii (genitive singular) of Octavius
Oculus, -i cati Eye of a cat, Cat's-eye, *Lamium amplexicaule* [25]
Ocymi facie (genitive and ablative singular) with the appearance of Basil (*Ocimum basilicum* L.)
oculus, -i an eye; *Oculus Christi* Christ's eye, *Salvia verbenaca* [292]
odor, -oris an odour, a scent, a smell; *odore gravi* with a heavy odour; *odore pulegii* with the scent of pennyroyal (*Mentha pulegium* [474]); *odore viroso* with a poisonous smell
odor[at]us, -a, -um scented, having a smell; *odorata flore luteo simplici* with a simple scented yellow flower
officina, -æ a shop (generally of an apothecary); *officinarum* of apothecaries' shops
olens, -ent- smelling
olidus, -a, -um smelly
olitoribus (dative plural) for the kitchen-gardeners
olus, oleris a vegetable, a potherb (see also *Album olus*); *Olus atrum* black vegetable (literally), *Smyrnium olusatrum* [285]
omninò altogether
omnium minimus, -a, -um smallest of all
orbicularis, -is, -e orbicular, disc-shaped
orbiculatus, -a, -um orbicular, disc-shaped
Origani facie (genitive and ablative singular) with the appearance of Marjoram (*Origanum* sp.)
Os, Oris leonis Mouth of a lion, Lion's-mouth, *Chaenorhinum minus* [50]
ovatus, -a, -um ovate, egg-shaped
ovillus/ovinus, -a, -um of a sheep, of sheep
ovum, -i an egg; *Ovum lupi* Wolf's-egg, a puff-ball fungus [207]
oxyschœnos, -i sharp rush (Greek ὀξυσχοινος, -ου)
pallido-luteus, -a, -um pale yellow
pallidus, -a, -um pale; *pallido flore [inodoro aut vix odoro]* with a pale [scentless or scarcely scented] flower
Palma, -æ Christi Palm of Christ, Christ's-palm, a marsh or spotted orchid (*Dactylorhiza* sp.)
palmatus, -a, -um palmate, like a palm (with outstretched fingers)
paludosus, -a, -um of marshes
palustris, -is, -e of marshes

VOCABULARY

Panax, -acis All-heal; *Panax coloni* Clown's (i.e. Countryman's) All-heal, *Stachys palustris* [540]
panicula, -æ a panicle; *paniculâ arundinaceâ* with a reed-like panicle; *paniculâ glomeratâ* with a clustered panicle; *paniculâ junceâ compactâ* with a compact rush-like panicle; *paniculâ multiplici* with a multiple panicle; *paniculâ spadiceâ delicatiore* with a rather delicate deep reddish-brown (date-coloured) panicle; *paniculâ sparsâ* with a sparse panicle; *paniculâ speciosâ* with a showy panicle; *paniculis plurimis sessilibus* with very many sessile panicles
paniculatus, -a, -um paniculate, with a panicle / panicles
Panis, -is cuculi Cuckoo's-bread, *Oxalis acetosella* [578]
Pannonicus, -a, -um Hungarian, of Hungary
Parnassi (genitive singular) of Mount Parnassus
parvus, -a, -um small, little; *parvis floribus* with small flowers; *parvo flore [sive Phœniceo]* with a small flower [or with a scarlet one]
pastor, -oris a shepherd
pastorius, -a, -um of a shepherd, of shepherds
Pauli (genitive singular) of Paulus (Paul of Aegina)
Pausaniæ (genitive singular) of Pausanias (Ancient Greek geographer)
Pecten Veneris Comb of Venus, Venus'-comb, *Scandix pecten-veneris* [425]
pediculus, -i a pedicel, a petiole, a footstalk, a stalk; *pediculo nigro* with a black petiole
penna, -æ - a feather, a wing
pennatus, -a, -um feathered
pentagonius, -a, -um pentagonal, five-angled (Greek πενταγωνος, -ος, -ον)
pentaphyllos, -os, -on five-leaved (Greek πενταφυλλος, -ος, -ον)
perennis, -is, -e perennial
perfoliatus, -a, -um perfoliate, with a leaf or pair of leaves completely surrounding the stem
perforatus, -a, -um perforated
perpetuò permanently
perpetuus, -a, -um perpetual, perennial
perpusillus, -a, -um very small, tiny
persicariæ foliis (genitive singular and ablative plural) with leaves [like those] of an arsmart (*Persicaria* sp. [433–435])
perticalis, -is, -e suitable for making poles or stakes
pervinca, -æ very-periwinkle (literally), applied to *Vinca minor* [138]
pes, pedis a foot; *Pes anserinus* Goose's-foot, Goosefoot, *Chenopodium* spp.; *Pes asini* Ass's-foot, *Alliaria petiolata* [17]; *Pes cati* Cat's-foot, *Antennaria dioica* [233]; *Pes columbinus* Dove's-foot, *Geranium molle* or *G. pusillum* [219]; *Pes Leonis* Lion's-foot, *Alchemilla filicaulis* subsp. *vestita* [B3]; *Pes leporinus / leporis* Hare's-foot, *Trifolium arvense* [316]; *Pes milvinus* Kite's-foot, *Lepidium coronopus* [153]; *Pesgruis* Crane's-foot, *Centaurea cyanus* [162]
petræus, -a, -um rock-loving, growing on rocks
Phœniceus, -a, -um Phoenician, of Phoenicia, Phoenician purple, scarlet
phragmites, -is growing in hedgerows (Greek φραγμιτης, -ου)
pilosus, -a, -um hairy
pinguis, -is, -e fat
pinnula, -æ a pinnule; *pinnulis angustis raris profunde dentatis* with spaced-out narrow pinnules [that are] deeply toothed; *pinnulis obtusis non dentatis* with obtuse untoothed pinnules
planta, -æ a plant

VOCABULARY

Plantaginis folio (genitive and ablative singular) with a leaf [like that] of a Plantain (*Plantago* sp.)
platyphyllos, -os, -on broad-leaved (Greek πλατυφυλλος, -ος, -ον)
Plauti (genitive singular) of Plautus (an early Roman poet)
plerisque (dative plural) for most [people, authors or botanists]
Plinianus, -a, -um of Plinius (Pliny the Elder)
Plinii, Plinio (genitive and dative singular) of / for Plinius (Pliny the Elder)
plurimus, -a, -um very much, very many, most; *plurimis* for most [people, authors or botanists]
poetarum (genitive plural) of the poets (of Greece and Rome)
Polyacanthos, -i many-prickled [plant] (Greek πολυακανθος, -ου)
polycarpos, -os, -on many-fruited (Greek πολυκαρπος, -ος, -ον)
Polygoni aut Helxines folio (genitive and ablative singular) with a leaf [like that] of *Polygonum aviculare sensu lato* [454] or *Parietaria judaica* [418]
polyspermos, -os, -on many-seeded (Greek πολυσπερμος, -ος, -ον)
polystachios, -os, on many-spiked (Greek πολυσταχυς, -εια, -υ)
Ponticus, -a, -um Pontic, of the Black Sea
porosus, -a, -um porous, with many pores
Portulaca, -æ aquatica, -æ Water Purslane, *Lythrum portula* [B8]; *Portulacæ aquaticæ similis* like Water Purslane (see last entry)
potamios, -os, -on of a river, of rivers (Greek ποταμιος, -α/ος, -ον)
Potamogeito affinis, -is, -e related to a Pondweed (*Potamogeton* sp.)
potiùs rather, preferably
præcox, præcoc- precocious, appearing or flowering early in the year
præmorsus, -a, -um premorse, [as if] bitten off; *præmorsâ radice* with the root [as if] bitten off
pratensis, -is, -e of meadows
pratum, -i a meadow; *pratorum* of meadows
primus, -a, -um first; *primum genus / prima species* (after a genitive) the first kind / species [of]
prior, prior, prius first (of two)
procerior, -ior, -ius taller, loftier, rather tall
procerus, -a, -um tall, lofty
proveniens, -ent- occurring, growing
provolutus, -a, -um rolled out (literally), prostrate
pubescens, -ent- pubescent, downy
pulcher, pulchra, pulchrum beautiful; *pulchro flore albo* with a beautiful white flower
pulchrior, -ior, -ius more beautiful; *pulchriore flore luteo* with a more beautiful yellow flower
pulegii odore (genitive and ablative singular) with the scent of pennyroyal (*Mentha pulegium* [474])
pulmonarius, -a, -um pulmonary, used to treat the lungs
pulverulentus, -a, -um powdered (as with dust)
pumilus, -a, -um dwarf
purpurascens, -ent- purplish
purpureus, -a, -um purple; *purpureo flore* with a purple flower
pusillus, -a, -um very small
putatus, -a, -um believed, considered [to be]
pyxidatus, -a, -um box-like (from Greek πυξις, -ιδος [pyxis, "a box-wood box"])

VOCABULARY

pyxiodes/pyxoides, -iod-/-oid- box-like (from Greek πυξις [pyxis])
quadratus, -a, -um quadrate, square-stemmed; *quadrato junco* with a square-stemmed [rush-like] twig
quadrifolius, -a, -um four-leaved
quæ (feminine singular) which; *quæ Ben album vulgò* which [is] commonly [called] white Ben, *Silene vulgaris* [80]; *quæ brevi est pediculo* which has (literally is with) a short petiole; *quæ Dactylon 2 vel 3 Plinii* which [is] the second or third Dactylon of Pliny; *quæ officinarum* which [is] of apothecaries' shops; *quam ovinam vocant* which they call sheep's
quartus, -a, -um fourth
quasi like, as if, so to speak, as it were
quercus, quercûs an oak (usually *Quercus robur* [476]); *Quercûs & cæterarum arborum* of the Oak & the rest of the trees; *quercûs floribus* with flowers [like those] of oak
quernus, -a, -um of oak (*Quercus* sp.)
quidam / quædam, cujusdam a certain [person]; *quibusdam, quorundam* (dative and genitive plural) for / of some [authors]
quinquefolius, -a, -um five-leaved
quinquenervius, -a, -um five-veined
quod (neuter singular) which; *quod aliquibus Hypericoides* which for some [authors] is] Hypericoides; *quod Amomum officinis nostris* which [is] Amomum for our apothecaries' shops; *quod est Pulmonaria Gallorum* which is Lungwort of the French; *quod in altitudinem ingentem adolescit* which grows up to an enormous height
racemosus, -a, -um racemose, with racemes
radicatus, -a, -um rooted
radix, radicis a root; *radice atrorubente* with a dark red root; *radice diuturnâ* with a long-lasting root; *radice fibrosa* with a fibrous root; *radice flavâ* with a yellow root; *radice granulosa* with a root full of granules (small tubers); *radice lutea [vel alba]* with a yellow [or white] root; *radice minùs intorta* with a less twisted root; *radice nigra* with a black root; *radice non nodosa* with a root that is not knotted; *radice oblonga* with an oblong root; *radice perenni* with a perennial root; *radice repente / reptrice* with a creeping root; *radice rubente / rubra* with a reddish / red root; *radice sanguinei coloris* with a root of a blood-red colour; *radice succisâ* with the root bitten off (or as if it was); *radice tuberosa* with a tuberous root
ramosior, -ior, -ius more branched
ramosus, -a, -um branched
ranunculi flore [& capitulo] (genitive and ablative singular) with a flower [& flower-head] [like that/those] of a buttercup / crowfoot (*Ranunculus* sp.)
Rapum, -i genistæ Turnip of broom (literally), Broomrape (*Orobanche* sp.)
rarior, -ior, -ius rather scattered
recentioribus, recentiorum (dative and genitive plural) for / of more recent [authors]
rectus, -a, -um erect, upright
recurvus, -a, -um recurved, curved backwards
Regina, -æ prati Queen of the meadow, *Filipendula ulmaria* [623]
regius, -a, -um royal, kingly
repens, repent- creeping
reptante cauliculo (ablative singular) with a creeping stem (literally little stem)
Resta, -æ bovis Arrester of the ox, Restharrow, *Ononis spinosa* [47]
reticulatus, -a, -um reticulate, netted

VOCABULARY

rhœas, rhœadis a kind of poppy, probably *Papaver rhoeas* [416] (Greek ῥοιας, ῥοιαδος)
risus, -ûs a laugh
rivulorum (genitive plural) of rivulets, of brooks
Robertianus, -a, -um of Robert (St Rupert or Ruprecht of Salzburg)
Romanus, -a, -um Roman; *Romanorum* of the Romans
Ros solis Dew of the sun, Sundew, *Drosera* spp. [500, 501]
roseus, -a, -um pink, rose-coloured, rose-like
rostrum, -i a snout, a beak; *Rostrum porcinum* Pig's-snout, *Taraxacum* sp. [167]
rotundifolius, -a, -um round-leaved
rotundior, -ior, -ius more rounded, rather rounded; *rotundiore folio* with a more / rather rounded leaf; *rotundioribus imis foliis* with the lowest leaves more / rather rounded
rotundus, -a, -um rounded
rubellus, -a, -um reddish
rubens, rubent- red, reddish
ruber, rubra, rubrum red; *rubro folio* with a red leaf
rubeus, -a, -um red, reddish
Rubiis accedens, -ent- resembling Madders (*Rubia* spp.)
Ruellii (genitive singular) of Ruellius (Jean Ruel)
rugosior, -ior, -ius more wrinkled, more crumpled; *rugosioribus foliis* with more wrinkled leaves
rugosus, -a, -um rugose, wrinkled, crumpled
rustic[an]us, -a, -um of the countryside
rutaceo folio (ablative singular) with a leaf like rue (*Ruta graveolens* L.), rue-leaved
Rutæ pratensis genus minus a smaller kind of Meadow Rue, *Thalictrum minus* [568]
Sabinæ ferè odore (genitive and ablative singular) almost with the smell of *Sabina* (the strongly fetid *Juniperus sabina* L.)
sabuletorum Zelandiæ & Bataviæ (genitive plural and singular) of the sandy places of Zeeland & Batavia
sacer, sacra, sacrum sacred, holy
sagitta, -æ an arrow; *sagittæ folio* with the leaf of an arrow, arrow-leaved
Salamanticus, -a, -um of Salamanca (in Spain)
salicis folio (genitive and ablative singular) with a leaf [like that] of a willow (*Salix* sp.), willow-leaved
salictarius, -a, -um growing in willow-beds
Salvia, -æ a salve, Sage (the herb *Salvia officinalis* L.); *Salvia vitæ* Salve of life, *Asplenium ruta-muraria* [11]
salvifolius, -a, -um sage-leaved, with leaves [like those] of sage (*Salvia officinalis* L.)
sanguinalis, -is, -e bloody, blood-red
sanguinarius/sanguineus, -a, -um bloody, blood-red
Saracenicus, -a, -um of the Saracens
Sardonius, -a, -um Sardinian
sativus, -a, -um sown, planted, cultivated
satorum (genitive plural) of sowers
saxatilis, -is, -e of rocks, of stones
saxifragus, -a, -um stone-breaking, rock-breaking
Scabiosæ capitulo cæruleo (genitive and ablative singular) with a blue head [like that] of Scabious (*Knautia arvensis* [526] or *Scabiosa columbaria* [527])
Scaligeri (genitive singular) of Scaliger (Giulio Cesare della Scala or Scaligero)

VOCABULARY

scandens, scandent- climbing
sceleratissimus, -a, -um most noxious, very harmful
scilicet evidently
scoparius, -a, -um twiggy
scorpioides, -oid- scorpioid, with a coiled inflorescence
Scribonii (genitive singular) of Scribonius (Scribonius Largus)
scutellatus, -a, -um bearing small dishes (referring to capsules)
Scythicus, -a, -um Scythian
se extendens, -ent- spreading itself
se spargens, -ent- sowing itself
secalinus, -a, -um like rye (*Secale cereale* [530])
secundus, -a, -um second; *secundum genus / secunda species* (after a genitive) the second kind / species [of]
seges, segetis a cornfield; *segetum* of cornfields, of standing corn
segetalis, -is, -e of cornfields, of standing corn
selinoides, -oid- like parsley (*Petroselinum crispum* (Mill.) Nyman ex A.W. Hill)
semen, seminis a seed; *semine albo vel russo* with a white or russet seed; *semine aspero [hispido]* with a rough [bristly] seed; *semine Echii* with a seed [like that] of Viper's Bugloss (*Echium vulgare* [171]); *semine etiam majore* with an even larger seed; *semine fœniculi* with a seed [like that] of fennel (*Foeniculum vulgare* Mill.); *semine læviore* with a smoother / rather smooth seed; *semine longo* with a long seed; *semine magno, nigro* with a large black seed; *semine multo* with much seed; *semine nigro* with a black seed; *semine parvo & nigro* with a small black seed; *semine rostrato* with a beaked seed; *semine rotundo nigro* with a round black seed; *semine russo* with a russet seed; *semine striato* with a striate / striped seed; *semine subnigro aut fusco* with a blackish or dark-coloured seed; *semine triangulo* with a triangular seed; *semine variegato* with a variegated seed; *seminibus asperis* with rough seeds; *seminibus lævibus* with smooth seeds
semper always; *semper virens* evergreen
Senecionis folio (genitive and ablative singular) with a leaf [like that] of Groundsel (*Senecio vulgaris* [535])
sepiarius, -a, -um of bushes
sepium (genitive plural) of bushes
septentrionalis, -is, -e northern; *Septentrionalium* of the Northern [authors / botanists]
serotinus, -a, -um late-flowering or late-fruiting
serpens, serpent- creeping
serpillifolius/serpyllifolius, -a, -um thyme-leaved (see next entry)
serpillum, -i thyme (*Thymus* sp.); *serpilli folio* with a leaf [like that] of thyme, thyme-leaved
serratifolius, -a, -um serrate-leaved
serratus, -a, -um serrate, saw-toothed; *serrato folio [glabro / hirsutiore]* with a [glabrous / hairier] serrate leaf
seseli pratensis facie (genitive and ablative singular) with the appearance of meadow Seseli (apparently *Silaum silaus* [524])
seu or
si non if not
Sii foliis (genitive singular and ablative plural) with leaves [like those] of Waterparsnip (*Sium latifolium* [543] or *Berula erecta* [544])
siligineus, -a, -um of white or winter wheat (siligo, siliginis)

VOCABULARY

siliqua, -æ a pod; *siliquâ angulosâ aut striatâ* with a strongly angled or striate / striped pod; *siliquâ cornutâ* with a horned pod; *siliquâ glabrâ articulatâ* with a glabrous jointed pod; *siliquâ hirsutâ [articulatâ]* with a hairy [jointed] pod; *siliquâ latiusculâ glabrâ* with a rather broad glabrous pod; *siliquâ longâ* with a long pod; *siliquis in summitate* with pods at the top; *siliquis longis* with long pods

siliquatus/siliquosus, -a, -um podded, with pods

simeterius, -a, -um of footpaths

similis, -is, -e (after a dative or genitive) like, similar [to]; *similis Silao* like Pepper Saxifrage (*Silaum silaus* [524])

simplex, simplic- simple, single-stemmed; *simplici flore* with a simple flower

simpliciter dictus, -a, -um simply called

sine pediculis without stalks

singulari flore (ablative singular) with a single flower

sinuatus, -a, -um sinuous, winding, wavy, wavy-leaved

siue/sive or

sol, solis the sun

Solanifolius, -a, -um Nightshade-leaved, with leaves [like those] of *Solanum* spp.

solisequus, -a, -um following the sun

solutivus, -a, -um loosening (of a purgative)

Sonchis cognatus, -a, -um related to the Sow-thistles (*Sonchus* spp.)

Soncho affinis, -is, -e related to a Sow-thistle (*Sonchus* sp.)

Spatula, -æ fœtida, -æ Stinking / Fetid Spatula, *Iris foetidissima* [630]

species, speciei (usually after a genitive) a kind, a species [of]; *species apud antiquos* a kind [known] among the ancients (Greeks / Romans)

speciosior, -ior, -ius showier, more handsome, more splendid; *speciosiore thyrso* with a showier / more handsome spike

speciosus, -a, -um showy, handsome, splendid

Speculum, -i Veneris Venus's Looking-glass, *Legousia hybrida* [556]

Spergulæ facie (genitive and ablative singular) with the appearance of Spurrey (*Spergula arvensis* [557])

sphærocephalus, -a, -um round-headed

sphegodes, -od- of a wasp (Greek σφηκωδης, -ης, -ες [sphecodes])

spica, -æ a spike, an ear of corn; *spicâ albicante* with a whitish ear; *spica angustiore* with a narrower spike; *spicâ asperâ [brevi]* with a [short] rough spike; *spicâ compactâ* with a compact spike; *spicâ congestâ pyramidali* with a compact pyramidal spike; *spicâ divisâ / divulsâ* with a divided spike; *spica et granis albicantibus* with the ear and the grains whitish; *spica et/& granis rufescentibus* with the ear and the grains reddish; *spicâ integrâ* with an undivided spike; *spicâ longâ* with a long spike; *spicâ longiore / longiusculâ* with a longer / longish spike; *spicâ multiplici rubrâ* with a multiple red spike; *spicâ multiplici sparsâ* with a sparse multiple spike; *spicâ muticâ [albicante]* with a [whitish] docked ear; *spicâ oblongâ rubrâ* with an oblong red spike; *spicâ pendulâ longiore [& angustiore]* with a longer / rather long [and narrower /rather narrow] pendulous spike; *spicâ plantaginis* with a spike [like that] of a plantain (*Plantago* sp.); *spicâ rufâ* with a reddish spike; *spicâ spadiceo-viridi* with a deep brownish-green spike; *spicis minoribus minúsq; (= minusque) compactis* with smaller and less compact spikes

spicatus, -a, -um spicate, bearing a spike

spina, -æ a spine, a thorn; *spinis carens, -ent-* lacking spines / thorns

Spinachiæ facie (genitive and ablative singular) with the appearance of Spinach (*Spinachia oleracea* L.)

VOCABULARY

spinosior, -ior, -ius spinier, more spiny
spinosissimus, -a, -um spiniest, very spiny
spinosus, -a, -um spiny
spiralis, -is, -e spiral
spongifer, -fera, -ferum sponge-bearing (i.e. bearing bedeguar galls or Robin's pincushions)
spontaneus, -a, -um spontaneous (i.e. uncultivated)
spumeus, -a, -um frothy, foaming
spurius, -a, -um spurious, false
sqammatis capitulis (ablative plural) with scaly heads
squammosus, -a, -um scaly
stellaris, -is, -e stellate, star-like
stellatus, -a, -um stellate, star-like
sterilis, -is, -e sterile, barren
stratiotes, -is a soldier (Greek στρατιωτης, -ου)
strumea, -æ a herb to cure a struma (a scrofulous tumour)
strumosâ radice (ablative singular) with a root with swellings on it
stylo longo (ablative singular) with a long style
sub (+ ablative) under; *sub aqua repens, -ent-* creeping under water
subrotundus, -a, -um subrotund, roundish
subterraneus, -a, -um subterranean
succo viroso (ablative singular) with poisonous juice
supinus, -a, -um supine, prostrate
surrectior, -ior, -ius more erect
surrectus, -a, -um erect; *surrectis cauliculis* with erect stems (literally little stems)
sylva, -æ a wood; *sylvarum, sylvis* of / in woods; *sylvis & umbrosis gaudens, -ent-* rejoicing in woods & shady [places]
sylvanus, -a, -um of woods, wild (?)
sylvaticus, -a, -um of woods
sylvestrior, -ior, -ius wilder
sylvestris, -tris, -tre wild
Tabernamontani/Tabernæmontani (genitive singular) of Tabernaemontanus (Jakob Dietrich of Bergzabern)
Tarsicus, -a, -um of Tarsus (St Paul's home town)
tectorum (genitive plural) of roofs
temulentus, -a, -um drunken (literally), making intoxicated
tenuifolius, -a, -um fine-leaved
tenuior, -ior, -ius finer, more slender, rather slender
tenuis, -is, -e fine, slender
tenuissimè divisus, -a, -um very finely divided
tenuissimus, -a, -um finest, slenderest, very slender; *tenuissimis glabris foliis* with very slender glabrous leaves
tenuiùs laciniatus, -a, -um more finely laciniate
tenuiùs spicatus, -a, -um more slenderly spiked, with a more / rather slender spike
tereti, rotundâ, oblongâ radice (ablative singular) with a terete / smooth, rounded, oblong root
teretifolius, -a, -um having terete / smooth leaves
terrestris, -tris, -tre terrestrial, growing on the ground (as opposed to in water or on rocks or trees)

VOCABULARY

tertius, -a, -um third
testiculatus, -a, -um with testicles (referring to testicle-like fruits)
testiculus, -i a testicle
tetrapetalos, -os, -on four-petalled (Greek τετραπεταλος, -ος, -ον)
tetraphyllos/us, -os/a, -on/um four-leaved (Greek τετραφυλλος, -ος, -ον)
Thelygono affinis herba, -æ a herb related to Thelygonum (Greek Θηλυγονον, -ου ["female-producer"], a herb believed to induce female births)
Theophrasti, Theophrasto (genitive and dative singular) of / for Theophrastus (Ancient Greek philosopher)
tinctorius, -a, -um of a dyer, of dyers, used in dyeing
tomentosus, -a, -um tomentose, covered with short matted hairs
tonsus, -a, -um shaven, beardless
topiarius, -a, -um of landscape-gardeners
Tormentillæ facie (genitive and ablative singular) with the appearance of Tormentil (*Potentilla erecta* [571])
torminalis, -is, -e used for treating colic
totus, -a, -um all, whole; *tota sua stirpe* on the whole of its stem / plant; *toto anno folia retinens, -ent-* retaining leaves for the whole year
Tragi (genitive singular) of Tragus (Jerome Bock)
tremulus, -a, -um trembling, quaking
triangularis, -is, -e triangular
trichophyllos, -os, -on hair-leaved (Greek τριχοφυλλος, -ος, -ον)
tricoccus, -a, -um three-berried, three-seeded (Greek τρικοκκος, -ος, -ον)
tricolor, -color- tricoloured, three-coloured
tridactylites, -it- three-fingered
trifidus, -a, -um trifid, three-cleft
Trifolii genus medicæ simile a kind of Trefoil like medick (*Medicago* sp.)
trifolius, -a, -um trifoliate, three-leaved
triglochin, -glochidis three-barbed
Trinitatis flos, floris Flower of the Trinity, *Viola arvensis* [620] or *V. tricolor* L.
triquetrus, -a, -um triquetrous, three-cornered
trisulcus, -a, -um trisulcate, three-pronged
tuberosus, -a, -um tuberous; *tuberosa radice* with a tuberous root
turbinatus, -a, -um turbinate, top-shaped
typhinus, -a, -um reedmace-like, like *Typha* spp. [604]
typhoides, -oid- reedmace-like, like *Typha* spp. [604]
uliginosus, -a, -um marshy, growing in marshes; *uliginosis proveniens, -ent- locis* occurring / growing in marshy places
ulophyllos, -os, -on with curling leaves (Greek οὐλοφυλλος, -ος, -ον)
umbellâ candidâ (ablative singular) with a pure white umbel
umbellatus, -a, -um umbellate, bearing an umbel
umbellifer, -fera, -ferum bearing an umbel (not necessarily an umbellifer)
umbilicatus, -a, -um navel-like, peltate, with the stalk in the middle; *umbilicato folio* with a peltate leaf
umbrosis (ablative plural) in shady places
unctuosus, -a, -um oily, greasy
Ungula, -æ caballina, -æ Horse's-hoof, Coltsfoot, *Tussilago farfara* [603]
urens, -ent- burning, stinging, acrid
ursinus, -a, -um of a bear, of bears, like a bear, eaten by bears

VOCABULARY

Urtica, -æ fatua, -æ foolish / insipid nettle; *Urticæ fœtidæ secunda species* the second kind of stinking nettle; *Urtica iners* stingless nettle, Deadnettle (*Lamium*, *Lamiastrum* or *Stachys* sp.); *Urticæ inerti affinis, -is, -e* related to a deadnettle
usitatior, -ior, -ius more useful
ut plurimum for the most part
utriculus, -i a small bottle, a bladder; *utriculis lanugine pubescentibus* with bladders pubescent with down
utrinque on both sides
Uva, -æ Taminia, -æ Taminian Grape (literally), properly Black Bryony, *Tamus communis* [92], but applied to White Bryony, *Bryonia dioica* [91] and Traveller's Joy, *Clematis vitalba* [621]
varietas, -atis (usually after a genitive) a variety; *varietas altera* (after a genitive) a second variety [of]; *varietas major* (after a genitive) a larger variety [of]; *varietas minor* (after a genitive) a smaller variety [of]
varius, -a, -um various, manifold; *variæ species* (plural) various kinds
vasculum, -i a seed-capsule; *vasculo tenui longo* with a slender long capsule
-ve / vel or
venenosus, -a, -um poisonous
ver, veris spring
verbenaceus, -a, -um like vervain (*Verbena officinalis* [608])
vermiculatus, -a, -um vermiculated, worm-like
vernaculus, -a, -um vernacular, native (i.e. Roman)
vernus, -a, -um of spring
veronicæ foliis with leaves [like those] of a speedwell (*Veronica* sp.)
verrucarius, -a, -um used for treating warts
verrucosus, -a, -um warty
verticillatus, -a, -um whorled
verus, -a, -um true
vervecinus, -a, -um of a sheep, of sheep
vescus, -a, -um providing food, edible
vesicarius, -a, -um bladder-like
vetus, veteris [an] old [man]; *veteribus, veterum* for / of the ancients (usually Greeks and / or Romans); *Veteribus Latinis* for the Ancient Latins (i.e. Romans)
viarum (genitive plural) of roads
villosus, -a, -um villous, woolly, shaggy-haired
vimen, viminis a withy; *vimine rubro* with a red withy; *viminibus albidioribus* with rather whitish withies; *viminibus rubris* with red withies
vinearum (genitive plural) of vineyards
violaceus, -a, -um violet, violet-coloured; *violaceo flore* with a violet-coloured flower
virga, -æ a rod; *Virga aurea* Golden Rod, *Solidago virgaurea* [A42/B102]; *Virga pastoris* Shepherd's-rod, *Dipsacus* spp. [168, 169] or *Polygonum aviculare sensu lato* [454]; *Virga sanguinea* Bloody Rod, *Cornus sanguinea* [151]; *Virgæ aureæ sive solidagini angustifoliæ affinis* related to Golden Rod or narrow-leaved solidago (*Solidago virgaurea* [A42/B102])
Virgilii, Virgilio (genitive and dative singular) of / for Virgilius (the Roman poet Virgil)
Virgineus, -a, -um of a Virgin, Maiden
viridis, -is, -e green; *viridi flore* with a green flower
viscosus, -a, -um viscous, sticky

vita, -æ life
vitis, -is a vine; *Vitis alba* White Vine, *Bryonia dioica* [91]; *Vitis Idea* Vine of Mount Ida, *Vaccinium* sp.; *Vitis nigra* Black Vine, *Tamus communis* [92]; *Vitis sylvestris* Wild Vine, *Clematis vitalba* [621] or *Solanum dulcamara* [547]
vitium, -i a blemish, a defect
vix scarcely, hardly
vulgarior, -ior, -ius commoner
vulgaris, -is, -e common; *vulgari mari non ramosæ accedens, -ent-* coming close to the common unbranched male [fern]; *vulgaria parva alba sive arvensia* (neuter plural) common small white or field [peas, *Pisum sativum* 444]
vulgatior, -ior, -ius commoner, more common
vulgatissimus, -a, -um commonest, very common
vulgi [Gallorum] (genitive singular [and plural]) of the common people [of the French]
vulgò commonly; *vulgò dictus, -a, -um* commonly called; *vulgo vera existimatus, -a, -um* commonly thought [to be] the true [one]
vulnerarius, -a, -um vulnerary, staunching wounds; *vulnerariorum* of surgeons (those who staunch wounds)
vulpinus, -a, -um of a fox, of foxes
Zelandia, -æ Zeeland or Zealand (a province of the Netherlands)

BIBLIOGRAPHY

In the following bibliography we have included as far as possible the published works directly cited by Ray, which are marked ‡ after the date, and those indirectly cited by Ray, including those only listed in the 'Explanation', which are marked †. For further details of this distinction see Chapter 5. Additional references cited in the 1685 appendix are marked ¶. We also include references we have cited in our introductory text and footnotes (which have no distinguishing mark). Ray rarely cited a specific edition of works which were published in two or more editions, and it is often impossible from the text to identify the edition he used. We have sometimes cited the first edition, sometimes a later edition, and sometimes more than one edition. All are marked by the appropriate symbol but this does not imply that we think Ray necessarily had access to them all. Further details of works with multiple editions are often provided in our footnotes in the *Catalogus*. We have not usually cited the printed editions of classical authors unless they include additional material from modern commentators and are cited by Ray.

The use by 16th- and 17th-century authors of latinised forms of their names – and to a lesser extent the variation in the spelling of some names in the native language – makes it difficult to construct a bibliography which is both easy to consult and faithful to the original texts. We have listed the works of an author under a single variant of his name, usually in the native language unless (as with Tabernaemontanus) the Latin version is much more familiar. If the variant on the title-page differs this is indicated in brackets after the date of publication, with Latin names converted where necessary to the nominative case. Variants of names that differ in their initial letters, and which might therefore be missed in the alphabetical list, are cross-referenced to the variant that we use. Long titles are often truncated, but we have sometimes included extended titles, especially if this provides a link with Ray's description of the work in his 'Explanation' or supports other statements in his or our text. We have altered i to j and u to v when they are used as consonants and similarly j to i and v to u when they are used as vowels, we have separated the letters in the ligatures æ and œ, and we have omitted all diacritical marks in Latin, unlike in our treatment of Ray's and other authors' original text. Publishers are omitted from titles published before 1800, but we have specified the formats of these books as 2° (folio), 4° (quarto), 8° (octavo), 12° (duodecimo) or 16° (sixteens). Places of publication are translated from Latin but not from other languages.

The on-line catalogues of the British Library, the Cambridge University libraries, the Oxford University libraries, Trinity College, Cambridge, and the Wellcome Library, London, were invaluable in tracing Ray's sources and constructing this bibliography, and we have generally followed their conventions when deciding on matters of format such as the capitalisation of words in the titles of books. We have also used a range of printed bibliographies including those of Adams (1967), Arber (1986), Henrey (1975), Jackson, Ferguson & Pantzer (1986), Lenger (1979), Pritzel (1872–1877), Voet (1980–1983), Wellcome Historical Medical Library (1962–2006) and Wing (1982–1998). We have consulted original copies or facsimiles of most but not all the pre-1660 works cited, mainly in Cambridge University Library but sometimes on-line or in the other libraries listed in our acknowledgements.

Aa, A.J. van der 1854. *Biographisch woordenboek der Nederlanden*. Volume 2(2). J.J. van Brederode, Haarlem.

BIBLIOGRAPHY

Acosta, C. 1582†. (as C. A Costa) *Aromatum & medicamentorum in Orientali India nascentium liber: ... Caroli Clusii Atrebatis opera ex Hispanico sermone Latinus factus, in epitomen contractus, & quibusdam notis illustratus*. Antverpiae [Antwerp]. 8°.

Adams, F., trans. 1856. *On the coeliac affection. The extant works of Aretaeus, the Cappadocian*. Sydenham Society, London.

Adams, H.M. 1967. *Catalogue of books printed on the continent of Europe, 1501–1600, in Cambridge libraries*. 2 vols. Cambridge University Press, London.

Ainsworth, G.C. 1976. *Introduction to the history of mycology*. Cambridge University Press, Cambridge.

Aldrovandi, U. 1599–1603‡. *Ornithologiae, hoc est, De avibus historiae libri XII*. 3 vols. Bononiae [Bologna]. 4°.

Allen, D.E. 1986. *The botanists*. St Paul's Bibliographies, Winchester.

Allen, D.E. 2010. *Books and naturalists*. Collins, London.

Allen, D.E. & Hatfield, G. 2004. *Medicinal plants in folk tradition. An ethnobotany of Britain & Ireland*. Timber Press, Portland, Oregon, & Swavesey, Cambridge.

Allgemeine Deutsche Biographie 1875–1912. 56 vols. Dunder & Humblot, Leipzig.

Amatus Lusitanus 1554†. *In Dioscoridis Anazarbei De materia medica libros quinque enarrationes eruditissimae*. Argentorati [Strasbourg]. 4°.

Amatus Lusitanus 1558†. *In Dioscoridis Anazarbei De medica materia libros quinque, Amati Lusitani ... enarrationes eruditissimae. Accesserunt huic operi praeter correctiones lemmatum, etiam adnotationes R. Constantini, necnon simplicium picturae ex Leonharto Fuchsio, Jacobo Dalechampio, atque aliis*. Lugduni [Lyon]. 8°.

Anderson, F.J. 1977. *An illustrated history of the herbals*. Columbia University Press, New York.

Anguillara, L. 1561†. *Semplici, liquali in piu Pareri à diversi nobili huomini scritti appaiono*. Vinegia. 8°.

Anon. 1523. Σχολια εἰς Θηριακα ["Commentaries on *Theriaca*"], in A.P. Manuzio (ed.), *Nicandri Theriaca ... Interpretatio innominati authoris in Theriaca*, pp. 27–63. [Venice]. 8°.

Anon. 1707. *A compleat history of Europe: or, a view of the affairs thereof, civil and military, for the year 1706*. London. 8°. A life of Ray appears on pp. 495–498 under the title of *Additions to the remarkables of the year 1705*.

Anon. 1866. *Nouvelle biographie générale depuis les temps les plus reculés jusqu'à nos jours ... publiée par MM. Firmin Didot Frères, sous la direction de M. le Dr Hoefer. Volume 41*. Firmin Didot Frères, Fils et Cie, Paris.

Anon. 1968–1982. *Oxford Latin Dictionary*. Clarendon Press, Oxford.

Anon. 1979. *Herbarium Apulei 1481: Herbolaro volgare 1522 con una introduzione di Erminio Caprotti e un saggio di William T. Stearn*. 2 vols. Edizioni il Polifilio, Milano.

Anon. 2002. *The Hartlib papers: a complete text and image database of the papers of Samuel Hartlib (c. 1600–1662)*, ed. 2. HROnline, Sheffield. [Booklet and CD.]

Arber, A. 1943. A seventeenth-century naturalist: John Ray. *Isis* 34: 319–324. [Review of Raven (1942).]

Arber, A. 1986. *Herbals: their origin and evolution*, ed. 3 with an introduction and annotations by W.T. Stearn. Cambridge University Press, Cambridge.

Armendáriz, J.-I.G. 2003. Lucius Junius Moderatus Columella. Addenda, in V. Brown (ed.), *Catalogus translationum et commentariorum, Volume VIII*, pp. 327–333. Catholic University of America Press, Washington D.C.

BIBLIOGRAPHY

Babington, C.C. 1860. *Flora of Cambridgeshire*. John van Voorst, London.

Backer, W. De, Buysscher, G. De, Depauw, C., Imhof, D., Lemli, J., Otte, E., Vandewiele, L.J. & Wille, H. 1993. Catalogue, in F. de Nave & D. Imhof (eds), *Botany in the Low Countries (end of the 15th century – ca. 1650). Plantin-Moretus Museum Exhibition*, pp. 81–146. Plantin-Moretus Museum and Stedelijk Prentenkabinet, Antwerp.

[Bacon], F., Lord Verulam, Viscount St Alban 1626‡. *Sylva Sylvarum: or a naturall historie in ten centuries*. London. 2°. [Title-page dated 1627.]

Baldwin, S.A. 1986. *John Ray (1627–1705) Essex naturalist*. Baldwin's Books, Witham.

Ball, W.W. Rouse & Venn, J.A., eds 1913. *Admissions to Trinity College, Cambridge. Volume II, 1546–1700*. Macmillan and Co., London.

Barbaro, E. 1530†. (as H. Barbarus) *In Dioscoridem corollariorum libri quinque*. Coloniae [Cologne]. 2°.

Barbaro, E. 1534†. (as H. Barbarus) *In C. Plinii Naturalis historiae libros castigationes*. Basileae [Basel]. 4°.

Barden, D.J. & Preston, C.D. 2009. Are *Polypodium interjectum* and *P. vulgare* increasing in Cambridgeshire (v.c. 29)? *Nature in Cambridgeshire* no. 51: 12–20.

Barker, N. 1994. *Hortus Eytstettensis: the Bishop's garden and Besler's magnificent book*. British Library, London.

Barr, C.B.L. & Pollard, M. 1962. The *Historia plantarum* of John Ray. *Transactions of the Cambridge Bibliographical Society* 3: 335–338.

Barrow, I. 1687. *Opuscula; viz. Determinationes, Conc[io] ad Clerum, Orationes, Poemata, &c*. [Volume 4 of *The works of the learned Isaac Barrow … Published by the Reverend Dr. Tillotson*.] Londini [London]. 2°.

Bauhin, C. [1596†]. (as C. Bauhinus) Φυτοπιναξ *[Phytopinax] seu Enumeratio plantarum ab herbariis nostro seculo descriptarum, cum earum differentiis: cui plurimarum hactenus ab iisdem non descriptarum succinctae descriptiones & denominationes accessere: additis aliquot hactenus non sculptarum plantarum vivis iconibus*. Basileae [Basel]. 4°.

Bauhin, C. 1598‡ see Mattioli 1598‡.

Bauhin, C. 1601†. (as C. Bauhinus) *Animadversiones in Historiam generalem plantarum Lugduni editam*. Francofurti [Frankfurt]. 4°.

Bauhin, C. 1613† see Tabernaemontanus 1613†.

Bauhin, C. 1614. (as C. Bauhinus) *De hermaphroditorum monstrosorumq[ue] partuum natura*. Oppenheimii [Oppenheim]. 8°.

Bauhin, C. 1620‡. (as C. Bauhinus) Προδρομος *[Prodromos] theatri botanici, in quo plantae supra sexcentae ab ipso primum descriptae cum plurimis figuris proponuntur*. Francofurti ad Moenum [Frankfurt]. 4°.

Bauhin, C. 1622†. (as C. Bauhinus) *Catalogus plantarum circa Basileam sponte nascentium: cum earundem synonymiis & locis in quibus reperiuntur*. Basileae [Basel]. 8°.

Bauhin, C. 1623‡. (as C. Bauhinus) Πιναξ *[Pinax] theatri botanici ... plantarum circiter sex millium ab ipsis exhibitarum nomina cum earundem synonymiis et differentiis methodice secundum earum et genera et species proponens*. Basileae [Basel]. 4°.

Bauhin, C. 1658‡. (as C. Bauhinus) *Theatri botanici: sive Historiae plantarum ex veterum et recentiorum placitis propriaq[ue] observatione concinnatae. Liber primus editus, opera & cura Io. Casp. Bauhini*. Basileae [Basel]. 2°.

Bauhin, J. 1591†. (as I. Bauhinus) *De plantis a divis sanctisve nomen habentibus. ... Additae sunt Conradi Gesneri … epistolae hactenus non editae*. Basileae [Basel]. 8°.

BIBLIOGRAPHY

Bauhin, J. 1593†. (as I. Bauhinus) *De plantis absynthii nomen habentibus.* Montisbeligardi [Montbéligard]. 8°.

Bauhin, J. 1598‡. (as I. Bauhinus) *Historia novi et admirabilis fontis balneique Bollensis in Ducatu Wirtembergico ad acidulas Goepingenses.* Montisbeligardi [Montbéligard]. 4°.

Bauhin, J. & Cherler, J.H. 1619. (as I. Bauhinus & I.H. Cherlerus) *Historiae plantarum generalis novae et absolutissimae quinquaginta annis elaboratae jam prelo commissae Prodromus.* Ebroduni [Yverdon]. 4°.

Bauhin, J. & Cherler, J.H. 1650–1651‡. (as I. Bauhinus & I.H. Cherlerus) *Historia plantarum universalis ... quam recensuit et auxit Dominicus Chabraeus. Volume 1* (1650); *Volumes 2–3* (1651). Ebroduni [Yverdon]. 2°. [The pagination of volumes 1 and 3 starts again in the middle and in citing page numbers in our references we distinguish the two halves as 1(1), 1(2) and 3(1), 3(2).]

Beaven, E.S. 1947. *Barley: fifty years of observation and experiment.* Duckworth, London.

Becanus, I.G. see Goropius, J.

Bech, S.C., ed. 1979–1984. *Dansk biografisk leksikon.* 16 vols. Gyldendal, København.

Beck, L.Y. 2005. *Pedanius Dioscorides of Anazarbus. De materia medica. Translated by Lily Y. Beck.* Altertumswissenschaftliche Texte und Studien 38. Olms-Weidman, Hildesheim.

Beier, A.L. 1974. Vagrants and the social order in Elizabethan England. *Past & present* no. 64: 3–29.

Belcher, H. & Swale, E. 1998. Moss that grows on skulls: a curious old remedy run to earth in Cambridge. *Nature in Cambridgeshire* no. 40: 74–75.

Belkin, J.S. & Caley, E.R. 1978. *Eucharius Rösslin the Younger On minerals and mineral products.* Walter de Gruyter, Berlin & New York.

Belon, P. 1553a. *Les observations de plusieurs singularitez et choses memorables, trouvées en Grèce, Asie, Iudée, Egypte, Arabie, & autres pays estranges.* Paris. 4°.

Belon, P. 1553b†. (as P. Bellonius) *De arboribus coniferis, resiniferis, aliis quoque nonnullis sempiterna fronde virentibus, cum earundem iconibus ad vivum expressis.* Parisiis [Paris]. 4°.

Belon, P. 1553c†. (as P. Bellonius) *De admirabili operum antiquorum et rerum suspiciendarum praestantia. Liber primus. De medicato funere, seu cadavere condito, & lugubri defunctorum ejulatione. Liber secundus. De medicamentis nonnullis, servandi cadaveris vim obtinentibus. Liber tertius.* Parisiis [Paris]. 4°.

Belon, P. 1589†. (as P. Bellonius) *Plurimarum singularium & memorabilium rerum in Graecia, Asia, Aegypto, Judaea, Arabia, aliisq[ue] exteris provinciis ab ipso conspectarum observationes ... Carolus Clusius Atrebas e gallicis latinas faciebat.* Antverpiae [Antwerp]. 8°.

Bendall, A.S. 1992. *Maps, land and society: a history, with a carto-bibliography of Cambridgeshire estate maps, c. 1600–1836.* Cambridge University Press, Cambridge.

Bendall, [A.]S. 2000. Merton College and the mapping of its estates 1601–1836. *Oxoniensia* 65: 79–100.

Bennet, C. 1654‡. (as C. Bennettus) *Theatri Tabidorum vestibulum: seu Exercitationes dianoeticae cum historiis et experimentis demonstrativis.* Londini [London]. 8°.

Bernard, J.P., Birch, T. & Lockman, J. 1739. *A general dictionary, historical and critical. Volume VIII.* London. 2°.

Besler, B. 1613†. (as B. Beslerus) *Hortus Eystettensis.* [Nuremberg]. 2°. [For a facsimile, see Littger, Lorenz & Menghini (2006).]

BIBLIOGRAPHY

Bietenholz, P.G., ed. 1985–1987. *Contemporaries of Erasmus: a biographical register of the Renaissance and Reformation.* 3 vols. University of Toronto Press, Toronto.

Bircham, P. & Jordan, P. 2000. Birds in Cambridgeshire – the past, present and future: a review for the Millennium. *Nature in Cambridgeshire* no. 42: 45–59.

Birkhead, T. 2008. *The wisdom of birds: an illustrated history of ornithology.* Bloomsbury, London.

Blockwitz, M. 1631†. (as M. Blochwitius) *Anatomia sambuci, quae non solum sambucum & hujusdem medicamenta singulatim delineat.* Lipsiae [Leipzig]. 12°.

Blockwitz, M. 1650†. (as M. Blochwitius) *Anatomia sambuci: quae non solum sambucum & hujusdem medicamenta singulatim delineat.* Londini [London]. 12°.

Blockwitz, M. 1655†. (as M. Blochwich) *Anatomia sambuci: or, The anatomie of the elder. Cutting out of it, plain, approved, and specifick remedies for most and chiefest maladies; confirmed and cleared by reason, experience, and history.* London. 12°.

Blunt, W. 1950. *The art of botanical illustration.* Collins, London.

Bock, H. 1539†. *New Kreütter Bůch von underscheydt, würckung und namen der kreütter so in Teütschen landen wachsen.* Strassburg. 2°.

Bock, H. 1552‡. (as H. Tragus) *De stirpium, maxime earum, quae in Germania nostra nascuntur, usitatis nomenclaturis ... commentariorum libri tres ... interprete Davide Kybero Argentinensi.* Argentorati [Strasbourg]. 4°.

Bock, H. 1595†. *Kreütterbuch, darin underscheidt Nammen, und Würcküng der Kreütter, Stauden, Hecken und Bäumen, mit ihren Früchten, so in Teutschen landen wachsen ... Alles durch H. Hieronymum Bock, auss langwüriger und gewisser Erfahrung, beschriben. Hernacher mit vilen nutzlichen Experimenten gemehret, durch den hochgelehrten Melchiorem Sebizium, der Artzney Doctorem und Physicum zu Strassburg.* Strassburg. 2°.

Bodaeus, J. 1644‡. *Theophrasti Eresii de historia plantarum libri decem, graece et latine: in quibus textum graecum variis lectionibus, emendationibus, hiulcorum supplementis, latinam Gazae versionem nova interpretatione ad margines: totum opus absolutissimis cum notis, tum commentariis: item rariorum plantarum iconibus illustravit Joannes Bodaeus à Stapel, medicus Amstelodamensis. Accesserunt Julii Caesaris Scaligeri, in eosdem libros animadversiones: et Roberti Constantini annotationes, cum indice locupletissimo.* Amstelodami [Amsterdam]. 2°.

Bond, D.A. 1995. Faba bean *Vicia faba* (Leguminosae – Papilionoideae), in J. Smartt & N.W. Simmonds (eds), *Evolution of crop plants*, ed. 2, pp. 312–316. Longman Scientific & Technical, Harlow.

Borel, P. 1656. (as P. Borellus) *Observationum microcospicarum* [sic] *centuria.* Hagae-comitis [The Hague]. 4°. [Bound in with Borel's *De vero telescopii inventore* of 1655 and covered by a common title-page.]

Boulger, G.S. 1892. The domestic life of John Ray at Black Notley. *Journal of Proceedings of the Essex Field Club* 4: clix–clxix.

Bowles, E.A. 1934. *A handbook of Narcissus.* Martin Hopkinson, London.

Boyle, R. 1744. *The works of the Honourable Robert Boyle. Volume V.* London. 2°.

Bradley, J.D., Tremewan, W.G. & Smith, A. 1979. *British tortricoid moths. Tortricidae: Olethreutinae.* Ray Society, London.

Brasavola, A.M. 1540†. (as A.M. Brasavolus) *Examen omnium syruporum, quorum publicus usus est.* Lugduni [Lyon]. 8°.

Brasavola, A.M. 1556†. (as A.M. Brasavolus) *Examen omnium simplicium, quorum usus in publicis est officinis.* Lugduni [Lyon]. 16°.

Braun, G. & Hogenberg, F. [1575]‡. (G. Braun as G. Bruin) *De praecipuis, totius universi urbibus, liber secundus.* [Cologne]. 2°.

Braunschweig, H. 1500. *Liber de arte distillandi, de simplicibus.* Strassburg. 2°.
Braunschweig, H. 1519†. (as I. Brunschwick) *Das buch zu distillieren die zusamen gethonen ding Composita genant.* Strassburg. 2°.
Breidbach, O. & Ghiselin, M.T. 2006. Baroque classification: a missing chapter in the history of systematics. *Annals of the history and philosophy of biology* 11: 1–30.
Briggs, T.R.A. 1880. *Flora of Plymouth.* John van Voorst, London.
Bristowe, W.S. 1946. Who was "Little Miss Muffet"? *Sunday Times* 3 March 1946: 6.
Britten, J., Jackson, B.D. & Stearn, W.T. 1965. *William Turner: Libellus de re herbaria 1538: The names of herbes 1548: Facsimiles with introductory matter.* Ray Society, London.
Bromfield, W.A. 1856. *Flora vectensis.* William Pamplin, London.
Brooke, J.H. 2000. 'Wise men nowadays think otherwise': John Ray, natural theology and the meanings of anthropocentrism. *Notes and records of the Royal Society* 54: 199–213.
Brown, V. 1976. Columella, Lucius Junius Moderatus, in F.E. Cranz (ed.), *Catalogus translationum et commentariorum, Volume III*, pp. 173–193. Catholic University of America Press, Washington D.C. [For addenda, see Armendáriz (2003).]
Brown, V. 1980a. Cato, Marcus Porcius, in F.E. Cranz (ed.), *Catalogus translationum et commentariorum, Volume IV*, pp. 223–247. Catholic University of America Press, Washington D.C.
Brown, V. 1980b. Varro, Marcus Terentius, in F.E. Cranz (ed.), *Catalogus translationum et commentariorum, Volume IV*, pp. 451–500. Catholic University of America Press, Washington D.C.
Browne, T. 1646. *Pseudodoxia epidemica: or, enquiries into very many received tenents, and commonly presumed truths.* London. 2°.
Browne, T. 1658‡. *Hydriotaphia, urne-buriall, or, A discourse of the sepulchrall urnes lately found in Norfolk. Together with The garden of Cyrus, or The quincunciall, lozenge, or net-work plantations of the ancients, artificially, naturally, mystically considered. With sundry observations.* London. 8°.
Bruce, N. & Sharpe, G. 2002. *Gamlingay, Cambridgeshire: portrait of an English village.* Publisher and place of publication not cited.
Brunf[els], O. 1530–1536†. *Herbarum vivae eicones ad naturae imitationem.* Volume I (1530); Volume II (1531); Volume III (1536). Argentorati [Strasbourg]. 2°.
Brunfels, O. 1532–1537. *Contrafayt Kreüterbuch. Nach rechter vollkommener art, vnud [sic] Beschreibungen der Alten, besst-berümpten ürtzt.* Volume I (1532); Volume II (1537). Strasszburg. 2°.
Brunfels, O. 1539†. *Kreüterbůch contrafayt, beyde Teyl vollkummen, nach rechter, warer beschreibung der Alten leerer und Artzt.* Strasszburg. 4°.
Burrows, M., Shaw, S.R. & Sutton, G.P. 2008. Resilin and chitinous cuticle form a composite structure for energy storage in jumping by froghopper insects. *BMC Biology* 6: article 41.
Bushell, W.D. 1938. *Hobson's Conduit: the New River at Cambridge commonly called Hobson's River.* Cambridge University Press, Cambridge.
Butler, C. 1634‡. *The Feminine Monarchie, or The Histori of Bees ... Written out of Experience.* Oxford. 4°.
Bynum, W.F. & Bynum, H., eds 2007. *Dictionary of medical biography.* 5 vols. Greenwood Press, Westport, Connecticut & London.
Cai, W., Yan, Y. & Li, L. 2005. The earliest records of insect parasitoids in China. *Biological control* 32: 8–11.
Cain, A.J. 1996. John Ray on 'accidents'. *Archives of natural history* 23: 343–368.

Cain, A.J. 1999a. Thomas Sydenham, John Ray, and some contemporaries on species. *Archives of natural history* 26: 55–83.
Cain, A.J. 1999b. John Ray on the species. *Archives of natural history* 26: 223–238.
Calepino, A. 1609‡. (as A. Calepinus) *Dictionarium octolingue in quo Latinis dictionibus Hebraeae, Graecae, Gallicae, Italicae, Germanicae, Hispanicae, atque Anglicae adjectae sunt.* 2 vols. [Paris]. 2°.
Camden, W. 1607. (as G. Camden) *Britannia.* Londini [London]. 4°.
Camerarius, J. 1588‡. (as I. Camerarius) *Hortus medicus et philosophicus: in quo plurimarum stirpium breves descriptiones, novae icones non paucae, indicationes locorum natalium, observationes de cultura earum peculiares atque insuper nonnulla remedia euporista, nec non philologica quaedam continentur. Cum Thalii Sylva Hercynia nunc primum editum.* Francofurti ad Moenum [Frankfurt]. 4°. [See also Thal 1588.]
Cancik, H. & Schneider, H. 2002–2010. *Brill's New Pauly: Encyclopaedia of the Ancient World. English edition. Antiquity.* 15 vols. Brill, Leiden.
Canini, A. 1555‡. (as A. Caninius) Ἑλληνισμος *[Hellenismos]. In quo quicquid vetustissimi scriptores de Graecae linguae ratione praecipiunt.* Parisiis [Paris]. 4°.
Canini, A. 1624‡. (as A. Caninius) Ἑλληνισμος *[Hellenismos], copiosissimi Graecarum Latinarumq[ue] vocum indicis accessione per Carolum Hauboesium locupletatus.* Londini [London]. 8°.
Cardano, G. 1557‡. (as H. Cardanus) *De rerum varietate libri XVII.* Basiliae [Basel]. 2°.
Cardano, G. 1559‡. (as H. Cardanus) *De subtilitate libri XXI.* Lugduni [Leiden]. 8°.
Castagna, L. 1996. *Quinti Ennii et Marci Pacuvii lexicon sermonis scaenici.* Olms-Weidmann, Hildersheim.
Cesalpino, A. 1583†. (as A. Caesalpinus) *De plantis libri XVI.* Florentiae [Florence]. 4°.
Cesalpino, A. 1603†. (as A. Caesalpinus) *Appendix ad libros de plantis et quaestiones peripateticas.* Romae [Rome]. 4°.
Chapman, G.T.L., McCombie, F. & Wesencraft, A.U., eds 1995. *A new herball by William Turner Parts II and III.* Cambridge University Press, Cambridge. [Facsimile reprint with transcription.]
Chapman, G.T.L. & Tweddle, M.N., eds 1995. *A new herball by William Turner Part I.* Cambridge University Press, Cambridge. [Facsimile reprint with transcription.]
Charleton, W. 1650a‡. (as G. Charleton) *Spiritus gorgonicus, vi sua saxipara exutus.* Lugd[uni] Batav[orum] [Leiden]. 8°.
Charleton, W. 1650b‡. *A ternary of paradoxes. The magnetick cure of wounds. The nativity of tartar in wine. The image of God in man. Written originally by Joh. Bapt. van Helmont, and translated, illustrated, and ampliated by Walter Charleton.* London. 4°.
Charleton, W. 1654. *Physiologia Epicuro-Gassendo-Charltoniana: or A fabrick of science natural, upon the hypothesis of atoms, founded by Epicurus, repaired by Petrus Gassendus, augmented by Walter Charleton.* London. 2°.
Chater, A.O. 1993. *Avena strigosa*, Bristle Oat, and other cereals as crops and casuals in Cardiganshire, VC 46. *Botanical Society of the British Isles Welsh Bulletin* no. 55: 7–14.
Clair, C. 1960. *Christopher Plantin.* Cassell & Co., London.
Clarke, W.A. 1900. *First records of British flowering plants*, ed. 2. West, Newman & Co., London.

BIBLIOGRAPHY

Clusius, C. see L'Écluse, C. de.
Cockayne, T.O. 1864–1866. *Leechdoms, wortcunning, and starcraft of early England.* Longman, Green, Longman, Roberts, and Green, London. 3 vols.
Cockayne, T.O. 1961. *Leechdoms, wortcunning, and starcraft of early England*, with a new introduction by Charles Singer. 3 vols. Holland Press, London.
Cokayne, G.E. 1902. *Complete baronetage. Volume II, 1625–1649.* William Pollard & Co., Exeter.
Colonna, F. 1592†. (as F. Columna) Φυτοβασανος *[Phytobasanos] sive plantarum aliquot historia.* Neapoli [Naples]. 4°.
Colonna, F. 1606†. (as F. Columna) *Minus cognitarum stirpium aliquot, ac etiam rariorum nostro coelo orientium* ἐκφρασις *[ecphrasis].* Romae [Rome]. 4°.
Colonna, F. 1616†. (as F. Columna) *Minus cognitarum rariorumque nostro coelo orientium stirpium* ἐκφρασις *[ecphrasis].* Romae [Rome]. 4°.
Considine, J.P. 2008. *Dictionaries in early modern Europe: lexicography and the making of heritage.* Cambridge University Press, Cambridge.
Cook, C.D.K. & Lüönd, R. 1982. A revision of the genus *Hydrocharis* (Hydrocharitaceae). *Aquatic Botany* 14: 177–204.
Coombe, D.E. 1987. Spiked speedwell, stone stripes and polygons, and the vanishing chalk heaths of Cambridgeshire. *Nature in Cambridgeshire* no. 29: 26–37.
Cooper, C.H. 1843. *Annals of Cambridge. Volume II.* Warwick and Co., Cambridge.
Cooper, C.H. 1845. *Annals of Cambridge. Volume III.* Warwick and Co., Cambridge.
Cooper, T. 1573. *Thesaurus linguae Romanae & Britannicae.* Londini [London]. 4°.
Cope, T. & Gray, A. 2009. *Grasses of the British Isles.* B.S.B.I. handbook no. 13. Botanical Society of the British Isles, London.
Cordus, E. 1534†. *Botanologicon.* Coloniae [Cologne]. 8°.
Cordus, V. 1561†. *Annotationes in Pedacii Dioscoridis Anazarbei De medica materia libros V. ... Ejusdem Val. Cordi Historiae stirpium Lib. IIII ... Item Conradi Gesneri De hortis Germaniae liber recens, una cum descriptione Tulipae Turcarum, Chamaecerasi montani, Chamaemespili, Chamaenerii, & Conizoides. Omnia summo studio atque industria doctilis ... Conr. Gesneri medici Tigurini collecta, & praefationius illustrata.* Argentorati [Strasbourg]. 2°.
Cordus, V. 1651†. *Valerii Cordi Dispensatorium, sive Pharmacorum conficiendorum ratio. Cum Petri Coudenbergii, & Matthiae Lobelii scholiis, emendationibus, & auctariis.* Lugduni Batavorum [Leiden]. 12°.
Cosenza, M.E. 1962. *Biographical and bibliographical dictionary of the Italian humanists and of the world of classical scholarship in Italy, 1300–1800*, ed. 2. *Volume 5, Synopsis and bibliography.* G.K. Hall & Co., Boston, Massachusetts.
Costeo, G. 1578‡. (as I. Costaeus) *De universali stirpium natura libri duo.* Augustae Taurinorum [Turin]. 8°.
Crolachius, H. [1563]†. *Isatis herba: De cultura herbae isatidis quam guadum vulgo vocant.* Tiguri [Zurich]. 8°.
Crompton, G. 1959. The Peat Holes of Triplow. *Nature in Cambridgeshire* no. 2: 24–34.
Crompton, G. 2001–2004. *Catalogue of Cambridgeshire flora records since 1538.* Part 1 (2001); Part 2 (2003); Part 3 (2004). Unpublished report. [For an updated version, see http://www.cambridgeshireflora.com]
Crompton, G. & Preston, C.D. 2000. Vascular plant records. *Nature in Cambridgeshire* no. 42: 93–94.
Crompton, G. & Whitehouse, H.L.K. 1983. *A checklist of the flora of Cambridgeshire.* Privately published, Cambridge.

BIBLIOGRAPHY

Cronk, Q.C.B. 1989. Some early notices of anthocyanin-rich variants of common British plants. *Watsonia* 17: 352–355.

Crossley, J., ed. 1847. *The diary and correspondence of Dr. John Worthington. Volume I.* Chetham Society, Manchester.

Dalby, A. 1998. *Cato: On farming: De agricultura. A modern translation with commentary.* Prospect Books, Totnes.

[Daléchamps, J.] 1586–1587‡. *Historia generalis plantarum: in libros XVIII.* 2 vols. Lugduni [Lyon]. 2°. [The first volume is dated 1587 and the second 1586, although their pagination is continuous so that it is difficult to imagine how the second could have preceded the first.]

Daléchamps, J. 1587‡. (as I. Dalechampsius) *C. Plinii Secundi Historiae mundi libri XXXVII ... Omnia quidem multorum antehac doctorum hominum, novissime vero laboriosis observationibus conquisita, & solerti judicio pensitata, Iacobi Dalecampii medici cadomensis.* Lugduni [Lyon]. 2°.

Dandy, J.E. & Taylor, G. 1938. Studies of British Potamogetons.– I. *Journal of Botany, British and Foreign* 76: 89–92.

Davis, P.H., ed. 1972. *Flora of Turkey and the East Aegean Islands. Volume 4.* Edinburgh University Press, Edinburgh.

Davis, P.V. 1969. *Macrobius: the Saturnalia.* Columbia University Press, New York & London.

De L'Écluse, see L'Écluse, C de.

De L'Obel, see L'Obel, M de.

Derham, W. 1718. *Philosophical letters between the late learned Mr. Ray and several of his ingenious correspondents, natives and foreigners.* London. 8°.

Derham, W. 1760. *Select remains of the learned John Ray, M.A. and F.R.S. with his life.* George Scott, London. 2°.

Derham, W. 1846. *Select remains and life of Ray*, in E. Lankester (ed.), *Memorials of John Ray*, pp. 7–52. Ray Society, London.

Dillenius, J.J. 1741. *Historia muscorum.* Oxonii [Oxford]. 2°.

Dizionario biografico degli Italiani 1960–. 73 vols (to date). Istituto della Enciclopedia Italiana fondata da Giovanni Treccani, Roma.

Dodoens, R. 1552†. (as R. Dodonaeus) *De frugum historia, liber unus.* Antverpiae [Antwerp]. 8°.

Dodoens, R. 1554. *Cruijde Boeck.* Antwerpen. 2°. [For a facsimile see Wittop Koning (1971).]

Dodoens, R. 1557†. *Histoire des plantes ... Nouvellement traduite de bas Aleman en François par Charles de l'Escluse.* Anvers. 4°. [For a facsimile see Opsomer (1978).]

Dodoens, R. 1559. (as R. Dodonaeus) *De stirpium historia commentariorum imagines, in duos tomos digestae.* Antverpiae [Antwerp]. 8°.

Dodoens, R. 1563†. *Cruijde Boeck.* Anvers. 2°.

Dodoens, R. 1568†. (as R. Dodonaeus) *Florum et coronariarum odoratarumque nonnullarum herbarum historia.* Antverpiae [Antwerp]. 8°.

Dodoens, R. 1574†. (as R. Dodonaeus) *Purgantium aliarumque eo facientium, tum et radicum, convolvulorum ac deleteriarum herbarum historiae libri IIII.* Antverpiae [Antwerp]. 8°.

Dodoens, R. 1578. *A niewe herball, or Historie of plantes ... First set foorth in the Doutche or Almaigne tongue, by that learned D. Rembert Dodoens ... And nowe first translated out of French into English, by Henry Lyte Esquyer.* Antwerpe. 2°.

Dodoens, R. 1583†. (as R. Dodonaeus) *Stirpium historiae pemptades sex. Sive libri XXX.* Antverpiae [Antwerp]. 2°.

Dodoens, R. 1608†. (as R. Dodonaeus) *Cruydt-Boeck van Rembertus Dodonaeus, volgens fijne laetste verbeteringe tot.* Leyden. 2°.
Dodoens, R. 1616†. (as R. Dodonaeus) *Stirpium historiae pemptades sex sive libri XXX. Varie ab auctore, paullo ante mortem, aucti & emendati.* Antverpiae [Antwerp]. 2°.
Dodoens, R. 1619. *A new herbal, or Historie of plants ... First set forth in the Dutch or Almaigne tongue, by ... Rembert Dodoens ... and now first translated out of French into English, by Henry Lyte ... Corrected and amended.* London. 2°.
Dodoens, R. 1644. (as R. Dodonaeus) *Cruydt-Boeck.* t'Antwerpen. 2°.
Druce, G.C. 1886. *The flora of Oxfordshire.* Parker and Co., Oxford.
Druce, G.C. 1912. Samuel Corbyn's catalogue of Cambridgeshire plants. *Journal of Botany, British and Foreign* 50: 76–79.
Druce, G.C. 1932. *The comital flora of the British Isles.* T. Buncle & Co., Arbroath.
Drury, P. 2010. *Audley End.* English Heritage, London.
DSB see Gillispie 1970–1980.
Du Chesne, J. 1607. (as I. Quercetanus) *Pharmacopoea dogmaticorum restituta. Pretiosis selectisque hermeticorum floribus abunde illustrata.* Parisiis [Paris]. 4°.
Du Choul, J. 1555†. (as I. du Choul) *De varia Quercus historia.* Lugduni [Lyon]. 8°.
Dunn, S.T. 1905. *Alien flora of Britain.* West, Newman and Co., London.
Durante, C. 1585†. *Herbario nuovo; con figure, che rappresentano le vive piante, che nascono in tutta Europa, & nell'Indie orientali, & occidentali. Con versi latini, che comprendono le facoltà de i semplici medicamenti.* Roma. 2°.
Edgington, J. 2007. A plant list of 1633: annotations in a copy of Thomas Johnson's *Iter plantarum. Archives of Natural History* 34: 272–292.
Egmond, F., Hoftijzer, P. & Visser, R., eds 2007. *Carolus Clusius. Towards a cultural history of a Renaissance naturalist.* Koninklijke Nederlandse Akademie van Wetenschappen, Amsterdam.
Eijk, P.J. van der 2000. *Diocles of Carystus: a collection of the fragments with translation and commentary. Volume 1, Text and translation.* Brill, Leiden.
Elrington, C.R., ed. 1973. *A history of the county of Cambridge and the Isle of Ely. Volume V.* Oxford University Press, London.
Elyot, T. 1539. *The castel of helth gathered and made by Syr Thomas Elyot knyghte.* London. 4°.
Eriksson, G. 2004. Flora writing in Norden, in B. Jonsell (ed.), *Flora nordica: general volume*, pp. 1–36. Bergius Foundation, Royal Swedish Academy of Sciences, Stockholm.
Evans, A.H. 1939. *A flora of Cambridgeshire.* Gurney and Jackson, London & Edinburgh.
Ewen, A.H. & Prime, C.T. [1975]. *Ray's Flora of Cambridgeshire.* Wheldon & Welsey, Hitchin.
Fawket, J. 1681. *An account of the late reverend and worthy Dr. George Seignior, fellow of Trinity College in Cambridge.* London. 12°.
Feingold, M. 1990a. Isaac Barrow: divine, scholar, mathematician, in M. Feingold (ed.), *Before Newton: the life and times of Isaac Barrow.* pp. 1–104. Cambridge University Press, Cambridge.
Feingold, M. 1990b. Isaac Barrow's library, in M. Feingold (ed.), *Before Newton: the life and times of Isaac Barrow*, pp. 333–372. Cambridge University Press, Cambridge.

Feingold, M. 1997. The mathematical sciences and new philosophies, in N. Tyacke (ed.), *The history of the University of Oxford. Volume IV, Seventeenth-century Oxford*, pp. 359–448. Clarendon Press, Oxford.

Feisenberger, H.A., ed. 1975. *Sale catalogues of the libraries of eminent persons. Volume 11, Scientists*. Mansell Information Publishing, London.

Fernel, J. 1644‡. (as J. Fernelius) *Consiliorum medicinalium liber: ex ejus adversariis quadringentarum consultationum selectus a Juliano Palmario editio praecentibus auctior*. Lugduni Batavorum [Leiden]. 8°.

Ferri, S., ed. 1997. *Pietro Andrea Mattioli Siena 1501 – Trento 1578. La vita le opere con l'identificazione delle piante*. Quattroenne, Perugia.

Findlay, W.M. 1956. *Oats: their cultivation and use from ancient times to the present day*. Oliver and Boyd, Edinburgh.

Fischer, H., Petit, G., Staedtke, J., Steiger, R. & Zoller, H. 1967. *Conrad Gessner, 1516–1565: Universalgelehrter, Naturforscher, Arzt*. Institut Orell Füssli, Zürich.

Fitter, A. & Smith, C. 1979. *A wood in Ascam: a study in wetland conservation*. William Sessions, York.

Fleitmann, S. 1986. *Walter Charleton (1620–1707), »Virtuoso«: Leben und Werk*. Aspekte der englischen Geistes- und Kulturgeschichte 7. Peter Lang, Frankfurt am Main.

Fragoso, J. 1601†. (as I. Fragoso) *Aromatum, fructuum, et simplicium aliquot medicamentorum ex India utraque, et Orientali et Occidentali, in Europam delatorum, quorum jam est usus plurimus, historia brevis, utilis, et jucunda. Conscripta primum Hispanice a Ioanne Fragoso ... nunc Latine edita opera ac studio Israelis Spachii*. Argentinae [Strasbourg]. 8°.

Fraser, A. 1992. *The gypsies*. Blackwell Publishers, Oxford.

Fraser, A. 1997. *Cromwell: our chief of men*. Weidenfeld and Nicolson, London.

Free, J.B. 1962. The behaviour of honeybees visiting field beans (*Vicia faba*). *Journal of Animal Ecology* 31: 497–502.

Friedman, A.T. 1989. *House and household in Elizabethan England: Wollaton Hall and the Willoughby family*. University of Chicago Press, Chicago & London.

Friedrich, E. 1986. *Breeding butterflies and moths: a practical handbook for British and European species*, edited by A. Maitland Emmet. Harley Books, Colchester.

Fuchs, L. 1542‡. (as L. Fuchsius) *De historia stirpium commentarii insignes*. Basileae [Basel]. 2°. [For a facsimile and commentary, see Meyer, Trueblood & Heller (1999).]

Fuchs, L. 1549†. (as L. Fuchsius) *Primi de stirpium historia commentariorum tomi vivae imagines, in exiguam angustioremq[ue] formam contractae*. Basileae [Basel]. 8°.

Fuller, T. 1662. *The history of the worthies of England*. London. 2°.

Füssel, S., ed. 2008. *Georg Braun and Franz Hogenberg. Civitates orbis terrarum. Cities of the world*. Taschen, Hong Kong. [Complete edition of the colour plates of 1572–1617.]

Garofalo, I. 1988. *Erasistrati fragmenta*. Biblioteca di studi antichi 62. Giardini Editori e Stampatori in Pisa, Pisa.

Gaskell, P. 1972. *A new introduction to bibliography*. Clarendon Press, Oxford.

Gaskell, P. 1980. *Trinity College library: the first 150 years*. Cambridge University Press, Cambridge.

Gassendi, P. 1658‡. *Opera omnia in sex tomos divisa. Tomus secundus quo continentur Syntagmatis philosophici partis secundae, seu physicae sectionis tertiae membra duo*. Lugduni [Lyon]. 2°.

BIBLIOGRAPHY

Gaza, T. 1529. *Theophrasti de historia, et causis plantaru[m], libri quindecim. Theodoro Gaza interprete.* Parisiis [Paris]. 8°.

Gerard, L. 1761. *Flora Gallo-Provincialis.* Parisiis [Paris]. 8°.

Gerarde, J. 1597‡. *The herball or generall historie of plantes.* London. 2°. [For the revised edition, see Johnson (1633, 1636).]

Gerli, E.M., ed. 2003. *Medieval Iberia: an encyclopedia.* Routledge, New York.

Gesner, C. 1541†. (as C. Gesnerus) *Historia plantarum et vires ex Dioscoride, Paulo Aegineta, Theophrasto, Plinio, & rece[n]tioribus Graecis, juxta elementoru[m] ordine[m].* Parisiis [Paris]. 8°.

Gesner, C. 1542†. (as C. Gesnerus) *Catalogus plantarum Latine, Graece, Germanice, & Gallice.* Tiguri [Zurich]. 4°.

Gesner, C. 1555†. (as C. Gesnerus) *De raris et admirandis herbis, quae sive quod noctu luceant, sive alias ob causas, lunariae nominantur, commentariolus: & obiter de aliis etiam rebus quae in tenebris lucent. ... Ejusdem descriptio Montis Fracti, sive Montis Pilati, juxta Lucernam in Helvetia.* Tiguri [Zurich]. 4°.

Gesner, C. 1565†. (as C. Gesnerus) *De rerum fossilium, lapidum et gemmarum maxime, figuris et similitudinibus liber.* [Published with seven other works in a volume entitled *De omni rerum fossilium genere, gemmis, lapidibus, metallis, et hujusmodi, libri aliquot, plerique nunc primum editi.*] Tiguri [Zurich]. 8°.

Gesner, C. 1577†. (as C. Gesnerus) *Epistolarum medicinalium ... libri III.* Tiguri [Zurich]. 4°.

Gesner, C. 1587†. (as C. Gesnerus) *De stirpium collectione tabulae tum generales, tum per duodecim menses, cum Germanicis nominibus ... olim per Conradum Gesnerum conscriptae ac aeditae: nunc authoris opera locupletatae, et de novo in usum Pharmacopolarum luci datae, per Casparum Wolphium.* Tiguri [Zurich]. 8°.

Gibson, E. 1695. *Camden's Britannia, newly translated into English: with large additions and improvements.* London. 2°.

Gibson, R.W. 1950. *Francis Bacon: a bibliography of his works and of Baconiana to the year 1750.* Scrivener Press, Oxford.

Gilbert, W. 1600‡. (as G. Gilbertus) *De magnete, magneticisque corporibus, et de magnete tellure; physiologia nova, plurimis & argumentis, & experimentis demonstrata.* Londini [London]. 2°.

Gilbert-Carter, H. 1964. *Glossary of the British flora*, ed. 3. Cambridge University Press, Cambridge.

Gillispie, C.C., ed. 1970–1980. *Dictionary of Scientific Biography.* 16 vols. Charles Scribner's Sons, New York.

Gilmour, J.S.L., ed. 1972. *Thomas Johnson: botanical journeys in Kent & Hampstead.* Hunt Botanical Library, Pittsburgh.

Godwin, H. 1967. Pollen-analytic evidence for the cultivation of *Cannabis* in England. *Review of Palaeobotany and Palynology* 4: 71–80.

Goebel, S. 1565†. (as S. Goebelius) *De succino libri duo.* [Published with seven other works in a volume of C. Gesner's entitled *De omni rerum fossilium genere, gemmis, lapidibus, metallis, et hujusmodi, libri aliquot, plerique nunc primum editi.*] Tiguri [Zurich]. 8°.

Gohlman, W.E. 1974. *The life of Ibn Sina: a critical edition and annotated translation.* State University of New York Press, Albany.

Goropius, J. 1580‡. (as I.G. Becanus) *Opera Ioan. Goropii Becani, hactenus in lucem non edita: nempe, Hermathena, Hieroglyphica, Vertumnus, Gallica, Francica, Hispanica.* Antwerpiae [Antwerp]. 2°.

BIBLIOGRAPHY

Gorris, J. de 1564†. (as J. Gorraeus) *Definitionum medicarum libri XXIIII.* Lutetiae Parisiorum [Paris]. 4°.
Gould, S.J. 1999. When fossils were young. *Natural History* 108(8): 24–26, 70–75.
Gouldman, F. 1669. *A copious dictionary in three parts*, ed. 2. Cambridge. 4°.
Gray, A. 1925. *The town of Cambridge: a history.* W. Heffer & Sons, Cambridge.
Greco, L. 1995. *From the Neolithic revolution to the gluten intolerance: benefits and problems associated to the cultivation of wheat.* http://www.enabling.org/ia/celiac/cul-wht.html
Green, P. & Miller, D. 2009. *The genus* Jasminum *in cultivation.* Kew Publishing, Kew.
Green, P.M. 1955. *Asclepiades: his life and writings.* Elizabeth Licht, New Haven.
Greene, E.L. 1905. The earliest local flora. *The Plant World* 8: 115–121.
Greene, E.L. 1983. *Landmarks of botanical history*, edited by F.N. Egerton. *Part I* (1983a), pp. 1–505, and *Part II* (1983b), pp. 506–1139. Stanford University Press, Stanford.
Grigson, G. 1955. *The Englishman's Flora.* J.M. Dent & Sons, London. [A facsimile edition was published by Phoenix House in 1987.]
Grigson, G. 1974. *A dictionary of English plant names (and some products of plants).* Allen Lane, London.
Guilandinus, M. 1558a†. *De stirpibus aliquot, epistolae V.* Patavii [Padua]. 4°.
Guilandinus, M. 1558b†. *Apologiae adversus Petr. Andream Matthaeolum: liber primus qui inscibitur Theon.* Patavii [Padua]. 4°.
Guilandinus, M. 1608†. *Hortus Patavinus. Cui accessere Melchioris Guilandini medici botanici cluentiss. Conjectanea synonymica plantarum eruditissima ... Publicante Joan. Georg. Schenckio à Grafenberg.* Francofurti [Frankfurt]. 8°. [The text of the title-page is dated 1608 and this is clearly the date of publication but the engraved border is misleadingly dated 1600, hence Bauhin's (1623) date of 1600 followed by Ray in his 'Explanation'.]
Guilandinus, M. 1613†. *Papyrus: Hoc est, commentarius in tria C. Plinii Majoris de papyro capita.* Ambergae [Amberg]. 8°.
Günther, H. 2009. Architectural theory/Vitruvianism, in M. Landfester (ed.), *Brill's New Pauly: Encyclopaedia of the Ancient World. English edition. Classical tradition, Volume 1 (A–Del)*, pp. 255–261. Brill, Leiden.
Gunther, R.T. 1922. *Early English botanists and their gardens.* Privately published, Oxford.
Gunther, R.T., ed. 1925. *The Herbal of Apuleius Barbarus from the early twelfth-century manuscript formerly in the Abbey of Bury St. Edmunds (MS. Bodley 130).* Privately published for the Roxburge Club, Oxford.
Gunther, R.T., ed. 1928. *Further correspondence of John Ray.* Ray Society, London.
Gunther, R.T. 1934. Letters from John Ray to Peter Courthope. *Journal of Botany, British and Foreign* 72: 217–223.
Hankinson, R.J. 2008. Philosophy of nature, in R.J. Hankinson (ed.), *The Cambridge companion to Galen*, pp. 210–241. Cambridge University Press, Cambridge.
Harderwijk, K.J.R. van 1860. *A.J. van der Aa, Biographisch woordenboek der Nederlanden.* Volume 9. J.J. van Brederode, Haarlem.
Harrap, A. & Harrap, S. 2005. *Orchids of Britain and Ireland: a field and site guide.* A. & C. Black, London.
Haslam, J. 1984. The development and topography of Saxon Cambridge. *Proceedings of the Cambridge Antiquarian Society* 72: 13–29.
Hawks, E. 1928. *Pioneers of plant study.* Sheldon Press, London.

BIBLIOGRAPHY

Heath, J. & Maitland Emmet, A. 1979. *The moths and butterflies of Great Britain and Ireland. Volume 9, Sphingidae – Noctuidae Noctuinae and Hadeninae.* Curwen Books, London.

Heer, H. de 1645‡. *Observationes medicae oppido rarae.* Lipsiae [Leipzig]. 12°. [Published with another work, *Spadacrene*; also published with this in Leiden in the same year.]

Helmont, J.B. van 1644‡. *Opuscula medica inaudita. I. De lithiasi. II. De febribus. III. De humoribus Galeni. IV. De peste.* Coloniae Agrippinae [Cologne]. 8°.

Helmont, J.B. van 1655. (as I.B. van Helmont) *Ortus medicinae, id est initia physicae inaudita progressus medicinae novus, in morborum ultionem ad vitam longam*, ed. 4. Lugduni [Lyon]. 2°.

Henning, B.D. 1983. *The history of Parliament. The House of Commons 1660–1690. III. Members M–Y.* Secker & Warburg, London.

Henrey, B. 1975. *British botanical and horticultural literature before 1800. I The sixteenth and seventeenth centuries. History and bibliography.* Oxford University Press, London.

Henrey, B. 1986. *No ordinary gardener: Thomas Knowlton, 1691–1781.* British Museum (Natural History), London.

Highmore, N. 1651‡. *The history of generation.* London. 8°.

Hirsch, A. 1884–1888. *Biographisches Lexikon der hervorragenden Aerzte aller Zeiten und Völker.* 6 vols. Urban & Schwarzenberg, Wien & Leipzig.

Hofmann, C. 1646‡. (as C. Hofmannus) *De medicamentis officinalibus, tam simplicibus quam compositis, libri duo ... Opus triginta annorum.* Parisiis [Paris]. 4°.

Holmes, C. 1974. *The Eastern Association in the English Civil War.* Cambridge University Press, London.

Holyoke, F. 1606. *Riders dictionarie corrected and augmented.* London. 4°.

Hooke, R. 1665. *Micrographia: or some physiological descriptions of minute bodies made by magnifying glasses.* London. 2°.

Hooker, W.J. 1842. *The British Flora. Volume 1, Comprising the phaenogamous or flowering plants and the ferns*, ed. 5. Longman, Brown, Green and Longmans, London.

Hooper, W.D. & Ash, H.B. 1967. *Marcus Porcius Cato: On agriculture. Marcus Terentius Varro: On agriculture.* Loeb Classical Library. William Heinemann, London.

Hornblower, S. & Spawforth, A. 2003. *The Oxford Classical Dictionary*, ed. 3 revised. Oxford University Press, Oxford.

[How, W.] 1650‡. *Phytologia britannica, natales exhibens indigenarum stirpium sponte emergentium.* Londini [London]. 8°.

Howald, E. & Sigerist, H.E., eds 1927. *Corpus medicorum latinorum. Volume IV.* B.G. Teubner, Lipsiae & Berolini [Leipzig & Berlin].

Howlett, C.P. & Moule, P. 1979. Field meetings in 1978. Sunday, 21 May, Elsworth Wood. *Nature in Cambridgeshire* no. 22: 9–10.

Hünemörder, C. 2009. Botany, in M. Landfester (ed.), *Brill's New Pauly: Encyclopaedia of the Ancient World. English edition. Classical tradition, Volume 1 (A–Del)*, pp. 569–574. Brill, Leiden.

Hunger, F.W.T. 1927. *Charles de L'Escluse (Carolus Clusius) Nederlandsch Kruidkundige 1526–1609.* Martinus Nijhoff, 's-Gravenhage.

Hurry, J.B. 1930. *The woad plant and its dye.* Oxford University Press, London.

Huss-Danell, K. 1997. Actinorhizal symbioses and their N_2 fixation. Tansley review no. 93. *New Phytologist* 136: 375–405.

Imperato, F. 1599†. *Dell'historia naturale di Ferrante Imperato Napolitano. Libri XXVIII.* Napoli. 2°.

Innes, H. McLeod, comp. 1941. *Fellows of Trinity College Cambridge.* Cambridge University Press, Cambridge.
Iunius, H. see Junius, A.
Jackson, W.A., Ferguson, F.S. & Pantzer, K.F. 1986. *A short-title catalogue of books printed in England, Scotland, & Ireland and of English books printed abroad, 1475–1640*, ed. 2. 2 vols. Bibliographical Society, London.
Jardine, L. 1975. Humanism and the sixteenth century Cambridge arts course. *History of Education* 4: 16–31.
Jarvis, C. 2007. *Order out of chaos: Linnaean plant names and their types.* Linnean Society of London, London.
Jayne, S. & Johnson, F.R. 1956. *The Lumley library: the catalogue of 1609.* Trustees of the British Museum, London.
Johnson, J. 1981. *Excellent Cassandra: the life and times of the Duchess of Chandos.* Alan Sutton Publishing, Gloucester.
Johnson, T. [1629]. (as T. Ionsonus) *Iter plantarum investigationis ergo susceptum a decem sociis, in Agrum Cantium. Anno Dom. 1629. Julii 13. Ericetum Hamstedianum sive Plantarum ibi crescentium observatio habita, Anno eodem 1. Augusti.* [London]. 4°. [For a facsimile and translation, see Gilmour (1972).]
Johnson, T. 1632. (as T. Iohnson) *Descriptio itineris plantarum investigationis ergo suscepti, in Agrum Cantianum Anno Dom. 1632. Et enumeratio plantarum in Ericeto Hampstediano locisq[ue] vicinis crescentium.* [London]. 12°. [For a facsimile and translation, see Gilmour (1972).]
Johnson, T. 1633‡. *The herball or generall historie of plantes. Gathered by John Gerarde ... Very much enlarged and amended by Thomas Johnson.* London. 2°.
Johnson, T. 1634. *Mercurius botanicus: sive, Plantarum gratia suscepti itineris anno M.DC.XXXIV descriptio.* Londini [London]. 8°.
Johnson, T. 1636‡. *The herball or generall historie of plantes. Gathered by John Gerarde ... Very much enlarged and amended by Thomas Johnson.* London. 2°. [A reprint of the 1633 edition without substantial alteration.]
Johnson, T. 1641. *Mercurii botanici pars altera: sive, Plantarum gratia suscepti itineris in Cambriam sive Walliam descriptio.* Londini [London]. 8°.
Jones, F. 1954. *The holy wells of Wales.* University of Wales Press, Cardiff.
Jones, H. 1981. *Pierre Gassendi 1592~1655: an intellectual biography.* Bibliotheca Humanista & Reformatorica 34. B. de Graaf, Nieuwkoop.
Jones, W.R.D. 1988. *William Turner: Tudor naturalist, physician and divine.* Routledge, London & New York.
Jung, J. 1678. (as J. Jungius) *Isagoge phytoscopica.* Hamburgi [Hamburg]. 4°. [The unpublished manuscript was cited by Ray.]
Junius, A. 1585‡. (as A. Iunius) *The Nomenclator, or Remembrancer of Adrianus Iunius Physician, divided into two tomes ... written by the said Ad. Iu. in Latine, Greeke, French and other forrein tongues: and now in English, by John Higins.* London. 8°.
Junius, A. 1602‡. (as H. Iunius) *Nomenclator, omnium rerum propria nomina septem diversis linguis explicata indicans.* Ursellis [Oberursel]. 8°.
Kangro, H. 1968. *Joachim Jungius' Experimente und Gedanken zur Begründung der Chemie als Wissenschaft.* Franz Steiner Verlag, Wiesbaden.
Kent, D.H. 1975. *The historical flora of Middlesex.* Ray Society, London.
Kent, D.H. 2000. *Flora of Middlesex: a supplement to The historical flora of Middlesex.* Ray Society, London.
Kentmann, J. 1565†. (as I. Kentmanus) *Nomenclaturae rerum fossilium qu[a]e in Misnia praecipue, & in aliis quoque regionibus inveniuntur.* [Published with seven

other works in a volume of C. Gesner's entitled *De omni rerum fossilium genere, gemmis, lapidibus, metallis, et hujusmodi, libri aliquot, plerique nunc primum editi.*] Tiguri [Zurich]. 8°.

Kessler, E., ed. 1994. *Girolamo Cardano: Philosoph, Naturforscher, Arzt.* Wolfenbütteler Abhandlungen zur Renaissanceforschung 15. Harrassowitz Verlag, Wiesbaden

Kew, H.W. & Powell, H.E. 1932. *Thomas Johnson: botanist and royalist.* Longmans, Green and Co., London.

Keynes, G. 1951. *John Ray: a bibliography.* Faber and Faber, London.

Keynes, G., ed. 1964. *The works of Sir Thomas Browne. Volume IV, Letters.* Faber & Faber, London.

Keynes, G. 1976. *John Ray, 1627–1705: a bibliography 1660–1970.* Gérard Th. van Heusden, Amsterdam.

Killy, W. & Vierhaus, R., eds 2001–2006. *Dictionary of German biography.* 10 vols. K.G. Saur, München.

Kircher, A. 1658‡. (as A. Kircherus) *Scrutinium physico-medicum contagiosae luis, quae pestis dicitur.* Romae [Rome]. 4°.

Krantz, A. 1580‡. (as A. Krantzius) *Saxonia. De Saxonicae gentis vetusta origine, longinquis expeditionibus susceptis, & bellis domi pro libertate diu fortiterq[ue] gestis.* Francofurti ad Moenum [Frankfurt]. 2°.

Kusukawa, S. 2000. The *Historia piscium* (1686). *Notes and records of the Royal Society* 54: 179–197.

Kusukawa, S. 2007. Uses of pictures in printed books: the case of Clusius' *Exoticorum libri decem*, in F. Egmond, P. Hoftijzer & R. Visser (eds), *Carolus Clusius. Towards a cultural history of a Renaissance naturalist*, pp. 221–246. Koninklijke Nederlandse Akademie van Wetenschappen, Amsterdam.

Laguna, A. de 1554†. (as A. Lacuna) *Annotationes in Dioscoridem Anazarbeum.* Lugduni [Lyon]. 16°.

Laguna, A. de 1563†. *Pedacio Dioscorides Anazarbeo, Acerca de la materia medicinal: y de los venenos mortiferos. Traduzido de lengua Griega en la vulgar Castellana, & illustrado con claras y substantiales annotationes, y con las figuras de innumeras plantas exquisitas y raras, por Andres de Laguna.* Salamanca. 2°.

Lankester, E., ed. 1846. *Memorials of John Ray.* Ray Society, London.

Lankester, E., ed. 1848. *The correspondence of John Ray.* Ray Society, London.

Lauremberg, P. 1632‡. (as P. Laurembergius) *Apparatus plantarius primus. Tributus in duos libros. I. De plantis bulbosis. II. De plantis tuberosis.* Francofurti ad Moenum [Frankfurt]. 4°.

Lauremberg, P. 1654a‡. (as P. Laurembergius) *Horticultura, libris II. comprehensa; huic nostro coelo & solo accommodata; regulis, observationibus, experimentis, & figuris novis instructa.* Francofurti ad Moenum [Frankfurt]. 4°.

Lauremberg, P. 1654b‡. (as P. Laurembergius) *Apparatus plantarius: Tributus in duos libros. I. De plantis bulbosis. II. De plantis tuberosis.* Francofurti ad Moenum [Frankfurt]. 4°.

L'Écluse, C. de 1557† see Dodoens 1557†.

L'Écluse, C. de 1574† see Monardes 1574†.

L'Écluse, C. de 1576†. (as C. Clusius) *Rariorum aliquot stirpium per Hispanias observatarum historia libris duobus expressa.* Antverpiae [Antwerp]. 8°.

L'Écluse, C. de 1582† see Acosta 1582†.

L'Écluse, C. de 1583†. (as C. Clusius) *Rariorum aliquot stirpium, per Pannoniam, Austrian et vicinas quasdam provincias observatarum historia: quatuor libris expressa.* Antverpiae [Antwerp]. 8°.

BIBLIOGRAPHY

L'Écluse, C. de 1601†. (as C. Clusius) *Rariorum plantarum historia.* Antverpiae [Antwerp]. 2°.

L'Écluse, C. de 1605†. (as C. Clusius) *Exoticorum libri decem: quibus animalium, plantarum, aromatum, aliorumque peregrinorum fructuum historiae describuntur: Item Petri Bellonii observationes, eodem Carolo Clusio interprete.* [Leiden]. 2°.

L'Écluse, C. de 1611†. (as C. Clusius) *Curae posteriores, seu plurimarum non ante cognitarum, aut descriptarum stirpium, peregrinorumque aliquot animalium novae descriptiones.* [Leiden]. 2°.

Leedham-Green, E.S. 1986. *Books in Cambridge inventories: book-lists from Vice-chancellor's Court Probate Inventories in the Tudor and Stuart periods.* 2 vols. Cambridge University Press, Cambridge.

Legré, L. 1897. La botanique en Provence au XVIe siècle; Mathias de Lobel et Pierre Pena. *Bulletin de la Société Botanique de France* 44: xi–xlvii.

Legré, L. 1899. La botanique en Provence au XVIe siècle. I. – Louis Anguillara. *Bulletin de la Société Botanique de France* 46: xxxiii–li.

Legré, L. 1901. *La botanique en Provence au XVIe siècle. Louis Anguillara, Pierre Belon, Charles de L'Escluse, Antoine Constantin.* H. Aubertin & G. Rolle, Marseille.

Legré, L. 1904. *La botanique en Provence au XVIe siècle. Les deux Bauhin, Jean-Henri Cherler et Valerand Dourez.* H. Aubertin & G. Rolle, Marseille.

Lenger, M.-T., ed. 1979. *Bibliotheca Belgica. Bibliographie générale des Pays-Bas.* 6 vols. Editions Culture et Civilisation, Bruxelles.

Lenteren, J.C. van & Godfray, H.C.J. 2005. European science in the Enlightenment and the discovery of the insect parasitoid life cycle in The Netherlands and Great Britain. *Biological control* 32: 12–24.

Leslie, A.[C.] 2007. Vascular plant records. *Nature in Cambridgeshire* no. 49: 89–95.

Leslie, A.C. 2011. An annotated check list of the flora of the Devil's Ditch, Cambridgeshire. *Nature in Cambridgeshire* no. 53: 3–24.

Leslie, A.C, Pannell, C.M. & Walters, S.M. 1982. Varieties of *Viola odorata* L. in Suffolk and Cambridgeshire. *Watsonia* 14: 73–74.

Letts, J.B. 2000. *Smoke blackened thatch.* English Heritage, London & University of Reading, Reading.

Lewis, C.T. & Short, C. 1890. *A Latin dictionary.* Clarendon Press, Oxford.

Liddell, H.G. & Scott, R. 1890. *A Greek-English lexicon*, ed. 7. Clarendon Press, Oxford.

Liddell, H.G., Scott, R. & Jones, H.S. 1940. *A Greek-English lexicon*, ed. 9. Clarendon Press, Oxford. Reprinted with a supplement in 1968.

Lind, H., Franzén, M., Pettersson, B. & Nilsson, L.A. 2007. Metapopulation pollination in the deceptive orchid *Anacamptis pyramidalis. Nordic Journal of Botany* 25: 176–182.

Linden, J.A. van der 1651‡. (as I.A. vander Linden) *De scriptis medicis libri duo*, ed. 2. Amstelredami [Amsterdam]. 8°.

Linnaeus, C. 1753. *Species plantarum.* 2 vols. Holmiae [Stockholm]. 8°. [For a facsimile, see Stearn (1957).]

Linnaeus, C. 1759. *Flora Anglica* [ed. 2], in *Amoenitates academicae.* Volume 4, pp. 88–111. Holmiae [Stockholm]. 8°. [For a facsimile, see Stearn (1973).]

Littger, K.W., Lorenz, G. & Menghini, A., eds 2006. *Basilius Besler Hortus Eystettensis commentarium.* Aboca Museum Edizioni, Sansepolcro.

L'Obel, M. de 1576‡. (as M. de Lobel) *Plantarum seu stirpium historia. Cui adnexum est Adversariorum volumen.* Antverpiae [Antwerp]. 2°. [Two volumes with a composite index, this one often called '*Observationes*' from the title on p. 9 and the running head, '*Stirpium observationes*'; for the second volume, see Pena & L'Obel (1576).]

L'Obel, M. de 1581†. *Kruydtboeck oft Beschrÿvinghe van allerleye ghewassen, kruyderen, hesteren, ende gheboomten.* t'Antwerpen [Antwerp]. 2°.

L'Obel, M. de 1581, 1591 see also Plantin 1581, 1591.

L'Obel, M. de 1655‡. *Stirpium illustrationes. Plurimas elaborantes inauditas plantas, subreptitis Joh: Parkinsoni rapsodiis (ex codice MS insalutato) sparsim gravatae.* Londini [London]. 4°.

Long, K.P. 2006. *Hermaphrodites in Renaissance Europe.* Ashgate Publishing, Aldershot.

Lonicer, A. 1551. (as A. Lonicerus) *Naturalis historiae opus novum.* Francofurti [Frankfurt]. 2°.

Lonicer, A. 1582†. (as A. Lonicerus) *Kreuterbuch: künstliche Conterfeytunge der Bäume, Stauden, Hecken, Kreuter, Getreyde, Gewürtze.* Franckfort am Meyn. 2°.

Louis, A. 1980. *Mathieu de L'Obel 1538–1616: episode de l'histoire de la botanique.* Story-Scientia, Ghent-Louvain.

Lynam, E. 1936. Maps of the Fenland, in W. Page, G. Proby & S.I. Ladds (eds), *The Victoria history of the county of Huntingdon. Volume III*, pp. 291–306. Saint Catherine Press, London.

Lyte 1578, 1619 see Dodoens 1578, 1619.

Maier, M. 1616‡. (as M. Majerus) *Lusus serius.* Oppenheimii [Oppenheim]. 4°.

Maier, M. 1654‡. (as M. Mayerus) *Lusus serius: or, Serious passe-time. A philosophicall discourse concerning the superiority of creatures under man.* London. 12°.

Malim, T. 1997. New evidence on the Cambridgeshire Dykes and Worsted Street Roman Road. *Proceedings of the Cambridge Antiquarian Society* 85: 27–122.

Manning, F. 1926. The works of Walter Charleton, in W. Charleton, *Epicurus's morals: collected, and faithfully Englished. With an introductory essay by Frederic Manning*, pp. xlii–xliii. Peter Davis, London.

Maranta, B. 1559‡. *Methodi cognoscendorum simplicium libri tres.* Venetiis [Venice]. 4°.

Marcellus Virgilius 1529†. (as Marcellus Vergilius) *Pedacii Dioscoridae Anazarbei ... Interprete Marcello Vergilio Secretario Florentino.* Coloniae [Cologne]. 4°.

Marchant, J.R.V. & Charles, J.F. 1955. *Cassell's Latin Dictionary*, revised edition. Cassell and Co., London.

Martini, M. 1623‡. (as M. Martinius) *Lexicon philologicum, praecipue etymologicum.* Bremae [Bremen]. 2°.

Martini, M. 1655‡. (as M. Martinius) *Lexicon philologicum: praecipue etymologicum et sacrum*, ed. 2. Francofurti ad Moenum [Frankfurt]. 2°.

Martyn, J. 1732. *Tournefort's History of plants growing about Paris ... Translated into English, with many additions.* 2 vols. London. 8°.

Martyn, T. 1763. *Plantae Cantabrigienses.* London. 8°.

Matthaeus Sylvaticus 1499†. (as Matthaeus Silvaticus) *Opus pandectarum.* Venetiis [Venice]. 2°.

Matthew, H.C.G. & Harrison, B.H., eds 2004. *Oxford Dictionary of National Biography.* 60 vols & unnumbered index volume. Oxford University Press, Oxford.

Mattioli, P.A. 1558†. (as P.A. Matthiolus) *Commentarii secundo aucti, in libros sex Pedacii Dioscoridis Anazarbei De materia medica. ... His accessit ejusdem Apologia adversus Amathum Lusitanum, quin & censura in ejusdem enarrationes.* Venetiis [Venice]. 2°.

Mattioli, P.A. 1563a†. (as P.A. Matthiolus) *New Kreüterbuch, mit den allerschönsten und artlichsten Figuren aller Gewechss, dergleichen vormals in keiner sprach* *nie an tag kommen. Von dem hochgelerten und weitberümbten Herrn Doctor Petro Andrea Matthiolo ... Erstlich in Latein gestellt. Folgendts durch Georgium Handsch ... verdeutscht und endtlich zu gemeinem nutz und wolfart Deutscher Nation in druck verfertigt.*[1] Prag. 2°.

Mattioli, P.A. 1563b†. (as P.A. Matthiolus) *I discorsi di Pietro And. Matthioli ... ne i sei libri di Pedacio Dioscoride Anazarbeo Della materia medicinale*. Venetia. 2°. [A facsimile was published by Editrice di Travel & Business, Todi, in 1989.]

Mattioli, P.A. 1568†. (as P.A. Matthiolus) *I discorsi di Pietro Andrea Matthioli ... nelli sei libri di Pedacio Dioscoride Anazarbeo della materia medicinale*. Vinegia. 2°.

Mattioli, P.A. 1571†. (as P.A. Matthiolus) *Compendium de plantis omnibus: una cum earum iconibus, de quibus scripsit suis in commentariis in Dioscoridem editis*. Venetiis [Venice]. 4°.

Mattioli, P.A. 1586†. (as P.A. Matthiolus) *De plantis Epitome utilissima, Petri Andreae Matthioli Senensis ... novis plane, et ad vivum expressis iconibus, descriptionesq[ue], longe & pluribus & accuratiorib[us] nunc primum diligenter aucta, & locupletata, a D. Joachimo Camerario*. Francofurti ad Moenum [Frankfurt]. 4°.

Mattioli, P.A. 1590†. (as P.A. Matthiolus) *Kreutterbuch dess ... Herrn D. Petri Andreae Matthioli jetzt widerumb mit viel schönen neuwen Figuren, auch nützlichen Artzeneyen, und andern guten stücken ... zum andern mal ... gemehret, und verfertigt durch Ioachimum Camerarium*. Franckfort am Mayn. 2°.

Mattioli, P.A. 1598‡. (as P.A. Matthiolus) *Opera quae extant omnia: hoc est, Commentarii in VI. libros Pedacii Dioscoridis Anazarbei De medica materia ... a Casparo Bauhino ... post diversarum editionum collationem infinitis locis aucti; ... Apologia in Amatum Lusitanum, cum censura in ejusdem enarrationes. Epistolarum medicinalium libri quinque*. Francofurti [Frankfurt]. 2°.

May, P. 1982. *Newmarket: medieval and tudor*. Privately published, [Newmarket].

May, P. 1984. *The changing face of Newmarket: a history from 1600 to 1760*. Peter May Publications, Newmarket.

McKitterick, D.J. 1978. *The library of Sir Thomas Knyvett of Ashwellthorpe c. 1539–1618*. University Library, Cambridge.

McKitterick, D.[J.] 1992. *A history of Cambridge University Press. Volume 1, Printing and the book trade in Cambridge, 1534–1698*. Cambridge University Press, Cambridge.

McKitterick, D.[J.], ed. 1995. *The making of the Wren Library*. Cambridge University Press, Cambridge.

McMahon, S. 2000. John Ray (1627–1705) and the Act of Uniformity 1662. *Notes and records of the Royal Society* 54: 153–178.

Meikle, R.D. 1984. *Willows and poplars of Great Britain and Ireland*. B.S.B.I. Handbook no. 4. Botanical Society of the British Isles, London.

Melich, G. 1601. (as G. Melichius) *Dispensatorium medicum sive De recta medicamentorum*. Francofurti [Frankfurt]. 12°.

Merrett, C. 1666. *Pinax rerum naturalium Britannicarum*. Londini [London]. 8°.

---

[1] The title can be translated thus: "New Herbal with the most beautiful and artistic figures of all plants, the like of which has never before seen the light of day in any language. Initially published in Latin by the learned and highly esteemed Doctor Petrus Andreas Matthiolus. Subsequently translated into German by George Handsch ... and finally published in print for the use and benefit of the German nation."

Meurs, J. van 1614. (as I. Meursius) *Glossarium Graeco-barbarum*, ed. 2. Lugduni Batavorum [Leiden]. 4°.

Meyer, F.G., Trueblood, E.E. & Heller, J.L. 1999. *The great herbal of Leonhart Fuchs: De historia stirpium commentarii insignes, 1542.* 2 vols. Stanford University Press, Stanford.

Monardes, N. 1574†. *De simplicibus medicamentis ex Occidentali India delatis, quorum in medicina usus est ... interprete Carolo Clusio Atrebate.* Antverpiae [Antwerp]. 8°. [Cited by Ray under Clusius in 'Explanation'.]

Mordue, J.E.M. & Ainsworth, G.C. 1984. *Ustilaginales of the British Isles.* Mycological Papers no. 154. Commonwealth Mycological Institute, Kew.

Morgan, V. 2004. *A history of the University of Cambridge. Volume II, 1546–1750.* Cambridge University Press, Cambridge.

Morison, R. 1680¶. *Plantarum historiae universalis Oxoniensis. Pars secunda.* Oxonii [Oxford]. 2°.

Mottelay, P.F., trans. 1990. On the loadstone and magnetic bodies and on the great magnet the Earth by William Gilbert, in M.J. Adler (ed.), *Great books of the western world. Volume 26, Gilbert, Galileo, Harvey*, ed. 2, pp. 1–121. Encyclopaedia Britannica, Chicago.

Moufet, T. 1634‡. (as T. Moufetus) *Insectorum sive minimorum animalium theatrum: olim ab Edoardo Wottono, Conrado Gesnero, Thomaque Pennio inchoatum.* Londini [London]. 2°. [For an English translation, see Topsell (1658).]

Moufet, T. 1655‡. (as T. Muffett) *Healths improvement: or, Rules comprizing and discovering the nature, method, and manner of preparing all sorts of food used in this nation. Corrected and enlarged by Christopher Bennet.* London. 4°.

Munk, W. 1878. *The roll of the Royal College of Physicians of London*, ed. 2. Volume I, 1518 to 1700. Royal College of Physicians, London.

Murrell, G. & Sell, P. 1990. *Ranunculus calcareus* and *R. pseudofluitans* in Cambridgeshire. *Nature in Cambridgeshire* no. 32: 14–16.

Nannius, P. 1559†. *In P. Virgilii Maronis Bucolica commentaria docta & accurata, nu[n]cque primum in lucem edita.* Basiliae [Basel]. 8°.

Napier, A., ed. 1859. *The theological works of Isaac Barrow, D.D. Volume IX.* Cambridge University Press, Cambridge.

Natale, A. De & Cellinese, N. 2009. Imperato, Cirillo, and a series of unfortunate events: a novel approach to assess the unknown provenance of historical herbarium specimens. *Taxon* 58: 963–970.

Nauert, C.G. 1980. Caius Plinius Secundus, in F.E. Cranz (ed.), *Catalogus translationum et commentariorum, Volume IV,* pp. 297–422. Catholic University of America Press, Washington D.C.

Netzloff, M. 2001. "Counterfeit Egyptians" and imagined borders: Jonson's *The gypsies metamorphosed*. *ELH* 68: 763–793.

Nutton, V. 2004. *Ancient medicine.* Routledge, London & New York.

Nutton, V. 2008. The fortunes of Galen, in R.J. Hankinson (ed.), *The Cambridge companion to Galen*, pp. 355–390. Cambridge University Press, Cambridge.

Oates, J.C.T. 1986. *Cambridge University Library: a history: from the beginnings to the Copyright Act of Queen Anne.* Cambridge University Press, Cambridge.

ODNB see Matthew & Harrison 2004.

OED see Simpson & Weiner 1989.

Ogilby, J. 1675. *Britannia, volume the first: or, An illustration of the kingdom of England and dominion of Wales: by a geographical and historical description of the principal roads thereof.* London. 2°.

OLD see Anon. 1968–1982.
Opie, I. & Opie, P. 1997. *The Oxford dictionary of nursery rhymes*, ed. 2. Oxford University Press, Oxford.
Opsomer, J.-E. 1978. *Histoire des plantes de Rembert Dodoens*. Centre national d'histoire de sciences, Bruxelles. [Facsimile of Dodoens (1557).]
Örstan, A. 2010. John Ray's hermaphrodite snails on their 350th anniversary. *Mollusc World* 23: 3–4.
Orta, Garcia de 1574†. (as Garcia ab Horto) *Aromatum, et simplicium aliquot medicamentorum apud Indos nascentium historia: primum quidem Lusitanica lingua per dialogos conscripta, D. Garcia ab Horto; Proregis Indiae medico, auctore. Nunc vero Latino sermone in epitomen contracta, & iconibus ad vivum expressis, locupletioribusq[ue] annotatiunculis illustrata a Carolo Clusio Atrebate*. Antverpiae [Antwerp]. 8°.
Oswald, P.H. 1992. Henry and other boggarts (or Heinz meanz elvz). *B.S.B.I. News* no. 60: 26–28.
Oswald, P.[H.] 1997. The quartercentenary of Gerarde's *Herball*. *Nature in Cambridgeshire* no. 39: 3–7.
Oswald, P.H. 2000. Historical records of *Lactuca serriola* L. and *Lactuca virosa* L. in Britain, with special reference to Cambridgeshire (v.c. 29). *Watsonia* 23: 149–159.
Oswald, P.H. & Preston, C.D. 1998. A mare's-nest of horsetails: John Ray's treatment of "Equisetum" in his *Cambridge Catalogue* (1660). *Nature in Cambridgeshire* no. 40: 2–18.
Padmore, P.A. 1957. The varieties of *Ranunculus flammula* L. and the status of *R. scoticus* E.S. Marshall and of *R. reptans* L. *Watsonia* 4: 19–27.
Parker, J. & Harley, B., eds 1992. *Martin Lister's English spiders 1678*. Harley Books, Colchester.
Parkinson, J. 1629†. *Paradisi in sole Paradisus terrestris*. London. 2°.
Parkinson, J. 1640‡. *Theatrum botanicum: The theater of plants: Or, an Herball of a large extent*. London. 2°.
Paulli, S. 1653‡. (as S. Paullus) *Viridaria varia regia & academica publica, in usum magnatum ac φιλοβοτανων [philobotanon], collecta ac recognita*. Hafniae [Copenhagen]. 12°. [For details of the component catalogues see Chapter 4.]
Pavord, A. 1999. *The tulip*. Bloomsbury Publishing, London.
Peglar, S.M. 1993. The development of the cultural landscape around Diss Mere, Norfolk, UK, during the past 7000 years. *Review of Palaeobotany and Palynology* 76: 1–47.
Pena, P. & L'Obel, M. de 1571‡. (as P. Pena & M. de Lobel) *Stirpium adversaria nova*. Londini [London]. 2°. [Title-page dated 1570 but colophon 1571.]
Pena, P. & L'Obel, M. de 1576‡. (as P. Pena & M. de Lobel) *Nova stirpium adversaria*. Antverpiae [Antwerp]. 2°. [A reprint of Pena & L'Obel (1571); the second volume of L'Obel (1576), q.v.]
Pena, P. & L'Obel, M. de 1605‡. *Dilucidae simplicium medicamenorum [sic] explicationes, & Stirpium adversaria. Second part of In G. Rondelletii ... methodicam pharmaceuticam officinam animadversiones*. Londini [London]. 2°. [Pages 458–542 comprise de L'Obel's *Adversariorum altera pars*.]
Percival, J. 1921. *The wheat plant: a monograph*. Duckworth & Co., London.
Percival, J. 1948. *Wheat in Great Britain*, ed. 2. Duckworth, London.
Perring, F.H., Sell, P.D., Walters, S.M. & Whitehouse, H.L.K. 1964. *A flora of Cambridgeshire*. Cambridge University Press, London.

BIBLIOGRAPHY

Petch, C.P. & Swann, E.L. 1962. West Norfolk plants today. *Proceedings of the Botanical Society of the British Isles* 4(4), supplement.

Pigott, C.D. 1955. Biological Flora of the British Isles. *Thymus* L. *Journal of Ecology* 43: 365–387.

Piso, W. 1658. (as G. Piso) *De Indiae utriusque re naturali et medica libri quatuordecim*. Amstelaedami [Amsterdam]. 2°.

Plantin, C. 1581†. *Plantarum seu stirpium icones*. Antverpiae [Antwerp]. 4°. [Sometimes attributed to M. de L'Obel.]

Plantin, C. 1591‡. *Icones stirpium, seu plantarum tam exoticarum, quam indigenarum*. Antverpiae [Antwerp]. 8°. [Sometimes attributed to M. de L'Obel.]

Pona, G. 1608†. (as J. Pona) *Plantae seu simplicia ut vocant, quae in Baldo Monte, et in via ab Verona ad Baldum reperiuntur ... Secunda editio. Cui additae sunt nonnullae stirpes insignes ab Honorio Bello Vicetino in Creta observatae*. Basileae [Basel]. 4°.

Pona, G. 1617†. *Monte Baldo descritto da Giovanni Pona Veronese: In cui si figurano et descrivono molte rare piante de gli antichi ... per Francesco Pona dal latino tradotti*. Venetia. 4°.

Porta, G. della 1558†. (as I.B. Porta) *Magiae naturalis: sive, De miraculis rerum naturalium libri IIII*. Neapoli [Naples]. 2°.

Porta, G. della 1589. (as I.B. Porta) *Magiae naturalis libri XX*. Neapoli [Naples]. 2°.

Porta, G. della 1658. (as J.B. Porta) *Natural magick*. London. 4°.

Portus, A., ed. 1619‡. Σουιδας. *Suidas, nunc primum integer Latinitate donatus, & ex collatione multorum manuscriptorum codicum infinitis mendis purgatus, pristinoque suo nitori redditus*. 2 vols. Coloniae Allobrogum [Geneva]. 2°.

Power, H. 1664. *Experimental philosophy: in three books; containing new experiments microscopical, mercurial, magnetical*. London. 4°.

Preston, C.D. 2008. The aquatic plants of the River Cam and its riparian commons, Cambridge, 1660–1999. *Nature in Cambridgeshire* no. 50: 18–37.

Preston, C.D. 2010. The first British records of *Potamogeton compressus* L. and *P. friesii* Rupr. *Watsonia* 28: 82–84.

Preston, C.D., Pearman, D.A. & Dines, T.D., eds 2002. *New Atlas of the British & Irish Flora*. Oxford University Press, Oxford.

Preston, C.D., Pearman, D.A. & Hall, A.R. 2004. Archaeophytes in Britain. *Botanical Journal of the Linnean Society* 145: 257–294.

Primrose, J. 1638‡. (as I. Primirosius) *De vulgi in medicina erroribus libri quatuor*. Londini [London]. 12°. [For an English translation, see Primrose (1651).]

Primrose, J. 1651. *Popular errours. Or the errours of the people in physick. First written in Latine by ... James Primrose ... Translated into English by Robert Wittie*. London. 8°.

Pritzel, G.A. 1872–1877. *Thesaurus literaturae botanicae omnium gentium. Editionem novam reformatam curavit G.A. Pritzel*. Lipsiae [Leipzig]. Reprinted in 1950 by G.G. Gorlich, Milan.

Proctor, M.C.F. 1956. A bryophyte flora of Cambridgeshire. *Transactions of the British Bryological Society* 3: 1–49.

Proctor, M.C.F. & Heywood, V.H. 1968. *Helianthemum* Miller, in T.G. Tutin, V.H. Heywood, N.A. Burges, D.M. Moore, D.H. Valentine, S.M. Walters & D.A. Webb (eds), *Flora Europaea. Volume 2 Rosaceae to Umbelliferae*, pp. 286–291. Cambridge University Press, Cambridge.

Pugh, R.B. ed. 1953. *The Victoria history of the county of Cambridge and the Isle of Ely. Volume Four*. Oxford University Press, Oxford.

Purchas, W.H. & Ley, A., eds [1889]. *A Flora of Herefordshire.* Jakeman and Carver, Hereford.

Quercetanus, I. see Du Chesne, J.

Quincy, J. 1718. *Pharmacopoeia officinalis & extemporanea: or, A compleat English dispensatory, in four parts.* London. 8°.

Rackham, O. 1975. *Hayley Wood: its history and ecology.* Cambridgeshire & Isle of Ely Naturalists' Trust, Cambridge.

Rackham, O. 1990. *Trees and woodland in the British landscape,* revised edition. J.M. Dent & Sons, London.

Rackham, O. 1992. Gamlingay Wood. *Nature in Cambridgeshire* no. 34: 3–15.

Rackham, O. 1994. *The illustrated history of the countryside.* Weidenfeld & Nicolson, London.

Rackham, O. 2003. *Ancient woodland: its history, vegetation and uses in England,* ed. 2. Castlepoint Press, Dalbeattie.

Rackham, O. & Coombe, D.E. 1996. Madingley Wood. *Nature in Cambridgeshire* no. 38: 27–54.

Rademaker, C.S.M. 1968. De Nomenclator van Hadrianus Junius. *Hermeneus* 7: 217–227.

Raven, C.E. 1942. *John Ray naturalist: his life and works.* Cambridge University Press, Cambridge.

Raven, C.E. 1947. *English naturalists from Neckam to Ray.* Cambridge University Press, Cambridge.

Raven, C.E. 1950. *John Ray naturalist: his life and works,* ed. 2. Cambridge University Press, Cambridge. [Reissued in the Cambridge Science Classics Series with an introduction by S.M. Walters in 1986.]

Raven, C.E. 1960. The debt of science to medicine. *British Medical Journal* 1960, 2: 169–171.

Ravensdale, J.R. 1974. *Liable to floods: village landscape on the edge of the fens, AD 450–1850.* Cambridge University Press, London.

[Ray, J.] 1660¶. *Catalogus plantarum circa Cantabrigiam nascentium.* Cantabrigiae [Cambridge]; also Londini [London]. 8°.

[Ray, J.] 1663. *Appendix ad Catalogum plantarum circa Cantabrigiam nascentium: continens addenda et emendanda.* Cantabrigiae [Cambridge]. 8°.

Ray, J. 1670¶. (as J. Raius) *Catalogus plantarum Angliae, et insularum adjacentium.* Londini [London]. 8°.

Ray, J. 1673. *Observations topographical, moral, & physiological: made in a journey through part of the Low-Countries, Germany, Italy, and France; with a catalogue of plants not native of England, found spontaneously growing in those parts, and their virtues.* London. 8°.

Ray, J. 1675. (as J. Raius) *Dictionariolum trilingue.* Londini [London]. 8°. [For a facsimile, see Stearn (1981).]

Ray, J. 1677¶. (as J. Raius) *Catalogus plantarum Angliae, et insularum adjacentium,* ed. 2. Londini [London]. 8°.

Ray, J. 1686–1704. (as J. Raius) *Historia plantarum. Volume 1* (1686); *Volume 2* (1688a); *Volume 3* (1704). Londini [London]. 2°. [Volumes 1 and 2 were reissued in 1693 (see Barr & Pollard 1962).]

Ray, J. 1688b. (as J. Raius) *Fasciculus stirpium Britannicarum, post editum Plantarum Angliae Catalogum observatarum.* Londini [London]. 8°.

Ray, J. 1690. (as J. Raius) *Synopsis methodica stirpium Britannicarum.* Londini [London]. 8°.

Ray, J. 1691. *The wisdom of God manifested in the works of the creation: being the substance of some common places delivered in the chappel of Trinity-College, in Cambridge.* London. 8°.

Ray, J. 1692. *The wisdom of God manifested in the works of the creation,* ed. 2. London. 8°.

Ray, J. 1693. (as J. Raius) *Synopsis methodica animalium quadrupedum et serpentini generis.* Londini [London]. 8°.

Ray, J. 1695. More rare plants growing wild in Cambridgeshire, in E. Gibson, *Camden's Britannia, newly translated into English: with large additions and improvements,* columns 415–420. London. 2°.

Ray, J. 1696. (as J. Raius) *Synopsis methodica stirpium Britannicarum,* ed. 2. Londini [London]. 8°.

Ray, J. 1710. (as J. Raius) *Historia insectorum.* Londini [London]. 4°.

Ray, J. 1724. (as J. Raius) *Synopsis methodica stirpium Britannicarum,* ed. 3 [edited by J.J. Dillenius]. Londini [London]. 8°. [For a facsimile, see Stearn (1973).]

[Ray, J. & Dent, P.] 1685. *Appendix ad Catalogum plantarum circa Cantabrigiam nascentium: continens addenda et emendanda,* ed. 2. Cantabrigiae [Cambridge]. 8°.

RCHM see Royal Commission on Historical Monuments, England.

Reaney, P.H. 1943. *The place-names of Cambridgeshire and the Isle of Ely.* English Place-name Society volume XIX. Cambridge University Press, Cambridge.

Redfern, M. & Askew, R.R. 1998. *Plant galls.* Naturalists' handbook 17, updated reprint. Richmond Publishing Co., Slough.

Redfern, M. & Shirley, P. 2002. *British plant galls: identification of galls on plants and fungi.* FSC Publications, Shrewsbury.

Redi, F. 1668. *Esperienze intorno alla generazione degl'insetti.* Firenze. 4°.

Redi, F. 1671. *Experimenta circa generationem insectorum.* Amstelodami [Amsterdam]. 12°.

Relhan, R. 1786. *Florae Cantabrigiensi supplementum.* Cantabrigiae [Cambridge]. 8°.

Reneaulme, P. de 1611‡. (as P. Renealmus) *Specimen historiae plantarum.* Parisiis [Paris]. 4°.

Reynolds, P. 2003. Kingston Wood. *Nature in Cambridgeshire* no. 45: 2–29.

Rice, E.F. 1980. Paulus Aegineta, in F.E. Cranz (ed.), *Catalogus translationum et commentariorum, Volume IV,* pp. 145–191. Catholic University of America Press, Washington D.C.

Riddle, J.M. 1980. Dioscorides, in F.E. Cranz (ed.), *Catalogus translationum et commentariorum, Volume IV,* pp. 1–143. Catholic University of America Press, Washington D.C.

Riggs, T.J. 1995. Carrot *Daucus carota* (Umbelliferae), in J. Smartt & N.W. Simmonds (eds), *Evolution of crop plants,* ed. 2, pp. 481–485. Longman Scientific & Technical, Harlow.

Roach, J.P.C., ed. 1959. *A history of the county of Cambridge and the Isle of Ely. Volume III. The City and University of Cambridge.* Oxford University Press, London.

Rossi, P. 2000. *The making of modern science,* translated by C. de Nardi Ipsen. Blackwell Publishing, Oxford.

Rösslin, E. 1533†. (as E. Rhodion) *Kreutterbůch von allem Erdtgewächs.* Franckfurt am Meyn. 2°.

Rösslin, E. 1569†. *Kreuterbuch künstliche Conterfeytunge der Bäume, Stauden, Hecken, Kreuter, Getreyde, Gewürtze.* Franckfort am Meyn. 2°.

BIBLIOGRAPHY

Royal Commission on Historical Monuments, England 1959. *An inventory of the historical monuments in the city of Cambridge.* 2 parts. Her Majesty's Stationery Office, London.

Royal Commission on Historical Monuments, England 1972. *An inventory of historical monuments in the county of Cambridge. Volume Two. North-east Cambridgeshire.* Her Majesty's Stationery Office, London.

Ruel, J. 1516†. (as I. Ruellius) *Pedacii Dioscoridis Anazarbei De medicinali materia libri quinque. De virule[n]tis animalibus, et venenis, cane rabioso, et eorum notis, ac remediis libri quattuor. Ioanne Ruellio Suessionensi interprete.* [Paris]. 2°.

Ruel, J. 1537‡. (as I. Ruellius) *De natura stirpium libri tres.* Basileae [Basel]. 2°.

Rydén, M., Helander, H. & Olsson, K., eds 1999. *William Turner* Libellus de re herbaria novus *1538. Edited with a translation into English.* Acta Societas Litterarum Humaniorum Regiae Upsaliensis 50. Uppsala.

Salmon, M.A. 2000. *The Aurelian legacy: British butterflies and their collectors.* Harley Books, Colchester.

Salter, J. 1970. *Madingley Hall: a short history and description of the hall and estate, their owners and occupiers*, ed. 3. University of Cambridge, [Cambridge].

Sandland, R. 1996. The real, the simulacrum, and the construction of the 'gypsy' in law. *Journal of Law and Society* 23: 383–405.

Sayle, C. 1921. The library of Thomas Lorkyn. *Annals of medical history* 3: 310–323.

Scaliger, G.C. 1557‡. (as I.C. Scaligerus) *Exotericarum exercitationum liber quintus decimus, De subtilitate, ad Hieronymum Cardanum.* Lutetiae [Paris]. 4°.

Scaliger, G.C. 1566‡. (as I.C. Scaligerus) *Commentarii, et animadversiones, in sex libros De causis plantarum Theophrasti.* Lugduni [Lyon; also Geneva]. 2°.

Scaliger, J.J. 1581‡. (as J. Scaligerus) Verborum etymologiae perperam a Varrone traditae, hic veris suis originibus redduntur: aut praetermissae explicantur, in *Conjectanea in M. Terentium Varronem De lingua Latina*, pp. 165–176, in *M. Terentii Varronis opera quae supersunt. Editio tertia, recognita & aucta.* [Paris]. 8°.

Schröder, J. 1648‡. (as J. Schröderus) *Quercetanus redivivus, hoc est, ars medica dogmatico-hermetica, ex scriptis Josephi Quercetani, chymiatri celeberrimi, consiliarii ac medici olim regii, tomis tribus digesta.* Francofurti [Frankfurt]. 4°.

Schröder, J. 1649‡. (as I. Schröderus) *Pharmacopëia medico-chymica.* Lugduni [Lyon]. 2°.

Schröder, J. 1669. *The compleat chymical dispensatory ... treating of all sorts of metals, precious stones, and minerals ... written in Latin, by Dr. John Schroder ... and Englished, by William Rowland.* London. 2°.

Schwenckfelt, K. 1600†. (as C. Schwenckfelt) *Stirpium et fossilium Silesiae catalogus.* Lipsiae [Leipzig]. 4°.

Scott, G.A.M. 1988. Studies in ancient bryology, II. Mosses to 1500 A.D. *Journal of Bryology* 15: 1–15.

S[cott], R.F. 1903. Notes from the college records. *The Eagle* 24: 153–176.

Sell, P.[D.] 1991. The cherries and plums of Cambridgeshire. *Nature in Cambridgeshire* no. 33: 29–39.

Sell, P.D. & Murrell, G. 1996. *Flora of Great Britain and Ireland. Volume 5, Butomaceae – Orchidaceae.* Cambridge University Press, Cambridge.

Sell, P.[D.] & Murrell, G. 2006. *Flora of Great Britain and Ireland. Volume 4, Campanulaceae – Asteraceae.* Cambridge University Press, Cambridge.

Sell, P.[D.] & Murrell, G. 2009. *Flora of Great Britain and Ireland. Volume 3, Mimosaceae – Lentibulariaceae.* Cambridge University Press, Cambridge.

Selosse, P. 2008. Traduire la nomenclature botanique néo-latine de la Renaisssance: la linguistique au secours de l'histoire des sciences, in P. Duris (ed.), *Traduire la science hier et aujourd'hui*, pp. 25–43. Maison des Sciences de l'Homme d'Aquitaine, Pessac.

Sennert, D. 1643‡. (as D. Sennertus) *Paralipomena: quibus praemittitur methodus discendi medicinam*. Lugduni [Lyon]. 8°.

Sennert, D. 1650. (as D. Sennertus) *Opera omnia in tres tomos divisa*. 3 vols. Lugduni [Lyon]. 8°.

Serrai, A. 1980. *Conrad Gesner*. Bulzoni Editore, Roma.

Sherrington, C. 1946. *The endeavour of Jean Fernel*. Cambridge University Press, Cambridge.

Simoons, F.J. 1982. Coeliac disease as a geographic problem. *Food, Nutrition and Evolution*, 1982: 179–199.

Simpson, J.A. & Weiner, E.S.C. 1989. *The Oxford English Dictionary*, ed. 2. 20 vols. Clarendon Press, Oxford. [A continuously updated version eventually to be published as the third edition has been consulted on-line.]

Siraisi, N.G. 1990. *Medieval and early Renaissance medicine: an introduction to knowledge and practice*. University of Chicago Press, Chicago & London.

Skelton, R.A. 1970. *County atlases of the British Isles, 1579–1850: a bibliography. 1579–1703*. Carta Press, London.

Skelton, R.A. & Vietor, A.O., eds 1965. *Mirror of the world. First series – Volumes I/III. Braun & Hogenberg: Civitates orbis terrarum 1572–1618*. 3 vols. Theatrum Orbis Terrarum, Amsterdam.

Slaughter, M.M. 1982. *Universal languages and scientific taxonomy in the seventeenth century*. Cambridge University Press, Cambridge.

Sloan, P.R. 1972. John Locke, John Ray, and the problem of the natural system. *Journal of the history of biology* 5: 1–53.

Smith, J.E. 1828. *The English Flora. Volume IV*. Longman, Rees, Orme, Brown, and Green, London.

[Smith, T.] 1659. *The Quaker disarm'd or A true relation of a late publick dispute held at Cambridge by three eminent Quakers, against one scholar of Cambridge*. London. 4°.

Speed, J. 1611. (as I. Speed) *The theatre of the empire of Great Britaine*. London. 2°.

Speed, J. 1627. *England, Wales, Scotland and Ireland described and abridged with ye historic relation of things worthy memory from a farr larger voulume*. London. 16°.

Spiegel, A. van der 1633‡. (as A. Spigelius) *Isagoges in rem herbariam libri duo*. Lugduni Batavorum [Leiden]. 4°.

Stace, C.[A.] 1997. *New flora of the British Isles*, ed. 2. Cambridge University Press, Cambridge.

Stace, C.A. 2010. *New flora of the British Isles*, ed. 3. Cambridge University Press, Cambridge.

Stafleu, F.A. & Cowan, R.S. 1976–1988. *Taxonomic literature: a selective guide to botanical publications and collections with dates, commentaries and types*, ed. 2. 7 vols. Bohn, Scheltema & Holkema, Utrecht.

Stannard, J. 1962. The plant called Moly. *Osiris* 14: 254–307.

Stannard, J. 1969. P.A. Mattioli: sixteenth century commentator on Dioscorides. *University of Kansas Bibliographical Contributions* 1: 59–81.

Stearn, W.T. 1957. *Carl Linnaeus: Species plantarum. A facsimile of the first edition 1753*. 2 vols. Ray Society, London.

Stearn, W.T. 1973. *John Ray: Synopsis methodica stirpium Britannicarum editio tertia 1724. Carl Linnaeus: Flora Anglica 1754 & 1759*. Facsimilies with an introduction by W.T. Stearn. Ray Society, London.

BIBLIOGRAPHY

Stearn, W.T. 1981. *John Ray: Dictionariolum trilingue editio prima 1675. Facsimile with an introduction by William T. Stearn.* Ray Society, London.

Stearn, W.T. 1984. Historical introduction, in W.T. Stearn & P.H. Davis, *Peonies of Greece: a taxonomic and historical survey of the genus* Paeonia *in Greece*, pp. 15–44. Goulandris Natural History Museum, Kifissia.

Stearn, W.T. 1986. The Wilkins Lecture, 1985: John Wilkins, John Ray and Carl Linnaeus. *Notes and records of the Royal Society* 40: 101–123.

Stearn, W.T. 1992. *Botanical Latin*, ed. 4. David & Charles, Newton Abbot.

Stearn, W.T. 1996. The tortuous tale of 'Bear's Breech', the puzzling bookname for *Acanthus mollis*. *Garden History* 24: 122–125.

Stephens, P. & Browne, W. 1658. (as P. Stephanus & G. Brouneus) *Catalogus Horti Botanici Oxoniensis*. Oxonii [Oxford]. 8°.

Stoner, J., trans. [1931]. *The book of my life (*De vita propria liber*) by Jerome Cardan*. J.M. Dent & Sons, London.

Sylbergius, F., ed. 1594‡. Ἐτυμολογικον το μεγα *[Etymologicon to mega]*. [Heidelberg]. 2°.

Tabernaemontanus, I.T. 1588–1591†. *Neuw Kreuterbuch*. 3 parts in 2 volumes (1588, 1591). Franckfurt am Mayn. 2°.

Tabernaemontanus, I.T. 1590†. *Eicones plantarum seu stirpium, arborum nempe, fructicum, herbarum, fructuum, lignorum, radicum, omnis generis*. Francofurti ad Moenum [Frankfurt]. 4°.

Tabernaemontanus, I.T. 1613†. *Neuw vollkommentlich Kreuterbuch ... Jetzt widerumb ... mit sonderm Fleisz gemehret durch Casparum Bauhinum Doctorem*. 3 parts in 2 volumes. Franckfurt am Mayn. 2°.

Tawney, R.H. & Power, E., eds 1951. *Tudor economic documents. Volume Two. Commerce, finance and the Poor Law.* Longmans, Green & Co., London.

Taylor, A. 1999. *Cambridge: the hidden history*. Tempus Publishing, Stroud.

Taylor, D.J. 1996. *Varro – De Lingua Latina X: a new critical text and English translation with prolegomena and commentary*. Studies in the History of Language Series 85. John Betjamins Publishing Company, Amsterdam & Philadelphia.

Tecusan, M. 2004. *The fragments of the Methodists. Volume one: Methodism outside Soranus*. Studies in Ancient Medicine 24(1). Brill, Leiden.

Thal, J. 1588†. (as I. Thalius) *Sylva Hercynia: sive Catalogus plantarum sponte nascentium in montibus et locis plerisque Hercyniae Sylvae*. Francofurti ad Moenum [Frankfurt]. 4°. [Published in the same volume as Camerarius 1588.]

Theodorus, J. see Tabernaemontanus, I.T.

Thirsk, J. 1967. Farming techniques, in J. Thirsk (ed.), *The agrarian history of England and Wales. Volume IV 1500–1640*, pp. 161–199. Cambridge University Press, Cambridge.

Thirsk, J. 1997. *Alternative agriculture: a history from the Black Death to the present day*. Oxford University Press, Oxford.

Thompson, R. 1974. Some newly discovered letters of John Ray. *Journal of the Society for the Bibliography of Natural History* 7: 111–123.

Thomson, G. 2000. *Insectorum sive minimorum animalium theatrum: the butterflies and moths*. Privately published, Lochmaben.

Thorne, G. 1961. *Principles of nematology*. McGraw-Hill Book Company, New York.

Tinniswood, A. 2007. *The Verneys*. Jonathan Cape, London.

Topsell, E. 1658. (as E. Topsel) *The history of four-footed beasts and serpents: ... Collected out of the writings of Conradus Gesner and other authors, by Edward*

BIBLIOGRAPHY

Topsel. Whereunto is now added, The theater of insects; or, Lesser living creatures: as bees, flies, caterpillars, spiders, worms, &c. A most elaborate work: by T. Muffet, Dr. of Physick. The whole revised, corrected, and inlarged ... by J[ohn] R[owland] M.D. London. 2°.

Tragus, H. see Bock, H.

Trevelyan, G.M. 1943. Undergraduate life under the Protectorate. *Cambridge Review* 64: 328–330.

Trist, P.J.O. 1988. Hildersham Furze Hills. *Nature in Cambridgeshire* no. 30: 4–12.

Turnèbe, A. 1564†. (as A. Turnebus) *Adversariorum, tomus primus duodecim libros continens.* Parisiis [Paris]. 4°.

Turner, R. 1664. Βοτανολογια *[Botanologia]. The Brittish physician: or, The nature and vertues of English plants.* London. 8°.

Turner, W. 1538. (as G. Turnerus) *Libellus de re herbaria novus.* Londini [London]. 4°. [For facsimiles, a translation and notes, see Britten, Jackson & Stearn (1965) and Rydén, Helander & Olsson (1999).]

Turner, W. 1548. *The names of herbes.* [London]. 8°. [For a facsimile, a transcription and notes, see Britten, Jackson & Stearn (1965).]

Turner, W. 1551†. *A new herball: wherin are conteyned the names of herbes in Greke, Latin, Englysh, Duch Frenche, and in the potecaries and herberies Latin.* London. 2°. [For a facsimile, see Chapman & Tweddle (1995).]

Turner, W. 1562. *The seconde parte of Vuilliam Turner's herball: wherein are conteyned the names of herbes in Greke, Latin, Duche, Frenche, and in the Apothecaries Latin, and sometyme in Italiane.* Collen [Cologne]. 2°. [For a facsimile, see Chapman, McCombie & Wesencraft (1995).]

Turner, W. 1568. *The thirde parte of Vuilliam Turners Herball, wherein are contained the herbes, trees, rootes and fruytes, whereof is no mention made of Dioscorides, Galene, Plinye, and other olde authores.* Collen. 2°. [For a facsimile, see Chapman, McCombie & Wesencraft (1995).]

Twigg, J. 1990. *The University of Cambridge and the English Revolution 1625–1688.* Boydell Press, Woodbridge.

Twigg, J. 1996. Student sports, and their context, in seventeenth-century Cambridge. *International Journal of the History of Sport* 13: 80–95.

Vánky, K. 1994. *European smut fungi.* Gustav Fischer Verlag, Stuttgart.

Venn, J., comp. 1897. *Biographical history of Gonville and Caius College 1349–1897. Volume I. 1349–1713.* Cambridge University Press, Cambridge.

Venn, J. & Venn, J.A., comp. 1913. *The book of matriculations and degrees: a catalogue of those who have matriculated or have been admitted to any degree in the University of Cambridge from 1544 to 1659.* Cambridge University Press, Cambridge.

Venn, J. & Venn, J.A., comp. 1922–1927. *Alumni Cantabrigienses. Part I. From the earliest times to 1751.* 4 vols. Cambridge University Press, Cambridge.

Verulam see [Bacon], F.

Vigo, G. da 1531. (as J. de Vigo) *Opera in chyrurgia.* [Lyons]. 8°.

Vigo, G. da 1571. (as J. Vigon) *The most excellent workes of chirurgerie ... translated into English.* London. 2°.

Voet, L. 1969–1972. *The golden compasses: a history and evaluation of the printing and publishing activities of the Officina Plantinianaat Antwerp in two volumes.* 2 vols. Vangendt & Co., Amsterdam.

Voet, L. 1980–1983. *The Plantin Press (1555–1589). A bibliography of the works printed and published by Christopher Plantin at Antwerp and Leiden.* 6 vols. Van Hoeve, Amsterdam.

Voet, L. 1993. Christopher Plantin as a promoter of the science of botany, in F. de Nave & D. Imhof (eds), *Botany in the Low Countries (end of the 15th century – ca. 1650). Plantin-Moretus Museum Exhibition*, pp. 39–45. Plantin-Moretus Museum and Stedelijk Prentenkabinet, Antwerp.

Vogt, S. 2008. Drugs and pharmacology, in R.J. Hankinson (ed.), *The Cambridge companion to Galen*, pp. 304–322. Cambridge University Press, Cambridge.

Walker, T.A. 1912. *Admissions to Peterhouse or S. Peter's College in the University of Cambridge: a biographical register.* Cambridge University Press, Cambridge.

Walker, W. 1958. *All the plants of the Bible.* Lutterworth Press, London.

Walters, S.M. 1946. Observations on varieties of *Viola odorata* L. *Report of the Botanical Society and Exchange Club of the British Isles* 12: 834–839.

Walters, S.M. 1954. Recent Cambridgeshire records. *Proceedings of the Botanical Society of the British Isles* 1: 89–90.

Walters, S.M. 1981. *The shaping of Cambridge botany.* Cambridge University Press, Cambridge.

Walters, S.M. 1986. *Introduction to reissue, 1986* in C.E. Raven, *John Ray naturalist: his life and works*, ed. 2, reissue, pp. xxi–xxv. Cambridge University Press, Cambridge.

Walters, S.M. & Stow, E.A. 2001. *Darwin's mentor: John Stevens Henslow 1796–1861.* Cambridge University Press, Cambridge.

Ward, L.K. 2010. Variation in ripening years of seed cones of *Juniperus communis* L. *Watsonia* 28: 11–19.

Wareham, A.F. & Wright, A.P.M., eds 2002. *A history of the county of Cambridge and the Isle of Ely. Volume X. North-eastern Cambridgeshire.* Oxford University Press, Oxford.

Warwick, S.I. & Briggs, D. 1979. The genecology of lawn weeds III. Cultivation experiments with *Achillea millefolium* L., *Bellis perennis* L., *Plantago lanceolata* L., *Plantago major* L. and *Prunella vulgaris* L. collected from lawns and contrasting grassland habitats. *New Phytologist* 83: 509–536.

Watkins, B., ed. 2002. *The book of saints*, ed. 7. A. & C. Black, London.

Webster, C. 2002. *The Great Instauration: science, medicine and reform 1626–1660*, ed. 2. Peter Lang, Oxford.

Webster, C. 2008. *Paracelsus: medicine, magic and mission at the end of time.* Yale University Press, New Haven & London.

Weeks, A. 1997. *Paracelsus: speculative theory and the crisis of the early Reformation.* State University of New York Press, Albany.

Welch, M.A. 1972. Francis Willoughby, F.R.S. (1635–1672). *Journal of the Society for the Bibliography of Natural History* 6: 71–85.

Welch, M.A. 1977. Francis Willughby of Middleton, Warwickshire and Wollaton, Nottinghamshire: a seventeenth century naturalist. *Transactions of the Thoroton Society of Nottinghamshire* 81: 33–40.

Wellcome Historical Medical Library 1962–2006. *A catalogue of printed books in the Wellcome Historical Medical Library.* 5 vols. Wellcome Historical Medical Library, London.

Wellisch, H.H. 1984. *Conrad Gessner: a bio-bibliography.* IDC, Zug.

Westfall, R.S. 1980. *Never at rest: a biography of Isaac Newton.* Cambridge University Press, Cambridge.

Wheatley, H.B., ed. 1867. Manipulus Vocabulorum: *a rhyming dictionary of the English language, by Peter Lewins (1570.).* Early English Text Society. Trübner & Co., London.

BIBLIOGRAPHY

Whewell, W. 1859. Barrow and his academical times as illustrated by his Latin works, in A. Napier (ed.), *The theological works of Isaac Barrow, D.D. Volume IX*, pp. i–lv. Cambridge University Press, Cambridge.

Wille, H. 1993. The botanical works of R. Dodoens, C. Clusius and M. Lobelius, in F. de Nave & D. Imhof (eds), *Botany in the Low Countries (end of the 15th century – ca. 1650). Plantin-Moretus Museum Exhibition*, pp. 33–37. Plantin-Moretus Museum and Stedelijk Prentenkabinet, Antwerp.

Williamson, T. 2000. *The origins of Hertfordshire.* Manchester University Press, Manchester.

Willis, R. & Clark, J.W. 1886. *The architectural history of the University of Cambridge, and of the colleges of Cambridge and Eton.* 4 vols. Cambridge University Press, Cambridge.

Wing, D. 1982–1998. *Short-title catalogue of books printed in England, Scotland, Ireland, Wales, and British America, and of English books printed in other countries, 1641–1700*, ed. 2. 4 vols. Modern Language Association of America, New York.

Wittop Koning, D.A. 1971. *Rembertus Dodonaeus Cruyde-boeck. Facsimile-herdruk van de oorspronkelijke uitgave van 1554.* De Forel N.V., Nieuwendijk.

Wood, A.C., ed. 1958. *The continuation of the history of the Willoughby family by Cassandra Duchess of Chandos.* Shakespeare Head Press, Windsor.

Worthington, J. 1664. The life of the Reverend and most learned Joseph Mede, B.D., in *The works of the pious and profoundly-learned Joseph Mede, B.D.*, pp. i–lxxvi. London. 2°.

Worthington, J. 1704. *Miscellanies ... also a collection of epistles, written to Mr Hartlib of pious memory.* London. 8°.

Wright, A.P.M., ed. 1978. *A history of the county of Cambridge and the Isle of Ely. Volume VI.* Oxford University Press, Oxford.

Wright, A.P.M., ed. 1982. *A history of the county of Cambridge and the Isle of Ely. Volume VIII. Armingford and Thriplow Hundreds.* Oxford University Press, Oxford.

Wright, A.P.M. & Lewis, C.P., eds 1989. *A history of the county of Cambridge and the Isle of Ely. Volume IX. Chesterton, Northstowe, and Papworth Hundreds (North and North-west Cambridgeshire).* Oxford University Press, Oxford.

Yeo, P.F. 1992. *Hardy geraniums*, ed. 2. Timber Press, Portland, Oregon.

Zoller, H. 1965. Konrad Gessner als Botaniker. *Gesnerus* 22: 216–227.

Zoller, H., Steinmann, M. & Schmid, K. 1972–1980. *Conradi Gesneri Historia plantarum. Faksimileausgabe.* 8 vols. Urs Graf-Verlag, Dietikon-Zürich.

INDEX

This index includes the current names of plant and animal taxa mentioned in the text and footnotes. Insignificant references (e.g. cross-references in Ray's text or our footnotes) are excluded. The current scientific names for plants in Ray's main catalogue and the appendices are given *in bold italic type*, followed by their English name and [in square brackets] their entry numbers. Page numbers **in bold type** indicate the location of these numbered species' entries. Simple page numbers are cited for index entries which refer to the main body of the text or to the text and its associated footnotes, but the page numbers are followed by f if the entries refer solely to footnotes. If an entry extends over more than one page, only the initial page number is given.

We have indexed the names of authors quoted by Ray if the quotation consists of more than just the attribution of a name or synonym. Significant references to earlier authors cited in the footnotes are also indexed. The annotation (biog.) indicates the page that includes the most detailed biographical summary of the person concerned. Topics covered by Ray in the text are also indexed, and there are some aggregate entries for topics of potential interest to current readers (e.g. Variation within species). We have not indexed Ray's own plant and animal names or his plant localities (but the latter are listed in the gazetteer).

Only a few synonyms of scientific names are given, cross-referenced to the later names; these synonyms include names accepted by Stace (1997) but not by Stace (2010). Current English names are given after many of the accepted scientific names and cross-referenced to them. These names are included for non-botanical readers but with the warning that there is no exact equivalence between scientific and English names; several English names may apply to one scientific name and the same English name may apply to more than one scientific name. English names are not given for some critical, hybrid or foreign vascular plants and for other groups for which they are not in common use (e.g. bryophytes and parasitic hymenoptera). Vertebrates are indexed under their English names.

Abrotănum, etymology 357
Absinthium, etymology 357
Acalyphe [Ἀκαλυφη], etymology 58, 413
Acanthium, etymology 358
Acanthus mollis (Bear's-breech) 367f
Acaria thomasi 294f
Acer, etymology 58, 358
Acer campestre (Field Maple) [6], **149**, 334
Acer platanoides (Norway Maple) 148f
Acer pseudoplatanus (Sycamore) [5], **148**, 334, 342
Acetōsa, etymology 358
Acetosella, etymology 358
Acheronia atropos (Death's-head Hawk-moth) 296
Achillēa, etymology 358, 363
Achillea ageratum 419f, 433f

Achillea millefolium (Yarrow) [371, 372], **243**, 334, 345
Achillea ptarmica (Sneezewort) [473], **270**, 337, 341
Achilles 160f, 179f, 201, 358, 363, 370, 409
Acĭnos, etymology 358
Aconitum, etymology 358
Aconitum napellus (Monk's-hood) 476
Acorus, etymology 359
Acorus calamus (Sweet Flag) 149f
Acosta, Cristóbal, works described in 'Explanation' 137; works in libraries 111
Act of Uniformity 6f, 16–18, 66
Actinorhizal symbioses 361f
Adder's Tongue *see Ophioglossum vulgatum*

INDEX

Adiantum, etymology 359
Adiantum capillus-veneris (Maidenhair Fern) 359f
Adonia 415f
Adonis 361f, 415f
Adoxa moschatellina (Moschatel) [477], **271**, 327, 347
Aegilops, etymology 359
Aegineta, Paulus (Paul of Aegina) 70 (biog.), 85, 135, 140; names cited by Ray 70; *see also* Paulus
Aegopodium podagraria (Ground Elder) [46], **159**, 323, 330, 383f
Aelianus, Claudius 238 (biog.); quoted in 'Etymology' 384
Aeneas 314
Aera [Ἀιρα], etymology 360
Aesculapius 179f, 363
Aethusa cynapium (Fool's Parsley) [133], **182**, 332, 336
Aëtius of Amida 35, 70 (biog.), 85, 135; names cited by Ray 70; quoted in 'Etymology' 404
Affinities of species discussed, *Brassica* 276; *Galium album* 452; *Montia fontana* 461; *Myosotis scorpioides* 245; *Persicaria amphibia* 266; *Petroselinum segetum* 448; *Potamogeton pectinatus* 243; *Scrophularia auriculata* 167; *Zannichellia palustris* 268; *see also* Classification of plants; Differences from closely related taxa outlined
Agamemnon 160f
Agerius, Nicolas 144
Aglais urticae (Small Tortoiseshell) 318
Agnos [Ἀγνος], etymology 374
Agrifolium, etymology 360
Agrimonia, etymology 360
Agrimonia eupatoria (Agrimony) [16], **151**, 197f, 323, 363
Agrimonia sylvestris, etymology 363
Agrimony *see Agrimonia eupatoria*
Agrimony, Hemp *see Eupatorium cannabinum*
Agrostemma githago (Corncockle) [472], **269**, 326
Agrostis [Ἀγρωστις], etymology 382
Aira caryophyllea (Silver Hair-grass) [258], **214**

Ajax, King of Salamis 384f, 409f
Ajuga chamaepitys (Ground-pine) [127], **180**, 331, 353
Ajuga reptans (Bugle) [95], **172**, 290, 325, 350
Albert, Saint 363
Alchemilla filicaulis subsp. **vestita** (Hairy Lady's-mantle) [B3], 115, **461**
Alchemists 288, 316, 360
Alchimilla, etymology 360
Alder *see Alnus glutinosa*
Aldrovandi, Ulisse 79 (biog.), 85, 186f, 318; source of observations 48; works in libraries 99, 100, 109
Alectorolŏphus, etymology 360
Algae, filamentous [140], **183**, 339
Alica 416
Alisma plantago-aquatica (Water-plantain) [445], **262**, 337
Alleluia, etymology 360
Allestree, James 24
Alliaria, etymology 360
Alliaria petiolata (Garlic Mustard) [17], **152**, 332, 340
Allium, etymology 360
Allium moly 152f
Allium sativum (Garlic) 152, 153, 360f, 417f
Allium ursinum (Ramsons) [18], **152**, 338, 347
Allium vineale (Crow Garlic) [19], **152**, 330, 451, 481
Alnus, etymology 361
Alnus glutinosa (Alder) [20], 44, **153**, 323
Alopecurus geniculatus (Marsh Foxtail) [237], **210**
Alopecurus myosuroides (Black-grass) [235], **210**, 330f
Alopecurus pratensis (Meadow Foxtail) [234], **210**, 330f
Alpago, Luigi 80
Alsine, etymology 361
Althaea cannabina 156f
Althaea officinalis (Marsh Mallow) [33], **156**, 334
Alyssum, etymology 361
Amaranthus albus (White Pigweed) 170f
Amaranthus blitum (Guernsey Pigweed) 170f

561

INDEX

Amatus Lusitanus 7, 80 (biog.), 92; names cited by Ray 36; works described in 'Explanation' 135, 138f, 142; works in libraries 99

Amellus, etymology 361; of Virgil 156, 173, 181

Ammophila arenaria (Marram Grass) 205f

Amphisbaena 192

Anacampsĕros, etymology 361

Anacamptis morio (Green-winged Orchid) [399], **249**, 347, 446

Anacamptis pyramidalis (Pyramidal Orchid) [406], 217f, **251**, 336, 347, 349, 352, 354

Anacyclus pyrethrum (Spanish Chamomile) 410f

Anagallis, etymology 361

Anagallis arvensis (Scarlet Pimpernel) [41], **158**, 337; subsp. ***arvensis*** (Scarlet Pimpernel) [42], **158**, 337; subsp. *arvensis* forma *azurea* 158f; subsp. *foemina* (Blue Pimpernel) 158f

Anagallis tenella (Bog Pimpernel) [389], 44, **247**, 334, 350, 353

Anaphalis margaritacea (Pearly Everlasting) 171f

Anchūsa, etymology 361

Anchusa arvensis (Bugloss) [93], **172**, 325

Androsaemum, etymology 361

Anemōne, etymology 361

Anemone coronaria 361f

Anemone nemorosa (Wood Anemone) [44], **158**, 323, 327, 330f, 350, 351, 361f

Anemone, Wood *see Anemone nemorosa*

Angelica, etymology 361

Angelica sylvestris (Wild Angelica) [45], **159**, 323

Angelica, Wild *see Angelica sylvestris*

Anguillara, Luigi 80 (biog.), 88, 92, 383f; names cited by Ray 36; works described in 'Explanation' 135; works in libraries 111

Anisantha sterilis (Barren Brome) [196], **198**, 336

Annual plants, definition 434; failure to regenerate 132, 156f; soil preferences 164, 257f

Anōnis, etymology 361

Anserina, etymology 362

Antennaria dioica (Mountain Everlasting) [233], 16, 66, **209**, 326, 327, 352

Anthemis arvensis (Corn Chamomile) [156], **188**, 325

Anthemis cotula (Stinking Chamomile) [155], **188**, 334

Anthony, Saint 363

Anthriscus caucalis (Bur Chervil) [384], **246**, 326

Anthriscus sylvestris (Cow Parsley) [135], **182**, 326

Anthyllis vulneraria (Kidney Vetch) [49], 115, **159**, 333, 344, 347, 348, 354

Antiphrasis, definition 357f; names coined by 357, 365, 373, 376, 382, 383, 391

Antirrhinum, etymology 362

Antirrhinum majus (Snapdragon) 289

Antonomasia, definition 417f; names coined by 417

Antony, Mark 138

Aparine, etymology 362

Aphanes arvensis *sensu lato* (Parsley-piert) [430], 114, 116, **258**, 336, 349

Aphanes arvensis sensu stricto 258f

Aphanes australis 258f

Apicii 303

Apicius, Marcus Gavius 303f (biog.), 377

Apium, etymology 362

Apium graveolens (Wild Celery) [53], **160**, 272f, 340, 362, 429f, 448

Apium nodiflorum (Fool's Watercress) [545], 272f, **296**, 337

Apollo 374, 384

Apuleius Platonicus ('Pseudo-Apuleius') 70 (biog.), 76, 168f; names cited by Ray 71; quoted in 'Etymology' 366; works described in 'Explanation' 135

Apuleius Platonicus, Lucius 135

Aquatic plants, life-spans 284; Ray's pioneer work 43

Aquilegia, etymology 161f, 362

Aquilegia vulgaris (Columbine) [54], 44, **160**, 326, 352

Arabidopsis thaliana (Thale Cress) [A30, B76], 66, 96, **446**, **474**

Arabis glabra see Turritis glabra

562

INDEX

Archangel, Yellow *see Lamiastrum galeobdolon*
Arctium lappa and/or ***A. minus*** (Greater/Lesser Burdock) [79], **166**, 325, 397f
Arctium lappa subsp. *pubens see Arctium pubens*
Arctium minus (Lesser Burdock) [B13], **462**; *see also A. lappa* and/or *A. minus*
Arctium nemorosum (Wood Burdock) 166f
Arctium pubens 166f, 463f
Arenaria serpyllifolia *sensu lato* (Thyme-leaved Sandwort) [30], **155**, 326
Aretaeus of Cappodocia 308f
Argemōne, etymology 362
Argentīna, etymology 362, 399
Aristis, etymology 383
Aristolochia clematitis (Birthwort) [B10], **462**
Aristotle, 9, 50, 71 (biog.), 73, 77, 78, 84, 131f, 143, 151f, 377, 396f, 398f; names cited by Ray 71; quoted: in 'Preface' 130, in main catalogue 180, 198, 281, 297, 312, 313f, 315, in 'Etymology' 368; source of observations 48
Aristoxenus of Tarentum, quoted in 'Etymology' 377
Armerii flores, etymology 363
Armoracia rusticana (Horse-radish) [489], 131, **274**, 338
Arnald of Villanova 71 (biog.), 96; names cited by Ray 35, 71
Arnoglosson [Ἀρνογλωσσον], etymology 398
Arnoseris minima (Lamb's Succory) [280], **219**, 331, 348
Arnulf, Emperor 153
Arrowgrass, Marsh *see Triglochin palustris*
Arrowhead *see Sagittaria sagittifolia*
Artemis *see* Diana
Artemisia, etymology 363
Artemisia II, Queen of Caria 363
Artemisia absinthium (Wormwood) [3], **147**, 345, 354
Artemisia campestris (Field Wormwood) [1, 2], 44, 65, **147**, 341, 352, 451, 481

Artemisia vulgaris (Mugwort) [62], 45, **162**, 335, 354
Arum maculatum (Lords-and-Ladies) [63], **163**, 327, 344
Asclepiades 71 (biog.), 78, 168
Ascyrum, etymology 364
Ash *see Fraxinus*
Asparagus, Garden *see Asparagus officinalis*
Asparagus officinalis (Garden Asparagus) 171, 195f, 236, 448
Aspen *see Populus tremula*
Asperugo procumbens (Madwort) [34], **156**, 325, 330, 333, 351
Asperula, etymology 364
Asperula cynanchica (Squinancywort) [523, A36, B86], 44, **290**, 291f, 324, 337, 348, 352, 354, **448**, 453, **476**, 486
Asphodel, Bog *see Narthecium ossifragum*
Asplenium adiantum-nigrum (Black Spleenwort) [B1], 200f, **461**
Asplenium ceterach (Rustyback) 200
Asplenium ruta-muraria (Wall-rue) [11], **150**, 200f, 339, 342, 349, 359f
Asplenium scolopendrium (Hart's-tongue) [A29, B75], 51, 200f, 331, 349, **446**, **474**
Asplenium trichomanes (Maidenhair Spleenwort) [577], 200f, **303**, 330f, 334
Aster amellus 156f
Aster tripolium (Sea Aster) [594], **306**, 341
Aster, Sea *see Aster tripolium*
Astragalus, etymology 55, 364
Astragalus danicus (Purple Milk-vetch) [227], 44, **209**, 342, 348, 350, 352, 381f
Astragalus glycyphyllos (Wild Liquorice) [228], **209**, 333, 349, 354, 381f
Athenaeus 71 (biog.); quoted: in main catalogue 230, in 'Etymology' 401, 402, 413
Athyrium filix-femina (Lady-fern) [B33], **466**
Atriplex, etymology 364; identification problems 67, 481
Atriplex patula (Common Orache) [68], **164**, 336

INDEX

Atriplex prostrata (Spear-leaved Orache) [71], **165**, 336, 481
Atropa belladonna (Deadly Nightshade) [548], **297**, 328, 335, 365f
Augustus, Emperor 73, 75, 76, 79, 154f, 168
Aurelianus, Caelius 58, 71 (biog.); quoted in 'Interpretation' 418
Auricularia auricula-judae (Jew's Ear) 288, 332
Avellana nux, etymology 364
Avena fatua (Wild Oat) [14], **150**, 331, 336
Avena sativa (Oat) [74, 75], **165**, 319, 335; germination 309; *see also* Cereals
Avena strigosa (Bristle Oat) 166f
Averroës (Ibn Rushd), quoted in 'Etymology' 365
Avicenna 50, 71 (biog.), 151f; names cited by Ray 72; quoted in 'Etymology' 409
Awns 165, 308

Babington, Charles Cardale, identification of Ray's plants: 113, 115, 116, *Anthemis* spp. 188f, *Arctium* 166f, *Asperula cynanchica* 290f, *Brassica* 246f, *Bromus hordeaceus* 199f, *Carduus tenuiflorus* 442f, *Carex* 211f, 212f, *Carex lepidocarpa* 215f, *Chenopodium polyspermum* 170f, *Crataegus* 253f, *Fumaria* 467f, *Geranium* 208f, 482f, *Hieracium umbellatum* 469f, *Juncus* 213f, *Mentha* 472f, *Mentha spicata* 472f, orchid 468f, *Orobanche elatior* 252f, *Osmunda regalis* 466f, *Potamogeton compressus* 267f, *Rubus fruticosus* agg. 282f, *Rumex* 229f, 484f, *Rumex conglomeratus* 230f, *Sagina procumbens* 291f, *Salix* 286f, *Salix alba* 283f, *Sedum rupestre* 292f, *Stellaria media* 155f, *Teucrium* 465f, *Viola arvensis* 314f
Bacon, Sir Francis, Baron Verulam, Viscount St Alban 18; quoted in main catalogue 165 (biog.), 207, 255, 259, 271, 281, 302, 313, 314; source of observations 48; works in libraries 99, 109

Bad Henry, etymology 366
Baianae, etymology 377
Baldellia ranunculoides (Lesser Waterplantain) [446], **262**, 337
Ballota nigra (Black Horehound) [77], **166**, 332, 349, 399f
Baptisecŭla, etymology 365
Barbara, Saint 364, 365
Barbarea, etymology 365
Barbarea vulgaris (Winter-cress) [78], **166**, 327, 339f
Barbaro, Ermolao 74, 80 (biog.); names cited by Ray 36; quoted in 'Etymology' 365, 373, 378; works described in 'Explanation' 140; works in libraries 99
Barberry *see Berberis vulgaris*
Bardana, etymology 365
Bark, inner discolours water 202, 285; taste, 374; uses: dyes 206, instead of paper 422, to induce vomiting 225; *see also* Trees; Wood
Barley *see Hordeum*
Barnard, Emma *see* Willughby, Emma
Barnard, Sir Henry 20
Barrow, Isaac 4, 6, 8, 9, 10, 12, 16, 18; oration to Trinity College 12; personal library 103, 109, 110
Bartsia, Red *see Odontites vernus*
Basilicum, etymology 394
Bassus, Cassianus 72 (biog.); names cited by Ray 72; quoted: in main catalogue 184, 185, 295, in 'Etymology' 387; source of observations 48
Bauhin, Caspar 1, 38, 58, 80 (biog.), 81, 88, 93, 95, 114, 139; Basel catalogue as model for *Catalogus* 26, 27, 33, 45; errors 452, 464; identification of plants 178, 199, 214, 220, 260, 485; names cited by Ray 35, 36, 40, 70–72, 87, 129, 130; pictures discussed 482; *Pinax* a source for 'Explanation' 35, 38, 97, 135, 139; quoted: in main catalogue 159, 162, 182, 200, 217, 220, 222, 226, 234, 235, 280, 307, 312, in 'Etymology' 361, 362, 376, 392, 399, 404, 407, 409, 411, in 'Interpretation' 416, in 1685 appendix 470, 482; source of observations 48; study of hermaphroditism 297f; works

564

INDEX

described in 'Explanation' 136, 142, 143; works in libraries 40f, 103–107, 109–111

Bauhin, Jean 1, 46, 58, 80, 81 (biog.), 85, 86, 197f, 363f; quoted in main catalogue 199, 233; Ray's tribute to 39, 141; source of observations 48; works described in 'Explanation' 39, 140; works in libraries 103, 110, 111; *see also* Bauhin, Jean & Cherler, Jean-Henri

Bauhin, Jean & Cherler, Jean-Henri, *Historia plantarum* (1650–1651) 39, 45, 47, 80, 97, 99, 188f, 444f; annotated copy at Trinity College 101; copies in libraries 99, 101, 103, 104, 107, 109, 111; identification of plants 176, 178, 179, 183, 196f, 199f, 211, 212, 213, 227, 234, 251, 260, 261, 298, 304–306, 446, 452, 464, 469f, 473; mentioned by Corbyn 16; names cited by Ray 36, 40, 129, 130; pictures discussed 156, 206, 210, 211, 213, 283, 452, 462, 482, 485; printing errors 16, 26f; quoted: in main catalogue 153, 159, 162, 167f, 172, 174, 179, 181, 182, 189, 193–195, 199, 200, 204, 206, 208, 209, 217, 220, 222, 231, 235, 240, 266, 270, 271, 280, 286, 287, 295, 299, 303, 312, 315–317, 319, in 'Etymology' 357f, 358, 360, 361, 363–368, 370, 371f, 373, 376, 377, 379, 381–383, 387–394, 398f, 399–403, 406, 408f, 410, in 'Interpretation' 419, 429, in 'More customary headings' 433, in appendices 442, 447, 448, 451, 452, 462, 476, 477, 481–483; Ray's assessment of 39, 141; source of Ray's classification 60; *see also* Bauhin, Jean

Bauhin, Jean Caspar 136

Bay-tree *see Laurus nobilis*

Beak-sedge, White *see Rhynchospora alba*

Beale, John 26

Bean, etymology 377

Beans, germination 308; Pythagorean views on 377; structure 198; *see also Vicia faba*

Bear's-breech *see Acanthus mollis*

Becabunga, beccabunga, etymology 157f, 365

Bechion [Βηχιον], etymology 411

Bedeguar 280; etymology 365

Bedstraw *see Galium*

Beer, brewing 199, 299; improvement of 147

Bees 154, 188, 191, 271, 288, 312

Beggars 273

Behen, etymology 365

Bella donna, etymology 297f, 365

Bellflower *see Campanula*

Belli, Onorio 35, 79, 81 (biog.); works described in 'Explanation' 135, 143

Bellis, etymology 365

Bellis perennis (Daisy) [82], **167**, 328

Belon, Pierre 35, 81 (biog.); works described in 'Explanation' 135; works in libraries 111

Bennett, Christopher 93; quoted in main catalogue 152 (biog.); source of observations 48; works in libraries 106, 109

Bents 215

Berberis, etymology 365

Berberis vulgaris (Barberry) [413], 65, 206, **253**, 324, 347

Berula, etymology 366

Berula erecta (Lesser Water-parsnip) [544], **296**, 337

Besler, Basileus 38f, 82 (biog.); names cited by Ray 36; works described in 'Explanation' 139; works in libraries 99f, 111

Betonica, etymology 366

Betonica officinalis (Betony) [84], **168**, 324, 350, 351, 412f

Betony *see Betonica officinalis*

Betula, etymology 366

Betula pendula (Silver Birch) 169f

Betula pubescens (Downy Birch) [85], **169**, 324

Betulla, etymology 366

Bidens cernua (Nodding Bur-marigold) [190], **197**, 323

Bidens tripartita (Trifid Bur-marigold) [189], **197**, 323

Bindweed, Black *see Fallopia convolvulus*

Bindweed, Field *see Convolvulus arvensis*

565

INDEX

Bindweed, Hedge *see Calystegia sepium*
Birch *see Betula*
Birdlime 151
Birds, cage-birds 175; dissection 12, 182; Great Bustard 182; partridges 308; predators of grain 222; predators of insects 255; swallows 180, 371; woodpeckers 270; *see also* Hens
Birdsfoot *see Ornithopus perpusillus*
Birdsfoot-trefoil *see Lotus*
Birthwort *see Aristolochia clematitis*
Biscutella 289
Bistort *see Persicaria*
Bitter plants 147, 148, 194, 209, 231, 270, 297, 357, 382; *see also* Taste of plants
Bittersweet *see Solanum dulcamara*
Black-grass *see Alopecurus myosuroides*
Blackstonia perfoliata (Yellow-wort) [122], 44, **179**, 326, 350, 354
Blackthorn *see Prunus spinosa*
Bladdernut *see Staphylea pinnata*
Bladderwort *see Utricularia*
Blechnum spicant (Hard-fern) [A23, B59], 66, **445**, **472**
Blechon [Βληχων], etymology 270, 400
Blinks *see Montia fontana*
Blisters 152
Blitum, etymology 366
Blockwitz, Martin 169; works in libraries 107
Blood, staunching flow 183, 205, 381, 398, 406
Bluebell *see Hyacinthoides non-scripta*
Boate, Gerard 26f
Bock, Jerome 82 (biog.), 83, 95, 139, 396f, 487f; names cited by Ray 36; plants described by 226f; quoted, often as *Trag.*: in main catalogue 165, 169, 200, 204, 206, 217, 259, in 'Etymology' 363, 380f, 395, 403, 407; source of observations 48; works described in 'Explanation' 138, 143; works in libraries 109
Bodaeus, Johannes 72, 74, 78, 82 (biog.), 151f, 203f, 309f, 366f, 396f; names cited by Ray 36; quoted: in main catalogue 161, 162, 175, 181, 184, 185, 199f, 208, 236, 238, 241, 261, 265, 292, 295, 297, 298, 310, in 'Etymology' 391, 393, 398, 406, 408; source of observations 48; works described in 'Explanation' 141; works in libraries 110
Bogbean *see Menyanthes trifoliata*
Bog-rush, Black *see Schoenus nigricans*
Böhme, Jakob 50
Bolboschoenus maritimus (Sea Club-rush) [246], **212**
Bombyx mori (Silkworm) 92, 187, 232, 279
Bonus Henricus, etymology 171f, 366
Boona, etymology 377
Borage *see Borago officinalis*
Borago, etymology 366
Borago officinalis (Borage) 245; cordial flower 367
Borel, Pierre 148f; works in libraries 109
Bos primigenius see Cows
Botrychium lunaria (Moonwort) [B60], **472**
Box *see Buxus sempervirens*
Boyle, Hon. Robert 14, 59f
Brabantic, spoken by Adam and Eve 58, 400f
Brachythecium rutabulum 244f, 245f
Bracken *see Pteridium aquilinum*
Bracket-fungi *see* Fungi, bracket
Braintree Grammar School 5
Bramble *see Rubus fruticosus* agg.
Branca ursina, etymology 367
Branches (boughs), definition 425; morphology 196; terminology: *comae* 418, *sarmentum* 425, *spadix* 425, *surculus* 427, *termes* 427, *thyrsus* 427, *turiones* 428, *virgultum* 428; tree rings 202–203
Brasavola, Antonio Musa 82 (biog.), 86; names cited by Ray 36; quoted in main catalogue 311; source of observations 48; works described in 'Explanation' 136; works in libraries 99
Brassica napus (Oil-seed Rape) [385, 495], **246**, **275**, 326, 335, 338, 343, 486
Brassica nigra (Black Mustard) [541], **295**, 335f
Brassica oleracea (Cabbage) 168, 246f, 275, 276

INDEX

Brassica rapa (Turnip, Wild Turnip) [385, 493, 494], 168, **246**, **275**, 326, 335, 343, 486f; subsp. *campestris* (Wild Turnip) 246f

Braun, Georg & Hogenberg, Franz, quoted in main catalogue 205; source of observations 48; works in libraries 101

Braunschweig, Hieronymus 82 (biog.), 181f; names cited by Ray 35, 36, 83

Bread, contamination 391; in ancient world 411; making 306f, 307, 420; reviving stale loaves 308

Britannica, etymology 367

Briza media (Quaking Grass) [266], **216**, 330, 354, 397f

Brome, Barren *see Anisantha sterilis*

Brome, Rye *see Bromus secalinus*

Brome, Soft *see Bromus hordeaceus*

Bromus hordeaceus (Soft Brome) [197], **198**, 328, 336, 391f

Bromus secalinus (Rye Brome) 199f

Brooklime *see Veronica beccabunga*

Brookweed *see Samolus valerandi*

Broom *see Cytisus scoparius*

Broomrape *see Orobanche*

Browne, Sir Thomas 45, 77; quoted in main catalogue 309 (biog.); source of observations 48; works in libraries 109

Browne, William 34

Brunfels, Otto 82, 83 (biog.), 87, 95; names cited by Ray 36; works described in 'Explanation' 136; works in libraries 111

Brydges, James, Duke of Chandos 22

Bryonia, etymology 367

Bryonia dioica (White Bryony) [91], **171**, 325, 347, 354

Bryony, Black *see Tamus communis*

Bryony, White *see Bryonia dioica*

Buck, Thomas 100

Buckthorn *see Rhamnus cathartica*

Buckthorn, Alder *see Frangula alnus*

Buckwheat *see Fagopyrum esculentum*

Buds, definitions 421, 423; eaten by bees 154; *Populus* 266; terminology: *alabastri* 416

Bugle *see Ajuga reptans*

Bugloss *see Anchusa arvensis*

Bugloss, Viper's *see Echium vulgare*

Buglossum, etymology 367

Bugula, etymology 367

Bulbocastanum, etymology 368

Bulbous plants, definition 417; growth from seed 223

Bullace *see Prunus domestica* subsp. *insititia*

Bulrush *see Typha latifolia*

Bunias, etymology 368

Bunt *see Tilletia tritici*

Bupleurum rotundifolium (Thorow-wax) [431], **258**, 342

Bupleurum tenuissimum (Slender Hare's-ear) [A6, B12], **442**, **462**

Burdock *see Arctium*

Burgoyne, Sir Roger 42, 164, 172, 206, 213, 223, 258, 280f, 298, 494 (biog.)

Bur-marigold *see Bidens*

Burnet, Great *see Sanguisorba officinalis*

Burnet, Salad *see Poterium sanguisorba* subsp. *sanguisorba*

Burnet-saxifrage *see Pimpinella*

Bur-parsley, Greater *see Turgenia latifolia*

Bur-parsley, Small *see Caucalis platycarpos*

Bur-reed *see Sparganium*

Burrell, Timothy 12, 22

Bustard, Great (*Otis tarda*) 182

Butcher's Broom *see Ruscus aculeatus*

Butler, Charles 46, 63; quoted in main catalogue 154, 188, 191, 271, 313; source of observations 48; works in libraries 99, 107

Butŏmos, etymology 368

Butomus umbellatus (Flowering Rush) [225], 44, 115, **208**, 330, 339, 352

Butter 152, 171, 303

Butterbur *see Petasites hybridus*

Buttercup *see Ranunculus*

Butterflies 261, 277, 288, 318; *see also* Larvae; Pupae

Butterwort, Common *see Pinguicula vulgaris*

Buxus sempervirens (Box) 284

Cabbage *see Brassica oleracea*

Caernarvonshire, Gods Well 445

Caesar, Gaius Julius 72 (biog.), 73, 75, 79, 154f, 274; quoted in main

INDEX

catalogue 209, 301; source of observations 48
Cage-birds *see* Birds
Calamint *see* Clinopodium
Calamintha, etymology 368
Calcitrapa, etymology 368
Calendula officinalis (Pot Marigold) 302
Calepino, Ambrogio 59, 72 (biog.); quoted in 'Interpretation' 417, 427; works in libraries 101, 107
Calliteara pudibunda (Pale Tussock) 187f
Callitriche obtusangula (Blunt-fruited Water-starwort) 300f
Callitriche platycarpa (Various-leaved Water-starwort) 300f
Callitriche spp. (Water-starworts) [562], 268, **300**, 326, 329, 334, 341
Callitriche stagnalis (Common Water-starwort) 300f
Calluna vulgaris (Heather) [184], 43, **196**, 331, 333, 348, 352
Caltha palustris (Marsh Marigold) [101], 18, **173**, 334
Calvatia gigantea (Giant Puff-ball) 204f
Calystegia sepium (Hedge Bindweed) [148], **185**, 324, 389f
Cambridge booksellers 24, 106, 459
Cambridge University Library 78, 96, 98, 101, 106f, 112
Camden, William 26, 101
Camellia galls 287
Camerarius, Joachim 46, 78, 82, 83 (biog.), 85, 88, 96, 97; identification of plants 148f; names cited by Ray 36; quoted: in main catalogue 147, 148, 150, 151, 160, 182, 183, 202, 225, 227, 239, 269, 271, 280, 288, 296, 300, 311, 316, in 'Etymology' 406; source of observations 48; works described in 'Explanation' 137, 142, 143; works in libraries 100, 107, 109
Campanula, etymology 368
Campanula glomerata (Clustered Bellflower) [573], **302**, 325, 342, 348, 352, 354
Campanula medium 433f
Campanula rapunculus 433f
Campanula rotundifolia (Harebell) [103], **173**, 324, 348

Campanula trachelium (Nettle-leaved Bellflower) [572], **302**, 325, 342, 350, 351
Campion *see Silene*
Campoplex difformis 192f
Canini, Angelo 56; quoted in 'Etymology' 386 (biog.), 406f, 413; works in libraries 103, 109
Cannăbis, etymology 368
Cannabis sativa (Hemp) [104], 45, **174**, 332
Capnos, etymology 369, 379
Capra hircus see Goats
Caprifolium, etymology 369
Capsella bursa-pastoris (Shepherd's Purse) [97], **172**, 289, 326, 340
Caraway *see Carum carvi*
Cardamīne, etymology 369
Cardamine pratensis (Cuckoo Flower) [106, B15], 68, **175**, 231f, 327, 333, **463**
Cardano, Giralamo 83 (biog.), 148f, 315f; names cited by Ray 35, 36; quoted in main catalogue 256; works in libraries 99, 109
Cardanus, Hieronymus *see* Cardano, Giralamo
Cardiospermum halicacabum 289
Carduus, etymology 369
Carduus acanthoides 442f
Carduus crispus (Welted Thistle) [A8, B17], 66, **442**, **463**
Carduus lacteus & Mariae, etymology 369
Carduus nutans (Musk Thistle) [111], **176**, 342, 451, 482
Carduus tenuiflorus (Slender Thistle) [A9, B18], 66, **442**, **463**
Carew, Richard 26
Carex, identification problems 42, 116, 211f, 212, 215f, 229f
Carex acuta (Slender Tufted Sedge) 211f
Carex acutiformis (Lesser Pond-sedge) 211f
Carex caryophyllea (Spring Sedge) 212f
Carex demissa (Common Yellow-sedge) 215f
Carex disticha (Brown Sedge) [249], **212**, 350f
Carex divisa (Divided Sedge) 212f

INDEX

Carex divulsa (Grey Sedge) 212f
Carex elata (Tufted Sedge) 211f
Carex flacca (Glaucous Sedge) 212f
Carex flava group (Yellow-sedge) 215f
Carex lepidocarpa (Long-stalked Yellow-sedge) [263], 44, **215**, 350
Carex muricata agg. [248], **212**
Carex muricata sensu stricto (Prickly Sedge) 212f
Carex nigra (Common Sedge) 212f
Carex oederi (Small-fruited Yellow-sedge) 215f
Carex otrubae (False Fox-sedge) [247], **212**, 350f
Carex pendula (Pendulous Sedge) [A16, B38], 44, 66, 213f, **444**, **467**
Carex pseudocyperus (Cyperus Sedge) 444f
Carex pulicaris (Flea Sedge) 69
Carex remota (Remote Sedge) [A17, B37], 213f, **444**, **467**
Carex riparia (Greater Pond-sedge) [243], **211**, 350f
Carex sp. [244, 245], **211**, **212**, 215f, 350f
Carex spicata (Spiked Sedge) 212f
Carex sylvatica (Wood Sedge) [250], **213**, 351
Carex viridula sensu lato 215f
Carex viridula subsp. *brachyrrhyncha* see *C. lepidocarpa*; subsp. *oedocarpa* see *C. demissa*; subsp. *viridula* see *C. oederi*
Carex, unidentified species 212
Carlina, etymology 369
Carlina vulgaris (Carline Thistle) [116], **177**, 342, 348, 352
Carpinus betulus (Hornbeam) [86], 39, 131, **170**, 332
Carrot *see Daucus carota* subsp. *sativus*
Carrot, Moon *see Seseli libanotis*
Carrot, Wild *see Daucus carota* subsp. *carota*
Carum carvi (Caraway) [B19], **464**
Caryophyllata, etymology 370
Caryophyllus, etymology 370
Cassia acutifolia 431f
Catchfly *see Silene*
Caterpillars *see* Larvae
Catharine Hall, Cambridge, Francis Willughby jnr attends 21; lack of wealth 5; Ray enters 5; Ray glad to leave 8
Cativulcus, King of the Eburones 301
Catkins, terminology: 283f, *catulus* 418, *julus* 422, *nucamentum* 423
Cat-mint *see Nepeta cataria*
Cato, Marcus Porcius 72 (biog.), 79, 108, 168, 394f, 429f
Cats (*Felix catus*) 241, 310, 393
Cat's Ear *see Hypochaeris*
Cat's-tail, Smaller *see Phleum bertolonii*
Caucalis, etymology 370
Caucalis platycarpos (Small Bur-parsley) [121], **178**, 336, 350
Cauda equina, etymology 375
Celandine, etymology 180f
Celandine, Greater *see Chelidonium majus*
Celandine, Lesser *see Ficaria verna*
Celery, Wild *see Apium graveolens*
Celsus, Aurelius Cornelius 58, 73 (biog.); names cited by Ray 73; quoted in 'Interpretation' 416
Centaurea calcitrapa (Red Star-thistle) [112], **176**, 342, 410f
Centaurea cyanus (Cornflower) [162], **189**, 324, 365f, 373f
Centaurea debeauxii 224f
Centaurea nigra *sensu lato* (Common Knapweed) [297], **224**, 333, 334, 387f
Centaurea nigra sensu stricto 224f
Centaurea scabiosa (Greater Knapweed) [298, 299], **224**, 252f, 313f, 333, 334, 354
Centaurea stoebe 433f
Centaurium, etymology 370
Centaurium erythraea (Common Centaury) [123], **179**, 326, 348, 349, 353
Centaury, Common *see Centaurium erythraea*
Cephalanthera damasonium (White Helleborine) 468f
Cerastium arvense (Field Mouse-ear) [76], **166**, 335, 348, 352, 464
Cerastium fontanum (Common Mouse-ear) [26], **155**, 326, 335, 441, 461
Cerastium glomeratum (Sticky Mouse-ear) [A3, B5], 66, **441**, **461**

INDEX

Ceratophyllum demersum (Rigid Hornwort) [183], **195**, 334, 482
Cercis, etymology 370
Cercis siliquastrum (Judas-tree) 169f, 265f, 370f, 431f
Cereals, crops 222, 230; germination 308; smut infection 318; terminology: *acus* 415, *alica* 416, *culmus* 419, *frit* 420, *gluma* 421, *granum* 421, *locustae* 422, *merges* 423, *mutica* 423, *spica* 425, *uruncus* 429
Ceres 374
Cervicaria, etymology 410
Cesalpino, Andrea 83 (biog.), 97; names cited by Ray 36, 84; quoted: in main catalogue 202, 217, in 'Etymology' 362, 399; source of observations 48; works described in 'Explanation' 136; works in libraries 111, 112
Ceterach officinarum see *Asplenium ceterach*
Chabrey, Dominic 81, 141
Chaenorhinum minus (Small Toadflax) [50], **159**, 325, 341, 348, 349, 352
Chaerophyllum temulentum see *C. temulum*
Chaerophyllum temulum (Rough Chervil) [124], **179**, 326, 354
Chafe-weed, etymology 381
Chaldean, words derived from 384f, 385, 386f
Chamaedrys, etymology 370, 411
Chamaemēlum, etymology 370
Chamaemelum nobile (Chamomile) 169f
Chamomile see *Chamaemelum nobile*
Chamomile, Corn see *Anthemis arvensis*
Chamomile, Spanish see *Anacyclus pyrethrum*
Chamomile, Stinking see *Anthemis cotula*
Chara spp. [179], **195**, 332
Charcoal 162, 227, 256
Charlemagne, Emperor 153f, 369
Charles the Fat, King of the East Franks 153f
Charleton, Walter 50, 63; quoted in main catalogue 169, 244; source of observations 48; works in libraries 106, 110, 111
Charlock see *Sinapis arvensis*
Cheese 241, 380f

Chelidonium, etymology 180f, 371
Chelidonium majus (Greater Celandine) [128], **180**, 326, 371f
Chenopodium album (Fat Hen) [70], **164**, 181f, 336, 481
Chenopodium bonus-henricus (Good King Henry) [90], **171**, 323, 334
Chenopodium murale (Nettle-leaved Goosefoot) [72], **165**, 181f, 330, 336, 341
Chenopodium polyspermum (Many-seeded Goosefoot) [88, 89], **170**, 324, 350, 451, 482
Chenopodium rubrum (Red Goosefoot) [73], **165**, 336
Chenopodium vulvaria (Stinking Goosefoot) [69], **164**, 336, 380f
Cherler, Jean-Henri 81; see also Bauhin, Jean & Cherler, Jean-Henri
Cherry see *Prunus*
Chervil, Bur see *Anthriscus caucalis*
Chervil, Rough see *Chaerophyllum temulum*
Chickweed, Common see *Stellaria media*
Chickweed, Greater see *Stellaria neglecta*
Chickweed, Upright see *Moenchia erecta*
Chickweed, Water see *Myosoton aquaticum*
Chicory see *Cichorium intybus*
Child, Sir Joshua 21, 22
Chippenham Fen 43, 44
Chiron 179f, 201f, 358, 363, 370
Chondrilla, etymology 370
Chrysalises see Pupae
Chrysanthemum, etymology 371
Chrysanthemum segetum see *Glebionis segetum*
Cicero, Marcus Tullius 9, 73 (biog.), 177f; model for Ray's prose in 'Preface' 125f–128f; quoted: in main catalogue 270, in Etymology 377, in 'Interpretation' 418; source of observations 48
Cichorium, etymology 371
Cichorium intybus (Chicory) [131], **181**, 342
Cicūta, etymology 371
Cinquefoil, Creeping see *Potentilla reptans*

570

INDEX

Cinquefoil, Hoary *see Potentilla argentea*
Cinquefoil, Marsh *see Comarum palustre*
Circaea, etymology 371
Circaea lutetiana (Enchanter's Nightshade) [136], **182**, 335, 349, 364, 371
Circe 152f, 371
Cirsium, etymology 372
Cirsium acaule (Dwarf Thistle) [107], **175**, 342
Cirsium arvense (Creeping Thistle) [113], **176**, 342
Cirsium dissectum (Meadow Thistle) [137], **183**, 237, 262, 342, 349
Cirsium eriophorum (Woolly Thistle) [115], **177**, 342, 351, 427f
Cirsium oleraceum 367f
Cirsium palustre (Marsh Thistle) [114], **176**, 342
Cirsium vulgare (Spear Thistle) [109, 110], **176**, 342, 451, 482
Cissos [Κισσος], etymology 382
Cissus 382
Citrullus vulgaris (Water Melon) 433f
Citrus aurantium (Orange) 227
Citrus limon (Lemon) 227
Cladium mariscus (Great Fen-sedge) [165], 44, **190**, 350
Cladonia spp. [379], **245**, 335, 349
Clary, Wild *see Salvia verbenaca*
Classification of plants 59, 431–434; *see also* Affinities of species discussed
Cleavers *see Galium*
Clematis, etymology 372
Clematis vitalba (Traveller's Joy) [621], **314**, 325, 326, 343, 350, 353
Cleopatra 138
Climbing plants 60, 61, 236, 418, 433
Clinopodium, etymology 372
Clinopodium acinos (Basil Thyme) [9], **149**, 324, 348, 352
Clinopodium ascendens (Common Calamint) [100], **173**, 325, 442, 463
Clinopodium calamintha (Lesser Calamint) [A7, B14], 66, **442**, **463**
Clinopodium vulgare (Wild Basil) [139], **183**, 252f, 324, 349, 354, 452, 485
Clover *see Trifolium*
Cloves, etymology 370
Clove-tree *see Syzygium aromaticum*

Club-rush, Bristle *see Isolepis setacea*
Club-rush, Common *see Schoenoplectus lacustris*
Club-rush, Grey *see Schoenoplectus tabernaemontani*
Club-rush, Sea *see Bolboschoenus maritimus*
Clusius *see* L'Écluse
Clymenos, etymology 363
Clymenos, King 363
Cnide [Κνιδη], etymology 413
Coastal protection 205
Cochlearia officinalis (Common Scurvygrass) 367f
Cochylis roseana 192f
Cock's-foot *see Dactylis glomerata*
Codlins and cream, etymology 238f
Coeloglossum viride (Frog Orchid) [404], 217f, **250**, 336, 347
Cole, William 51
Collins, John 102, 103f, 111, 112
Collins, Samuel 5
Colocasia coryli (Nut-tree Tussock) 187f
Colonna, Fabio 84 (biog.); names cited by Ray 36; quoted: in 'Etymology' 364, in 1663 Appendix 443; Ray's attempt to buy his books 111f; works described in 'Explanation' 138; works in libraries 111
Coltsfoot *see Tussilago farfara*
Columbine *see Aquilegia vulgaris*
Columbines, etymology 362
Columella, Lucius Junius Moderatus 58, 72, 73 (biog.), 79, 108, 429f; quoted: in main catalogue 236, 276, in 'Etymology' 362, 375, 395, 405, in 'Interpretation' 415, 418, 425; source of observations 48
Comarum palustre (Marsh Cinquefoil) [428], **258**, 326, 348
Comfrey, Common *see Symphytum officinale*
Coneion [Κωνειον], etymology 371
Conferva, etymology 372
Confusion of earlier authors, *Atriplex* 67, 481; *Carex* 213; *Crepis capillaris* 220; *Epipactis* 218; germination 309; grasses 67, 199f, 210f; *Helianthemum* 180; insect life-histories 278; *Lactuca* 227; *Rumex* 67, 485; *Salix* 282; thistles

571

INDEX

176; yellow composites (hawkweeds) 67; yellow crucifers 275; *see also* Differences not or only doubtfully constituting specific distinctions
Conium maculatum (Hemlock) [132], **182**, 332
Conopodium majus (Pignut) [96], **172**, 328, 348
Consolida, etymology 372
Constantin, Robert 82; quoted in 'Etymology' 366 (biog.)
Constantine VII Porphyrogenitus, Emperor 72
Contarenus, Nicolas 143
Convallaria majalis (Lily-of-the-Valley) 252
Convolvulus, etymology 372
Convolvulus arvensis (Field Bindweed) [149, B28], 68, **185**, 324, **465**
Convolvulus scammonia 218f
Conyza, etymology 372
Cooking, with mustard 295; peas 261; *see also* Edible plants and plant products
Coot (*Fulica atra*) 397
Corbyn, Samuel 15
Cordials 367f
Cordus, Euricius 82, 84 (biog.); names cited by Ray 36; works described in 'Explanation' 139; works in libraries 111
Cordus, Valerius 55, 56, 74, 81, 84 (biog.), 88, 139, 375f; names cited by Ray 36; quoted in 'Etymology' 365, 378, 389; works described in 'Explanation' 138, 140; works in libraries 103, 104
Coris monspeliensis 224f
Cornarius, Janus 85 (biog.); names cited by Ray 35, 36, 85
Corncockle *see Agrostemma githago*
Cornflower *see Centaurea cyanus*
Cornsalad, Common *see Valerianella locusta*
Cornus sanguinea (Dogwood) [151], **186**, 327, 328, 330
Coronŏpus, etymology 372
Coronopus squamatus see Lepidium coronopus
Corylus, etymology 372

Corylus avellana (Hazel) [154], 45, **186**, 271, 335, 364, 372f, 422
Costeo, Giovanni 46, 85 (biog.), 92; names cited by Ray 35, 36, 85; quoted in main catalogue 183, 270; source of observations 48
Costmary *see Tanacetum balsamita*
Cotesia glomerata 277f
Cotton, Sir John 302, 492 (biog.)
Cottongrass, Common *see Eriophorum angustifolium*
Cotylēdon, etymology 373
Couch-grass *see Elytrigia repens*
Courthope, Peter (1639–1725) 10–12 (biog.), 19, 22, 62, 63, 110; acknowledged by Ray 34, 129; collects plants for Ray's garden 15; correspondence with Ray 11f, 15–20, 66, 106, 213f, 349f, 445f, 449f; residence at Trinity College 1658–1661 64
Courthope, Peter (1676–1695) 12
Cows (*Bos primigenius*) 152
Cowslip *see Primula veris*
Cow-wheat, etymology 391; *see also Melampyrum*
Crab Apple *see Malus sylvestris*
Cranberry *see Vaccinium oxycoccos*
Cranesbill *see Geranium*
Cranmer, Thomas 98, 99
Crataegus heldreichii 478f
Crataegus laevigata (Midland Hawthorn) [414], 116, **253**, 315, 331, 342
Crataegus monogyna (Hawthorn) [414], 116, **253**, 315, 331, 342
Crataegus pentagyna 478f
Cratevas 363
Creeping Jenny *see Lysimachia nummularia*
Crepis capillaris (Smooth Hawk's-beard) [282], **219**, 331, 354f; var. *agrestis* 220f; var. *capillaris* 220f
Crepis foetida (Stinking Hawk's-beard) [283], **220**, 331
Crepis rubra (Pink Hawk's-beard) 220
Crepis vesicaria (Beaked Hawk's-beard) 220f
Crespinus, etymology 373
Cress, Garden *see Lepidium sativum*
Cress, Shepherd's *see Teesdalia nudicaulis*

572

INDEX

Cress, Thale *see Arabidopsis thaliana*
Crista Galli, etymology 396
Crocos 407
Crocus, etymology 373
Crocus sativus (Saffron Crocus) [159], **188**, 206, 339
Crocus, Saffron *see Crocus sativus*
Crolach, Henry 169
Cross, Maltese *see Tribulus terrestris*
Crosswort *see Cruciata laevipes*
Crowfoot, Ivy-leaved *see Ranunculus hederaceus*
Cruciāta, etymology 373
Cruciata laevipes (Crosswort) [160], **189**, 327, 335, 351, 354
Crucifer, unidentifiable [602], 44, **309**
Cuckoo Flower *see Cardamine pratensis*
Cuckoo-spit 255, 396
Cuckoo-spit Insect *see Philaenus spumarius*
Cucurbita pepo 433f
Cudweed, etymology 381
Cudweed, Common *see Filago vulgaris*
Cudweed, Heath *see Gnaphalium sylvaticum*
Cudweed, Marsh *see Gnaphalium uliginosum*
Cudweed, Small *see Filago minima*
Cultivation, difficulties 240; effects of sowing time 222, 261; effects on plants 174, 181, 186, 208, 235, 260, 269, 270; of saffron 189; *see also* Gardens; Grafted trees and shrubs; Transplantation
Curcuma longa (Turmeric) 206
Currant, Black *see Ribes nigrum*
Curtius, quoted in 'Etymology' 373
Cuscuta europaea (Greater Dodder) [161], **189**, 200f, 328
Cuttings 427, 428
Cyamoe [Κυαμοι], etymology 377
Cyănus, etymology 373
Cynoglossum officinale (Hound's-tongue) [164], **190**, 332
Cynosurus cristatus (Crested Dog's-tail) [236], **210**, 215f, 330f, 452, 483
Cypērus, etymology 373
Cytisus scoparius (Broom) [213], **206**, 252f, 253, 325, 348, 389f

Dactylis glomerata (Cock's-foot) [241], **211**
Dactylorhiza fuchsii (Common Spotted Orchid) [402], **250**, 336, 340, 347, 354
Dactylorhiza incarnata and/or ***D. praetermissa*** (Early/Southern Marsh Orchid) [401], **249**, 336, 340, 347
Dactylorhiza praetermissa (Southern Marsh Orchid) *see D. incarnata* and/or *D. praetermissa*
Daffodil *see Narcissus*
Dagger-flower *see Mantisalca salmantica*
Daisy *see Bellis perennis*
Daisy, Oxeye *see Leucanthemum vulgare*
Dale, Samuel 15, 17, 22, 23, 69, 107, 226f, 231f, 254f, 463f, 478f
Daléchamps, Jacques 85 (biog.), 86, 443f; names cited by Ray 36; quoted, sometimes as *Lugd*.: in main catalogue 231, 287, 296, in 'Etymology' 376, 377, 388, 391; source of observations 48; works described in 'Explanation' 136, 138, 141; works in libraries 99, 102, 111
Damocrates Servilius 73 (biog.), 185f; names cited by Ray 73
Dandelion, etymology 191f; *see also Taraxacum*
Daphne laureola (Spurge Laurel) [327], 101, **230**, 333, 341, 352, 388f
Darnel *see Lolium temulentum*
Date Palm *see Phoenix dactylifera*
Dates *see Phoenix dactylifera*
Daucus, etymology 373
Daucus carota subsp. ***carota*** (Wild Carrot) [166], **190**, 324, 325, 354; subsp. ***sativus*** (Carrot) [423, 424], **257**, 303, 325
De La Brosse, Guy 51
Dead-nettle *see Lamium*
Deciduous trees and shrubs, classification 431
Democritus 374f; quoted in 'Etymology' 403
Dens leonis, etymology 373
Dent, Peter 254f, 286f; authorship of 1685 Appendix 67–69, 459; records: in 1663 Appendix 443, 445, in 1685 Appendix 68, 463f, 464, 465f, 466, 468f, 469f, 472f, 473f, 475f, in Ray's

INDEX

English catalogue 68; *see also* Ray, John (1685, with P. Dent)
Deschampsia cespitosa (Tufted Hairgrass) [262], **214**
Descriptions of plants, *Anacamptis pyramidalis* 251; *Atriplex prostrata* 481; *Caucalis platycarpos* 178; *Ceratophyllum demersum* 195; *Chenopodium album* 481; *Coeloglossum viride* 250; *Crepis foetida* 220; *Dianthus deltoides* 464; *Epipactis palustris* 217; *Euphorbia platyphyllos* 478; *Isolepis setacea* 470; *Juncus squarrosus* 471; *Liparis loeselii* 248; *Moenchia erecta* 470; *Montia fontana* 461; orchids 217f; *Potamogeton berchtoldii* and/or *P. pusillus* 267; *P. compressus* 267; *Ranunculus parviflorus* 447, 475; *Raphanus raphanistrum* subsp. *raphanistrum* 274; *Rumex conglomeratus* 484; *R. crispus* 484; *R. obtusifolius* 484; *Salix* spp. 283–287; *Sinapis alba* 275; *Trifolium ochroleucon* 305; *T. striatum* 306; *T. subterraneum* 479; *Zannichellia palustris* 268; see also Differences from closely related taxa outlined
Descurainia sophia (Flixweed) [552], **298**, 329
Desmoulins, Jean 85, 141
Dewberry *see Rubus caesius*
Diana (Artemis) 363, 374, 384f
Dianthus deltoides (Maiden Pink) [B21], 68, **464**
Dicas, Thomas 24
Dieter, Jakob *see* Tabernaemontanus
Differences between segregates not resolved by Ray, *Gentianella amarella* 207; *Nasturtium officinale sensu lato* 247; *Plantago lanceolata* 263; *Schoenoplectus lacustris* 226
Differences from closely related taxa outlined, *Arctium minus* 463; *Carduus crispus* 463; *Cerastium glomeratum* 441, 461; *Chenopodium album* 481; *Convolvulus arvensis* 465; *Euphorbia* spp. 302; *Fumaria* sp. 467; *Galium saxatile* 473; *G. tricornutum* 462; *Hieracium sabaudum* 469; *Hordeum distichon* 222; *Hypericum hirsutum* 470; *Lactuca serriola* 484; *Luzula multiflora* 213; *Lythrum salicaria* 237; *Moehringia trinervia* 441, 462; *Oreopteris limbosperma* 467; *Raphanus raphanistrum* subsp. *raphanistrum* 275; *Rumex maritimus* 471; *Rumex* spp. 484; *Scrophularia auriculata* 167; *Sinapis alba* 275; *S. arvensis* 274; *Torilis arvensis* 464; *Trifolium fragiferum* 304; *T. ochroleucon* 305; *Triticum turgidum* 308; *Utricularia minor* 473; see also Affinities of species discussed; Descriptions of plants
Differences maintained when plants growing together, *Convolvulus arvensis* 465
Differences not or only doubtfully constituting specific distinctions, *Achillea millefolium* 243; *Arabidopsis thaliana* 447, 474; *Artemisia vulgaris* 162; *Carex* sp. or spp. 212; *Chenopodium polyspermum* 451, 482; *Cirsium dissectum* 183; *C. vulgare* 451, 482; *Echium vulgare* 194, 451, 482; *Equisetum fluviatile* 195; *Hedera helix* 217; *Hordeum vulgare* 222; *Humulus lupulus* 235; *Juncus acutiflorus* and/or *J. subnodulosus* 214; *Lithospermum officinale* 235; *Luzula multiflora* 452, 483; *Poterium sanguisorba* 260; *Salix alba* 486; *S. rosea* 287; *Salix* sp. 486; *Stellaria media* 481; *Teesdalia nudicaulis* 451, 482; *Veronica* spp. 451, 481; see also Confusion of earlier authors; Differences between segregates not resolved by Ray; Variation within species
Digitalis purpurea (Foxglove) 131
Dillenius, Johann Jakob 114, 115, 150f, 231f, 244f, 245f, 446f, 462f, 483f
Dillingham, Theophilus 99
Dilsius, I. 167
Diocles 73 (biog.), 168
Diodorus Siculus, quoted in 'Etymology' 374
Dioecious plants, *Cannabis sativa* 174; *Humulus lupulus* 235; *Marchantia*

polymorpha 445f, 468f; *Salix alba* 486; *Valeriana dioica* 311f, 479f
Dioscorides, Pedanius 12, 47, 58, 70, 73 (biog.), 75, 83, 84, 142, 443f; commentaries on 47, 74, 80, 83–86, 88, 90, 92, 95, 103, 135, 138, 140, 141, 371f, 391f, 398f, 408f; names cited by Ray 74; quoted: in main catalogue 163, 186, 190, 200, 203, 240, 241, 270, in 'Etymology' 358, 362, 370, 374, 375, 380–383, 387, 389, 393, 395, 403–405, 412; source of observations 48; works described in 'Explanation' 138, 143; works in libraries 99, 100
Diplolepis rosae 281f
Dipsăcus, etymology 373
Dipsacus fullonum (Wild Teasel) [168], **191**, 342, 343
Dipsacus pilosus (Small Teasel) [169], **193**, 340, 347
Dipsacus sativus (Fuller's Teasel) 369f
Dissection, birds 11, 12, 182; described facetiously by Barrow 12; Frog 12, 400; Great Bustard 182; *Liparis* bulbs 248; seeds 148
Dittander *see Lepidium latifolium*
Dock *see Rumex*
Doctrine of Signatures 46, 50, 181f, 288, 398; *see also* Medicine
Dodder, Greater *see Cuscuta europaea*
Dodecatheon, etymology 373
Dodoens, Rembert 51, 85 (biog.), 90, 91, 94; identification of plants 452, 485; names cited by Ray 36; pictures discussed 452, 485; pictures used by Johnson (1633) 89; quoted: in main catalogue 200, 299, in 'Etymology' 366, 377, 380, 396, 407, 411; source of observations 48; text of Gerarde (1597) based on 87, 139; works described in 'Explanation' 137, 138; works in libraries 99; *see also* Lyte
Dog's-tail, Crested *see Cynosurus cristatus*
Dogwood *see Cornus sanguinea*
Doody, Samuel 463f
Dorsten, Theodor 141f
Dough 307
Dourez, Valerand 86 (biog.); names cited by Ray 35, 36, 86

Dropwort, etymology 394; *see also Filipendula vulgaris*
Drosera intermedia (Oblong-leaved Sundew) [500], 44, **280**, 339, 342, 350
Drosera rotundifolia (Round-leaved Sundew) [501], **280**, 339, 342, 348
Druce, George Claridge 15, 16
Dryden, John 18
Dryocopus martius see Woodpecker, Black
Dryopteris 131
Dryopteris filix-mas (Male Fern) [199], **199**, 329, 348, 466, 467
Du Chesne, Joseph 50, 185f
Du Choul, Jean 168
Duckweed *see Lemna*
Dulcamāra, etymology 374
Duport, James 8 (biog.), 10–12, 24, 56, 110; personal library 109; rules for pupils 9, 103f, 131f
Duport, John 8
Durante, Castore 86 (biog.), 164f; names cited by Ray 36; works described in 'Explanation' 137; works in libraries 111
Dye-plants 45, 206; *Berberis vulgaris* 206; *Crocus sativus* 189, 206; *Curcuma longa* 206; *Indigofera* spp. 207f; *Isatis tinctoria* 169, 206, 209; *Juglans regia* 225; *Lycopus europaeus* 239; *Quercus coccifera* 206; *Rhamnus cathartica* 280; *Rubia tinctorum* 206

Ebulus, etymology 374
Echium, etymology 374
Echium vulgare (Viper's Bugloss) [171, 172], **193**, 325, 348, 451, 482; cordial flower 367f
Edible plants and plant products 45; *Arctium* 167; *Artemisia absinthium* 147; *Arum* 163; *Avena sativa* 165; *Chenopodium bonus-henricus* 171; *Humulus lupulus* 236; *Lactuca* 302; *Ononis spinosa* 159; *Onopordum acanthium* 148; *Pisum sativum* 261; pot-herbs 423; *Tamus communis* 236; terminology: *fruges* 420, *frumenta* 420; *Tragopogon* 303; *Triticum* spp. 307–308; *Tulipa* 289; *see also* Cereals; Cooking; Fruits; Leguminous plants; Sap

INDEX

Egenolph, Christian 141f
Eggs, hens' 152, 175, 378; insects' 279f, 288
Egyptians *see* Gypsies
Eichstätt, Bishops of 82, 139
Elaeagnus, etymology 374
Elaeagnus angustifolia (Oleaster, Russian Olive) 375f
Elaphoboscum, etymology 375
Elder *see Sambucus nigra*
Elder, Dwarf *see Sambucus ebulus*
Elder, Ground *see Aegopodium podagraria*
Elecampane *see Inula helenium*
Eleocharis palustris (Common Spike-rush) [305], **226**, 339
Ellistone, John 50
Elm *see Ulmus*
Elyot, Thomas 366f
Elytrigia repens (Couch-grass) [242], **211**, 267, 452, 483
Empedocles, quoted in 'Etymology' 378
Empiricus, Marcellus 74 (biog.), 366f, 367f; names cited by Ray 74
Endothenia gentianaeana 192f
Endothenia marginana 192f
Ennius, Quintus, quoted in 'Etymology' 373
Enŭla, etymology 375
Epilepsy 84, 162
Epilobium, length of species' names 40
Epilobium hirsutum (Great Willowherb) [352], **237**, 344
Epilobium montanum (Broad-leaved Willowherb) [349], **237**, 345
Epilobium palustre (Marsh Willowherb) [351], 44, **237**, 345, 353
Epilobium parviflorum (Hoary Willowherb) [353], **238**, 345
Epilobium tetragonum (Square-stalked Willowherb) [350], **237**, 345
Epipactis helleborine (Broad-leaved Helleborine) [275], **218**, 331, 335f, 350
Epipactis palustris (Marsh Helleborine) [274], 44, **217**, 331, 335f, 350, 353, 468f, 483
Equisētum, etymology 375
Equisetum arvense (Field Horsetail) [178], **195**, 332

Equisetum fluviatile (Water Horsetail) [180], **195**, 332
Equisetum hyemale (Rough Horsetail) 195f
Equisetum sp. [182], **195**, 332
Equisetum telmateia (Great Horsetail) 195f
Equus ferus see Horses
Erasistratus 74 (biog.), 168
Erica, etymology 58, 375
Erica cinerea (Bell Heather) 131
Erica tetralix (Cross-leaved Heath) [B30], **466**
Erigeron ['Ηριγερων], etymology 375, 405
Erigeron acris (Blue Fleabane) [144], 44, **184**, 329, 350
Eriophorum angustifolium (Common Cottongrass) [265], 44, **216**, 350, 353, 427f
Erodium cicutarium (Common Storksbill) [217], **207**, 327, 348
Erophila glabrescens 256f
Erophila majuscula 256f
Erophila verna *sensu lato* (Whitlowgrass) [419], 155f, **256**, 344, 349, 396f
Erophila verna sensu stricto 256f
Erūca, etymology 58, 375
Eruca vesicaria subsp. *sativa* (Garden Rocket) 375f
Erysĭmum, etymology 376
Erysimum cheiranthoides (Treacle Mustard) [102], **173**, 345
Erysimum cheiri (Wallflower) [330], **231**, 274, 326, 344
Essex 34, 231f, 286; Black Notley 4, 21, 444f; Maldon 442
Esŭla, etymology 376
Etymologicon Magnum, quoted in 'Etymology' 371, 382, 411
Eugalacton, etymology 381
Euonymus, etymology 376
Euonymus europaeus (Spindle) [188], **196**, 338, 341, 350, 351
Eupatorium, etymology 58, 363, 376
Eupatorium cannabinum (Hemp Agrimony) [191], **197**, 323
Euphorbia (Spurges) 168, 289
Euphorbia amygdaloides (Wood Spurge) [570], **302**, 341, 410f

Euphorbia characias (Large Mediterranean Spurge) 410f
Euphorbia dendroides (Tree Spurge) 410f
Euphorbia exigua (Dwarf Spurge) [187], **196**, 341
Euphorbia helioscopia (Sun Spurge) [569], **302**, 341, 453, 487
Euphorbia peplus (Petty Spurge) [569, A28, B74], **302**, 341, **446**, 453, **474**, 487
Euphorbia platyphyllos (Broad-leaved Spurge) [B97], 69, **478**
Euphorbia resinifera 168f
Euphorbia tirucalli 410f
Euphorbium, etymology 363
Euphorbus 76, 168f
Euphrasia, etymology 377
Euphrasia anglica see *E. officinalis* subsp. *anglica*
Euphrasia confusa 197f
Euphrasia nemorosa 197f
Euphrasia officinalis subsp. *anglica* 197f
Euphrasia pseudokerneri 197f
Euphrasia spp. (Eyebright) [192], 44, 188f, **197**, 329, 348
Eustathius, quoted in 'Etymology' 392, 405
Evergreens, classification 431; leaf phenology 151; soil preferences 164
Everlasting, Mountain see *Antennaria dioica*
Everlasting, Pearly see *Anaphalis margaritacea*
Everlasting, Shrubby see *Helichrysum stoechas*
Ewen & Prime, identification of Ray's plants: 113, 116, *Asperula cynanchica* 290f, *Brassica* 246f, *Chenopodium polyspermum* 170f, fruticose lichens 244f, *Juncus* 213f, *Orobanche elatior* 252f, *Potamogeton compressus* 267f, *Rumex* 229f, *Sagina procumbens* 291f, *Salix* 286f, *Sedum rupestre* 293f, *Stellaria media* 155f, *Veronica anagallis-aquatica sensu lato* 157f, *Viola arvensis* 314f; translation of *Catalogus* 2, 3f, 52, 117, 125f, 129f, 192f, 193f, 307f, 316f, 354f

Excrement, *Crocus sativus* stains 189; *Fragaria vesca* affects 201
Extrafloral nectaries 312
Eyebright see *Euphrasia*
Eyes, abscesses in 359; improving vision with hawkweed 383; restoration of 180, 316; strengthening of 310

Faba, etymology 377
Faba inversa, etymology 378
Fabaria, etymology 378
Fagopyrum esculentum (Buckwheat) [195], **198**, 324, 325, 379f
Fagotriticum, etymology 379
Fallopia convolvulus (Black Bindweed) [150], **185**, 324
Falloppio, Gabriele 78, 86 (biog.), 88; names cited by Ray 35, 36, 86
False-helleborine see *Veratrum*
Farfara, etymology 378
Farfarella, etymology 378
Farfarus, etymology 378
Farmers 174, 246, 261, 308
Fasciola hepatica (Liver fluke) 280f
Fasēlus, etymology 378
Fat Hen see *Chenopodium album*
Favagello, etymology 181f, 379
Fegopyron, etymology 379
Felix catus see Cats
Felwort see *Gentianella amarella*
Fennel see *Foeniculum vulgare*
Fen-sedge, Great see *Cladium mariscus*
Fenugreek see *Trigonella foenum-graecum*
Fern, '*Dryopteris*' absent from Cambridge 131
Fern, Lemon-scented see *Oreopteris limbosperma*
Fern, Limestone see *Gymnocarpium robertianum*
Fern, Maidenhair see *Adiantum capillus-veneris*
Fern, Male see *Dryopteris filix-mas*
Fern, Royal see *Osmunda regalis*
Fern, unidentifiable [A13], [B29], **443**, **466**
Ferne, Henry 19
Fernel, Jean François, quoted in main catalogue 162; source of observations 48; works in libraries 106, 108

INDEX

Ferns, 'seed' 200; use to catch flies 199
Ferula communis 297f, 375f
Festus, Sextus Pompeius, quoted in 'Etymology' 366, 382, 390 (biog.), 402
Feverfew *see Tanacetum parthenium*
Ficaria, etymology 379
Ficaria verna (Lesser Celandine) [129], **181**, 337, 371f, 379f
Ficus carica (Fig) 200, 421
Field, John 24, 25, 67, 123, 437
Fig *see Ficus carica*
Figwort *see Scrophularia*
Filago minima (Small Cudweed) [232], **209**, 328, 348
Filago vulgaris (Common Cudweed) [230], **209**, 237f, 327, 381f
Filamentous algae *see* Algae
Filipendula, etymology 394
Filipendula ulmaria (Meadow-sweet) [623], **315**, 334, 351
Filipendula vulgaris (Dropwort) [198], **199**, 328, 348, 394f
Filix, etymology 379
Fire-lighters *see* Tinder
Flaccus, Marcus Verrius 382f
Flag, Sweet *see Acorus calamus*
Flag, Yellow *see Iris pseudacorus*
Flammula, etymology 400
Flax, manufacture and use 233–234; *see also* Linum
Fleabane, Blue *see Erigeron acris*
Fleabane, Common *see Pulicaria dysenterica*
Fleabane, Small *see Pulicaria vulgaris*
Fleawort *see Tephroseris*
Flies 199, 259, 288; *see also* Insects
Flint 233, 310
Flixweed *see Descurainia sophia*
Flour, 306f, 307f, 381f; in ancient world 411
Flower, Eternal *see Helichrysum stoechas*
Flower festivals 241
Flowers, absence in figs 200; appearance in winter 270; colour 151, 157f, 181, 224, 252, 269, 312, 373, 451, 453, 473, 482, 486; cordial 367; definition 420; diurnal opening 185; morphology 155, 220, 233, 256, 261, 291, 300; preceding leaves 237f, 259, 310; recovery from removal 292; relative size 156; resemblance to animals 161, 288, 362, 374, 380, 381, 405, 410; smell 247, 280, 313, 382, 451, 482; terminology: *anthera* 416, *calyx* 417, *capillamenta* 417, *fibrae* 420, *pediculus* 424, petal 84, 118f; *perianthium* 424, *stamina* 427, *stylus* 427, *umbilicus* 429, *unguis* 429; uses 311; visited by bees 188, 191, 312; wilting 185; *see also* Descriptions of plants; Dioecious plants; Variation within species
Foeniculum vulgare (Fennel) 415f, 448
Foliose lichens *see* Lichens, foliose
Fontinalis antipyretica [A25, B68], 66, **445**, **473**
Fontinalis squamosa 446, 473
Forget-me-not *see Myosotis*
Fortho, John 101
Foundations of buildings 153
Foxglove *see Digitalis purpurea*
Fox-sedge, False *see Carex otrubae*
Foxtail *see Alopecurus*
Foxton, Richard 98, 99f
Fracastoro, Girolamo, quoted in 'Etymology' 380
Fragaria, etymology 379
Fragaria vesca (Wild Strawberry) [201], **201**, 341
Fragoso, Juan 86 (biog.); names cited by Ray 36, 86; works described in 'Explanation' 139; works in libraries 111
Frangula alnus (Alder Buckthorn) [B4], 432f, **461**
Frankia 361f
Fraxinus, etymology 379
Fraxinus excelsior (Ash) [203], 114, 148, **201**, 323, 345
Fraxinus ornus (Manna Ash) 201f, 386f
Frog (*Rana temporaria*), copulation studied 11; diet revealed by dissection 400; plants resembling 288; predatory habits 255, 400
Frogbit *see Hydrocharis morsus-ranae*
Froghopper, Common *see Philaenus spumarius*
Fruits, absence in jasmine 200; colour 196, 280; definition 420; dyes and paints from 280; edibility 186, 301;

INDEX

effects of cultivation on 269, 288; maturation 227, 269, 387; morphology 196, 239, 255, 303, 312; of evergreens 151; resemblance to animals 362; size in relation to seeds 239; terminology: *acinus* 415, *baccae* 417, *caro* 418, *conus* 418, *echinus* 420, *folliculus* 420, *glans* 421, *loculamentum* 422, *nucleus* 423, *nux* 423, *ossiculum* 424, *pappus* 424, *pediculus* 424, *pericarpium* 424, *pomum* 424, *pulpa* 424, *racemus* 424; *samera* 425, *siliqua* 425, *umbilicus* 429, *volva* 429; trees 271; *see also* Hazelnuts; Seeds; Variation within species

Fruticose lichens, *see* Lichens, fruticose

Fuchs, Leonhart 81, 83, 85, 86, 87 (biog.), 96; identification of plants 182, 305; names cited by Ray 36; pictures discussed 305; quoted in 'Etymology' 393; works described in 'Explanation' 139, 144; works in libraries 100, 101, 109

Fuga daemonum, etymology 397

Fulica atra see Coot

Fumaria, etymology 379

Fumaria capreolata (White Ramping Fumitory) 467f

Fumaria muralis (Common Ramping Fumitory) 467f

Fumaria officinalis (Common Fumitory) [204], 101, **203**, 329, 369f, 467

Fumaria sp. [B35], **467**

Fumitory *see Fumaria*

Funaria hygrometrica 150f

Fungi 69, 90, 203–205, 288

Fungi, bracket [206], **204**, 289

Fungi, puff-ball [207], **204**, 335, 338, 348, 352

Fungus, etymology 379

Fusanus, etymology 376

Galanthus nivalis (Snowdrop) 231f

Galega, etymology 380

Galega officinalis (Goat's Rue) [208], 44, 93, **205**, 339

Galen 47, 49, 50, 71, 73, 74 (biog.), 75, 84, 85, 88, 256f, 411f; names cited by Ray 74; quoted: in main catalogue 200, 302, in 'Etymology' 371, 380, 390; source of observations 48; works described in 'Explanation' 139; works in libraries 100; *see also* Medicine, Galenic

Galeopsis, etymology 380

Galeopsis angustifolia (Red Hemp-nettle) [315], 101, **228**, 332, 388f

Galeopsis bifida 175f

Galeopsis tetrahit *sensu lato* (Common Hemp-nettle) [105], **175**, 332

Galeopsis tetrahit sensu stricto 175f

Galium album (Hedge Bedstraw) [373], **243**, 333, 452, 485

Galium aparine (Cleavers) [51], **160**, 326, 330, 362f, 382f, 462

Galium mollugo see Galium album

Galium odoratum (Woodruff) [67], **164**, 345

Galium palustre (Marsh Bedstraw) [210], 44, **205**, 333, 353, 473

Galium saxatile (Heath Bedstraw) [B66], **473**

Galium tricornutum (Corn Cleavers) [A5, B9], 66, **441**, **462**

Galium verum (Lady's Bedstraw) [211], **205**, 326, 333

Gallium, etymology 380

Gallus domesticus see Hens

Galls 158, 271, 280, 287, 294f, 316

Garden plant catalogues 16, 34, 40, 93, 140; Oxford catalogue a possible model for *Catalogus* 34; Ray's lost Cambridge catalogue 15, 201f; Ray's proposed English catalogue 15, 215f; *see also* Paulli, Simon

Gardens, cultivation of plants in 14, 15, 126, 186, 240, 241, 269, 271, 393f, 443, 463f, 464; of Adonis 415

Garlands, floral 240, 362, 406, 425, 433

Garlic *see Allium*

Garosmum, etymology 380

Garum 164f

Gassendi, Pierre 50, 187 (biog.); names cited by Ray 36; quoted: in main catalogue 151, 202, 219, 284, in 'Etymology' 360, 398; source of observations 48; works described in 'Explanation' 142; works in libraries 99, 101

579

INDEX

Gaza, Theodorus 78 (biog.), 82, 169f, 313f; quoted in 'Etymology' 396; source for main catalogue 151f, 222f, 254f, 262f, 271f, 308f; works in libraries 99

Gellius, Aulus 59, 74 (biog.); quoted: in 'Etymology' 377, 406, in 'Interpretation' 425f, 427

Gemmingen, Johann Konrad von, Bishop of Eichstätt 82

Genista, etymology 380

Genista anglica (Petty Whin) [214], **206**, 330, 344, 348, 483

Genista tinctoria (Dyer's Greenweed) [215], **206**, 328, 331, 345, 351

Genserich, etymology 362

Gentian, Great Yellow *see Gentiana lutea*

Gentiana, etymology 363

Gentiana lutea (Great Yellow Gentian) 380f

Gentianella, etymology 380

Gentianella amarella (Felwort) [216], **207**, 329, 330, 348, 483

Gentius, King of the Illyrians 363, 380

Geophilus electricus 187f

Geranium, etymology 381

Geranium dissectum (Cut-leaved Cranesbill) [220], **207**, 328

Geranium lucidum (Shining Cranesbill) [224], **208**, 328, 347

Geranium molle and/or ***G. pusillum*** (Dove's-foot/Small-flowered Cranesbill) [219, 221], **207**, **208**, 327, 349, 452, 482

Geranium pratense (Meadow Cranesbill) [218], **207**, 327, 347, 349

Geranium pusillum (Small-flowered Cranesbill) *see Geranium molle* and/or *G. pusillum*

Geranium robertianum (Herb Robert) [223], **208**, 363, 381

Geranium sanguineum (Bloody Cranesbill) [222], **208**, 327, 351, 352, 381

Gerard, Saint 159f

Gerarde, John 34, 44, 45, 87 (biog.), 89, 91, 95, 152f, 170f, 171f, 180f, 188f, 204f, 224f, 238f, 248f, 288f, 367f, 473f, 476f; errors 180, 182, 260, 264, 286, 295f, 300, 303, 487; identification of plants 162, 177, 178, 190, 199f, 210f, 211, 235, 240, 264f, 282f, 295f, 309f, 317, 446, 451, 452, 472, 482, 485; mentioned by Corbyn 16; names cited by Ray 36, 39, 40, 129; pictures discussed 16, 180, 209, 211, 240, 451, 452, 464, 472, 482, 485; qualities of plants 47; quoted: in main catalogue 165, 222, 240, 243, 263, 266, 280, 297, in 'Etymology' 395; source of observations 48; Ray's assessment of 139; records omitted from *Catalogus* 44; works described in 'Explanation' 139; works in libraries 100, 101, 103, 107, 110; *see also* Johnson, Thomas

Germander *see Teucrium*

Germination *see* Seeds

Gesner, Conrad, 74, 81, 84–87, 88 (biog.), 92, 96; names cited by Ray 36; quoted: in main catalogue 153, 161, 197, in 'Etymology' 378; source of observations 48; works described in 'Explanation' 138, 139; works in libraries 98, 99

Geum rivale × ***G. urbanum*** [A10, B20], 15, **443**, **464**

Geum urbanum (Herb Bennet) [117], **177**, 324, 370

Giardia sanamunda 177f

Gilbert, William, quoted in 'Etymology' 360; works in libraries 99, 108, 109

Gladdon *see Iris foetidissima*

Gladiolus 289

Glass-making 201

Glaux, etymology 381

Glaux maritima (Sea Milkwort) 381f

Glebionis segetum (Corn Marigold) [130], **181**, 334

Glechoma hederacea (Ground Ivy) [272], 101, **217**, 323, 330, 332, 343, 477f

Glow-worms 187

Gluten intolerance 308f

Glyceria fluitans (Floating Sweet-grass) [238], **210**, 330

Glyceria maxima (Reed Sweet-grass) [239], **211**, 338f

Glycyrrhiza glabra (Liquorice) [229], **209**, 333

Gnaphalium, etymology 381

INDEX

Gnaphalium sylvaticum (Heath Cudweed) [A15, B36], 66, **444**, **467**
Gnaphalium uliginosum (Marsh Cudweed) [231], **209**, 327
Goats (*Capra hircus*) 193, 270, 359
Goat's-beard *see Tragopogon pratensis*
Goat's Rue *see Galega officinalis*
Gods, eponyms listed 363
Goebel, Severin, quoted in main catalogue 296; source of observations 48
Golden bough 314
Goldenrod *see Solidago virgaurea*
Goldilocks *see Helichrysum stoechas* and *Ranunculus auricomus*
Good King Henry, etymology 171f; *see also Chenopodium bonus-henricus*
Good, Thomas 5
Goodyer, John 74, 89; personal library 40f, 62, 89, 104f, 111f; quoted in 1663 Appendix 443 (biog.)
Goosefoot *see Chenopodium*
Goropius, Johannes 58; quoted: in main catalogue 234, in 'Etymology' 399; source of observations 48; works in libraries 99, 101
Gorris, Jean de 357f
Gorse *see Ulex europaeus*
Graffenried, Franz Ludwig von 81, 141
Grafted trees and shrubs 151, 288
Grain crops *see* Cereals
Gramen, etymology 382
Gramen Parnassi, etymology 382
Grapes *see Vitis vinifera*
Grass of Parnassus *see Parnassia palustris*
Grass Poly *see Lythrum hyssopifolia*
Grass, Canary *see Phalaris canariensis*
Grass, Marram *see Ammophila arenaria*
Grass, Quaking *see Briza media*
Grasses, construction of bents 215; flowering behaviour 292; identification problems 67, 69, 116, 199f, 210f, 215; L'Obel's discoveries appropriated by Parkinson 213; mowing 214; Ray urges Lister to study 67; unidentified 215; *see also* Cereals; Confusion of earlier authors
Graves, William 24
Graymill, etymology 392

Greenfly (plant lice) 161, 274
Greenweed, Dyer's *see Genista tinctoria*
Groenlandia densa (Opposite-leaved Pondweed) [575], **303**, 325, 329f
Gromwell *see Lithospermum*
Ground Ivy *see Glechoma hederacea*
Ground-pine *see Ajuga chamaepitys*
Groundsel *see Senecio*
Gruel 308
Grummell, etymology 235, 392
Guelder Rose *see Viburnum opulus*
Guilandinus, Melchior 88 (biog.), 92; names cited by Ray 36; works described in 'Explanation' 140; works in libraries 111
Gymnadenia conopsea sensu lato (Fragrant Orchid) [403], **250**, 336, 347, 349, 352, 354
Gymnadenia conopsea sensu stricto 250f
Gymnadenia densiflora 250f
Gymnadenia rhellicani 252f
Gymnetron villosulum 158
Gymnocarpium robertianum (Limestone Fern) 443f
Gypsies 239
Gypsy Moth *see Lymantria dispar*
Gypsywort, etymology 239f; *see also Lycopus europaeus*

Hacket, John 99f
Hair-grass, Silver *see Aira caryophyllea*
Hair-grass, Tufted *see Deschampsia cespitosa*
Handsch, Georg 137
Hard-fern *see Blechnum spicant*
Harebell *see Campanula rotundifolia*
Hare's-ear, Slender *see Bupleurum tenuissimum*
Hartlib, Samuel 1, 2, 14, 24, 26f, 45, 59 (biog.), 418
Hart's-tongue *see Asplenium scolopendrium*
Hawkbit, Autumn *see Scorzoneroides autumnalis*
Hawkbit, Rough *see Leontodon hispidus*
Hawk-moth, Death's-head *see Acheronia atropos*
Hawk-moth, Humming-bird *see Macroglossum stellatarum*
Hawk-moth, Privet *see Sphinx ligustri*

INDEX

Hawk's-beard *see Crepis*
Hawkweed, Mouse-ear *see Pilosella officinarum*
Hawthorn *see Crataegus*
Hayes, John 67, 455
Hazel *see Corylus avellana*
Hazelnuts, classical names 186
Heath, Cross-leaved *see Erica tetralix*
Heather *see Calluna vulgaris*
Heather, Bell *see Erica cinerea*
Heberden, William 244f
Hebrew, words derived from 52, 56, 57, 384
Hector 409f
Hedera, etymology 382
Hedera helix (Ivy) [270, 271], **216**, **217**, 332, 433f
Hedge-parsley *see Torilis*
Hedypnois, etymology 382
Hedysarum, etymology 382
Heer, Henry from, quoted in main catalogue 316; source of observations 48; works in libraries 110, 111
Helen 363, 383
Helenium, etymology 363, 383
Helianthemum nummularium (Common Rockrose) [125], **179**, 326, 332, 342, 348, 352
Helichrysum stoechas (Eternal Flower, Goldilocks, Shrubby Everlasting) 419f
Heliotropic movements 302
Hellebore *see Helleborus*
Helleborine, Broad-leaved *see Epipactis helleborine*
Helleborine, Marsh *see Epipactis palustris*
Helleborine, White *see Cephalanthera damasonium*
Hellebŏrus, etymology 382
Helleborus foetidus (Stinking Hellebore) [A18, B41], 66, **444**, **468**
Helleborus viridis (Green Hellebore) [276], 43, 131, **218**, 260f, 324, 332, 347
Helminthotheca echioides (Bristly Oxtongue) [94], **172**, 333, 354f; cordial flower 367f
Helmont, Jean Baptiste van 50, 169, 244f; source of observations 48; works in libraries 99, 107, 108, 110

Helxīne, etymology 382
Hemlock *see Conium maculatum*
Hemp *see Cannabis sativa*
Hemp-nettle *see Galeopsis*
Hemp-retting 174
Henbane *see Hyoscyamus niger*
Henbit *see Lamium amplexicaule*
Hens (*Gallus domesticus*) 152, 175, 202, 316, 318
Henslow, John Stevens 52
Heracles *see Hercules*
Heracleum sphondylium (Hogweed) [560], **299**, 337, 367f; forma **angustisectum** and/or forma **stenophyllum** [B96], **478**
Heraclion, etymology 393
Herb Bennet *see Geum urbanum*
Herb Paris *see Paris quadrifolia*
Herb Robert *see Geranium robertianum*
Herba benedicta, etymology 370
Herba Gerardi, etymology 159, 383
Herba haemorhoidum, etymology 379
Herba impia, etymology 381
Herba Paralysis, etymology 268f
Herba Ruperti, etymology 363f
Herba Sardoa, etymology 272f
Herba venti, etymology 400
Herbae capillares, etymology 398
Herbe impious, etymology 237f
Herbs, classification 60, 431, 432; definition 421; life-forms 434
Herb-selling women 149
Hercules 265f, 363, 393, 402f
Hermaphrodites, Renaissance interest in 297f; slugs and snails 297
Hermaphroditus 297f
Hermes *see Mercury*
Herminium monorchis (Musk Orchid) [A27, B71], 66, **446**, **474**
Hermolaus *see Barbaro, Ermolao*
Hermu botanion [Ἑρμου βοτανιον], etymology 392
Hermupoa, etymology 392
Herodotus, quoted in 'Etymology' 374
Heroes, eponyms listed 363
Hesperis matronalis (Dame's Violet) 231f
Hesychius 394f, 399f
Hieracium, etymology 383
Hieracium murorum 447f

582

INDEX

Hieracium sabaudum forma ***bladonii*** [B46], **469**; forma ***sabaudum*** [B45], **469**
***Hieracium* sp.** [A32, B81], 66, **447**, 469, **475**
Hieracium umbellatum [B47], **469**
Hieracium vagum 469f
Hierobotane [Ἱεροβοτανη], etymology 412
Highmore, Nathaniel, quoted in main catalogue 148; source of observations 48; works in libraries 110, 111
Hill, Thomas 6
Hills, Henry 25
Hippocrates 47, 71, 75 (biog.), 85, 190; identification of plants 152f, 228f; names cited by Ray 75; quoted in main catalogue 168, 186; source of observations 48
Hippocrepis comosa (Horseshoe Vetch) [452, A14, B31], 44, **263**, 334, 348, 350, 352, 354, **444**, 453, **466**, 485
Hippolapathum, etymology 388
Hippuris, etymology 375
Hippuris vulgaris (Mare's-tail) [181], **195**, 332
Hirundo rustica see Swallows
Hobbes, Thomas (of Gray's Inn) 5
Hofmann, Caspar 46, 50, 85, 88 (biog.), 92; names cited by Ray 36, 88; quoted: in main catalogue 147, 148, 168, 186, 189, 190, 201, 207, 209, 216, 227, 230, 241, 270, 273, 299, 302, in 'Etymology' 357, 366, 368, 371, 397, 399, 402, 405, 412; source of observations 48; works described in 'Explanation' 136; works in libraries 106
Hogenberg, Franz *see* Braun, Georg & Hogenberg, Franz
Hogweed *see Heracleum sphondylium*
Holcus, etymology 383
Holcus lanatus (Yorkshire Fog) [261], **214**
Holdsworth, Richard, personal library 99
Holland, Philemon 77
Holler, Thomas 99
Holly *see Ilex aquifolium*
Holosteum, etymology 383
Holy Cross 364

Homalothecium sericeum 244f
Homer 9, 75 (biog.), 152f, 160f, 201, 264f, 270f, 272f, 371f, 392f, 419f; names cited by Ray 75; quoted in 'Etymology' 403
Honeydew 255, 271
Honeysuckle *see Lonicera periclymenum*
Honoratus, Servius Maurus, quoted in 'Etymology' 364, 371 (biog.), 387, 412
Hop *see Humulus lupulus*
Horace (Quintus Horatius Flaccus) 75 (biog.); quoted: in 'Preface' 128, 131, in 'Etymology' 385
Hordeum (Barley) 168, 230, 253, 318, 381, 415f, 421; germination 308, 309; *see also* Cereals
Hordeum distichon (Two-rowed Barley) [289, 290], **222**, 324
Hordeum murinum (Wall Barley) [288], **221**, 324, 330, 339, 344, 383f; etymology 222f
Hordeum vulgare (Six-rowed Barley) [291], **222**, 324
Hordeum zeocriton 222f
Horehound, Black *see Ballota nigra*
Horehound, White *see Marrubium vulgare*
Hormīnum, etymology 383
Hornbeam *see Carpinus betulus*
Horned-poppy, Violet *see Papaver bivalve*
Hornwort, Rigid *see Ceratophyllum demersum*
Horse-radish *see Armoracia rusticana*
Horses (*Equus ferus*) 159, 240
Horseshoe Vetch *see Hippocrepis comosa*
Horsetail *see Equisetum*
Hottonia palustris (Water Violet) [365], **242**, 334, 344, 434f
Hound's-tongue *see Cynoglossum officinale*
How, William 44, 65, 88 (biog.), 91; errors in *Phytologia* 14; identification of plants 306; names cited by Ray 36, 89; *Phytologia* as model for *Catalogus* 26, 27, 30, 34, 45; quoted in main catalogue 147; works described in 'Explanation' 39, 142; works in libraries 107, 109

INDEX

Humulus lupulus (Hop) [342], 148, 153, 174, **235**, 332, 347; substitutes 199, 299
Huntingdonshire, Stilton 442
Hyacinthoides non-scripta (Bluebell) [293], **223**, 331, 350, 351
Hyacinthos 384
Hyacinthus, etymology 384
Hydrocharis morsus-ranae (Frogbit) [375, 376], **244**, 329
Hydrocotyle vulgaris (Marsh Pennywort) [157], 44, **188**, 337, 339, 350, 353
Hydropiper, etymology 397
Hyoscyămus, etymology 384
Hyoscyamus niger (Henbane) [294], 188, **223**, 332
Hypericum, etymology 386
Hypericum androsaemum (Tutsan) 361f
Hypericum coris 224
Hypericum elodes (Marsh St John's-wort) [66], **164**, 280, 339, 348
Hypericum hirsutum (Hairy St John's-wort) [A22, B50], 66, 68, **445**, **470**
Hypericum humifusum (Trailing St John's-wort) [B51], **470**, 472
Hypericum perforatum (Perforate St John's-wort) [296], **223**, 339, 354, 364f, 397f, 445f, 470
Hypericum pulchrum (Slender St John's-wort) [295], **223**, 339, 348
Hypericum tetrapterum (Square-stalked St John's-wort) [65], **163**, 339, 364f
Hypochaeris maculata (Spotted Cat's Ear) [A21, B48], 66, **445**, **469**
Hypochaeris radicata (Cat's Ear) [279], **219**, 331, 354f
Hyssop *see Hyssopus officinalis*
Hyssōpus, etymology 56, 57, 384
Hyssopus officinalis (Hyssop) 384f

Ibn Rushd *see* Averroës
Ilex aquifolium (Holly) [15], 45, **150**, 216, 332
Imperato, Ferrante 89 (biog.); names cited by Ray 37, 89; works described in 'Explanation' 141; works in libraries 109
Indexing, 17th century 35f
Indigo *see Indigofera*
Indigofera suffruticosa (Indigo) 207f

Indigofera tinctoria (Indigo) 207f
Inflorescences, terminology: *capitulum* 417, *corymbus* 419, *juba* 421, *panicula* 424, *spica* 425, *thyrsus* 427, *umbella* 428, *verticillatae plantae* 428; *see also* Catkins
Ink 234
Insects, eaten by frogs 400; indicators of plant relationships 276; parasitoids 192, 193, 276, 281; 'sphondyle' 407; *see also* Butterflies; Flies; Galls; Glow-worms; Larvae; Maggots; Moths; Pupae
Introduced plants 131, 156f, 218f, 473f–475f
Inula, etymology 375
Inula conyzae (Ploughman's Spikenard) [141], **184**, 329, 337, 341
Inula helenium (Elecampane) [177], **194**, 328, 350, 351
Iolaus, quoted in 'Etymology' 402
Ipomoea batatas 432f
Irio, etymology 387
Iris 289; etymology 387
Iris foetidissima (Gladdon) [630], **319**, 330, 352, 407f
Iris pseudacorus (Yellow Flag) [10], 114, **149**, 329
Isatis tinctoria (Woad) [226], 169, 206, **208**, 345
Isidore 75 (biog.), 78; quoted: in 'Etymology' 388, in 'Interpretation' 419
Isolepis setacea (Bristle Club-rush) [B53], **470**
Itea ['Ἰτέα], etymology 58, 403, 428f
Ivy *see Hedera helix*

Jacea, etymology 387
Jacea nigra, etymology 387
James, Saint 364
Jasione montana (Sheep's-bit) [496], **279**, 338, 340, 348
Jasmine *see Jasminum*
Jasminum officinale (Common or Cottage Jasmine) 200
Jason 179f
Jerome, Saint 368
Jessop, Francis 20
Jew's Ear, etymology 288f; *see also* Auricularia auricula-judae

INDEX

John the Baptist, 364; collection of plants on his eve/day 47, 162, 200, 223f, 292

Johnson, Thomas 45, 87, 89 (biog.), 93, 96, 171f, 177f, 178f, 256f, 262f, 263f, 291f, 295f; errors 293, 312, 467; *Herball* as source for 'Explanation' 38; herb-sellers and 149f; identification of plants 170, 183, 227, 293, 295f, 296, 442f, 453, 464, 478; local Floras 51; *Mercurius* as indirect model for *Catalogus* 26–29, 45; names cited by Ray 37, 40; pictures discusssed 293, 478; quoted, often as *Ger. emac.*: in 'Explanation' 135, 139, 142, in main catalogue 162, 170, 183, 200, 215, 226, 227, 234, 295, 296, in 'Etymology' 395, in appendices 442, 453, 463, 468f, 484, 486, 487; Ray's assessment of 39, 139; source of observations 48; works described in 'Explanation' 139; works in libraries 100, 103f, 107

Jove *see* Jupiter

Juba II of Mauretania 75 (biog.), 168, 363

Judas-tree *see Cercis siliquastrum*

Juglans, etymology 387

Juglans regia (Walnut) [303], 131, 169f, 204, **225**, 335, 372f

Julian, Emperor 76

Juncus, etymology 387

Juncus acutiflorus and/or
  J. subnodulosus (Sharp-flowered/ Blunt-flowered Rush) [255], **213**, 350f

Juncus ambiguus see J. ranarius

Juncus articulatus (Jointed Rush) [253], **213**, 350f

Juncus bufonius (Toad Rush) [256], **214**, 350f, 467

Juncus bulbosus (Bulbous Rush) [B39], **467**

Juncus compressus (Round-fruited Rush) [254], **213**

Juncus conglomeratus (Compact Rush) [308], **226**, 339

Juncus effusus (Soft Rush) [307], **226**, 339

Juncus inflexus (Hard Rush) [304], **225**, 339

Juncus ranarius (Frog Rush) 468f

Juncus squarrosus (Heath Rush) [B54], **471**

Juncus subnodulosus (Blunt-flowered Rush) *see J. acutiflorus* and/or *J. subnodulosus*

Jung, Joachim 59, 84; quoted: in 'Interpretation' 415, 417, 418, 420–422, 424, 425, 427, 429, in 'More customary headings' 433

Juniper *see Juniperus communis*

Junipĕrus, etymology 387

Juniperus communis (Juniper) [310], **227**, 332, 387f

Junius, Adrianus 59, 90 (biog.); names cited by Ray 35, 37; quoted in 'Interpretation' 415, 416, 418, 421–425, 427; works in libraries 106, 109

Juno 374

Jupiter 225f, 374, 387, 411f, 412

Juvenal (Decimus Junius Juvenalis) 9, 76 (biog.); quoted in 'Preface' 133

Kentmann, Johannes, quoted in main catalogue 153; source of observations 48

Kermes, etymology 206f

Kickxia elatine (Sharp-leaved Fluellen) [175], **194**, 329

Kickxia spuria (Round-leaved Fluellen) [176], **194**, 329, 341f

Kidneys 201; *see also* Stones, bladder and kidney

Kircher, Athanasius, quoted in main catalogue 182, 204, 296; source of observations 48; works in libraries 107

Knapweed *see Centaurea*

Knautia arvensis (Field Scabious) [526, A38, B90], 66, **291**, 340, 354, **448**, **476**

Knawel, Annual *see Scleranthus annuus*

Knotgrass *see Polygonum aviculare sensu lato*

Krantz, Albert, quoted in main catalogue 153; source of observations 48; works in libraries 99, 101

Knyvet, Edmund 278f

Knyvet, Sir Thomas 277

Kunigunde, Saint 197f, 364

Kyber, David 143

INDEX

Labrum Veneris, etymology 191, 373
Lactūca, etymology 387
Lactuca saligna (Least Lettuce) [314], **228**, 333, 484
Lactuca sativa (Garden Lettuce) 302, 415f
Lactuca serriola (Prickly Lettuce) forma ***integrifolia*** [313], **227**, 333, 484; forma ***serriola*** [312], **227**, 333, 484
Lactuca virosa (Great Lettuce) 228f
Ladănum segetum, etymology 388
Lady-fern *see Athyrium filix-femina*
Lady's-mantle, Hairy *see Alchemilla filicaulis* subsp. *vestita*
Lady's Tresses, Autumn *see Spiranthes spiralis*
Lagōpus, etymology 228f, 388
Laguna, Andreas 74, 90 (biog.); names cited by Ray 37; works described in 'Explanation' 141
Lambert, John 14
Lambeth Palace Library 98, 100
Lamiastrum galeobdolon (Yellow Archangel) [318], **228**, 323, 350, 351
Lamium, etymology 388
Lamium album (White Dead-nettle) [317], **228**, 323, 335
Lamium amplexicaule (Henbit) [25], **155**, 332, 399f, 477f
Lamium purpureum (Red Dead-nettle) [319], **229**, 323, 335, 477f
Lampsāna, etymology 388
Lantana, etymology 58, 312, 412
Lapăthum, etymology 388
Lappa, etymology 388
Lapsana communis (Nipplewort) [320], **229**, 327, 335, 342, 388f
Largus, Scribonius *see* Scribonius Largus
Larvae 187, 191, 232, 270f, 276, 281, 287, 296, 318; *see also* Maggots
Latex *see* Sap
Lathyrus aphaca (Yellow Vetchling) [52], **160**, 344
Lathyrus nissolia (Grass Vetchling) [A11, B22], 66, **443**, **464**
Lathyrus pratensis (Meadow Vetchling) [326], **230**, 342, 351
Lathyrus sylvestris (Narrow-leaved Everlasting Pea) [325], **230**, 337, 350, 351, 433f

Lathyrus tuberosus (Tuberous Pea) 289
Latin words, coined by changing Greek breathings: rough to 's' 56, 405, smooth to 'v' 56, 413
Laurel *see Laurus nobilis*
Laurel, Spurge *see Daphne laureola*
Lauremberg, Peter 46, 90 (biog.); names cited by Ray 35, 37, 90; quoted: in 'Preface' 132, in main catalogue 152, 163, 167, 170, 185, 188, 189, 214, 223, 232, 239, 241, 254, 261, 262, 271, 302; source of observations 49; works in libraries 104, 105, 108, 109
Laureola, etymology 388
Laurus, etymology 388
Laurus nobilis (Bay-tree, Laurel) 388f
Lavandula stoechas 432f
Laver, etymology 407
Lawson, Thomas 463f
Leaves, absence 189, 200; colour 265; definition 420; evergreens 151; on square stems 189; response to summer solstice 265; shape 316, 476; terminology: *crenae* 419, *lacinia* 422, *nervus* 423, *pediculus* 424; uses 241; wilting 185; *see also* Smell of plants; Variation within species
L'Écluse, Charles de (Clusius) 81, 86, 90 (biog.), 91, 94, 111f, 149f; identification of plants 447, 452, 475, 483, 487; names cited by Ray 37; pictures discussed 180, 452; quoted in 'Etymology' 391; works described in 'Explanation' 135, 137–139, 143; works in libraries 102, 107
Leeuwenhoek, Anthony van 279f
Legousia hybrida (Venus's Looking-glass) [556], **298**, 343, 344
Legumen, etymology 422
Leguminous plants, bitterness 209; effects on soil 230; flowers 261; germination 308; growth 230; parasitised by *Orobanche* 389; terminology: *fruges* 420, *frumenta*, *frumentacea* 420, *legumen*, *legumentum* 422
Lemna minor (Common Duckweed) [329], **230**, 328, 389f
Lemna trisulca (Ivy-leaved Duckweed) [273], **217**, 328, 332

INDEX

Lemon *see Citrus limon*
Lens 269, 276
Lens, etymology 58, 389
Lens culinaris (Lentil) [328], **230**, 389f, 433f
Lens palustris, etymology 389
Lenticula palustris, etymology 389
Lentil *see Lens culinaris*
Lentil, Bastard *see Vicia ervilia*
Leontodon autumnalis see Scorzoneroides autumnalis
Leontodon hispidus (Rough Hawkbit) [278], **219**, 331, 354f
Leonurus cardiaca (Motherwort) [B16], 432f, **463**
Lepidium coronopus (Swine-cress) [153], **186**, 327, 381f
Lepidium latifolium (Dittander) [B58], **471**
Lepidium sativum (Garden Cress) 369f
Lettuce *see Lactuca*
Leucanthemum vulgare (Oxeye Daisy) [81], **167**, 328, 336, 354
Leucographis, etymology 369
Leucoium, etymology 231f
Lhwyd, Edward 23
Liber 382
Lice 189, 396; *see also* Greenfly (plant lice)
Lichen, etymology 389
Lichens 232, 244, 245, 333, 335
Lichens, foliose [332], **232**, 333
Lichens, fruticose [377], **244**, 335
Life-forms, terminology: *stirps*, 427; *see also* Annual plants; Aquatic plants; Herbs; Perennial plants; Shrubs; Subshrubs; Trees
Life-spans, aquatic plants 284; trees 202–203
Ligustrum, etymology 389
Ligustrum vulgare (Wild Privet) [333], **232**, 287f, 338, 350, 351, 354
Lily-of-the-Valley *see Convallaria majalis*
Limodōron, etymology 389
Limodorum abortivum 389f
Limonium, etymology 389
Limosella aquatica (Mudwort) [A31, B77], 66, **447**, **475**
Linaria vulgaris (Toadflax) [335], **233**, 342, 348, 354

Linden, Jan Antonides van der, quoted in main catalogue 169; source of observations 49; works in libraries 106, 108, 109
Linen 233
Linum catharticum (Fairy Flax) [338], 156, 233f, **234**, 329f, 334, 347–349, 353, 354
Linum perenne (Perennial Flax) [337], **234**, 329, 348, 349, 354, 485
Linum usitatissimum (Flax) [336], 45, **233**, 329f
Liparis loeselii (Fen Orchid) [397], 44, 217f, **248**, 336, 350, 353
Liquorice *see Glycyrrhiza glabra*
Liquorice, Wild *see Astragalus glycyphyllos*
Lister, Martin 19, 20, 34, 67–69
Listera ovata see Neottia ovata
Lithospermon [Λιθοσπερμον], etymology 392
Lithospermum arvense (Field Gromwell) [43], **158**, 331, 340
Lithospermum officinale (Common Gromwell) [339], 158f, **234**, 331, 350, 352, 392f
Littorella uniflora (Shoreweed) 69
Liver 206
Liver fluke *see Fasciola hepatica*
Lobaria pulmonaria 232f
L'Obel, Mathias de (Lobelius) 58, 86, 87, 89, 90, 91 (biog.), 94, 160f, 180f; identification of plants 485; names cited by Ray 37; pictures discussed 209, 249, 485; quoted: in main catalogue 211, 214, 215, 280, in 'Etymology' 404; works described in 'Explanation' 135, 139, 141, 142; works in libraries 99f, 103; works now attributed to Plantin 91, 94, 141f; *see also* Pena, Pierre & L'Obel, Mathias de
Lolium, etymology 58, 389
Lolium perenne (Perennial Rye-grass) [341], **235**, 328f, 397f
Lolium temulentum (Darnel) [340], **235**, 328f, 332, 338, 360f, 389f, 391f
London, botanists 452, 483; herb-sellers 149
Lonicer, Adam 91 (biog.); names cited by Ray 37; work annotated by Nidd 101;

587

INDEX

works described in 'Explanation' 141
Lonicera periclymenum (Honeysuckle) [432], **259**, 332, 345
Loosestrife *see Lysimachia*
Loosestrife, Purple *see Lythrum salicaria*
Loosestrife, Yellow *see Lysimachia vulgaris*
Lords-and-Ladies *see Arum maculatum*
Lorkyn, Thomas 98, 99
Lotus, etymology 390
Lotus corniculatus (Birdsfoot-trefoil) [579], **304**, 343
Lotus pedunculatus (Greater Birdsfoot-trefoil) [580], **304**, 343, 348
Lotus tenuis (Narrow-leaved Birdsfoot-trefoil) [B98], **478**
Lousewort *see Pedicularis sylvatica*
Lucilius, Caius Ennius, quoted in 'Etymology' 402 (biog.)
Lucretius (Titus Lucretius Carus), quoted in 'Etymology' 357 (biog.)
Lumley, John, Lord 98, 99
Lunularia cruciata see Marchantia polymorpha and/or *Lunularia cruciata*
Lupulus, etymology 174f, 389
Lutzius, Master 233
Luzula campestris (Field Woodrush) [251], **213**, 452, 483
Luzula multiflora (Heath Woodrush) [252], **213**, 348, 452, 483
Lychnis, etymology 390
Lychnis flos-cuculi see Silene flos-cuculi
Lycopus europaeus (Gypsywort) [358], **239**, 432f
Lymantria dispar (Gypsy Moth) 187f
Lynceans 127, 138
Lynnett, William 103, 106
Lysimachia 168; etymology 363, 390
Lysimachia nemorum (Yellow Pimpernel) 16, 51, 337
Lysimachia nummularia (Creeping Jenny) [388], **247**, 334, 343, 350
Lysimachia vulgaris (Yellow Loosestrife) [346], **236**, 272f, 345, 349, 390f, 425f
Lysimachus, King of Thrace 168f, 363, 390

Lyte, Henry, translation of Dodoens 86, 138f, 238f, 239f, 428f; works in libraries 109
Lythrum hyssopifolia (Grass Poly) [269], 44, **216**, 332, 350
Lythrum portula (Water Purslane) [B8], 266f, 441, **462**
Lythrum salicaria (Purple Loosestrife) [347, 348], **237**, 272f, 345, 390f, 425f, 433f

Macrobius, Ambrosius Theodosius 59, 76 (biog.); quoted: in 'Etymology' 364, 387, 405, in 'Interpretation' 423
Macroglossum stellatarum (Humming-bird Hawk-moth) 290
Madder *see Rubia tinctorum*
Madder, Field *see Sherardia arvensis*
Madwort *see Asperugo procumbens*
Maggots, eating into peas 261; generated in flesh 153; *see also* Larvae
Magnetism 360
Maier, Michael, quoted in main catalogue 234; source of observations 49; works in libraries 110, 111
Malache [Μαλαχη], etymology 390
Malaria 191, 233, 236f, 295, 409
Mallow, Common *see Malva sylvestris*
Mallow, Dwarf *see Malva neglecta*
Mallow, Marsh *see Althaea officinalis*
Mallow, Musk *see Malva moschata*
Mallows *see Malva*
Malus domestica see M. pumila
Malus pumila (Apple) 26f, 315, 323f, 362f
Malus sylvestris (Crab Apple) [356], **238**, 323f, 344
Malva (Mallows) 238f; etymology 390
Malva moschata (Musk Mallow) [A1, B2], 66, **441**, **461**
Malva neglecta (Dwarf Mallow) [354], **238**, 334
Malva sylvestris (Common Mallow) [355], **238**, 334
Manchester, Edward Montagu, 2nd Earl of 6
Mandragora officinarum (Mandrake) 297f
Mandrake *see Mandragora officinarum*
Manna 255

INDEX

Mantisalca salmantica (Dagger-flower) 220
Maple *see* Acer
Mapletoft, John 62, 103
Maranta, Bartolomeo, quoted in main catalogue 153, 190, 218, 240, 265, 269, 317; source of observations 49; works in libraries 107
Marcellus, Nonius, quoted in 'Etymology' 402 (biog.)
Marchantia polymorpha [A19, A20, B43, B44], 44, 66, **445**, **468**
Marchantia polymorpha and/or ***Lunularia cruciata*** [331], **231**, 333, 445f
Marcion *see* Moschion
Mare's-tail *see* Hippuris vulgaris
Marigold, Corn *see* Glebionis segetum
Marigold, Marsh *see* Caltha palustris
Marigold, Pot *see* Calendula officinalis
Marinellus, Joannes 135
Marjoram, Wild *see* Origanum vulgare
Marrubium, etymology 390
Marrubium vulgare (White Horehound) [357], **239**, 332, 349, 399, 432f
Mars 127f, 270, 374
Marsilius, Otto 279f
Martial 9; quoted in 'Etymology' 375
Martin, feast of Saint 280
Martin, John 24
Martini, Matthias 58; quoted in 'Etymology' 369 (biog.), 371–375, 377–380, 382f, 384f, 389, 392, 393, 394f, 395, 396f, 399, 402f, 407, 408, 411, 412f, 413; works in libraries 101, 107, 110
Martyn, Thomas 52; identification of Ray's plants 150f, 245f, 286f
Mary, Saint 364, 369
Mat-grass *see* Nardus stricta
Matricaria, etymology 390
Matricaria chamomilla (Scented Mayweed) 188f
Matricaria recutita see M. chamomilla
Matthaeus Sylvaticus 76 (biog.); names cited by Ray 76; works described in 'Explanation' 142; works in libraries 99
Mattioli, Pietro Andrea 47, 74, 80, 83, 85, 88, 90, 92 (biog.), 95, 140; names cited by Ray 37, 76; quoted in 'Etymology' 380, 392; works described in 'Explanation' 136, 137, 142; works in libraries 100, 103, 107, 111
Mausolus, King of Caria 363
Maximilian II, Emperor 137
Mayerne, Sir Theodore 92
Mayerus, Michael *see* Majerus, Michael
Mayweed 168
Mayweed, Scented *see* Matricaria chamomilla
Mayweed, Scentless *see* Tripleurospermum inodorum
Mead, Joseph 56
Meadow-grass *see* Poa
Meadow Rue *see* Thalictrum
Meadowsweet *see* Filipendula ulmaria
Meat pies 148, 307
Mecon [Μηκων], etymology 395
Medicago (Medicks) 39, 239f, 289
Medicago arabica (Spotted Medick) [581], 289f, **304**, 343
Medicago lupulina (Black Medick) [586], **304**, 343
Medicago minima (Bur Medick) [582], 65, **304**, 343, 351
Medicago sativa subsp. ***falcata*** (Sickle Medick) [587], **305**, 343
Medicinal plants, *Aegopodium podagraria* 159f; *Ajuga reptans* 290; *Betonica officinalis* 168; *Centaurium erythraea* 179; *Marrubium vulgare* 239; moss 244; *Persicaria hydropiper* 259; *Petasites hybridus* 259; *Primula veris* 268f; *Sanicula europaea* 290, *Ulmus procera* 316; *Veratrum* spp. 218f; *see also* Medicine; Qualities of plants
Medicine, 47; Galenic 47, 49, 50, 265f, 360f, 367f; Paracelsian 50, 92, 167f, 185f, 244f; *see also* Doctrine of Signatures; Epilepsy; Malaria; Medicinal plants; Stones, bladder and kidney
Medick *see* Medicago
Medlar *see* Mespilus germanicus
Mela, Pomponius 76 (biog.), 78; names cited by Ray 76
Melampodion, etymology 363
Melampus, King 363

INDEX

Melampyrum, etymology 390
Melampyrum arvense (Field Cow-wheat) 56, 391f
Melampyrum cristatum (Crested Cow-wheat) [359], 16, 101, **240**, 327, 350, 351
Melampyrum pratense (Common Cow-wheat) [360], 15, 101, **240**, 327
Melanorrhizon [Μελανορριζον], etymology 413
Melich, Georg 92 (biog.); names cited by Ray 35, 37, 92
Melilot, Tall *see Melilotus altissimus*
Melilōtus, etymology 391
Melilotus altissimus (Tall Melilot) [361], **240**, 334, 354, 406f
Melon, Water *see Citrullus vulgaris*
Melothron, etymology 367
Menelaus, King of Sparta 363, 383
Mentha, etymology 391
Mentha aquatica (Water Mint) [362], 231, **241**, 334
Mentha aquatica × *spicata* (*M.* × *piperita*) (Peppermint) 241f
Mentha arvensis (Corn Mint) [99], **173**, 325
Mentha longifolia 472f
Mentha longifolia × *suaveolens* (*M.* × *rotundifolia*) 472f
Mentha pulegium (Pennyroyal) [474], **270**, 337, 338, 400f
Mentha rotundifolia auct. 472f
Mentha spicata (Spearmint) [B63], 65, **472**; hybrids 241f
Mentha ? *spicata* × *suaveolens* (*M.* × *villosa*) [B62], **472**
Mentha suaveolens (Round-leaved Mint) 472f
Menyanthes trifoliata (Bogbean) [592], **306**, 325, 343
Mercurialis, etymology 363, 392
Mercurialis annua (Annual Mercury) [B64], 174, **472**
Mercurialis perennis (Dog's Mercury) [163], 174, **190**, 334, 351, 366f
Mercury *see Mercurialis*
Mercury (Hermes) 152f, 297f, 363, 374, 392
Mercury (quicksilver), use to kill trees 271

Merrett, Christopher 67, 89
Mespilus germanicus (Medlar) 289, 478f
Metamorphosis, of insects *see* Pupae; of plants 230, 231, 266, 275
Meum, etymology 55, 369f, 392
Meurs, Johannes van 396f (biog.); works in libraries 100
Mexicans, love of floral garlands 241
Microscope 47, 148f, 270f, 296
Midsummer Day/Eve *see* John the Baptist
Mignonette, Wild *see Reseda lutea*
Milium Solis, etymology 234, 392
Milk 152, 241, 289, 308, 311, 380, 381, 387, 395f
Milk-vetch, Purple *see Astragalus danicus*
Milkwort, Common *see Polygala vulgaris*
Milkwort, Sea *see Glaux maritima*
Millet, Broomcorn *see Panicum mileaceum*
Millet, Common *see Panicum mileaceum*
Millet, Foxtail *see Setaria italica*
Millet, Italian *see Setaria italica*
Minerva 374
Mint *see Mentha*
Mintha (Pluto's concubine) 391
Minuartia hybrida (Fine-leaved Sandwort) [32], **156**, 326, 353
Mistletoe *see Viscum album*
Mithridates VI Eupator, King of Pontus 363, 376
Mithridation, etymology 363
Moehringia trinervia (Three-nerved Sandwort) [A4, B7], 66, **441**, **462**
Moenchia erecta (Upright Chickweed) [B49], **470**
Moly, identification 152f
Monardes, Nicholaus, works described in 'Explanation' 137; works in libraries 111
Monk's-hood *see Aconitum napellus*
Monographs 140, 168
Montane plants 165, 173
Monte, Jo. de 193
Montia fontana (Blinks) [A2, B6], 66, 266f, **441**, **461**
Moon, harvesting in relation to 163, 170, 214

590

INDEX

Moonwort *see Botrychium lunaria*
Moors near Cambridge 43, 44, 173f
Morchella esculenta (Morel) 204f
Morchella sp. [205], **204**, 335
Morden, William 24, 437
More, Henry 56
Morel *see Morchella esculenta*
Morison, Robert 68; quoted in 1685 Appendix 479
Moschatel *see Adoxa moschatellina*
Moschion 76 (biog.), 168
Mosses, unidentifiable [13, 378, 380], **150**, **244**, **245**, 330f, 334, 398f
Motherwort *see Leonurus cardiaca*
Moths 187, 192, 193, 232, 288, 290, 296f; *see also* Larvae; Pupae
Moufet, Thomas 46, 86, 92 (biog.), 152f, 367f; names cited by Ray 35, 37, 93; quoted in main catalogue 148, 161, 165, 167, 187, 191–193, 232, 277, 278, 281, 287, 290, 318; source of observations 49; works in libraries 106, 107, 109
Mouse-ear *see Cerastium*
Mousetail *see Myosurus minimus*
Mudwort *see Limosella aquatica*
Mugwort *see Artemisia vulgaris*
Mullein *see Verbascum*
Muralium, etymology 382
Musa, Antonius 76 (biog.), 168, 363; source of observations 49
Muscus, etymology 392
Mustard 152, 295, 406
Mustard, Black *see Brassica nigra*
Mustard, Garlic *see Alliaria petiolata*
Mustard, Hedge *see Sisymbrium officinale*
Mustard, Tower *see Turritis glabra*
Mustard, White *see Sinapis alba*
Mustarda, etymology 406
Mycenae, etymology 379
Myces [Μυκης], etymology 379
Myosōtis, etymology 392
Myosotis arvensis (Field Forget-me-not) [381], **245**, 335, 340
Myosotis scorpioides (Water Forget-me-not) [382], **245**, 340
Myosoton aquaticum (Water Chickweed) [21], **154**, 155f, 326
Myosurus, etymology 245f

Myosurus minimus (Mousetail) [383], **245**, 335
Myriapods 187
Myrica gale (Sweet Gale) [173], 187f, **194**, 330, 335, 344, 375f
Myriophyllum spicatum (Spiked Water-milfoil) [368], **243**, 334, 352, 434f
Myriophyllum verticillatum (Whorled Water-milfoil) [367], **242**, 334, 434f

Naaman 361f
Nannius, Petrus, quoted in 'Etymology' 384 (biog.)
Napus, etymology 392
Narcissus (mythical youth) 392
Narcissus 289; etymology 392
Narcissus poeticus (Pheasant's-eye Daffodil) 393f
Narcissus poeticus × N. tazetta (*N. × medioluteus*) (Primrose Peerless) [B69], 68, **473**, 474f
Narcissus pseudonarcissus (Daffodil) [386], 43, 218, **246**, 260, 328
Nardostachys grandiflora (Spikenard) 397f
Nardus stricta (Mat-grass) [555], **298**, 334, 348
Narthecium ossifragum (Bog Asphodel) [B11], **462**, 476
Nasturtium *see Tropaeolum majus*
Nasturtium, etymology 393
Nasturtium microphyllum 246f
Nasturtium microphyllum × officinale (*N. × sterile*) 246f
Nasturtium officinale *sensu lato* (Watercress) [387], 116, 160, **246**, 272f, 327
Nasturtium officinale sensu stricto 246f
Nealand, William 24, 123
Needham, Walter 10, 12
Nematodes ("little worms"), generated in nose and teeth 223; in vinegar 148
Nenufar, etymology 393
Nenuphar, etymology 393
Neotinea ustulata (Burnt Orchid) [405], **251**, 336, 348, 352
Neottia ovata (Twayblade) [396], **248**, 251, 343, 347, 351, 354
Nepĕta, etymology 393

INDEX

Nepeta cataria (Cat-mint) [363], 15, **241**, 326, 335, 347, 393f
Neptune 374
Nettle *see Urtica*
Newton, Sir Isaac 7, 18
Nicander of Colophon 357f (biog.); his interpreter quoted in 'Etymology' 357
Nicholas, Friar 51
Nidd, Gervase 11, 103f
Nidd, John 10, 11 (biog.), 59f, 112; acknowledged by Ray 34, 128; books annotated by 93, 101; date of death 11f; mentioned in *Catalogus* 11, 63, 203, 451, 481; personal library 26, 40, 63, 103–107, 109–111, 498; possible co-author of *Catalogus* 62–64; residence at Trinity College 1658–1659 11f, 64; will 63, 101, 103, 104, 107
Nightshade, Black *see Solanum nigrum*
Nightshade, Deadly *see Atropa belladonna*
Nightshade, Enchanter's, etymology 371; *see also Circaea lutetiana*
Nipplewort, etymology 396; *see also Lapsana communis*
Nonius *see* Marcellus, Nonius
Norton, John 26
Norwich 241
Nufar, etymology 393
Nummularia, etymology 393
Nuphar lutea (Yellow Water-lily) [391], **247**, 333
Nymphaea, etymology 393
Nymphaea alba (White Water-lily) [390], 44, **247**, 333, 352
Nymphoides peltata (Fringed Water-lily) [392], **247**, 333, 352

Oak *see Quercus*
Oat *see Avena*
Oath of Engagement 6, 13
Oats, Black 165f
Ocymum, etymology 393
Ocymum garyophyllatum, etymology 394
Odontites vernus (Red Bartsia) [158], **188**, 327
Odontītis, etymology 394
Odysseus 152f, 272f, 358f, 371f
Oenanthe, etymology 394

Oenanthe aquatica (Fine-leaved Water-dropwort) 296f
Oenanthe aquatica and/or *O. fluviatilis* (Fine-leaved/River Water-dropwort) [134], 116, **182**, 332
Oenanthe fistulosa (Tubular Water-dropwort) [393], **248**, 296, 328
Oenanthe fluviatilis see O. aquatica and/or *O. fluviatilis*
Oil, flax 233, 234; rape 246, 275, 486
Olea europaea (Oleaster, Olive) 362f, 374f
Oleaster *see Elaeagnus angustifolia* and *Olea europaea*
Olesicarpos ['Ωλεσικαρπος], etymology 403
Olive *see Olea europaea*
Olive, Russian *see Elaeagnus angustifolia*
Olusatrum, etymology 221f
Onobrychis, etymology 394
Onobrychis viciifolia (Sainfoin) [394], **248**, 326, 329, 348, 354; subsp. *collina* 248f; subsp. *decumbens* 248f
Ononis, etymology 361
Ononis repens (Common Restharrow) [48], **159**, 339, 347, 354
Ononis spinosa (Spiny Restharrow) [47], **159**, 325, 339
Onopordum acanthium (Cotton Thistle) [4], **148**, 281f, 342
Ophioglossum vulgatum (Adder's Tongue) [395], **248**, 323, 347
Ophioscorōdon, etymology 394
Ophrys apifera (Bee Orchid) [408], 65, **252**, 336, 340, 348, 349
Ophrys insectifera (Fly Orchid) [400], **249**, 336, 340, 349, 352, 353
Ophrys sphegodes (Early Spider Orchid) [A26, B70], **446**, **474**
Opŭlus, etymology 395
Opuntia ficus-indica (Prickly Pear) 410f
Orache *see Atriplex*
Orange *see Citrus aurantium*
Orchid, Bee *see Ophrys apifera*
Orchid, Burnt *see Neotinea ustulata*
Orchid, Common Spotted *see Dactylorhiza fuchsii*
Orchid, Early Marsh *see Dactylorhiza incarnata*
Orchid, Early Purple *see Orchis mascula*

INDEX

Orchid, Early Spider *see Ophrys sphegodes*
Orchid, Fen *see Liparis loeselii*
Orchid, Fly *see Ophrys insectifera*
Orchid, Fragrant *see Gymnadenia conopsea sensu lato*
Orchid, Frog *see Coeloglossum viride*
Orchid, Greater Butterfly *see Platanthera chlorantha*
Orchid, Green-winged *see Anacamptis morio*
Orchid, Musk *see Herminium monorchis*
Orchid, Pyramidal *see Anacamptis pyramidalis*
Orchid, Southern Marsh *see Dactylorhiza praetermissa*
Orchid, unidentifiable [B42], 68, **468**
Orchids, growth of tubers 408; Ray's descriptions 217f
Orchis, etymology 395, 404
Orchis mascula (Early Purple Orchid) [398], **249**, 350, 351
Orchis morio see Anacamptis morio
Orchis ustulata see Neotinea ustulata
Oreopteris limbosperma (Lemon-scented Fern) [B34], **467**
Oribasius 76 (biog.); names cited by Ray 35, 76
Origanum, etymology 395
Origanum syriacum 384f
Origanum vulgare (Wild Marjoram) [410], **252**, 334, 349, 452, 485
Orléans, Gaston Jean-Baptiste de France, Duke of 479
Ornithogalum pyrenaicum (Spiked Star-of-Bethlehem) [B72], 68, 236f, **474**
Ornithopus perpusillus (Birdsfoot) [411], **252**, 324, 348
Orobanche, etymology 389
Orobanche elatior (Knapweed Broomrape) [412], **252**, 325, 349, 354
Orobanche minor (Common Broomrape) 253f
Orobanche rapum-genistae (Greater Broomrape) [412], 206, **252**, 325, 348, 389f
Orpine *see Sedum telephium*
Orta, Garcia de, works described in 'Explanation' 137; works in libraries 99, 105
Orthopelma mediator 281f

Osier *see Salix viminalis*
Osiris, etymology 395
Osmunda regalis (Royal Fern) [B32], **466**, 477
Osyris, etymology 395
Otis tarda see Bustard, Great
Ovid (Publius Ovidius Naso) 76 (biog.), 391f; quoted: in Preface 128, in 'Etymology' 358, 375, 384, 392, 407
Oxalis, etymology 358, 402
Oxalis acetosella (Wood-sorrel) [578], **303**, 341, 360f, 402f
Oxlip *see Primula elatior*
Oxtongue, Bristly *see Helminthotheca echioides*
Oxtongue, Hawkweed *see Picris hieracioides*
Oxyacantha, etymology 365
Oxyacanthus, etymology 365

Paints 280
Paliurus aculeatus 432f
Palm, Date *see Phoenix dactylifera*
Palm, leaves 151
Pamplin, William 15
Panax, etymology 179, 395
Panax Asclepium, etymology 363
Panax Centaurion, etymology 363
Panax Chironium, etymology 363
Panax coloni, etymology 295f, 395
Panax Heraclion, etymology 363
Panicum mileaceum (Common or Broomcorn Millet) 421f
Pansy *see Viola*
Papaver (Poppy) 131
Papāver, etymology 395
Papaver argemone (Prickly Poppy) [58], 115, **161**, 338
Papaver bivalve (Violet Horned-poppy) [415], **254**, 338
Papaver dubium sensu lato (Long-headed Poppy) 254f
Papaver hybridum (Rough Poppy) [59], 115, **161**, 338
Papaver lecoqii (Yellow-juiced Poppy) 254f
Papaver rhoeas (Common Poppy) [416], **254**, 338
Papaver somniferum (Opium Poppy) [417], 132f, **255**, 338, 396f

INDEX

Papaver spumeum, etymology 396
Papillaris, etymology 396
Paracelsus 49 (biog.), 245f; *see also* Medicine, Paracelsian
Paralysis, etymology 396
Parasitoid insects *see* Insects
Parietaria, etymology 382
Parietaria judaica (Pellitory-of-the-Wall) [418], **256**, 337, 382f
Paris 409f
Paris quadrifolia (Herb Paris) [277], **219**, 335, 336, 343, 350
Parkinson, John 34, 44, 45, 62, 89, 91, 93 (biog.), 96, 114, 188f, 204f, 206f, 220f, 224f, 262f, 263f, 289f, 339f, 444f, 475f; errors 156, 260, 286, 300, 303, 304, 487; identification of plants 177, 178, 190, 199f, 211, 212f, 235, 240, 242f, 251f, 264f, 317, 446, 452, 462, 470, 472, 485; mentioned by Corbyn 16; names cited by Ray 37, 40, 129; pictures discussed 16, 209, 211, 240, 452, 472, 485; quoted: in main catalogue 165, 189, 200, 205, 213, 222, 226, 240, 263, 264, 287, 293, 299, in 1685 Appendix 469; Ray's assessment of 39, 41, 142; source of observations 49; suspected plagiarism 213, 477; works described in 'Explanation' 142; works in libraries 99, 104, 107, 109, 110
Parnassia palustris (Grass of Parnassus) [264], 44, **216**, 331, 350, 353, 382f
Paronychia, etymology 396
Parsley, Corn *see Petroselinum segetum*
Parsley, Cow *see Anthriscus sylvestris*
Parsley, Fool's *see Aethusa cynapium*
Parsley, Garden *see Petroselinum crispum*
Parsley, Stone *see Sison amomum*
Parsley-piert *see Aphanes arvensis sensu lato*
Parsnip *see Pastinaca sativa* subsp. *sativa*
Parsnip, Wild *see Pastinaca sativa* subsp. *sylvestris*
Partridge (*Perdix perdix*) 308
Pasque Flower *see Pulsatilla vulgaris*
Pastināca, etymology 396

Pastinaca sativa subsp. ***sativa*** (Parsnip) [422], **257**, 303, 337; subsp. ***sylvestris*** (Wild Parsnip) [174], **194**, 337, 354
Paul of Aegina *see* Aegineta
Paulinus, Saint 368 (biog.)
Paulli, Simon 40, 93 (biog.); names cited by Ray 37; quoted in main catalogue 266; titles of published garden catalogues 93; work annotated by Nidd 93, 101; works described in 'Explanation' 140; works in libraries 104–106
Paulus (?Paulus Aegineta) 236; source of observations 49; *see also* Aegineta
Pausanias 265f (biog.)
Pea, Garden *see Pisum sativum*
Pea, Narrow-leaved Everlasting *see Lathyrus sylvestris*
Pea, Tuberous *see Lathyrus tuberosus*
Pear, Prickly *see Opuntia ficus-indica*
Pear, Wild *see Pyrus pyraster*
Pearlwort *see Sagina*
Pedicularis, etymology 396
Pedicularis sylvatica (Lousewort) [427], **258**, 333, 338
Peganon [Πηγανον], etymology 402
Peleus 201f
Pellitory-of-the-Wall *see Parietaria judaica*
Pena, Pierre 86, 91, 94 (biog.); *see also* Pena, Pierre & L'Obel, Mathias de
Pena, Pierre & L'Obel, Mathias de (1571, 1576, 1605) 2, 91, 188f; identification of plants 317; names cited by Ray 37; pictures discussed 210; quoted: in main catalogue 173, 174, 180, 255, 297, 300, 486, in 'Etymology' 358, 360, 368–370, 380, 381, 383, 389, 392, 393f, 396, 399, 405, 408, 410; source of observations 49; works described in 'Explanation' 135, 139; works in libraries 99f
Penny, Thomas 86, 92, 248f
Pennyroyal *see Mentha pulegium*
Pennywort, Marsh *see Hydrocotyle vulgaris*
Peppermint *see Mentha aquatica* × *spicata*
Perdicium, etymology 383
Perdix perdix see Partridge

INDEX

Perennial plants, definition 434; soil preferences 164
Perforāta, etymology 397
Periclyměnum, etymology 397
Periwinkle, Lesser *see Vinca minor*
Perne, Andrew 98
Perottus, Nicolaus 379f, 382f
Persicaria, etymology 397
Persicaria amphibia (Amphibious Bistort) [461], **266**, 338
Persicaria bistorta (Common Bistort) [87], 43, 131, **170**, 218f, 324, 340
Persicaria hydropiper (Water-pepper) [435], **259**, 323, 397f
Persicaria maculosa (Redshank) [433], **259**, 323
Persicaria minor (Small Water-pepper) [434], **259**, 323
Persicaria vivipara (Alpine Bistort) 170f
Personāta, etymology 397
Petasītes, etymology 397
Petasites hybridus (Butterbur) [436], 237f, **259**, 325, 337
Peter, Saint 364
Petroselinum crispum (Garden Parsley) 182f, 264f, 296f, 448
Petroselinum segetum (Corn Parsley) [A39, B91], **448**, **477**
Phaenias 77 (biog.), 168
Phalăris, etymology 397
Phalaris arundinacea (Reed-grass) [240], **211**, 339f, 397f
Phalaris canariensis (Canary Grass) 397f
Phanias *see* Phaenias
Pharnaceon, etymology 363
Pharnaces, King of Pontus 363
Philaenus spumarius (Cuckoo-spit Insect) 255f
Philanthropos, etymology 362
Philetaeria, etymology 398
Philoponus, Joannes 368f
Phleum bertolonii (Smaller Cat's-tail) [268], **216**, 330
Phleum pratense (Timothy) [267], **216**, 330
Phlomos [Φλομος], etymology 58, 412
Phoenix, etymology 397
Phoenix dactylifera (Date Palm), seeds 198; terminology: 58, *dactylus* 420, *palmula* 424, *spadix* 425, *termes* 427

Phragmites australis (Common Reed) [64], **163**, 338
Phu, etymology 397
Phyllitis scolopendrium see Asplenium scolopendrium
Picris echioides see Helminthotheca echioides
Picris hieracioides (Hawkweed Oxtongue) [284], **221**, 331, 354f
Pieris brassicae (Large White) 276
Pignut *see Conopodium majus*
Pigs (*Sus scrofa*) 165, 300, 404, 407
Pigweed *see Amaranthus*
Piles 181f, 379
Pilewort, etymology 181f
Pilosella, etymology 392
Pilosella officinarum (Mouse-ear Hawkweed) [437], **260**, 335, 392f, 469
Pimpernel, Blue *see Anagallis arvensis* subsp. *foemina*
Pimpernel, Bog *see Anagallis tenella*
Pimpernel, Scarlet *see Anagallis arvensis*
Pimpernel, Yellow *see Lysimachia nemorum*
Pimpinella, etymology 398
Pimpinella anisum 428f, 432f
Pimpinella major (Greater Burnet-saxifrage) [441], 65, **260**, 340, 348
Pimpinella saxifraga (Burnet-saxifrage) subsp. **saxifraga** var. *dissecta* [442], **260**, 340; var. *intercedes* 261f; var. *saxifraga* [440], **260**, 340
Pine *see Pinus*
Pinguicula vulgaris (Common Butterwort) [443], 44, **261**, 325, 340, 350
Pink, Maiden *see Dianthus deltoides*
Pinus (Pine) 284
Pinus pinea (Stone or Umbrella Pine) 362f, 386f
Piso, Gulielmus 106
Pistachia terebinthus 167f
Pisum, etymology 398
Pisum sativum (Garden Pea) [444], 45, 188, **261**, 337, 433f
Plane, Oriental *see Platanus orientalis*
Plantago (Plantain) 168, 408; etymology 398
Plantago coronopus (Buck's-horn Plantain) [152], 15, **186**, 348, 351

INDEX

Plantago lanceolata (Ribwort Plantain) [450], **263**, 337, 339; var. *pusilla* 263f
Plantago major (Greater Plantain) [447], **262**, 337, 344; teratological variants [449, B78], 44, 68, **262**, 337, **475**
Plantago media (Hoary Plantain) [448], **262**, 333, 337
Plantain *see Plantago*
Plantin, Christopher 86, 89, 91, 94 (biog.), 141f, 188f, 296f; names cited by Ray 37, 94; pictures discussed 158, 210; works in libraries 104
Plants, uses of 132
Platanaria, etymology 398
Platanthera chlorantha (Greater Butterfly Orchid) [407], **252**, 336, 340, 351
Platanus orientalis (Oriental Plane) 298, 398f
Plautus, Titus Maccius 265f (biog.), 366f, 418f
Pliny the Elder (Gaius Plinius Secundus) 46, 47, 58, 71, 73–76, 77 (biog.), 78, 80 (biog.), 151f, 157f, 163f, 174f, 181f, 182f, 185f, 187f, 221f, 250f, 268f–270f, 272f, 283f, 292f, 303f, 309f, 408f; names cited by Ray 77; plants described by 160f, 167f–169f, 171f, 191f, 222f, 226f, 244f, 247f, 273f, 274f, 299f, 359f, 367f, 369f–371f, 380f, 381f, 383f, 388f; quoted, in main catalogue: 149, 154, 162, 168f, 186, 189, 200–203, 209, 225, 233, 265, 270, 280, 301, 310, 315, in 'Etymology' 357–363, 366, 367, 369–376, 379–381, 383, 388–390, 392–395, 398, 401, 403, 405, 406, 409, 412, in 'Interpretation' 415, 416, 418, 419, 421; source of observations 49; works described in 'Explanation' 138, 140, 142; works in libraries 99, 106
Plot, Robert 291f
Ploughman's Spikenard *see Inula conyzae*
Plume, Joseph 5
Plutarch 9, 75, 77 (biog.); quoted in 'Etymology' 377, 393, 401, 402
Pluto 391
Poa angustifolia (Narrow-leaved Meadow-grass) 214f

Poa annua (Annual Meadow-grass) [260], **214**
Poa humilis (Spreading Meadow-grass) 214f
Poa pratensis *sensu lato* [259], **214**
Poa pratensis sensu stricto (Smooth Meadow-grass) 214f
Pockley, Thomas 62, 103
Poisonous plants or fungi 188; *Atropa bellabonna* 297; *Cannabis sativa* 174; *Conium maculatum* 182; *Fraxinus excelsior* 202; fungi 203; *Lactuca sativa* sap 302; *Taxus baccata* 301
Polemonium, etymology 398
Polishing 233
Polygala vulgaris (Common Milkwort) [451], 44, **263**, 334, 347, 348, 350, 354
Polygonum arenastrum 264f
Polygonum aviculare *sensu lato* (Knotgrass) [454], **264**, 333
Polygonum aviculare sensu stricto 264f
Polygonum rurivagum 264f
Polypodium, etymology 398
Polypodium interjectum 264f
Polypodium vulgare *sensu lato* (Polypody) [455], **264**, 337, 351
Polypodium vulgare sensu stricto 264f
Polypody *see Polypodium vulgare sensu lato*
Polytrichum, etymology 398
Polytrichum commune [12], 44, **150**, 330f, 334, 350, 398f
Polytrichum juniperinum 150f
Pomegranate *see Punica granatum*
Pona, Francesco 143
Pona, Giovanni 94 (biog.); names cited by Ray 37, 94; works described in 'Explanation' 143; works in libraries 109
Pond-sedge *see Carex*
Pondweed, Broad-leaved *see Potamogeton natans*
Pondweed, Curled *see Potamogeton crispus*
Pondweed, Fennel *see Potamogeton pectinatus*
Pondweed, Flat-stalked *see Potamogeton friesii*
Pondweed, Grass-wrack *see Potamogeton compressus*

596

INDEX

Pondweed, Horned *see Zannichellia palustris*
Pondweed, Lesser *see Potamogeton pusillus*
Pondweed, Opposite-leaved *see Groenlandia densa*
Pondweed, Perfoliate *see Potamogeton perfoliatus*
Pondweed, Sharp-leaved *see Potamogeton acutifolius*
Pondweed, Shining *see Potamogeton lucens*
Pondweed, Small *see Potamogeton berchtoldii*
Poplar *see Populus*
Poppy *see Papaver*
Population density of plants, effects 174, 222, 317
Populus (Poplar) 265, 284; etymology 399
Populus alba and/or ***P. alba × tremula*** (White/Grey Poplar) [456], **264**, 338, 347, 353, 378f
Populus alba × tremula (*P. × canescens*) (Grey Poplar) *see P. alba* and/or *P. alba × tremula*
Populus nigra (Black Poplar) [458], **266**, 338
Populus tremula (Aspen) [457], **265**, 324, 338, 347, 353, 399
Porphyrius 77 (biog.); quoted in 'Interpretation' 427
Porta, Giambattista della 50, 318; source of observations 49
Portulāca, etymology 399
Portulaca oleracea (Common Purslane) [459], 44, 65, **266**
Potamogeton acutifolius (Sharp-leaved Pondweed) 267f
Potamogeton berchtoldii and/or ***P. pusillus*** (Small/Lesser Pondweed) [465], 44, **267**, 338, 350
Potamogeton compressus (Grass-wrack Pondweed) [464], **267**, 338
Potamogeton crispus (Curled Pondweed) [576], **303**, 325, 329
Potamogeton friesii (Flat-stalked Pondweed) 267f
Potamogeton lucens (Shining Pondweed) [462], **267**, 338, 485

Potamogeton natans (Broad-leaved Pondweed) [460], 44, **266**, 337, 350
Potamogeton pectinatus (Fennel Pondweed) [370], **243**, 329f, 334, 352
Potamogeton perfoliatus (Perfoliate Pondweed) [463], **267**, 338
Potamogeton pusillus (Lesser Pondweed) *see P. berchtoldii* and/or *P. pusillus*
Potentilla, etymology 363, 399
Potentilla anserina (Silverweed) [60], **161**, 340, 342
Potentilla argentea (Hoary Cinquefoil) [B73], **474**
Potentilla erecta (Tormentil) [571], 44, **302**, 342, 350
Potentilla palustris see Comarum palustre
Potentilla reptans (Creeping Cinquefoil) [429], **258**, 326, 329
Potentilla sterilis (Barren Strawberry) [202], **201**, 341
Poterium sanguisorba subsp. ***balearicum*** 260f; subsp. ***sanguisorba*** (Salad Burnet) [439], **260**, 325, 347, 348, 354
Power, Henry 45
Prassium, etymology 399
Priam, King of Troy 409f
Priest, Robert 87, 139
Primrose *see Primula vulgaris*
Primrose Peerless *see Narcissus poeticus × N. tazetta* (*N. × medioluteus*)
Primrose, James, quoted in main catalogue 244; works in libraries 106
Primula 188; etymology 399f
Primula elatior (Oxlip) [468], **268**, 336, 350, 351, 396f
Primula veris (Cowslip) [467], **268**, 327, 336, 354, 396f; etymology 396, 399
Primula vulgaris (Primrose) [469], **268**, 338, 354, 396f
Printing, declining standards 1550–1700 25
Priscianus Caesariensis, quoted in 'Etymology' 372 (biog.)
Privet, Wild *see Ligustrum vulgare*
Proserpina 315f, 319
Pruinnen, etymology 399
Prune [Προυνη], etymology 400
Prunella, etymology 399

INDEX

Prunella vulgaris (Selfheal) [470], **269**, 340
Prunos [Προυνος], etymology 400
Prunus, etymology 399
Prunus avium (Wild Cherry) [B24], **465**
Prunus cerasus (Dwarf Cherry) [B25], **465**
Prunus domestica subsp. *insititia* var. *nigra* (Black Bullace) [B80], 68, **475**; var. *syriaca* (White Bullace) [B79], 68, **475**
Prunus spinosa (Blackthorn) [471], **269**, 324, 340, 342
Pseudacorus, etymology 149f
Psora [Ψωρα], etymology 404
Ptarmĭca, etymology 400
Pteridium aquilinum (Bracken) [200], **199**, 324, 329, 348
Pteris [Πτερις], etymology 379
Ptisan 168
Pubescence, terminology: *lanugo* 422, *pappus* 424, *tomentum* 427
Puff-ball fungi *see* Fungi, puff-ball
Puff-ball, Giant *see Calvatia gigantia*
Pulegium, etymology 400
Pulicaria, etymology 372
Pulicaria dysenterica (Common Fleabane) [142], **184**, 329, 372f
Pulicaria vulgaris (Small Fleabane) [143], **184**, 329, 372f
Pulsatilla, etymology 400
Pulsatilla vulgaris (Pasque Flower) [475], 15, 44, 115, **270**, 272f, 337, 348
Pulses *see* Leguminous plants
Punic, words derived from 385, 386
Punica granatum (Pomegranate), terminology: 58, *balaustium* 417, *ciccum, cicus* 418, *cytinus* 419, *malicorium* 422
Pupae (chrysalises) 187, 192, 232, 276–279, 287, 318
Purgatives 169, 219f, 313, 448
Purslane, Common *see Portulaca oleracea*
Purslane, Water *see Lythrum portula*
Putrefaction, animals generated by 147, 296
Pyros [Πυρος], etymology 411
Pyrus pyraster (Wild Pear) 51, 337

Pythagoras 238 (biog.); quoted in 'Etymology' 377, 378
Pythagoreans 133; regarded mallows as sacred 238f; views on beans variously explained 377

Qualities of plants 47; aquatic plants 273; *Heracleum sphondylium* 299; *Humulus lupulus* 299; *Marchantia polymorpha* and/or *Lunularia cruciata* 231; *Papaver* 255; *see also* Medicine
Quartan fever *see* Malaria
Quercus (Oak) 169, 225, 284
Quercus coccifera (Kermes Oak) 206, 269f
Quercus ilex (Holm Oak) 150f
Quercus robur (Pedunculate Oak) [476], 45, 116, **271**, 335, 351
Quicksilver *see* Mercury

Rabdophaga cinerearum 287f
Rabdophaga strobilina 287f
Radish *see Raphanus*
Radix, etymology 401
Ragged Robin *see Silene flos-cuculi*
Ragwort *see Senecio*
Ramsons *see Allium ursinum*
Rana temporaria see Frog
Rantzius, Master 199
Ranunculus, etymology 272f, 400
Ranunculus acris (Meadow Buttercup) [A35, B84], 66, **447**, **476**
Ranunculus aquatilis (Common Water-crowfoot) [479], 160f, 263f, **272**, 327
Ranunculus arvensis (Corn Buttercup) [480], **272**, 327
Ranunculus auricomus (Goldilocks) [486], **273**, 327, 330f, 347
Ranunculus bulbosus (Bulbous Buttercup) [481], **272**, 327
Ranunculus ficaria see Ficaria verna
Ranunculus flammeus, etymology 400
Ranunculus flammula (Lesser Spearwort) [483, 484], **273**, 341
Ranunculus fluitans (River Water-crowfoot) 113
Ranunculus hederaceus (Ivy-leaved Crowfoot) [485], **273**, 327, 453, 486
Ranunculus lingua (Greater Spearwort) [482], 44, **273**, 341, 352

INDEX

Ranunculus parviflorus (Small-flowered Buttercup) [A33, B83], 66, 113, **447**, **475**
Ranunculus penicillatus subsp. *pseudofluitans* (Stream Water-crowfoot) [B82], **475**
Ranunculus repens (Creeping Buttercup) [487], **274**, 325, 327
Ranunculus sardous (Hairy Buttercup) [A34, B85], 66, 272f, **447**, **476**
Ranunculus sceleratus (Celery-leaved Buttercup) [478], 160f, 263f, **272**, 327, 341
Ranunculus trichophyllus (Thread-leaved Water-crowfoot) [366], **242**, 327f, 329f, 434f
Rape, Oil-seed *see Brassica napus*
Raphănus, etymology 401
Raphanus raphanistrum subsp. ***raphanistrum*** (Wild Radish) [491], **274**, 326, 486
Raphanus sativus (Garden Radish) 168
Rapum, etymology 401
Rapum Genistae, etymology 389
Ravelingen, Françoys van 138f
Raven, Charles Earle, life of Ray 3; identification of Ray's insects: *Campoplex difformis* 192f, *Cotesia glomerata* 277f, moth on *Corylus*187f; identification of Ray's plants: 113, 116, *Alopecurus pratensis* 210f, *Asperula cynanchica* 290f, *Carex disticha* 212f, *Orobanche elatior* 252f, *Sagina procumbens* 291f, *Stellaria media* 155f; interpretation of Ray's motives in 1662 18; understates Nidd's contribution to *Catalogus* 63
Ray, Catharine 22
Ray, Elizabeth (John's mother) 4, 21, 22, 24
Ray, Elizabeth (John's sister) 4
Ray, Jane 22, 23
RAY, JOHN, biography, early life 4; enters University of Cambridge 5; enters Trinity College 5; returns to Trinity 8; pupil of James Duport 8, 10; illness 13, 117f; positions held 1649–1662 8; residence at Trinity: 1651–1654 14, 1658–1661 64; bequest from Nidd 103, 104, 107; ordination 8; position at Restoration 17; leaves Cambridge 17, 439, 457; at Middleton Hall 19, 20; elected F.R.S. 20; changes name from Wray 4f; Willughby's executor 20; annuity from Willughby 20; marries 20; leaves Middleton Hall 20; returns to Black Notley 21; daughters born 22; supports Wren Library 18; later life 22; anticipates eternity 13; commemorated by Trinity College 18

~~scientific work, begins to study botany 13, 34, 125; book purchases 106; botanical tours: Britain 15, 18, 20, Europe 19; catalogue of Cambridge garden plants 15; personal library 22, 26f, 103f, 106, 498; plans for work after Cambridge catalogue 1, 15, 18, 131, 215, 439, 457; studies plants in small garden 15, 126; work with Willughby 19, 20

~~(1660) *Catalogus plantarum circa Cantabrigiam nascentium*, aims 129; arrangement 25, 129; authorship 62; Babington's annotated copy 113; chosen names 39; classification of plants in 'More customary headings' 59; concise format 2; coverage of county 65, 489; date of completion 59f; delimitation of species 41; description of localities 41, 42; digressions 56; direct and indirect citation of sources 97; experiments reported in 47; fieldwork 42, 43, first County Flora 1, 64; habitats 42, 44, 118; identification of plants 113; imperfect alphabetical order 39, 51, 58; length of names 40; list of common plants not found in county 15, 131; models 26; modern plant names 54; names coined by Ray 37, 41; names treated in 'Etymology' 52; notes or observations 45, 48; plans described by Hartlib 14; preparation 13, 14, 59f; printing 24; publication 24; records outside county 65; significance 1; sources: 'Etymology' 58, 'Explanation' 35, 38, in libraries 97, 'Interpretation' 58, names 36, 40, observations 46, 48; style of 'Preface' 34, 117f; target audience 133; treatment of earlier

INDEX

records 44; treatment of garden escapes 43; variant copies 24
~~(1663) *Appendix ad Catalogus plantarum circa Cantabrigiam nascentium* 15, 17, 24, 51, 63, 65, 67–69, 96, 111; Babington's annotated copy 113; fieldwork, 1660–1662 66; reasons for publication 66, 439, 457; scope outlined 439, 457
~~(1670) *Catalogus plantarum Angliae* 41, 43, 45, 67–69, 115, 150f, 158f, 176f, 199f, 210f, 217f, 225f, 227f–229f, 237f, 243f, 244f, 273f, 282f, 283f, 286f, 294f, 313f, 314f, 337f, 441f, 443f, 444f, 446f–448f, 459, 461f–473f, 476f, 478f, 479f, 481f, 482, 483f, 484f, 485, 486, 487f; taxonomic problems posed by commonest species 67; treatment of Cambridgeshire species 68
~~(1674) *Of the specifick differences of plants* 41, 237f, 257f, 275f
~~(1675) *Dictionariolum trilingue* 20, 153f, 229f, 270f
~~(1677) *Catalogus plantarum Angliae* 2nd ed. 43, 67, 69, 115, 158f, 228f, 243f, 244f, 264f, 266f, 273f, 282f, 294f, 313f, 337f, 444f, 459, 461f–465f, 467f–471f, 473f–476f, 479f, 481f, 482, 483f, 484f, 485, 486, 487f; excluded records from Dent 68; treatment of garden escapes 68
~~(1685, with P. Dent) *Appendix ad Catalogus plantarum circa Cantabrigiam nascentium* 2nd ed., authorship 67, 69; omission of some published records 69; reasons for publication 459; scope outlined 459; sources 67, 68
~~(1686–1688) *Historia plantarum* 11, 22, 67, 69, 115, 158f, 189f, 200f, 211f, 217f, 218f, 220f, 229f, 231f, 243f, 244f, 254f, 260f, 282f, 289f, 291f, 293f, 298f, 313f, 444f, 447f, 463f, 465f, 469f, 473f, 476f–478f
~~(1688) *Fasciculus* 22, 64, 115, 261f, 465f, 473f, 478f
~~(1690) *Synopsis methodica stirpium Britannicarum* 22, 115, 176f, 196f, 228f, 229f, 244f, 261f, 282f, 293f, 313f, 441f, 446f, 447f, 451f, 452f, 462f–465f, 473f, 476f–478f, 482f, 483f
~~(1691, 1692) *The wisdom of God* 11, 13, 18, 22, 23, 26, 47, 125f, 288f
~~(1695) in *Camden's Britannia* 65, 156f, 177f, 194f, 220f, 233f, 294f
~~(1696) *Synopsis methodica stirpium Britannicarum*, 2nd ed. 22, 115, 166f, 196f, 212f, 226f, 228f, 229f, 261f, 282f, 293f, 313f, 441f, 444f, 446f, 451f, 452f, 463f
~~other works 12, 20, 22, 23, 47, 161f, 279f
Ray, Margaret (John's daughter) 22
Ray, Margaret (John's wife, née Oakeley) 20, 22
Ray, Mary 22
Ray, Roger (John's brother) 4
Ray, Roger (John's father) 4
Redshank *see Persicaria maculosa*
Reed, Common *see Phragmites australis*
Reed-grass *see Phalaris arundinacea*
Reed-pipes 163
Regional studies before 1660 26
Reneaulme, Paul de 94 (biog.); names cited by Ray 37, 94; quoted in main catalogue 172, 179; works described in 'Explanation' 142; works in libraries 109
Resēda, etymology 401
Reseda lutea (Wild Mignonette) [497], **279**, 339, 347, 348, 354, 401f, 425f
Reseda luteola (Weld) [343], **236**, 328, 345, 354
Restharrow *see Ononis*
Revolting things, healing virtues of 51f
Rhamnus cathartica (Buckthorn) [498], **279**, 325, 342, 351, 353
Rhaphanus, etymology 401
Rhinanthus minor (Yellow-rattle) [426], **258**, 326, 338
Rhododendron *see Rhododendron ponticum*
Rhododendron ponticum (Rhododendron) 273f
Rhodon ['Ροδον], etymology 401
Rhœas, etymology 396
Rhynchospora alba (White Beak-sedge) [B40], **468**

INDEX

Ribes nigrum (Black Currant) [499], **280**, 328, 341
Robinson, Tancred 23
Rocket, Garden *see Eruca vesicaria* subsp. *sativa*
Rockrose, Common *see Helianthemum nummularium*
Rodigues, João *see* Amatus Lusitanus
Roemeria hybrida see Papaver bivalve
Rogga, etymology 401
Rondelet, Guillaume 81, 85, 90, 91, 94, 103f; works described in 'Explanation' 135
Root-collectors, extirpation of plants by 132, 274
Roots, absence 189, 200; culms with 211; definition 425; depth 257; downward movement 219; growth on *Lemna* 231; medicinal 218; shape 258, 408; size 163; terminology: *fibrae* 420, *fundus* 421, *tubera* 428; use as tinder 310; *see also* Edible plants and plant products; Orchids, growth of tubers; Variation within species
Ropalon, etymology 393
Rorippa amphibia (Great Yellow-cress) [488], **274**, 338
Rorippa microphylla see Nasturtium microphyllum
Rorippa nasturtium-aquaticum see Nasturtium officinale
Rorippa sylvestris (Creeping Yellow-cress) [185], **196**, 339
Rorippa × *sterilis see Nasturtium microphyllum* × *officinale* (*N.* × *sterile*)
Rosa (Roses), 188; cordial flower 367f; cultivation 151; etymology 401
Rosa canina (Dog Rose) [502], 192, **280**, 324, 339; etymology 401
Rose, Dog *see Rosa canina*
Rosemary *see Rosmarinus officinalis*
Roses *see Rosa*
Rosmarinus officinalis (Rosemary) 384f
Ross, Alexander 98, 99
Rösslin, Eucharius 91; works described in 'Explanation' 141
Roubiliac, Louis François 18
Rough breathing, changed to 's' to form Latin words 56, 405
Rowan *see Sorbus aucuparia*

Rubia tinctorum (Madder) 189, 206, 247f
Rubus, etymology 402
Rubus caesius (Dewberry) [504], **281**, 324, 328
Rubus fruticosus agg. (Bramble) [505], **282**, 324
Rubus ulmifolius 282f
Ruel, Jean 74, 78, 90, 92, 94 (biog.); names cited by Ray 37; quoted in 'Etymology' 368; works described in 'Explanation' 138, 141, 143; works in libraries 99, 112
Rumex, etymology 402; identification problems 42, 67, 116, 229f, 230f, 484
Rumex acetosa (Common Sorrel) [7], 115, **149**, 341, 402f
Rumex acetosella (Sheep's Sorrel) [8], 115, **149**, 341, 348, 352, 402f
Rumex conglomeratus (Clustered Dock) [324], 229f, **230**, 328, 484
Rumex crispus (Curled Dock) [B55], 229f, **471**, 484
Rumex hydrolapathum (Water Dock) [323], **230**, 328, 484
Rumex maritimus (Golden Dock) [B56], **471**, 484
Rumex obtusifolius (Broad-leaved Dock) [322], **229**, 328, 484
Rumex palustris (Marsh Dock) 230f, 484f
Rumex patientia (Patience Dock) 170f
Rumex pulcher (Fiddle Dock) [B57], **471**, 485
Rumex sanguineus (Wood Dock) 229f; var. *viridis* 229f
Rumex spp. [321], **229**, 328, 484
Rupert, Saint 363, 381
Ruscus aculeatus (Butcher's Broom) [A37, B87], 66, 131, 254, **448**, 476
Rush, Blunt-flowered *see Juncus subnodulosus*
Rush, Bulbous *see Juncus bulbosus*
Rush, Compact *see Juncus conglomeratus*
Rush, Flowering *see Butomus umbellatus*
Rush, Frog *see Juncus ranarius*
Rush, Hard *see Juncus inflexus*
Rush, Heath *see Juncus squarrosus*
Rush, Jointed *see Juncus articulatus*
Rush, Round-fruited *see Juncus compressus*

INDEX

Rush, Sharp-flowered *see Juncus acutiflorus*
Rush, Soft *see Juncus effusus*
Rush, Toad *see Juncus bufonius*
Russell, Sir Francis 169, 492 (biog.)
Rustyback *see Asplenium ceterach*
Ruta, etymology 402
Rye *see Secale cereale*
Rye-grass, Perennial *see Lolium perenne*

Sage, Wood *see Teucrium scorodonia*
Sagina apetala sensu lato (Annual Pearlwort) 291f
Sagina filicaulis 291f
Sagina nodosa (Knotted Pearlwort) [31], 44, **156**, 340, 350, 353
Sagina procumbens (Procumbent Pearlwort) [525], 290f, **291**, 340, 352, 453, 487
Sagitta, etymology 403
Sagittaria sagittifolia (Arrowhead) [506, 507], **282**, 323, 465
Sainfoin *see Onobrychis viciifolia*
St John's College, Cambridge, Library 40f, 102, 110, 111, 112
St John's-wort *see Hypericum*
Saints, eponyms listed 363
Salicastrum, etymology 403
Salix, etymology 403; identification problems 282
Salix alba (White Willow) [508, 509], **283**, 287, 344, 486; cv. 'Britzensis' 283f; var. *vitellina* 283f, 286f
Salix caprea (Goat Willow) [514], **285**, 287f, 375f
Salix caprea × *viminalis* (*S.* × *smithiana*) 286f
Salix cinerea (Grey Willow) [513], **285**, 286f, 287f, 340
Salix cinerea × *viminalis* (*S.* × *holosericea*) 286f
Salix fragilis (Crack Willow) [510], **284**, 344
Salix purpurea (Purple Willow) [512], **285**, 286f, 287f
Salix repens (Creeping Willow) [B88, B89], 187f, **476**
Salix 'rosea' see Camellia galls
Salix sp. [516, 517], **286**, 486

Salix triandra (Almond Willow) [511], **284**, 344
Salix viminalis (Osier) [515], 34f, **286**, 336
Salix × *sericans see S. caprea* × *viminalis* (*S.* × *smithiana*)
Salix × *smithiana* auct. *see S. cinerea* × *viminalis* (*S.* × *holosericea*)
Salsify *see Tragopogon porrifolius*
Salt 147, 159, 167, 171, 261
Salvia, etymology 403
Salvia verbenaca (Wild Clary) [292], **223**, 326
Sambucus, etymology 403
Sambucus ebulus (Dwarf Elder) [170], **193**, 328, 344, 350, 351
Sambucus nigra (Elder) [519], 169, **288**, 328; heartwood 202
Samolus valerandi (Brookweed) [39], 44, 86, **157**, 337, 350, 353
Sanamunda, etymology 177f, 370
Sandwort, Fine-leaved *see Minuartia hybrida*
Sandwort, Three-nerved *see Moehringia trinervia*
Sandwort, Thyme-leaved *see Arenaria serpyllifolia*
Sanguisorba, etymology 398
Sanguisorba minor subsp. *minor see Poterium sanguisorba* subsp. *sanguisorba*; subsp. *muricata see Poterium sanguisorba* subsp. *balearicum*
Sanguisorba officinalis (Great Burnet) [438], 43, 218f, **260**, 295f, 325
Sanicle *see Sanicula europaea*
Sanicula, etymology 404
Sanicula europaea (Sanicle) [520], **289**, 340, 351
Sap 148, 193, 202, 209, 225, 289, 302, 358, 402, 407, 422; potable 169
Saponaria, etymology 404
Saponaria officinalis (Soapwort) [521], **290**, 341, 349
Sardoa, etymology 272f
Sare, Mr 147
Satyrium, etymology 404
Saw-wort *see Serratula tinctoria*
Saxifraga granulata (Meadow Saxifrage) [522], **290**, 340, 349

INDEX

Saxifraga tridactylites (Rue-leaved Saxifrage) [420], **256**, 344, 349
Saxifrage, Meadow *see Saxifraga granulata*
Saxifrage, Pepper *see Silaum silaus*
Saxifrage, Rue-leaved *see Saxifraga tridactylites*
Scabiosa, etymology 404
Scabiosa columbaria (Small Scabious) [527], **291**, 340, 349, 354
Scabious, Devil's-bit *see Succisa pratensis*
Scabious, Field *see Knautia arvensis*
Scabious, Small *see Scabiosa columbaria*
Scaliger, Joseph Justus 59, 399f, 400f, 423f; quoted in 'Interpretation' 423
Scaliger, Julius Caesar 59, 82; quoted: in main catalogue 315 (biog.), in 'Etymology' 360, in 'Interpretation' 423, 427; source of observations 49; works in libraries 99, 107
Scandix pecten-veneris (Shepherd's Needle) [425], **258**, 285, 340, 343
Scented plants 188, 194, 313, 432; *see also* Smell of plants
Schenck, Johann Georg 140
Schoenoplectus lacustris (Common Club-rush) [306], **226**, 339, 471
Schoenoplectus tabernaemontani (Grey Club-rush) 226f
Schoenus nigricans (Black Bog-rush) [309], 44, 115, **227**, 339, 350, 353
Schola Salernitana 390; in libraries 109
Schröder, Johannes 59, 95 (biog.); names cited by Ray 37, 95; quoted: in main catalogue 159, 162, 179, 185, 205, 269, 310, 313, in 'Interpretation' 418, 422, 423, 425, 428; source of observations 49; works described in 'Explanation' 143; works in libraries 105, 106, 108, 109
Schwenckfelt, Kaspar 95 (biog.); names cited by Ray 37, 95; works described in 'Explanation' 143
Scirpus, etymology 404
Scleranthus annuus (Annual Knawel) [453], **263**, 333, 348
Scribonius Largus 59, 78 (biog.), 86; names cited by Ray 35, 78; quoted in 'Interpretation' 429

Scordium, etymology 404
Scorzoneroides autumnalis (Autumn Hawkbit) [281], **219**, 331, 354f, 408f
Scrophularia, etymology 379, 404
Scrophularia auriculata (Water Figwort) [83], **167**, 324
Scrophularia nodosa (Common Figwort) [529], 167, **292**, 329, 350, 351
Scurvygrass, Common *see Cochlearia officinalis*
Scutellaria galericulata (Skullcap) [345], **236**, 332, 345, 409f
Sea-spurrey *see Spergularia*
Sebisch, Melchior 144
Secāle, etymology 405
Secale cereale (Rye) [530], 199, **292**, 318, 339, 401f; germination 309; *see also* Cereals
Sedge *see Carex*
Sedum acre (Biting Stonecrop) [533], **293**, 341, 344
Sedum album (White Stonecrop) [532], **293**, 338, 341
Sedum rupestre (Reflexed Stonecrop) [531], **292**, 338, 341
Sedum telephium (Orpine) [566], 181f, **301**, 333, 336
Seeds, absence in ferns 200; beans 198; collection 182; dates 198; dormancy 160, 254, 262; eaten by birds 175, 182, 222; emetic and purgative 313; fertility 241, 254; forceful expulsion from fruit 303; germination 148, 160, 240, 308, 419, 427; late developing 358; morphology 148, 198, 209, 269; rarity in some mints 241; regeneration of *Salix* 403; regeneration of *Viscum* 315; relationship to fruit size 239, terminology: *cor*, *corculum* 419, *semen* 425; time to maturity 223; vermifugous 223; *see also* Oil; Fruits
Seignior, George 17f
Selfheal *see Prunella vulgaris*
Selīnum, etymology 405
Semecarpus anacardium 289
Semitic languages, words derived from 52, 56, 57, 384
Senecio, etymology 375, 405
Senecio aquaticus (Marsh Ragwort) [301], **225**, 338, 470

INDEX

Senecio erucifolius (Hoary Ragwort) [B52], **470**
Senecio jacobaea (Common Ragwort) [300], **225**, 338, 340, 364f
Senecio paludosus (Fen Ragwort) [146], **184**, 329, 352
Senecio ?sylvaticus (Heath Groundsel) [534], **293**, 331
Senecio viscosus (Sticky Groundsel) 293f
Senecio vulgaris (Groundsel) [535], **294**, 331, 340
Sennert, Daniel 50, 167f; quoted in main catalogue 316; source of observations 49; works in libraries 99, 103, 107
Serapias, etymology 405
Serapio, Joannes, quoted in 'Etymology' 392 (biog.), 409
Serapis 405
Serpyllum, etymology 405
Serratula, etymology 406
Serratula tinctoria (Saw-wort) [538], **294**, 340, 351
Sertula campana, etymology 406
Service Tree, Wild *see Sorbus torminalis*
Servilius, Damocrates *see* Damocrates Servilius
Servius *see* Honoratus, Servius Maurus
Seseli libanotis (Moon Carrot) 261
Setaria italica (Foxtail or Italian Millet) 421f
Shade, effects on *Chenopodium polyspermum* 451, 482
Sharpe, Samuel 103
Sheep 270, 280
Sheep rot 280
Sheep's-bit *see Jasione montana*
Shepherd's Needle *see Scandix pecten-veneris*
Shepherd's Purse *see Capsella bursa-pastoris*
Sherardia arvensis (Field Madder) [503], **281**, 333
Shoreweed *see Littorella uniflora*
Shropshire 199
Shrubs 428; classification 60, 431, 432; definition 421; *see also* Branches; Subshrubs; Topiary
Sideritis, etymology 406

Silaum silaus (Pepper Saxifrage) [524], 290f, **291**, 340
Silene dioica (Red Campion) 131f
Silene flos-cuculi (Ragged Robin) [61], **162**, 327, 337, 344, 347
Silene gallica (Small-flowered Catchfly) [B61], 68, 69, **472**
Silene latifolia (White Campion) [344], **236**, 325
Silene otites (Spanish Catchfly) [539], 65, **294**, 301, 304, 326, 351
Silene vulgaris (Bladder Campion) [80], **167**, 338, 396f
Silkworm *see Bombyx mori*
Silverweed *see Potentilla anserina*
Silybum marianum (Milk Thistle) [108], **176**, 342, 369f
Simeon, Saint 364
Simpson, Thomas 23
Sināpi, etymology 406
Sinapis alba (White Mustard) [492], **275**, 326
Sinapis arvensis (Charlock) [490], **274**, 326, 335f
Sison amomum (Stone Parsley) [542], **295**, 296f, 336, 349, 448
Sisymbrium officinale (Hedge Mustard) [186], **196**, 327, 335
Sium, etymology 406
Sium latifolium (Greater Water-parsnip) [543], 217, **296**, 337
Skin, application of plants to 152, 229f, 239, 270, 273
Skippon, Philip (d. 1660) 19
Skippon, Sir Philip (1641–1692) 19
Skullcap *see Scutellaria galericulata*
Sloane, Sir Hans 23
Slugs, hermaphrodite 297
Smell of plants, *Allium sativum* 153, *A. vineale* 152, *Carduus nutans* 451, 482, *Cynoglossum officinale* 190, *Epilobium hirsutum* 238, *Lactuca* 227, *Melilotus altissimus* 240, *Nepeta cataria* 241, 393f, *Nuphar lutea* 247, *Onopordum acanthium* 148, *Populus* 266, *Pulicaria dysenterica* 184, *Rosa canina* 280, *Thalictrum minus* 301, *Valeriana officinalis* 310; *see also* Scented plants
Smilax [Σμιλαξ], etymology 407

INDEX

Smith, Thomas 102
Smooth breathing, changed to 'v' to form Latin words 56, 413
Smut *see Ustilago segetum sensu lato*
Smyrnium creticum 216
Smyrnium olusatrum (Alexanders) [285], **221**, 323, 349, 375f
Smyth, William 101
Snagge, John 62
Snails, eaten by frogs 400; hermaphrodite 297
Snapdragon *see Antirrhinum majus*
Sneezewort *see Achillea ptarmica*
Snowdrop *see Galanthus nivalis*
Soapwort *see Saponaria officinalis*
Soil, effects of plants on 230; effects on: cultivated plants 174, flower colour 225f, 313f, plant size 306; preference of different life-forms for 164; relationship to root structure 257
Solānum, etymology 407
Solanum dulcamara (Bittersweet) [547], **296**, 324, 335, 372f, 374f
Solanum nigrum (Black Nightshade) [546], **296**, 335
Solemn League and Covenant 6, 17
Solidago virgaurea (Goldenrod) [A42, B102], 66, **449**, **479**
Solinus, Gaius Julius 78 (biog.); quoted in 'Etymology' 358
Solstice, summer 151, 200, 265, 283; winter 270
Son before father, etymology 237f, 309
Sonchus [Σογχος], etymology 58, 407
Sonchus arvensis (Perennial Sowthistle) [551], **298**, 341, 354f
Sonchus asper (Prickly Sowthistle) [549], **298**, 341
Sonchus oleraceus (Smooth Sowthistle) [550], **298**, 331, 341
Sorbus aucuparia (Rowan) [B94], **477**
Sorbus torminalis (Wild Service Tree) [B95], **478**
Sorrel *see Rumex*
Sorrell, etymology 358
Sowthistle *see Sonchus*
Spachius, Israel 139
Sparganium, etymology 407
Sparganium emersum (Unbranched Bur-reed) [554], 115, **298**, 339

Sparganium erectum (Branched Bur-reed) [553], 115, **298**, 325, 339, 398f
Spatŭla foetida, etymology 407
Spearmint *see Mentha spicata*
Spearwort *see Ranunculus*
Speed, John 26, 498
Speedwell *see Veronica*
Spelt *see Triticum spelta*
Spergula, etymology 407
Spergula arvensis (Corn Spurrey) [557], **299**, 341, 348
Spergularia marina and/or ***S. media*** (Lesser/Greater Sea-spurrey) [559], **299**, 341
Spergularia media (Greater Sea-spurrey) *see S. marina* and/or *S. media*
Spergularia rubra (Sand Spurrey) [558], **299**, 341, 348
Sphagnum 245f
Sphinx ligustri (Privet Hawk-moth) 232, 287
Sphondylium, etymology 407
Spiders 288
Spiegel, Adriaan van der 35, 59, 95 (biog.); names cited by Ray 35, 37, 95; quoted: in main catalogue 153, 189, 196, 198, 200, 224, 230, 255, 261, 281, 306, in 'Etymology' 395, 404, in 'Interpretation' 416, 418, 421, 422, 427; source of observations 49; works in libraries 109
Spigelius *see* Spiegel
Spikenard *see Nardostachys grandiflora*
Spike-rush, Common *see Eleocharis palustris*
Spindle *see Euonymus europaeus*
Spiraea, etymology 408
Spiranthes spiralis (Autumn Lady's Tresses) [409], 44, **252**, 333, 352
Spleenwort *see Asplenium*
Spontaneous generation 47, 147, 153, 223, 296
Spurge *see Euphorbia*
Spurrey, Corn *see Spergula arvensis*
Spurrey, Sand *see Spergularia rubra*
Squalermo, Luigi *see* Anguillara, Luigi
Squinancywort *see Asperula cynanchica*
Stachys, etymology 408
Stachys arvensis (Field Woundwort) [B93], **477**

605

INDEX

Stachys officinalis see Betonica officinalis
Stachys palustris (Marsh Woundwort) [540], **295**, 323, 395f
Stachys sylvatica (Hedge Woundwort) [209], **205**, 335, 354, 380f
Stanhope, Sir Michael 101
Staphylea pinnata (Bladdernut) [561], 44, 93, **299**, 324, 424f
Staphylodendron, etymology 408
Starch 307
Star-of-Bethlehem, Spiked *see Ornithogalum pyrenaicum*
Star-thistle, Red *see Centaurea calcitrapa*
Stellaria alsine (Bog Stitchwort) [27], **155**, 326, 348
Stellaria aquatica, etymology 408
Stellaria graminea (Lesser Stitchwort) [287], **221**, 341, 348
Stellaria holostea (Greater Stitchwort) [286], **221**, 341, 350, 351
Stellaria media (Common Chickweed) [28, 29], **155**, 256, 326, 441, 462, 481
Stellaria neglecta (Greater Chickweed) 155f
Stellaria uliginosa see S. alsine
Stems, absence 190; climbing clockwise 236; definition 418; morphology 157, 159, 172, 179, 189, 190, 195, 196; pubescence 295; terminology: *ala* 415, *asparagus* 195f, 416, *cyma* 419, *fundus* 421, *geniculum* 421, *internodium* 421, *scapus* 425, *stolones* 427; *see also* Branches; Edible plants and plant products; Variation within species
Stephen, Saint 182f, 364
Stephens, Philip 34
Stitchwort *see Stellaria*
Stomach 152
Stonecrop *see Sedum*
Stones, bladder and kidney 169, 375, 399, 406
Stones (testicles) 249f, 251f
Storksbill, Common *see Erodium cicutarium*
Strabo, quoted in 'Etymology' 405 (biog.)
Strachey, Robert 44, 262
Stratiotes aloides (Water Soldier) [364], **242**, 340, 341, 352, 434f

Strawberrry, Barren *see Potentilla sterilis*
Strawberry, Wild *see Fragaria vesca*
Struthium, etymology 404
Subshrubs, classification 60, 431, 432; definition ('*suffrutex*') 427
Succisa, etymology 408
Succisa pratensis (Devil's-bit Scabious) [374], **243**, 328, 354, 387f, 448, 476
Succory, Lamb's *see Arnoseris minima*
Suda 138; in libraries 101
Sundew *see Drosera*
Sura, Mamilius 394f
Sus scrofa see Pigs
Swallows (*Hirundo rustica*), restoration of sight 180, 371
Swammerdam, Jan Jacob 279f
Sweet Gale *see Myrica gale*
Sweet-grass *see Glyceria*
Swine-cress *see Lepidium coronopus*
Sybil of Cumae 315f
Sycamore *see Acer pseudoplatanus*
Symphyton [Συμφυτον], etymology 372, 409f
Symphytum, etymology 408
Symphytum officinale (Common Comfrey) [147], **185**, 328
Syriac, words derived from 385, 386
Syzygium aromaticum (Clove-tree) 394f

Tabernaemontanus, Jakob Theodor 87, 95 (biog.); names cited by Ray 37; quoted in 'Etymology' 381; works described in 'Explanation' 136, 139, 143; works in libraries 99, 111
Tammuz 361f
Tamus communis (Black Bryony) [92], **171**, 236, 325, 347, 354
Tanacētum, etymology 409
Tanacetum balsamita (Costmary) 364f
Tanacetum parthenium (Feverfew) [421], 114, **256**, 329
Tanacetum vulgare (Tansy) [563], **300**, 342
Tansy *see Tanacetum vulgare*
Tapsus, etymology 409
Taraxacon, etymology 409
Taraxacum spp. (Dandelion) [167], **191**, 328, 354f, 469f
Tare *see Vicia*

606

INDEX

Taste of plants, *Allium vineale* 152; *Artemisia absinthium* 147; *Caucalis platycarpos* 178; *Marchantia polymorpha* and/or *Lunularia cruciata* 231; *Ranunculus acris* 447, 476; *Solanum dulcamara* 297, *Tragopogon* 178; *see also* Bitter plants
Taxus, etymology 409
Taxus baccata (Yew) [565], **300**, 345
Teasel *see Dipsacus*
Teesdalia nudicaulis (Shepherd's Cress) [98], **172**, 340, 348, 451, 482
Telamon, King of Salamis 409f
Telephium, etymology 363, 409
Telephus, King of Mysia 358, 363, 409
Tephroseris integrifolia (Field Fleawort) [302], 16, **225**, 338, 348, 352
Tephroseris palustris (Marsh Fleawort) [145], **184**, 329
Tertianaria, etymology 409
Tetragonium, etymology 376
Tetter 229f
Teucer, King of Troas 363, 409
Teucrium, etymology 363, 409
Teucrium chamaedrys (Wall Germander) 154f, 465f
Teucrium polium 432f
Teucrium scordium (Water Germander) [528], 89, **292**, 330, 465f
Teucrium scorodonia (Wood Sage) 131
Teucrium sp. [B27], **465**
Thal, Johannes 83, 95 (biog.); names cited by Ray 37; works described in 'Explanation' 143; works in libraries 109
Thalictrum, etymology 409
Thalictrum flavum (Common Meadow Rue) [567], **301**, 339
Thalictrum minus (Lesser Meadow Rue) [568], 65, **301**, 339, 351
Themison 71, 78 (biog.), 168
Theophrastus 46, 58, 71, 72, 74, 77, 78 (biog.), 82, 85, 87, 148f, 370f, 390f; identification of plants 150f, 163f, 169f, 191f, 201f, 206f, 247f, 253f, 297f, 359f, 375f, 400f, 443f, 478f; names cited by Ray 78; quoted: in main catalogue 151, 154, 160, 163, 165, 186, 189, 194, 200, 203, 217, 222, 230, 239, 240, 254, 257, 258, 262, 265, 266, 271, 284, 301, 308, 313, 317, 319, in 'Etymology' 357, 358, 376, 389, 393, 400, 403, 410; source of observations 49; works described in 'Explanation' 140, 141, 143; works in libraries 99, 100, 107, 110
Thesium humifusum (Bastard Toadflax) [334], **233**, 342, 348, 352
Thesmophoria 375
Thistle, Carline *see Carlina vulgaris*
Thistle, Cotton *see Onopordum acanthium*
Thistle, Creeping *see Cirsium arvense*
Thistle, Dwarf *see Cirsium acaule*
Thistle, Marsh *see Cirsium palustre*
Thistle, Meadow *see Cirsium dissectum*
Thistle, Milk *see Silybum marianum*
Thistle, Musk *see Carduus nutans*
Thistle, Slender *see Carduus tenuiflorus*
Thistle, Spear *see Cirsium vulgare*
Thistle, Welted *see Carduus crispus*
Thistle, Woolly *see Cirsium eriophorum*
Thistles, identification problems 176
Thornton, Thomas 62
Thorow-wax *see Bupleurum rotundifolium*
Thuidium tamariscinum [A24, B67], 66, **445**, **473**
Thyme, Basil *see Clinopodium acinos*
Thyme, Breckland *see Thymus serpyllum*
Thyme, Large *see Thymus pulegioides*
Thyme, Wild *see Thymus polytrichus*
Thymus polytrichus (Wild Thyme) [536, 537], 66, **294**, 343, 348, 354, 448, 477; var. *britannicus* 294f; var. *neglectus* 294f
Thymus pulegioides (Large Thyme) 477f
Thymus serpyllum (Breckland Thyme) 477f
Thymus sp. [A40, B92], 66, **448**, **477**
Tilletia tritici and/or **Ustilago segetum** *sensu lato* (Bunt/Smut) [629], **318**, 325
Timothy *see Phleum pratense*
Tinder 204, 310
Tiredness, prevention of 162
Tithymālus, etymology 410
Tithymalus characias, etymology 410
Toadflax *see Linaria vulgaris*
Toadflax, Bastard *see Thesium humifusum*

INDEX

Toadflax, Small *see Chaenorhinum minus*
Tomentaria, etymology 381
Topiary 232, 428
Topsell, Edward 92
Torilis arvensis (Spreading Hedge-parsley) [B23], **464**
Torilis japonica (Upright Hedge-parsley) [119], **178**, 336, 464
Torilis nodosa (Knotted Hedge-parsley) [120], **178**, 336
Tormentil *see Potentilla erecta*
Tormentilla, etymology 410
Tortoiseshell, Small *see Aglais urticae*
Trachelium, etymology 410
Tragopōgon, etymology 410
Tragopogon porrifolius (Salsify) 303f
Tragopogon pratensis (Goat's-beard) [574], 289, **303**, 330, 354
Tragus, Hieronymus *see* Bock, Jerome
Transplantation 15, 202, 241, 269
Traveller's Joy *see Clematis vitalba*
Treacle Mustard *see Erysimum cheiranthoides*
Tree rings 63, 202–203
Trees, classification 60, 431; definition 416; fruit and seed size 239; growth rate 284; improvement of fruit 26f; killing 271; relationship of habit to density 317; relationship of size to fruit size 271; roots 257; shaped by wind 317; terminology: *caudex* 418, *comae* 418, *cortex* 419, *liber* 422, *matrix* 422; *termes* 427, *truncus* 428; time of fruiting 151; *see also* Bark; Branches; Buds; Tree rings; Wood
Trefoil *see Trifolium*
Tremellius 405
Tribŭlus, etymology 410
Tribulus terrestris (Maltese Cross) 410f
Trifolium arvense (Hare's-foot Clover) [316], **228**, 331, 349
Trifolium campestre (Hop Trefoil) [584], **304**, 343, 347, 348, 354
Trifolium dubium (Lesser Trefoil) [585], **304**, 343, 347
Trifolium fragiferum (Strawberry Clover) [583], **304**, 343
Trifolium medium (Zigzag Clover) [591], **305**, 326, 343, 487

Trifolium ochroleucon (Sulphur Clover) [589], **305**, 343, 349, 354
Trifolium pratense (Red Clover) [590], **305**, 332, 343, 487
Trifolium repens (White Clover) [588], **305**, 343
Trifolium striatum (Knotted Clover) [593], **306**, 343, 347
Trifolium subterraneum (Subterranean Clover) [A41, B99], 66, 68, **449**, **478**
Triglochin palustris (Marsh Arrowgrass) [257], 44, **214**, 331, 350, 352
Trigonella foenum-graecum (Fenugreek) 289
Trinity College, Cambridge, Admissions Books 8; appeal for new library building 18; Edit/Redit Books 14; Library 11, 16, 18, 26, 40, 90, 93, 96, 100, 102–106, 109, 110, 112; religious differences within 16, 17; scientific community 10–12; Upper Buttery Books 8f, 11f, 12, 14, 63; wealth 5
Tripleurospermum inodorum (Scentless Mayweed) 188f
Trisāgo, etymology 55, 411
Triticum (Wheat) 45, 292, 319, 390, 411, 415f, 416, 421; etymology 411; germination 309; *see also* Cereals
Triticum aestivum (Bread Wheat) [595–599], **306**, **307**, 344
Triticum spelta (Spelt) 416
Triticum turgidum (Rivet Wheat) [600, 601], **307**, **308**, 344; var. *dinurum* 307f; var. *iodurum* 307f
Tropaeolum majus (Nasturtium) 467
Tulipa (Tulips) 289
Tulips *see Tulipa*
Turbatrix aceti (Vinegar Eel) 148
Turgenia latifolia (Greater Bur-parsley) [118], **177**, 336, 350
Turmeric *see Curcuma longa*
Turnèbe, Adrien, quoted in 'Etymology' 399
Turner, William 44, 96 (biog.), 174f, 222f, 274f, 381f, 428f; names cited by Ray 37; quoted in 'Etymology' 392; records omitted from *Catalogus* 44, 96; works described in 'Explanation' 144; works in libraries 99, 100
Turnip *see Brassica rapa*

INDEX

Turnip, Wild *see Brassica rapa*
Turritis glabra (Tower Mustard) 309f
Tussilāgo, etymology 411
Tussilago farfara (Coltsfoot) [603], 237f, **309**, 326, 349, 378f
Tussock, Nut-tree *see Colocasia coryli*
Tussock, Pale *see Calliteara pudibunda*
Tutisan, etymology 361
Tutsan *see Hypericum androsaemum*
Twayblade *see Neottia ovata*
Typha, etymology 411
Typha latifolia (Bulrush) [604], 44, **310**, 326, 353
Typhon 411

Ulcers 152, 259
Ulex europaeus (Gorse) [212], **206**, 330, 344, 348, 352
Ulmus glabra (Wych Elm) [625], **315**, 329
Ulmus minor (Small-leaved Elm) [624], **315**, 329
Ulmus procera (English Elm) [626], 45, **315**, 329
Uncertain identifications discussed by Ray, *Anacamptis pyramidalis* 251, *Apium nodiflorum* 296, *Carduus nutans* 176, *Carex* sp. or spp. 212, *Caucalis platycarpos* 178, *Cirsium palustre* 176, *Cladium mariscus* 190, *Crepis capillaris* 220, *Cynosurus cristatus* 210, *Elytrigia repens* 211, *Epipactis helleborine* 218, *Geranium* 452, 482, *Helianthemum nummularium* 180, *Hippocrepis comosa* 263, *Juncus compressus* 213, *Limosella aquatica* 447, 475, *Ophrys sphegodes* 446, *Pimpinella saxifraga* 260, *Plantago major* 475, *Portulaca oleracea* 266, *Potamogeton lucens* 267, *Salix* spp. 282, *Schoenus nigricans* 227, *Seseli libanotis* 261, *Sinapis alba* 275, *Sonchus arvensis* 298, *Trifolium dubium* 304, *T. medium* 305, *T. ochroleucon* 305, *T. striatum* 306, *Turgenia latifolia* 177, *Veronica scutellata* 157
University of Cambridge, Act of Uniformity 6f, 18; aristocratic element 7; effects of Civil War 5; fellow-commoners 7; growth 1564–1621 5; Parliamentary purges 6; pensioners 7; poor students 5, 7; Ray bemoans lack of botanical expertise 126; Ray urges men to study science 13, 132; recreations 9, 110f; sizars & subsizars 7; syllabus 6; tutorial system 7; *see also* Cambridge University Library; Catharine Hall; St John's College; Trinity College
Urceolaria, etymology 383
Urination 159, 169, 236
Urine, stained by *Crocus sativus* 189
Urtica (Nettles) 168; etymology 413
Urtica dioica (Common Nettle) [628], **317**, 335
Urtica urens (Small Nettle) [627], **317**, 335
Usnea spp. or other fruticose lichens [377], **244**, 335
Usnea subfloridana 244f
Ustilago avenae 318f
Ustilago hordei 318f
Ustilago segetum *sensu lato* (Smut) *see Tilletia tritici* and/or *Ustilago segetum sensu lato*
Ustilago tritici 318f
Utricularia minor (Lesser Bladderwort) [B65], 44, **473**
Utricularia vulgaris (Greater Bladderwort) [369], **243**, 334, 473
Uvularia, etymology 413

Vaccinium oxycoccos (Cranberry) [B100], **479**
Valerian *see Valeriana*
Valeriāna, etymology 412
Valeriana dioica (Marsh Valerian) [606, B101], 44, **310**, 343, 397f, **479**
Valeriana officinalis (Common Valerian) [605], **310**, 343, 351, 397f
Valerianella locusta (Common Cornsalad) [311], **227**, 326, 348
Variation within species, awns: *Triticum turgidum* 308; colour: *Artemisia vulgaris* 162, *Chenopodium polyspermum* 451, 482, *Salix* sp. 486; flore pleno: *Cardamine pratensis* 68, 463, *Hydrocharis morsus-ranae* 41, 244f, *Knautia arvensis*

448, 476; flower colour: *Achillea millefolium* 41, *Anacamptis morio* 249, *A. pyramidalis* 251, *Anagallis arvensis* 158f, *Antennaria dioica* 210, *Artemisia vulgaris* 162, *Carduus nutans* 176, *Centaurea cyanus* 189, *C. scabiosa* 225f, 313f, *Centaurium erythraea* 179, *Cichorium intybus* 181, *Dactylorhiza* spp. 250, *Galeopsis tetrahit sensu lato* 175, *Geranium molle* and/or *G. pusillum* 208, *Gymnadenia conopsea sensu lato* 250, *Knautia arvensis* 291, *Linum perenne* 234, *Myosotis arvensis* 245, *Odontites vernus* 188, *Ononis spinosa* 159, *Onopordon acanthium* 148, *Orchis mascula* 249, *Polygala vulgaris* 263, *Prunella vulgaris* 269, *Raphanus raphanistrum* 275f, *Viola tricolor* agg. 314; fruiting time: *Hordeum distichon* 222; fruits: *Sinapis arvensis* 274; leaves of mature plants: *Capsella bursa-pastoris* 172, *Lythrum salicaria* 41, *Persicaria amphibia* 266, *P. maculosa* 259; leaves, ontogenetic changes: *Hedera helix* 41, 216, 265, *Ilex aquifolium* 151, 216, *Populus* 265, *Sium latifolium* 217, 296, *Smyrnium creticum* 216; roots: *Blackstonia perfoliata* 179; rows of grains: *Hordeum vulgare* 222; size: *Arabidopsis thaliana* 447, 474, *Aster tripolium* 306, *Capsella bursa-pastoris* 41, 172, *Crepis capillaris* 220, *Eleocharis palustris* 226, *Plantago coronopus* 15, 186, *Stellaria media* 481, *Torilis japonica* 178; stem pubescence: *Brassica nigra* 295; teratological: *Plantago major* 44, 68, 262f, 475f; treatment in Ray's later works: *Achillea millefolium* 243f, *Anagallis arvensis* 158f, *Brassica rapa* 275f, *Centaurea scabiosa* 225f, *Daucus carota* subsp. *sativus* 257f, *Knautia arvensis* 448f, *Hedera helix* 217f, *Heracleum sphondylium* 478f, *Hydrocharis morsus-ranae* 244f, *Lythrum salicaria* 237f, *Pimpinella saxifraga* 260f, *Ranunculus flammula* 273f, *Sagittaria sagittifolia* 282f, *Salix alba* 283f, *S. repens* 476f, *Salix* sp. 286f, *Thymus polytrichus* 294f, *Viola odorata* 313f; *see also* Differences from closely related taxa outlined; Differences not or only doubtfully constituting specific distinctions

Varro, Marcus Terentius 58, 59, 72, 78 (biog.), 423, 429f; names cited by Ray 79; quoted: in main catalogue 275, in 'Etymology' 365, 394, 401, 411, in 'Interpretation' 418, 420, 422; source of observations 49; works in libraries 108

Venus 297f, 315f, 373, 374, 403

Venus's Looking-glass *see Legousia hybrida*

Verātrum, etymology 58, 382, 413

Veratrum album (White False-helleborine) 218f

Veratrum nigrum (Black False-helleborine) 218f

Verbascum, etymology 412

Verbascum nigrum (Dark Mullein) [607], **311**, 335, 348

Verbascum thapsus (Great Mullein) [564], **300**, 332, 333, 335, 409f

Verbēna, etymology 58, 412

Verbena officinalis (Vervain) [608], **311**, 344, 354

Veronica, etymology 58, 412

Veronica agrestis (Field Speedwell) [22], **154**, 326

Veronica anagallis-aquatica *sensu lato* (Water Speedwell) [37, 38], **157**, 158f, 325, 451, 481

Veronica anagallis-aquatica sensu stricto 157f

Veronica arvensis (Wall Speedwell) [23], **154**, 326, 348

Veronica beccabunga (Brooklime) [35, 36], **157**, 158f, 451, 481

Veronica catenata 157f

Veronica chamaedrys (Germander Speedwell) [126], **180**, 330

Veronica hederifolia (Ivy-leaved Speedwell) [24], **154**, 326, 332, 348, 477f

Veronica montana (Wood Speedwell) [A12, B26], 66, 111f, **443**, **465**

Veronica officinalis (Heath Speedwell) [609], **311**, 329, 341, 352, 433f

INDEX

Veronica scutellata (Marsh Speedwell) [40], 44, **157**, 325, 352
Veronica serpyllifolia (Thyme-leaved Speedwell) [610], **311**, 329, 341, 351, 412f
Veronica spicata (Spiked Speedwell) [611], **311**, 329, 341, 352
Verulam, Baron *see* Bacon, Sir Francis
Verus, Cloatius, quoted in 'Etymology' 387 (biog.)
Vervain *see Verbena officinalis*
Vesta 374
Vetch, Bitter *see Vicia ervilia*
Vetch, Bush *see Vicia sepium*
Vetch, Common *see Vicia sativa*
Vetch, Kidney *see Anthyllis vulneraria*
Vetch, Tufted *see Vicia cracca*
Vetchling *see Lathyrus*
Viburnum, etymology 412
Viburnum lantana (Wayfaring Tree) [612], 101, **312**, 327, 337, 344, 350, 351, 412f
Viburnum opulus (Guelder Rose) [518], **287**, 329, 351, 433f; cv. 'Roseum' 287f
Vicia, etymology 412
Vicia cracca (Tufted Vetch) [55], **161**, 342, 344, 354
Vicia ervilia (Bastard Lentil, Bitter Vetch) 433f
Vicia faba (Broad Bean) 189, 313f; var. *equina* and/or var. *minor* [194], **198**, 324; var. *faba* [193], **198**, 324; var. *minor see* var. *equina* and/or var. *minor*; *see also* Beans
Vicia hirsuta (Hairy Tare) [57, 613], **161**, **312**, 342, 453, 481
Vicia sativa (Common Vetch) subsp. *nigra* and/or subsp. *segetalis* [56], **161**, 312f, 341, 413f; subsp. *sativa* [615], **312**, 342; subsp. *segetalis see* subsp. *nigra* and/or subsp. *segetalis*
Vicia sepium (Bush Vetch) [616], **313**, 344
Vicia tetrasperma (Smooth Tare) [614], **312**, 342
Vigo, Giovanni da 96 (biog.); names cited by Ray 35, 37, 96
Vinca minor (Lesser Periwinkle) [138], **183**, 337, 350
Vine, Grape *see Vitis vinifera*
Vinegar Eel *see Turbatrix aceti*
Viola (Violets) 231f; cordial flower 367f; etymology 413
Viola arvensis (Field Pansy) [620], **314**, 331, 336, 349, 38'/f, 487
Viola canina (Heath Dog Violet) 314f
Viola odorata (Sweet Violet) var. *dumetorum* and/or var. *imberbis* [618], **313**, 344; var. *imberbis see* var. *dumetorum* and/or var. *imberbis*; var. *odorata* [617], **313**, 344, 387f
Viola reichenbachiana and/or *V. riviniana* (Early/Common Dog Violet) [619], **314**, 344, 350
Viola riviniana (Common Dog Violet) *see V. reichenbachiana* and/or *V. riviniana*
Viola tricolor (Wild Pansy) 314f, 487f
Violet, Common Dog *see Viola riviniana*
Violet, Dame's, etymology 231f; *see also Hesperis matronalis*
Violet, Early Dog *see Viola reichenbachiana*
Violet, Heath Dog *see Viola canina*
Violet, Sweet *see Viola odorata*
Violet, Water *see Hottonia palustris*
Violets *see Viola*
Virgil 9, 79 (biog.), 150f, 173, 181, 198f, 269f, 292f, 316f, 384f, 423f; identification of plants 156f, 226f, 315f; names cited by Ray 35, 79
Virgilius, Marcellus, quoted in 'Etymology' 371 (biog.), 391, 398, 408; works in libraries 99
Viscum album (Mistletoe) [622], 200, **314**
Viscus, etymology 413
Vitex, etymology 374
Vitex agnus-castus 374f, 432f
Vitis vinifera (Grape Vine), terminology: *acinus* 415, *arilli* 416, *capreolus* 418, *palmes* 427, *racemus* 424, *sarmentum* 425, *vitilia* 428, *vinaceum* 428
Vitrearia, etymology 383
Vitruvius 79 (biog.); quoted in main catalogue 153; source of observations 49
Volatile salts 185, 274
Vomiting, induction of 197, 225, 289, 313

INDEX

Vulcan 374
Vulvaria, etymology 164f

Wallflower *see Erysimum cheiri*
Wall-rue *see Asplenium ruta-muraria*
Walnut *see Juglans regia*
Warts 191, 373
Watercress *see Nasturtium officinale sensu lato*
Watercress, Fool's *see Apium nodiflorum*
Water-crowfoot *see Ranunculus*
Water-dropwort *see Oenanthe*
Water finding 184, 310
Water-lily, Fringed *see Nymphoides peltata*
Water-lily, White *see Nymphaea alba*
Water-lily, Yellow *see Nuphar lutea*
Water-milfoil *see Myriophyllum*
Water-parsnip, Greater *see Sium latifolium*
Water-parsnip, Lesser *see Berula erecta*
Water-pepper *see Persicaria*
Water-plantain *see Alisma plantago-aquatica*
Water-plantain, Lesser *see Baldellia ranunculoides*
Water Soldier *see Stratiotes aloides*
Water-starwort *see Callitriche*
Wayfaring Tree *see Viburnum lantana*
Weeds, arable land 224f, 253f, 281f
Weiditz, Hans 83
Weld *see Reseda luteola*
Wendy, Lady Lettice 21, 22
Wheat *see Triticum*
Whelock, Abraham 98
Whetstones 153
Whin, Petty *see Genista anglica*
White things, weakness of 262
White, Large *see Pieris brassicae*
Whitlowgrass *see Erophila verna sensu lato*
Wieland, Melchior *see Guilandinus, Melchior*
Wild Basil *see Clinopodium vulgare*
Wilkins, John 19
Willow *see Salix*
Willowherb *see Epilobium*
Willughby, Cassandra, later Duchess of Chandos (1670–1735) 20, 21, 22f
Willughby, Emma (née Barnard) 20, 21

Willughby, Francis (1635–1672) 10, 11 (biog.), 15, 18, 19, 24, 62, 63, 66; acknowledged by Ray 34, 129; death 20, 22; family library 40, 63, 104, 106f, 107–109, 110f; marriage and children 20; mentioned in *Catalogus* 63, 277; residence at Trinity College 1658–1661 64; will set aside 21; *see also* Ray, John, biography
Willughby, Lady Cassandra (d. 1675) 19, 20, 22f
Willughby, Sir Francis (1668–1688) 20; escapes wicked stepfather 21
Willughby, Sir Thomas, Lord Middleton 20, 21, 107
Wind, effect on trees 317
Winter-cress *see Barbarea vulgaris*
Woad *see Isatis tinctoria*
Wolf, Caspar 140
Wood 149, 153, 202–203, 227; terminology: *pecten* 424, *scobs* 425; wood-turning 149; *see also* Tree rings, Trees
Woodlice, eaten by frogs 400
Woodpecker, Black (*Dryocopus martius*) 270f
Woodruff *see Galium odoratum*
Woodrush *see Luzula*
Wood-sorrel *see Oxalis acetosella*
Worms, little *see* Nematodes
Wormwood *see Artemisia*
Worthington, John 1, 2, 14, 24, 45, 56, 62, 110
Wotton, Edward 92
Woundwort *see Stachys*
Wray see Ray

Xyris [Ξυρις], etymology 407

Yarrow *see Achillea millefolium*
Yellow-cress *see Rorippa*
Yellow-rattle *see Rhinanthus minor*
Yellow-sedge *see Carex*
Yellow-wort *see Blackstonia perfoliata*
Yew *see Taxus baccata*
Yorkshire Fog *see Holcus lanatus*

Zacharias, Saint 364
Zannichellia palustris (Horned Pondweed) [466], 115, **268**, 331
Zeus *see* Jupiter